An Introduction to Quantum Optics

Series in Optics and Optoelectronics

For more information about this series, please visit:
https://www.crcpress.com/Series-in-Optics-and-Optoelectronics/book-series/TFOPTICSOPT

An Introduction to Quantum Optics

Optics

Photon and Biphoton Physics

Yanhua Shih

CRC Press
Taylor & Francis Group
Boca Raton London New York

CRC Press is an imprint of the
Taylor & Francis Group, an **informa** business

Second edition published 2021

by CRC Press

6000 Broken Sound Parkway NW, Suite 300, Boca Raton, FL 33487-2742

and by CRC Press

2 Park Square, Milton Park, Abingdon, Oxon, OX14 4RN

© 2021 Taylor & Francis Group, LLC

First edition published by CRC Press 2011

CRC Press is an imprint of Taylor & Francis Group, LLC

ISBN: 978-1-138-60125-3 (hbk)
ISBN: 978-0-367-67359-8 (pbk)
ISBN: 978-1-003-13060-4 (ebk)

Typeset in Computer Modern font
by KnowledgeWorks Global Ltd.

Visit the [companion website/eResources]: [insert CW/eResources URL]

Contents

Preface

Thanks to CRC Express and many students and colleagues for continuously encouraging me to complete the second edition of this book.

After the publication of the first edition, I received a considerable number of encouragements, comments, suggestions, as well as critiques and objections from readers all around world. These comments, suggestions, encouragements, and especially the critiques and objections, have helped me to significantly improve this book for its second edition. Compared with the first edition, one significant improvement is the introduction and emphasize of Einstein's granularity picture of light. Einstein's granularity picture and the quantum mechanical picture, together, provide us a modern view of light. In the second edition, in addition to updating recent experimental results and observations in the field of quantum optics, readers will find significant changes in the structure of the book, which should be especially helpful to graduate students and young researchers to delve into this attractive but challenging subject step-by-step.

This book is a selected collection of my thirty-year lecture notes on quantum optics for my Ph.D. thesis students and for an introductory-level graduate course at the University of Maryland.

My students and colleagues encouraged me to publish these lecture notes as a textbook or reference book that might be helpful in understanding the quantum theory of light from a relatively elementary introductory level. The successful introduction of the concept of photon, or quantum of light, stimulated a new foundation of physics, namely, the quantum theory. Today, quantum theory has turned out to be the overarching principle of modern physics. To quote Wheeler: It would be difficult to find a single subject among the physical science which is not affected in its foundations or in its applications by quantum theory. After a century of wondering, what do we know about the photon itself? The photon is a wave: it has no mass, it travels at the highest speed in the universe, and it interferes with itself. The photon is a particle: it has well-defined value of momentum and energy, and it even spins like a particle. The photon is neither a wave nor a particle, because whichever we think it is, we would have difficulty in explaining the other part of its behavior. The photon is a wave-like particle and/or a particle-like wave: a photon can never be divided into parts, but interference of a single photon can be easily observed in a modern optics laboratory. It seems that a photon passes both paths of an interferometer when interference patterns are observed; however, if the interferometer is set in such a way that its two paths are distinguishable, the photon knows which path to follow and never passes through both paths. Apparently, a photon has to make a choice when facing an interferometer: a choice of both-path like a wave or which-path like a particle. Surprisingly, the choice is not necessary before passing through the interferometer. It has been experimentally demonstrated that the choice of which-path and/or both-path can be delayed until after the photon has passed through the interferometer. More surprisingly, the which-path information can even be erased (Scullys quantum eraser) after the annihilation of the photon itself. In light of new technology, many historical gedankenexperiments became testable. In the past two decades, we have at least experimentally proven the existence of a biphoton system that behaves exactly as Einstein-Podolsky-Rosen (EPR) expected in 1935. In this biphoton system, the value of the momentum and the position for neither single subsystem is determinate. However, if one of the subsystems is measured with a certain momentum and/or position, the momentum and/or position of the other one is determined with certainty despite the distance between them. More exciting results have come from recent multiphoton interference experiments dealing with classical thermal light. Similar nonlocal superpositions between two-photon amplitudes and three-photon amplitudes have been observed in the joint photodetection of thermal radiation, and the resulting nonlocal point-to-point correlation of a randomly created and randomly distributed photon pair has been utilized in reproducing turbulence-free ghost images and to observing turbulence-free interferences.

The behavior of a photon, or a pair of photons, apparently does not follow any of the basic criteria, reality, causality, or locality of our everyday life. From the point of view of quantum theory, however, all of these surprises are predictable and explainable. The nonclassical behaviors of light quanta are the results of quantum interference, involving the superposition of single-photon or multi-photon amplitudes, a nonclassical entity corresponding to different yet indistinguishable alternative ways of producing a photodetection event or a joint photodetection event. The superposition of quantum amplitudes is a common phenomenon in the quantum world. It occurs between either single-photon amplitudes or multi-photon amplitudes in the measurement of either quantum light or classical light. Although questions regarding the fundamental issues about the concept of photon still exist, the quantum theory of light has contributed perhaps the most influential and successful, yet controversial, part to quantum mechanics.

This book aims to introduce these and many other exciting developments in quantum optics together with the basic theory and concepts of quantum optics to students and scientists in a simple and straightforward way. Differing from most traditional textbooks on this subject, it places greater emphasis on the experimental part of the analysis. All fundamental concepts are introduced in the process of analyzing typical experimental measurements and observations. The basic methods of classical and quantum mechanical treatments for optical measurements are naturally and gradually explored in the analysis of typical experiments. This attempt is aimed at: (1) helping students and young scientists analyze, summarize, and resolve quantum optical problems; and (2) encouraging students and young researchers to be more open minded in looking for the truth and improving their ability in making new discoveries in the field of physics.

In this regard, this book attempts to provide a number of nontraditional treatments and interpretations in certain historical and recent experimental discoveries in the field of quantum optics. The reader may find the following differences between this book and other traditional books on this subject: (1) This book introduces Einstein's granularity picture of light in terms of quantized bundle of rays or subfields and relates atomic transition to sub-sources in the early chapters. This attempt is aimed at preparing a general physical picture and background for introducing the concept of photon and the quantum mechanical theory of light. (2) It connects the interference phenomenon among a large number of sub-radiations with the concept of statistical ensemble average in the classical treatments of optical measurements and optical coherence. What is the physical cause of the measured intensity fluctuations? This book gives a nontraditional but perhaps more reasonable answer. (3) It distinguishes quantum mechanical multi-photon interference from classical statistical correlation of intensities. From the point of view of classical theory, the joint-detection by two or more photodetectors measures the statistical correlation of intensities. Any nontrivial second-order or higher-order correlation of light is caused by the nontrivial correlation of intensity fluctuations. From the point of view of quantum theory, the joint-detection by two or more photodetectors measures the probability of jointly having two or more photons contribute to the joint-photodetection event. If more than one multi-photon amplitude contributes to the event, the superposition between these quantum amplitudes results in a multi-photon constructive or destructive interference effect, which may not be considered or may not be explainable in the classical statistical theory of intensity fluctuation correlation. It does not seem difficult to distinguish multi-photon interference from statistical correlation in the measurement of entangled photon pairs. However, it is definitely not easy to appreciate the quantum interference picture in the measurement of classical thermal light, even if the measurement is at the single photon level. This book gives much experimental evidence and theoretical analysis in supporting the viewpoint of quantum mechanics. I hope that this effort would help readers see a general quantum interference picture of light, which perhaps reflects the physical truth behind all of the optical observations. To this end, the last chapter provides a detailed analysis of Bells theorem and Bells inequality as well as Scullys quantum eraser and Poppers experiment to distinguish quantum interference, especially multi-photon interference, from classical statistical correlation.

Introducing the basic concepts, tools, and the exciting developments of quantum optics to the reader, this book starts from Maxwell's equations, which explain the electromagnetic wave nature of light. We then review the historic block body radiation problem and photoelectric effect to address a critical question of why the quantization of light is necessary. Einstein's granularity picture of light is followed immediately as a semi-quantum theory. Although Einstein's concept of quantized subfield is still in the framework of electromagnetic theory, the physical behavior of his subfields, especially the self-interference of a single-subfield or a group of subfields, consistent with that of quantum mechanical concept of photon or a group of photons, except for a serious problem of locality. Having such a picture in mind, it would be natural to introduce the concept of field quantization and the concepts and tools of quantum optics based on the principles and rules of quantum mechanics. In this book, we introduce the concept of effective wavefunction of a photon or a group of photons and compare it with Einstein's subfield or subfields. There is no surprise that the effective wavefunction of a photon or a group of identical photons in thermal state is mathematically the same as that of Einstein's subfield. Based on these basic understandings we introduce the concepts of coherent and incoherent radiations in terms of coherently and incoherently radiated subfields and Fourier modes. After the introduction of typical quantum states of quantized field or photon, we study the first-order coherence and the second-order coherence or correlation of light in terms of the self-interference of a single-photon and the self-interference of a pair of photons in different quantum states. We then generalize the concept to Nth-order coherence or correlation in terms of N-photon interference: a group of N photons interfering with the group itself. Standing solidly on these fundamental concepts and theories, we then concentrate on the recent research topics of quantum optics, including quantum entanglement, multi-photon interferometry, quantum imaging, and optical tests of foundations of quantum theory. I hope to not only introduce these exciting developments to the reader, but also to give the reader an opportunity to practice the concepts and tools learned from this book through the analysis of these fascinating observations.

I hope that this book, which has been written to the best of my ability and knowledge, can be helpful to students, researchers, and all readers in general, in their efforts to understand, develop, and advance the field of optical science, which is currently undergoing tremendous change.

Acknowledgments

I would first like to thank my distinguished teachers Carroll O. Alley, David N. Klyshko, Morton H. Rubin, and John A. Wheeler, not only for teaching me physics but also for passing the ethos of physicists to me: seeking the truth and only the truth. My thanks also go to my students, postdoctors, and coworkers for their impressive experimental and theoretical research work, which have provided great support to this book. At the same time, my apologizes to them for not having included all of their research findings due to limited space and time, only those experiments and theories that are directly relevant to the contents and discussion points have been selected. For the same reason, I would like to apologize to all other researchers in the field of quantum optics for not including their works. I would also like to thank Ling-An Wu, Jason Simon, and Liang Shih for helping me with language editing issues. Finally, I would like to thank Ping Xu for translating and updating this book into Chinese as a text book for graduate and post-graduate level courses in China.

Author

Yanhua Shih, Professor of Physics, received his Ph.D. in 1987 from the Department of Physics, University of Maryland College Park. He started the Quantum Optics Laboratory at the University of Maryland Baltimore County (UMBC) in the fall of 1989. His group has been recognized as one of the leading groups in the field of quantum optics that attempts to probe the foundation of quantum theory. His pioneering research on multi-photon entanglement, multi-photon interferometry, quantum imaging, and optical tests of foundations of quantum theory has attracted a great deal of attention from the physics and engineering community all around the world. In thirty years of teaching and research, he published hundreds of experimental and theoretical works in leading refereed journals on the subject of quantum optics and has given hundreds of invited lectures and presentations at national and international professional conferences and workshops. Yanhua Shih received the Willis Lamb Medal in 2002 for his pioneer contributions to quantum optics and especially the study of coherence effects of multi-photon entangled states.

CHAPTER 1

Electromagnetic Wave Theory and Measurement of Light

What is light? This question had been asked from the very beginning of human civilization. Particle? Wave? Perhaps, this is one of the longest debated scientific topics in human history. In this chapter, we review the Maxwell continuum electromagnetic wave theory of light. Starting from Maxwell's equations, we derive the electromagnetic (EM) wave equation and obtain its plane-wave solution; we then introduce the classical superposition "principle" and the wavepacket concept of the electromagnetic field, which is a mathematical result of the Fourier transform. The electromagnetic field of light is never easy to measure directly. Most of the observations of light are based on the measurement of its energy flow or intensity. The measurement of light is another focus of this chapter.

1.1 ELECTROMAGNETIC WAVE THEORY OF LIGHT

To introduce the basic concepts on the coherence property of light we begin with the Maxwell's equations—the foundation of the classical electromagnetic wave theory of light. The set of four Maxwell's equations forms the basis of the theory for classical electromagnetic phenomena and electromagnetic wave phenomena. In free space, the Maxwell's equations have the form

$$\nabla \times \mathbf{E} = -\frac{\partial \mathbf{B}}{\partial t}, \tag{1.1.1}$$

$$\nabla \times \mathbf{H} = \frac{\partial \mathbf{D}}{\partial t}, \tag{1.1.2}$$

$$\nabla \cdot \mathbf{D} = 0, \tag{1.1.3}$$

$$\nabla \cdot \mathbf{B} = 0, \tag{1.1.4}$$

where $\mathbf{E}$ and $\mathbf{H}$ are the electric and magnetic field vectors, $\mathbf{D}$ and $\mathbf{B}$ are the electric displacement and magnetic induction vectors, respectively. We also have the relations

$$\mathbf{D} = \epsilon_0 \mathbf{E},$$
$$\mathbf{B} = \mu_0 \mathbf{H}, \tag{1.1.5}$$

where ϵ_0 and μ_0 are the free space electric permittivity and magnetic permeability, respectively.

Taking the curl of Eq.(1.1.1), using Eqs.(1.1.2), (1.1.3), and (1.1.5), as well as the identity

$$\nabla \times \nabla \times \mathbf{E} = \nabla(\nabla \cdot \mathbf{E}) - \nabla^2 \mathbf{E}, \tag{1.1.6}$$

the electric field vector $\mathbf{E}(\mathbf{r}, t)$ can be shown to satisfy the wave equation

$$\nabla^2 \mathbf{E} - \frac{1}{c^2} \frac{\partial^2 \mathbf{E}}{\partial t^2} = 0. \tag{1.1.7}$$

Similarly, the magnetic field vector $\mathbf{H}(\mathbf{r}, t)$, or the magnetic induction vector $\mathbf{B}(\mathbf{r}, t)$, can be shown to satisfy the same wave equation

$$\nabla^2 \mathbf{H} - \frac{1}{c^2} \frac{\partial^2 \mathbf{H}}{\partial t^2} = 0 \tag{1.1.8}$$

where $c \equiv 1/\sqrt{\epsilon_0 \mu_0}$ is the speed of light in free space.

Eqs. (1.1.7) and (1.1.8), namely the Maxwell wave equation or electromagnetic (EM) wave equation, both contain the basic wave equation structure (for a variable v)

$$\nabla^2 v - \frac{1}{c^2} \frac{\partial^2 v}{\partial t^2} = 0. \tag{1.1.9}$$

Now, suppose $v(\mathbf{r}, t)$ has a Fourier integral representation

$$v(\mathbf{r}, t) = \int_{-\infty}^{\infty} d\omega \, v(\mathbf{r}, \omega) \, e^{-i\omega t} \tag{1.1.10}$$

with the inverse transform

$$v(\mathbf{r}, \omega) = \frac{1}{2\pi} \int_{-\infty}^{\infty} dt \, v(\mathbf{r}, t) \, e^{i\omega t}. \tag{1.1.11}$$

Substituting Eq. (1.1.10) into Eq. (1.1.9) it is straightforward to find that the Fourier transform $v(\mathbf{r}, \omega)$ also satisfies the Helmholtz wave equation

$$\nabla^2 v(\mathbf{r}, \omega) + k^2 v(\mathbf{r}, \omega) = 0 \tag{1.1.12}$$

where $k = \omega/c$ is the wave number. The wave number takes either a set of discrete or continuous values determined by the boundary conditions. Eq. (1.1.12) has well-known plane-wave solutions

$$v_k(\mathbf{r}) = v_k \, e^{i\mathbf{k} \cdot \mathbf{r}} \tag{1.1.13}$$

and thus, we have

$$v_k(\mathbf{r}, t) = v_k \, e^{-i(\omega t - \mathbf{k} \cdot \mathbf{r})} + c.c. \tag{1.1.14}$$

where v_k is the complex amplitude associated with mode $\mathbf{k}$, and c.c. stands for the complex conjugate. To simplify the mathematics we will neglect writing c.c. in the following discussions unless otherwise specified.

Due to the linear nature of the differential equation, any linear superposition of the plane-wave is also a solution of Eq.(1.1.9). Therefore, we can, in principle, find a set of transforms of $v_k(\mathbf{r}, t)$ for the wave function $v(\mathbf{r}, t)$. This property may be considered as the "classical superposition principle".

With the convention that the physical fields are the real parts of the complex solutions, we may write the plane-wave electric and magnetic fields in the form

$$\mathbf{E}(\mathbf{r}, t) = \mathbf{E} \, e^{-i(\omega t - \mathbf{k} \cdot \mathbf{r})} \tag{1.1.15}$$

$$\mathbf{H}(\mathbf{r}, t) = \mathbf{H} \, e^{-i(\omega t - \mathbf{k} \cdot \mathbf{r})},$$

where $\mathbf{E}$ and $\mathbf{H}$ are vectors that are constant in space-time, usually named field strengths.

To satisfy the divergence Eqs. (1.1.3) and (1.1.4) of the Maxwell's equations, requires that both the $\mathbf{E}$ and $\mathbf{H}$ vector fields be purely transverse, i.e., perpendicular to the wave vector $\mathbf{k}$. The curl equations (1.1.1) and (1.1.2) require even further restrictions on the $\mathbf{E}$ and $\mathbf{H}$ vector fields: (1) $\mathbf{E} \perp \mathbf{H}$; (2) in free space, $\mathbf{E}$ and $\mathbf{H}$ are in phase at all points of space-time, with their magnitudes related by $|\mathbf{E}| = \sqrt{\mu_0/\epsilon_0}|\mathbf{H}|$.

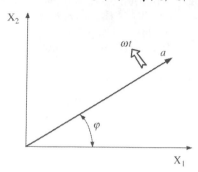

FIGURE 1.1.1 A harmonic oscillator picture of a plane-wave solution with frequency ω. The real and positive amplitude a rotates at angular speed ω starting from initial phase φ. The X_1 and X_2 axis indicate the two quadratures.

It is then useful to define a set of orthogonal unit vectors $(\hat{\mathbf{e}}_1, \hat{\mathbf{e}}_2, \hat{\mathbf{k}})$ and rewrite the field strengths as:

$$\mathbf{E} = \hat{\mathbf{e}}_1 E_0, \qquad \mathbf{H} = \hat{\mathbf{e}}_2 \sqrt{\frac{\epsilon_0}{\mu_0}} E_0 \qquad (1.1.16)$$

or

$$\mathbf{E} = \hat{\mathbf{e}}_2 E_0', \qquad \mathbf{H} = -\hat{\mathbf{e}}_1 \sqrt{\frac{\epsilon_0}{\mu_0}} E_0', \qquad (1.1.17)$$

where E_0 and E_0' are complex constants. It is easy to see that there are two independent polarization directions of $\hat{\mathbf{e}}_1$ and $\hat{\mathbf{e}}_2$ for the plane-wave field of frequency $(\omega, \mathbf{k})$. In certain types of problems, it is convenient to write the field strength E_0 as a product of the amplitude a and phase term $e^{i\varphi}$, or as quadratures in the complex amplitude plane:

$$\begin{aligned} E_0 &= a\, e^{i\varphi} \\ &= a \cos\varphi + i\, a \sin\varphi, \end{aligned} \qquad (1.1.18)$$

where the amplitude a is real and positive. The physical picture of the amplitude-phase-quadrature concept is schematically shown in Fig. 1.1.1. It is interesting to see that the plane-wave solution of the electromagnetic field corresponds to a classical harmonic oscillator.

1.2 CLASSICAL SUPERPOSITION

Since the electromagnetic wave equation is a linear differential equation, it is elementary to draw the following two corollaries:

(1) If each $\mathbf{E}_j(\mathbf{r}, t)$ is a solution of the wave equation, then the linear superposition

$$\mathbf{E}(\mathbf{r}, t) = \sum_j \mathbf{E}_j(\mathbf{r}, t) \qquad (1.2.1)$$

is also a solution of the wave equation. We may refer to Eq. (1.2.1) as the principle of classical superposition. A wide range of optical coherent phenomena can be formulated in the form of Eq. (1.2.1). As a simple example, consider a number of independent sub-sources, if each sub-source gives rise to a subfield $\mathbf{E}_j(\mathbf{r}, t)$, which is a solution of the EM wave equation, at space-time point $(\mathbf{r}, t)$, the measured field $\mathbf{E}(\mathbf{r}, t)$ must be the result of the superposition of all of the subfields.

(2) It is possible, in principle, to decompose a radiation field $\mathbf{E}(\mathbf{r}, t)$ into an appropriate linear superposition of a set of solutions of the wave equation. This is a different approach, or a different view point, to define classical superposition. For example, we may find a set of plane-waves to write the field $\mathbf{E}(\mathbf{r}, t)$ as the following superposition

$$\mathbf{E}(\mathbf{r}, t) = \sum_{\mathbf{k}} \hat{\mathbf{e}}_{\mathbf{k}}\, E(\mathbf{k})\, e^{-i(\omega t - \mathbf{k} \cdot \mathbf{r})} \tag{1.2.2}$$

or

$$\mathbf{E}(\mathbf{r}, t) = \int d\mathbf{k}\, \hat{\mathbf{e}}_{\mathbf{k}}\, E(\mathbf{k})\, e^{-i(\omega t - \mathbf{k} \cdot \mathbf{r})}. \tag{1.2.3}$$

Note, as we have discussed in section 1.1, there are two independent polarizations of $\hat{\mathbf{e}}_{\mathbf{k}}$ for each plane-wave solution of $\mathbf{k}$. A coherent superposition of the two independent polarizations results in a polarized field (vector), otherwise, the field is unpolarized with an random relationship between the two independent polarizations. We will focus our attention onto one of the polarizations of the field in the following discussions unless otherwise specified. The polarization state of the field will be discussed in later chapters.

To further simplify the mathematics we consider a 1-D approximation of Eq. (1.2.3) and focus our discussion on the longitudinal behavior of the field along one selected propagation direction. Our goal is to learn the temporal behavior of the field through the measurement at a point-like photodetector. For a point-like radiation source at coordinate $\mathbf{r}_0 = 0$ that give rise to radiation at $t_0 = 0$, the field observed at $(\mathbf{r}, t)$ field can be written as

$$E(r, t) = \int_0^\infty d\omega\, E(\omega)\, e^{-i[\omega t - k(\omega) r]} \tag{1.2.4}$$

where $r = |\mathbf{r} - \mathbf{r}_0|$ is the optical length along the path between the point source at $\mathbf{r}_0 = 0$ and the point of observation at $\mathbf{r}$, and $E(\omega) = a(\omega) e^{i\varphi(\omega)}$ is the complex spectral amplitude density (or spectral amplitude for short) for the plane-wave mode of frequency ω. The dispersion relation $k = k(\omega)$ allows us to express the wavenumber through the frequency detuning ν,

$$\omega \equiv \omega_0 + \nu.$$

Therefore, Eq. (1.2.4) can be formally integrated:

$$\begin{aligned} E(r, t) &= e^{-i[\omega_0 t - k(\omega_0) r]} \int_{-\infty}^\infty d\nu\, E(\nu)\, e^{-i\nu\tau} \\ &= e^{-i[\omega_0 t - k(\omega_0) r]}\, \mathcal{F}_\tau\{E(\nu)\} \end{aligned} \tag{1.2.5}$$

where $\mathcal{F}_\tau\{E(\nu)\}$ is the Fourier transform of the complex spectral amplitude $E(\nu)$. In Eq. (1.2.5), a first order approximation of dispersion has been applied:

$$k(\omega) \simeq k(\omega_0) + \frac{dk}{d\omega}\bigg|_{\omega_0} \nu.$$

We have also defined

$$\tau \equiv t - \frac{r}{u(\omega_0)}$$

where the inverse of the first order dispersion

$$u(\omega_0) = \frac{1}{\frac{dk}{d\omega}\big|_{\omega_0}}$$

is named the group velocity of the wavepacket if there exists a wavepacket. In vacuum, the phase speed of the carrier frequency and the group speed of the wavepacket (envelope) are both equal to c,

$$\frac{\omega}{k} = \frac{d\omega}{dk} = c,$$

since $\omega = kc$. Eq. (1.2.5) is then simplified to

$$E(r,t) = e^{-i\omega_0\tau} \mathcal{F}_\tau\{E(\nu)\} \tag{1.2.6}$$

where $\tau \equiv t - r/c$. In most of the following discussions, we will assume light propagates in a vacuum and use the simplified notation of Eq. (1.2.6).

The field $E(r,t)$ is now formally written in terms of the Fourier transform of $E(\omega)$ in Eq. (1.2.6). The resulting function of the Fourier integral $\mathcal{F}_\tau\{E(\omega)\}$ is determined by the complex spectral amplitude $E(\omega) = a(\omega)e^{i\varphi(\omega)}$. Both the real-positive amplitude $a(\omega)$ and the phase $\varphi(\omega)$ play important roles. If $a(\omega)$ and $\varphi(\omega)$ are well-defined functions of ω, the resulting field $E(r,t)$ will be a well-defined function of space-time, and vice versa. On the other hand, if $a(\omega)$ and/or $\varphi(\omega)$ vary rapidly and randomly from time to time, no deterministic function of $E(r,t)$ is expected.

In classical optics, both $a(\omega)$ and $\varphi(\omega)$ can be simultaneously defined precisely. Consequently, $E(r,t)$ will also be a precisely defined function in space-time. The uncertainty relation between $\Delta a(\omega)$ and $\Delta\varphi(\omega)$ is not the subject of this chapter. In the following discussions, we will assume a well-defined distribution function of $a(\omega)$ and leave freedom to the phase $\varphi(\omega)$. The variations or uncertainties of the phase $\varphi(\omega)$ will determine the coherent property of the radiation.

We discuss two extreme cases in the following: (I) Coherent Superposition and (II) Incoherent Superposition.

Case (I): Coherent Superposition

(1) $\varphi(\omega) = \varphi_0 =$ constant.

With a constant phase of $\varphi(\omega) = \varphi_0$, if $a(\nu)$ is a well-defined function of the detuning frequency ν then the Fourier integral of Eq.(1.2.5) defines a wavepacket in space-time given by

$$E(r,t) = e^{-i[\omega_0 t - k(\omega_0)r - \varphi_0]} \mathcal{F}_\tau\{a(\nu)\}. \tag{1.2.7}$$

The envelope of the wavepacket, which is the Fourier transform of the spectral amplitude $a(\nu)$, propagates with group velocity $u(\omega_0)$. Under the envelope is the carrier harmonic wave of frequency ω_0, which propagates at phase velocity $\omega_0/k(\omega_0)$. We may imagine the wavepacket is formed in a coherent radiation source located at $\mathbf{r} = 0$ in which a set of excited plane-waves of $\omega = \omega_0 + \nu$, each defined as a "Fourier-mode", are created at $t = 0$ with a common initial phase φ_0. The coherent superposition of these Fourier-modes results in the wavepacket of Eq. (1.2.7) propagating in space-time. The "center" of the wavepacket, in which all Fourier-modes added constructively, is always located at $\tau = t - r/u(\omega_0) = 0$. In vacuum, the group velocity $u(\omega_0)$ of the envelope and the phase velocity $\omega_0/k(\omega_0)$ of the carrier wave have the same value of c. The group velocity, however, could be quite different from the phase velocity in dispersive media. In a dispersive medium, each plane-wave mode of $\omega = \omega_0 + \nu$ may have different phase velocities $\omega/k(\omega)$ depending on the dispersion of the medium. In special circumstances, superposition of these Fourier-modes results in an interesting effect wherein the group velocity could be greater than the phase velocity or greater than c, even if all the phase velocities of the modes are less than c. The value of the group velocity, whether less than or greater than the phase velocity, is determined by the first order dispersion of the medium. The wavepacket defined in Eq.(1.2.7) is the result of a coherent superposition of the electromagnetic field.

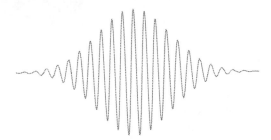

FIGURE 1.2.1 Schematic of a Gaussian wavepacket. The envelope of the wavepacket propagates with group velocity $u(\omega_0)$. Under the envelope is the carrier wave of frequency ω_0. The carrier propagates with phase velocity $\omega_0/k(\omega_0)$. The wavepacket is centered at space-time point $\tau = t - r/u(\omega_0) = 0$. At $\tau = 0$, all of its "Fourier-modes" superpose coherently and constructively due to their common initial phase $\varphi(\omega) = \varphi_0$.

For certain amplitude distribution functions of $a(\nu)$, the Fourier integral in Eq.(1.2.5) can be easily evaluated. For example, a Gaussian distribution function in terms of the detuning frequency ν

$$E(\nu) = a_0 \, e^{-\sigma^2 \nu^2} \, e^{i\varphi_0}, \tag{1.2.8}$$

where σ is a constant, results in a Gaussian wavepacket in space-time

$$E(r, t) = E_0 \, e^{-\tau^2/4\sigma^2} \, e^{-i[\omega_0 t - k(\omega_0)r - \varphi_0]} \tag{1.2.9}$$

where all constants have been absorbed into E_0. Figure 1.2.1 illustrates a classical Gaussion wavepacket. The wavepacket has a well-defined envelope, which propagates with group velocity $u(\omega_0)$. Under the envelope is the carrier wave of frequency ω_0, which propagates at phase velocity $\omega_0/k(\omega_0)$.

Another useful example is for a constant distribution

$$E(\nu) = a_0 \, e^{i\varphi_0}, \tag{1.2.10}$$

where a_0 is a constant. The envelope of the wavepacket in space-time turns out to be a δ-function

$$\mathcal{F}_\tau \big\{ E(\nu) \big\} = E_0 \, \delta(\tau), \tag{1.2.11}$$

where, again, all constants have been absorbed into E_0. In reality, we may assume a constant $E(\nu)$ within a certain bandwidth $\Delta\nu$

$$E(\nu) = \begin{cases} a_0 \, e^{i\varphi_0} & -\Delta\nu/2 \leq \nu \leq \Delta\nu/2 \\ 0 & \text{otherwise.} \end{cases}$$

where $a_0 \sim E_0/\Delta\nu$. In this case, the Fourier transform gives a sinc-function, which is defined as $\mathrm{sinc}(x) \equiv \sin(x)/x$,

$$E(r, t) = E_0 \, \mathrm{sinc}\Big(\frac{\Delta\nu \, \tau}{2}\Big) \, e^{-i[\omega_0 t - k(\omega_0)r - \varphi_0]}. \tag{1.2.12}$$

(2) $\varphi(\omega) = \omega t_0$ with $t_0 = $ constant.

With $\varphi(\omega) = \omega t_0$, if $a(\nu)$ is a well-defined function of the detuning frequency ν, the Fourier integral of Eq.(1.2.5) also defines a wavepacket in space-time,

$$E(r,t) = \mathcal{F}_{(\tau - t_0)}\{a(\nu)\} e^{-i[\omega_0(t-t_0)-k(\omega_0)r]}. \qquad (1.2.13)$$

This wavepacket is the same as that of Eq. (1.2.7) except all its Fourier-modes exhibit a common initial phase $\varphi_0 = 0$ when the wavepacket is created at $r = 0$, $t = t_0$.

Case (II): Incoherent superposition, $\varphi(\omega) = $ random number.

The phases of the complex spectral amplitude are completely random. There is no defined phase relationship between different harmonic modes. The Fourier transform of Eq.(1.2.5) consequently results in a random function of space-time even if the real-positive amplitude $a(\omega)$ has a well defined distribution. In this situation we may model each Fourier component or each plane-wave mode of ω as an independent harmonic oscillator.

1.3   INTENSITY OF LIGHT: A MEASURABLE QUANTITY

The field $E(r,t)$ is not the physical quantity directly measurable by a photodetector. Roughly speaking, the measurement process of a photodetector involves the annihilation of a photon and the release of a photoelectron contributing to the output current of the photodetector. The output current of the photodetector is proportional to the energy carried by the annihilated photons per unit time. The detailed quantum process of photodetection and the quantization of the electromagnetic field will be discussed later. Classically, the photoelectron current of a photodetector measures the intensity of light, which is defined as the amount of energy crossing a unit area per unit of time, and is given by the Poynting vector of the electromagnetic field

$$\mathbf{S} = \frac{1}{\mu_0}\mathbf{E} \times \mathbf{B},$$

where $\mathbf{E}$ and $\mathbf{B}$ are the electric and magnetic fields (real functions of space-time). At optical frequencies, the Poynting vector is an extremely rapid varying function of time, twice as rapid as the field. Unfortunately, there is no photodetector that is able to resolve these fast vibrations. This suggests that at least practically, a time average is occurring in the process of photodetection. The photocurrent is a measure of the magnitude of the cycle-averaged Poynting vector. Thus, we define the "instantaneous" intensity, or simply the intensity, of light as:

$$I(\mathbf{r},t) = \frac{\epsilon_0 c}{2} |\mathbf{E}(\mathbf{r},t)|^2, \qquad (1.3.1)$$

where we have considered the effect that $\mathbf{B}$ is at right angles to $\mathbf{E}$. The "cycle-average theorem" is also applied:

$$\overline{[Re\,\mathbf{E}(\mathbf{r},t)]^2} = \frac{1}{2}Re\,|\mathbf{E}(\mathbf{r},t)|^2. \qquad (1.3.2)$$

For convenience, we will absorb the constant $\epsilon_0 c/2$ of Eq. (1.3.1) into the field, except for certain necessary quantitative discussions.

Substituting Eq.(1.2.3) into Eq.(1.3.1) the measured intensity is calculated as

$$I(\mathbf{r},t) = \int d\mathbf{k}\, \hat{\mathbf{e}}_\mathbf{k}\, E^*(\mathbf{k})\, e^{i(\omega t - \mathbf{k}\cdot\mathbf{r})} \cdot \int d\mathbf{k}'\, \hat{\mathbf{e}}_{\mathbf{k}'}\, E(\mathbf{k}')\, e^{-i(\omega' t - \mathbf{k}'\cdot\mathbf{r})}$$

$$= \int d\mathbf{k} \int d\mathbf{k}'\, (\hat{\mathbf{e}}_\mathbf{k} \cdot \hat{\mathbf{e}}_{\mathbf{k}'})\, E^*(\mathbf{k}) E(\mathbf{k}') e^{i[(\omega-\omega')t - (\mathbf{k}-\mathbf{k}')\cdot\mathbf{r}]}. \tag{1.3.3}$$

where the modes are treated continuously, and the $\mathbf{k}$ vectors may take any value and may propagate to any direction. Simplifying the mathematics to 1-D, again, we assume a point-like radiation source at $\mathbf{r}_0 = 0$, a point-like photodetector at $\mathbf{r}$, and free propagation along the 1-D path, r, in vacuum. The point source excites a set of Fourier-modes of ω at $t_0 = 0$ and the point-like photodetector measures the instantaneous intensity $I(r,t)$,

$$I(r,t) = E^*(r,t)E(r,t)$$

$$= \int d\omega\, E^*(\omega)\, e^{i(\omega t - kr)} \int d\omega'\, E(\omega')\, e^{-i(\omega' t - k'r)}. \tag{1.3.4}$$

We rewrite Eq. (1.3.4) in the following form for easy discussion,

$$I(r,t) = \int_{\omega=\omega'} d\omega |E(\omega)|^2$$

$$+ \int_{\omega\neq\omega'} d\omega\, d\omega'\, E^*(\omega)E(\omega')\, e^{i[(\omega-\omega')t - (k-k')r]} \tag{1.3.5}$$

where, again, $E(\omega) = a(\omega)e^{i\varphi(\omega)}$ and $E(\omega') = a(\omega')e^{i\varphi(\omega')}$ are the complex amplitudes of the ω-mode and the ω'-mode, respectively, and $\tau \equiv t - r/c$. The first term of Eq.(1.3.5) is a constant. The second term of Eq.(1.3.5) is a superposition of the frequency beats $\omega - \omega'$. This term may contribute significantly or trivially depending on the complex amplitudes of the conjugated amplitudes $E(\omega) = a(\omega)e^{i\varphi(\omega)}$ and $E(\omega') = a(\omega')e^{i\varphi(\omega')}$.

We discuss two extreme cases in the following: (I) Coherent Superposition and (II) Incoherent Superposition.

Case (I): Coherent Superposition, $\varphi(\omega) = \varphi_0 = $ constant.

With a constant phase $\varphi(\omega) = \varphi_0$, if $a(\nu)$ is a well-defined function of the detuning frequency ν then the intensity becomes

$$I(r,t) = E^*(r,t)E(r,t) = \left| \mathcal{F}_\tau\{a(\nu)\} \right|^2 \tag{1.3.6}$$

which is the normal square of the Fourier transform or the "envelope" of the EM wavepacket. For example, a Gaussian distribution function of the detuning frequency

$$a(\nu) = a_0\, e^{-\sigma^2\nu^2} \tag{1.3.7}$$

yields a Gaussian pulse

$$I(r,t) = E_0^2\, e^{-\tau^2/2\sigma^2}. \tag{1.3.8}$$

Case (II): Incoherent Superposition, $\varphi(\omega) = $ random number.

In this case, if $\varphi(\nu) - \varphi(\nu')$ takes all possible values in the superposition of Eq.(1.3.5), the second term of Eq.(1.3.5) superposes destructively and vanishes. The intensity $I(r,t)$ becomes a constant

$$I(r,t) = \int_{\omega=\omega'} d\omega\, |E(\omega)|^2 = \int_{\omega=\omega'} d\omega\, I(\omega;r,t). \tag{1.3.9}$$

where $I(\omega;r,t)$ indicates the intensity of the ω mode. In the case of weak light, if $\varphi(\nu) - \varphi(\nu')$ takes a particular set of values in the superposition of Eq.(1.3.5), the second term may not vanish completely and contribute an intensity variation or fluctuation.

1.4 INTENSITY OF LIGHT: EXPECTATION AND FLUCTUATION

In the classical electromagnetic wave theory of light, the expectation value of intensity at a space-time coordinate $(\mathbf{r}, t)$ is defined as

$$\langle I(\mathbf{r}, t) \rangle \equiv \langle E^*(\mathbf{r}, t) E(\mathbf{r}, t) \rangle. \tag{1.4.1}$$

The notation $\langle ... \rangle$, which is adapted from statistics, denotes the mathematical expectation of the intensity measurement by *taking into account all possible realizations* of the EM field. In probability theory, the expectation value of a measurement equals the mean value of measurements on an ensemble of similarly prepared systems. The expectation value $\langle I(\mathbf{r}, t) \rangle$ can be a well-defined function of the space-time coordinates of the measurement events. To emphasize the dependence or independence as either a nontrivial distribution or a constant distribution of space-time coordinates, $\langle I(\mathbf{r}, t) \rangle$ is also called expectation function.

Substitute Eq.(1.2.3) into Eq.(1.4.1), the expectation value of intensity is calculated from

$$\langle I(\mathbf{r}, t) \rangle = \langle \int d\mathbf{k}\, \hat{\mathbf{e}}_{\mathbf{k}}\, E^*(\mathbf{k})\, e^{i(\omega t - \mathbf{k} \cdot \mathbf{r})} \cdot \int d\mathbf{k}'\, \hat{\mathbf{e}}_{\mathbf{k}'}\, E(\mathbf{k}')\, e^{-i(\omega' t - \mathbf{k}' \cdot \mathbf{r})} \rangle$$

$$= \langle \int d\mathbf{k} \int d\mathbf{k}'\, (\hat{\mathbf{e}}_{\mathbf{k}} \cdot \hat{\mathbf{e}}_{\mathbf{k}'})\, E^*(\mathbf{k}) E(\mathbf{k}')\, e^{i[(\omega - \omega')t - (\mathbf{k} - \mathbf{k}') \cdot \mathbf{r}]} \rangle \tag{1.4.2}$$

by *taking into account all possible complex amplitudes of the EM fields in the superposition*. The statistical result of *taking into account all possible complex amplitudes of the EM fields in the superposition* is equivalent to an ensemble average. A single measurement of intensity at a space-time point yields the expectation value, if and only if the measured fields in the superposition take *all possible values* of their complex amplitudes $E(\mathbf{k}) = a(\mathbf{k}) e^{i\varphi(\mathbf{k})}$ and $E(\mathbf{k}') = a(\mathbf{k}') e^{i\varphi(\mathbf{k}')}$, especially, the phase factor $e^{i\varphi(\mathbf{k})}$ and $e^{i\varphi(\mathbf{k}')}$ must take all possible values. This is equivalent to having $\varphi \equiv \varphi(\mathbf{k}) - \varphi(\mathbf{k}')$ taking all possible values from 0 to 2π in a random manner. Realistically, the fields may not be able to take *all possible* realizations in a measurement and therefore the measured intensity at a space-time coordinate or the observed intensity as a function of space-time, $I(\mathbf{r}, t)$, may differ from $\langle I(\mathbf{r}, t) \rangle$. We may write $I(\mathbf{r}, t)$ into the sum of its expectation value or function $\langle I(\mathbf{r}, t) \rangle$ and its fluctuation $\Delta I(\mathbf{r}, t)$

$$I(\mathbf{r}, t) = \langle\, I(\mathbf{r}, t)\, \rangle + \Delta I(\mathbf{r}, t). \tag{1.4.3}$$

The expectation value or expectation function, $\langle I(r, t) \rangle$, for the simple 1-D measurement, in which the radiation propagates from a point-like source to a point-like photodetector in vacuum, is easy to calculate from Eq.(1.3.4),

$$\langle I(r, t) \rangle = \langle \int d\omega\, E^*(\omega)\, e^{i(\omega t - kr)} \int d\omega'\, E(\omega')\, e^{-i(\omega' t - k'r)} \rangle$$

$$= \langle \int_{\omega = \omega'} d\omega |E(\omega)|^2 \rangle$$

$$+ \langle \int_{\omega \neq \omega'} d\omega\, d\omega'\, E^*(\omega) E(\omega')\, e^{i[(\omega - \omega')t - (k - k')r]} \rangle$$

$$= \langle \int_{\nu = \nu'} d\nu |E(\nu)|^2 \rangle$$

$$+ \langle \int_{\nu \neq \nu'} d\nu\, d\nu'\, E^*(\nu) E(\nu')\, e^{i(\nu - \nu')\tau} \rangle \tag{1.4.4}$$

by *taking into account all possible values of the complex amplitudes (a and φ)* in terms of the Fourier-modes. In Eq.(1.4.4), again, we have treated the Fourier-modes as continuous.

It is easy to find that the first term of Eq. (1.4.4), corresponding to the *self-interference of a Fourier-mode: a Fourier-mode interferences with itself*, always takes a constant value despite the coherent or incoherent nature of the radiation, while the outcome of the second term, corresponding to the *cross-interference of two different Fourier-modes*, is determined by the coherent or incoherent property of the measured radiation.

Nowadays, we usually divide light into roughly two categories: incoherent and coherent. A discharge tube radiates incoherent light and a laser beam is considered coherent. In Maxwell's continuum picture of light, the Fourier-modes are superposed incoherently if emitted from an incoherent radiation source and coherently if emitted from a coherent radiation source. Whether incoherent or coherent, the property of light is determined intrinsically by the radiation source.

(I) Natural light

In the early days the only light sources available for optical observations and measurements were natural light prepared in the stochastic radiation processes. Natural light has the popular name "thermal light" or "thermal radiation". Following the Maxwell EM wave theory, we model the complex amplitude of each of the Fourier-mode with random initial phases because thermal radiation is created from stochastic processes. For natural thermal light, the expectation value of intensity at coordinate (r, t), by *taking into account all possible values of the random phases of the Fourier-modes*, is thus

$$
\begin{aligned}
\langle I(r,t) \rangle &= \langle \int_{\omega=\omega'} d\omega |E(\omega)|^2 \rangle + \langle \int_{\omega \neq \omega'} d\omega\, d\omega'\, E^*(\omega) E(\omega')\, e^{i[(\omega-\omega')t-(k-k')r]} \rangle \\
&= \int d\omega\, |E(\omega)|^2 + \langle \Delta I(r,t) \rangle \\
&= \int d\omega\, I(\omega; r, t).
\end{aligned}
\tag{1.4.5}
$$

The second term of the superposition in Eq.(1.4.5) added destructively, by *taken into account all possible values of the random phases of the Fourier-modes*, results in a null contribution to $\langle I(r,t) \rangle$. The above expectation value evaluation is equivalent to a partial ensemble average, which has only *taken into account all possible values of the random phases of the Fourier-modes*.

The intensity variation or fluctuation, resulted from two-mode interferences, can be defined as

$$
\Delta I(r,t) \equiv \int_{\omega \neq \omega'} d\omega\, d\omega'\, E^*(\omega) E(\omega')\, e^{i[(\omega-\omega')t-(k-k')r]}.
\tag{1.4.6}
$$

In a realistic intensity measurement that cannot be treated as an ensemble average, the Fourier-modes may not be able to take all possible random phases and the incomplete destructive superposition of the second term of Eq.(1.4.4) may have a nonzero contribution. In this case, the second integral in Eq.(1.4.4) gives raise to a nondeterministic intensity variation, or fluctuation, in the neighborhood the expectation value of intensity:

$$
\begin{aligned}
I(r,t) &= \int_{\omega=\omega'} d\omega |E(\omega)|^2 + \int_{\omega \neq \omega'} d\omega\, d\omega'\, E^*(\omega) E(\omega')\, e^{i[(\omega-\omega')t-(k-k')r]} \\
&= \langle I(r,t) \rangle + \Delta I(r,t)
\end{aligned}
\tag{1.4.7}
$$

which is consistent with Eq.(1.4.3). The two-mode interference induced nondeterministic intensity fluctuation

$$\Delta I(r,t) = \int_{\omega \neq \omega'} d\omega \, d\omega' \, E^*(\omega) E(\omega') \, e^{i[(\omega - \omega')t - (k-k')r]}$$

$$= \int_{\nu \neq \nu'} d\nu \, d\nu' \, E^*(\nu) E(\nu') \, e^{i[(\nu - \nu')t - (k-k')r]}$$

contains all possible temporal and spatial frequency "beats" in terms of $\omega - \omega' = \nu - \nu'$ and $k - k'$.

(II) Coherent light.

Since the invention of the laser, coherent light has become the most popular light sources in modern optical measurements. These sources not only produce light coherently, but are also, in certain cases, non-stationary. For example, a pulsed laser, either Q-switched or mode-locked, generates well-defined wavepackets in $E(\mathbf{r},t)$ or pulses in $I(\mathbf{r},t)$. The intensity, $I(\mathbf{r},t)$, no longer fluctuates randomly in the neighborhood of a constant value from time to time nondeterministically, but becomes a well-defined function of time deterministically. In this extreme case, the Fourier-modes in Eq. (1.4.4) are excited coherently at the source and superposed coherently at the observation point.

$$\langle I(r,t) \rangle = \left\langle \int d\omega \, E^*(\omega) \, e^{i(\omega t - kr)} \int d\omega' \, E(\omega') \, e^{-i(\omega' t - k'r)} \right\rangle$$

$$= \int d\nu d\nu' \, E^*(\nu) \, E(\nu') \, e^{i[(\nu - \nu')t - (k-k')r]}$$

$$= \left| \mathcal{F}_\tau \{ E(\nu) \} \right|^2 \qquad (1.4.8)$$

where we have used the formally integrated Fourier transform of Eq.(1.2.5) by assuming a well-defined function of $E(\nu)$. For example, a Gaussian distribution function in terms of the detuning frequency

$$E(\nu) = E_0 \, e^{-\sigma^2 \nu^2} \qquad (1.4.9)$$

where the complex constant E_0 contains a constant phase factor $e^{i\varphi_0}$, yields a Gaussian pulse in space-time

$$\langle I(r,t) \rangle = |E_0|^2 \, e^{-\tau^2/2\sigma^2}. \qquad (1.4.10)$$

1.5 MEASUREMENT OF INTENSITY: ENSEMBLE AVERAGE AND TIME AVERAGE

In classical theory of light, an idealized point photodetector measures the instantaneous intensity of radiation $I(\mathbf{r},t)$ at space-time coordinate $(\mathbf{r},t)$. Assuming a point photodetector that is placed at a chosen coordinate $\mathbf{r}$ reads a value of I at time t, we may find the measured value I slightly differs from the expectation value $\langle I \rangle$

$$I = \langle I \rangle + \Delta I, \qquad (1.5.1)$$

where ΔI, which may take a positive value or a negative value, denotes the difference. Note, the coordinate $(\mathbf{r},t)$ has been dropped from $I(\mathbf{r},t)$. Suppose we are not making one measurement but a large number of independent measurements simultaneously on a set of identical radiation fields under the same experimental condition, we may find each measured value I_j slightly differs from $\langle I \rangle$ and from each other. The statistical mean intensity is defined as

$$\bar{I} = \frac{1}{N} \sum_{j=1}^{N} I_j = \langle I \rangle + \frac{1}{N} \sum_{j=1}^{N} \Delta I_j, \tag{1.5.2}$$

where I_j is the measured instantaneous intensity of the jth member of the ensemble. The mean value $\bar{I}$ equals the expectation value $\langle I \rangle$ when $N \sim \infty$,

$$\langle I \rangle = \bar{I} = \lim_{N \sim \infty} \frac{1}{N} \sum_{j=1}^{N} I_j, \tag{1.5.3}$$

since statistically

$$\lim_{N \sim \infty} \frac{1}{N} \sum_{j=1}^{N} \Delta I_j = 0 \tag{1.5.4}$$

for a large set of randomly distributed values of ΔI_j. Besides, Eq. (1.5.3) is physically reasonable. The expectation value is calculated by taking into account all possible realizations of the fields. In the measurement of a member of the ensemble, the field may not be able to take all possible realizations but only a particular set of complex amplitudes in the superposition. When a large number of measurements contribute to the statistical averaging, especially when $N \sim \infty$, the field will definitely have a chance to take all possible complex amplitudes and consequently give an averaged value equal to $\langle I \rangle$. Eq. (1.5.3), in general, connects the expectation value $\langle I \rangle$ with the concept of ensemble average. In certain observations, such as the measurement of a bright thermal source, a measurement may be approximated as if the measurement has taken into account all possible relative phases of the Fourier-modes that is randomly distributed between 0 to 2π. In this case, the measured instantaneous intensity could be indistinguishable from its expectation value.

We now introduce the concept of time averaged intensity $\langle I(t) \rangle_T$.

$$\langle I(t) \rangle_T \equiv \frac{1}{T} \int_{t-\frac{T}{2}}^{t+\frac{T}{2}} dt\, I(t) \tag{1.5.5}$$

where T is the integration period. What is the relationship between $\langle I(t) \rangle$ and $\langle I(t) \rangle_T$? For randomly radiated and randomly distributed thermal radiation, it is easy to show

$$\langle I(t) \rangle_{T \sim \infty} = \langle I(t) \rangle. \tag{1.5.6}$$

Since the ensemble averaged intensity equals to the time averaged intensity when $T \sim \infty$, randomly radiated and randomly distributed thermal radiation is considered as stationary and ergodic. In statistics, ergodic implies that the ensemble average is equivalent to the time average of a typical member of the ensemble; stationary implies that the ensemble averaged mean value is independent of time. Since time average and ensemble average are equivalent for randomly radiated thermal light, we may take a large number of $I(t_j)$, each at a different time t_j, to evaluate the statistical mean intensity of thermal light

$$\bar{I} = \frac{1}{N} \sum_{j}^{N} I(t_j). \tag{1.5.7}$$

It is obviously that $\bar{I} = \langle I(t) \rangle_T$ when $N \sim \infty$.

Eq. (1.5.6) is valid for randomly radiated thermal light. The situation is different in the measurement of coherent radiation. For instance, in a pulsed laser the coherent superposition of the cavity modes produces a wavepacket which is a function of t deterministically. The result of the time average will be different from that of the ensemble average.

We discuss two types of time average in the following.

(I) Unavoidable time average caused by the finite response time of the measurement device.

In a real measurement, the finite response time of the photodetector and the associated electronics may physically impose a time average on $I(\mathbf{r}, t)$. The resolving time, or response time, t_c, of a measurement device is usually much longer than a few cycles of the light wave. For example, a fast photodetector may have a response time on the order of nanoseconds, which differs from the femtosecond cycle period of a visible light wave by a factor of 10^6. The output current of a photodetector, $i(t)$, cannot follow any fast variations of the intensity beyond the time scale of t_c. Unavoidable time averaging within that time scale occurs during the detection process. Hence, a much longer time average other than the "cycle-average" is always present during an experimental measurement. The output current of the photodetector with finite response time t_c is thus

$$i(\mathbf{r}, \tilde{t}) \propto \langle I(\mathbf{r}, t) \rangle_{t_c} = \int_{t_c} dt\, | E(\mathbf{r}, t) |^2, \qquad (1.5.8)$$

where we have introduced a "mean" time $\tilde{t}$ (or "slow" time) in $i(\mathbf{r}, \tilde{t})$, which has a minimum basic time scale of t_c. Any meaningful physics we learn from the measurement cannot go beyond that time scale. This unavoidable resolving time limit, t_c, is usually referred to as the characteristic time of the measurement device. Considering a particular response function of the detection system, we usually write the time average as a convolution

$$i(\mathbf{r}, \tilde{t}) \propto \int dt\, | E(\mathbf{r}, t) |^2 \, \mathcal{D}(\tilde{t} - t). \qquad (1.5.9)$$

We have used a generic normalized function $\mathcal{D}(\tilde{t} - t)$ to simulate the response distribution function of the photodetector, where $\tilde{t}$ represents the mean time of a photodetection event. $\mathcal{D}(\tilde{t} - t)$ is usually taken to be a Gaussian. To simplify the mathematics, it is also common to use a rectangular-function, which turns Eq. (1.5.9) into Eq. (1.5.8). With either the use of a Gaussian or a rectangular-function, the result of the convolution of Eqs. (1.5.9) will be different for $t_c <$ pulse-width and for $t_c >$ pulse-width, if $I(\mathbf{r}, t)$ is a well-defined function of time, such as a Gaussian-like pulse. When $t_c \ll$ pulse-width, $\mathcal{D}(\tilde{t} - t)$ can be approximated as a δ-function. The output of the convolution will be the Gaussian-like pulse itself, perhaps broadened slightly. On the other hand, if $t_c \gg$ pulse-width, for instance in the measurement of femtosecond laser pulse, the pulse itself can be approximated as a δ-function. The output of the convolution will be the response function of the photodetector,

$$i(\tilde{t}) \propto \int dt\, \delta(t - t_0) \, \mathcal{D}(\tilde{t} - t) = \mathcal{D}(\tilde{t} - t_0). \qquad (1.5.10)$$

In either case, the convolution yields a time averaged function of $\tilde{t}$.

Assuming a rectangular temporal response function, following Eq. (1.5.8), the output photocurrent of a photodetector is formally calculated as follows:

$$i(r, \tilde{t}) \propto \langle I(r,t) \rangle_{t_c}$$

$$= \int_{t_c} dt \left[\int d\omega \, E^*(\omega) \, e^{i(\omega t - kr)} \int d\omega' \, E(\omega') e^{-i(\omega' t - k'r)} \right]$$

$$= \int_{t_c} dt \left[\int d\nu \, E^*(\nu) \, e^{i\nu\tau} \int d\nu' \, E(\nu') e^{-i\nu'\tau} \right]$$

$$= \int \int d\nu \, d\nu' \, E^*(\nu) \, E(\nu') \left[\int_{t_c} dt \, e^{i(\nu - \nu')\tau} \right]$$

$$= \int \int d\nu \, d\nu' \, E^*(\nu) \, E(\nu') \, \text{sinc} \, \frac{(\nu - \nu')t_c}{2} \, e^{i[(\nu - \nu')\tilde{\tau}]}. \tag{1.5.11}$$

Comparing with the instantaneous intensity $I(r,t)$, the output photocurrent $i(r,\tilde{t})$ has lost all of its beat frequencies that slow photodetectors may not be able to follow. The superposition of the surviving harmonic beats of $\omega - \omega' = \nu - \nu' < 2\pi/t_c$ yields a much smoother function for the measured $i(r, \tilde{t})$ than that of the instantaneous intensity $I(r,t)$. The response time, or characteristic time of the photodetector, t_c, is a critical physical parameter, which must be carefully chosen for certain experimental expectations. For instance, a nanosecond photodetector will never be able to resolve the temporal profile of a femtosecond pulse-like $I(r,t)$. This phenomenon can also be seen from another perspective: the sinc-function in Eq. (1.5.11), namely the non-causal impulse response of a low-pass filter, will narrow the bandwidth of the spectrum. When the original spectrum of the radiation, determined by the amplitude $E(\nu)$, is much wider than $2\pi/t_c$, the sinc-function, $\text{sinc} \, (\nu - \nu')t_c/2$, effectively narrow the spectrum of the radiation to $\Delta\omega = \Delta\nu < 2\pi/t_c$ and consequently broaden the temporal profile of the wavepacket.

Further assuming a measurement of wave packets with broad enough bandwidth of spectrum that can be approximated as a constant within the spectrum that a photodetector can respond to, the integral of Eq. (1.5.11) yields a rectangular low-pass filter function that is the Fourier transform of its non-causal impulse response sinc-function

$$i(r, \tilde{t}) \propto \langle I(r,t) \rangle_{t_c} \propto \mathcal{F}_\tau \left\{ \text{sinc} \, \frac{(\nu - \nu')t_c}{2} \right\} = \frac{1}{|t_c|} \, \text{rect} \frac{\tau}{\pi t_c} \tag{1.5.12}$$

where $\text{rect}(x)$ is a rectangular function

$$\text{rect}(x) = \begin{cases} 0 & |x| > 1/2 \\ 1/2 & |x| = 1/2 \\ 1 & |x| < 1/2. \end{cases}$$

It is interesting to find that although the measurement of a slow photodetector is on ultra-short pulsed wavepackets, its output current is a rectangular-like pulse with much broader temporal width.

In the early days, the most popular measurement device for light was the human eye, which has a response time of $\sim 1/15$ second. For a time average of $t_c \sim 1/15$ second, the sinc-function in Eq. (1.5.11) can be approximately treated as a delta function $\delta(\omega - \omega')$. We thus effectively have a time averaged intensity of constant, or a pulse with infinite temporal width, by taking $T \sim \infty$:

$$\langle I(r,t) \rangle_{T \sim \infty} = \int \int d\omega \, d\omega' \, E^*(\omega) \, E(\omega') \, \delta(\omega - \omega') \, e^{i[(\omega - \omega')\tilde{t} - (k - k')r]}$$

$$= \int_0^\infty d\omega \, |E(\omega)|^2. \tag{1.5.13}$$

This result is consistent with the Parseval theorem:

$$\int_{T\sim\infty} dt \, | \, E(\omega) \, |^2 = \int_0^\infty d\omega \, | \, E(\omega) \, |^2. \tag{1.5.14}$$

In the measurement of coherent light, especially when $\varphi_j - \varphi_k$ takes a constant value, the time average has a null effect on the second term of Eq. (1.4.4).

The unavoidable time average caused by the slow response time of a measurement device may yield the same constant value as the expectation value of thermal radiation, however, the physics behind these two types of "averaging" is very different. The time average is physically imposed by the slow time resolving ability of the measurement device. The constant obtained from the expectation operation is the result of a superposition, which may not be simply treated as an "averaging", especially in the case of coherent superposition. In a coherent superposition, the expectation calculation yields a well-defined pulse. The time average will broaden the pulse significantly if the response time of the photodetector is much greater than the pulse width, $t_c \gg \Delta t$. For instance, as we have mentioned earlier, a photodetector with nanosecond response time will broaden a femtosecond laser pulse to nanosecond in $i(\tilde{t})$. When $t_c \sim \infty$, the time average yields a constant photocurrent in any circumstances.

(II) Timely accumulative measurement.

Another type of time integral may apply if a measurement has to be taken accumulatively in time. The time integrated expectation value of intensity is defined as

$$Q(\mathbf{r}) = \int_{T_1}^{T_2} dt \, \langle \, I(\mathbf{r}, t) \, \rangle, \tag{1.5.15}$$

where we have assumed that the accumulative measurement starts from $t = T_1$ and ends at $t = T_2$. It is easy to see that $Q(\mathbf{r})$ will be a constant in time for the above simple measurement of intensity, if the accumulative time period is long enough to be treated as infinity, $T_2 - T_1 \sim \infty$, even if the expectation function is a well-defined pulse.

The time averaged intensity measured in time accumulative measurements is defined as follows

$$\langle \, I(\mathbf{r}) \, \rangle_T = \frac{1}{T_2 - T_1} \int_{T_1}^{T_2} dt \, I(\mathbf{r}, t). \tag{1.5.16}$$

It is easy to find that

$$\langle \, I(\mathbf{r}) \, \rangle_T = \frac{Q(\mathbf{r})}{T_2 - T_1}$$

when $T_2 - T_1 \sim \infty$.

1.6 MEASUREMENT OF INTENSITY: TEMPORAL FLUCTUATION AND SPATIAL FLUCTUATION

The physics of intensity fluctuation is a complex subject. Intensity fluctuation may come from the light source and may be caused by propagation. In Maxwell's continuum picture of natural light, an incomplete destructive interference between different Fourier modes may also cause intensity fluctuation,

$$\Delta I(r, t) = \int_{\omega \neq \omega'} d\omega \, d\omega' \, E^*(\omega) E(\omega') \, e^{i[(\omega-\omega')t - (k-k')r]}$$

$$= \int_{\omega \neq \omega'} d\omega \, d\omega' \, E^*(\omega) E(\omega') \, e^{i(\omega-\omega')\tau}.$$

Even if a measurement deals with strong light and guarantees a complete destructive interference between different Fourier modes, the measured mean intensity, corresponding to its expectation value, may also fluctuate from time to time and from point to point. Dealing with intensity fluctuations, we may need to specify the type of fluctuations we are discussing. In this book we follow the tradition to consider intensity fluctuation *nondeterministic* quantity whether in terms of temporal variable or spatial variable. Any deterministic temporal and spatial intensity variations, such as laser pules (temporal) and "manmade speckles" (spatial), will be treated differently.

Despite all these complications, the intensity fluctuation of the radiation is measurable. For instance, we may setup a point-like photodetector at a chosen coordinate $\mathbf{r}$ to monitor the intensity $I(\mathbf{r}, t)$ of a natural light from time to time. The measured mean intensity mean at coordinate $\mathbf{r}$ is calculated from

$$\bar{I}(\mathbf{r}) = \frac{1}{N} \sum_j I(\mathbf{r}, t_j) = \frac{1}{T} \int_T dt \, I(\mathbf{r}, t) = \langle I(\mathbf{r}, t) \rangle_T. \qquad (1.6.1)$$

The temporal intensity fluctuation, i.e., the nondeterministic random intensity fluctuation from time to time, is usually defined as

$$\Delta I(\mathbf{r}, t) = I(\mathbf{r}, t) - \langle I(\mathbf{r}, t) \rangle_T, \qquad (1.6.2)$$

indicating the temporal variations or fluctuations of the intensity of a chosen coordinate $\mathbf{r}$ from its time averaged value. $\Delta I(\mathbf{r}, t)$ may take a positive or negative value from time to time. The courses of the intensity fluctuation could be completed. The incomplete destructive interference between different Fourier modes will definitely produce intensity fluctuations.

Similarly, we may set up a 2-D array of Charged Couple Device (CCD) or Complementary Metal-Oxide Semiconductor (CMOS) in the transverse plane of a light beam to monitor the intensity distribution on an observation plane of $z = z_o$, i.e., the transverse spatial distribution function $I(\vec{\rho}, z = z_o, t)$, where $\vec{\rho}$ is the transverse coordinate vector. The concept of "spatial fluctuation" usually refers to an nondeterministic random transverse spatial distribution of intensity $I(\vec{\rho}, z = z_o, t)$. Spatial fluctuation of intensity is usually defined in terms of the variations of the transverse spatial distribution of the intensity comparing with its spatially-averaged value

$$\Delta I(\vec{\rho}, t) = I(\vec{\rho}, t) - \frac{1}{\sigma} \int_\sigma d\vec{\rho} \, I(\vec{\rho}, t)$$
$$= I(\vec{\rho}, t) - \langle I(\vec{\rho}, t) \rangle_\sigma, \qquad (1.6.3)$$

where σ is the transverse area of the integral (average). $\Delta I(\vec{\rho}, t)$ may take a positive or negative value from speckle to speckle. In a realistic measurement of $\Delta I(\vec{\rho}, t)$, a time average is also involved due to the slow measurement devices, such as a CCD or CMOS,

$$\Delta I(\vec{\rho}, \tilde{t}) = \frac{1}{\Delta t_c} \int_{\tilde{t} - \Delta t_c/2}^{\tilde{t} + \Delta t_c/2} dt \left[I(\vec{\rho}, t) - \frac{1}{\sigma} \int_\sigma d\vec{\rho} \, I(\vec{\rho}, t) \right], \qquad (1.6.4)$$

where Δt_c is the characteristic time of the measurement device, and $\tilde{t}$ is defined as the effective detection time which has a minimum unit of Δt_c. If the spatial fluctuation of intensity is "pictured" by a CCD or CMOS, Δt_c is usually determined by the frame rate of the CCD or CMOS.

1.7 BLACKBODY RADIATION UNDER MAXWELL'S CONTINUUM ELECTRODYNAMICS

In Maxwell's continuum picture of light, a natural light source, such as the sun or a distant star, emits EM waves in a continuous manner. The light source may emit any amount of energy in the form of EM waves. There exists neither a minimum nor a maximum restriction. Although Maxwell's continuum picture of light does not have any "quantization" related concepts, it received great success almost immediately after the discovery of Maxwell. Especially, the theory of classical superposition, which is a mathematical consequence of the Maxwell linear differential wave equation, is able to provide accurate predictions and reasonable explanations to all optical coherence phenomena interested in the time of Maxwell.

Since the discovery of light bulb, searching for long lasting materials with higher emitting efficiency in the visible spectrum became interesting and useful. A typical research of this kind involves the measurement on "blackbody radiation". A blackbody is an idealized object which absorbs and emits all frequencies, ν, of light. The best approximation to a blackbody radiation is the radiation coming from a tiny pinhole in the wall of a hollow enclosure, or cavity that is in thermal equilibrium with a heat reservoir at temperature T. The hollow enclosure is usually made by a chosen metal or other material. The intensity, $I(\nu)$, of radiation per unit solid angle, coming from the pinhole, in the frequency range between ν and $\nu + d\nu$ can be accurately measured to compare with the prediction of Maxwell's continuum theory of light. The following analysis on blackbody radiation follows the historical treatment of Rayleigh-Jean.

(1) Number of modes

Suppose we have a rectangular cavity of dimension L_x, L_y, and L_z. Applying the boundary condition required by electromagnetic theory, the allowed wavevector $\mathbf{k}$ is thus:

$$k_x = \frac{2\pi l}{L_x} \qquad k_y = \frac{2\pi m}{L_y} \qquad k_z = \frac{2\pi n}{L_z} \tag{1.7.1}$$

where l, m, and n are integers taking values from $-\infty$ to ∞. The number of allowed modes is therefore

$$\Delta N = \Delta l \, \Delta m \, \Delta n = \frac{V}{(2\pi)^3} \, dk_x \, dk_y \, dk_z, \tag{1.7.2}$$

where $V = L_x L_y L_z$. To estimate the number of permissible modes in the frequency range between ν and $\nu + d\nu$, it is more convenient to adopt polar coordinates in $\mathbf{k}$ space by considering the volume element $dk_x \, dk_y \, dk_z = k^2 \, dk \, d\Omega$, where $d\Omega$ is the element of solid angle. The number of permissible modes in the frequency range between ν and $\nu + d\nu$ is thus

$$\Delta N = \frac{4\pi V}{(2\pi)^3} \, k^2 \, dk = \frac{4\pi V}{c^3} \, \nu^2 \, d\nu$$

where we have integrated over $d\Omega$, since we are not interested in the direction of the wavevector $\mathbf{k}$. Taking into consideration the polarization for each mode, the number of permissible modes in the frequency range between ν and $\nu + d\nu$ will be doubled,

$$\Delta N = \frac{8\pi V}{c^3} \, \nu^2 \, d\nu. \tag{1.7.3}$$

(2) Mean energy of each mode

To calculate the mean energy of each mode, we will adopt the results of classical statistical mechanics. We consider each mode of radiation to be in thermodynamic equilibrium with the walls of the enclosure which is treated as a heat reservoir. A heat reservoir is defined as a very large system with constant temperature. Physically, this means that the temperature of the enclosure remains unaffected by whatever small amount of energy it gives to the radiation mode. Under the condition of equilibrium, the probability of finding the radiation mode between E and $E + dE$ follows the canonical distribution:

$$P(E)\, dE = \frac{e^{-E/kT}\, dE}{\int_0^\infty e^{-E/kT}\, dE}. \tag{1.7.4}$$

The mean value of the energy $\bar{E}$ is calculated by weighting each possible energy according to its probability:

$$\bar{E} = \frac{\int_0^\infty E\, e^{-E/kT}\, dE}{\int_0^\infty e^{-E/kT}\, dE} = kT \frac{\int_0^\infty \epsilon\, e^{-\epsilon}\, d\epsilon}{\int_0^\infty e^{-\epsilon}\, d\epsilon} = kT \tag{1.7.5}$$

where $\epsilon = E/kT$. In Eq. (1.7.5), we have used the solution of the Γ-function, $\Gamma(n+1) = n!$ and $\Gamma(1) = 1$.

The mean energy of each radiation mode is kT. This is a good example of the theorem of equipartition of energy. The amount of radiation energy in the frequency range between ν and $\nu + d\nu$ is thus

$$u(\nu)\, d\nu = \bar{E} \Delta N = \frac{8\pi V}{c^3} kT \nu^2\, d\nu \tag{1.7.6}$$

which is called the Rayleigh-Jean's law.

As we know, however, the blackbody radiation measurements disagree with the Rayleigh-Jeans law. We have to face the truth and accept that Maxwell's continuum electrodynamic approach may not be adequate for blackbody radiation.

SUMMARY

In this chapter we introduced the following theory and concepts:

(1) The Maxwell's equations and the electromagnetic wave equation: the foundation of the classical EM wave theory of light.

(2) The classical superposition and the Fourier transform of classical EM field.

(3) The measurement of light: the Poynting vector and the concept of intensity; the expectation value and fluctuations of intensity; the expectation value and the ensemble measurement; the time averaged intensity,

(4) We have built a simple classical model of light with multi-Fourier-modes,

$$\langle I(r,t)\rangle = \Big\langle \int d\omega\, E^*(\omega)\, e^{i(\omega t - kr)} \int d\omega'\, E(\omega')\, e^{-i(\omega' t - k'r)} \Big\rangle$$

$$= \Big\langle \int_{\omega=\omega'} d\omega |E(\omega)|^2 \Big\rangle$$

$$+ \Big\langle \int_{\omega\neq\omega'} d\omega\, d\omega'\, E^*(\omega) E(\omega')\, e^{i[(\omega-\omega')t - (k-k')r]} \Big\rangle$$

(5) The expectation value or expectation function of intensity $\langle I(\mathbf{r},t)\rangle$ is defined as

$$\langle I(\mathbf{r},t)\rangle = \langle E^*(\mathbf{r},t) E(\mathbf{r},t)\rangle$$

by means of *taking into account all possible realizations of the field.*

(6) The mean value of the measured intensity $\bar{I}(\mathbf{r}, t)$ based on a set of measurements is defined as

$$\bar{I}(\mathbf{r}, t) = \frac{1}{N} \sum_{j=1}^{N} I_j(\mathbf{r}, t),$$

where $I_j(\mathbf{r}, t)$ is the measured intensity of the jth measurement.

(7) The time averaged intensity $\langle I(\mathbf{r}, t) \rangle_T$ is defined as

$$\langle I(\mathbf{r}, t) \rangle_T = \frac{1}{T} \int_{t-\frac{T}{2}}^{t+\frac{T}{2}} dt\, I(\mathbf{r}, t) \equiv \int_T dt\, I(\mathbf{r}, t),$$

where $I(\mathbf{r}, t)$ is the measured intensity at time t.

(8) The intensity fluctuation (introduced in four different ways so far):

  (a) Two-mode interference induced intensity fluctuation at space-time $(\mathbf{r}, t)$;

  (b) Intensity fluctuation from measurement to measurement;

  (c) Intensity fluctuation from time to time (temporal fluctuation);

  (d) Intensity fluctuation from point to point (spatial fluctuation).

We especially labeled two-mode interference induced intensity fluctuation to distinguish it from all other types of intensity fluctuations.

(9) Two additional points about intensity fluctuation:

  (a) A well-defined wavepacket or pulse that is a function of time t, such as a Gaussian, cannot be considered as intensity fluctuation. The intensity fluctuations should be statistically random and nondeterministic.

  (b) Even if the instantaneous intensity $I(\mathbf{r}, t)$ is observed to be $I(\mathbf{r}, t) = I_0 + I \cos \omega \tau$, the deterministic modulation $I \cos \omega \tau$ cannot be treated as intensity fluctuation, although the measured intensity sinusoidally "fluctuates" in the neighborhood of a constant I_0.

REFERENCES AND SUGGESTIONS FOR READING

[1] J.D. Jackson, *Classical Electrodynamics*, John Wiley & Sons, New York, 1998.

[2] M. Born and E. Wolf, *Principle of Optics*, Cambridge, 2002.

[3] J.W.S. Rayleigh, Phil. Mag. **49**, 539 (1900); Nature, **72**, 54, 243 (1905).

[4] J.H. Jeans, Phil. Mag. **10**, 91 (1905).

CHAPTER 2

Quantum Theory of Light: Field Quantization and Photon

The quantum theory of light began in the early 20th century. The successful introduction of the concept of photon, or light quantum, stimulated a new foundation of physics, namely, the quantum theory. Today, quantum theory has turned out to be the overarching principle of modern physics. It would be difficult to find a single subject among the physical sciences which is not affected in its foundations or in its applications by quantum theory.

After 100 years of studies, how much do we know about the photon? The photon is a wave: it has no mass, it travels at the highest speed in the universe, and it interferes with itself. The photon is a particle: it has well-defined values of momentum and energy, and it even "spins" like a particle. The photon is neither a wave nor a particle, because whichever we think it is, we would be tripped into difficulties in explaining the other part of its behavior. The photon is a wave-like-particle and/or a particle-like-wave: a photon can never be divided into parts, but interference of a single photon can be easily observed by modern technologies. It seems that a photon passes both paths of an interferometer when interference patterns are observed; however, if the interferometer is set in such a way that its two paths are "distinguishable", the photon "knows" which path to follow and never passes through both paths. Apparently, a photon has to make a choice in its behavior when facing an interferometer: a choice of "both-path" like a wave or "which-path" like a particle. Surprisingly, the choice is not necessary before passing through the interferometer. It has been experimentally demonstrated that the choice of "which-path" and/or "both-path" can be delayed until after the photon has passed through the interferometer. More surprisingly, the which-path information can even be "erased" after the annihilation of the photon itself. The behavior of a photon apparently does not follow any of the basic criterion: reality, causality, and locality, of our everyday life. Of course, the peculiarity of wave-particle duality is not only the property of photons, it belongs to all quanta in the quantum world. Perhaps, it is easy to accept the particle picture of an electron with mass, m_e, and charge, e; it is definitely not easy to accept the particle nature of a photon. On the other hand, perhaps, it is easy to accept the wave picture of a photon with frequency, ω, and wavevector, $\mathbf{k}$; it is definitely not easy to accept the wave nature of an electron.

Although questions regarding the fundamental issues about the concept of a photon still exist, the quantum theory of light has contributed perhaps the most influential and successful, yet controversial part to quantum mechanics.

In this chapter, we will constrain ourselves to the following basic questions about the quantum theory of light: (1) why quantization of light radiation is necessary; (2) how to quantize the radiation field; (3) how to describe the state of the quantized field; and (4)

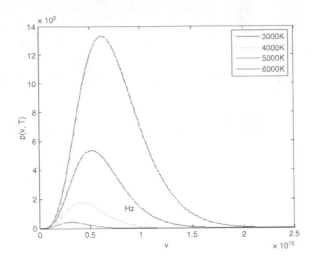

FIGURE 2.1.1 Blackbody radiation curves. $I(\nu)$ depends only on the temperature of the walls of the enclosure. At a particular temperature, $I(\nu) \propto \nu^2$ for low frequencies, while at high frequencies $I(\nu)$ drops off exponentially. Another interesting feature of the spectrum is that the maximum $I(\nu)$ is shifted toward higher frequencies while the temperature of the blackbody is raised.

how to physically model and mathematically formulate a photodetection event or a joint photodetection event.

2.1   THE EXPERIMENTAL FOUNDATION—I: BLACKBODY RADIATION

Around the year of 1900, an unexpected observation happened in experimental physics. It was found that the experimentally observed spectral distribution of blackbody radiation disagreed with the theoretical predications of classical physics.

We have briefly discussed the classical treatment of blackbody radiation in Chapter 1. A "blackbody" is a perfect absorber that absorbs all incident radiation. The best approximation to a blackbody is a tiny pinhole in the wall of a hollow enclosure, or cavity. The intensity, $I(\nu)$, of radiation per unit solid angle, coming from the hole, in the frequency range between ν and $\nu + d\nu$ can be accurately measured. It was found experimentally, under the condition of thermodynamic equilibrium, that any blackbody has the same characteristic emission function $I(\nu)$. Typical curves of $I(\nu)$ are shown in Fig. 2.1.1. $I(\nu)$ depends only on the temperature of the walls of the enclosure, and not on the material of which the enclosure is made nor on the shape of the cavity. At a particular temperature, $I(\nu) \propto \nu^2$ for low frequencies, while at high frequencies $I(\nu)$ drops off exponentially. Another interesting feature of the spectrum distribution is that the maximum $I(\nu)$ is shifted toward higher frequencies while the temperature of the blackbody is raised. The four curves in Fig. 2.1.1 clearly indicate these characteristics of the blackbody radiation.

It was indeed a surprise at that time that the theory of classical mechanics, electrodynamics, and thermodynamics all together failed to explain this simple observation.

In the frame work of classical electrodynamics, the intensity of the blackbody radiation per unit solid angle is related to the energy density of the radiation in the cavity, or the enclosure:

$$I(\nu) = \frac{c}{4\pi}\, u(\nu) \tag{2.1.1}$$

where c is the speed of light and $u(\nu)$ is the energy density of frequency ν. Therefore, the measured intensity of blackbody radiation $I(\nu)$ is determined by $u(\nu)$ of the cavity. The source of the radiation energy in the cavity is obviously the walls of the enclosure, which continually emit waves of every possible frequency and wave vector, or say all possible modes. In thermodynamic equilibrium, the amount of energy $u(\nu)d\nu$, in the frequency range between ν and $\nu + d\nu$ is easily calculated. What we need is to (1) estimate the number of permissible modes under the boundary condition of the cavity and (2) calculate the mean energy of the permissible modes.

In Chapter 1, we have given a simple analysis of blackbody radiation by following the historical treatment of Rayleigh-Jeans. The amount of radiation energy in the frequency range between ν and $\nu + d\nu$ is thus

$$u(\nu)\, d\nu = \bar{E}\Delta N = \frac{8\pi V}{c^3}\, kT\, \nu^2\, d\nu \tag{2.1.2}$$

which is called the Rayleigh-Jean's law. Comparing with the blackbody radiation curves shown in Fig. (2.1.1), Rayleigh-Jean's law only agrees with the experimental observations at low frequencies; it gives too much power of radiation for high frequencies. In addition, if we integrate over all frequencies to calculate the total energy, the result diverges, meaning an infinite amount of energy contained in the cavity.

Wien attempted a different classical approach. Based on classical thermodynamic arguments, Wien showed that the blackbody radiation distribution must be of the form $u(\nu) = \nu^3 f(\nu/T)$. The function $f(\nu/T)$, however, cannot be determined from thermodynamics alone. Wien obtained a distribution function that was later named as Wien's law:

$$u(\nu)\, d\nu \sim \nu^3\, e^{-h\nu/kT}\, d\nu \tag{2.1.3}$$

where h is a constant determined experimentally by data fitting. This constant later turned out to be the symbolic constant of quantum theory and was named Planck constant. Wien's distribution improved the high frequency spectrum fitting, but got worse at low frequencies.

In history, all classical attempts, whether the electrodynamic approaches or the thermodynamic treatments failed to give an accurate distribution function to fit the observation curves of blackbody radiation, within experimental error. We thus conclude the concepts we have used to derive these laws, or distributions, may not be adequate to describe the behavior of blackbody radiation.

In the year 1900, Planck decided to abandon the classical tradition and in doing so he succeed in fitting the experimentally measured blackbody radiation spectrum. Planck's hypothesis was very simple. He assumed that a radiation mode can only take energy values of an integer multiple of a basic unit, $E = nh\nu$, where n is an integer running from 0 to ∞. This basic unit $h\nu$ is not the same for all modes, but rather is proportional to the frequency of the mode. With this assumption, phenomenologically, Planck explained the blackbody radiation by accurately fitting the experimentally measured distribution curves, within experimental error.

Planck's assumption is truly inconsistent with classical concepts. According to classical mechanics and classical electrodynamics, there are no restrictions on the energy of a radiation mode. The only "restriction" regarding the energy of a radiation mode is the mean value, kT, which is independent of the frequency of the radiation. Planck's assumption also seems inconsistent with many of our everyday experiences. For instance, the output power of an AM or FM radio oscillator may have any value. There is no experimental evidence that the energy of a radio oscillator must be quantized to $E = nh\nu$. Does it mean Planck's theory is inadequate to describe the behavior of radio waves? The answer is NO. The apparent inconsistency arises from the fact that h is a very small quantity, $h \sim 6.6 \times 10^{-34}$ joule-second. In the AM and FM radio frequencies, for example $\nu \sim 10^6$ Hz and $\nu \sim 10^8$ Hz, the basic unit of energy $h\nu$ is on the order of 6.6×10^{-28} joule and 6.6×10^{-26} joule which are not detectable by any available sensitive detection apparatus. With light waves, however, the values of $h\nu$ increase significantly, for $\nu \sim 10^{15}$ Hz, $h\nu \sim 10^{-19}$ joule. This value is measurable by modern measurement devices. Therefore, as we go to higher frequencies, Planck's quantization hypothesis is easier to verify.

We now derive the spectrum distribution function for blackbody radiation by following Planck's energy quantization. Similar to what we did in the early classical analysis, we recalculate the mean energy per mode by applying the canonical distribution. The probability for a mode to be in a given energy $E_n = nh\nu$ is then,

$$P(n) \propto e^{-E_n/kT} = e^{-nh\nu/kT} \tag{2.1.4}$$

Normalizing Eq. (2.1.4), we obtain

$$P(n) = \frac{e^{-nh\nu/kT}}{\sum_{n=0}^{\infty} e^{-nh\nu/kT}} = \frac{e^{-nh\nu/kT}}{1 - e^{-h\nu/kT}}. \tag{2.1.5}$$

The mean energy per mode is then

$$\bar{E} = \sum_{n=0}^{\infty} E_n P(n) = (1 - e^{-h\nu/kT})^{-1} \sum_{n=0}^{\infty} nh\nu \, e^{-nh\nu/kT}$$

$$= h\nu \, (1 - e^{-h\nu/kT})^{-1} \sum_{n=0}^{\infty} n \, e^{-nh\nu/kT}. \tag{2.1.6}$$

To evaluate Eq. (2.1.6), we can write

$$\sum_{n=0}^{\infty} n \, e^{-nh\nu/kT} = -\frac{d}{d\alpha} \sum_{n=0}^{\infty} e^{-n\alpha} = -\frac{d}{d\alpha} \left(\frac{1}{1 - e^{-\alpha}} \right) = \frac{e^{-\alpha}}{(1 - e^{-\alpha})^2}$$

where $\alpha = h\nu/kT$. The mean energy per mode, $\bar{E}$, is then

$$\bar{E} = \frac{h\nu \, e^{-h\nu/kT}}{1 - e^{-h\nu/kT}}. \tag{2.1.7}$$

The amount of radiation energy in the frequency range between ν and $\nu + d\nu$ is found to be

$$u(\nu) \, d\nu = \bar{E}\Delta N = \frac{8\pi V}{c^3} \, h\nu^3 \, \frac{e^{-h\nu/kT}}{1 - e^{-h\nu/kT}} \, d\nu.$$

We thus obtain the Planck distribution

$$u(\nu) = \frac{8\pi V}{c^3} \, \frac{h\nu^3 \, e^{-h\nu/kT}}{1 - e^{-h\nu/kT}} \tag{2.1.8}$$

which is an exact fit, within experimental error, to the distribution of blackbody radiation.

2.2   THE EXPERIMENTAL FOUNDATION—II: PHOTOELECTRIC EFFECT

The study of blackbody radiation concluded indirectly that electromagnetic waves may increase or decrease energy only in the units of $h\nu$. The discovery of the photoelectric effect confirms this surprising conclusion in a more direct way. In fact, the photoelectric effect was first reported by Hertz in 1887, more than ten years prior to Planck's work. The quantum explanation of the effect was given later by Einstein in 1905 as a result of

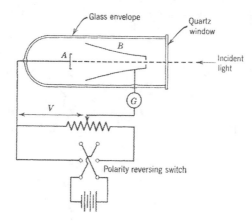

FIGURE 2.2.1 Typical schematic experimental setup for observing the photoelectric effect.

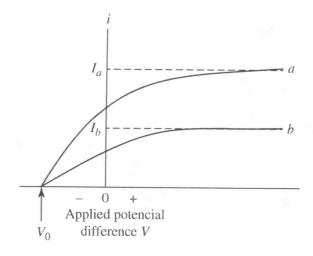

FIGURE 2.2.2 Typical measured photoelectric current, i, as a function of the applied voltage, V, between the metal plate, A, and the anode, B. Note, the voltage can be switched to negative or positive. The two curves, a and b, correspond to two different incident light intensities: $I_a = 2I_b$. It is a surprise to find that V_0 is independent of the intensity of the incident light.

five years of thinking about Planck's hypothesis. Figure 2.2.1 shows a typical experimental setup for observing the photoelectric effect. A simple vacuum tube, containing a metal plate, A, and an anode, B, is used for the experimental observation. Monochromatic light is incident through the quartz window of the vacuum tube on the metal A. The photoelectrons liberated from the surface of the metal A are collected by the anode B. An adjustable potential difference V is applied between the metal plate A and the anode B. The output photocurrent of the anode is monitored by a sensitive ammeter G.

Figure 2.2.2 is a typical observation of the photoelectric current, i, as a function of the applied potential difference, V. When V is positive and takes large enough values, the photocurrent i reaches a saturated value, which means all liberated photoelectrons are collected by the anode A. The saturated value of the photoelectric current i is proportional to the intensity of the incident light. This result is reasonable because a large intensity should indeed eject more photoelectrons. The surprise, however, comes when V is reversed, making

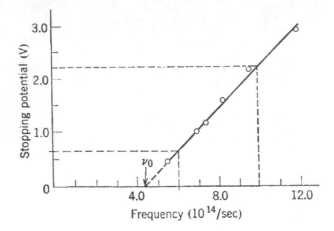

FIGURE 2.2.3 Typical measurement of V_0 as a linear function of the frequency of the incident light. The slope of the experimental curve is on the order of 3.9×10^{-15} V-sec. There exists, for each different metal plate, a characteristic cutoff frequency ν_0. For any frequency less than ν_0, photoelectric effect stops occurring, no matter how intense the incident light.

it negative, and adjusted to reach the stopping potential V_0 where the photoelectric current drops to zero, $i \sim 0$. It was found that V_0 is independent of the intensity of the incident light, as shown in Fig. 2.2.2. When a negative potential, V, is applied, the photoelectric current does not immediately drop to zero. This suggests the electron escapes from the surface of the metal with a certain kinetic energy. Some of the escaping electrons can still reach the anode, B, if their kinetic energies are large enough, $K_{max} > eV_0$, to overcome the applied electric potential against their motion. If the negative potential V is made large enough to be equal or greater than the maximum kinetic energy of the escaping electrons, $eV_0 \geq K_{max}$, no electrons can reach the anode, B, and consequently the photoelectric current i drops to zero. It is a surprise from the classical point of view that the stopping potential V_0 and consequently the kinetic energy of a liberated electron does not dependent on the intensity of the incident light. In classical electromagnetic wave theory, $I \propto |E|^2$. Since the force applied to an electron is eE, the kinetic energy of the photoelectron should increase as the intensity of the incident light increases.

More surprises came later from Millikan's work. Millikan's experiment showed that the stopping potential, V_0, is linearly dependent on the frequency of the incident light and there exists, for each different metal plate, a characteristic cutoff frequency ν_0. For any frequency lower than ν_0, photoelectric effect stops occurring, no matter how intense the incident light. Figure 2.2.3 shows a typical measurement of V_0 as a linear function of the incident light frequency. The slope of the experimental curve is on the order of 3.9×10^{-15} V-sec. The existence of the cutoff frequency, ν_0, is inconsistent with classical concepts of the electromagnetic wave theory. According to classical electrodynamics, the photoelectric effect should be observable for any frequency, provided the incident light is intense enough to give the necessary amount of kinetic energy to the photoelectron.

In 1905, Einstein proposed a theory which successfully explained the photoelectric effect. In his theory, Einstein quantized the radiation energy into localized "bundles", $E = h\nu$, which were later named "photons" (1926). Einstein assumed that one photon, individually, is completely absorbed by one excited electron in the process of a photoelectron ejection. When the electron is ejected from the surface of the metal, its kinetic energy is given by

$$K = h\nu - W \tag{2.2.1}$$

where W is the the work required to overcome the attractive forces of the atoms that bind the electron to the metal. W is called the work function.

How does Einstein's hypothesis explain the photoelectric effect? This question is assigned as an excise at the end of the chapter. Here, we provide a very brief discussion regarding the cutoff frequency, ν_0, which is the observation that conflicts the most with classical theory. Although Einstein's theory was published before Millkan's experiment, it provided a perfect explanation to Millkan's cutoff frequency ν_0. When the kinetic energy of the escaped electron equals zero, we have $h\nu_0 = W$, which asserts that a photon of frequency ν_0 has just enough energy to overcome the work function. If the frequency of light is reduced below ν_0, a photon will not have enough energy, individually, to eject a photoelectron, no matter how many photons are incident on the surface of the metal.

Einstein was the first physicist to relate the photoelectric effect with Plank's hypothesis. Einstein's Eq. (2.2.1) can be rewritten as

$$V_0 = \frac{h}{e}\nu - \frac{W}{e}, \qquad (2.2.2)$$

where we have substituted eV_0 for K_{max}, and V_0 is the applied stopping potential at which the photoelectric current drops to zero. Eq. (2.2.2) indicates a linear relationship between the stopping potential V_0 and the frequency of the incident light, in agreement with Millikan's experimental results, see Fig. 2.2.3. The measured slop of the experimental curve in Millkan's experiment is

$$\frac{h}{e} \sim \frac{2.20V - 0.65V}{(10.0 - 6.0) \times 10^{14}/\text{sec}} \simeq 3.9 \times 10^{-15} V\text{-sec}.$$

Multiplying the measured slope by the electronic charge, e, yields $h \sim 6.2 \times 10^{-34}$ joule-sec which is close to the value $h \sim 6.6 \times 10^{-34}$ joule-sec, appearing in Plank's distribution function of 1900. Later, more accurate photoelectric experiments measured $h \sim 6.6262 \times 10^{-34}$ joule-sec. The agreement between the two constants, h, appearing in the photoelectric experiments and the blackbody radiation observation is a strong confirmation that h is a universal constant and the radiation field can exchange energy only in units of $h\nu$.

2.3 EINSTEIN'S GRANULARITY PICTURE OF LIGHT

After five years of analyzing about the blackbody radiation hypothesis of Planck and the experimental studies of Hertz on the photoelectric effect, in 1905, Einstein introduced a granularity to radiation, abandoning the continuum interpretation of Maxwell. This led to a microscopic picture of radiation and a statistical view of light. Although Einstein did not name his "bundle of ray", which is an English translation from German "strahlenbündel", a photon, at the heart of Einstein's theory is the particle picture of radiation. Einstein assumed that the radiation energy is quantized into localized bundles as light quanta. The energy of the bundle or light quantum is initially localized in a small volume of space and remains localized as it moves away from the radiation source with velocity c. The energy of the bundle or light quantum is related to its frequency by multiplication of a universal constant h. In the photoelectric process, one bundle of energy $E = h\nu$, or one photon, is completely absorbed by one electron originally bound with the metal. How can one bounded electron, which is a particle localized within a very small volume, completely absorb a photon to become an photoelectron? The simplest physical picture is that the particle-like photon transferred all its energy and momentum to the electron during a *collision*. Is the photon a localized object? Yes and no. When facing its particle-like behavior, it must be localized. When facing its wave-like behavior, it cannot be localized. The biggest confusion, perhaps, happens when the photon is treated as a particle in the interpretation, but is treated as a wave in the calculation.

Standing on the shoulders of giants, now we have a better picture on Einstein's concept of granularity. The following picture may not be exactly the same as the original idea of Einstein, we have enough reason to name it Einstein's picture of light.

In Einstein's picture, a natural light source, such as the sun or a distant star, consists of many point-like sub-sources ("atomic transitions" in modern language), each of which emits their own subfields (originally labeled in German by "strahlenbündel", translated to "bundle of ray" in English, and now named as "photon" in modern language) that carries energy $E = h\nu$ in a random manner: the mth subfield (photon) that is emitted from the mth point-like sub-source (an atomic transition in which the atom changes its state from higher energy level E_2 to lower energy level E_1, and releases a photon with energy $h\nu = E_2 - E_1$) may propagate in all possible directions with all possible random phase. Einstein's concept of "bundle of ray" or "subfield" refers a quantized microscopic realistic substance of electromagnetic field, corresponding to the quantum mechanical concept of photon, precisely, it has the same mathematical expression as the effective wavefunction of photon in thermal state. Later in this book we will introduce a quantum mechanical concept of "effective wavefunction" from the quantum theory of optical coherence. It is interesting to find that the effective wavefunction of a photon plays the same role as Einstein's subfield; and the effective wavefunction of a photon in the thermal state is mathematically the same function of space-time as that of Einstein's subfield of natural light. It should be emphasized that the physical means of the two are fundamentally different: Einstein's subfield $E_m(\mathbf{r}, t)$ is a quantized microscopic realistic entity of electromagnetic wave propagating in space-time. The quantum mechanical effective wavefunction is the probability amplitude for a photon to produce a photodetection event at space-time coordinate $(\mathbf{r}, t)$.

In Einstein's picture, the radiation measured at coordinate $(\mathbf{r}, t)$ is the result of a super-position among a large number of subfields. We need to modify Eq.(1.2.3) accordingly:

$$
\begin{aligned}
\mathbf{E}(\mathbf{r}, t) &= \sum_{m=1}^{M} \int d\mathbf{k}\, \hat{\mathbf{e}}_{m,\mathbf{k}}\, E_m(\mathbf{k})\, e^{-i[\omega(t-t_m) - \mathbf{k}\cdot(\mathbf{r}-\mathbf{r}_m)]} \\
&= \sum_{m=1}^{M} \mathbf{E}_m(\mathbf{r}, t)
\end{aligned}
\tag{2.3.1}
$$

where we have treated mode $\mathbf{k}$ continuous and the sub-sources and the subfield discontinu-ous, and the subscript index m, $m = 1, 2, ..., M$, labels the mth subfield that is created from the mth point-like sub-source (atomic transition) at space-time coordinate $(\mathbf{r}_m, t_m)$ with energy $\hbar\omega = E_2 - E_1$; M is the total number of subfields contributed to the superposition; $E_m(\mathbf{k}) = a_m(\mathbf{k})e^{\varphi_m(\mathbf{k})}$ is the complex amplitude of the $\mathbf{k}$-mode of the mth subfield, $\mathbf{E}_m(\mathbf{r}, t)$ is the observed mth subfield at space-time $(\mathbf{r}, t)$. In the following discussion, we will simplify the expression of sum from $\sum_{m=1}^{M}$ to $\sum_m$. In Eq.(2.3.1), we have considered the energy-time uncertainty of the atomic transition $\Delta E \Delta t \geq h$: when the life time of $\Delta t \neq \infty$, $\Delta E_2 \neq 0$. The mth subfield emitted from the mth atomic transition may carry its energy with any value between $\hbar\omega = (E_2 - \Delta E_2/2) - E_1$ and $\hbar\omega = (E_2 + \Delta E_2/2) - E_1$, where we have chosen E_1 the ground level. In Eq.(2.3.1), we have also considered the momentum-position uncertainty: when the mth subfield is emitted from a point-like sub-source, it may propa-gate to any direction within $\Omega = 4\pi$ solid angle. The uncertainty in energy and momentum restrict the integrals of $d\omega$ (frequency) and $d\Omega$ (direction) and thus $d\mathbf{k}$.

Einstein's picture of light introduced an additional freedom of superposition: the ob-served field at a space-time point is not only the result of the superposition among a large number of coherent or incoherent harmonic modes but also the result of the superposition among a large number of coherent or incoherent subfields created from a large number of point-like sub-sources. For a thermal radiation source, such as the sun, it is reasonable to consider incoherent superposition between different subfields $m \neq n$ due to the stochastic nature of the creation process of the photons. It is also reasonable to assume coherent superposition among the continuous harmonic modes within a subfield, the harmonic modes excited from the mth atomic transition must have the same initial phase φ_m and the same emission time t_m with a certain uncertainty.

In terms of the stochastically created subfields $E_m(\mathbf{r}_m, t_m)$ at their creation space-time points $(\mathbf{r}_m, t_m)$, we may relate the observed field $E(\mathbf{r}, t)$ at their observation space-time point $(\mathbf{r}, t)$ with $E_m(\mathbf{r}_m, t_m)$ by considering a free propagation:

$$E(\mathbf{r}, t) = \sum_m \int d\mathbf{k}\, \hat{\mathbf{e}}_{m,\mathbf{k}} \frac{E_m(\mathbf{k}; \mathbf{r}_m, t_m)}{|\mathbf{r} - \mathbf{r}_m|} e^{-i[\omega(t-t_m) - \mathbf{k}\cdot(\mathbf{r}-\mathbf{r}_m)]}$$

$$= \sum_m \int d\mathbf{k}\, \hat{\mathbf{e}}_{m,\mathbf{k}}\, E_m(\mathbf{k}; \mathbf{r}_m, t_m)\, G_\mathbf{k}(\mathbf{r}_m, t_m; \mathbf{r}, t) \qquad (2.3.2)$$

where

$$G_\mathbf{k}(\mathbf{r}_m, t_m; \mathbf{r}, t) = \frac{e^{-i[\omega(t-t_m) - \mathbf{k}\cdot(\mathbf{r}-\mathbf{r}_m)]}}{|\mathbf{r} - \mathbf{r}_m|}$$

is the Green's function of free propagation that propagates the $\mathbf{k}$-mode of the mth subfield $E_m(\mathbf{k}; \mathbf{r}_m, t_m)$ from $(\mathbf{r}_m, t_m)$ to $(\mathbf{r}, t)$. Formally, Eq.(2.3.2) can be written as:

$$E(\mathbf{r}, t) = \sum_m E_m(\mathbf{r}_m, t_m)\, G(\mathbf{r}_m, t_m; \mathbf{r}, t), \qquad (2.3.3)$$

with a common statement: the Green's function $G(\mathbf{r}_m, t_m; \mathbf{r}, t)$ propagates the mth subfield from $(\mathbf{r}_m, t_m)$ to $(\mathbf{r}, t)$.

In most optical measurements, it is not the real-positive amplitude $a_m(\mathbf{k}; \mathbf{r}_m, t_m)$, but the phase $\varphi_m(\mathbf{k})$ playing the significant role determining the coherent or incoherent nature of the radiation. To simplify the notation, we write Eq.(2.3.2) as

$$E(\mathbf{r}, t) = \sum_m \int d\mathbf{k}\, \hat{\mathbf{e}}_{m,\mathbf{k}}\, E_m(\mathbf{k})\, g(\mathbf{k}; \mathbf{r}_m, t_m; \mathbf{r}, t)$$

$$= \sum_m \int d\mathbf{k}\, \hat{\mathbf{e}}_{m,\mathbf{k}}\, a_m(\mathbf{k})\, e^{i\varphi_m(\mathbf{k})}\, g(\mathbf{k}; \mathbf{r}_m, t_m; \mathbf{r}, t) \qquad (2.3.4)$$

For point-like sub-sources and free propagation,

$$E_m(\mathbf{k}) = \frac{a_m(\mathbf{k}; \mathbf{r}_m, t_m)}{|\mathbf{r} - \mathbf{r}_m|} e^{i\varphi_m(\mathbf{k})} = a_m(\mathbf{k})\, e^{i\varphi_m(\mathbf{k})}$$

$$g(\mathbf{k}; \mathbf{r}_m, t_m; \mathbf{r}, t) = e^{-i[\omega(t-t_m) - \mathbf{k}\cdot(\mathbf{r}-\mathbf{r}_m)]} \qquad (2.3.5)$$

where $a_m(\mathbf{k}; \mathbf{r}_m, t_m)$ is the real-positive amplitude of the $\mathbf{k}$-mode of the mth subfield at its creation point $(\mathbf{r}_m, t_m)$, $a_m(\mathbf{k})$ is the real-positive amplitude of the $\mathbf{k}$-mode of the mth subfield at its observation point $(\mathbf{r}, t)$; $g(\mathbf{k}; \mathbf{r}_m, t_m; \mathbf{r}, t)$ is now named the Green's function, or the phase propagator, that propagates the $\mathbf{k}$-mode of the mth subfield from $(\mathbf{r}_m, t_m)$ to $(\mathbf{r}, t)$. Shorten the notation, we will write $g(\mathbf{k}; \mathbf{r}_m, t_m; \mathbf{r}, t)$ as $g_m(\mathbf{k}; \mathbf{r}, t)$.

Simplifying the mathematics to 1-D, we assume a simple setup of a point-like light source at $\mathbf{r}_0 = 0$, such as a distant star, and a point-like photodetector at $\mathbf{r}$. The observed radiation comes from the point-like source to the point-like photodetector along the 1-D path $r = |\mathbf{r} - \mathbf{r}_0|$. The observed field $E(\mathbf{r}, t)$ is the result of a superposition among a large number of subfields each is created from a point-like sub-source at time t_m, such as trillions of atomic transitions in a point-like distant star, the resultant field can be modeled as a superposition of subfields in terms of the sub-sources and their harmonic modes of frequency ω:

$$E(r,t) = \sum_m \int d\omega \, E_m(\omega) g_m(\omega; r, t)$$
$$= \sum_m \int d\omega \, a_m(\omega) e^{i\varphi_m(\omega)} \, e^{-i\omega[(t-t_m)-r/c]}, \qquad (2.3.6)$$

where $E_m(\omega) = a_m(\omega) e^{i\varphi_m(\omega)}$ is the complex amplitude of the Fourier-mode, ω, of the mth subfield. Eq. (2.3.6) can be formally written into the following form of Fourier transform,

$$E(r,t) = \sum_m \left\{ \int d\nu \, E_m(\nu) e^{-i\nu\tau_m} \right\} e^{-i\omega_0\tau_m}$$
$$= \sum_m \mathcal{F}_{\tau_m} \{E_m(\nu)\} \, e^{-i\omega_0\tau_m} \qquad (2.3.7)$$

where $\tau_m \equiv (t - t_m) - r/c = \tau - t_m$ and $\tau \equiv (t - r/c)$.

In a special case in which all the Fourier-modes of the mth subfield are excited coherently with a constant common phase φ_{0m} at $(0, t_m)$, the above formally written Fourier transform yields a well-defined wavepacket with a constant initial phase φ_{0m}:

$$E(r,t) = \sum_m \left\{ \int d\nu \, a_m(\nu) e^{-i\nu\tau_m} \right\} e^{-i(\omega_0\tau_m - \varphi_{0m})}$$
$$= \sum_m \mathcal{F}_{\tau_m} \{a_m(\nu)\} \, e^{-i(\omega_0\tau_m - \varphi_{0m})}. \qquad (2.3.8)$$

Eq. (2.3.8) indicates that the total field $E(r,t)$ is the superposition of a certain number of wavepackets, each is independently and randomly created from a point-like sub-source at its own creation time. The mth wavepacket has a well-defined envelope which is the Fourier transform of its amplitude and a harmonic carrier of frequency ω_0. Based on Einstein's granularity, we thus have the following physical picture of the above simple 1-D measurement: a large number of wavepackets excited at spatial coordinate $\mathbf{r} = 0$ but different time t_m and initial phase φ_{0m}, are propagated and superposed at space-time point $(\mathbf{r}, t)$. Eq.(2.3.8) is commonly used to model thermal light. We will have more discussions about this model in later sections.

Eq.(2.3.6) can also be formally written in the following form:

$$E(r,t) = \int d\omega \sum_m E_m(\omega) g_m(\omega; r, t)$$
$$= \int d\omega \left\{ \sum_m a_m(\omega) e^{i[\omega t_m + \varphi_m(\omega)]} \right\} e^{-i\omega\tau}. \qquad (2.3.9)$$

In the same special case in which all the Fourier-modes of the mth subfield are excited coherently with a constant initial phase φ_{0m} at $(0, t_m)$, Eq.(2.3.9) becomes

$$E(r,t) = \int d\omega \left\{ \sum_m a_m(\omega) \, e^{i(\omega t_m + \varphi_{0m})} \right\} e^{-i\omega\tau}$$
$$= \mathcal{F}_\tau \{A(\nu)\} \, e^{-i\omega_0\tau} \qquad (2.3.10)$$

where $A(\nu)$ is the complex amplitude of the ω mode. Eq. (2.3.10) is consistent with the formally written classical field in Maxwell's continuum picture of light. We then have a slightly different physical picture of the above simple 1-D measurement: a set of continuously distributed Fourier-modes are superposed at space-time point $(\mathbf{r}, t)$, however, the complex amplitude of each Fourier-mode is the sum of a set of discrete complex amplitudes of subfields that are created from stochastic process.

In the case of broad spectrum natural light, comparing with the broad spectrum of the total field, each quantized subfield, or bundle of ray, can be treated as a single mode light quantum with energy $\hbar\omega$, as Einstein did in 1905. The indexes of m and n can be replaced by ω. In this case, the radiation field $E(r, t)$ can be approximated as the incoherent superposition among these single mode subfields:

$$E(r, t) = \sum_{\omega} E(\omega) g_m(\omega; r, t)$$
$$\simeq \sum_{\omega} a(\omega) e^{i\varphi(\omega)} e^{-i\omega[(t-t_m)-r/c]}, \qquad (2.3.11)$$

where $\varphi(\omega)$ is the random phase of the complex amplitude. The sum of ω is usually approximated as an integral of ω for continuous broad spectrum.

Einstein's picture of light is still in the frame work of Maxwell electromagnetic wave theory, except Einstein introduced the concept of granularity, or quantized subfield to radiation and abandoned the continuum interpretation of Maxwell. Einstein's bundle of ray, light quantum, or subfield, is a quantized microscopic realistic entity of electromagnetic wave propagating in space-time. Einstein may have never felt comfortable to this picture. In the last a few years of his life, Einstein posed his students a question: suppose a photon of energy $\hbar\omega$ is created from a point source, such as an atomic transition; how big is the photon after propagating one year? This question seems easy to answer for his students. Since the photon is created from a point source, it would propagate in the form of a spherical wave and its wavefront must be a sphere with a diameter of two lightyears after one year propagating. We may find a "consensus" or "common sense" behind the above statement: the total energy of the photon $\hbar\omega$ must be carried by the wavefront of the subfield and thus uniformly distributed on the big sphere. Einstein then asked again: suppose that photon is annihilated by a point-like photon counting detector located on the surface of the big sphere, how long does it take for the energy on the other side of the big sphere to arrive at the detector? Two years? Bohr provided a famous answer to this question: the "wavefunction collapses" instantaneously! Why does the wavefunction need to "collapse"? Bohr did not explain. Nevertheless, Bohr has passed an important message to us: quantum mechanical picture of photon is different from Einstein's granularity, or EM subfield. Although we still have questions regarding the wave-particle duality of a photon, we have to accept the experimental effect that the energy of the electromagnetic field is quantized in nature. We may have to face the truth that a new theory of radiation with quantization is necessary.

2.4 FIELD QUANTIZATION AND THE LIGHT QUANTUM

In blackbody radiation, the atoms on the walls of the cavity box continuously radiate electromagnetic waves into the cavity. In general, there are two fundamental principles governing the physical process of the radiation and determining the physical properties of the radiation field. The Schrödinger equation determines the quantized atomic energy level, and the govern the behavior of the radiation field. The interaction between the field and the atom results in a quantized electromagnetic field. The energy and the frequency of the emitted photon are determined by the quantized energy levels of the atom, $\hbar\omega = E_2 - E_1$.

On the other hand, any excited electromagnetic field must satisfy the Maxwell equations which determine the harmonic mode structure and the superposition.

In the quantum theory of light, the radiation field is treated as a set of harmonic oscillators. The energy of each mode is quantized in a similar way as that of a harmonic oscillator. To quantize the field, we will follow the standard procedure. First, we proceed to link the Hamiltonian of the free electromagnetic field to a set of independent harmonic oscillators. The quantum mechanical results of harmonic oscillators are then adapted to the quantized radiation field. Notice, here, free field means no "sources" or "drains" of the radiation field in the chosen volume of $V = L^3$ that covers the field of interest. The energy of the free field is given by

$$H = \frac{1}{2} \int_V d^3 r \, [\, \epsilon_0 \mathbf{E}^2(\mathbf{r}, t) + \frac{1}{\mu_0} \mathbf{B}^2(\mathbf{r}, t) \,], \qquad (2.4.1)$$

where V is the total volume of the field of interest. The volume is usually, but not necessarily, treated as a large finite cubic cavity of L^3 to simplify the mathematics. To link the energy of the free field with the Hamiltonian of a set of independent harmonic oscillators, we first need to turn Eq. (2.4.1) into the following form

$$H = \frac{1}{2} \sum_{\mathbf{k}} [\, p_{\mathbf{k}}^2(t) + \omega^2 q_{\mathbf{k}}^2(t) \,], \qquad (2.4.2)$$

where $q_{\mathbf{k}}(t)$ and $p_{\mathbf{k}}(t)$ are a pair of real canonical variables.

To achieve our goal, we start to construct a classical solution of the vector potential $\mathbf{A}(\mathbf{r}, t)$ of the field. The vector potential of the free electromagnetic field satisfies the EM wave equation

$$\nabla^2 \mathbf{A} - \frac{1}{c^2} \frac{\partial^2 \mathbf{A}}{\partial t^2} = 0 \qquad (2.4.3)$$

with the Coulomb gauge

$$\nabla \cdot \mathbf{A} = 0.$$

The electric and magnetic fields, $\mathbf{E}(\mathbf{r}, t)$ and $\mathbf{B}(\mathbf{r}, t)$, are thus given in terms of $\mathbf{A}(\mathbf{r}, t)$:

$$\mathbf{E}(\mathbf{r}, t) = -\frac{\partial}{\partial t} \mathbf{A}(\mathbf{r}, t)$$
$$\mathbf{B}(\mathbf{r}, t) = \nabla \times \mathbf{A}(\mathbf{r}, t). \qquad (2.4.4)$$

The most convenient way for analyzing the field structure is to begin with a very large but finite cubic cavity. Applying the periodic boundary condition, we write $\mathbf{A}(\mathbf{r}, t)$ in terms of the Fourier expansion of plane-wave modes

$$\mathbf{A}(\mathbf{r}, t) = \sum_{\mathbf{k}} \mathbf{A}_{\mathbf{k}}(t) \, e^{i \mathbf{k} \cdot \mathbf{r}} \qquad (2.4.5)$$

with

$$\mathbf{A}_{\mathbf{k}}(t) = \mathbf{A}_{-\mathbf{k}}^*(t).$$

Where we have introduced a wave vector

$$\mathbf{k} = \left\{ \frac{2\pi l}{L_x}, \, \frac{2\pi m}{L_y}, \, \frac{2\pi n}{L_z} \right\}, \qquad (2.4.6)$$

with

$$l = 0, \pm 1, \pm 2, \ldots$$
$$m = 0, \pm 1, \pm 2, \ldots$$
$$n = 0, \pm 1, \pm 2, \ldots$$

forming a discrete sum of $\mathbf{k}$ in the Fourier expansion of Eq. (2.4.5).

Considering the Coulomb gauge, we have

$$\sum_{\mathbf{k}} \mathbf{k} \cdot \mathbf{A_k}(t)\, e^{i\mathbf{k}\cdot\mathbf{r}} = 0 \tag{2.4.7}$$

for all values of $\mathbf{r}$, which requires that

$$\mathbf{k} \cdot \mathbf{A_k}(t) = 0. \tag{2.4.8}$$

Substituting Eq. (2.4.5) into Eq. (2.4.3), we have

$$\sum_{\mathbf{k}} \left(-k^2 - \frac{1}{c^2}\frac{\partial^2}{\partial t^2} \right) \mathbf{A_k}(t)\, e^{i\mathbf{k}\cdot\mathbf{r}} = 0 \tag{2.4.9}$$

for all values of $\mathbf{r}$. Thus, we have an equation to determine each of the amplitudes, $\mathbf{A_k}(t)$, of the Fourier expansion

$$\left(\frac{\partial^2}{\partial t^2} + \omega^2 \right) \mathbf{A_k}(t) = 0, \tag{2.4.10}$$

where $\omega = ck$ is the angular frequency of the mode. A solution of Eq. (2.4.10) is given by

$$\mathbf{A_k}(t) \tag{2.4.11}$$
$$= \hat{\mathbf{e}}_1\, \mathcal{A}_{\mathbf{k},1}\big(a^*_{-\mathbf{k},1}\, e^{i\omega t} + a_{\mathbf{k},1}\, e^{-i\omega t}\big) + \hat{\mathbf{e}}_2\, \mathcal{A}_{\mathbf{k},2}\big(a^*_{-\mathbf{k},2}\, e^{i\omega t} + a_{\mathbf{k},2}\, e^{-i\omega t}\big),$$

where $a_{\mathbf{k},s}$ $(a^*_{\mathbf{k},s})$, $s = 1, 2$ is the amplitude for the mode $\mathbf{k}$ and the polarization s. $\mathcal{A}_{\mathbf{k},s}$ is determined by the initial conditions of the electromagnetic field. In Eq. (2.4.11) we have assigned two orthogonal polarization, $\hat{\mathbf{e}}_1$ and $\hat{\mathbf{e}}_2$, by considering the transverse condition of Eq. (2.4.8). The unit vectors $\hat{\mathbf{e}}_1$, $\hat{\mathbf{e}}_2$ and the unit vector $\hat{\mathbf{k}}$ in the direction of the wave vector $\mathbf{k}$, together, form a right-hand, orthogonal, Cartesian basis.

To simplify the mathematical expression, we focus on one of the polarizations

$$\mathbf{A}(\mathbf{r}, t) = \sum_{\mathbf{k}} \hat{\mathbf{e}}\, \mathcal{A}_{\mathbf{k}} \big[a_{\mathbf{k}}(t)\, e^{i\mathbf{k}\cdot\mathbf{r}} + a^*_{\mathbf{k}}(t)\, e^{-i\mathbf{k}\cdot\mathbf{r}} \big] \tag{2.4.12}$$

where $a_{\mathbf{k}}(t) = a_{\mathbf{k}} e^{-i\omega t}$, $a^*_{\mathbf{k}}(t) = a^*_{\mathbf{k}} e^{i\omega t}$. The vector potential $\mathbf{A}(\mathbf{r}, t)$ is written as a superposition of the orthogonal harmonic modes

$$q_{\mathbf{k}}(\mathbf{r}, t) = a_{\mathbf{k}}(t)\, e^{i\mathbf{k}\cdot\mathbf{r}} + a^*_{\mathbf{k}}(t)\, e^{-i\mathbf{k}\cdot\mathbf{r}} \tag{2.4.13}$$

These orthogonal (independent) harmonic modes are equivalent to the normal mode of a harmonic oscillator system. In the following, we will treat these orthogonal harmonic modes of the electromagnetic field as independent harmonic oscillators, and treat $a_{\mathbf{k}}(t)$ $(a^*_{\mathbf{k}}(t))$ as normal mode amplitudes. In terms of the vector potential $\mathbf{A}(\mathbf{r}, t)$ of Eq. (2.4.12), the electric field $\mathbf{E}(\mathbf{r}, t)$ and the magnetic field $\mathbf{B}(\mathbf{r}, t)$ are calculated from Eq. (2.4.4)

$$\mathbf{E}(\mathbf{r}, t) = \sum_{\mathbf{k}} \omega \hat{\mathbf{e}}\, \mathcal{A}_{\mathbf{k}} \big[i\, a_{\mathbf{k}}(t)\, e^{i\mathbf{k}\cdot\mathbf{r}} - i\, a^*_{\mathbf{k}}(t)\, e^{-i\mathbf{k}\cdot\mathbf{r}} \big] \tag{2.4.14}$$

$$\mathbf{B}(\mathbf{r}, t) = \sum_{\mathbf{k}} (\mathbf{k} \times \hat{\mathbf{e}})\, \mathcal{A}_{\mathbf{k}} \big[i\, a_{\mathbf{k}}(t)\, e^{i\mathbf{k}\cdot\mathbf{r}} - i\, a^*_{\mathbf{k}}(t)\, e^{-i\mathbf{k}\cdot\mathbf{r}} \big].$$

Now we examine the energy of the electromagnetic field. We can either treat the electromagnetic field as a system of independent harmonic oscillators or calculate the Hamiltonian of the field from Eq. (2.4.1). Viewing the field as a set of independent harmonic oscillators, the Hamiltonian of the system is readily given in classical mechanics

$$H = 2 \int_V d^3r \sum_{\mathbf{k}} \omega^2 \left| q_{\mathbf{k}}(\mathbf{r}, t) \right|^2 = 2 \sum_{\mathbf{k}} \omega^2 \left| a_{\mathbf{k}}(t) \right|^2, \qquad (2.4.15)$$

where $a_{\mathbf{k}}(t)$ is treated as the normal mode amplitude. To calculate the Hamiltonian from Eq. (2.4.1), substitute Eq. (2.4.14) into Eq. (2.4.1), the integral gives

$$H = 2 \sum_{\mathbf{k}} |\mathcal{A}_{\mathbf{k}}|^2 \omega^2 \left| a_{\mathbf{k}}(t) \right|^2. \qquad (2.4.16)$$

To be consistent with the result of Eq. (2.4.15), we may take a constant mode distribution of $\mathcal{A}_{\mathbf{k}} = \mathcal{A}_0 = 1/\epsilon_0^{1/2} L^{3/2}$ to "normalize" the energy of the field to $H = 2 \sum_{\mathbf{k}} \omega^2 \left| a_{\mathbf{k}}(t) \right|^2$. It is quite reasonable to treat all the independent and "free" harmonic oscillators equally. Although, in reality, non-constant mode distribution may have to be taken into consideration, it would not affect the quantization of each independent mode of the electromagnetic field.

Next, we introduce a pair of canonical variables $q_{\mathbf{k}}(t)$ and $p_{\mathbf{k}}(t)$

$$\begin{aligned} q_{\mathbf{k}}(t) &= a_{\mathbf{k}}(t) + a_{\mathbf{k}}^*(t) \\ p_{\mathbf{k}}(t) &= -i\omega \left[a_{\mathbf{k}}(t) - a_{\mathbf{k}}^*(t) \right]. \end{aligned} \qquad (2.4.17)$$

The Hamiltonian of the field in Eq. (2.4.15) is then rewritten in terms of $q_{\mathbf{k}}(t)$ and $p_{\mathbf{k}}(t)$ as

$$H = \sum_{\mathbf{k}} \frac{1}{2} \left[p_{\mathbf{k}}^2(t) + \omega^2 q_{\mathbf{k}}^2(t) \right] = \sum_{\mathbf{k}} H_{\mathbf{k}}, \qquad (2.4.18)$$

which is recognized as the Hamiltonian of a harmonic oscillator system. Each harmonic oscillator of the system corresponds to a mode of the field specified by wave vector $\mathbf{k}$. In Eq. (2.4.18) $H_{\mathbf{k}}$ indicates the Hamiltonian of the $\mathbf{k}$-th harmonic oscillator. The Hamiltonian of a classical harmonic oscillator is allowed to take any non-negative values, because $p_{\mathbf{k}}$ and $q_{\mathbf{k}}$ can take any values in classical theory. It is easy to derive from Eq. (2.4.17) and Eq. (2.4.18) that $q_{\mathbf{k}}(t)$ and $p_{\mathbf{k}}(t)$ satisfy the classical canonic equations of motion

$$\begin{aligned} \frac{\partial q_{\mathbf{k}}}{\partial t} &= \frac{\partial H}{\partial p_{\mathbf{k}}} \\ \frac{\partial p_{\mathbf{k}}}{\partial t} &= -\frac{\partial H}{\partial q_{\mathbf{k}}}. \end{aligned} \qquad (2.4.19)$$

It is also easy to find the results of the following Poisson bracket

$$\begin{aligned} \left[q_{\mathbf{k}}(t), \, p_{\mathbf{k}'}(t) \right] &= \delta_{\mathbf{k}\mathbf{k}'} \\ \left[q_{\mathbf{k}}(t), \, q_{\mathbf{k}'}(t) \right] &= 0 \\ \left[p_{\mathbf{k}}(t), \, p_{\mathbf{k}'}(t) \right] &= 0. \end{aligned} \qquad (2.4.20)$$

Now, we follow the quantum theory of harmonic oscillator to formulate and quantize the electromagnetic field. As a standard procedure, we first replace the canonical variables of $q_{\mathbf{k}}(t)$ and $p_{\mathbf{k}}(t)$ with quantum mechanical canonical conjugate operators $\hat{q}_{\mathbf{k}}(t)$ and $\hat{p}_{\mathbf{k}}(t)$

with the following commutation relations:

$$\left[\hat{q}_{\mathbf{k}}(t),\ \hat{p}_{\mathbf{k}'}(t)\right] = i\hbar\,\delta_{\mathbf{k}\mathbf{k}'}$$
$$\left[\hat{q}_{\mathbf{k}}(t),\ \hat{q}_{\mathbf{k}'}(t)\right] = 0$$
$$\left[\hat{p}_{\mathbf{k}}(t),\ \hat{p}_{\mathbf{k}'}(t)\right] = 0. \qquad (2.4.21)$$

The quantum mechanical Hamiltonian of the harmonic oscillator system is thus written in terms of the operators $\hat{q}_{\mathbf{k}}(t)$ and $\hat{p}_{\mathbf{k}}(t)$

$$\hat{H} = \sum_{\mathbf{k}} \frac{1}{2}\left[\hat{p}_{\mathbf{k}}^2(t) + \omega^2\,\hat{q}_{\mathbf{k}}^2(t)\right] = \sum_{\mathbf{k}} \hat{H}_{\mathbf{k}} \qquad (2.4.22)$$

where $\hat{H}_{\mathbf{k}}$ is the Hamiltonian operator of the $\mathbf{k}$-th harmonic oscillator. Differing from the Hamiltonian, $H_{\mathbf{k}}$, of a classical oscillator in Eq. (2.4.18), which may take any values, $\hat{H}_{\mathbf{k}}$ in Eq. (2.4.22) can only admit quantized energy.

To show the energy quantization of a quantum harmonic oscillator, we introduce a pair of non-Hermitian operators to replace the $\hat{p}_{\mathbf{k}}$ and $\hat{q}_{\mathbf{k}}$ in Eq. (2.4.22)

$$\hat{a}_{\mathbf{k}}(t) = \frac{1}{\sqrt{2\hbar\omega}}\left[\omega\,\hat{q}_{\mathbf{k}}(t) + i\hat{p}_{\mathbf{k}}(t)\right]$$
$$\hat{a}_{\mathbf{k}}^{\dagger}(t) = \frac{1}{\sqrt{2\hbar\omega}}\left[\omega\,\hat{q}_{\mathbf{k}}(t) - i\hat{p}_{\mathbf{k}}(t)\right], \qquad (2.4.23)$$

and a set of commutation relations derived from Eq. (2.4.21):

$$\left[\hat{a}_{\mathbf{k}}(t),\ \hat{a}_{\mathbf{k}'}^{\dagger}(t)\right] = \delta_{\mathbf{k}\mathbf{k}'}$$
$$\left[\hat{a}_{\mathbf{k}}(t),\ \hat{a}_{\mathbf{k}'}(t)\right] = 0,$$
$$\left[\hat{a}_{\mathbf{k}}^{\dagger}(t),\ \hat{a}_{\mathbf{k}'}^{\dagger}(t)\right] = 0. \qquad (2.4.24)$$

The Hamiltonian of Eq. (2.4.22) is thus written in terms of $\hat{a}_{\mathbf{k}}(t)$ and $\hat{a}_{\mathbf{k}}^{\dagger}(t)$

$$
\begin{aligned}
\hat{H} &= \frac{1}{2}\sum_{\mathbf{k}} \hbar\omega\left[\hat{a}_{\mathbf{k}}(t)\,\hat{a}_{\mathbf{k}}^{\dagger}(t) + \hat{a}_{\mathbf{k}}^{\dagger}(t)\,\hat{a}_{\mathbf{k}}(t)\right] \\
&= \sum_{\mathbf{k}} \hbar\omega\left[\hat{a}_{\mathbf{k}}^{\dagger}(t)\,\hat{a}_{\mathbf{k}}(t) + \frac{1}{2}\right] \\
&= \sum_{\mathbf{k}} \hbar\omega\left[\hat{n}_{\mathbf{k}} + \frac{1}{2}\right] \\
&= \sum_{\mathbf{k}} \hat{H}_{\mathbf{k}},
\end{aligned}
\qquad (2.4.25)
$$

where we have used the commutation relations of Eq. (2.4.24) and introduced the number operator

$$\hat{n}_{\mathbf{k}} = \hat{a}_{\mathbf{k}}^{\dagger}(t)\,\hat{a}_{\mathbf{k}}(t) \qquad (2.4.26)$$

for the mode $\mathbf{k}$. It is clear that $\hat{n}_{\mathbf{k}}$ is Hermitian and is independent of time.

We now show that the Hamiltonian, $\hat{H}_{\mathbf{k}}$, of the $\mathbf{k}$-th harmonic oscillator, or mode, in Eq. (2.4.25) only admits quantized energies. To simplify the notation, we will drop the subscript $\mathbf{k}$ in the following discussion by considering a single harmonic oscillator, or mode.

Assume $|n\rangle$ is an eigenstate of $\hat{n}$ with eigenvalue of n,

$$\hat{n}\,|\,n\,\rangle = n\,|\,n\,\rangle \tag{2.4.27}$$

where n must be real, since $\hat{n}$ is Hermitian. It is easy to show that $|n\rangle$ is also an eigenstate of $\hat{H}$ with eigenvalue of $E_n = \hbar\omega(n + 1/2)$,

$$\hat{H}\,|\,n\,\rangle = \hbar\omega\left[\hat{n} + \frac{1}{2}\right]|\,n\,\rangle = \hbar\omega\left(n + \frac{1}{2}\right)|\,n\,\rangle. \tag{2.4.28}$$

We will show that the Hamiltonian in Eq. (2.4.28) is quantized because the commutations of Eq. (2.4.24) limit n to integer values. To prove this, we show that $\hat{a}\,|\,n\,\rangle$ is an eigenstate of $\hat{n}$ with eigenvalue of $n-1$, since

$$\hat{n}\,\hat{a}|\,n\,\rangle = \hat{a}\,(n-1)\,|\,n\,\rangle = (n-1)\,\hat{a}\,|\,n\,\rangle \tag{2.4.29}$$

where we have used the commutation relations of Eq. (2.4.24). If $\langle\,n\,|\,n\,\rangle = 1$, then $\hat{a}\,|\,n\,\rangle$ has normalization

$$\langle\,n\,|\,\hat{a}^\dagger\,\hat{a}\,|\,n\,\rangle = \langle\,n\,|\,\hat{n}\,|\,n\,\rangle = n. \tag{2.4.30}$$

Therefore,

$$\hat{a}\,|\,n\,\rangle = \sqrt{n}\,|\,n-1\,\rangle \tag{2.4.31}$$

where state $|\,n-1\,\rangle$ is normalized $\langle\,n-1\,|\,n-1\,\rangle = 1$.

Similarly, $(\hat{a})^2\,|\,n\,\rangle$ is an eigenstate of $\hat{n}$ with eigenvalue of $n-2$, and

$$(\hat{a})^2\,|\,n\,\rangle = \sqrt{n(n-1)}\,|\,n-2\,\rangle.$$

Repeatedly applying the operator $\hat{a}$, we have all the normalized eigenstates in terms of $|\,n\,\rangle$

$$(\hat{a})^m\,|\,n\,\rangle = \sqrt{n(n-1)...(n-m+1)}\,|\,n-m\,\rangle \tag{2.4.32}$$

Thus, if n is an eigenvalue of the number operator $\hat{n}$, then $n-1$, $n-2$, ... are all eigenvalues of $\hat{n}$. Since

$$n = \langle\,n\,|\,\hat{n}\,|\,n\,\rangle = \langle\,n\,|\,\hat{a}^\dagger\,\hat{a}\,|\,n\,\rangle = \langle\,\Phi\,|\,\Phi\,\rangle \geq 0,$$

where $|\,\Phi\,\rangle = \hat{a}\,|\,n\,\rangle$, all the eigenvalues of $\hat{n}$ must be positive. Therefore n must be an integer and bounded by the lowest eigenvalue of zero. The null eigenstate $|\,0\,\rangle$, or the vacuum state, is defined as

$$\hat{a}\,|\,1\,\rangle = |\,0\,\rangle, \quad \text{and} \quad \hat{a}\,|\,0\,\rangle = 0. \tag{2.4.33}$$

The Hermitian operator $\hat{n}$, therefore, takes eigenvalues from a set of integer $0, 1, 2, ...$, and is called the number operator for mode **k**. Regarding Eq. (2.4.32), it is interesting that we can only apply n times the operator $\hat{a}$ to reach the lowest permissible vacuum state $|\,0\,\rangle$ with a non-zero lowest energy $E_0 = \hbar\omega/2$:

$$(\hat{a})^n\,|\,n\,\rangle = \sqrt{n!}\,|\,0\,\rangle. \tag{2.4.34}$$

So far we have explored two important differences between quantum theory and classical theory regarding the energy of a radiation mode: (1) the energy of a radiation mode is quantized; (2) the lowest energy of a radiation mode, the vacuum state $|0\rangle$, takes a non-zero value $E_0 = \hbar\omega/2$ which is called the zero-point energy. It should be emphasized that both properties (1) and (2) follow from the commutation relation $[\hat{a}, \hat{a}^\dagger] = 1$. We will come back to discuss the physics associated with these two special features later.

Similar to $\hat{a}|n\rangle$, $\hat{a}^\dagger|n\rangle$ is also an eigenstate of $\hat{n}$ with eigenvalue $n + 1$, since

$$\hat{n}\,\hat{a}^\dagger|n\rangle = \hat{a}^\dagger\,(n+1)|n\rangle = (n+1)\,\hat{a}^\dagger|n\rangle, \tag{2.4.35}$$

where we have again used the commutation relations of Eq. (2.4.24). It is easy to show that

$$\hat{a}^\dagger|n\rangle = \sqrt{n+1}\,|n+1\rangle, \tag{2.4.36}$$

where the state $|n+1\rangle$ is normalized, $\langle n+1|n+1\rangle = 1$. Unlike $\hat{a}$ the eigenvalues of $\hat{a}^\dagger$ are not bounded. by repeatedly applying $\hat{a}^\dagger$ the eigenvalues of $\hat{n}$ go to infinity and therefore are integers from zero to infinity. Using Eq. (2.4.35,) we have all the normalized eigenstates in terms of $|0\rangle$.

$$|1\rangle = \hat{a}^\dagger|0\rangle$$

$$|2\rangle = \frac{1}{\sqrt{2}}\,\hat{a}^\dagger|1\rangle = \frac{1}{\sqrt{2}}\,(\hat{a}^\dagger)^2|0\rangle$$

$$|3\rangle = \frac{1}{\sqrt{3}}\,\hat{a}^\dagger|2\rangle = \frac{1}{\sqrt{6}}\,(\hat{a}^\dagger)^3|0\rangle$$

$$\vdots$$

$$|n\rangle = \frac{1}{\sqrt{n!}}\,(\hat{a}^\dagger)^n|0\rangle, \tag{2.4.37}$$

where the state $|n\rangle$ is normalized by means of $\langle n|n\rangle = 1$.

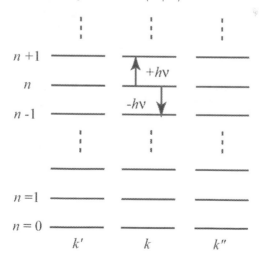

FIGURE 2.4.1 Quantized energy levels for a radiation mode of an electromagnetic field. The creation operator $\hat{a}_\mathbf{k}^\dagger$ adds a quantum of energy $\hbar\omega$, or a photon, to the mode $\mathbf{k}$ to excite the mode to a higher energy level. The annihilation operator $\hat{a}_\mathbf{k}$ subtract the same amount of energy, or annihilate a photon, from the radiation mode. In connection with the concept of photon, the illustrated energy eigenstates of $\hat{H}_\mathbf{k}$ are also named as photon number states, or Fock states.

By applying $\hat{a}^\dagger$, we have created a set of normalized number states $|n_\mathbf{k}\rangle$ which are eigenstates of the Hamiltonian $\hat{H}_{\mathbf{k}}$ for mode $\mathbf{k}$ and for polarization $\hat{\mathbf{e}}_{\mathbf{k}}$. These energy eigenstates form a complete, orthonormal vector space for characterizing the radiation field. The eigenvalues of the Hamiltonian, $\hat{H}_{\mathbf{k}}$, are quantized with discrete values of $n\hbar\omega$, or $nh\nu$, in contrast to classical electromagnetic theory where the energy of a radiation mode can have any value. Figure 2.4.1 illustrates the quantized energy levels for a radiation mode of the electromagnetic field.

Connecting $\hat{a}^{\dagger}$ and $\hat{a}$ with Einstein's concept of an energy bundle, or a photon, $\hat{a}^{\dagger}$ adds a quantum of energy, $\hbar\omega$, or a photon, to the $\mathbf{k}$-th mode of the radiation field. Consequently its called the *photon creation operator*. Similarly, $\hat{a}$ subtracts a quantum of energy, $\hbar\omega$, from the $\mathbf{k}$-th mode of the radiation field. Therefore, its called the *photon annihilation operator*. The excited state $|n\rangle$ of a radiation mode contains n quanta of energy along with the zero-point energy, and n is called the *occupation number* or the number of photons. Accordingly, the set of energy eigenstates in Eq. (2.4.37) are called *photon number states*. The physical process of photon creation and annihilation in an atomic transition will be addressed later, in the discussions of the quantum state and the field operator.

We are now ready to characterize the quantized radiation field in terms of the energy eigenstates of its Hamiltonian. In general, the state of any radiation field can be described as the superposition of the energy eigenstates,

$$|\Psi\rangle = \sum_n c_n |n\rangle, \tag{2.4.38}$$

where c_n is the complex amplitude associated with the photon number state $|n\rangle$. In principle, we are not restricted to using the energy eigenstates to characterize the state vector of a radiation mode. However, it has advantage when calculating the time evolution of the radiation field. It is readily shown that an eigenstate of the Hamiltonian $\hat{H}$ simply develops a phase factor $e^{-iE_n(t-t_0)/\hbar}$, from time t_0 to t,

$$|\Psi_n(t)\rangle = e^{-iE_n(t-t_0)/\hbar} |\Psi_n(t_0)\rangle, \tag{2.4.39}$$

since

$$i\hbar \frac{\partial}{\partial t} |\Psi_n\rangle = \hat{H} |\Psi_n\rangle = E_n |\Psi_n\rangle.$$

With the radiation field quantized, we have simultaneously turned the electromagnetic field into operators in terms of the creation operator and the annihilation operator,

$$\hat{\mathbf{A}}(\mathbf{r},t) = \sum_{\mathbf{k}} \hat{\mathbf{e}}_k \mathcal{A}_{\mathbf{k}} \left[\hat{a}_{\mathbf{k}}(t)\, e^{i\mathbf{k}\cdot\mathbf{r}} + \hat{a}_{\mathbf{k}}^{\dagger}(t)\, e^{-i\mathbf{k}\cdot\mathbf{r}} \right],$$

$$\hat{\mathbf{E}}(\mathbf{r},t) = \sum_{\mathbf{k}} \hat{\mathbf{e}}_k \mathcal{E}_{\mathbf{k}} \left[i\,\hat{a}_{\mathbf{k}}(t)\, e^{i\mathbf{k}\cdot\mathbf{r}} - i\,\hat{a}_{\mathbf{k}}^{\dagger}(t)\, e^{-i\mathbf{k}\cdot\mathbf{r}} \right],$$

$$\hat{\mathbf{B}}(\mathbf{r},t) = \sum_{\mathbf{k}} (\mathbf{k} \times \hat{\mathbf{e}}_k)\, \mathcal{B}_{\mathbf{k}} \left[i\,\hat{a}_{\mathbf{k}}(t)\, e^{i\mathbf{k}\cdot\mathbf{r}} - i\,\hat{a}_{\mathbf{k}}^{\dagger}(t)\, e^{-i\mathbf{k}\cdot\mathbf{r}}) \right]. \tag{2.4.40}$$

We usually write the field operator into two parts:

$$\hat{\mathbf{E}}(\mathbf{r},t) = \sum_{\mathbf{k}} \hat{\mathbf{e}}_k\, i\, \mathcal{E}_{\mathbf{k}}\, \hat{a}_{\mathbf{k}}(t)\, e^{i\mathbf{k}\cdot\mathbf{r}} + \sum_{\mathbf{k}} \hat{\mathbf{e}}_k (-i) \mathcal{E}_{\mathbf{k}}\, \hat{a}_{\mathbf{k}}^{\dagger}(t)\, e^{-i\mathbf{k}\cdot\mathbf{r}}$$

$$= \hat{\mathbf{E}}^{(+)}(\mathbf{r},t) + \hat{\mathbf{E}}^{(-)}(\mathbf{r},t) \tag{2.4.41}$$

where $\hat{\mathbf{E}}^{(+)}(\mathbf{r},t)$ and $\hat{\mathbf{E}}^{(-)}(\mathbf{r},t)$ contain the annihilation and creation operators, respectively.

Notice in the above analysis we have simplified the mathematics by assuming a large but finite cubic cavity and by applying the plane-wave solutions. In certain experimental situations, the plane-wave solutions may need to be generalized for different size, shape, and nature of boundaries. We then replace the plane-wave solutions with the generalized spatial mode functions $\mathbf{u}_k(\mathbf{r})$ in Eq. (2.4.5)

$$\mathbf{A}(\mathbf{r}, t) = A_0 \sum_{\mathbf{k}} \mathbf{A}_{\mathbf{k}}(t)\, \mathbf{u}_k(\mathbf{r}), \tag{2.4.42}$$

corresponding to the excited frequency ω_k. $\mathbf{u}_k(\mathbf{r})$ must be a solution to the Helmholtz equation

$$\nabla^2 \mathbf{u}_k + \frac{\omega_k^2}{c^2}\, \mathbf{u}_k = 0 \tag{2.4.43}$$

subject to the corresponding boundary conditions. We further require that the mode functions form a complete orthonormal set

$$\int dr^3\, \mathbf{u}_l^*(\mathbf{r})\, \mathbf{u}_m(\mathbf{r}) = \delta_{lm}, \tag{2.4.44}$$

with the condition

$$\nabla \cdot \mathbf{u}_k(\mathbf{r}) = 0.$$

Using the above three equations and the boundary conditions, we find a suitable set of mode functions for the field quantization. The non-plane-wave solutions will be used in later chapters.

SUMMARY

In this chapter, we quantized the radiation field of light and introduced a number of key concepts of quantum optics. Why do we need to quantize the radiation field? This chapter reviewed two important discoveries in the history of physics to address this question. Based on the experimental observations, Einstein introduced his granularity picture of light. Einstein assumed that the radiation energy is quantized into localized bundles. A bundle of energy is initially localized in a small volume of space and remains localized as it moves away from the radiation source with velocity c. The energy of the bundle is related to its frequency by multiplication of a universal constant h. In the photoelectric process, one bundle of energy $E = h\nu$, or one photon, is completely absorbed by one electron originally bound with the metal. How could one bounded electron, which is a particle localized within a very small volume, completely absorb a photon to become an photoelectron? Einstein treated his bundle of ray as a particle. A particle-like bundle of ray transfers all its energy and momentum to an electron during a *collision*. Accordingly, Einstein proposed a quantized radiation with particle-like bundles (each is localized) and wave-like rays (nonlocalized). According to Einstein, a natural light source, such as the sun or a distant star, consists of a large number of point-like sub-sources ("atomic transitions" in modern language), each of which emits their own subfields (originally labeled in German by "strahlenbündel", translated to "bundle of ray" in English, and now named as "photon" in modern language) that carries energy $E = h\nu$ in a random manner: the mth subfield (photon) that is emitted from the mth point-like sub-source (an atomic transition in which the atom changes its state from higher energy level E_2 to lower energy level E_1, and releases a photon with energy $h\nu = E_2 - E_1$) may propagate in all possible directions with all possible random phase. Each quantized subfield propagates independently and may interference with itself and with others at a local space-time point,

$$E(\mathbf{r}, t) = \sum_m E_m(\mathbf{r}, t).$$

Although his concept of subfield is still under the framework of electromagnetic field theory, Einstein abandoned the continuum interpretation of Maxwell. His concept of "bundle of ray" or "subfield" refers a quantized microscopic realistic substance of electromagnetic field, corresponding to the quantum mechanical concept of photon, precisely, it has the same mathematical expression as the effective wavefunction of photon in thermal state. The introduction of electromagnetic field quantization and the introduction of light quantum in this chapter are standard, inherited from Dirac. Similar to most of the textbooks, we started from classical Maxwell's equations and ended with a quantized electromagnetic field. The Maxwell's equations require wave solutions of the electromagnetic field with a discrete or continual mode structure according, subject to certain boundary conditions. Quantum mechanics, in addition, quantized the energy of each mode of the radiation field as harmonic oscillator. In the quantum theory of light, the quantum *state* characterizes the state of the quantized field, a photon or a group of photons (either identical or distinguishable). The field *operators*, in terms of the annihilation and creation operators, characterize the generation, annihilation, and propagation of the quantized field, a photon or a group of photons (either identical or distinguishable). Later in this book we will introduce a quantum mechanical concept of "effective wavefunction" from the quantum theory of optical coherence. It is interesting to find that the effective wavefunction of a photon plays the same role as Einstein's subfield; and the effective wavefunction of a photon in the thermal state is mathematically the same function of space-time as that of Einstein's subfield of natural light. It should be emphasized that the physical means of the two are fundamentally different: Einstein's subfield $E_m(\mathbf{r}, t)$ is a quantized microscopic realistic entity of electromagnetic wave propagating in space-time. The quantum mechanical effective wavefunction is the probability amplitude for a photon to produce a photodetection event at space-time coordinate $(\mathbf{r}, t)$.

REFERENCES AND SUGGESTIONS FOR READING

[1] M. Plank, Verh. dt. phys., Ges. **2**, 202, 237 (1900).

[2] H.R. Hertz, Annalen der Physik, **267**, 421, 983 (1987).

[3] J.W.S. Rayleigh, Phil. Mag. **49**, 539 (1900); Nature, **72**, 54, 243 (1905).

[4] J.H. Jeans, Phil. Mag. **10**, 91 (1905).

[5] W. Wien, Annln. Phys., **58**, 662 (1896).

[6] A. Einstein, Annln. Phys., **17**, 132 (1905).

[7] P.A.M. Dirac, Proc. Roy., Soc., A **114**, 243 (1927).

[8] L.D. Landau and E.M Lifshitz, *Quantum Mechanics*, Pergamon Press, 1965.

CHAPTER 3

Quantum Theory of Light: The State of Quantized Field and Photon

In this chapter, we study the quantum state of quantized field. To analyze and to solve a problem of quantum optics, we need to know the state of the quantized field, or the state of a photon, or the state of a group of photons. We first study the photon number state and coherent state. Photon number state representation, especially the single-photon state representation, and the coherent state representation are widely used to specify the state of a radiation field. The concept of density operator or density matrix is introduced in the process of calculating the expectation value of a quantum observable. We then introduce and distinguish the pure state and mixed state. The quantum states of composite system, especially the state of two-photon field, are studied by introducing an 2-D Hilbert space constructed as the direct or tensor product of the Hilbert spaces of two subsystems. A simple model for the creation of single-photon state and multi-photon state is given with details. The concepts of product state, entangled state, and mixed state are introduced and distinguished in the last section of this chapter.

3.1 PHOTON NUMBER STATE OF RADIATION FIELD

By applying the creation operators, we have generated a set of normalized photon number states which are eigenstates of the Hamiltonian in Eq. (2.4.37) for mode $\mathbf{k}$ and polarization $\hat{\mathbf{e}}_{\mathbf{k}}$. These energy eigenstates form a complete and orthonormal vector space for characterizing the radiation field, or for characterizing the state of light quanta. We now proceed to generalize our discussion to multi-mode radiation. We will show that the generalized Fock state of Eq. (3.1.1) is an eigenstate of the multi-mode Hamiltonian in Eq. (2.4.25). In addition, we introduce a useful operator, namely the total photon number operator, $\hat{n}$, in Eq. (3.1.3).

The generalized multi-mode photon number state, or Fock state of the radiation field is written as:

$$| \Psi \rangle = \prod_{\mathbf{k},s} | n_{\mathbf{k},s} \rangle = \prod_{\mathbf{k},s} \frac{1}{\sqrt{n_{\mathbf{k},s}!}} (\hat{a}_{\mathbf{k},s}^{\dagger})^n | 0 \rangle \tag{3.1.1}$$

where $\mathbf{k}$ and s indicate the mode and the polarization. Eq. (3.1.1) defines the state for all and for each radiation mode and polarization.

It follows immediately that the multi-mode Fock state of Eq. (3.1.1) is an eigenstate of the single-mode photon number operator $\hat{n}_{\mathbf{k},s}$

$$\hat{n}_{\mathbf{k},s} \left(\prod_{\mathbf{k},s} |n_{\mathbf{k},s}\rangle \right) = n_{\mathbf{k},s} \prod_{\mathbf{k},s} |n_{\mathbf{k},s}\rangle, \tag{3.1.2}$$

where $n_{\mathbf{k},s}$ is the occupation number of the radiation mode, $\mathbf{k}$, and polarization, s. We now define a total photon number operator, $\hat{n}$, by summing $\hat{n}_{\mathbf{k},s}$ over all the radiation modes, $\mathbf{k}$, and all the polarizations, s,

$$\hat{n} = \sum_{\mathbf{k},s} \hat{n}_{\mathbf{k},s}. \tag{3.1.3}$$

It is easy to find that the multi-mode Fock state of Eq. (3.1.1) is an eigenstate of the total photon number operator,

$$\hat{n} \left(\prod_{\mathbf{k},s} |n_{\mathbf{k},s}\rangle \right) = \left(\sum_{\mathbf{k},s} n_{\mathbf{k},s} \right) \prod_{\mathbf{k},s} |n_{\mathbf{k},s}\rangle = n \prod_{\mathbf{k},s} |n_{\mathbf{k},s}\rangle, \tag{3.1.4}$$

where $n = \sum n_{\mathbf{k},s}$ is the total number of photons in the radiation field.

It is easy to show that the multi-mode Fock state of Eq. (3.1.1) is an eigenstate of the total Hamiltonian (multi-mode) in Eq. (2.4.25),

$$\begin{aligned}
\hat{H}|\Psi\rangle &= \left[\sum_{\mathbf{k},s} \hbar\omega_{\mathbf{k},s} \left(\hat{n}_{\mathbf{k},s} + \frac{1}{2} \right) \right] \left(\prod_{\mathbf{k},s} |n_{\mathbf{k},s}\rangle \right) \\
&= \left[\sum_{\mathbf{k},s} \hbar\omega_{\mathbf{k},s} \left(n_{\mathbf{k},s} + \frac{1}{2} \right) \right] \left(\prod_{\mathbf{k},s} |n_{\mathbf{k},s}\rangle \right) \\
&= E|\Psi\rangle,
\end{aligned} \tag{3.1.5}$$

where $E = \sum \hbar\omega_{\mathbf{k},s} \left(n_{\mathbf{k},s} + \frac{1}{2} \right)$ is the total energy of the radiation.

For convenience, we define a short-hand notation $\{n\}$ and rewrite the multi-mode Fock state of Eq. (3.1.1) as

$$|\{n\}\rangle \equiv \prod_{\mathbf{k},s} |n_{\mathbf{k},s}\rangle. \tag{3.1.6}$$

Eq. (3.1.4) is then rewritten, in short-hand form, as

$$\hat{n}|\{n\}\rangle = n|\{n\}\rangle, \tag{3.1.7}$$

which simply indicates that the multi-mode Fock state $|\{n\}\rangle$ is an eigenstate of the total number operator, $\hat{n}$, defined in Eq. (3.1.3) with an eigenvalue of n, which is the total number of photons in the radiation field.

The eigenvalue, n, of the total number operator, is commonly used to name the Fock state, either single-mode or multi-mode, as an n-photon state. For instance, the state

$$\begin{aligned}
|\Psi\rangle &= \hat{a}_{\mathbf{k},s}^{\dagger} |0\rangle = |0\rangle \dots |0\rangle |1_{\mathbf{k},s}\rangle |0\rangle \dots |0\rangle \\
&= |0, \dots 0, 1_{\mathbf{k},s}, 0, \dots 0\rangle
\end{aligned} \tag{3.1.8}$$

is a *single-photon* state ($n = 1$) of the mode $\mathbf{k}$ and polarization s. Shorten the notation, it is common to write the above single-photon state as $|\Psi\rangle = |1_{\mathbf{k},s}\rangle$, we should not forget the ground states of the other modes. Referring Fig. 2.4.1, $|\Psi\rangle = |1_{\mathbf{k},s}\rangle$ means the occupation number of the mode $(\mathbf{k}, s)$ takes a value of one, while all the other modes stay on their ground state (vacuum) with occupation number of zeros. The states

$$|\Psi\rangle = \hat{a}_{\mathbf{k},s}^{\dagger} \hat{a}_{\mathbf{k}',s'}^{\dagger} |0\rangle = |0, \dots 0, 1_{\mathbf{k},s}, 0, \dots 1_{\mathbf{k}',s'}, 0, \dots 0\rangle \qquad (3.1.9)$$

and

$$|\Psi\rangle = \frac{1}{\sqrt{2}} (\hat{a}_{\mathbf{k},s}^{\dagger})^2 |0\rangle = |0, \dots 0, 2_{\mathbf{k},s}, 0, \dots 0\rangle \qquad (3.1.10)$$

are both *two-photon states* $(n = 2)$ but characterize different physics. The state in Eq. (3.1.9) indicates the excitation of two different radiation modes with occupation numbers $n_{\mathbf{k},s} = 1$ and $n_{\mathbf{k}',s'} = 1$. The state in Eq. (3.1.10) indicates the excitation of a radiation mode with occupation number $n_{\mathbf{k},s} = 2$. Such states in Eq. (3.1.8), Eq. (3.1.9), and Eq. (3.1.10) are all known as Fock states, or photon number states. Again, shorten the notation, it is also common to write the two-photon state as $|\Psi\rangle = |1_{\mathbf{k},s}, 1_{\mathbf{k}',s'}\rangle$ and $|\Psi\rangle = |2_{\mathbf{k},s}\rangle$, one should not forget the vacuum states of the other modes.

In connection with the concept of photons, Eq. (3.1.8) indicates the excitation of one light quantum, or a photon, of energy $\hbar\omega_{\mathbf{k},s}$ to the radiation mode $\mathbf{k}$ and polarization s. Eq.(3.1.9) means the excitation of two photons with energies $\hbar\omega_{\mathbf{k},s}$ and $\hbar\omega_{\mathbf{k}',s'}$, respectively, to the radiation mode $\mathbf{k}$ - polarization s and the radiation mode $\mathbf{k}'$-polarization s'. Eq.(3.1.10) corresponds to the excitation of two photons both with energy of $\hbar\omega_{\mathbf{k},s}$ to the same radiation mode $\mathbf{k}$ and polarization s. Eq. (3.1.8), Eq. (3.1.9), and Eq. (3.1.10) are also popular to be considered as the state of a photon with wavevector $\mathbf{k}$ and polarization s, the state of two photons, one with wavevector $\mathbf{k}$ and polarization s and the other one with wavevector $\mathbf{k}'$ and polarization s', and the state of two photons both with wavevector $\mathbf{k}$ and polarization s. The two photons in state Eq. (3.1.10) are "indistinguishable" or "degenerate".

Since Fock states form a complete, orthonormal vector space, following the rules of quantum mechanics, we are able to express any state vector of a radiation field in terms of a superposition of Fock states, or photon number states,

$$|\Psi\rangle = \sum_j c_j |\Psi_j\rangle = \sum_{\{n\}} f(\{n\}) |\{n\}\rangle, \qquad (3.1.11)$$

where $f(\{n\})$ is the normalized probability amplitude to find the radiation field in the state $|\{n\}\rangle$.

As an example, Eq. (3.1.12) represents a single-photon state:

$$|\Psi\rangle = \sum_j c_j |\Psi_j\rangle = \sum_{\mathbf{k},s} f(\mathbf{k}, s) \hat{a}_{\mathbf{k},s}^{\dagger} |0\rangle$$

$$= \sum_s \int d\mathbf{k} \, f(\mathbf{k}, s) \hat{a}_{\mathbf{k},s}^{\dagger} |0\rangle \qquad (3.1.12)$$

where $c_j = f(\mathbf{k}, s) = \langle \Psi_j |\Psi\rangle$ is the normalized probability amplitude for the radiation field to be in the Fock state $|\Psi_j\rangle = \hat{a}_{\mathbf{k},s}^{\dagger} |0\rangle = |1_{\mathbf{k},s}\rangle$. In the last line of Eq.(3.1.12), we have treated mode $\mathbf{k}$ continuously as usual, we have also formally written the two polarizations, $s = 1, 2$, for each mode $\mathbf{k}$ as a superposition. If the two polarizations have a constant phase relationship, the two polarizations superpose coherently resulting in a pure polarization state, or a vector in the polarization space. However, if the two polarizations are completely independent with indeterministic random relative phases, the two polarizations superpose incoherently resulting in a mixed polarization state. Even if we decide to simplify the mathematics to one polarization by ignoring the sum of s, the state of Eq. (3.1.12) can still be a pure state or a mixed state, depending on the nature of the superposition of $|\Psi_j\rangle = \hat{a}_{\mathbf{k}}^\dagger |0\rangle$: a coherent superposition gives a pure state, an incoherent superposition gives a mixed state. We focus our discussion to pure state in this section and leave the mixed state after introducing the concept of density operator.

The single-photon state of Eq.(3.1.12) can be generated from a two level atomic transition. Fig.3.1.1 illustrates a simple model of the process.

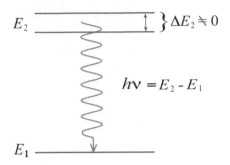

E_2 $\left.\rule{0pt}{12pt}\right\}\Delta E_2 \neq 0$

$h\nu = E_2 - E_1$

E_1

FIGURE 3.1.1 Schematic model of a possible atomic transition that generates the single-photon state, or wavepacket, characterized by Eq. (3.1.12). A two level atom is placed inside a large but finite sized cubic cavity. The upper energy level E_2 has a finite width of $\Delta E_2 \neq 0$. The emitted photon, or the radiation field excited by this atomic transition, may be in any, or in all permissible states, of $|\Psi_j\rangle = \hat{a}_{\mathbf{k},s}^\dagger |0\rangle$ with probability amplitude $f(\mathbf{k}, s)$.

A two level atom is placed inside a large but finite size cubic cavity. The upper energy level E_2 has a finite width of $\Delta E_2 \neq 0$. A single-photon, or wavepacket, is emitted from the atomic transition from E_2 to E_1. The created photon may excite any or excite all permissible states $|\Psi_j\rangle = \hat{a}_{\mathbf{k},s}^\dagger |0\rangle$ with probability amplitude $f(\mathbf{k}, s) = |f(\mathbf{k}, s)|e^{i\varphi(\mathbf{k},s)}$. The space-time behavior of each permissible radiation mode is determined by the boundary condition in solving the Maxwell equations. The probability amplitude distribution is determined by the property of the atomic transition and the property of the cavity, mainly the energy uncertainty ΔE_2. Because all superposed states are generated from the same atomic transition, it is reasonable to consider $\varphi(\mathbf{k}, s) = \varphi_0$. Mathematically, the coherent superposition of a special set of Fock states $|\Psi_j\rangle = |0, ...0, 1_{\mathbf{k},s}, 0, ...0\rangle$ of total occupation number $n = 1$ represents a vector in Hilbert space. Physically, it characters the state of a radiation field that is excited by an atomic transition.

What can we learn from Eq. (3.1.12) about the radiation field?

(1) The radiation field is characterized by a pure state, a vector in Hilbert space. The vector basis of this Hilbert space is Fock state of $n = 1$. In general, a pure state can be used to characterize the state of a quantum system or an ensemble of identical quantum systems.

(2) The radiation contains a set of possible excited modes $(\mathbf{k}, s)$ of occupation number $n_{\mathbf{k},s} = 1$ with probability amplitude $f(\mathbf{k}, s)$, while all other modes stay in their ground state with occupation number $n = 0$. The radiation field is expressed as the superposition of a special set of Fock states of total photon number, $n = \sum n_{\mathbf{k},s} = 1$.

(3) The state of the field is not an eigenstate of the number operator $\hat{n}_{\mathbf{k},s}$. The mean occupation number of the mode $(\mathbf{k}, s)$ is calculated as

$$\langle \Psi | \hat{n}_{\mathbf{k},s} | \Psi \rangle = |f(\mathbf{k}, s)|^2,$$

which equals the probability of exciting the mode $(\mathbf{k}, s)$.

(4) The state of the field is an eigenstate of the total number operator $\hat{n} = \sum_{\mathbf{k},s} \hat{n}_{\mathbf{k},s}$

with eigenvalue $n = 1$,

$$\hat{n}\,|\Psi\rangle = \sum_{\mathbf{k}',s'} \hat{n}_{\mathbf{k}',s'} \sum_{\mathbf{k},s} f(\mathbf{k},s)\,\hat{a}^{\dagger}_{\mathbf{k},s}\,|0\rangle = 1 \sum_{\mathbf{k},s} f(\mathbf{k},s)\,\hat{a}^{\dagger}_{\mathbf{k},s}\,|0\rangle.$$

(5) The state of the field is not an eigenstate of the Hamiltonian $\hat{H}_{\mathbf{k},s}$. The mean energy of the mode $(\mathbf{k},s)$ is calculated as

$$\langle\,\Psi\,|\,\hat{H}_{\mathbf{k},s}\,|\,\Psi\,\rangle = |f(\mathbf{k},s)|^2\,\hbar\omega_{\mathbf{k},s}.$$

(6) The state of the field is not an eigenstate of the total Hamiltonian $\hat{H} = \sum_{\mathbf{k},s}\hat{H}_{\mathbf{k},s}$. The expectation value of the total Hamiltonian, or the mean energy of the radiation, is calculated as

$$\langle\Psi|\hat{H}\,|\Psi\rangle = \sum_{\mathbf{k},s} |f(\mathbf{k},s)|^2\,\hbar\omega_{\mathbf{k},s},$$

which is the mean value of all possible quantized energy $\hbar\omega_{\mathbf{k},s}$, averaged statistically with a weighting function $|f(\mathbf{k},s)|^2$.

What can we say about the state of Eq. (3.1.12) in terms of the concept of photon?

(1) Eq. (3.1.12) is a pure single-photon state thus characterizes the state of a photon that is created from an atomic transition. The state can also be used to characterize an homogenous ensemble of identical photons.

(2) The photon is not in any defined single-mode Fock state but has a certain probability to be in any or in all single-mode Fock state of $n = 1$ within the superposition.

(3) The photon does not have any defined energy of $\hbar\omega$, but may take any or all possible values of $\hbar\omega_{\mathbf{k},s}$ within the superposition.

(4) The photon is localized within a wavepacket which is a vector in the Hilbert space.[1]

(5) The mean energy of the wavepacket carries is $\sum_{\mathbf{k},s}|f(\mathbf{k},s)|^2\hbar\omega_{\mathbf{k},s}$, however, the energy of the localized photon is allowed only to take values of $\hbar\omega_{\mathbf{k},s}$. More precisely, the photon has a probability $|f(\mathbf{k},s)|^2$ of carrying energy $\hbar\omega_{\mathbf{k},s}$.

3.2 COHERENT STATE OF RADIATION FIELD

Coherent state is defined as the eigenstate of the annihilation operator

$$\hat{a}|\alpha\rangle = \alpha|\alpha\rangle. \qquad (3.2.1)$$

It is convenient to write the eigenvalue α in terms of an amplitude and a phase

$$\alpha = a\,e^{i\varphi} \quad \text{with} \quad a = |\alpha|.$$

Coherent states form a complete set of vector space. Similar to number states, coherent states can be used as a vector basis for characterizing radiation field, except coherent states are nonorthogonal vectors in general. It is straightforward to obtain an expression of $|\alpha\rangle$ in terms of the number state $|n\rangle$ from Eq. (3.2.1)

$$|\alpha\rangle = e^{-|\alpha|^2/2} \sum_{n=0}^{\infty} \frac{\alpha^n}{\sqrt{n!}}\,|n\rangle, \qquad (3.2.2)$$

---

[1]This statement is different from the statement of "photon is a wavepacket". Perhaps, a more careful statement should be the follows: the created photon excites radiation field in the form of a localized wavepacket. Moreover, in certain measurements, the state of a photon may be described by a set of wavepackets, which means that a photon can be localized within a set of wavepackets.

by applying

$$\hat{a}|n\rangle = \sqrt{n}\,|n-1\rangle.$$

Since

$$|n\rangle = \frac{1}{\sqrt{n!}}(\hat{a}^\dagger)^n\,|0\rangle,$$

Eq. (3.2.2) can be written as

$$|\alpha\rangle = e^{-|\alpha|^2/2}\,e^{\alpha\hat{a}^\dagger}\,|0\rangle. \tag{3.2.3}$$

Eq. (3.2.3) is thus formally rewritten as

$$|\alpha\rangle = e^{-|\alpha|^2/2}\,e^{\alpha\hat{a}^\dagger}e^{-\alpha^*\hat{a}}\,|0\rangle = e^{\alpha\hat{a}^\dagger - \alpha^*\hat{a}}\,|0\rangle, \tag{3.2.4}$$

since

$$e^{-\alpha^*\hat{a}}\,|0\rangle = 0,$$

and

$$e^{\hat{A}+\hat{B}} = e^{-[\hat{A},\hat{B}]/2}\,e^{\hat{A}}e^{\hat{B}},$$

where $\hat{A}$ and $\hat{B}$ are any operators that satisfy

$$[[\hat{A},\hat{B}],\hat{A}] = [[\hat{A},\hat{B}],\hat{B}] = 0.$$

Here, we have taken $\hat{A} = \alpha\hat{a}^\dagger$, $\hat{B} = -\alpha^*\hat{a}$. By writing $|\alpha\rangle$ in the form of Eq. (3.2.4), we thus introduced a unitary operator, namely the displacement operator

$$\hat{D}(\alpha) = e^{\alpha\hat{a}^\dagger - \alpha^*\hat{a}} = e^{-|\alpha|^2/2}\,e^{\alpha\hat{a}^\dagger}e^{-\alpha^*\hat{a}} = e^{|\alpha|^2/2}\,e^{-\alpha^*\hat{a}}e^{\alpha\hat{a}^\dagger},$$

for the purpose of expressing $|\alpha\rangle$ as a unitary transformation of $|0\rangle$:

$$|\alpha\rangle = \hat{D}(\alpha)\,|0\rangle. \tag{3.2.5}$$

Eq. (3.2.5) is useful for further theoretical concerns and discussions about coherent state.

Some important properties of the coherent states are listed as follows:

(1) The mean number of photons in the coherent state $|\alpha\rangle$ is $|\alpha|^2$,

$$\bar{n} = \langle\hat{n}\rangle = \langle\alpha|\,\hat{n}\,|\alpha\rangle = |\alpha|^2, \tag{3.2.6}$$

and the probability of finding n photons in $|\alpha\rangle$ is given by a Poisson distribution,

$$P(n) = \langle n|\alpha\rangle\langle\alpha|n\rangle = \frac{\bar{n}^n}{n!}e^{-\bar{n}}. \tag{3.2.7}$$

The photon number distribution of laser light is close to the Poisson distribution. The Poisson distribution is useful for simulating single-photon state. For instance, reducing the intensity of a laser field to mean number $\bar{n} = 0.01$, we find that the field has $\sim 99\%$ probability to be in its ground state $|0\rangle$, the probability of being in $|1\rangle$ and in $|2\rangle$ are $\sim 1\%$ and $\sim 0.01\%$, respectively. For a photon counting type experiment with 1% error, the contributions from $|2\rangle$ and higher numbers are ignorable. Taking first-order approximation, the state of the measured field can be approximated as

$$|\Psi\rangle \simeq |0\rangle + \epsilon\,|1\rangle + \epsilon^2...$$

where $\epsilon \ll 1$. The Poisson distribution is also useful for simulating number state of $n \gg 1$.

For instance, achieving mean number of $\bar{n} = 10^6$, its photon number distribution has a peak at $n = 10^6$ with $\Delta n \sim 10^3$ which is three-orders smaller than $\bar{n}$.

(2) The vector set of coherent states $|\alpha\rangle$ is a complete set in Hilbert space

$$\frac{1}{\pi} \int d^2\alpha \, |\alpha\rangle\langle\alpha| = 1. \tag{3.2.8}$$

The completeness relation of Eq. (3.2.8) indicates that the coherent states can be used as a vector basis for expanding any quantum state of radiation. To prove this, we substitute the number state expansion of the coherent state into the integral, obtaining

$$\int d^2\alpha \, |\alpha\rangle\langle\alpha| = \int d^2\alpha \, e^{-|\alpha|^2} \frac{\alpha^n (\alpha^*)^n}{n!} \sum_n |n\rangle\langle n| = \pi \sum_n |n\rangle\langle n|, \tag{3.2.9}$$

where we have applied the result of the following integral[2]

$$\int d^2\alpha \, e^{-|\alpha|^2} \frac{\alpha^n (\alpha^*)^n}{n!} = \pi.$$

Since the Fock states $|n\rangle$ form a complete orthonormal set of vector basis, the sum in Eq. (3.2.9) gives the unit operator.

(3) Two coherent states $|\alpha\rangle$ and $|\alpha'\rangle$ are not orthogonal unless $|\alpha - \alpha'| \gg 1$,

$$\langle\alpha|\alpha'\rangle = e^{-\frac{1}{2}(|\alpha|^2 - 2\alpha^*\alpha' + |\alpha'|^2)}, \tag{3.2.10}$$

and

$$|\langle\alpha|\alpha'\rangle|^2 = e^{-|\alpha - \alpha'|^2}. \tag{3.2.11}$$

This means that the vector basis of coherent states is in principle overcomplete. However, if the experimental condition achieves $|\alpha - \alpha'| \gg 1$ the set of coherent states can be approximated as orthonormal.

(4) Similar to photon number state, we can define a multi-mode coherent state, which is written as a product of single-mode coherent state $|\alpha_{\mathbf{k},s}\rangle$

$$|\{\alpha\}\rangle = \prod_{\mathbf{k},s} |\alpha_{\mathbf{k},s}\rangle, \tag{3.2.12}$$

where $\mathbf{k}$ and s indicate the wavenumber vector and the polarization of the mode, respectively. $|\{\alpha\}\rangle$ is an eigenstate of the annihilation operator with an eigenvalue $\alpha_{\mathbf{k},s}$,

$$\hat{a}_{\mathbf{k},s} |\{\alpha\}\rangle = \alpha_{\mathbf{k},s} |\{\alpha\}\rangle. \tag{3.2.13}$$

Similar to the discussion for number state, we may also define a multi-mode annihilation operator

$$\hat{a} = \sum_{\mathbf{k},s} \hat{a}_{\mathbf{k},s}. \tag{3.2.14}$$

---

[2]

$$\int d^2\alpha \, e^{-|\alpha|^2} \alpha^m (\alpha^*)^n = \int_0^\infty d|\alpha| \, e^{-|\alpha|^2} |\alpha|^{m+n+1} \int_0^{2\pi} d\varphi \, e^{i(m-n)\varphi} = \pi n! \, \delta_{mn}$$

It is easy to find that the multi-mode coherent state defined in Eq. (3.2.12) is an eigenstate of the multi-mode annihilation operator with eigenvalue $\sum_{\mathbf{k},s} \alpha_{\mathbf{k},s}$

$$\hat{a}\,|\{\alpha\}\rangle = \left(\sum_{\mathbf{k},s}\hat{a}_{\mathbf{k},s}\right)\prod_{\mathbf{k},s}|\alpha_{\mathbf{k},s}\rangle = \left(\sum_{\mathbf{k},s}\alpha_{\mathbf{k},s}\right)|\{\alpha\}\rangle. \tag{3.2.15}$$

For example, similar to the number state, we may construct a two-mode coherent state

$$|\Psi\rangle = |0,...,0,\alpha_{\mathbf{k},s},0,...\alpha_{\mathbf{k}',s'},0,...0\rangle, \tag{3.2.16}$$

which is an eigenstate of $\left(\sum_{\mathbf{k},s}\hat{a}_{\mathbf{k},s}\right)$ with eigenvalue $(\alpha_{\mathbf{k},s} + \alpha_{\mathbf{k}',s'})$,

$$\left(\sum_{\mathbf{k},s}\hat{a}_{\mathbf{k},s}\right)|0,...,0,\alpha_{\mathbf{k},s},0,...\alpha_{\mathbf{k}',s'},0,...\rangle$$
$$= (\alpha_{\mathbf{k},s} + \alpha_{\mathbf{k}',s'})|0,...,0,\alpha_{\mathbf{k},s},0,...\alpha_{\mathbf{k}',s'},0,...\rangle.$$

3.3 DENSITY OPERATOR, DENSITY MATRIX, AND THE EXPECTATION VALUE OF AN OBSERVABLE

For a pure state, the expectation value of an observable (operator) is easily calculated

$$\langle\hat{A}\rangle = \langle\Psi|\hat{A}|\Psi\rangle = \langle\hat{A}\rangle_{\text{QM}} \equiv \langle\hat{A}\rangle_{|\Psi\rangle}. \tag{3.3.1}$$

Unfortunately, we are not always dealing with pure states. In certain measurements, the radiation field is in mixed state and can only be described statistically. In this case, a density operator will be defined to characterize the field. To calculate the expectation value of an observable for mixed state, in addition to the quantum average, a statistical ensemble average is necessary. The density operator is helpful in calculating the expectation value of an observable for mixed states.

The density operator is defined in the following. Assume a measurement in which we have to deal with radiation in a mixed state. The field can be described by N randomly distributed state vectors $|\Psi_j\rangle$ in Hilbert space, where N can be any integer from two to infinity, $N = 2\ldots\infty$. The best knowledge we can have about this radiation is that it has a certain probability P_j of being in the state $|\Psi_j\rangle$. The expectation value of an observable $\hat{A}$ is calculated as

$$\langle\hat{A}\rangle = \sum_j P_j\langle\Psi_j|\hat{A}|\Psi_j\rangle = \langle\langle\hat{A}\rangle_{\text{QM}}\rangle_{\text{En}}. \tag{3.3.2}$$

In Eq.(3.3.2), we first calculate the expectation values of $\hat{A}$ for each individual state $|\Psi_j\rangle$, and than evaluate the statistical averaged value for the ensemble. Applying completeness $\sum_n |n\rangle\langle n| = 1$,

$$\langle\langle\hat{A}\rangle_{\text{QM}}\rangle_{\text{Ensamble}} = \sum_n\sum_j P_j\langle\Psi_j|\hat{A}|n\rangle\langle n|\Psi_j\rangle$$
$$= \sum_n\sum_j P_j\langle n|\Psi_j\rangle\langle\Psi_j|\hat{A}|n\rangle$$
$$= \sum_n\langle n|\hat{\rho}\hat{A}|n\rangle$$
$$= tr\,\hat{\rho}\hat{A}$$
$$\equiv \langle\hat{A}\rangle_{\hat{\rho}} \tag{3.3.3}$$

where we have introduced the density operator

$$\hat{\rho} = \sum_j P_j |\Psi_j\rangle\langle\Psi_j| \qquad (3.3.4)$$

to specify the radiation field in mixed state. By tracing operators $\hat{\rho}\hat{A}$ we obtain the expectation value of an observable from a radiation field in mixed state.

Following the same procedure, we can formally define a density operator for pure state

$$\hat{\rho} \equiv |\Psi\rangle\langle\Psi|. \qquad (3.3.5)$$

The expectation value of an observable (operator) can be calculated accordingly,

$$\langle\hat{A}\rangle = tr\,\hat{\rho}\hat{A} = \langle\Psi|\hat{\rho}\hat{A}|\Psi\rangle = \langle\Psi|\Psi\rangle\langle\Psi|\hat{A}|\Psi\rangle = \langle\Psi|\hat{A}|\Psi\rangle \qquad (3.3.6)$$

Since any state $|\Psi_j\rangle$ can be expanded in terms of a chosen vector basis, the density operator $\hat{\rho}$ is also able to be expanded in terms of a chosen vector basis such as photon number states, coherent states, etc. For the density operator of a pure state,

$$\hat{\rho} = \sum_m \sum_n |m\rangle\langle m|\Psi\rangle\langle\Psi|n\rangle\langle n| \equiv \sum_m \sum_n \rho_{mn} |m\rangle\langle n|, \qquad (3.3.7)$$

which defines the density matrix for pure state

$$\rho_{mn} \equiv \langle m|\Psi\rangle\langle\Psi|n\rangle.$$

For the density operator of a mixed state,

$$\hat{\rho} = \sum_m \sum_n \sum_j P_j |m\rangle\langle m|\Psi_j\rangle\langle\Psi_j|n\rangle\langle n| \equiv \sum_m \sum_n \rho_{mn} |m\rangle\langle n|. \qquad (3.3.8)$$

which defines the density matrix for mixed state

$$\rho_{mn} \equiv P_j \langle m|\Psi_j\rangle\langle\Psi_j|n\rangle.$$

Some useful properties of the density matrix are listed in the following:

(1) $\rho_{mm} \geq 0$.

The diagonal elements of the density matrix are real, and non-negative. This follows immediately from $\rho_{mm} = \langle m|\Psi_j\rangle\langle\Psi_j|m\rangle = |\langle m|\Psi_j\rangle|^2 \geq 0$.

(2) $\sum_m \rho_{mm} = 1$ or $tr\,\hat{\rho} = 1$.

The density matrix is defined in terms of the normalized states, $|\Psi_j\rangle$, and the probability distribution of the field, P_j. This makes the interpretation of the diagonal elements as probabilities valid. It is obvious $\rho_{mm} < 1$ for a mixed state. Pure state can be treated as a special case of mixed state in which the field is in the state $|\Psi\rangle$ with certainty ($P = 1$).

(3) $\rho_{mn}^* = \rho_{nm}$ or $\hat{\rho}^\dagger = \hat{\rho}$.

This means that the density matrix is Hermitian. Since the density matrix is Hermitian it can be diagonalized by a unitary transformation. In this book, we will choose photon number state as the basis. Photon number states are the eigen states of the Hamiltonian. The time evolution of photon number states involves phase propagation only in optical measurements. This property is useful for the propagation of the state or the operator.

We will show in the next a few sections that the density matrix of thermal field only have diagonal elements in the basis of photon number states.

(4) $\hat{\rho}^2 = \hat{\rho}$ for pure state.

This property is easy to prove and is useful for distinguishing pure states from mixed states.

We have mentioned earlier that the single-photon state of Eq.(3.1.12) can be a pure state or a mixed state. The mixed state is usually represented by the density operator

$$\hat{\rho} = \sum_{\mathbf{k},s} |f(\mathbf{k},s)|^2 \, \hat{a}^{\dagger}_{\mathbf{k},s} \, |\, 0 \,\rangle\langle 0|\hat{a}_{\mathbf{k},s}. \tag{3.3.9}$$

In some discussion, it is convenient to write Eq.(3.3.9) in 1-D :

$$\hat{\rho} = \int d\omega \, |f(\omega)|^2 \, \hat{a}^{\dagger}(\omega) \, |\, 0 \,\rangle\langle 0| \, \hat{a}(\omega) \tag{3.3.10}$$

where we have also ignored the polarization of the field. In this representation, although we have chosen the same single-photon Fock state of $n = 1$

$$\hat{a}^{\dagger}(\omega) \, |\, 0 \,\rangle = |0, \, ... \, 0, \, 1(\omega), \, 0, \, ... \, 0\rangle \tag{3.3.11}$$

as the vector basis of the Hilbert space that was used for specifying a pure state, for the mixed state of Eq.(3.3.10), we do not have enough knowledge on the superposition between these Fock states. These single-photon state vector are randomly distributed in Hilbert space. The state of the radiation can only be described statistically. What we know is the probability, $|f(\omega)|^2$, for the radiation field to be in a Fock state of Eq.(3.3.11).

In the following, we introduce another freedom to the single-photon state in addition to the wave vector $\mathbf{k}$ and polarization s. Suppose we are evaluating $\langle \hat{n} \rangle$ for a weak natural light source by assuming a photodetection device that is able to measure the mean photon number of the radiation. The photons are randomly created from a large number of spontaneous atomic transitions, and we know each atomic transition, such as the mth atomic transition, produces a single photon state of Eq.(3.1.12),

$$|\Psi_m\rangle = \sum_s \int d\mathbf{k} \, f_m(\mathbf{k},s) \, \hat{a}^{\dagger}_m(\mathbf{k},s) \, |\, 0 \,\rangle, \tag{3.3.12}$$

as a coherent superposition of Fock states $n = 1$. To simplify the mathematics, we write Eq.(3.3.12) in 1-D and focus on one polarization, the single-photon state created from the mth atomic transition is thus

$$|\Psi_m\rangle = \int d\omega \, f_m(\omega) \, \hat{a}^{\dagger}_m(\omega) \, |\, 0 \,\rangle. \tag{3.3.13}$$

We have enough reason to consider $|\Psi_m\rangle$ a pure state and treat it as a vector in the space of single-photon Fock state of Eq.(3.3.11). Mathematically, it is easy to prove $|\Psi_m\rangle$, $m = 1$ to $\sim \infty$, form a complete orthogonal vector basis of a new Hilbert space. The following superposition in the new vector space may represent the state of a radiation excited from these atomic transitions,

$$|\Psi\rangle = \sum_m c_m |\Psi_m\rangle = \sum_m c_m \int d\omega \, f_m(\omega) \, \hat{a}^{\dagger}_m(\omega) \, |\, 0 \,\rangle \tag{3.3.14}$$

where $c_m = |c_m|e^{-i\varphi_m}$ is the normalized complex probability amplitude of observing a photon from the mth atomic transition, or finding the radiation field in state of Eq.(3.3.13). Similar to the superposition in the polarization space of $|\hat{R}\rangle$ and $|\hat{L}\rangle$, and in the cat space of $|+\text{cat}\rangle$ and $|-\text{cat}\rangle$, when $\varphi_m = \text{constant}$, this coherent superposition forms a vector in the space of $|\Psi_m\rangle$, $m = 1$ to $\sim \infty$, representing a pure state of the radiation. When $\varphi_m = \text{random}$ number, this incoherent superposition represents a set of randomly distributed vectors in the space of $|\Psi_m\rangle$, which is usually written as a density operator

$$\hat{\rho} = \sum_m P_m |\Psi_m\rangle\langle\Psi_m|$$

$$= \sum_m P_m \int\int d\omega\, d\omega'\, f_m(\omega)f_m^*(\omega')\, \hat{a}_m^\dagger(\omega)|0\rangle\langle 0|\hat{a}_m(\omega'). \tag{3.3.15}$$

where $P_m = |c_m|^2$ is the probability of observing a photon from the mth atomic transition, or finding the radiation field in the state of Eq.(3.3.13). Eq.(3.3.15) indicates a mixed state of the radiation.

In general, we usually use Eq.(3.3.14) representing a radiation field in pure state and use Eq.(3.3.15) representing a radiation field in mixed state.

The mean photon number $\langle\hat{n}\rangle$ for the pure state of Eq.(3.3.14) is straightforward,

$$\langle\hat{n}\rangle_{|\Psi\rangle} = \langle\Psi|\hat{n}|\Psi\rangle = 1. \tag{3.3.16}$$

The mean photon number $\langle\hat{n}\rangle$ for the mixed state of Eq.(3.3.15) is calculated as follows

$$\langle\hat{n}\rangle_{\hat{\rho}} = tr\,\hat{\rho}\hat{A} = \sum_m P_m\langle\Psi_m|\hat{n}|\Psi_m\rangle = 1. \tag{3.3.17}$$

There should be no surprise that both Eq.(3.3.14) and Eq.(3.3.15) represent single-photon states (pure and mixed). We have observed similar situation earlier in the discussions of polarization and Schrödinger cat. The polarization state represented in the Hilbert space of $|\hat{R}\rangle$ and $|\hat{L}\rangle$, and the cat state represented in the Hilbert space of $|+\text{cat}\rangle$ and $|-\text{cat}\rangle$, are still single-photon state and single-cat state. In this sense, we can consider the Hilbert space of $|\Psi_m\rangle$, $m = 1$ to $\sim \infty$, as a "rotation" in Hilbert space of single-photon Fock state $n = 1$.

In the above mean photon number calculation for mixed state, we have used the density operator of Eq.(3.3.15). To calculate the expectation value of an observable for mixed state, it is unnecessary to start from tracing operators. For example, we may write a mixed single-photon state formally as an incoherent superposition of $|\Psi_m\rangle$,

$$|\Psi\rangle = \sum_m c_m|\Psi_m\rangle = \sum_m |c_m|e^{i\varphi_m}\int d\omega\, f_m(\omega)\,\hat{a}_m^\dagger(\omega)\,|0\rangle \tag{3.3.18}$$

$$\varphi_m = \text{random number}$$

where $c_m = |c_m|e^{i\varphi_m}$ is the normalized complex probability amplitude for the system to be in the state $|\Psi_m\rangle$, and φ_m takes any indeterministic random value. The expectation value of an arbitrary observable $\langle\hat{A}\rangle$ is calculated accordingly:

$$\langle\langle\hat{A}\rangle_{\text{QM}}\rangle_{\text{En}} = \Big\langle\sum_{m,n}|c_m||c_n|\,e^{-i(\varphi_m-\varphi_n)}\langle\Psi_m|\hat{A}|\Psi_n\rangle\Big\rangle_{\text{En}}$$

$$= \sum_m P_m\langle\Psi_m|\hat{A}|\Psi_m\rangle \tag{3.3.19}$$

where $P_m = |c_m|^2$, and we have completed the ensemble average by taking into account all possible values of the relative phases of $(\varphi_m - \varphi_n)$. The ensemble average sums all the $m \neq n$ terms to zero. Replace the arbitrary operator $\hat{A}$ with the photon number operator $\hat{n}$, we obtain

$$\langle\langle\hat{n}\rangle_{\mathrm{QM}}\rangle_{\mathrm{En}} = \langle \sum_{m,n} |c_m||c_n|\, e^{-i(\varphi_m-\varphi_n)}\langle\Psi_m|\hat{n}|\Psi_n\rangle\rangle_{\mathrm{En}}$$

$$= \sum_m P_m \langle\Psi_m|\hat{n}|\Psi_m\rangle$$

$$= 1, \tag{3.3.20}$$

agreeing with Eq.(3.3.17).

3.4 PURE STATE AND MIXED STATE

The concepts of pure states and mixed states include two important aspects of physics: the state of the individual quantum and the state of the measured ensemble.

(1) In terms of the individual quantum system: a pure state represents a vector in Hilbert space. If the state of a quantum can be described by a vector, the state of the quantum is said to be *pure*. On the contrary, if the state of a quantum cannot be described as a vector but rather a mixture of vectors, the quantum is said to be in a *mixed state*.

(2) In terms of the measured ensemble: if the state of all the measured quanta of the ensemble can be described by the same vector, the state of the ensemble, or the measured systems of quanta, is said to be *homogeneous* or in a statistically *pure state*. If the measured ensemble cannot be described by the same vector, or the measured ensemble has to be described by several, or by many different vectors, the ensemble is referred to as a statistical *mixture*.

The second aspect is easy to understand. We usually have more questions about aspect (1). In the following, we give three simple examples: (I) polarization measurement of a single-photon state; (II) Schrödinger cat; (III) general measurement of single-photon state. These three simple examples might be helpful for exploring the important physics behind the concept, especially for aspect (1).

Example (I): Polarization measurement of single-photon state.

Consider the set up of a polarization measurement of Fig. 14.0.2, which consists of a far-field single-photon radiation source of frequency ω, a uniaxial crystal, and two photon counting detectors. The point-like photodetectors D_1 and D_2 are used for measuring the ordinary-ray (o-ray) and the extraordinary-ray (e-ray) of the uniaxial crystal, respectively.

After a large number of measurements, we find: (1) no joint photodetection event happens, i.e., D_1 and D_2 are never triggered simultaneously; and (2) the ratio between the single-photon counting rates of D_1 and D_2 is 1, i.e., 50% of the photodetection events are registered by D_1 and another 50% of the photodetection events are recorded by D_2. What is the state of the radiation field? From observation (1), the state can be approximated a single-photon state. Observation (2) suggests at least three possible states:

(1) Possibility one: The state can be a linearly polarized single-photon state with polarization either $45°$ or $-45°$ relative to the polarization of the ordinary-ray,

$$|\Psi_{45°}\rangle = \frac{1}{\sqrt{2}}\left[\hat{a}^\dagger(\omega,\hat{o})|0\rangle + \hat{a}^\dagger(\omega,\hat{e})|0\rangle\right]$$

$$|\Psi_{-45°}\rangle = \frac{1}{\sqrt{2}}\left[\hat{a}^\dagger(\omega,\hat{o})|0\rangle - \hat{a}^\dagger(\omega,\hat{e})|0\rangle\right], \tag{3.4.1}$$

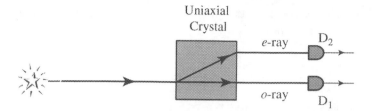

FIGURE 3.4.1 Polarization measurement of a photon. The measurement system consists of a uniaxial crystal and two photon counting detectors, D_1 and D_2. D_1 and D_2 are used for measuring the o-ray and the e-ray of the uniaxial crystal, respectively. The o-ray (e-ray) is polarized horizontally (vertically) before any "rotation" of the crystal-detector system.

where the unit vectors $\hat{o}$ and $\hat{e}$ indicate the polarization of the o-ray and the e-ray, respectively, and the $\pm$ sign indicates that the field is either polarized at $45°$ or $-45°$ relative to the unit vector $\hat{o}$. If the measurements only focus on polarization, we may rewrite Eq. (3.4.1) in the following simplified form

$$|\Psi_{45°}\rangle = |\hat{o}\rangle\langle\hat{o}|\Psi_{45°}\rangle + |\hat{e}\rangle\langle\hat{e}|\Psi_{45°}\rangle = \frac{1}{\sqrt{2}}\left[\,|\hat{o}\rangle + |\hat{e}\rangle\,\right]$$

$$|\Psi_{-45°}\rangle = |\hat{o}\rangle\langle\hat{o}|\Psi_{-45°}\rangle + |\hat{e}\rangle\langle\hat{e}|\Psi_{-45°}\rangle = \frac{1}{\sqrt{2}}\left[\,|\hat{o}\rangle - |\hat{e}\rangle\,\right], \tag{3.4.2}$$

where $|\hat{o}\rangle$ and $|\hat{e}\rangle$ form a complete orthonormal 2-D vector basis. We regard the linear polarization states of $|\Psi_{45°}\rangle$ and $|\Psi_{-45°}\rangle$ as coherent *superposition* of $|\hat{o}\rangle$ and $|\hat{e}\rangle$. The state is a pure state. In the chosen 2-D vector space of $|\hat{o}\rangle$ and $|\hat{e}\rangle$, the density matrix of the density operator $\hat{\rho} = |\Psi\rangle\langle\Psi|$ is thus

$$\hat{\rho}_{45°} = \begin{pmatrix} 1/2 & 1/2 \\ 1/2 & 1/2 \end{pmatrix}, \qquad \hat{\rho}_{-45°} = \begin{pmatrix} 1/2 & -1/2 \\ -1/2 & 1/2 \end{pmatrix}. \tag{3.4.3}$$

(2) Possibility two: The state can be a circularly polarized single-photon state with right-hand polarization or left-hand polarization,

$$|\Psi_R\rangle = \frac{1}{\sqrt{2}}\left[\hat{a}^\dagger(\omega,\hat{o})|0\rangle + i\,\hat{a}^\dagger(\omega,\hat{e})\,|0\rangle\right]$$

$$|\Psi_L\rangle = \frac{1}{\sqrt{2}}\left[-\hat{a}^\dagger(\omega,\hat{o})|0\rangle + i\,\hat{a}^\dagger(\omega,\hat{e})\,|0\rangle\right], \tag{3.4.4}$$

where, again, the unit vectors $\hat{o}$ and $\hat{e}$ indicate the polarization of the o-ray and the e-ray, respectively. If the measurements only focus on polarization, we may rewrite Eq. (3.4.4) in the following simplified form

$$|\Psi_R\rangle = |\hat{o}\rangle\langle\hat{o}|\Psi_R\rangle + |\hat{e}\rangle\langle\hat{e}|\Psi_R\rangle = \frac{1}{\sqrt{2}}\left[\,|\hat{o}\rangle + i\,|\hat{e}\rangle\,\right]$$

$$|\Psi_L\rangle = |\hat{o}\rangle\langle\hat{o}|\Psi_L\rangle + |\hat{e}\rangle\langle\hat{e}|\Psi_L\rangle = \frac{1}{\sqrt{2}}\left[-|\hat{o}\rangle + i\,|\hat{e}\rangle\,\right]. \tag{3.4.5}$$

We regard the circular polarization states of $|\Psi_R\rangle$ and $|\Psi_L\rangle$ as coherent *superposition* of $|\hat{o}\rangle$ and $|\hat{e}\rangle$. The state is a pure state. In the chosen 2-D vector space of $|\hat{o}\rangle$ and $|\hat{e}\rangle$, the density matrix of the density operator is written as

$$\hat{\rho}_R = \begin{pmatrix} 1/2 & -i/2 \\ i/2 & 1/2 \end{pmatrix}, \qquad \hat{\rho}_L = \begin{pmatrix} 1/2 & i/2 \\ -i/2 & 1/2 \end{pmatrix}. \tag{3.4.6}$$

(3) Possibility three: The radiation field could be unpolarized. We regard a mixed polarization state in terms of the ordinary-ray (o-ray) and the extraodinary-ray (e-ray) of the uniaxial crystal

$$\hat{\rho} = \frac{1}{2} \left[\hat{a}^\dagger(\omega, \hat{o}) \, |0\rangle\langle 0 | \, \hat{a}(\omega, \hat{o}) + \hat{a}^\dagger(\omega, \hat{e}) \, |0\rangle\langle 0 | \, \hat{a}(\omega, \hat{e}) \right], \qquad (3.4.7)$$

or in the simplified form

$$\hat{\rho} = P_o \, |\hat{o}\rangle\langle\hat{o}| + P_e \, |\hat{e}\rangle\langle\hat{e}| = \frac{1}{2} \left[|\hat{o}\rangle\langle\hat{o}| + |\hat{e}\rangle\langle\hat{e}| \right], \qquad (3.4.8)$$

where P_o and P_e are the probabilities for the system to be in the polarization states $|\hat{o}\rangle$ and $|\hat{e}\rangle$, respectively. In the chosen vector space in which $|\hat{o}\rangle$ and $|\hat{e}\rangle$ form a complete orthonormal 2-D vector basis, the matrix form of the density operator is written as

$$\hat{\rho} = \begin{pmatrix} 1/2 & 0 \\ 0 & 1/2 \end{pmatrix}. \qquad (3.4.9)$$

It is not difficult to see the above three states (possibilities) yield the same results for the experimental set up of Fig. 14.0.2, i.e., 50%:50% chance to register an photon at D_1 or D_2.

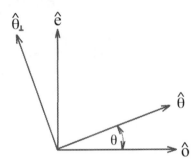

FIGURE 3.4.2   A polarization projection operator $\hat{\theta} = |\hat{\theta}\rangle\langle\hat{\theta}|$ corresponds to a polarization analyzer oriented at angle θ relative to vector $\hat{o}$. Note, $\hat{o}$ is defined by the uniaxial crystal in Fig.14.0.2 without any "rotation".

For a more general measurement, we introduce projection operators $\hat{\theta} = |\hat{\theta}\rangle\langle\hat{\theta}|$ and $\hat{\theta}_\perp = |\hat{\theta}_\perp\rangle\langle\hat{\theta}_\perp|$, corresponding to a polarization analyzer oriented at angle θ relative to vector $\hat{o}$. Note, $\hat{o}$ is defined by the uniaxial crystal of Fig.14.0.2 without any "rotation", see Fig. 3.4.2, in the natural right-hand coordinate system. In the chosen vector space of $|\hat{o}\rangle$ and $|\hat{e}\rangle$, the density matrixes of the projection operators $\hat{\theta}$ and $\hat{\theta}_\perp$ are

$$\hat{\theta} = \begin{pmatrix} \cos^2\theta & \cos\theta\sin\theta \\ \sin\theta\cos\theta & \sin^2\theta \end{pmatrix}, \qquad (3.4.10)$$

and

$$\hat{\theta}_\perp = \begin{pmatrix} \sin^2\theta & -\sin\theta\cos\theta \\ -\cos\theta\sin\theta & \cos^2\theta \end{pmatrix} \qquad (3.4.11)$$

respectively, where the matrix elements $\langle\hat{i}|\hat{\theta}\rangle\langle\hat{\theta}|\hat{j}\rangle = \cos(\hat{i}, \hat{\theta}) \cos(\hat{\theta}, \hat{j})$, and $\langle\hat{i}|\hat{\theta}_\perp\rangle\langle\hat{\theta}_\perp|\hat{j}\rangle = \cos(\hat{i}, \hat{\theta}_\perp) \cos(\hat{\theta}_\perp, \hat{j})$, with $\hat{i}, \hat{j} = \hat{o}, \hat{e}$, and θ is the angle between the unit vectors $\hat{\theta}$ and $\hat{o}$.

In general, the expectation value of the polarization measurement is calculated by taking the trace of the matrix $\hat{\rho}\hat{\theta}$ and $\hat{\rho}\hat{\theta}_\perp$

$$\langle \hat{\theta} \rangle = tr\, \hat{\rho}\hat{\theta}, \quad \langle \hat{\theta}_\perp \rangle = tr\, \hat{\rho}\hat{\theta}_\perp. \tag{3.4.12}$$

For pure states,

$$\langle \hat{\theta} \rangle = \langle \Psi | \hat{\theta} \rangle \langle \hat{\theta} | \Psi \rangle = |\langle \hat{\theta} | \Psi \rangle|^2$$
$$\langle \hat{\theta}_\perp \rangle = \langle \Psi | \hat{\theta}_\perp \rangle \langle \hat{\theta}_\perp | \Psi \rangle = |\langle \hat{\theta}_\perp | \Psi \rangle|^2. \tag{3.4.13}$$

For mixed states,

$$\langle \hat{\theta} \rangle = \sum_j P_j \langle \Psi_j | \hat{\theta} \rangle \langle \hat{\theta} | \Psi_j \rangle = \sum_j P_j |\langle \hat{\theta} | \Psi_j \rangle|^2$$
$$\langle \hat{\theta}_\perp \rangle = \sum_j P_j \langle \Psi_j | \hat{\theta}_\perp \rangle \langle \hat{\theta}_\perp | \Psi_j \rangle = \sum_j P_j |\langle \hat{\theta}_\perp | \Psi_j \rangle|^2. \tag{3.4.14}$$

In the measurement of Fig. 14.0.2, the projection angle θ of the polarization analyzer (the uniaxial crystal) is chosen to be either $\theta = 0$ (o-ray) ($\theta_\perp = 90°$) or $\theta = 90°$ (e-ray) ($\theta_\perp = 180°$). Definitely, these measurements cannot distinguish the above three different states.

One may immediately ask the following question: Can we distinguish the pure states Eq. (3.4.1) and Eq. (3.4.4) from the mixed state of Eq. (3.4.7), in principle and in practice? For this kind of simple polarization measurement, the answer is positive. For example, one may consider to rotate the polarization analyzer to either the $45°$ or the $-45°$) directions, If D_1 (D_2) shows a maximum (minimum) counting rate at $45°$ and a minimum (maximum) counting rate at $-45°$, the field is polarized at $45°$

$$| \Psi \rangle = \hat{a}^\dagger(\omega, 45°) | 0 \rangle \equiv | \widehat{45°} \rangle$$

If D_1 (D_2) has a minimum (maximum) counting rate at $45°$ and a maximum (minimum) counting rate at $-45°$, the field is polarized at $-45°$,

$$| \Psi_\perp \rangle = \hat{a}^\dagger(\omega, -45°) | 0 \rangle \equiv | -\widehat{45°} \rangle.$$

The $45°$ polarization and $-45°$ polarization are distinguished from the mixed state and the circular polarization.

In general, a linearly polarized single-photon state of $|\hat{\theta}\rangle$ or $|\hat{\theta}\rangle_\perp$ can be easily distinguished by rotating the polarization analyzer to the directions of θ and $\theta_\perp$. In fact, the state of

$$| \Psi \rangle = \hat{a}^\dagger(\omega, \theta) | 0 \rangle \equiv | \hat{\theta} \rangle \tag{3.4.15}$$

and

$$| \Psi_\perp \rangle = \hat{a}^\dagger(\omega, \theta_\perp) | 0 \rangle \equiv | \hat{\theta}_\perp \rangle \tag{3.4.16}$$

form a new complete orthogonal 2-D vector basis for characterizing any polarization state. For example, the states of horizontal polarization, which was originally labeled as $|\hat{o}\rangle$, and vertical polarization, which was originally labeled as $|\hat{e}\rangle$, can be written as the superposition of $|\hat{\theta}\rangle$ and $|\hat{\theta}_\perp\rangle$

$$| \Psi_o \rangle = \frac{1}{\sqrt{2}} [\hat{a}^\dagger(\omega, \theta) | 0 \rangle + \hat{a}^\dagger(\omega, \theta_\perp) | 0 \rangle] \tag{3.4.17}$$

$$| \Psi_e \rangle = \frac{1}{\sqrt{2}} [\hat{a}^\dagger(\omega, \theta) | 0 \rangle - \hat{a}^\dagger(\omega, \theta_\perp) | 0 \rangle],$$

or in the simplified form

$$|\hat{o}\rangle = \frac{1}{\sqrt{2}}\left[|\hat{\theta}\rangle + |\hat{\theta}_\perp\rangle\right] \qquad (3.4.18)$$

$$|\hat{e}\rangle = \frac{1}{\sqrt{2}}\left[|\hat{\theta}\rangle - |\hat{\theta}_\perp\rangle\right].$$

Of course, θ and $\theta_\perp$ can be used to characterize a mixed state of polarization too:

$$\hat{\rho} = P(\theta)|\hat{\theta}\rangle\langle\hat{\theta}| + P(\theta_\perp)|\hat{\theta}_\perp\rangle\langle\hat{\theta}_\perp|, \qquad (3.4.19)$$

indicating a 50%:50% chance to be in the states $|\hat{\theta}\rangle$ and $|\hat{\theta}_\perp\rangle$.

Similarly, if we build a polarization analyzer to distinguish the single-photon state of spin +1 from that of spin -1, the circularly polarized single-photon state with right-hand polarization or left-hand polarization can be distinguished easily. In fact, the state of

$$|\Psi\rangle = \hat{a}^\dagger(\omega, +1)|0\rangle \equiv |\hat{R}\rangle \qquad (3.4.20)$$

and

$$|\Psi_\perp\rangle = \hat{a}^\dagger(\omega, -1)|0\rangle \equiv |\hat{L}\rangle \qquad (3.4.21)$$

form another new complete orthogonal 2-D vector basis for characterizing any polarization state. For example, the states of horizontal polarization, which was originally labeled as $|\hat{o}\rangle$, and vertical polarization, which was originally labeled as $|\hat{e}\rangle$ can be written as the coherent superposition of $|\hat{R}\rangle$ and $|\hat{L}\rangle$

$$|\Psi_o\rangle = \frac{1}{\sqrt{2}}\left[\hat{a}^\dagger(\omega, +1)|0\rangle - \hat{a}^\dagger(\omega, -1)|0\rangle\right] \qquad (3.4.22)$$

$$|\Psi_e\rangle = \frac{-i}{\sqrt{2}}\left[\hat{a}^\dagger(\omega, +1)|0\rangle + \hat{a}^\dagger(\omega, -1)|0\rangle\right],$$

or in the simplified form

$$|\hat{o}\rangle = \frac{1}{\sqrt{2}}\left[|\hat{R}\rangle - |\hat{L}\rangle\right] \qquad (3.4.23)$$

$$|\hat{e}\rangle = \frac{-i}{\sqrt{2}}\left[|\hat{R}\rangle + |\hat{L}\rangle\right],$$

Of course, the right-hand and left-hand polarization states can be used to characterize a mixed polarization too:

$$\hat{\rho} = P(+1)|\hat{R}\rangle\langle\hat{R}| + P(-1)|\hat{L}\rangle\langle\hat{L}|, \qquad (3.4.24)$$

indicating a 50%:50% chance to be in the states of $|\hat{R}\rangle$ and $|\hat{L}\rangle$.

A pure state represents a vector in Hilbert space. If we could "rotate" our measurement device to identify the vector and its orthogonal conjugates, it is not surprising to have maximum and minimum counting rates at different "directions" in the chosen Hilbert space. Thus, in principle, it is possible, and "easy", to distinguish a pure state from a mixed state as well as from its orthogonal states, such as the state with the "+" sign and the state with the "−" sign in Eq. (3.4.1), by a simple rotation. If we consider the measurement operation a projection, to distinguish a pure state from mixed states as well as from its orthogonal states, we can project the pure state onto itself, or project it onto its orthogonal states.

Mathematically, we may define a projection operator $\hat{P} = |\Psi\rangle\langle\Psi|$, which is nothing but the density operator, to write these projections as

$$\hat{P}|\Psi\rangle = |\Psi\rangle\langle\Psi|\Psi\rangle = |\Psi\rangle$$
$$\hat{P}|\Psi_\perp\rangle = |\Psi\rangle\langle\Psi|\Psi_\perp\rangle = 0. \tag{3.4.25}$$

Eq. (3.4.25) can be expressed in a general form in terms of the density operator

$$\hat{\rho}^2 = |\Psi\rangle\langle\Psi|\Psi\rangle\langle\Psi| = \hat{\rho}, \tag{3.4.26}$$

where the state $|\Psi\rangle$ (and $|\Psi\rangle_\perp$) is normalized as is in Eq. (3.4.1). Any pure state must satisfy Eq. (3.4.26).

It is easy to find that the states $|45°\rangle$ and $|-45°\rangle$ (as well as $|\hat{R}\rangle$ and $|\hat{L}\rangle$) form a complete orthonormal vector basis and satisfy the completeness relation

$$|\Psi\rangle\langle\Psi| + |\Psi_\perp\rangle\langle\Psi_\perp| = 1. \tag{3.4.27}$$

In measurements other than polarization, however, the situation is not that simple. In many cases, we cannot "rotate" our measurement-projection device in the chosen vector space. We may have to keep the state vector as a superposition of the chosen set of eigenvectors of a known operator such as the Hamiltonian. In this case, we may be confused again in understanding the physics of a quantum superposition. For example, if one insists that a photon has to be polarized as either an o-ray or an e-ray when propagated in the uniaxial crystal, it is very reasonable to ask: how could a $45°$ or $-45°$ polarized photon be an o-ray and an e-ray simultaneously? Since we know that an energy bundle of $\hbar\omega$, or a photon, can never be divided into parts. A similar question naturally follows: Has the $45°$ polarized single-photon been "which ray or both ray" when passing the uniaxial crystal? This is equivalent to the question we have been asking since the beginning of quantum theory: Has the observed single-photon passed "which slit or both slit" in a Young's double-slit interferometer? Regarding this historically long standing problem, the Schrödinger cat may be considered the most vivid cartoon model.

Before turning to the physics of the Schrödinger cat, we need a brief discussion about the measurement of a single-photon. Talking about the measurement of a single-photon, it does not mean we only deal with a photon. In fact, one measurement of one photon cannot provide us any meaningful information about the radiation system. In quantum theory, any meaningful physics has to be learned from a large number of measurements, or from the measurement of an ensemble. There are two types of ensembles in quantum theory. The first type is said to be homogeneous or in a statistically *pure* state. In this case, each of the measured N systems in the ensemble is in the same pure state, and thus the ensemble itself is in a pure state. The second type is referred to as a *mixture*. In this case, the measured ensemble of N systems are not necessarily in the same pure state. The measured N systems can be in several, or in many, different states with certain probabilities, and thus the ensemble itself is inhomogeneous or in a statistically mixed state.

We have two pictures of a pure state of a single-photon. The pure state can represent (1) an individual photon or (2) an ensemble of photons. For example, in the polarization measurement, the pure states of Eq. (3.4.1) with "+" sign

$$|\Psi_{45°}\rangle = \frac{1}{\sqrt{2}} \left[\hat{a}^\dagger(\omega, \hat{o})|0\rangle + \hat{a}^\dagger(\omega, \hat{e})|0\rangle \right]$$

means (1) each measured photon (individual) is polarized at $45°$; (2) all measured photons (ensemble) are polarized at 45°, or simply, the measured radiation field is polarized at 45°. The pure states of Eq. (3.4.1) with "-" sign

$$|\Psi_{-45°}\rangle = \frac{1}{\sqrt{2}}[\hat{a}^\dagger(\omega, \hat{o})|0\rangle - \hat{a}^\dagger(\omega, \hat{e})|0\rangle],$$

means (1) each measured photon (individual) is polarized at $-45°$; (2) all measured photons (ensemble) are polarized at $-45°$, or simply, the measured radiation field is polarized at $-45°$. Mathematically, the only difference between $|\Psi_{45°}\rangle$ and $|\Psi_{-45°}\rangle$ is the "$\pm$" sign between the orthogonal basis $\hat{a}^\dagger(\omega, \hat{o})|0\rangle$, or $|\hat{o}\rangle$ in the simplified form, and $\hat{a}^\dagger(\omega, \hat{e})|0\rangle$, or $|\hat{e}\rangle$ in the simplified form. The "+" sign means the $|\hat{o}\rangle$ and $|\hat{e}\rangle$ are "in-phase", and the "-" sign means $|\hat{o}\rangle$ and $|\hat{e}\rangle$ are "out-of-phase".

In fact, we can formally write any polarization state into the following superposition in the 2-D Hilbert space of single-photon Fock state $\hat{a}^\dagger(\omega, \hat{o})|0\rangle$ (or $|\hat{o}\rangle$) and $\hat{a}^\dagger(\omega, \hat{e})|0\rangle$ (or $|\hat{e}\rangle$):

$$|\Psi\rangle = c_o\,\hat{a}^\dagger(\omega, \hat{o})|0\rangle + c_e\,\hat{a}^\dagger(\omega, \hat{e})|0\rangle \tag{3.4.28}$$

where $c_o = |c_o|e^{i\varphi_o}$ and $c_e = |c_e|e^{i\varphi_e}$ are the probability amplitudes of observing a photon in the state $|\hat{o}\rangle$ and/or in the state $|\hat{e}\rangle$, respectively. It is easy to find, when $|c_o| = |c_e|$ and $\varphi_e - \varphi_o = 0$, $|\Psi\rangle = |\Psi_{45°}\rangle$; while $|c_o| = |c_e|$ and $\varphi_e - \varphi_o = \pi$, $|\Psi\rangle = |\Psi_{-45°}\rangle$. It is also easy to find, when $|c_o| = |c_e|$ and $\varphi_e - \varphi_o = \pi/2$, $|\Psi\rangle = |\Psi_R\rangle$; while $|c_o| = |c_e|$ and $\varphi_e - \varphi_o = -\pi/2$, $|\Psi\rangle = |\Psi_L\rangle$.

Because $|\Psi_{45°}\rangle$ and $|\Psi_{45°}\rangle$ ($|\Psi_R\rangle$ and $|\Psi_L\rangle$ as well) form a complete orthogonal 2-D Hilbert space, we can also formally write any polarization state into the following superposition in the 2-D Hilbert space of single-photon Fock state $\hat{a}^\dagger(\omega, 45°)|0\rangle$ (or $|\Psi_{45°}\rangle$) and $\hat{a}^\dagger(\omega, -45°)|0\rangle$ (or $|\Psi_{-45°}\rangle$)

$$|\Psi\rangle = c_{45°}\,\hat{a}^\dagger(\omega, 45°)|0\rangle + c_{-45°}\,\hat{a}^\dagger(\omega, -45°)|0\rangle \tag{3.4.29}$$

where $c_{45°} = |c_{45°}|e^{i\varphi_{45°}}$ and $c_{-45°} = |c_{-45°}|e^{i\varphi_{-45°}}$ are the probability amplitudes of observing a photon in the state $|\Psi_{45°}\rangle$ and/or in the state $|\Psi_{-45°}\rangle$, respectively. It is easy to find, when $|c_{45°}| = |c_{-45°}|$ and $\varphi_{-45°} - \varphi_{45°} = 0$, $|\Psi\rangle = |\hat{o}\rangle$; while $|c_{45°}| = |c_{-45°}|$ and $\varphi_{-45°} - \varphi_{45°} = \pi$, $|\Psi\rangle = |\hat{e}\rangle$.

In terms of the 2-D Hilbert space of $|\Psi_{45°}\rangle$ and $|\Psi_{-45°}\rangle$, the mixed state of Eq. (3.4.19) may have the following meanings: (1) Half of the measured photons in the ensemble are polarized at 45° and the other half of the measured photons in the ensemble are polarized at $-45°$. (2) Each measured photon is neither polarized at 45° nor at $-45°$, but a random polarization in the form of an incoherent superposition

$$|\Psi\rangle = \frac{e^{i\varphi_{45°}}}{\sqrt{2}}\hat{a}^\dagger(\omega, 45°)|0\rangle + \frac{e^{i\varphi_{-45°}}}{\sqrt{2}}\hat{a}^\dagger(\omega, -45°)|0\rangle$$

with $\varphi_{45°}$ = random number and $\varphi_{-45°}$ = random number, or $\varphi_{-45°} - \varphi_{45°}$ = random number. Each photon is polarized with a set of special values of $\varphi_{45°}$ and $\varphi_{-45°}$, and the set takes different values from one to another. The ensemble is represented by a large number of these randomly distributed vectors in the 2-D polarization space of vector basis $|\hat{o}\rangle$ and $|\hat{e}\rangle$, and is thus unpolarized.

Example (II): Schrödinger cat.

There has been a famous *cat* in the history of physics, namely the Schrödinger cat. Schrödinger's cat is in the superposition state of $|\,\mathrm{alive}\,\rangle$ and $|\,\mathrm{dead}\,\rangle$:

$$|+\mathrm{cat}\,\rangle = \frac{1}{\sqrt{2}}[\,|\,\mathrm{alive}\,\rangle + |\,\mathrm{dead}\,\rangle\,], \tag{3.4.30}$$

where $|\,\text{alive}\,\rangle$ and $|\,\text{dead}\,\rangle$ are the eigenvectors of the "life status" operator of a cat. Similar to the linear polarization of a radiation field, the vector space defined by the "life status" operator has a dimension of two. The eigenvectors $|\,\text{alive}\,\rangle$ and $|\,\text{dead}\,\rangle$ form a complete, orthonormal vector space. The Schrödinger cat is not in any of the eigenstates, but rather a superposition of the eigenstates. Of course, there is no such cat in our everyday life. In the quantum world, however, a Schrödinger cat is allowed to be alive and dead simultaneously. Besides the pure state of Eq. (3.4.30), quantum theory permits another pure state which is orthogonal to the state of Eq. (3.4.30)

$$|-\text{cat}\,\rangle = \frac{1}{\sqrt{2}}\left[\,|\,\text{alive}\,\rangle - |\,\text{dead}\,\rangle\,\right]. \tag{3.4.31}$$

First, both states $|+\text{cat}\,\rangle$ and $|-\text{cat}\,\rangle$ in Eq. (3.4.30) and Eq. (3.4.31) are pure states. It is easy to show that

$$\hat{\rho}_{+\text{cat}}^2 = \hat{\rho}_{+\text{cat}}, \qquad \hat{\rho}_{-\text{cat}}^2 = \hat{\rho}_{-\text{cat}}.$$

Second, the states of $|+\text{cat}\,\rangle$ and $|-\text{cat}\,\rangle$ form a complete orthonormal set of vector basis,

$$|+\text{cat}\,\rangle\langle+\text{cat}\,| + |-\text{cat}\,\rangle\langle-\text{cat}\,| = \mathbf{1}. \tag{3.4.32}$$

Therefore, any cat-states can be written as the superposition of $|+\text{cat}\,\rangle$ and $|-\text{cat}\,\rangle$, such as

$$|\,\text{alive}\,\rangle = \frac{1}{\sqrt{2}}\left[\,|+\text{cat}\,\rangle + |-\text{cat}\,\rangle\right. \tag{3.4.33}$$

$$|\,\text{dead}\,\rangle = \frac{1}{\sqrt{2}}\left[+|\,\text{cat}\,\rangle - |-\text{cat}\,\rangle.\right.$$

If we have a measurement-projection device capable of identifying $|+\text{cat}\,\rangle$ and $|-\text{cat}\,\rangle$, we are going to find that an alive-cat (dead-cat as well) would be a $|+\text{cat}\,\rangle$-cat (Schrödinger cat) and $|-\text{cat}\,\rangle$-cat (conjugate Schrödinger cat) simultaneously. Should we ask: How could an alive-cat to be a Schrödinger cat and its conjugate-cat simultaneously?

Third, the pure states of Eq. (3.4.30) and Eq. (3.4.31) are very different from the mixed state

$$\hat{\rho} = \frac{1}{2}\left[\,|\,\text{alive}\,\rangle\langle\,\text{alive}\,| + |\,\text{dead}\,\rangle\langle\,\text{dead}\,|\,\right]. \tag{3.4.34}$$

Comparing with the linear polarization measurement in example one, the states of Eq. (3.4.30) and Eq. (3.4.31) are equivalent to the linear polarization states of $45°$ and $-45°$. We have made a similar rotation to achieve Eq. (3.4.30) and Eq. (3.4.31). Theoretically, this is a simple rotation in the 2-D "cat-space". Unfortunately, no one knows how to achieve that rotation in practice. The only measurement we can do is to identify if a cat is alive or dead. If our measurement-projection device can only project a cat state onto the eigenstate $|\,\text{alive}\,\rangle$ or $|\,\text{dead}\,\rangle$, we may never be able to distinguish the pure states of Eq. (3.4.30) and Eq. (3.4.31) from the mixed state of Eq. (3.4.34). Although we do not know how to achieve a rotation from $|\,\text{alive}\,\rangle$ or $|\,\text{dead}\,\rangle$ to $|\,\text{alive}\,\rangle \pm |\,\text{dead}\,\rangle$ in the cat space, it does not mean the non-existence of this kind of superposition states, especially if we know how to generate such a state.

Similar to the polarization measurement of a single-photon, one measurement of a Schrödinger cat does not give us any meaningful information about the state of the cat. In the theory of Schrödinger's cat, any meaningful physics has to be learned from the measurement of an ensemble of cats. If all measured cats are in the same state, such as that of Eq. (3.4.30) or Eq. (3.4.31), we say that each measured cat is in a pure state and the measured cat-system is in a pure state. If the measured cats are in mixed state of Eq. (3.4.34), it may have the following meaning: (1) half of the cats in the ensemble are alive and the other half of cats in the ensemble are dead. (2) Each measured cat is neither in |alive⟩ nor |dead⟩, but in an incoherent superposition

$$|\Psi\rangle = \frac{e^{i\varphi_{\text{alive}}}}{\sqrt{2}}|\text{alive}\rangle + \frac{e^{i\varphi_{\text{dead}}}}{\sqrt{2}}\hat{a}^\dagger|\text{dead}\rangle \tag{3.4.35}$$

with φ_{alive} = random number and φ_{dead} = random number, or $\varphi_{\text{dead}} - \varphi_{\text{alive}}$ = random number. Each cat is characterized by a set of special values of φ_{alive} and φ_{dead}; and the set takes different values from one cat to another cat. The ensemble of cats is represented by a large number of these randomly distributed vectors in the 2-D cat space.

Eq.(3.4.35) can be formally used to characterize a pure state of cat too. For example, when $\varphi_{\text{dead}} - \varphi_{\text{alive}} = 0$, the coherent superposition between |alive⟩ and |dead⟩ results in the cat state $|+\text{cat}\rangle$ of Eq.(3.4.30); when $\varphi_{\text{dead}} - \varphi_{\text{alive}} = \pi$, the coherent superposition between |alive⟩ and |dead⟩ results in the cat state $|-\text{cat}\rangle$ of Eq.(3.4.31).

Example (III): general measurement of single-photon state.

We now discuss aspect (1) and (2) based on the measurements of single-photon states.

(1) What is the difference between the pure single-photon state of Eq.(3.1.12)

$$|\Psi\rangle = \sum_{\mathbf{k},s} f(\mathbf{k},s)\,\hat{a}^\dagger_{\mathbf{k},s}\,|\,0\,\rangle$$

and the mixed single-photon state of Eq.(3.3.9)

$$\hat{\rho} = \sum_{\mathbf{k},s} |f(\mathbf{k},s)|^2\,\hat{a}^\dagger_{\mathbf{k},s}\,|\,0\,\rangle\langle\,0\,|\,\hat{a}_{\mathbf{k},s} \quad ?$$

In terms of the state of a photon, Eq.(3.1.12) represents a pure state by means of a vector in the Hilbert space of single-photon Fock state $n = 1$. On the contrary, Eq.(3.3.9) represents a mixed state by means of a set of randomly distributed vectors in the same Hilbert space of single-photon Fock state $n = 1$.

Eq.(3.1.12) and Eq.(3.3.9) can be formally written as

$$|\Psi\rangle = \sum_{\mathbf{k},s} |f(\mathbf{k},s)|e^{i\varphi_{\mathbf{k},s}}\,\hat{a}^\dagger_{\mathbf{k},s}\,|\,0\,\rangle \tag{3.4.36}$$

where $\varphi_{\mathbf{k},s}$ represents the phase of the probability amplitude for the photon to be in the single-photon Fock state $\hat{a}^\dagger_{\mathbf{k},s}|0\rangle$. When $\varphi_{\mathbf{k},s}$ = constant, the coherent superposition among the single-photon Fock state $\hat{a}^\dagger_{\mathbf{k},s}|0\rangle$s results in a pure state of Eq.(3.1.12); when $\varphi_{\mathbf{k},s}$ = random number, the incoherent superposition among the single-photon Fock state $\hat{a}^\dagger_{\mathbf{k},s}|0\rangle$s results in a mixed state of Eq.(3.3.9).

Assume both states are generated from an atomic transition, the photon in state of Eq.(3.3.9) must be created from a coherent process in which all radiation modes of $\mathbf{k}$ with polarization s are excited with constant phase and superposed coherently. The probability amplitude for the photon to be in the single-photon Fock state $\hat{a}^\dagger_{\mathbf{k},s}|0\rangle$ is $f(\mathbf{k},s)$, which has a real and positive amplitude $|f(\mathbf{k},s)|$ and a phase $\varphi_{\mathbf{k},s}$, indicating precise magnitude and phase for all projections of the state vector onto each vector basis of the single-photon Fock state $n = 1$. On the contrary, the photon in state Eq.(3.3.9) must be created from a stochastic process in which all radiation modes $\mathbf{k}$ with polarization s are excited with random phases and thus superposed incoherently. The only knowledge we can have about this photon is that it takes probability $|f(\mathbf{k}, s)|^2$ to be in the single-photon Fock state $\hat{a}_{\mathbf{k},s}^\dagger |0\rangle$. The mathematical picture is interesting: although we have precise knowledge on the magnitude of the projection $|f(\mathbf{k}, s)|$ or $|f(\mathbf{k}, s)|^2$, however, the phase takes random value indeterministically.

Can a simple measurement distinguish the pure state of Eq.(3.1.12) from the mixed state of Eq.(3.3.9)? If we can find a "rotation" (measurement, operation) in terms of the Hilbert space of single-photon Fock state of $n = 1$ to project the pure state of Eq.(3.1.12) onto itself and onto its orthogonal states, certainly, we can. This "rotation" is similar to the rotations we have discussed in earlier discussions on the measurement of polarization and Schrödinger's cat. Do we have such a measurement corresponding to such a "simple rotation"? Unfortunately, we do not.

In terms of the ensemble, similar to the physics of the cat system, a pure state requires all photons of the ensemble in the same state, such as the state of Eq.(3.1.12). A mixed state means the photons in the ensemble are not in the same state, but rather in all possible different states. We have two slightly different physical pictures of the ensemble: (1) if the ensemble has N photons, the number of photons in the single-photon Fock state $\hat{a}_{\mathbf{k},s}^\dagger |0\rangle$ is proportional to $|f(\mathbf{k}, s)|^2$; (2) the state of each photon in the ensemble is represented by an incoherent superposition of single-photon Fock state $\hat{a}_{\mathbf{k},s}^\dagger |0\rangle$, such as Eq.(3.4.36), with a special set of $\varphi(\mathbf{k}, s)$s. The special set takes different values from one to another. The ensemble is represented by a large number of these randomly distributed vectors in the Hilbert space of single-photon Fock state $n = 1$. Therefore, an ensemble average on $\varphi(\mathbf{k}, s)$ is necessary for calculating the expectation value of an observable.

(2) What is the difference between the pure single-photon state of Eq.(3.3.14)

$$|\Psi\rangle = \sum_m c_m |\Psi_m\rangle = \sum_m c_m \int d\omega\, f_m(\omega)\, \hat{a}_m^\dagger(\omega)\, |0\rangle$$

and the mixed state of Eq.(3.3.15)

$$\hat{\rho} = \sum_m P_m |\Psi_m\rangle\langle\Psi_m|$$
$$= \sum_m P_m \int\int d\omega\, d\omega'\, f_m(\omega) f_m^*(\omega')\, \hat{a}_m^\dagger(\omega) |0\rangle\langle 0| \hat{a}_m(\omega') \quad ?$$

In terms of the state of a photon, Eq.(3.3.14) represents a pure state by means of a vector in the multi-dimensional Hilbert space of $|\Psi_m\rangle$. On the contrary, Eq.(3.3.15) represents a mixed state by means of a set of randomly distributed vectors in the same multi-dimension Hilbert space of $|\Psi_m\rangle$.

Eq.(3.3.14) and Eq.(3.3.15) can be formally written as

$$|\Psi\rangle = \sum_m |c_m| e^{i\varphi_m} |\Psi_m\rangle = \sum_m |c_m| e^{i\varphi_m} \int d\omega\, f_m(\omega)\, \hat{a}_m^\dagger(\omega)\, |0\rangle \qquad (3.4.37)$$

where φ_m represents the phase of the probability amplitude for the photon to be in the single-photon state of $|\Psi_m\rangle$. When $\varphi_m = $ constant, the coherent superposition among these $|\Psi_m\rangle$s results in a pure state of Eq.(3.3.14); when $\varphi_m = $ random number, the incoherent superposition among these $|\Psi_m\rangle$s results in a mixed state of Eq.(3.3.15).

Assume both states are generated from a large number of atomic transitions, the photon in state of Eq.(3.3.14) must be created from a coherent process in which all atomic transitions are excited with constant phase relation and superposed coherently. The probability amplitude for the photon to be created from the mth atomic transition is c_m, which has a real and positive amplitude $|c_m|$ and a phase φ_m, indicating precise magnitude and phase for all projections of the state vector onto each vector basis of $|\Psi_m\rangle$. On the contrary, the photon in state Eq.(3.3.15) must be created from a stochastic process in which the mth subfield is excited with a random phase φ_m from the mth atomic transition and thus superposed incoherently with others. The only knowledge we can have about this photon is that it has probability $|c_m|^2$ to be created from the mth atomic transition. The mathematical picture also is interesting: although we have precise knowledge on the magnitude of the projection $|c_m|$ or $|c_m|^2$ on the vector basis $|\Psi_m\rangle$, however, the phase takes random value indeterministically.

Can a simple measurement distinguish the pure state of Eq.(3.3.14) from the mixed state of Eq.(3.3.15)? If we can find a "rotation" (measurement, operation) in the Hilbert space of $|\Psi_m\rangle$ to project the pure state of Eq.(3.3.14) onto itself and onto its orthogonal states, certainly, we can. This "rotation", again, is similar to the rotations we have discussed in earlier discussions on the measurement of polarization and Schrödinger's cat. Do we have such a measurement corresponding to such a "simple rotation"? Unfortunately, we do not.

In terms of the ensemble, similar to the physics of the cat system, a pure state requires all photons created from the atomic system in the same state, such as the state of Eq.(3.3.14). A mixed state means the photons in the ensemble are not in the same state, but rather in all possible different states. We have two slightly different physical pictures of the ensemble: (1) if the ensemble has N photons, the number of photons in the single-photon state $|\Psi_m\rangle$, or it is created from the mth atomic transition, is proportional to $|c_m|^2$; (2) the state of each photon in the ensemble is represented by an incoherent superposition of single-photon state $|\Psi_m\rangle$, such as Eq.(3.4.37), with a special set of φ_ms. The special set takes different values from one to another. The ensemble is represented by a large number of these randomly distributed vectors in the Hilbert space of single-photon state $|\Psi_m\rangle$. Therefore, an ensemble average on φ_m is necessary for calculating the expectation value.

3.5 COMPOSITE SYSTEM AND TWO-PHOTON STATE OF RADIATION FIELD

In certain measurements, we need to deal with two-photon states, which describe the state of a composite system of two photons. The subsystems may be spatially separated in large distance. Despite the distance between the subsystems, in quantum theory, a composite system composed of two subsystems is described by a Hilbert space constructed as the direct or tensor product of the Hilbert spaces of the two subsystems

$$H = H_1 \otimes H_2. \tag{3.5.1}$$

If state $|\Psi_1\rangle \in H_1$ and state $|\Psi_2\rangle \in H_2$, then we denote the direct product of these states by $|\Psi\rangle = |\Psi_1\rangle \otimes |\Psi_2\rangle$, or simply $|\Psi\rangle = |\Psi_1\rangle|\Psi_2\rangle$. The inner product on H is defined in terms of the inner product on H_1 and H_2 by

$$\langle\Psi|\Psi'\rangle = \langle\Psi_1|\Psi_1'\rangle\langle\Psi_2|\Psi_2'\rangle. \tag{3.5.2}$$

If $\{|m\rangle\}$ is an orthonormal base of H_1 and $\{|n\rangle\}$ is an orthonormal basis of H_2, then $\{|m\rangle \otimes |n\rangle\}$ or simple $\{|m\rangle|n\rangle\}$ is an orthonormal basis of H. This basis is usually called the Schmidt basis.

In a composite system, the subsystems may be independent, correlated or entangled. For independent subsystems the state of the composite system can be written as a product state

$$|\Psi\rangle = \sum_{m,n} c_m c_n |m\rangle|n\rangle = \sum_m c_m|m\rangle \sum_n c_n |n\rangle = |\Psi_1\rangle \otimes |\Psi_2\rangle, \tag{3.5.3}$$

where $c_{mn} = c_m c_n$ is factorizable. When the two subsystems are correlated or entangled, in general, $|\Psi\rangle$ cannot be written in the form of Eq. (3.5.3), i.e., c_{mn} is non-factorizable and consequently the state is non-factorizable

$$|\Psi\rangle = \sum_{m,n} c_{mn} |m\rangle|n\rangle. \tag{3.5.4}$$

We say the two subsystems are in a correlated state or in an entangled state, depending on the form of c_{mn}.

The simplest two-photon states are the Fock states of Eq. (3.1.9). To simplify the mathematics, we rewrite Eq. (3.1.9) in 1-D and ignore the polarization

$$\begin{aligned}|\Psi\rangle &= \hat{a}_1^\dagger(\omega)\,\hat{a}_2^\dagger(\omega')\,|\,0\,\rangle \\ &= |0, ...0, 1_1(\omega), 0, ...0\rangle|0, ...0, 1_2(\omega'), 0, ...0\rangle,\end{aligned} \tag{3.5.5}$$

where we have also rewritten the state of Eq.(3.1.9) explicitly as that of a composite two subsystems. In general, we may deal with a set of two-photon Fock states either in a coherent superposition

$$|\Psi\rangle = \sum_\omega \sum_{\omega'} f(\omega,\omega')\,\hat{a}_1^\dagger(\omega)\,\hat{a}_2^\dagger(\omega')\,|\,0\,\rangle, \tag{3.5.6}$$

or in a statistical mixture

$$\hat{\rho} = \sum_\omega \sum_{\omega'} |f(\omega,\omega')|^2\,\hat{a}_1^\dagger(\omega)\,\hat{a}_2^\dagger(\omega')\,|\,0\,\rangle\langle 0|\hat{a}_2(\omega')\,\hat{a}_1(\omega), \tag{3.5.7}$$

where $f(\omega,\omega')$ is the probability amplitude for the quantized field to be in the two-photon Fock state $|0, ...0, 1_\omega, 0, ...0\rangle|0, ...0, 1_{\omega'}, 0, ...0\rangle$. The two-photon probability amplitude $f(\omega,\omega')$ may be factorizable into a product of $f_1(\omega) \times f_2(\omega')$, or non-factorizable at all. The physical properties of the states are very different between these two cases. If $f(\omega,\omega')$ can be factorized into a product of $f_1(\omega) \times f_2(\omega')$, the state itself is also factorizable into a product state of two independent single-photons. If $f(\omega,\omega')$ is non-factorizable and consequently the state itself cannot be factorized into a product state, we name the state a correlated or an entangled two-photon state. In terms of the concept of a photon, the product states describe the behavior of two independent photons, the correlated states describe the behavior of two correlated photons, and the entangled states characterize the behavior of an entangled pair of photons. Below are three examples showing factorizable or non-factorizable probability amplitudes. We will have a detailed study of entangled states in later chapters.

Example (I): Product state.

Eq. (3.5.8) is a two-photon state which can be factorized into a product of two single-photon states:

$$\begin{aligned}|\Psi\rangle &= \sum_\omega \sum_{\omega'} f(\omega)f'(\omega')\,\hat{a}_1^\dagger(\omega)\,\hat{a}_2^\dagger(\omega')\,|\,0\,\rangle \tag{3.5.8} \\ &= \sum_\omega f(\omega)\,\hat{a}_1^\dagger(\omega)\,|\,0\,\rangle \times \sum_{\omega'} f'(\omega')\,\hat{a}_2^\dagger(\omega')\,|\,0\,\rangle \\ &= |\Psi_1\rangle|\Psi_2\rangle, \tag{3.5.9}\end{aligned}$$

where we have assumed a factorizable amplitude: $f(\omega,\omega') = f(\omega) \times f'(\omega')$.

Example (II): EPR state.

Eq. (3.5.10) is a non-factorizable two-photon state with total photon number $n = 2$

$$|\Psi\rangle = \sum_{\omega,\omega'} \Psi_0 \, \delta \left[\omega + \omega' - \omega_0\right] \hat{a}_1^\dagger(\omega) \, \hat{a}_2^\dagger(\omega') \, |0\rangle, \qquad (3.5.10)$$

where Ψ_0 is a normalization complex constant, the delta function and the constant ω_0 together indicates the conservation of energy. In the state of Eq. (3.5.10), a pair of modes are always excited together and the frequencies (energies) of the pair remains constant, although each mode may take any value within the superposition. The two-photon superposition state of Eq. (3.5.10) has certain properties that may never be understood classically. For example, in terms of the concept of a photon, the energy of neither photon is determined; however, if one of the photons is measured with a certain value the energy of the other photon is determined with certainty, despite the distance between the two photons. These kind of states are called entangled states by Schrödinger following the 1935 paper of Einstein, Podolsky, and Rosen. Quantum entanglement will be intensively discussed in later chapters.

Example (III): Number state of $n_\omega = 2$.

Eq. (3.5.11) defines another type of two-photon state with total photon number $n = 2$

$$|\Psi\rangle = \sum_\omega f(\omega) \left[\hat{a}^\dagger(\omega)\right]^2 |0\rangle. \qquad (3.5.11)$$

Eq. (3.5.11) indicates the excitation of one mode with occupation number $n_\omega = 2$.

It should be emphasized that the states in Eqs. (3.5.8), (3.5.10), (3.5.11) are all pure states. They are very different from the following corresponding mixed states,

$$\hat{\rho} = \sum_{\omega,\omega'} P(\omega) \, P'(\omega') \hat{a}^\dagger(\omega) \, \hat{a}^\dagger(\omega') \, |0\rangle\langle 0| \, \hat{a}(\omega) \, \hat{a}(\omega')$$

$$\hat{\rho} = \sum_{\omega,\omega'} \delta \left[\omega + \omega' - \omega_0\right] \hat{a}^\dagger(\omega) \, \hat{a}^\dagger(\omega') \, |0\rangle\langle 0| \, \hat{a}(\omega) \, \hat{a}(\omega')$$

$$\hat{\rho} = \sum_\omega P(\omega) \left[\hat{a}^\dagger(\omega)\right]^2 |0\rangle\langle 0| \left[\hat{a}(\omega)\right]^2.$$

3.6 A SIMPLE MODEL OF SINGLE-PHOTON AND MULTI-PHOTON STATE CREATION

To simplify the discussion and mathematics, we assume a point light source contains a large number of atoms that are ready for two-level atomic transitions at any time t. For a point source, each atomic transition excites a sub-field in the form of a symmetrical spherical wave propagating to all 4π directions. The excited radiations are monitored by a point-like photodetector or a set of independent N point-like photodetectors that are placed at distance, such as a light year, for single-photon counting measurement or for joint N-photon counting measurement. We assume the measured light is weak at such a distance.

Although the chance to have a spontaneous emission is very small, there is indeed a small probability for an atom to create a photon whenever the atom decays from its higher energy level E_2 ($\Delta E_2 \neq 0$) down to its ground energy state of E_1. It is reasonable to approximate the mth atomic transition excites a sub-field in the following state:

$$|\Psi\rangle = c_0|0\rangle + c_1|\Psi_m\rangle \simeq |0\rangle + \epsilon|\Psi_m\rangle, \qquad (3.6.1)$$

where $|c_0| \sim 1$ is the probability amplitude for no-field-excitation and $|c_1| = |\epsilon| \ll 1$ is the probability amplitude for the creation of a photon from the mth atomic transition, and

$$|\Psi_m\rangle = \sum_s \int d\mathbf{k}\, f_m(\mathbf{k}, s)\, \hat{a}_m^\dagger(\mathbf{k}, s)|0\rangle$$

or

$$|\Psi_m\rangle = \int d\omega\, f_m(\omega)\, \hat{a}_m^\dagger(\omega)\,|0\rangle \qquad \text{(1-D)}$$

as shown earlier in Eq.(3.3.12) and Eq.(3.3.13). The generalized state of the radiation field that is excited by the light source, which contains such a large number atomic transitions, can be formally written as

$$
\begin{aligned}
|\Psi\rangle &= \prod_m \left\{ |0\rangle + \epsilon\, c_m\, |\Psi_m\rangle \right\} \\
&\simeq |0\rangle + \epsilon \Big[\sum_m |c_m|\, e^{i\varphi_m}\, |\Psi_m\rangle \Big] \\
&\quad + \epsilon^2 \Big[\sum_{m<n} |c_m||c_n|\, e^{i(\varphi_m + \varphi_n)}\, |\Psi_m\rangle|\Psi_n\rangle \Big] + \dots
\end{aligned} \qquad (3.6.2)
$$

where, again, we have defined a phase factor $e^{i\varphi_m}$ associated with the mth atomic transition. In principle, this phase factor belongs to the amplitude of each excited mode associated with the creation operator. We assume a common phase for each wavepacket, although it can still take arbitrary values from transition to transition. This common phase is determined by a common physical parameter in the creation process, such as the creation time of the mth wavepacket. Since $|\epsilon| \ll 1$, In Equation (3.6.2) we listed the first-order and the second-order approximations on ϵ. For a particular measurement, only a certain lower-order approximations are necessary. In the following we give four types of approximations corresponding to different types of measurements or different mechanisms of light generation.

(I): Mixed single-photon state

We assume all atomic transitions in the source are radiated randomly and independently with $\varphi_m =$ random number and $\varphi_n =$ random number for all m and n. In certain measurements, a first-order expansion on ϵ in Equation (3.6.2) is a good approximation. The higher-order approximations may contribute to the measurement, however, these contributes are usually small enough under the weak light condition. A radiation is said to be at single-photon's level when the higher-order approximations are ignorable, and the radiation field is characterized by the following formally written single-photon state:

$$|\Psi\rangle \simeq \sum_m |c_m|\, e^{i\varphi_m}\, |\Psi_m\rangle, \qquad (3.6.3)$$

where each coherently superposed wavepacket

$$|\Psi_m\rangle = \int d\omega\, f_m(\omega)\, \hat{a}_m^\dagger(\omega)\,|0\rangle$$

is associated with a random phase φ_m which may take any arbitrary value. The first-order approximation is thus written in the form of an incoherent superposition of a set of single-photon state vectors or wavepacket $|\Psi_m\rangle$. The incoherently superposed state of Eq.(3.6.3) can be considered as: (1) the state of the measured radiation which may be created from any or from all possible randomly radiated independent atomic transitions with probability $|c_m|^2$; (2) the state of an ensemble of measured photons, each photon is created from a randomly radiated atomic transition, as an inhomogeneous ensemble.

The expectation value of an operator $\hat{A}$ corresponding to the measurement is calculated from

$$
\begin{aligned}
\langle\langle \Psi | \hat{A} | \Psi \rangle\rangle_{\mathrm{En}} &= \Big\langle \sum_m |c_m| \, e^{-i\varphi_m} \langle \Psi_m | \, \hat{A} \sum_n |c_n| e^{i\varphi_n} | \Psi_n \rangle \Big\rangle_{\mathrm{En}} \\
&= \Big\langle \sum_{m,n} |c_m| |c_n| \, e^{i(\varphi_n - \varphi_m)} \langle \Psi_m | \hat{A} | \Psi_n \rangle \Big\rangle_{\mathrm{En}} \\
&= \sum_m |c_m|^2 \langle \Psi_m | \hat{A} | \Psi_m \rangle \\
&= tr \, \hat{\rho}^{(1)} \hat{A}.
\end{aligned}
\tag{3.6.4}
$$

where we have defined a density operator to characterize the first-order approximated mixed single-photon state

$$
\hat{\rho}^{(1)} \equiv \sum_m P_m |\Psi_m\rangle\langle\Psi_m|
\tag{3.6.5}
$$

with $P_m = |c_m|^2$ the probability for the radiation field to be in the state $|\Psi_m\rangle$. The ensemble average in Eq. (3.6.4) has taken into account all possible values of the relative phases $\varphi_n - \varphi_m$.

Under the above approximation, traditionally, we say the measured field is in a mixed single-photon state:

$$
\begin{aligned}
\hat{\rho} &= \sum_m P_m \, |\Psi_m\rangle\langle\Psi_m| \\
&= \sum_m P_m \int d\omega \, d\omega' \, f_m(\omega) f_m^*(\omega') \, \hat{a}_m^\dagger(\omega) \, |0\rangle\langle 0| \, \hat{a}_m(\omega').
\end{aligned}
\tag{3.6.6}
$$

If we further assume all the atomic transitions have the same mode distribution $f_m(\omega) = f(\omega)$ and each mode in the mth wavepacket has a phase $\varphi_m(\omega)$ that is determined by the transition time t_{0m} of the mth atomic transition, $\varphi_m(\omega) = \omega t_{0m}$, and consider a random and continuous distribution on t_{0m}, the sum on m turns into an integral of t_0,

$$
\begin{aligned}
\hat{\rho} &= \int dt_0 \, e^{-i(\omega-\omega')t_0} \int d\omega \, d\omega' \, f(\omega) f^*(\omega') \, \hat{a}^\dagger(\omega) \, |0\rangle\langle 0| \, \hat{a}(\omega') \\
&\cong \int d\omega \, |f(\omega)|^2 \, \hat{a}^\dagger(\omega) |0\rangle\langle 0| \, \hat{a}(\omega),
\end{aligned}
\tag{3.6.7}
$$

where the result of the integral on t_0 is approximated $\delta(\omega - \omega')$ when taking t_0 to ∞. Although it is popular to use the density operator of Eq.(3.6.7) to represent mixed single-photon state, we should not forget it requires $t_0 \sim \infty$. In a realistic stochastic process of finite time, or equivalently, in a realistic measurement of finite time, however, $\Delta\omega \equiv (\omega - \omega') \neq 0$. In this case, Eq.(3.6.6) is a better representation.

(II): Mixed two-photon state

Now we consider two independent point-like photodetectors at distance for joint photon counting measurement, i.e., to count the number of events in which each of the two photodetectors observes a photoelectron, respectively and simultaneously, or within a short time window. For this type of measurement, a second-order approximation is necessary. The third-order and higher-order approximations may also contribute to the measurement, however, these contributions are usually ignorable under weak light condition. In the above given simple model, the second-order approximation, or the two-photon state, can be approximated as:

$$|\Psi\rangle \simeq \sum_{m<n} c_m c_n \, e^{i(\varphi_m+\varphi_n)} |\Psi_m\rangle |\Psi_n\rangle. \tag{3.6.8}$$

where, again,

$$|\Psi_m\rangle = \int d\omega f_m(\omega) \, \hat{a}_m^\dagger(\omega) \, |0\rangle.$$

When $\varphi_m + \varphi_n$ = random number for all m and n, the incoherent superposition turns Eq.(3.6.8) a mixed two-photon state. The incoherently superposed two-photon state can be considered as: (1) the state of a jointly measured pair of photons which may be created from any or from all possible randomly radiated and randomly paired atomic transitions with probability $P_{m,n} = |c_m|^2|c_n|^2$; (2) the state of an ensemble of jointly measured photon pairs, each pair is created from a randomly radiated and randomly paired atomic transitions, as an inhomogeneous ensemble.

In the case of φ_m = random number, φ_n = random number, and $\varphi_m + \varphi_n$ = random number for all m and n, the incoherent superpositions turn both first-order and second-order approximations into mixed states. Natural light, such as the sunlight, is a typical example of this type of radiation. This kind of light has a popular name of thermal light.

For the mixed two-photon approximation, the expectation value of an operator is calculated from Eq.(3.6.8)

$$\begin{aligned}
&\langle\langle\Psi|\hat{A}|\Psi\rangle\rangle_{\text{En}} \\
&= \Big\langle \sum_{m<n} c_m^* c_n^* \, e^{i(\varphi_m+\varphi_n)} \langle\Psi_m|\langle\Psi_n| \, \hat{A} \sum_{p<q} c_p c_q e^{-i(\varphi_p+\varphi_q)} |\Psi_p\rangle|\Psi_q\rangle \Big\rangle_{\text{En}} \\
&= \Big\langle \sum_{m<n,p<q} c_m^* c_n^* c_p c_q \, e^{i[(\varphi_m+\varphi_n)-(\varphi_p+\varphi_q)]} \langle\Psi_m|\langle\Psi_n|\hat{A}|\Psi_p\rangle|\Psi_q\rangle \Big\rangle_{\text{En}} \\
&\simeq \sum_{m<n} |c_m|^2|c_n|^2 \langle\Psi_m|\langle\Psi_n|\hat{A}|\Psi_n\rangle|\Psi_m\rangle \\
&= tr \, \hat{\rho}^{(2)} \hat{A}, \tag{3.6.9}
\end{aligned}$$

where we have defined a density operator to characterize the second-order approximated mixed two-photon state

$$\hat{\rho}^{(2)} = \sum_{m<n} |c_m|^2|c_n|^2 |\Psi_n\rangle|\Psi_m\rangle\langle\Psi_m|\langle\Psi_n|. \tag{3.6.10}$$

The ensemble average in Equation (3.6.9) has taken into account all possible values of the relative phases $(\varphi_m + \varphi_n) - (\varphi_p - \varphi_q)$. If we further assume the atoms have the same mode distribution $f(\omega)$ and each mode in the mth wavepacket has a phase $\varphi_m(\omega)$ that is determined by the transition time t_{0m} of the mth atomic transition, $\varphi_m(\omega) = \omega t_{0m}$, and consider a random and continuous distribution on t_{0m} and t_{0n}, Similar to the single-photon state, the integrals on t_{0m} ($\sim \infty$) and t_{0n} ($\sim \infty$) turn the density operator into the following form:

$$\hat{\rho}^{(2)} \cong \int d\omega \, d\omega' \, |f(\omega)|^2 |f(\omega')|^2 \, \hat{a}^\dagger(\omega) \, \hat{a}^\dagger(\omega') \, |0\rangle\langle0| \, \hat{a}(\omega')\hat{a}(\omega), \tag{3.6.11}$$

where $f(\omega)$ is renormalized.

Eq.(3.6.10) and Eq.(3.6.11) are both commonly used to represent the mixed two-photon state, such as in the calculation of second-order coherence of thermal light.

(III): Entangled two-photon state

In the second-order approximation, if a "peculiar" photon creation mechanism is able to force $\varphi_m + \varphi_n = $ constant, but keeps φ_m and φ_n random number for all m and n, the two-photon state turns to be a coherently superposed pure state, however, the two subsystems, $\hat{\rho}_m$ and $\hat{\rho}_n$, will still keep their nature of mixed single-photon state with incoherent superposition. In this case, the second-order approximation is written as,

$$|\Psi\rangle \simeq \sum_{m<n} |c_m||c_n|\, e^{i(\varphi_m+\varphi_n)}|\Psi_m\rangle|\Psi_n\rangle_{\varphi_m+\varphi_n=\varphi_0}$$
$$= \sum_{m<n} |c_m||c_n|\, \delta(\varphi_m + \varphi_n - \varphi_0)|\Psi_m\rangle|\Psi_n\rangle \qquad (3.6.12)$$

where, again,

$$|\Psi_m\rangle = \int d\omega f_m(\omega)\, \hat{a}_m^\dagger(\omega)\,|0\rangle.$$

In Eq.(3.6.12), we have used a δ-function, $\delta(\varphi_m + \varphi_n - \varphi_0)$, instead of the face factor $e^{i\varphi_0}$ representing the nonfactorizable two-photon amplitude, to emphasize freedom for both φ_m and φ_n when taking sum of m and n. Equation (3.6.12) defines a two-photon state which cannot be factorized into the product of two single-photon states. We name this type state entangled two-photon state. The coherently superposed two-photon state of Eq.(3.6.12) is a pure state which characterizes (1) the state of an entangled photon pair, created from any or from all possible paired atomic transitions in which $\varphi_m + \varphi_n = $ constant but $\varphi_m = $ random number and $\varphi_n = $ random number; and (2) the state of an ensemble of entangled photon pairs, all in the same two-photon state, as a homogenous ensemble. Although the two-photon state of Eq.(3.6.12) is a pure state, the two subsystems are both in mixed single-photon states

$$\hat{\rho}_m = \sum_m P_m |\Psi_m\rangle\langle\Psi_m|$$
$$\hat{\rho}_n = \sum_n P_n |\Psi_n\rangle\langle\Psi_n|. \qquad (3.6.13)$$

The above mechanism of creating entangled two-photon state is based on the "pairing" of two individual atomic transitions by achieving $\varphi_m + \varphi_n = $ constant. Forcing $\varphi_m + \varphi_n = $ constant for all m and n, but keeping $\varphi_m = $ random number and $\varphi_n = $ random number for all m and n is not easy but possible.

There exist other mechanisms to create entangled two-photon state, such as atomic cascade decay and spontaneous parametric down-conversion (SPDC). The atomic cascade decay creates a pair of entangled photons from one cascade atomic transition. The entangled two-photon state of SPDC is generated from nonlinear optical process that involves a laser beam of coherent "pump". The two-photon states created from atomic cascade decay and SPDC are different from that of Eq.(3.6.12). A typical 1-D entangled two-photon state of SPDC is given as follows:

$$|\Psi\rangle = \Psi_0 \int d\omega_s\, d\omega_i\, \delta(\omega_s + \omega_i - \omega_p)\, \hat{a}^\dagger(\omega_s)\hat{a}^\dagger(\omega_i)\,|0\rangle, \qquad (3.6.14)$$

where Ψ_0 is the normalization constant, and ω_p is the frequency of the pump laser, which is approximated as constant. For historical reason, the created pair of photons are named signal and idler. The signal photon and the idler photon are labeled by frequency ω_s and ω_i, respectively. The entangled two-photon state of the signal-idler photon pair is a pure state; however, the state of the signal photon and the idler photon, respectively, is a mixed single-photon state,

$$\hat{\rho}_s = \int d\omega_s \hat{a}^\dagger(\omega_s)|0\rangle\langle 0|\hat{a}(\omega_s)$$

$$\hat{\rho}_i = \int d\omega_i \hat{a}^\dagger(\omega_i)|0\rangle\langle 0|\hat{a}(\omega_i). \tag{3.6.15}$$

The entangled two-photon state of Eq.(3.6.14) was named by Klyshko "biphoton" state to distinguish the entangled two-photon state of Eq.(3.6.12).

The two-photon states of Eq.(3.6.12) and Eq.(3.6.14) are different mathematically and physically. It might be easier to find the difference by rewriting Eq.(3.6.12) and Eq.(3.6.13) to the following forms to compare with Eq.(3.6.14) and Eq.(3.6.15):

$$|\Psi\rangle = \sum_{m<n} |c_m||c_n|\,\delta(\varphi_m + \varphi_n - \varphi_0)\,|\Psi_m\rangle|\Psi_n\rangle \tag{3.6.16}$$

$$= \sum_{m<n} |c_m||c_n|\,\delta(\varphi_m + \varphi_n - \varphi_0)\int d\omega\, d\omega'\, f_m(\omega) f_n(\omega')\, \hat{a}_m^\dagger(\omega)\, \hat{a}_n^\dagger(\omega')\,|0\rangle$$

and

$$\hat{\rho}_m = \sum_m P_m \int d\omega\, d\omega'\, f_m(\omega) f_m^*(\omega')\, \hat{a}_m^\dagger(\omega)\,|0\rangle\langle 0|\,\hat{a}_m(\omega')$$

$$\hat{\rho}_n = \sum_n P_n \int d\omega\, d\omega'\, f_n(\omega) f_n^*(\omega')\, \hat{a}_n^\dagger(\omega)\,|0\rangle\langle 0|\,\hat{a}_n(\omega'). \tag{3.6.17}$$

(IV): Coherent radiation

Assuming a photon creation mechanism that makes all atomic transitions in phase and indistinguishable, the state of Equation (3.6.2) is rewritten in the following form

$$|\Psi\rangle = \left\{|0\rangle + \epsilon \int d\omega\, f(\omega)\, \hat{a}^\dagger(\omega)\,|0\rangle\right\}^n, \tag{3.6.18}$$

where n is the number of atomic transitions participate to the coherent excitation of the radiation field. The expansion of Eq.(3.6.18) is thus

$$|\Psi\rangle \simeq |0\rangle + \epsilon\,\frac{n}{1!}|\Psi\rangle + \epsilon^2\,\frac{n(n-1)}{2!}\big(|\Psi\rangle\big)^2 + \epsilon^3\,\frac{n(n-1)(n-2)}{3!}\big(|\Psi\rangle\big)^3 + \dots$$

$$+ \epsilon^m\,\frac{n(n-1)(n-2)\dots(n-m+1)}{m!}\big(|\Psi\rangle\big)^m + \dots, \tag{3.6.19}$$

with

$$|\Psi\rangle = \int d\omega\, f(\omega)\,\hat{a}^\dagger(\omega)\,|0\rangle.$$

In single-mode approximation, the state $|\Psi\rangle$ turns to be

$$|\Psi\rangle \simeq |0\rangle + \epsilon\,\frac{n}{1!}\,\hat{a}^\dagger(\omega)|0\rangle$$

$$+ \epsilon^2\,\frac{n(n-1)}{2!}\,[\hat{a}^\dagger(\omega)]^2|0\rangle + \epsilon^3\,\frac{n(n-1)(n-2)}{3!}\,[\hat{a}^\dagger(\omega)]^3|0\rangle$$

$$+ \epsilon^m\,\frac{n(n-1)(n-2)\dots(n-m+1)}{m!}\,[\hat{a}^\dagger(\omega)]^m|0\rangle + \dots. \tag{3.6.20}$$

A laser system has the ability to force all atomic transitions in phase and indistinguishable. Laser light is therefore named coherent light.

3.7 PRODUCT STATE, ENTANGLED STATE, AND MIXED STATE OF PHOTON PAIRS

The concepts of product photon states and entangled photon states involve multi-photon systems. In this section, we restrict ourselves to two-photon states. We will give two simple discussions before introducing the general entangled muti-mode two-photon states: (I) Product, entangled and mixed states of Schrödinger's cats; (II) Product, entangled and mixed polarization states of a two-photon system.

Example (I): Product, entangled and mixed states of Schrödinger's cats.

(1) Product state of two Schrödinger's cats:

Suppose two Schrödinger's cats, cat-one and cat-two, are created from a cat source independently, cat-one is prepared in the state $|+\text{cat}\rangle$ and cat-two is prepared in the state $|-\text{cat}\rangle$. Due to the independent creation process, there is no interaction between the two cats; the two created Schrödinger's cats can be treated as a cat system in product states

$$
\begin{aligned}
|\Psi\rangle &= |+\text{cat}_1\rangle \times |-\text{cat}_2\rangle \\
&= \frac{1}{2}\big[\,|\text{alive}_1\rangle|\text{alive}_2\rangle + |\text{dead}_1\rangle|\text{dead}_2\rangle \\
&\quad - |\text{alive}_1\rangle|\text{dead}_2\rangle - |\text{dead}_1\rangle|\text{alive}_2\rangle\big].
\end{aligned} \tag{3.7.1}
$$

We further assume the two cats "propagate" freely in opposite directions with no interaction between them. If the propagation, or the time evolution, of the system simply develops an overall phase but keeps the relative phases between the four terms in the superposition, the state of the two Schrödinger's cats will remain the same as that of Eq. (3.7.1), except the trivial overall phase. At a large distance, such as a light year, two observers decide to measure the live status of the two cats simultaneously, or the joint-live-status of the two cats. It is not surprising that the measurements show four equal possibilities: (1) cat-one and cat-two both alive, (2) cat-one and cat-two both dead, (3) cat-one alive and cat-two dead, and (4) cat-one dead and cat-two alive.

Here, "not surprising" means that there is no difference between the quantum predication and the classical convention.

(2) Entangled state of two Schrödinger's cats:

Suppose a pair of Schrödinger's cats is created at the cat-source with one of the following two-cat states

$$
\begin{aligned}
|\Psi^{(+)}\rangle &= \frac{1}{\sqrt{2}}\big[\,|\text{alive}_1\rangle|\text{alive}_2\rangle + |\text{dead}_1\rangle|\text{dead}_2\rangle\big] \\
|\Psi^{(-)}\rangle &= \frac{1}{\sqrt{2}}\big[\,|\text{alive}_1\rangle|\text{alive}_2\rangle - |\text{dead}_1\rangle|\text{dead}_2\rangle\big] \\
|\Phi^{(+)}\rangle &= \frac{1}{\sqrt{2}}\big[\,|\text{alive}_1\rangle|\text{dead}_2\rangle + |\text{dead}_1\rangle|\text{alive}_2\rangle\big] \\
|\Phi^{(-)}\rangle &= \frac{1}{\sqrt{2}}\big[\,|\text{alive}_1\rangle|\text{dead}_2\rangle - |\text{dead}_1\rangle|\text{alive}_2\rangle\big],
\end{aligned} \tag{3.7.2}
$$

and neither can be written in the form of a product state, $|\Psi\rangle = |\Psi_1\rangle|\Psi_2\rangle$. The pair of cats is in an entangled two-cat state. All four states in Eq. (3.7.2) are defined as entangled two-cat states. We may name them Bell's cat-states. The Bell's cat-states form a complete orthogonal vector basis in two-cat space.

If the same criterion for free propagation and time evolution of the system is applicable, i.e., the time evolution simply develops an overall phase but keeps the relative phase between the two terms in the superposition, the state of the two Schrödinger's cats will remain the same as that shown in Eq. (3.7.2), except for a trivial phase factor. Therefore, the outcome of any "correlation" measurement of the two cats will remain the same despite the distance between them.

Now, we consider a joint measurement of the cat system. Suppose two observers, at equal distances from the cat-source, measure the live-status of the cat pair simultaneously, equivalent to having a set of joint projection operators:

$$\langle \Psi | \, \text{alive}_1 \rangle | \, \text{alive}_2 \rangle \langle \text{alive}_2 \, | \langle \text{alive}_1 \, | \Psi \rangle = | \langle \text{alive}_2 \, | \langle \text{alive}_1 \, | \Psi \rangle |^2,$$

$$\langle \Psi | \, \text{alive}_1 \rangle | \, \text{dead}_2 \rangle \langle \text{dead}_2 \, | \langle \text{alive}_1 \, | \Psi \rangle = | \langle \text{dead}_2 \, | \langle \text{alive}_1 \, | \Psi \rangle |^2,$$

$$\langle \Psi | \, \text{dead}_1 \rangle | \, \text{alive}_2 \rangle \langle \text{alive}_2 \, | \langle \text{dead}_1 \, | \Psi \rangle = | \langle \text{alive}_2 \, | \langle \text{dead}_1 \, | \Psi \rangle |^2,$$

$$\langle \Psi | \, \text{dead}_1 \rangle | \, \text{dead}_2 \rangle \langle \text{dead}_2 \, | \langle \text{dead}_1 \, | \Psi \rangle = | \langle \text{dead}_2 \, | \langle \text{dead}_1 \, | \Psi \rangle |^2 \qquad (3.7.3)$$

for the four Bell's cat-states in Eq. (3.7.2). The following table lists the expectation values of the joint projection operators:

	$|\Psi^{(+)}\rangle$	$|\Psi^{(-)}\rangle$	$|\Phi^{(+)}\rangle$	$|\Phi^{(-)}\rangle$
$\langle \text{alive}_2 \, | \langle \text{alive}_1 \, |$	$1/2$	$1/2$	0	0
$\langle \text{dead}_2 \, | \langle \text{alive}_1 \, |$	0	0	$1/2$	$1/2$
$\langle \text{alive}_2 \, | \langle \text{dead}_1 \, |$	0	0	$1/2$	$1/2$
$\langle \text{dead}_2 \, | \langle \text{dead}_1 \, |$	$1/2$	$1/2$	0	0

Again these results are not surprising. The quantum expectations are consistent with the classical convention. If the ensemble is prepared with 50% "alive-alive" and "dead-dead" cat-pairs ($|\Psi^{(+)}\rangle$ and $|\Psi^{(-)}\rangle$), the two observers should have a 50% chance of observing an "alive-alive" or "dead-dead" cat-pair. If the ensemble is prepared with 50% "alive-dead" and "dead-alive" cat-pairs ($|\Phi^{(+)}\rangle$ and $|\Phi^{(-)}\rangle$), the two observers should have a 50% chance of observing an "alive-dead" or "dead-alive" cat-pair.

Although this measurement distinguishes $|\Psi^{(+)}\rangle$ and $|\Psi^{(-)}\rangle$ from $|\Phi^{(+)}\rangle$ and $|\Phi^{(-)}\rangle$, choosing the set of projection measurements in Eq. (3.7.3) may not be the best choice. More importantly, this measurement cannot distinguish the entangled two-cat pure states from the mixed states of cat-pairs defined in Eq. (3.7.5).

We now consider a different set of joint projection measurements defined by the vectors $| + \text{cat} \rangle$ and $| - \text{cat} \rangle$,

$$\langle \Psi | + \text{cat}_1 \rangle | + \text{cat}_2 \rangle \langle +\text{cat}_2 \, | \langle +\text{cat}_1 \, | \Psi \rangle = | \langle +\text{cat}_2 \, | \langle +\text{cat}_1 \, | \Psi \rangle |^2,$$

$$\langle \Psi | + \text{cat}_1 \rangle | - \text{cat}_2 \rangle \langle -\text{cat}_2 \, | \langle +\text{cat}_1 \, | \Psi \rangle = | \langle -\text{cat}_2 \, | \langle +\text{cat}_1 \, | \Psi \rangle |^2,$$

$$\langle \Psi | - \text{cat}_1 \rangle | + \text{cat}_2 \rangle \langle +\text{cat}_2 \, | \langle -\text{cat}_1 \, | \Psi \rangle = | \langle +\text{cat}_2 \, | \langle -\text{cat}_1 \, | \Psi \rangle |^2,$$

$$\langle \Psi | - \text{cat}_1 \rangle | - \text{cat}_2 \rangle \langle -\text{cat}_2 \, | \langle -\text{cat}_1 \, | \Psi \rangle = | \langle -\text{cat}_2 \, | \langle -\text{cat}_1 \, | \Psi \rangle |^2 \qquad (3.7.4)$$

for the four Bell's cat-states in Eq. (3.7.2). The following table lists the expectation values of the joint projection operators:

	$|\Psi^{(+)}\rangle$	$|\Psi^{(-)}\rangle$	$|\Phi^{(+)}\rangle$	$|\Phi^{(-)}\rangle$
$\langle +\text{cat}_2 \, | \langle +\text{cat}_1 \, |$	$1/2$	0	$1/2$	0
$\langle -\text{cat}_2 \, | \langle +\text{cat}_1 \, |$	0	$1/2$	0	$1/2$
$\langle +\text{cat}_2 \, | \langle -\text{cat}_1 \, |$	0	$1/2$	0	$1/2$
$\langle -\text{cat}_2 \, | \langle -\text{cat}_1 \, |$	$1/2$	0	$1/2$	0

For states $|\Psi^{(+)}\rangle$ and $|\Phi^{(+)}\rangle$, quantum mechanics predicts that the two observers both have a 50% chance of registering a $|+\text{cat}\rangle$ or a $|-\text{cat}\rangle$ state. However, if one of them has found his cat in the state $|+\text{cat}\rangle$ or the state $|-\text{cat}\rangle$, the other observer must find his cat in the same state. For states $|\Psi^{(-)}\rangle$ and $|\Phi^{(-)}\rangle$, quantum mechanics predicts that the two observers both have a 50% chance of registering a $|+\text{cat}\rangle$ or a $|-\text{cat}\rangle$ state. However, if one of them has found his cat in the state $|+\text{cat}\rangle$ or the state $|-\text{cat}\rangle$, the other observer must find his cat in the orthogonal state. How can this be? This is indeed a surprising. Remember, the ensemble is neither prepared with the "$|+\text{cat}\rangle \,\&\, |+\text{cat}\rangle$" and "$|-\text{cat}\rangle \,\&\, |-\text{cat}\rangle$" pairs for the $|\Psi^{(+)}\rangle$ and $|\Phi^{(+)}\rangle$ states nor with the "$|+\text{cat}\rangle \,\&\, |-\text{cat}\rangle$" and "$|-\text{cat}\rangle \,\&\, |+\text{cat}\rangle$" pairs for the $|\Psi^{(-)}\rangle$ and $|\Phi^{(-)}\rangle$ states. In the classical point of view, the ensemble is prepared with the "$|\text{alive}\rangle \,\&\, |\text{alive}\rangle$" and "$|\text{dead}\rangle \,\&\, |\text{dead}\rangle$" pairs for the states $|\Psi^{(+)}\rangle$ and $|\Psi^{(-)}\rangle$ and with the "$|\text{alive}\rangle \,\&\, |\text{dead}\rangle$" and "$|\text{dead}\rangle \,\&\, |\text{alive}\rangle$" pairs for the states $|\Phi^{(+)}\rangle$ and $|\Phi^{(-)}\rangle$. According to Eq. (3.4.33), the states $|\text{alive}\rangle$ and $|\text{dead}\rangle$ both have equal chances of being in the state $|+\text{cat}\rangle$ and in the state $|-\text{cat}\rangle$. Classically, the ensemble should have an equal number of four possible pair combinations: "$|+\text{cat}\rangle \,\&\, |+\text{cat}\rangle$", "$|+\text{cat}\rangle \,\&\, |-\text{cat}\rangle$", "$|-\text{cat}\rangle \,\&\, |+\text{cat}\rangle$" and "$|-\text{cat}\rangle \,\&\, |-\text{cat}\rangle$".

We will learn in later chapters that this surprise is the result of an interference called "two-cat" interference, a unique phenomenon in the quantum world.

(3) Mixed state:

To simulate the entangled states of Eq. (3.7.2), historically, the following mixed states were proposed:

$$\hat{\rho}^{(S)} = \frac{1}{2}\left[|\,\text{alive}_1\,\rangle\langle\,\text{alive}_2\,| + |\,\text{dead}_1\,\rangle\langle\,\text{dead}_2\,|\right]$$

$$\hat{\rho}^{(A)} = \frac{1}{2}\left[|\,\text{alive}_1\,\rangle\langle\,\text{dead}_2\,| + |\,\text{dead}_1\,\rangle\langle\,\text{alive}_2\,|\right], \qquad (3.7.5)$$

where the density operator is used to specify the mixed state of the ensemble of a cat system. The density operator $\hat{\rho}^{(S)}$ indicates that the ensemble is prepared with one-half "alive & alive" cats and one-half "dead & dead" cats. The density operator $\hat{\rho}^{(A)}$ indicates that the ensemble is prepared with one-half "alive & dead" cats and one-half "dead & alive" cats. There is no surprise that the two distant observers will have a 50%:50% chance of observing two alive cats or two dead cats simultaneously, if the ensemble is prepared in $\hat{\rho}^{(S)}$, or a 50% - 50% chance of observing "cat one alive & cat two dead" or "cat one dead & cat two alive" simultaneously, if the ensemble is prepared in $\hat{\rho}^{(A)}$. However, neither $\hat{\rho}^{(S)}$ nor $\hat{\rho}^{(A)}$ can achieve the correlation or expectations of the Bell's cat-states for the measurements defined in Eq. (3.7.4). In fact, these measurements on either $\hat{\rho}^{(S)}$ or $\hat{\rho}^{(A)}$ will have equal probabilities of observing the pairs "$+\text{cat}_1 \,\&\, +\text{cat}_2$", "$+\text{cat}_1 \,\&\, -\text{cat}_2$", "$-\text{cat}_1 \,\&\, +\text{cat}_2$", and "$-\text{cat}_1 \,\&\, -\text{cat}_2$", in the joint detection of the distant observers.

Example (II): Product, entangled and mixed polarization state of photons.

(1) Product polarization state of two photons:

Suppose two independent photons are created from a radiation source at time $t = 0$ with photon-one polarized at $45°$ and photon-two polarized at $-45°$. Due to the independent creation process, or non-interaction between the subsystems, the two created photons can be treated as a photon system of a product state

$$|\Psi\rangle = |\,45°_1\,\rangle \times |-45°_2\,\rangle$$

$$= \frac{1}{2}\left[|\,\hat{o}_1\,\rangle|\,\hat{o}_2\,\rangle + |\,\hat{e}_1\,\rangle|\,\hat{e}_2\,\rangle - |\,\hat{o}_1\,\rangle|\,\hat{e}_2\,\rangle - |\,\hat{e}_1\,\rangle|\,\hat{o}_2\,\rangle\right], \qquad (3.7.6)$$

where $\hat{o}$ and $\hat{e}$ indicate the ordinary-ray and the extraordinary-ray of the uniaxial crystal that we defined in early section for polarization measurements.

We apply the same criterion for the free propagation and time evolution for the two photon system as that for the cat system. It is not surprising that the two observers at large distances will observe four equal possibilities for the polarization correlation: (1) photon-one and photon-two are both polarized in the $\hat{o}$ direction, (2) photon-one and photon-two are both polarized in the $\hat{e}$ direction, (3) photon-one is polarized in the $\hat{o}$ direction and photon-two is polarized in the $\hat{e}$ direction, and (4) photon-one is polarized in the $\hat{e}$ direction and photon-two is polarized in the $\hat{o}$ direction.

Again, "not surprising" means that there is no difference between the quantum prediction and the classical convention.

(2) Entangled polarization state of two photons:

Suppose a pair of photons is created at the radiation source in one of the following states

$$|\Psi^{(+)}\rangle = \frac{1}{\sqrt{2}}\left[|\hat{o}_1\rangle|\hat{o}_2\rangle + |\hat{e}_1\rangle|\hat{e}_2\rangle\right]$$

$$|\Psi^{(-)}\rangle = \frac{1}{\sqrt{2}}\left[|\hat{o}_1\rangle|\hat{o}_2\rangle - |\hat{e}_1\rangle|\hat{e}_2\rangle\right]$$

$$|\Phi^{(+)}\rangle = \frac{1}{\sqrt{2}}\left[|\hat{o}_1\rangle|\hat{e}_2\rangle + |\hat{e}_1\rangle|\hat{o}_2\rangle\right]$$

$$|\Phi^{(-)}\rangle = \frac{1}{\sqrt{2}}\left[|\hat{o}_1\rangle|\hat{e}_2\rangle - |\hat{e}\rangle_1|\hat{o}\rangle_2\right] \tag{3.7.7}$$

where the state cannot be written in the form of a product $|\Psi\rangle = |\Psi_1\rangle|\Psi_2\rangle$, the state of the photon pair is defined as an entangled state. The above four states are called Bell's states. The Bell's states form a complete, orthogonal two-photon polarization vector basis. It is worth emphasizing that although the Bell's states of the pair are pure states the state of photon one and photon two are both mixed states. The state of a subsystem can easily be found by taking a partial trace of its twin

$$\hat{\rho}_j = \frac{1}{2}\left[|\hat{o}_j\rangle\langle\hat{o}_j| + |\hat{e}_j\rangle\langle\hat{e}_j|\right] \tag{3.7.8}$$

where $j = 1, 2$. Eq. (3.7.8) indicates that the subsystems are unpolarized.

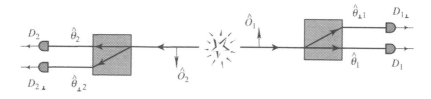

FIGURE 3.7.1 Polarization correlation measurement of a photon pair. Two independent polarization analyzers are oriented at angle θ_1 and θ_2 relative to the unit vector $\hat{o}_1$ and $\hat{o}_2$, respectively. Each polarization analyzer has two output ports: $\hat{\theta}_j$ and $\hat{\theta}_{\perp j}$. The four possible joint detection events correspond to four projection operations defined in Eq. (3.7.9). Remark: all unit vectors are defined in the natural right-hand coordinate system.

Again, we apply the same criterion for free propagation and time evolution of the system. The state of the photon pair will remain the same as that of Eq. (3.7.7), except for a trivial overall phase factor. Two distant observers decide to measure the polarization of the photon pair with two independent polarization analyzers oriented at angles θ_1 and θ_2 relative to the unit vectors $\hat{o}_1$ and $\hat{o}_2$, respectively, as shown in Fig. 3.7.1. Each polarization analyzer has two output ports: θ_j and $\theta_{\perp j}$, $j = 1, 2$. The four possible joint detection events correspond to four joint-projections of the two-photon state defined as follows,

$$\langle\Psi|\hat{\theta}_1\,\hat{\theta}_2\rangle\langle\hat{\theta}_2\,\hat{\theta}_1|\Psi\rangle = |\langle\hat{\theta}_2\,\hat{\theta}_1|\Psi\rangle|^2,$$

$$\langle\Psi|\hat{\theta}_{\perp 1}\,\hat{\theta}_{\perp 2}\rangle\langle\hat{\theta}_{\perp 2}\,\hat{\theta}_{\perp 1}|\Psi\rangle = |\langle\hat{\theta}_{\perp 2}\,\hat{\theta}_{\perp 1}|\Psi\rangle|^2,$$

$$\langle\Psi|\hat{\theta}_1\,\hat{\theta}_{\perp 2}\rangle\langle\hat{\theta}_{\perp 2}\,\hat{\theta}_1|\Psi\rangle = |\langle\hat{\theta}_{\perp 2}\,\hat{\theta}_1|\Psi\rangle|^2,$$

$$\langle\Psi|\hat{\theta}_{\perp 1}\,\hat{\theta}_2\rangle\langle\hat{\theta}_2\,\hat{\theta}_{\perp 1}|\Psi\rangle = |\langle\hat{\theta}_2\,\hat{\theta}_{\perp 1}|\Psi\rangle|^2, \tag{3.7.9}$$

where we have written $|\hat{\theta}_1\rangle\,|\hat{\theta}_2\rangle$ as $|\hat{\theta}_1\,\hat{\theta}_2\rangle$. The Bell's states have the following polarization correlation:

$$\langle\Psi^{(+)}|\hat{\theta}_1\,\hat{\theta}_2\rangle\langle\hat{\theta}_2\,\hat{\theta}_1|\Psi^{(+)}\rangle = \frac{1}{2}\cos^2(\theta_1 - \theta_2)$$

$$\langle\Psi^{(+)}|\hat{\theta}_{\perp 1}\,\hat{\theta}_{\perp 2}\rangle\langle\hat{\theta}_{\perp 2}\,\hat{\theta}_{\perp 1}|\Psi^{(+)}\rangle = \frac{1}{2}\cos^2(\theta_1 - \theta_2)$$

$$\langle\Psi^{(+)}|\hat{\theta}_1\,\hat{\theta}_{\perp 2}\rangle\langle\hat{\theta}_{\perp 2}\,\hat{\theta}_1|\Psi^{(+)}\rangle = \frac{1}{2}\sin^2(\theta_1 - \theta_2)$$

$$\langle\Psi^{(+)}|\hat{\theta}_{\perp 1}\,\hat{\theta}_2\rangle\langle\hat{\theta}_2\,\hat{\theta}_{\perp 1}|\Psi^{(+)}\rangle = \frac{1}{2}\sin^2(\theta_1 - \theta_2) \tag{3.7.10}$$

$$\langle\Psi^{(-)}|\hat{\theta}_1\,\hat{\theta}_2\rangle\langle\hat{\theta}_2\,\hat{\theta}_1|\Psi^{(-)}\rangle = \frac{1}{2}\cos^2(\theta_1 + \theta_2)$$

$$\langle\Psi^{(-)}|\hat{\theta}_{\perp 1}\,\hat{\theta}_{\perp 2}\rangle\langle\hat{\theta}_{\perp 2}\,\hat{\theta}_{\perp 1}|\Psi^{(-)}\rangle = \frac{1}{2}\cos^2(\theta_1 + \theta_2)$$

$$\langle\Psi^{(-)}|\hat{\theta}_1\,\hat{\theta}_{\perp 2}\rangle\langle\hat{\theta}_{\perp 2}\,\hat{\theta}_1|\Psi^{(-)}\rangle = \frac{1}{2}\sin^2(\theta_1 + \theta_2)$$

$$\langle\Psi^{(-)}|\hat{\theta}_{\perp 1}\,\hat{\theta}_2\rangle\langle\hat{\theta}_2\,\hat{\theta}_{\perp 1}|\Psi^{(-)}\rangle = \frac{1}{2}\sin^2(\theta_1 + \theta_2) \tag{3.7.11}$$

$$\langle\Phi^{(+)}|\hat{\theta}_1\,\hat{\theta}_2\rangle\langle\hat{\theta}_2\,\hat{\theta}_1|\Phi^{(+)}\rangle = \frac{1}{2}\sin^2(\theta_1 + \theta_2)$$

$$\langle\Phi^{(+)}|\hat{\theta}_{\perp 1}\,\hat{\theta}_{\perp 2}\rangle\langle\hat{\theta}_{\perp 2}\,\hat{\theta}_{\perp 1}|\Phi^{(+)}\rangle = \frac{1}{2}\sin^2(\theta_1 + \theta_2)$$

$$\langle\Phi^{(+)}|\hat{\theta}_1\,\hat{\theta}_{\perp 2}\rangle\langle\hat{\theta}_{\perp 2}\,\hat{\theta}_1|\Phi^{(+)}\rangle = \frac{1}{2}\cos^2(\theta_1 + \theta_2)$$

$$\langle\Phi^{(+)}|\hat{\theta}_{\perp 1}\,\hat{\theta}_2\rangle\langle\hat{\theta}_2\,\hat{\theta}_{\perp 1}|\Phi^{(+)}\rangle = \frac{1}{2}\cos^2(\theta_1 + \theta_2) \tag{3.7.12}$$

$$\langle\Phi^{(-)}|\hat{\theta}_1\,\hat{\theta}_2\rangle\langle\hat{\theta}_2\,\hat{\theta}_1|\Phi^{(-)}\rangle = \frac{1}{2}\sin^2(\theta_1 - \theta_2)$$

$$\langle\Phi^{(-)}|\hat{\theta}_{\perp 1}\,\hat{\theta}_{\perp 2}\rangle\langle\hat{\theta}_{\perp 2}\,\hat{\theta}_{\perp 1}|\Phi^{(-)}\rangle = \frac{1}{2}\sin^2(\theta_1 - \theta_2)$$

$$\langle\Phi^{(-)}|\hat{\theta}_1\,\hat{\theta}_{\perp 2}\rangle\langle\hat{\theta}_{\perp 2}\,\hat{\theta}_1|\Phi^{(-)}\rangle = \frac{1}{2}\cos^2(\theta_1 - \theta_2)$$

$$\langle\Phi^{(-)}|\hat{\theta}_{\perp 1}\,\hat{\theta}_2\rangle\langle\hat{\theta}_2\,\hat{\theta}_{\perp 1}|\Phi^{(-)}\rangle = \frac{1}{2}\cos^2(\theta_1 - \theta_2). \tag{3.7.13}$$

The correlation functions in Eq. (3.7.10) to Eq. (3.7.13) specify the probabilities for the photodetector at output port θ_1 (θ_2) to register a photon while its twin is registered at output port θ_2 (θ_1), for each of the Bell's states. It is easy to see that for any value of θ_1 (θ_2) there is a unique orientation, θ_2 (θ_1), for which we expect a maximum probability of $1/2$ to register the pair simultaneously, while the pair is prepared in any one of the Bell's states. This is truly a surprise. For each photon pair, coming from the two-photon source, each polarization analyzer has a 50%:50% chance to register its photon at θ_j or $\theta_{\perp j}$, as if the photon is unpolarized. However, whenever observer one (two) finds a photon polarized along θ_1 (θ_2) or $\theta_{\perp 1}$ ($\theta_{\perp 2}$), the photon measured by observer two (one) must be polarized at a unique orientation θ_2 (θ_1) or $\theta_{\perp 2}$ ($\theta_{\perp 1}$). Surprisingly, after a large number of measurements, the two distant observers conclude that although the polarization of neither photon one nor photon two is defined, if one of the photons is measured with a polarization along a certain orientation the polarization of its twin is uniquely determined, despite the distance between the two observers. Remember, (1) there is no interaction between the two photons during their propagation and there is no communication between the two observers either; (2) the ensemble is neither prepared with photon pairs of $|\hat{\theta}_1\rangle$ & $|\hat{\theta}_2\rangle$ and $|\hat{\theta}_{\perp 1}\rangle$ & $|\hat{\theta}_{\perp 2}\rangle$, nor with photon pairs of $|\hat{\theta}_1\rangle$ & $|\hat{\theta}_{\perp 2}\rangle$ and $|\hat{\theta}_{\perp 1}\rangle$ & $|\hat{\theta}_2\rangle$, except when $\hat{\theta} = \hat{o}$, or $\hat{\theta} = \hat{e}$.

Here, again, a "surprise" means that the quantum predication is different from the classical convention and it seems impossible to understand it under the framework of classical physics.

(3) Mixed state:

It is very clear that the Bell's states in Eq. (3.7.7) are very different from the following mixed states

$$\hat{\rho}^{(S)} = \frac{1}{2} \left[|\hat{o}_1\rangle\langle\hat{o}_2| + |\hat{e}_1\rangle\langle\hat{e}_2| \right]$$

$$\hat{\rho}^{(A)} = \frac{1}{2} \left[|\hat{o}_1\rangle\langle\hat{e}_2| + |\hat{e}_1\rangle\langle\hat{o}_2| \right], \tag{3.7.14}$$

where the ensemble is prepared with one-half of the photon-pairs with polarization $\hat{o}_1$ & $\hat{o}_2$ ($\hat{o}_1$ & $\hat{e}_2$) and the other one-half photon-pairs with polarization $\hat{e}_1$ & $\hat{e}_2$ ($\hat{e}_1$ & $\hat{o}_2$) for $\hat{\rho}^{(S)}$ ($\hat{\rho}^{(A)}$). Although in $\hat{\rho}^{(S)}$ ($\hat{\rho}^{(A)}$) the two distant observers will receive half of the pairs in $\hat{o}_1$ & $\hat{o}_2$ ($\hat{o}_1$ & $\hat{e}_2$) and another half of the pairs in $\hat{e}_1$ & $\hat{e}_2$ ($\hat{e}_1$ & $\hat{o}_2$), neither $\hat{\rho}^{(S)}$ nor $\hat{\rho}^{(A)}$ can achieve any correlation similar to that in Eq. (3.7.10) to Eq. (3.7.13), if θ_1 and θ_2 are chosen at orientations other then $\hat{o}$ and $\hat{e}$.

SUMMARY

In this chapter we have introduced and studied a number of critical concepts of quantum optics:

(I) Photon number state representation and coherent state representation

The generalized multi-mode photon number state, or Fock state of the radiation field is written as:

$$|\Psi\rangle = \prod_{\mathbf{k},s} |n_{\mathbf{k},s}\rangle = \prod_{\mathbf{k},s} \frac{1}{\sqrt{n_{\mathbf{k},s}!}} (\hat{a}^{\dagger}_{\mathbf{k},s})^n |0\rangle$$

which is an eigenstate of the photon number operator $n = \sum n_{\mathbf{k},s}$,

$$\hat{n} \left(\prod_{\mathbf{k},s} |n_{\mathbf{k},s}\rangle \right) = \left(\sum_{\mathbf{k},s} n_{\mathbf{k},s} \right) \prod_{\mathbf{k},s} |n_{\mathbf{k},s}\rangle = n \prod_{\mathbf{k},s} |n_{\mathbf{k},s}\rangle,$$

where $n = \sum n_{\mathbf{k},s}$ is the total number of photons in the radiation field. The multi-mode coherent state is defined as the product of single-mode coherent state $|\alpha_{\mathbf{k},s}\rangle$

$$|\{\alpha\}\rangle = \prod_{\mathbf{k},s} |\alpha_{\mathbf{k},s}\rangle,$$

which is an eigenstate of the multi-mode annihilation operator $\hat{a} = \sum_{\mathbf{k},s} \hat{a}_{\mathbf{k},s}$ with eigenvalue $\sum_{\mathbf{k},s} \alpha_{\mathbf{k},s}$:

$$\hat{a}\,|\{\alpha\}\rangle = \Big(\sum_{\mathbf{k},s} \hat{a}_{\mathbf{k},s}\Big) \prod_{\mathbf{k},s} |\alpha_{\mathbf{k},s}\rangle = \Big(\sum_{\mathbf{k},s} \alpha_{\mathbf{k},s}\Big)|\{\alpha\}\rangle.$$

The photon number state representation, especially single-photon state representation, and coherent state representation are widely used to specify the quantum state of radiation.

(II) Density operator or density matrix

The density operator (density matrix) is defined from the calculation of the expectation value of an observable:

$$
\begin{aligned}
\big\langle\langle\hat{A}\rangle_{\mathrm{QM}}\big\rangle_{\mathrm{En}} &= \sum_j P_j \langle\Psi_j|\hat{A}|\Psi_j\rangle \\
&= \sum_j P_j \langle\Psi_j|\hat{A}\big[\sum_n |\Psi_n\rangle\langle\Psi_n|\big]|\Psi_j\rangle \\
&= \sum_n \langle\Psi_n|\big[\sum_j P_j|\Psi_j\rangle\langle\Psi_j|\big]\hat{A}|\Psi_n\rangle \\
&= \sum_n \langle\Psi_n|\hat{\rho}\hat{A}|\Psi_n\rangle \\
&= tr\,\hat{\rho}\hat{A}
\end{aligned}
$$

where

$$\hat{\rho} \equiv \sum_j P_j\,|\Psi_j\rangle\langle\Psi_j|.$$

In principle, the trice can be taken in any vector base:

$$
\begin{aligned}
\big\langle\langle\hat{A}\rangle_{\mathrm{QM}}\big\rangle_{\mathrm{Ensamble}} &= \sum_n \sum_j P_j \langle\Psi_j|\hat{A}|n\rangle\langle n|\Psi_j\rangle \\
&= \sum_n \langle n|\big[\sum_j P_j|\Psi_j\rangle\langle\Psi_j|\big]\hat{A}|n\rangle \\
&= \sum_n \langle n|\hat{\rho}\hat{A}|n\rangle \\
&= tr\,\hat{\rho}\hat{A}
\end{aligned}
$$

Obviously, the density operator for pure state is defined as

$$\hat{\rho} \equiv |\Psi\rangle\langle\Psi|.$$

(III) Pure state and mixed state

The concepts of pure states and mixed states include two important aspects of physics: the state of the individual quantum and the state of the measured ensemble.

(1) In terms of the individual quantum system: a pure state represents a vector in Hilbert space. If the state of a quantum can be described by a vector, the state of the quantum is said to be *pure*. On the contrary, if the state of a quantum cannot be described as a vector but rather a mixture of vectors, the quantum is said to be in a *mixed state*.

(2) In terms of the measured ensemble: if the state of all the measured quanta of the ensemble can be described by the same vector, the state of the ensemble, or the measured systems of quanta, is said to be *homogeneous* or in a statistically *pure state*. If the measured ensemble cannot be described by the same vector, or the measured ensemble has to be described by several, or by many different vectors, the ensemble is referred to as a statistical *mixture*.

We have given three examples to explore the physics behind, especially for aspect (1): (I) polarization measurement of a single-photon state; (II) Schrödinger cat; and (III) general measurement of single-photon state.

(IV) Composite system and two-photon state of radiation field

In certain measurements, we need to deal with multi-photon states, which describe the state of a composite system of two or more photons. In a composite system, the subsystems may be independent, correlated, or entangled. The state of a composite system with two independent photons can be written as a product state

$$|\Psi\rangle = \sum_{m,n} c_m\, c_n\, |m\rangle|n\rangle = \sum_m c_m|m\rangle \sum_n c_n\, |n\rangle = |\Psi_1\rangle \otimes |\Psi_2\rangle,$$

where $c_{mn} = c_m\, c_n$ is factorizable. When the two subsystems are correlated or entangled, in general, $|\Psi\rangle$ cannot be written as a product state, i.e., c_{mn} is non-factorizable and consequently the state is non-factorizable

$$|\Psi\rangle = \sum_{m,n} c_{mn}\, |m\rangle|n\rangle.$$

Depending on the form of c_{mn}, we name the two subsystems in a correlated state or in an entangled state.

(V) A simple model of single-photon and multi-photon source

We have given a simple model of single-photon and multi-photon source for the creation of single-photon and multi-photon state in photon number representation and in coherent state representation.

(VI) Product state, Entangled state and mixed state of photon pairs

The concepts of product states and entangled states involve multi-photon systems. As examples, we studied and distinguished product state, entangled state and mixed states of two-photon systems. To simply the mathematical picture, we have also reduced the dimension to 2-D product, entangled and mixed states of Schrödinger's cats as well as the 2-D polarization state of photons.

REFERENCES AND SUGGESTIONS FOR READING

[1] R.J. Glauber, Phys. Rev. **130**, 2529 (1963); Phys. Rev. **131**, 2766 (1963).
[2] G. Baym, *Lectures on Quantum Mechanics*, Benjamin/Cummings, 1981.
[3] R. Loudon, *The Quantum Theory of Light*, Oxford Science, 2000.
[4] M.O. Scully and M.S. Zubairy, *Quantum Optics*, Cambridge, 1997.
[5] L. Mandel and E. Wolf, *Optical Coherence and Quantum Optics*, Cambridge, London, 1995.

CHAPTER 4

Measurement of Quantized Field and Photon

In the Maxwell continuum wave theory of light, a photodetector measures the intensity of the EM wave. The output current of the photodetector is proportional to the intensity of input light. Since the introduction of the concept of photon, or the concept of field quantization, physicists started to premeditate single-photon detection, theoretically and experimentally. Not until the 1950's, did we have the ability to observe a single light quantum, not theoretically but experimentally. Is this ability important? In fact, it is not only important experimentally but also theoretically. As we know, for a certain period of time, we used to consider a quantum state of radiation only meaningful in terms of an ensemble. For example, in the early experiments of Young's double-slit interferometer, the interference was judged by human eyes. Human eyes do not work well at weak light conditions. In terms of the measurement of an ensemble, the interpretation of Young's interference is reasonable and simple: 50% of the photons passed slit-A and another 50% passed slit-B; the photons that passed slit-A "meet" with the photons that passed slit-B at different locations in the observation plane. Due to slight differences in their optical paths, at certain locations, the radiations superposed constructively, at other locations the field superposed destructively. A sinusoidal modulation is thus observable. However, we found this interpretation may not be appropriate when the experimental physicists improved the sensitivity of their photodetector to the single-photon level. Almost everyone was surprised that the interference was still observable in the observation plane, and the sinusoidal modulation observed from single-photon was exactly the same as before. A single-photon counting detector is able to count photons one by one. How does a photon pass the double-slit then? It passes one slit or both slits? One may not care which slit a photon passes, but we do need to understand the measurement process: How does a photoelectron or a pair of photoelectrons physically relate to the quantized radiation field? Nowadays, we may easily find a statement like: "... the observed photon pair is in an entangled state ... ". What does it mean?

4.1   MEASUREMENT OF EINSTEIN'S BUNDLE OF RAY

In section 2.3, we concluded that Einstein's granularity picture of light is still in the framework of Maxwell's electromagnetic wave theory, except Einstein introduced the concept of bundle of ray, or subfield, to radiation and abandoned the continuum interpretation of Maxwell. Einstein's bundle of ray, light quantum, or subfield is a realistic quantized microscopic entity of electromagnetic wavepacket propagating in space-time. In the view of electromagnetic field theory, Einstein introduced an additional freedom of superposition in terms of quantized subfields: in addition to the superposition of a large number of coherent or incoherent harmonic radiation modes, Einstein introduced the superposition of a large number of coherent or incoherent bundle of rays or subfields. In Einstein's granularity picture, we have written the radiation field measured at coordinate $(\mathbf{r}, t)$ as the result of a superposition among a large number of subfields, each created from a point-like sub-source at $(\mathbf{r}_m, t_m)$,

$$E(\mathbf{r}, t) = \sum_m E_m(\mathbf{r}, t) = \sum_m E_m(\mathbf{r}_m, t_m) \, g_m(\mathbf{r}, t),$$

where $E_m(\mathbf{r}_m, t_m)$ labels the subfield emitted from the mth sub-source at coordinate $(\mathbf{r}_m, t_m)$, and $g_m(\mathbf{r}, t)$ represents the field propagator or Green's function that propagates the mth subfield from coordinate $(\mathbf{r}_m, t_m)$ to coordinate $(\mathbf{r}, t)$. If there exist more than one different yet indistinguishable paths or alternatives for the mth subfield to propagate from $(\mathbf{r}_m, t_m)$ to $(\mathbf{r}, t)$, $g_m(\mathbf{r}, t)$ must be written as a superposition

$$g_m(\mathbf{r}, t) = \frac{1}{\sqrt{N}} \sum_{s=1}^{N} g_{ms}(\mathbf{r}, t),$$

where $g_{ms}(\mathbf{r}, t)$ indicates the sth path of the mth subfield.

(I) Measurement of $\langle I(\mathbf{r}, t) \rangle$

Assume a point-like photodetector D, placed at space-time coordinate $(\mathbf{r}, t)$, is measuring the intensity of a natural thermal radiation, the expectation value of its measured intensity at $(\mathbf{r}, t)$ is

$$\begin{aligned}
\langle I(\mathbf{r}, t) \rangle &= \langle E^*(\mathbf{r}, t) E(\mathbf{r}, t) \rangle \\
&= \Big\langle \sum_m E_m^*(\mathbf{r}, t) \sum_n E_n(\mathbf{r}, t) \Big\rangle \\
&= \Big\langle \sum_m |E_m(\mathbf{r}, t)|^2 \Big\rangle + \Big\langle \sum_{m \neq n} E_m^*(\mathbf{r}, t) E_n(\mathbf{r}, t) \Big\rangle \\
&= \sum_m |E_m(\mathbf{r}, t)|^2 + 0 \\
&= \sum_m I_m(\mathbf{r}, t).
\end{aligned} \tag{4.1.1}$$

In Eq.(4.1.1) the ensemble average makes the second term vanishing. Eq.(4.1.1) connects the expectation of the measured intensity, which is proportional to the mean current of the photodetector, $\langle i(\mathbf{r}, t) \rangle \propto \langle I(\mathbf{r}, t) \rangle$, with Einstein's concept of subfield or "bundle of ray". Eq. (4.1.1) indicates $\langle I(\mathbf{r}, t) \rangle$ is the sum of sub-intensities, $\sum_m I_m(\mathbf{r}, t)$, and each $I_m(\mathbf{r}, t)$ is the result of the mth subfield interfering with the mth subfield itself. The interference between different subfields, corresponding to the $m \neq n$ term, vanishes when taking into account all possible random phases of the subfields,

$$\langle \Delta I(\mathbf{r}, t) \rangle = \Big\langle \sum_{m \neq n} E_m^*(\mathbf{r}, t) \, E_n(\mathbf{r}, t) \Big\rangle = 0 \tag{4.1.2}$$

However, if the measurement is under weak light condition, or if the measurement happens within a short time window in which only a few photons or a limited number of photons contribute to the measurement, so that the measurement is unable to take into account all possible random phases of the subfields, the $m \neq n$ term may not vanish and is observable from the measurement of a photodetector. In this case, the $m \neq n$ terms contribute to the intensity a fluctuation. Similar to the two-mode interference induced intensity fluctuation under the frame of Maxwell, in Einstein's picture we name it two-photon interference induced intensity fluctuation,

$$\Delta I(\mathbf{r}, t) = \sum_{m \neq n} E_m^*(\mathbf{r}, t) E_n(\mathbf{r}, t) \neq 0. \tag{4.1.3}$$

Due to the random phases associated with the subfields, $\Delta I(\mathbf{r}, t)$ takes random positive or negative values in the neighborhood of $\langle I(\mathbf{r}, t)\rangle$ from time to time or from measurement to measurement. $\Delta I(\mathbf{r}, t)$ is usually considered "noise" in the measurement of $\langle I(\mathbf{r}, t)\rangle$.

(II) Measurement of $\langle I(\mathbf{r}_1, t_1) I(\mathbf{r}_2, t_2)\rangle$

Now, we setup a correlation measurement of thermal field by means of joint-photodetection between two point-like photodtectors D_1 and D_2 placed at space-time coordinates $(\mathbf{r}_1, t_1)$ and $(\mathbf{r}_2, t_2)$, respectively. Figure 4.1.1 schematically illustrated such a measurement. The output currents of D_1 and D_2 are amplified by two individual amplifiers, respectively. The output currents, $i_1(t)$ and $i_2(t)$, of the two amplifiers are multiplied by an electronic linear multiplier, or RF mixer at time t. The correlation of $i_1(t)$ and $i_2(t)$ is proportional to the correlation of the intensities measured by D_1 and D_2 at early times t_1 and t_2, respectively, $\langle i_1(t) i_2(t)\rangle \propto \langle I(\mathbf{r}_1, t_1) I(\mathbf{r}_2, t_2)\rangle$, where $t_1 = t - \tau_1^e, t_2 = t - \tau_2^2$ are the times of the two photodetection events of D_1 and D_2. τ_1^e and τ_2^e are the electronic time delays, including the delays of the detectors, the amplifiers, and the delay-line cables. The expectation value of $\langle I(\mathbf{r}_1, t_1) I(\mathbf{r}_2, t_2)\rangle$ is calculated in Einstein's picture

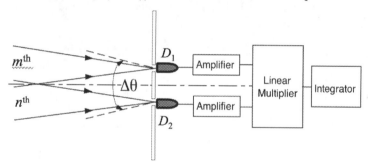

FIGURE 4.1.1 Schematic setup for the intensity correlation measurement of $\langle I(\mathbf{r}_1, t_1) I(\mathbf{r}_2, t_2)\rangle$ by two point-like photodetectors placed at space-time coordinates $(\mathbf{r}_1, t_1)$ and $(\mathbf{r}_2, t_2)$, respectively. An electronic linear multiplier, or RF mixer, is employed to multiply the two currents $i_1(t)$ and $i_2(t)$ at time t. Note, the times of the two photodetection events t_1 and t_2, respectively, as well as the relative time delay $t_1 - t_2 = \tau_1^e - \tau_2^e$ are all defined by the electronics.

$$
\begin{aligned}
\langle I(\mathbf{r}_1, t_1) I(\mathbf{r}_2, t_2)\rangle &= \langle E^*(\mathbf{r}_1, t_1) E(\mathbf{r}_1, t_1) E^*(\mathbf{r}_2, t_2) E(\mathbf{r}_2, t_2)\rangle \\
&= \Big\langle \sum_m E_m^*(\mathbf{r}_1, t_1) \sum_n E_n(\mathbf{r}_1, t_1) \sum_p E_p^*(\mathbf{r}_2, t_2) \sum_q E_q(\mathbf{r}_2, t_2) \Big\rangle \\
&= \sum_m E_m^*(\mathbf{r}_1, t_1) E_m(\mathbf{r}_1, t_1) \sum_n E_n^*(\mathbf{r}_2, t_2) E_n(\mathbf{r}_2, t_2) \\
&\quad + \sum_{m \neq n} E_m^*(\mathbf{r}_1, t_1) E_n(\mathbf{r}_1, t_1) E_n^*(\mathbf{r}_2, t_2) E_m(\mathbf{r}_2, t_2) \\
&= \sum_{m \neq n} \Big| \frac{1}{\sqrt{2}} \big[E_m(\mathbf{r}_1, t_1) E_n(\mathbf{r}_2, t_2) + E_n(\mathbf{r}_1, t_1) E_m(\mathbf{r}_2, t_2) \big] \Big|^2 \\
&= \langle I(\mathbf{r}_1, t_1)\rangle \langle I(\mathbf{r}_2, t_2)\rangle + \langle \Delta I(\mathbf{r}_1, t_1) \Delta I(\mathbf{r}_2, t_2)\rangle. \tag{4.1.4}
\end{aligned}
$$

where m, n, p, q label the randomly created and randomly distributed subfields and their corresponding point-like sub-sources. Considering the random phases of the subfields, as the result of a destructive interference cancelation when taking into account all possible phases of the subfields, the only surviving terms from the ensemble sum are the following two groups: (1) $m = q, n = p$, (2) $m = p, n = q$. $\langle I(\mathbf{r}_1, t_1) I(\mathbf{r}_2, t_2) \rangle$ can be written into the form of Eq. (4.1.4). The first terms of Eq. (4.1.4) is a product of two mean intensities measured by D_1 and D_2, respectively. The second term of Eq. (4.1.4) is the intensity fluctuation correlation measured by D_1 and D_2, jointly. For randomly created thermal radiations, the first term usually contributes a constant to $\langle I(\mathbf{r}_1, t_1) I(\mathbf{r}_2, t_2) \rangle$, and the second term contributes a nontrivial function of space-time $(\mathbf{r}_1, t_1)$ and $(\mathbf{r}_2, t_2)$ to $\langle I(\mathbf{r}_1, t_1) I(\mathbf{r}_2, t_2) \rangle$ in terms of either temporal delay or spatial separation between the two photodetection events of D_1 and D_2. Mathematically, it is straightforward to find that Eq. (4.1.4) indicates an interference phenomenon: a randomly created and randomly paired subfields, or photons in thermal state, interferes with the pair itself. Physically, it is easy to find that the mth-nth pair of subfields or photons has two different yet indistinguishable alternatives to produce a joint-photodetection event: (1) the mth subfield, or photon, is annihilated at D_1 and the nth subfield, or photon, is annihilated at D_2; (2) the nth subfield, or photon, is annihilated at D_1 and the mth subfield, or photon, is annihilated at D_2. We name this interference or superposition "two-photon interference".

(III) Measurement of $\langle \Delta I(\mathbf{r}_1, t_1) \Delta I(\mathbf{r}_2, t_2) \rangle$

It is also easy to find from Eq. (4.1.4), the two-photon, or two-subfield, interference induced intensity fluctuation correlation of randomly created and randomly distributed thermal radiation is observable when

$$\langle \Delta I(\mathbf{r}_1, t_1) \Delta I(\mathbf{r}_2, t_2) \rangle = \sum_{m \neq n} E_m^*(\mathbf{r}_1, t_1) E_n(\mathbf{r}_1, t_1) E_n^*(\mathbf{r}_2, t_2) E_m(\mathbf{r}_2, t_2) \neq 0. \qquad (4.1.5)$$

Indeed, it is the cross-interference term of the two-subfield or two-photon superposition of Eq. (4.1.4) results in the nontrivial intensity fluctuation correlation; however, to observe the intensity fluctuation correlation, certain experimental conditions must be achieved to yield the above non-zero cross-interference term.

How does a randomly created thermal radiation have its nontrivial intensity fluctuation correlation in terms of its temporal or spatial variables? To estimate the correlation, we may rewrite Eq. (4.1.5) in terms of the superposition of a large number of subfields and Fourier modes. To simplify the math, we calculate the temporal correlation only by taking the Fourier transform in 1-D as usual,

$$\langle \Delta I(\mathbf{r}_1, t_1) \Delta I(\mathbf{r}_2, t_2) \rangle$$
$$= \sum_{m \neq n} \int d\omega\, E_m^*(\omega) e^{i(\omega t_1 - k r_1)} \int d\omega'\, E_n(\omega') e^{-i(\omega' t_1 - k' r_1)}$$
$$\int d\omega''\, E_n^*(\omega'') e^{i(\omega'' t_2 - k'' r_2)} \int d\omega'''\, E_m(\omega''') e^{-i(\omega''' t_2 - k''' r_2)}$$
$$= \sum_{m \neq n} \mathcal{F}_{\tau_1}^* \{ E_m(\nu) \} \mathcal{F}_{\tau_1} \{ E_n(\nu) \} \mathcal{F}_{\tau_2}^* \{ E_n(\nu) \} \mathcal{F}_{\tau_2} \{ E_m(\nu) \} \qquad (4.1.6)$$

indicating the product of two wavepackets and their conjugates. Producing a non-zero cross-interference term, the above two wavepackets and their conjugates must "overlap" in space-time τ_1 and τ_2. This means: (1) in order for the mth wavepacket and the conjugate of the nth wavepacket to be both non-zero at space-time τ_1 and τ_2, $t_m - t_n$ cannot be greater than the temporal width of the wavepacket; (2) in order for the mth (or nth) wavepacket and its conjugate to be both non-zero at τ_1 and τ_2, $\tau_1 - \tau_2$ cannot be greater than the temporal width of the wavepacket. We will have more detailed discussions on the measurement of intensity fluctuation correlation in later chapters.

In the case of broad spectrum thermal field, comparing with the broad spectrum of the total field, each quantized subfield, or bundle of ray, can be treated as a single mode light quantum with energy $\hbar\omega$, as Einstein did in 1905. The indexes of m and n can be replaced by ω. In this case, the radiation field $E(r,t)$ can be approximated as an incoherent superposition among these single mode subfields. Eq. (4.1.4) (1-D) can be simplified as the follows:

$$\langle I(r_1,t_1)I(r_2,t_2)\rangle$$
$$= \langle \sum_\omega E^*(\omega)e^{i\omega\tau_1} \sum_{\omega'} E(\omega')e^{-i\omega'\tau_1} \sum_{\omega''} E^*(\omega'')e^{i\omega''\tau_2} \sum_{\omega'''} E(\omega''')e^{-i\omega'''\tau_2} \rangle$$
$$= \sum_{\omega=\omega'} |E(\omega)|^2 \sum_{\omega''=\omega'''} |E(\omega'')|^2$$
$$+ \sum_{\omega=\omega'''} |E(\omega)|^2 e^{i\omega(\tau_1-\tau_2)} \sum_{\omega'=\omega''} |E(\omega')|^2 e^{-i\omega'(\tau_1-\tau_2)}$$
$$= \langle I(r_1,t_1)\rangle\langle I(r_1,t_1)\rangle + \langle \Delta I(r_1,t_1)\Delta I(r_2,t_2)\rangle \tag{4.1.7}$$

Regarding the intensity fluctuation and the intensity fluctuation correlation, we should emphasize:

(1) The two-photon interference induced intensity fluctuation is not the same as the intensity fluctuation associated with $\sum_m |E_m(\mathbf{r},t)|^2$, although the value of $\sum_m |E_m(\mathbf{r},t)|^2$ may varie from time to time or from measurement to measurement when the total number of subfields varies from time to time or from measurement to measurement. For instance, we may observe a Poisson distribution in terms of photon numbers in a photon counting measurements, the concept of "Poisson distribution" is related to $\sum_m |E_m(\mathbf{r},t)|^2$, but not to $\sum_{m\neq n} E_m^*(\mathbf{r},t)E_n(\mathbf{r},t)$.

(2) The two-photon interference induced intensity fluctuation of thermal radiation $\sum_{m\neq n} E_m^*(\mathbf{r},t)E_n(\mathbf{r},t)$ is the one that contributes to the intensity fluctuation correlation measurement of $\langle \Delta I(\mathbf{r}_1,t_1)\Delta I(\mathbf{r}_2,t_2)\rangle$. We will discuss this in detail in Chapter 8.

(3) The two-photon interference induced $\Delta I(\mathbf{r}_j,t_j)$, $j = 1,2$, is usually a nondeterministic functions of space-time variables, even if the two-photon interference induced intensity fluctuation correlation is a deterministic function of the *temporal delay* or the *spatial separation* of the two photodetection events of D_1 and D_2. For example, we usually emphasize the spatial correlation of thermal field in the language of Einstein-Podolsky-Rosen: although the measured intensities fluctuate randomly in the observation planes of D_1 and D_2, if D_1 observed a positive (negative) intensity fluctuation at space-time coordinate $(\mathbf{r}_1,t_1)$, D_2 must observe a corresponding positive (negative) intensity fluctuation at an unique space-time coordinate $(\mathbf{r}_2,t_2)$.

4.2 TIME DEPENDENT PERTURBATION THEORY

In certain problems of practice, Schrödinger equation may not be solved with exact solutions, but only with a good approximation. In quantum optics, we usually face a problem in finding the eigenstates and the eigenvalues of a Hamiltonian in the form of $H_0 + V(t)$, where $V(t)$ is time-dependent and H_0 has exact solutions. Photodetection is one of the typical problems. This problem is formulated in general as time-dependent perturbation theory in quantum mechanics. We will have a very brief review of the subject. A model and calculation of the photon counting process will follow as an exercise of time-dependent perturbation theory.

Suppose we shine light on an atomic system at time t_0 and ask what are the chances that the light ionizes the atoms, or releases one photoelectron, two photoelectrons, three photoelectrons, ..., or N photoelectrons from the atomic system. Alternatively, the problem may be stated in a more general way: we begin with a system of Hamiltonian H_0 and then add a time-dependent perturbation $V(t)$ at time t_0. What changes of the eigenstates and eigenvalues of the system occurred due to the perturbation?

It is assumed that, before time t_0, the system is in state $|\Psi^{(0)}\rangle$ which is an exact solution of the Schrödinger equation

$$i\hbar \frac{\partial}{\partial t} |\Psi^{(0)}\rangle = H_0 |\Psi^{(0)}\rangle \qquad \text{for } t < t_0 \tag{4.2.1}$$

and the time dependent perturbation $V(t)$ is applied at t_0. To find the new state $|\Psi^{(t)}\rangle$ of the system we need to solve the Schrödinger equation

$$i\hbar \frac{\partial}{\partial t} |\Psi^{(t)}\rangle = [H_0 + V(t)] |\Psi^{(t)}\rangle \qquad \text{for } t > t_0 \tag{4.2.2}$$

subject to the boundary condition $|\Psi^{(0)}\rangle = |\Psi^{(t)}\rangle$ at $t = t_0$. Although we may not be able to find exact solutions for Eq. (4.2.2), a good approximation of $|\Psi^{(t)}\rangle$ is permissible in terms of the superposition

$$|\Psi^{(t)}\rangle = \sum_j c_j |j\rangle \tag{4.2.3}$$

where c_j is the probability amplitude for the system staying in the jth eigenstate, $|j\rangle$, of H_0, and $|j\rangle$ forms a complete orthonormal vector basis. Furthermore, if $V(t)$ is much smaller than H_0, only a few lower-order terms of the expansion are necessary for a good approximation. In this case, $V(t)$ is treated as a perturbation. In the following discussion, we will assume $V(t) \ll H_0$. Thus, $|\Psi^{(t)}\rangle$ can be approximated as

$$|\Psi^{(t)}\rangle = e^{-iH_0 t/\hbar} |\Psi(t)\rangle, \tag{4.2.4}$$

where $e^{-iH_0 t/\hbar}$ dominates the time-dependence of $|\Psi^{(t)}\rangle$. It is easy to see that this term contributes to the jth eigenstate of H_0 a phase factor $e^{-iE_j t/\hbar} = e^{-i\omega_j t}$.

Substituting Eq. (4.2.4) into Eq. (4.2.2), we obtain the Schrödinger equation that $|\Psi(t)\rangle$ must obey

$$i\hbar \frac{\partial}{\partial t} |\Psi(t)\rangle = H_I(t) |\Psi(t)\rangle, \tag{4.2.5}$$

where

$$H_I(t) = e^{iH_0 t/\hbar} V(t) e^{-iH_0 t/\hbar}. \tag{4.2.6}$$

The state $|\Psi(t)\rangle$ and the operator $H_I(t)$ in Eq. (4.2.5) are said to be in the interaction representation. In quantum mechanics, Eq. (4.2.6) defines the interaction representation for any operator.

Eq. (4.2.5) can be formally integrated

$$|\Psi(t)\rangle = \left[e^{\frac{1}{i\hbar} \int_{t_0}^{t} dt'\, H_I(t')} \right]_T |\Psi(t_0)\rangle, \tag{4.2.7}$$

where T stands for time order. The first three leading terms of the state, i.e., the zero-order, the first-order and the second-order changes of the state are thus

$$|\Psi(t)\rangle \cong |\Psi(t)\rangle^{(0)} + |\Psi(t)\rangle^{(1)} + |\Psi(t)\rangle^{(2)}$$

$$= |\Psi(t_0)\rangle - \frac{i}{\hbar} \int_{t_0}^{t} dt'\, H_I(t')\, |\Psi(t_0)\rangle$$

$$+ \frac{1}{(i\hbar)^2} \int_{t_0}^{t} dt' \int_{t_0}^{t'} dt''\, H_I(t')\, H_I(t'')\, |\Psi(t_0)\rangle. \qquad (4.2.8)$$

The notation $|\Psi(t)\rangle^{(s)}$, $s = 0, 1, 2...$, indicates the sth order changes of the state. It should be emphasized that the operators in $|\Psi(t)\rangle^{(2)}$ are in time-order. Care has to be taken when the operators do not commute and, in general, this is the more likely case.

It is interesting to find, term by term, the physical correspondence of Eq. (4.2.8), in connection with the photodetection process. The zero-order term $|\Psi(t)\rangle^{(0)}$, of course, is the no-action term. The state of the atomic system and the state of the field have no changes. The second term, $|\Psi(t)\rangle^{(1)}$, and the third term, $|\Psi(t)\rangle^{(2)}$, in Eq. (4.2.8) are the first-order and the second-order changes of the state corresponding to the annihilation of one photon and two photons from the field, respectively.

4.3 MEASUREMENT OF LIGHT: PHOTON COUNTING

In classical theory, a photodetector measures the intensity of a light wave and thus relates to the electromagnetic fields in a phenomenological way. It is quantum theory that provides a successful solution to the physical process of atomic transition—photon annihilation. Although a detailed study of the theory is not the goal of the following discussion, a brief review of the concepts is necessary for understanding the physics of photon counting detection and the measurement of quantum correlations. The following review follows Glauber's theory of photodetection.

We start by defining an idealized photon counting detector as a point detector. One may imagine it as an atom with all concerned transition probabilities in terms of the frequencies of the light field. H_0 has two parts,

$$H_0 = H_0^A + H_0^F, \qquad (4.3.1)$$

where H_0^A and H_0^F are the Hamiltonian for the free atom and the free field, with complete sets of orthonormal eigenstates $|j\rangle$ and $|n\rangle$, respectively. To simplify the analysis, we assume the atom is in its ground state $|0\rangle$ before the interaction. The initial state of the field, $|\Psi_F^{(0)}\rangle$, will be treated more generally as a superposition, in terms of the eigenstates, $|n\rangle$, of $H_0^F = \hbar\omega\hat{N}$, where $\hat{N}$ is the number operator. To simplify the discussion, we will restrict ourselves to single mode. The result can be easily generalized to the multimode case. Assume the atom is ionized from its ground state due to the dipole-field interaction, $V(t) = -\hat{\mu} \cdot \mathbf{E}^{(+)}$, where $\hat{\mu}$ is the dipole moment of the atom. The interaction Hamiltonian is thus

$$H_I(t) = e^{iH_0 t/\hbar}\, V(t)\, e^{-iH_0 t/\hbar}$$

$$= -e^{iH_0^A t/\hbar}\, \hat{\mu}\, e^{-iH_0^A t/\hbar}\, e^{iH_0^F t/\hbar}\, E^{(+)}\, e^{-iH_0^F t/\hbar}, \qquad (4.3.2)$$

where we have simplified the problem to 1-D. Following Eq. (4.2.8), the state $|\Psi(t)\rangle$ in first order approximation is thus

$$|\Psi(t)\rangle \cong |0\rangle|\Psi_F^{(0)}\rangle - \frac{i}{\hbar} \int_{t_0}^{t} dt'\, e^{iH_0 t'/\hbar}\, (-\hat{\mu}\, E^{(+)})\, e^{-iH_0 t'/\hbar}\, |0\rangle|\Psi_F^{(0)}\rangle. \qquad (4.3.3)$$

The total probability of ionizing the atom from its ground state to a set of possible final states by absorbing a photon from the field is thus

$$
P = \sum_f |\langle f | \Psi(t) \rangle|^2 \cong \frac{1}{\hbar^2} \int_{t_0}^{t} dt' \int_{t_0}^{t'} dt''
$$

$$
\times \sum_{j,n} \langle 0 | e^{-iH_0^A t''/\hbar} \, \hat{\mu}^* \, e^{iH_0^A t''/\hbar} | j \rangle \langle j | e^{iH_0^A t'/\hbar} \, \hat{\mu} \, e^{-iH_0^A t'/\hbar} | 0 \rangle
$$

$$
\times \langle \Psi_F^{(0)} | e^{-iH_0^F t''/\hbar} E^{(-)}(t'') e^{iH_0^F t''/\hbar} | n \rangle \langle n | e^{iH_0^F t'/\hbar} E^{(+)}(t') e^{-iH_0^F t'/\hbar} | \Psi_F^{(0)} \rangle
$$

$$
\cong \frac{1}{\hbar^2} \int_{t_0}^{t} dt' \int_{t_0}^{t'} dt'' \sum_{j,n} e^{-i(\frac{E_j - E_0}{\hbar} - \omega)(t'' - t')} \langle 0 | \hat{\mu}^* | j \rangle \langle j | \hat{\mu} | 0 \rangle
$$

$$
\times \langle \Psi_F^{(0)} | E^{(-)}(t'') | n \rangle \langle n | E^{(+)}(t') | \Psi_F^{(0)} \rangle. \tag{4.3.4}
$$

To complete the calculation we first consider summing Eq. (4.3.4) over all final states. Suppose $\langle j | \hat{\mu} | 0 \rangle$ and $\langle n | E^{(+)} | \Psi_F^{(0)} \rangle$ do not change violently within the group of final states. The sum turns into an integral over the energy E_j with a certain density distribution function. We further assume a smooth density distribution function. The total transition probability is thus proportional to

$$
P \propto \int_{t_0}^{t} dt' \, \langle \Psi_F^{(0)} | E^{(-)}(t') E^{(+)}(t') | \Psi_F^{(0)} \rangle, \tag{4.3.5}
$$

where we have used the completeness relation

$$
\sum_m | m \rangle \langle m | = 1 \tag{4.3.6}
$$

and the mathematical approximation for broadband integral

$$
\int dE_j \, e^{-i(\frac{E_j - E_0}{\hbar} - \omega)(t'' - t')} \sim \delta(t'' - t'). \tag{4.3.7}
$$

The transition rate of ionizing the atom by absorbing a photon from the field at coordinate $\mathbf{r}$ between time t and $t + dt$ is thus

$$
G^{(1)}(\mathbf{r}, t) \propto \langle \Psi_F^{(0)} | E^{(-)}(\mathbf{r}, t) E^{(+)}(\mathbf{r}, t) | \Psi_F^{(0)} \rangle. \tag{4.3.8}
$$

In summary, we have concluded that the probability of annihilating a photon from the radiation field as a photodetection event at space-time point $(\mathbf{r}, t)$, is proportional to

$$
G^{(1)}(\mathbf{r}, t) = \langle \Psi | E^{(-)}(\mathbf{r}, t) E^{(+)}(\mathbf{r}, t) | \Psi \rangle = | E^{(+)}(\mathbf{r}, t) | \Psi \rangle |^2.
$$

Notice that a pure state has been used in the above discussion. Eq. (4.3.8) can be easily generalized to a mixed state

$$
G^{(1)}(\mathbf{r}, t) = \sum_j P_j \, \langle \Psi_j | E^{(-)}(\mathbf{r}, t) E^{(+)}(\mathbf{r}, t) | \Psi_j \rangle
$$

$$
= Tr \left[\hat{\rho} \, E^{(-)}(\mathbf{r}, t) E^{(+)}(\mathbf{r}, t) \right] \tag{4.3.9}
$$

where P_j is the probability of being in the jth state $|\Psi_j\rangle$, and

$$\hat{\rho} = \sum_j P_j |\Psi_j\rangle\langle\Psi_j| \tag{4.3.10}$$

is the density operator representing the mixed state of the radiation field or the system of photons. Notice that we have dropped the superscript from the state $|\Psi_j^{(0)}\rangle$.

If the mixed state can be formally written into an incoherent superposition

$$|\Psi\rangle = \sum_j e^{-i\varphi_j} |\Psi_j\rangle$$

where φ_j is a random number. The probability of annihilating a photon from the radiation field can be calculated, in general, from

$$G^{(1)}(\mathbf{r}, t) = \left\langle \langle\Psi| E^{(-)}(\mathbf{r}, t) E^{(+)}(\mathbf{r}, t) |\Psi\rangle \right\rangle_{\mathrm{En}}. \tag{4.3.11}$$

The time averaged photon counting rate is thus

$$R \propto \frac{1}{T} \int_0^T dt\, G^{(1)}(\mathbf{r}, t) = \int_T dt\, \langle E^{(-)}(\mathbf{r}, t) E^{(+)}(\mathbf{r}, t) \rangle, \tag{4.3.12}$$

where T is the total time of averaging. The time averaging here is no different than that given in chapter 1.

4.4   MEASUREMENT OF LIGHT: JOINT DETECTION OF PHOTONS

In the following, we calculate the probability of the joint detection of photon pairs by two independent photon counting detectors. Again, we assume idealized point detectors. To simplify the analysis, we assume the two independent atoms are both in their ground state, $|0_1\rangle$ and $|0_2\rangle$, before the interaction, where the subscripts label the individual atoms, or the point photon counting detectors. The Hamiltonian H_0 has three parts

$$H_0 = H_{01}^A + H_{02}^A + H_0^F, \tag{4.4.1}$$

where H_{0j}^A, $j = 1, 2$, is the Hamiltonian of the jth free atom. Assume the two atoms are both ionized from their ground state due to the dipole-field interaction $V_1(t) = -\hat{\mu}_1 \cdot \mathbf{E}_1^{(+)}$ and $V_2(t) = -\hat{\mu}_2 \cdot \mathbf{E}_2^{(+)}$, where $\hat{\mu}_j$, $j = 1, 2$, is the dipole moment of the jth atom. The interaction Hamiltonian is thus

$$\begin{aligned}
H_I(t) &= e^{iH_0 t/\hbar} \left[V_1(t) + V_2(t)\right] e^{-iH_0 t/\hbar} \\
&= -e^{iH_{01}^A t/\hbar} \hat{\mu}_1 e^{-iH_{01}^A t/\hbar} e^{iH_0^F t/\hbar} E_1^{(+)} e^{-iH_0^F t/\hbar} \\
&\quad - e^{iH_{02}^A t/\hbar} \hat{\mu}_2 e^{-iH_{02}^A t/\hbar} e^{iH_0^F t/\hbar} E_2^{(+)} e^{-iH_0^F t/\hbar} \\
&= H_{I1}(t) + H_{I2}(t).
\end{aligned} \tag{4.4.2}$$

Following Eq. (4.2.8), the first-order change of the state, $|\Psi(t)\rangle^{(1)}$ is thus

$$|\Psi(t)\rangle^{(1)} = -\frac{i}{\hbar} \int_{t_0}^t dt'\, H_{I1}(t') |\Psi(t_0)\rangle - \frac{i}{\hbar} \int_{t_0}^t dt'\, H_{I2}(t') |\Psi(t_0)\rangle.$$

The two terms correspond to the single-photon transition of atoms one and two, respectively. The second order change of the state, $|\Psi(t)\rangle^{(2)}$, has four terms

$$
|\Psi(t)\rangle^{(2)} = \frac{1}{(i\hbar)^2} \int_{t_0}^{t} dt' \int_{t_0}^{t'} dt'' \, H_{I1}(t') \, H_{I1}(t'') \, |\Psi(t_0)\rangle
$$

$$
+ \frac{1}{(i\hbar)^2} \int_{t_0}^{t} dt' \int_{t_0}^{t'} dt'' \, H_{I2}(t') \, H_{I2}(t'') \, |\Psi(t_0)\rangle
$$

$$
+ \frac{1}{(i\hbar)^2} \int_{t_0}^{t} dt' \int_{t_0}^{t'} dt'' \, H_{I1}(t') \, H_{I2}(t'') \, |\Psi(t_0)\rangle
$$

$$
+ \frac{1}{(i\hbar)^2} \int_{t_0}^{t} dt' \int_{t_0}^{t'} dt'' \, H_{I2}(t') \, H_{I1}(t'') \, |\Psi(t_0)\rangle.
$$

All four terms correspond to the joint annihilation of two photons from the field. The first two terms correspond to the two-photon transition associated with an atom that contributes to the single detector counting rate. For the joint detection events of two photodetectors, we are interested in the last two terms of this group. Assume the coincidence detection circuit only selects one of these joint detection events by time ordering. The total probability of jointly ionizing atom one and atom two from their ground states to a set of possible final states is thus

$$
\sum_f |\langle f | \Psi(t) \rangle|^2
$$

$$
\cong \frac{1}{\hbar^4} \int_{t_0}^{t_1} dt'_1 \int_{t_0}^{t'_1} dt''_1 \int_{t_0}^{t_2} dt'_2 \int_{t_0}^{t'_2} dt''_2
$$

$$
\times \sum_{j_1, j_2, n,} \langle 0_1 | e^{-iH_{01}^A t''_1/\hbar} \hat{\mu}_1^* e^{iH_{01}^A t''_1/\hbar} | j_1 \rangle \langle j_1 | e^{iH_{01}^A t'_1/\hbar} \hat{\mu}_1 e^{-iH_{01}^A t'_1/\hbar} | 0_1 \rangle
$$

$$
\times \langle 0_2 | e^{-iH_{02}^A t''_2/\hbar} \hat{\mu}_2^* e^{iH_{02}^A t''_2/\hbar} | j_2 \rangle \langle j_2 | e^{iH_{02}^A t'_2/\hbar} \hat{\mu}_2 e^{-iH_{02}^A t'_2/\hbar} | 0_2 \rangle
$$

$$
\times \langle \Psi_F^{(0)} | e^{-iH_{01}^F t''_1/\hbar} e^{-iH_{02}^F t''_2/\hbar} E_1^{(-)}(t''_1) E_2^{(-)}(t''_2) e^{iH_{02}^F t''_2/\hbar} e^{iH_{01}^F t''_1/\hbar} | n \rangle
$$

$$
\times \langle n | e^{iH_{01}^F t'_1/\hbar} e^{iH_{02}^F t'_2/\hbar} E_2^{(+)}(t'_2) E_1^{(+)}(t'_1) e^{-iH_{02}^F t'_2/\hbar} e^{-iH_{01}^F t'_1/\hbar} | \Psi_F^{(0)} \rangle. \quad (4.4.3)
$$

Similar to the single detection case, to complete the calculation, we first consider summing Eq. (4.4.3) over all final states and then apply the completeness relation and the delta function approximation. The total joint transition probability is thus proportional to

$$
P \propto \int_{t_0}^{t_1} dt'_1 \int_{t_0}^{t_2} dt'_2 \langle \Psi_F^{(0)} | E_1^{(-)}(t'_1) E_2^{(-)}(t'_2) E_2^{(+)}(t'_2) E_1^{(+)}(t'_1) | \Psi_F^{(0)} \rangle. \quad (4.4.4)
$$

The joint transition rate of ionizing the two atoms by absorbing a photon from the field at coordinate $\mathbf{r}_1$ between time t_1 and $t_1 + dt_1$ and another photon from the field at coordinate $\mathbf{r}_2$ between time t_2 and $t_2 + dt_2$ is thus

$$
G^{(2)}(\mathbf{r}_1, t_1; \mathbf{r}_2, t_2)
$$

$$
\propto \langle \Psi_F^{(0)} | E_1^{(-)}(\mathbf{r}_1, t_1) E_2^{(-)}(\mathbf{r}_2, t_2) E_2^{(+)}(\mathbf{r}_2, t_2) E_1^{(+)}(\mathbf{r}_1, t_1) | \Psi_F^{(0)} \rangle. \quad (4.4.5)
$$

In summary, we have concluded that the joint atomic transition rate, or the probability of jointly annihilating a photon as a photodetection event occurs at space-time point $(\mathbf{r}_1, t_1)$, and a second photodetection event occurs at space-time point $(\mathbf{r}_2, t_2)$, is proportional to the expectation value of the Hermitian operator $E_1^{(-)} E_2^{(-)} E_2^{(+)} E_1^{(+)}$ on the initial state of the light field $|\Psi_F^{(0)}\rangle$. This Hermitian operator is a fourth order operator of the field, and thus, is also called the fourth order moment. It should be emphasized that the operator $E_1^{(-)} E_2^{(-)} E_2^{(+)} E_1^{(+)}$, in general, cannot be defined as the product of two intensity operators due to the commutation property of the field operators. Although $G(\mathbf{r}_1, t_1; \mathbf{r}_2, t_2)$ is sometimes written in the following form,

$$G^{(2)}(\mathbf{r}_1, t_1; \mathbf{r}_2, t_2) = \langle\, : \hat{I}(\mathbf{r}_1, t_1)\, \hat{I}(\mathbf{r}_2, t_2) : \,\rangle, \qquad (4.4.6)$$

where $\langle\, : \dots : \,\rangle$ denotes normal-order. We should not be confused with the notation. The notation means nothing except writing Eq. (4.4.5) in a different form. The commutation rule of the field operators distinguishes Eq. (4.4.5) from its classical analogy.

Notice again that we have been using a pure state in the above calculation. The result of Eq. (4.4.5) can easily be generalized to a mixed state

$$
\begin{aligned}
G^{(2)}&(\mathbf{r}_1, t_1; \mathbf{r}_2, t_2) \\
&\propto \sum_j P_j \langle \Psi_j | E^{(-)}(\mathbf{r}_1, t_1)\, E^{(-)}(\mathbf{r}_2, t_2)\, E^{(+)}(\mathbf{r}_2, t_2)\, E^{(+)}(\mathbf{r}_1, t_1) | \Psi_j \rangle \\
&= tr\,[\,\hat{\rho}\, E^{(-)}(\mathbf{r}_1, t_1)\, E^{(-)}(\mathbf{r}_2, t_2)\, E^{(+)}(\mathbf{r}_2, t_2)\, E^{(+)}(\mathbf{r}_1, t_1)\,], \qquad (4.4.7)
\end{aligned}
$$

where P_j is the probability of being in the jth initial pure state $|\Psi_j\rangle$, and $\hat{\rho}$ is the density operator representing the mixed state of the radiation field or the system of photons, as defined earlier.

If the mixed state can be formally written into an incoherent superposition

$$|\Psi\rangle = \sum_j e^{-i\varphi_j} |\Psi_j\rangle \;\; \text{or} \;\; |\Psi\rangle = \sum_{j,k} e^{-i(\varphi_j + \varphi_k)} |\Psi_j\rangle |\Psi_k\rangle$$

where $(\varphi_j + \varphi_k)$ is a random number. The probability of annihilating a photon from the radiation field can be calculated, in general, from

$$
\begin{aligned}
G^{(2)}&(\mathbf{r}_1, t_1; \mathbf{r}_2, t_2) \qquad\qquad\qquad\qquad\qquad\qquad\qquad (4.4.8) \\
&= \Big\langle\langle \Psi | E^{(-)}(\mathbf{r}_1, t_1)\, E^{(-)}(\mathbf{r}_2, t_2)\, E^{(+)}(\mathbf{r}_2, t_2)\, E^{(+)}(\mathbf{r}_1, t_1) | \Psi \rangle\Big\rangle_{\text{En}}.
\end{aligned}
$$

The averaged joint detection counting rate is thus

$$
\begin{aligned}
R &\propto \int_T dt_1 \int_T dt_2\, G^{(2)}(\mathbf{r}_1, t_1; \mathbf{r}_2, t_2) \\
&= \int_T dt_1 \int_T dt_2\, \langle\, E^{(-)}(\mathbf{r}_1, t_1)\, E^{(-)}(\mathbf{r}_2, t_2)\, E^{(+)}(\mathbf{r}_2, t_2)\, E^{(+)}(\mathbf{r}_1, t_1)\, \rangle \qquad (4.4.9)
\end{aligned}
$$

where T is the total time of averaging.

4.5 FIELD PROPAGATION IN SPACE-TIME

The quantum theory of photodetection is a straightforward application of the quantum theory of time dependent perturbation. The first-order approximation calculates the probability that a photon is annihilated at space-time point $(\mathbf{r}, t)$ of the photo-dection event, or the photon counting rate of a photodetector at position $\mathbf{r}$ and time t,

$$G^{(1)}(\mathbf{r}, t) \propto \langle \Psi_F^{(0)} | E^{(-)}(\mathbf{r}, t)\, E^{(+)}(\mathbf{r}, t) | \Psi_F^{(0)} \rangle.$$

The second-order approximation gives the probability that a pair of photons is annihilated jointly at the space-time points $(\mathbf{r}_1, t_1)$ and $(\mathbf{r}_2, t_2)$, or the joint-detection counting rate of a pair of photodetectors at positions $(\mathbf{r}_1, \mathbf{r}_2)$, and times (t_1, t_2),

$$G^{(2)}(\mathbf{r}_1, t_1; \mathbf{r}_2, t_2)$$
$$\propto \langle \Psi_F^{(0)} | E^{(-)}(\mathbf{r}_1, t_1) \, E^{(-)}(\mathbf{r}_2, t_2) \, E^{(+)}(\mathbf{r}_2, t_2) \, E^{(+)}(\mathbf{r}_1, t_1) | \Psi_F^{(0)} \rangle.$$

The joint-detection counting rate is also called "coincidence" counting rate, although t_1 is not necessarily equal to t_2. In these formulas, $|\Psi_F^{(0)}\rangle$ is the state of the field at t_0, which is the time when the perturbation $V(t)$ is applied, i.e., at the very beginning of the photodetection event. In typical optical measurements, $|\Psi_F^{(0)}\rangle$ is often evaluated from the state $|\Psi(0)\rangle$ that is prepared at the radiation source. The state $|\Psi(0)\rangle$ of the field that is excited at the radiation source is usually known from early measurements or predictable through the study of the radiation mechanism of the light source.

This is a typical quantum mechanical dynamical problem: we are interested in knowing the expectation value of an operator $\hat{A}$ at time t

$$\langle \hat{A} \rangle_t = \langle \Psi(t) | \hat{A} | \Psi(t) \rangle, \tag{4.5.1}$$

with the knowledge of the quantum state of the system at an early time, such as $|\Psi(0)\rangle$ at time $t = 0$.

Since we have shown that any state of a radiation field can be described by the superposition of the eigenstates of the Hamiltonian, such as Eq. (2.4.38), and proved that an energy eigenstate simply develops a phase factor, $e^{-iE_n t/\hbar}$, from time $t = 0$ to t. The quantum state of the field at time t is thus written as

$$|\Psi(t)\rangle = \sum_E e^{-iE_n t/\hbar} | E \rangle \langle E | \Psi(0) \rangle$$
$$= e^{-i\hat{H}t/\hbar} | \Psi(0) \rangle, \tag{4.5.2}$$

where $| E \rangle$ represents the eigenstate of the Hamiltonian. To obtain the result of Eq. (4.5.2), we have assumed a complete orthogonal set of energy eigenstates

$$\sum_E | E \rangle \langle E | = \mathbf{1}.$$

Eq. (4.5.1) is thus rewritten as

$$\langle \hat{A} \rangle_t = \langle \Psi(0) | e^{i\hat{H}t/\hbar} \, \hat{A} \, e^{-i\hat{H}t/\hbar} | \Psi(0) \rangle. \tag{4.5.3}$$

We are ready to work in the Heisenberg representation to propagate the field operator from $t = 0$ at the radiation source to t_0 of the photodetection event, rather then working in the Schrödinger representation to develop the state of $| \Psi(0) \rangle$. If we define a time-dependent operator $\hat{A}(t)$ by

$$\hat{A}(t) = e^{i\hat{H}t/\hbar} \, \hat{A} \, e^{-i\hat{H}t/\hbar}, \tag{4.5.4}$$

then we can rewrite Eq. (4.5.3) as

$$\langle \hat{A} \rangle_t = \langle \Psi(0) | \hat{A}(t) | \Psi(0) \rangle. \tag{4.5.5}$$

Differentiating Eq. (4.5.4) with respect to time t, we find

$$i\hbar \frac{d\hat{A}(t)}{dt} = [\, \hat{A}(t), \, \hat{H} \,], \tag{4.5.6}$$

if $\hat{A}(t)$ does not depend explicitly on time t. In Eq. (4.5.6), we have assumed that $\hat{H}$ commutes with $e^{-i\hat{H}t/\hbar}$. Eq. (4.5.6) is called Heisenberg equation of motion for operators.

It is easy to show that for any product of two or more operators, such as $\hat{A} = \hat{B}\hat{C}$, we can define a product of two or more time-dependent operators, such as $\hat{A}(t) = \hat{B}(t)\hat{C}(t)$,

$$
\begin{aligned}
\langle\Psi(t)|\,\hat{B}\,\hat{C}\,|\Psi(t)\rangle &= \langle\Psi(0)|\,e^{i\hat{H}t/\hbar}\,\hat{B}\,\hat{C}\,e^{-i\hat{H}t/\hbar}\,|\Psi(0)\rangle \\
&= \langle\Psi(0)|\,e^{i\hat{H}t/\hbar}\,\hat{B}\,e^{-i\hat{H}t/\hbar}\,e^{i\hat{H}t/\hbar}\,\hat{C}\,e^{-i\hat{H}t/\hbar}\,|\Psi(0)\rangle \\
&= \langle\Psi(0)|\,\hat{B}(t)\,\hat{C}(t)\,|\Psi(0)\rangle,
\end{aligned} \tag{4.5.7}
$$

where $e^{-i\hat{H}t/\hbar}e^{i\hat{H}t/\hbar} = 1$ has been used. It is also easy to show that $\hat{B}(t)$ and $\hat{C}(t)$ obey the Heisenberg equation of motion, respectively, and thus can be propagated, individually,

$$
i\hbar\,\frac{d\hat{B}(t)}{dt} = [\,\hat{B}(t),\,\hat{H}\,], \qquad i\hbar\,\frac{d\hat{C}(t)}{dt} = [\,\hat{C}(t),\,\hat{H}\,]. \tag{4.5.8}
$$

Based on the above results, in the Heisenberg representation, we have the counting rate of a photodetector, or the probability for a photon to be annihilated at space-time point $(\mathbf{r}, t)$,

$$
\begin{aligned}
G^{(1)}(\mathbf{r}, t) &\propto \langle\,\Psi(0)\,|\,E^{(-)}(\mathbf{r}, t)\,E^{(+)}(\mathbf{r}, t)\,|\,\Psi(0)\,\rangle \\
&= \big|\,E^{(+)}(\mathbf{r}, t)\,|\,\Psi(0)\,\rangle\,\big|^{2},
\end{aligned} \tag{4.5.9}
$$

and the joint-detection counting rate of a pair of photodetectors, or the probability for a pair of photons to be annihilated jointly at space-time points $(\mathbf{r}_1, t_1)$ and $(\mathbf{r}_2, t_2)$

$$
\begin{aligned}
&G^{(2)}(\mathbf{r}_1, t_1; \mathbf{r}_2, t_2) \\
&\propto \langle\Psi(0)|\,E^{(-)}(\mathbf{r}_1, t_1)\,E^{(-)}(\mathbf{r}_2, t_2)E^{(+)}(\mathbf{r}_2, t_2)\,E^{(+)}(\mathbf{r}_1, t_1)\,|\Psi(0)\rangle \\
&= \big|\,E^{(+)}(\mathbf{r}_2, t_2)\,E^{(+)}(\mathbf{r}_1, t_1)\,|\Psi(0)\rangle\,\big|^{2}
\end{aligned} \tag{4.5.10}
$$

with the understanding that the field operators $E^{(\pm)}(\mathbf{r}_j, t_j)$ are propagated from $(\mathbf{r}^0, t^0)$ of the radiation source. Notice, we have used a different notation, t^0, to distinguish the perturbation time, t_0, of the photodetection event. To simplify the notation, we usually choose $t^0 = 0$ and write the state $|\Psi(0)\rangle$ as $|\Psi\rangle$. If the radiation source can be treated as a point source, we may also choose $\mathbf{r}^0 = 0$. Care has to be taken in the cases when the finite size of the source has to be taken into account, especially in "near-field" measurements. Eqs. (4.5.9) and (4.5.10) are the basic formalisms for photon counting measurements.

Regarding the propagation of the field operator in space-time, in general, we need to obtain the time evolution of the annihilation and creation operator and to find the solution of the Helmholtz equation subject to the corresponding boundary condition of the field. The mathematical tools for defining the spatial mode function of the field are well developed in classical optics and there is no difference in quantum optics. The time evolution of the annihilation operator, $\hat{a}_{\mathbf{k}}(t)$, and the creation operator, $\hat{a}_{\mathbf{k}}^{\dagger}(t)$, can be easily obtained for certain types of Hamiltonians, such as the free field Hamiltonian of Eq. (2.4.25), by solving the Heisenberg equation of motion

$$
\begin{aligned}
\frac{d}{dt}\,\hat{a}_{\mathbf{k}}(t) &= \frac{1}{i\hbar}\big[\,\hat{a}_{\mathbf{k}}(t),\,\hat{H}\,\big] = -i\omega\,\hat{a}_{\mathbf{k}}(t) \\
\frac{d}{dt}\,\hat{a}_{\mathbf{k}}^{\dagger}(t) &= \frac{1}{i\hbar}\big[\,\hat{a}_{\mathbf{k}}^{\dagger}(t),\,\hat{H}\,\big] = i\omega\,\hat{a}_{\mathbf{k}}^{\dagger}(t)
\end{aligned} \tag{4.5.11}
$$

with the solution

$$\hat{a}_{\mathbf{k}}(t) = \hat{a}_{\mathbf{k}}(0) \, e^{-i\omega t}$$
$$\hat{a}_{\mathbf{k}}^{\dagger}(t) = \hat{a}_{\mathbf{k}}^{\dagger}(0) \, e^{i\omega t}, \tag{4.5.12}$$

where we have assigned $t^0 = 0$. We thus write the free field operator in the Heisenberg representation

$$\hat{\mathbf{E}}(\mathbf{r}, t)$$
$$= \sum_{\mathbf{k}} \hat{\mathbf{e}}_k \, i \, \mathcal{E}_{\mathbf{k}} \, \hat{a}_{\mathbf{k}}(0) \, e^{-i\omega t} \, u_k(\mathbf{r}) + \sum_{\mathbf{k}} \hat{\mathbf{e}}_k \, (-i) \mathcal{E}_{\mathbf{k}} \, \hat{a}_{\mathbf{k}}^{\dagger}(0) \, e^{i\omega t} \, u_k^*(\mathbf{r})$$
$$= \hat{\mathbf{E}}^{(+)}(\mathbf{r}, t) + \hat{\mathbf{E}}^{(-)}(\mathbf{r}, t), \tag{4.5.13}$$

where $u_k(\mathbf{r})$ is the solution of the Helmholtz equation of Eq. (2.4.43) subject to the required boundary conditions. In Eq. (4.5.13), $\hat{\mathbf{E}}^{(+)}(\mathbf{r}, t)$ and $\hat{\mathbf{E}}^{(-)}(\mathbf{r}, t)$ correspond to the physical processes of photon annihilation and creation at the space-time point $(\mathbf{r}, t)$, respectively.

Under certain experimental conditions, such as far-field measurements, the plane-wave is a good approximation. The positive and the negative field operators in Eq. (4.5.13) can then be written as

$$\hat{\mathbf{E}}^{(+)}(\mathbf{r}, t) = \sum_{\mathbf{k}} \hat{\mathbf{e}}_k \, i \, \mathcal{E}_{\mathbf{k}} \, \hat{a}_{\mathbf{k}}(0) \, e^{-i(\omega t - \mathbf{k} \cdot \mathbf{r})}$$
$$\hat{\mathbf{E}}^{(-)}(\mathbf{r}, t) = \sum_{\mathbf{k}} \hat{\mathbf{e}}_k \, (-i) \mathcal{E}_{\mathbf{k}} \, \hat{a}_{\mathbf{k}}^{\dagger}(0) \, e^{i(\omega t - \mathbf{k} \cdot \mathbf{r})} \tag{4.5.14}$$

It is common to formally write $\hat{\mathbf{E}}^{(+)}(\mathbf{r}, t)$ and $\hat{\mathbf{E}}^{(-)}(\mathbf{r}, t)$ as follows

$$\hat{\mathbf{E}}^{(+)}(\mathbf{r}, t) = \sum_{\mathbf{k}} \hat{\mathbf{e}}_k \, \hat{a}_{\mathbf{k}}(0) \, g(\mathbf{k}; \mathbf{r}, t)$$
$$\hat{\mathbf{E}}^{(-)}(\mathbf{r}, t) = \sum_{\mathbf{k}} \hat{\mathbf{e}}_k \, \hat{a}_{\mathbf{k}}^{\dagger}(0) \, g^*(\mathbf{k}; \mathbf{r}, t), \tag{4.5.15}$$

with $g(\mathbf{k}; \mathbf{r}, t)$ the propagator, or the Green's function, propagating the $\mathbf{k}$th mode of the field from the source to $(\mathbf{r}, t)$; we have also formally absorbed the constants into the Green's function as usual.

4.6 QUANTIZED SUBFIELD AND EFFECTIVE WAVEFUNCTION OF PHO-TON

Einstein's bundle of ray, or subfield, is a quantized microscopic realistic entity of electro-magnetic wavepacket propagating in space-time. This concept is still in the frame work of Maxwell electromagnetic wave theory, except Einstein introduced a concept of granular-ity and abandoned the continuum interpretation of Maxwell. It is interesting to find that an "effective wavefunction", $\psi_m(\mathbf{r}, t)$, of the mth photon, or the mth group of identical photons, among an ensemble of photons, or groups of identical photons, can be defined from the quantum photodetection theory, precisely, from the calculations of $G^{(1)}(\mathbf{r}, t)$ and $G^{(2)}(\mathbf{r}_1, t_1; \mathbf{r}_2, t_2)$. It is also interesting to find that $\psi_m(\mathbf{r}, t)$ has the same mathematical representation and play the same role as that of Einstein's subfield $E_m(\mathbf{r}, t)$. It should be emphasized even before introducing the concept, however, $\psi_m(\mathbf{r}, t)$ represents the proba-bility amplitude for the mth photon, or the mth groip of identical photons, to produce a photodetection event at space-time $(\mathbf{r}, t)$, which is different in nature from a quantized realistic entity of electromagnetic wavepacket. In the following, we introduce the "effective wavefunction" of a photon, or a group of identical photons, in terms of the single-photon state representation and the coherent state representation.

(I) Single-photon state representation.

Assuming a natural weak point-like light source and a far-field measurement of a point-like photodetector. In the light source, a large number of randomly distributed spontaneous atomic transitions excite a large number of subfields. Although the chance to have a spontaneous emission is very small, there is indeed a tiny probability for an atom to create a photon whenever the atom decays from its higher energy level E_2 ($\Delta E_2 \neq 0$) down to its ground energy state of E_1. We have approximated the state of the mth subfield, which is excited by the mth atomic transition, as

$$|\Psi\rangle = c_0|0\rangle + c_1|\Psi_m\rangle \simeq |0\rangle + \epsilon|\Psi_m\rangle,$$

where $|c_0| \sim 1$ is the probability amplitude for no-field-excitation and $|c_1| = |\epsilon| \ll 1$ is the probability amplitude for the creation of a photon from the mth atomic transition. In a 1-D approximation,

$$|\Psi_m\rangle = e^{i\varphi_m} \int d\omega\, a_m(\omega)\, \hat{a}_m^\dagger(\omega)\, |0\rangle,$$

here we have intentionally written the complex amplitudes $a_m(\omega)e^{i\varphi_m(\omega)}$, which is determined by the creation and annihilation operators, into the state and assumed that all radiation modes ω excited by the mth atomic transition have a common phase $\varphi_m(\omega) = \varphi_m$, so that it can be moved outside of the integral. The generalized state of the radiation field, created from such a large number atomic transitions, can be formally written as

$$|\Psi\rangle = \prod_m \left\{ |0\rangle + \epsilon\, c_m\, |\Psi_m\rangle \right\}$$
$$\simeq |0\rangle + \epsilon\Big[\sum_m c_m\, |\Psi_m\rangle \Big] + \epsilon^2\Big[\sum_{m<n} c_m\, c_n|\Psi_m\rangle|\Psi_n\rangle \Big] + \dots \qquad (4.6.1)$$

where, again, we have assumed a common complex amplitude for each wavepacket, although it can still take arbitrary phases from transition to transition. Since $|\epsilon| \ll 1$, In Equation (4.6.1) we listed the first-order and the second-order approximations on ϵ.

Considering a simple measurement in which a point-like photodetecter at $(\mathbf{r}, t)$ is facing the point-like light source, the field operator at $(\mathbf{r}, t)$ can be approximated in 1-D:

$$\hat{E}^{(+)}(\mathbf{r}, t) = \sum_p \int d\omega\, \hat{a}_p(\omega)\, g_p(\omega; \mathbf{r}, t)$$
$$\hat{E}^{(-)}(\mathbf{r}, t) = \sum_q \int d\omega'\, \hat{a}_q^\dagger(\omega')\, g_q^*(\omega'; \mathbf{r}, t) \qquad (4.6.2)$$

where $g_p(\omega; \mathbf{r}, t)$ is the Green's function that propagates the ω-mode of the pth subfield from the point-like source to space-time coordinate $(\mathbf{r}, t)$. We are ready to calculate $G^{(1)}(\mathbf{r}, t)$:

$$
\begin{aligned}
G^{(1)}&(\mathbf{r}, t) \\
&= \left\langle \langle \Psi | E^{(-)}(\mathbf{r}, t) E^{(+)}(\mathbf{r}, t) | \Psi \rangle \right\rangle_{\text{En}} \\
&= \left\langle \langle \Psi | \Big[\sum_p \int d\omega \, \hat{a}_p^\dagger(\omega) \, g_p^*(\omega; \mathbf{r}, t) \Big] \Big[\sum_q \int d\omega' \hat{a}_q(\omega') g_q(\omega'; \mathbf{r}, t) \Big] | \Psi \rangle \right\rangle_{\text{En}} \\
&= \left\langle \langle \Psi | \Big[\sum_p \int d\omega \, \hat{a}_p^\dagger(\omega) \, g_p^*(\omega; \mathbf{r}, t) \Big] |0\rangle\langle 0| \Big[\sum_q \int d\omega' \hat{a}_q(\omega') g_q(\omega'; \mathbf{r}, t) \Big] | \Psi \rangle \right\rangle_{\text{En}} \\
&= \left\langle \Big[\sum_m \psi_m^*(\mathbf{r}, t) \Big] \Big[\sum_n \psi_n(\mathbf{r}, t) \Big] \right\rangle_{\text{En}} \\
&= \left\langle \sum_m |\psi_m(\mathbf{r}, t)|^2 \right\rangle_{\text{En}} + \left\langle \sum_{m \neq n} \psi_m^*(\mathbf{r}, t) \psi_n(\mathbf{r}, t) \right\rangle_{\text{En}} \\
&\propto \langle n(\mathbf{r}, t) \rangle
\end{aligned}
\tag{4.6.3}
$$

where we have applied the completeness relation

$$
\sum_n | n \rangle \langle n | = 1.
\tag{4.6.4}
$$

In Eq.(4.6.3) we have defined the effective wavefunction of the mth photon,

$$
\psi_m(\mathbf{r}, t) \equiv \langle 0| E^{(+)}(\mathbf{r}, t) | \Psi_m \rangle = \int d\omega \, a_m(\omega) e^{i\varphi_m} \, g_m(\omega; \mathbf{r}, t).
\tag{4.6.5}
$$

where, again, we have assumed a common phase φ_m for the mth atomic transition, we have also absorbed $|c_m|$ into $a_m(\omega)$ as usual. In the point-to-point propagation, or the far-field approximation, $\psi_m(\mathbf{r}, t)$ can be written as

$$
\begin{aligned}
\psi_m(\mathbf{r}, t) &= \Big[\int d\nu \, a_m(\nu) e^{-i\nu\tau} \Big] e^{-i(\omega_0 \tau - \varphi_m)} \\
&= \mathcal{F}_\tau \big\{ a_m(\nu) \big\} e^{-i(\omega_0 \tau - \varphi_m)}
\end{aligned}
\tag{4.6.6}
$$

where we have chosen the mth atomic transition as the origin of the space-time coordinate system. It is easy to find that the effective wavefunction $\psi_m(\mathbf{r}, t)$ has the same mathematical representation and plays the same role as that of Einstein's subfield $E_m(\mathbf{r}, t)$.

It is easy to find that $\langle \Delta n(\mathbf{r}, t) \rangle = 0$ for thermal state:

$$
\begin{aligned}
\langle \Delta n(\mathbf{r}, t) \rangle &= \left\langle \sum_{m \neq n} \psi_m^*(\mathbf{r}, t) \psi_n(\mathbf{r}, t) \right\rangle_{\text{En}} \\
&= \left\langle \sum_{m \neq n} e^{-i(\varphi_m - \varphi_n)} \mathcal{F}_\tau^* \big\{ a_m(\nu) \big\} \mathcal{F}_\tau \big\{ a_n(\nu) \big\} \right\rangle_{\text{En}} = 0
\end{aligned}
$$

when taking into account all possible values of $(\varphi_m - \varphi_n)$, agreeing with that of Einstein's picture, which is calculated from the same mathematics of ensemble average.

(II) Coherent state representation.

We have assumed a weak thermal radiation in the above calculation, in which only the lower order of Eq.(4.6.1) contributes. In the case of arbitrary brightness, especially bright light condition, we may need to keep all orders of the expansion. In this case, quantum coherent state representation is a better choice. In the quantum coherent state representation, the state of a thermal radiation field can be written as

$$|\Psi\rangle = \prod_m |\Psi_m\rangle = \prod_m |\{\alpha_m\}\rangle = \prod_{m,\mathbf{k}} |\alpha_m(\mathbf{k})\rangle. \tag{4.6.7}$$

In this model, the radiation field contains a large number of randomly created and randomly distributed subfields, or wavepackets, each is created from an atomic transition, or a group of identical atomic transitions, or other creation mechanisms, and each subfield contains a large number of radiation modes that can be represented by coherent state. In Eq.(4.6.7), m labels the mth subfield, and $\mathbf{k}$ labels the $\mathbf{k}$th wavevector. $|\alpha_m(\mathbf{k})\rangle$ is an eigenstate of the annihilation operator with an eigenvalue $\alpha_m(\mathbf{k})$,

$$\hat{a}_m(\mathbf{k})|\alpha_m(\mathbf{k})\rangle = \alpha_m(\mathbf{k})|\alpha_m(\mathbf{k})\rangle. \tag{4.6.8}$$

Thus, we have,

$$\hat{a}_m(\mathbf{k})|\Psi\rangle = \alpha_m(\mathbf{k})|\Psi\rangle. \tag{4.6.9}$$

Considering the same simple measurement with the same field operators of Eq. (4.6.2), we have

$$\langle\Psi|\hat{E}^{(-)}(\mathbf{r},t)|\Psi\rangle = \langle\Psi|\sum_m \int d\mathbf{k}\, \hat{a}_m^\dagger(\mathbf{k})\, g_m^*(\mathbf{k};\mathbf{r},t)|\Psi\rangle$$

$$= \sum_m \int d\mathbf{k}\, \alpha_m^*(\mathbf{k})\, g_m^*(\mathbf{k};\mathbf{r},t)]$$

$$\langle\Psi|\hat{E}^{(+)}(\mathbf{r},t)|\Psi\rangle = \langle\Psi|\big[\sum_n \int d\mathbf{k}\, \hat{a}_n(\mathbf{k})\, g_n(\mathbf{k};\mathbf{r},t)\big]|\Psi\rangle$$

$$= \sum_n \int d\mathbf{k}\, \alpha_n(\mathbf{k})\, g_n(\mathbf{k};\mathbf{r},t) \tag{4.6.10}$$

Following Glauber's theory, the probability to produce a photoelectron event at $(\mathbf{r},t)$ is proportional to $G^{(1)}(\mathbf{r},t) = \langle\langle E^{(-)}(\mathbf{r},t)E^{(+)}(\mathbf{r},t)\rangle_{\mathrm{QM}}\rangle_{\mathrm{En}}$. Substitute the quantum state and the field operators into $G^{(1)}(\mathbf{r},t)$, we have

$$G^{(1)}(\mathbf{r},t) = \langle\langle\Psi|E^{(-)}(\mathbf{r},t)E^{(+)}(\mathbf{r},t)|\Psi\rangle\rangle_{\mathrm{En}}$$

$$= \langle\langle\Psi|\big[\sum_m \int d\mathbf{k}\, \hat{a}_m^\dagger(\mathbf{k})\, g_m^*(\mathbf{k};\mathbf{r},t)\big]$$

$$\times \big[\sum_n \int d\mathbf{k}\, \hat{a}_n(\mathbf{k}) g_n(\mathbf{k};\mathbf{r},t)\big]|\Psi\rangle\rangle_{\mathrm{En}}$$

$$= \langle\langle\Psi|\big[\sum_m \int d\mathbf{k}\, \hat{a}_m^\dagger(\mathbf{k})\, g_m^*(\mathbf{k};\mathbf{r},t)\big]$$

$$\times |\Psi\rangle\langle\Psi|\big[\sum_n \int d\mathbf{k}\, \hat{a}_n(\mathbf{k}) g_n(\mathbf{k};\mathbf{r},t)\big]|\Psi\rangle\rangle_{\mathrm{En}}$$

$$= \langle\big[\sum_m \psi_m^*(\mathbf{r},t)\big]\big[\sum_n \psi_n(\mathbf{r},t)\big]\rangle_{\mathrm{En}}$$

$$= \langle\sum_m |\psi_m(\mathbf{r},t)|^2\rangle_{\mathrm{En}} + \langle\sum_{m\neq n}\psi_m^*(\mathbf{r},t)\psi_n(\mathbf{r},t)\rangle_{\mathrm{En}}$$

$$\propto \langle n(\mathbf{r},t)\rangle \tag{4.6.11}$$

where we have defined $\psi_m(\mathbf{r}, t)$ the effective wavefunction of the mth subfield:

$$\psi_m(\mathbf{r}, t) \equiv \langle \Psi_m | E^{(+)}(\mathbf{r}, t) | \Psi_m \rangle$$

$$= \int d\mathbf{k}\, \alpha_m(\mathbf{k}) e^{i\varphi_m}\, g_m(\mathbf{k}; \mathbf{r}, t), \qquad (4.6.12)$$

We have shown earlier $\langle \Delta n(\mathbf{r}, t) \rangle = 0$ when taking into account all possible values of $(\varphi_m - \varphi_n)$ in the ensemble average.

It is interesting to find that the quantum mechanical concept of effective wavefunction has the same mathematical representation and plays the same role as Einstein's concept of subfield. However, the physical means of the two are different in nature: Einstein's subfield $E_m(\mathbf{r}, t)$ is a classical real entity of electromagnetic field. The quantum mechanical effective wavefunction $\psi_m(\mathbf{r}, t)$ is the probability amplitude for the mth subfield to produce a photodetection event at space-time $(\mathbf{r}, t)$.

If there exists two different yet indistinguishable alternatives, such as in the Young's double-slit interferometer, to produce a photodetection event at space-time coordinate $(\mathbf{r}, t)$, the effective wavefunction is easily calculated to be the following linear superposition

$$\psi_m(\mathbf{r}, t) = \int d\mathbf{k}\, a_m(\mathbf{k}) \frac{1}{\sqrt{2}} \big[g_{mA}(\mathbf{k}; \mathbf{r}, t) + g_{mB}(\mathbf{k}; \mathbf{r}, t) \big]$$

$$= \frac{1}{\sqrt{2}} \Big[\int d\mathbf{k}\, a_m(\mathbf{k}) g_{mA}(\mathbf{k}; \mathbf{r}, t) + \int d\mathbf{k}\, a_m(\mathbf{k}) g_{mB}(\mathbf{k}; \mathbf{r}, t) \Big]$$

$$= \frac{1}{\sqrt{2}} \big[\psi_{mA}(\mathbf{r}, t) + \psi_{mB}(\mathbf{r}, t) \big] \qquad (4.6.13)$$

which is, again, the same superposition as that in Einstein's picture, except the electromagnetic subfield $E_m(\mathbf{r}, t)$ is replaced with the effective wavefunction $\psi_m(\mathbf{r}, t)$. However, the "interpretation" of the superposition between $\psi_{mA}(\mathbf{r}, t)$ and $\psi_{mB}(\mathbf{r}, t)$ is different from that of Einstein's picture. In quantum theory, $\psi_{mA}(\mathbf{r}, t)$ and $\psi_{mB}(\mathbf{r}, t)$ represent two different yet indistinguishable probability amplitudes for a photon or a group of identical photons to produce a photodetection event at $(\mathbf{r}, t)$. It is a principle of quantum mechanics: the probability of producing a photodetection event at $(\mathbf{r}, t)$ is proportional to $|\psi_{mA}(\mathbf{r}, t) + \psi_{mB}(\mathbf{r}, t)|^2$.

4.7   JOINT MEASUREMENT OF COMPOSITE RADIATION SYSTEMS

Following Glauber's theory, the probability to produce a joint photodetection event at space-time coordinates $(\mathbf{r}_1, t_1)$ and $(\mathbf{r}_2, t_2)$ from thermal radiation field in single-photon state representation is calculated as follows:

$$G^{(2)}(\mathbf{r}_1, t_1; \mathbf{r}_2, t_2)$$

$$= \Big\langle \big\langle E^{(-)}(\mathbf{r}_1, t_1) E^{(-)}(\mathbf{r}_2, t_2) E^{(+)}(\mathbf{r}_2, t_2) E^{(+)}(\mathbf{r}_1, t_1) \big\rangle_{\text{QM}} \Big\rangle_{\text{En}}$$

$$= \Big\langle \Big\langle \Psi \Big| \sum_m E_m^{(-)}(\mathbf{r}_1, t_1) \sum_n E_n^{(-)}(\mathbf{r}_2, t_2)$$

$$\times |0\rangle\langle 0| \sum_q E_q^{(+)}(\mathbf{r}_2, t_2) \sum_p E_p^{(+)}(\mathbf{r}_1, t_1) \Big| \Psi \Big\rangle \Big\rangle_{\text{En}}$$

$$= \Big\langle \sum_m \psi_m^*(\mathbf{r}_1, t_1) \sum_n \psi_n^*(\mathbf{r}_2, t_2) \sum_p \psi_p(\mathbf{r}_2, t_2) \sum_q \psi_q(\mathbf{r}_1, t_1) \Big\rangle_{\text{En}}$$

$$= \sum_m \psi_m^*(\mathbf{r}_1, t_1) \psi_m(\mathbf{r}_1, t_1) \sum_n \psi_n^*(\mathbf{r}_2, t_2) \psi_n(\mathbf{r}_2, t_2)$$

$$+ \sum_{m \neq n} \psi_m^*(\mathbf{r}_1, t_1) \psi_n(\mathbf{r}_1, t_1) \psi_n^*(\mathbf{r}_2, t_2) \psi_m(\mathbf{r}_2, t_2)$$

$$= \sum_{m,n} \left| \frac{1}{\sqrt{2}} [\psi_m(\mathbf{r}_1, t_1) \psi_n(\mathbf{r}_2, t_2) + \psi_m(\mathbf{r}_2, t_2) \psi_n(\mathbf{r}_1, t_1)] \right|^2$$

$$= \langle n(\mathbf{r}_1, t_1) \rangle \langle n(\mathbf{r}_2, t_2) \rangle + \langle \Delta n(\mathbf{r}_1, t_1) \Delta n(\mathbf{r}_2, t_2) \rangle \tag{4.7.1}$$

where we have defined the effective wavefunction of the m-nth pair of randomly created and randomly distributed photons

$$\psi_{mn}(\mathbf{r}_1, t_1; \mathbf{r}_2, t_2) \equiv \langle 0 | E^{(+)}(\mathbf{r}_2, t_2) E^{(+)}(\mathbf{r}_1, t_1) | \Psi_{mn} \rangle \tag{4.7.2}$$

with

$$|\Psi_{mn}\rangle = |\Psi_m\rangle |\Psi_n\rangle.$$

$G^{(2)}(\mathbf{r}_1, t_1; \mathbf{r}_2, t_2)$ is also named the second-order coherence function, which is the result of a superposition between two different yet indistinguishable two-photon amplitudes: (1) the mth wavepacket is annihilated at $(\mathbf{r}_1, t_1)$ and the nth wavepacket is annihilated at $(\mathbf{r}_2, t_2)$; and (2) the mth wavepacket is annihilated at $(\mathbf{r}_2, t_2)$ and the nth wavepacket is annihilated at $(\mathbf{r}_1, t_1)$. We name this superposition two-photon interference: two randomly created and randomly paired wavepackets, or photons in thermal state interfering with the pair itself.

Examine the cross-interference term of Eq.(14.5.9) which contributes to the photon number fluctuation correlation measurement,

$$\langle \Delta n(\mathbf{r}_1, t_1) \Delta n(\mathbf{r}_2, t_2) \rangle = \sum_{m \neq n} \psi_m^*(\mathbf{r}_1, t_1) \psi_n(\mathbf{r}_1, t_1) \psi_n^*(\mathbf{r}_2, t_2) \psi_m(\mathbf{r}_2, t_2) \tag{4.7.3}$$

we may find that this term is the same as these we have calculated in previous sections, except Einstein's subfield is replaced by the effective wavefunction.

A similar effective wavefunction of a pair of randomly created and randomly distributed photons can be also calculated from coherent state representation

$$\psi_{mn}(\mathbf{r}_1, t_1; \mathbf{r}_2, t_2) = \langle \Psi_{mn} | E^{(+)}(\mathbf{r}_2, t_2) E^{(+)}(\mathbf{r}_1, t_1) | \Psi_{mn} \rangle. \tag{4.7.4}$$

We will have more discussions on the effective wavefunctions of photons and photon pairs in Chapters 6 and 7.

SUMMARY

In this chapter, we discuss the measurement of quantized light. We start from the measurement of Einstein's bundle of ray

$$I(\mathbf{r}, t) = \sum_m |E_m(\mathbf{r}, t)|^2 + \sum_{m \neq n} E_m^*(\mathbf{r}, t) E_n(\mathbf{r}, t)$$

$$= \langle I(\mathbf{r}, t) \rangle + \Delta I(\mathbf{r}, t),$$

and then introduced the Glauber photodetection theory in first-order photon counting and in second-order coincidence photon counting based on the time dependent perturbation theory of quantum mechanics,

$$G^{(1)}(\mathbf{r}, t) = \left\langle \langle \Psi | E^{(-)}(\mathbf{r}, t) E^{(+)}(\mathbf{r}, t) | \Psi \rangle \right\rangle_{\text{En}}$$

and

$$G^{(2)}(\mathbf{r}_1, t_1; \mathbf{r}_2, t_2)$$
$$= \langle\langle \Psi | E^{(-)}(\mathbf{r}_1, t_1) E^{(-)}(\mathbf{r}_2, t_2) E^{(+)}(\mathbf{r}_2, t_2) E^{(+)}(\mathbf{r}_1, t_1) | \Psi \rangle\rangle_{\text{En}}.$$

During the calculations of $G^{(1)}(\mathbf{r}, t)$ and $G^{(2)}(\mathbf{r}_1, t_1; \mathbf{r}_2, t_2)$, we introduced the concept of effective wavefunction for a photon and the concept of effective wavefunction for a pair of photons in the single-photon state representation and in the coherent state representation.

REFERENCES AND SUGGESTIONS FOR READING

[1] R.J. Glauber, Phys. Rev. **130**, 2529 (1963); Phys. Rev. **131**, 2766 (1963).
[2] M.O. Scully and M.S. Zubairy, *Quantum Optics*, Cambridge, 1997.

Coherent and Incoherent Radiation

What do we mean when name a radiation coherent or incoherent? For instance, what is the physical reason for us to consider a discharge tube radiates incoherent light, but a laser radiates coherent light? In this chapter we introduce the concept of coherence and discuss the coherent property of radiation. We start from Einstein's granularity picture of light to introduce two category of superpositions in terms of Einstein's subfields and Maxwell's Fourier-modes. We then show this simple model is consistent with that of quantum mechanics. The concepts of temporal coherence and spatial coherence are introduced and discussed at the end of this chapter.

5.1 COHERENT AND INCOHERENT RADIATION—EINSTEIN'S PICTURE

Einstein's picture of light is still in the framework of electromagnetic theory of radiation, except the concept of "bundle of ray" (field quantization) added an additional freedom to the superposition of Fourier mode.

Assume a point-like light source at $\mathbf{r}_0 = 0$, such as a distant star, and a point-like photodetector at $\mathbf{r}$, the observed radiation propagates from the point-like source to the point-like photodetector along the 1-D path $r = |\mathbf{r} - \mathbf{r}_0|$ in vacuum. In Einstein's granularity picture, the observed field $E(\mathbf{r}, t)$ is the result of a superposition among a large number of subfields. Each subfield is randomly created from a point-like sub-source at time t_m, such as trillions of atomic transitions in a point-like distant star, the resultant field can be modeled as the superposition of subfields in terms of the sub-sources and their harmonic modes of frequency ω:

$$E(r, t) = \sum_m \int d\omega \, E_m(\omega) g_m(\omega; r, t)$$

$$\simeq \sum_m \int d\omega \, a_m(\omega) e^{i\varphi_m(\omega)} \, e^{-i\omega[(t-t_m)-r/c]}, \qquad (5.1.1)$$

The expectation value of the measured intensity is thus

$$\langle I(r,t)\rangle = \langle E^*(r,t)E(r,t)\rangle$$

$$= \langle \sum_m \int d\omega\, a_m(\omega)e^{-i\varphi_m(\omega)}\, e^{i\omega(\tau-t_m)}$$

$$\times \sum_n \int d\omega'\, a_n(\omega')e^{i\varphi_n(\omega')}\, e^{-i\omega(\tau-t_n)}\rangle \qquad (5.1.2)$$

where $\tau \equiv t - r/c$. Eq.(5.1.2) can be written in the following form

$$\langle I(r,t)\rangle$$

$$= \langle \sum_{m=n} \int_{\omega=\omega'} d\omega\, a_m^2(\omega)\rangle$$

$$+ \langle \sum_{m\neq n} \int_{\omega=\omega'} d\omega \left\{ a_m(\omega)a_n(\omega)e^{-i[\varphi_m(\omega)-\varphi_n(\omega)]}e^{-i\omega(t_m-t_n)}\right\}\rangle \qquad (5.1.3)$$

$$+ \langle \sum_{m=n} \int_{\omega\neq\omega'} d\omega d\omega' \left\{ a_m(\omega)a_m(\omega')e^{-i[\varphi_m(\omega)-\varphi_m(\omega')]}\right\}e^{i[\omega(\tau-t_m)-\omega'(\tau-t_n)]}\rangle$$

$$+ \langle \sum_{m\neq n} \int_{\omega\neq\omega'} d\omega d\omega' \left\{ a_m(\omega)a_n(\omega')e^{-i[\varphi_m(\omega)-\varphi_n(\omega')]}\right\}e^{i[\omega(\tau-t_m)-\omega'(\tau-t_n)]}\rangle$$

Focusing on the coherent and incoherent superposition in terms of the subfields and Fourier-modes, we analyze the following four extreme cases:

(I) Incoherent subfields   &   Incoherent Fourier-modes;
(II) Coherent subfields   &   Coherent Fourier-mode;
(III) Incoherent subfields   &   Coherent Fourier-mode; and
(IV) Coherent subfields   &   Incoherent Fourier-mode.

Eq.(5.1.3) indicates two-freedom of superposition in terms of Einstein's subfield, m, and the Fourier-mode, ω. As we have learned in Chapter 2, Einstein's "bundle of ray", or subfield, corresponds to the quantum mechanical concept of photon: the mth subfield is created from the mth sub-source, or the mth photon is created from the mth atomic transition; within each atomic transition, the Fourier-modes may be excited coherently or incoherently. In Eq.(5.1.3), we have treated the Fourier-modes continuously, which is reasonable as we have discussed in Chapter 2. The physics may become complicated when a cavity is involved. In this case, the modes ω may have to be treated discontinuously.

(I) Incoherent subfield and incoherent Fourier-modes.

An extreme case of superposition in Eq. (5.1.3) involves incoherently superposed subfields and incoherently superposed Fourier-modes. Imagine a radiation source in which all its subfields and Fourier-modes are created randomly, and thus exhibiting random relative phases $\varphi_m(\omega) - \varphi_n(\omega')$. The observed radiation at a space-time point $(\mathbf{r}, t)$ is the result of an incoherent superposition among its incoherent subfields and incoherent Fourier-modes. As a result of the incoherent superposition, the only surviving terms in Eq. (5.1.3) are those amplitudes (and their conjugates) that belong to the same sub-source and the same mode of frequency ω, i.e., $m = n$ and $\omega = \omega'$, when taking into account *all possible* values of $\varphi_m(\omega) - \varphi_n(\omega')$. These surviving terms are known as the "diagonal-terms" of the matrix elements

$$\langle I(r,t)\rangle = \langle E^*(r,t)E(r,t)\rangle = \sum_{m=n} \int_{\omega=\omega'} d\omega\, a_m^2(\omega) \qquad (5.1.4)$$

corresponding to the first term of Eq. (5.1.3). Eq. (5.1.4) can also be written in the following form

$$\langle I(r,t) \rangle = \sum_m \int d\omega \, |E_m(\omega)|^2 = \sum_m \int d\omega \, I_m(\omega). \tag{5.1.5}$$

Eq.(5.1.5) indicates that the expectation value of the total intensity is the sum of the sub-intensities in terms of the subfields and the Fourier-modes. Notice that the above evaluation of expectation is only partially completed, we have left out the evaluation for the real-positive amplitudes. The coherence property of light is determined by the relative phases between the sub-radiations[1]. In certain measurements, we may need to take into account *all possible* values of $a_m^2(\omega)$ in terms of the subfields and the Fourier-modes.

Examining Eq. (5.1.3), the second, third, and fourth integrals may all contribute to the variation of the intensity δI, if the interference cancelation is incomplete. These contributions may vary from time to time causing random fluctuations of the intensity in the neighborhood of its expectation value. The incomplete cancellations are the major contributions to δI compared with the variations of the real-positive amplitude.

We have given a simply model of incoherent light, which is the result of incoherent superposition between its subfields and Fourier modes by means of the randomly distributed relative phases $\varphi_m(\omega) - \varphi_n(\omega')$. It is the randomness of the phases that distinguishes an *incoherent sum* from a *coherent superposition*. The incoherent superposition of the subfields and Fourier modes, however, does not exclude the following possibilities: (1) the real and positive amplitude $a_m(\omega)$ may be a well-defined function of the spectral frequency ω; (2) the intensity itself may have a certain distribution function $p(I)$, such as a Gaussian, in the neighborhood of its mean value $\bar{I}$.

(II) Coherent subfields and coherent Fourier-modes.

Another extreme case of superposition in Eq. (5.1.3), obviously, involves coherently superposed subfields and coherently superposed Fourier-modes. A Q-switched laser pulse is a good example. In this extreme case, all terms of the superposition in Eq. (5.1.3) survive. The expectation function of intensity is thus

$$\langle I(r,t) \rangle = \left\langle \int d\nu \sum_m a_m(\nu) \, e^{i\nu\tau} \int d\nu' \sum_n a_n(\nu') \, e^{-i\nu'\tau} \right\rangle$$
$$= \left| \mathcal{F}_\tau \{ A(\nu) \} \right|^2. \tag{5.1.6}$$

where we have defined $A(\nu) = \sum_m a_m(\nu)$ as the total amplitude of the mode of frequency ω. The intensity expectation turns out to be a well-defined pulse in Eq. (5.1.6).

Differing from incoherent subfields, here, we assume a constant phase relationship between the subfields and Fourier modes. The calculated wavepacket, or pulse, is the result of constructive interference or coherent superposition of the subfields and the Fourier modes. In regard to the fluctuations, incoherent field and coherent wavepacket take two extremes. In incoherent radiation, the subfields may not take all possible relative phases, and in coherent light the subfields may not take an identical constant phase. The incomplete destructive interference in incoherent radiation and the incomplete constructive interference in coherent light between sub-sources are the major causes of the intensity fluctuation.

Fig. 5.1.1 shows a laser pulse train measured by a fast photodetector. The pulse is a well-defined function of time t deterministically, but fluctuate from pulse to pulse in a nondeterministic manner. The expectation value of the intensity can be calculated from

---

[1]Note, "sub-radiation" refers a Fourier-mode ω within the mth atomic transition. Here, "sub-radiation" is different from "subfield". "Subfield", labeled by m, refers the mth "bundle of ray".

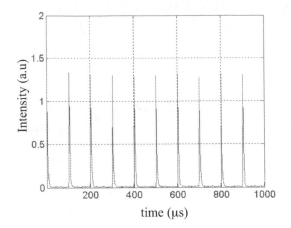

FIGURE 5.1.1 Measured laser pulse train. The intensity of each pulse is a function of time t deterministically, but fluctuates from pulse to pulse.

Eq. (5.1.3) by taking a constant phase relationship in terms of the subfield and Fourier-mode. In a measurement, however, the radiations may not be able to take the same constant phase relationship and thus produce a wavepacket slightly differ from its expectation. Consequently, the measured intensity $I(\mathbf{r}, t)$ may fluctuate in the neighborhood of its expectation function $\langle I(\mathbf{r}, t)\rangle$ from pulse to pulse randomly in a nondeterministic manor.

(III) Incoherent subfields and coherent Fourier-modes.

In the third simplified model we assume each sub-source randomly emits its own subfield with random relative phases. The Fourier-modes of the mth subfield, however, are coherently excited at time t_m. This model is very close to Einstein's picture of natural light. We will use Einstein's picture in the next a few chapters to model randomly created and randomly distributed thermal radiation.

Under the assumption of incoherent sub-sources, the only surviving terms in Eq. (1.4.4) are those with $m = n$, which includes the first and third sums of Eq. (5.1.3),

$$
\begin{aligned}
&\langle I(r, t)\rangle \\
&= \int d\omega \, d\omega' \sum_m a_m(\omega) e^{-i\omega t_m} \, a_m(\omega') e^{i\omega' t_m} \, e^{i(\omega - \omega')\tau} \\
&= \sum_m \left\{ \int d\nu \left[a_m(\nu) e^{-i\nu t_m} \right] e^{i\nu\tau} \right\} \left\{ \int d\nu' \left[a_m(\nu') e^{i\nu' t_m} \right] e^{-i\nu'\tau} \right\} \\
&= \sum_m \left| \mathcal{F}_{(\tau - t_m)} \left\{ a_m(\nu) \right\} \right|^2.
\end{aligned}
\tag{5.1.7}
$$

This result reflects explicitly the incoherent nature of the subfields and the coherent nature of the Fourier-modes associated with the mth subfield. The expectation value of intensity, $\langle I(r, t)\rangle$, is the sum of a set of sub-pulses excited by all possible independent randomly radiated sub-sources. Each sub-pulse corresponds to the Fourier transform of $a_m(\nu)$ of the mth subfield. Each Fourier transform yields a well-defined wavepacket in space-time if $a_m(\nu)$ is well-defined. The expectation value, or expectation function, $\langle I(r, t)\rangle$ is now determined by the summation of these well-defined sub-pulses. We examine the following simplified cases:

(1) Single wavepacket or pulse.

We assume the field is weak enough so only one wavepacket, or one pulse, is involved in the measurement of $\langle I(r,t)\rangle$, we may keep Eq. (5.1.7) as the expectation value, except only one nonzero a_m. For identical sub-sources, such as a large number of identical atomic transitions, we may simplify Eq. (5.1.7) to

$$\langle I(r,t)\rangle = \big|\mathcal{F}_{(\tau-t_m)}\{a(\nu)\}\big|^2. \tag{5.1.8}$$

The expectation value, or expectation function, is obviously a well-defined function of time.

(2) A large number of overlapped-partially-overlapped wavepackets or pulses.

After the ensemble average by taking into account all possible random phases of the subfields,

$$\begin{aligned}
\langle I(r,t)\rangle &= \sum_m \big|\mathcal{F}_{(\tau-t_m)}\{a_m(\nu)\}\big|^2 \\
&\cong \int dt_m\,\big|\mathcal{F}_{(\tau-t_m)}\{a_m(\nu)\}\big|^2 \\
&= \int d\nu\, d\nu'\, a(\nu)\, a(\nu')\,\Big[\int_\infty dt_m\, e^{i(\nu-\nu')t_m}\Big]\, e^{-i(\nu-\nu')\tau} \\
&\cong N\int d\nu\, a^2(\nu)
\end{aligned} \tag{5.1.9}$$

agreeing with the Parseval theorem. Here, we have assumed the wavepackets or pulses all have the same function of $\tau - t_m$, and are distributed randomly and continuously, so that the summation can be approximated to an integral over t_m from $-\infty$ to $+\infty$. To compare with the earlier results, we rewrite Eq. (5.1.9) as

$$\langle I(r,t)\rangle = \int d\nu \sum_m a_m^2(\nu).$$

It is interesting to see that the expectation value of the intensity of a large number of *randomly* distributed overlapped-partially-overlapped wavepackets is the same as that of the radiation with incoherent subfields and incoherent Fourier-modes. It should be emphasized that this result is under the approximation of the integral over t_m from $-\infty$ to $+\infty$ as a convolution[2]. This approximation may not apply to certain measurements, for instance, when the integral over t_m is limited within a time window of t_c,

$$\begin{aligned}
\langle I(r,t)\rangle &= \int d\nu\, d\nu'\, a(\nu)\, a(\nu')\,\Big[\int_{t_c} dt_m\, e^{i(\nu-\nu')t_m}\Big]\, e^{-i(\nu-\nu')\tau} \\
&\simeq \int d\nu\, d\nu'\, a(\nu)\, a(\nu')\,\Big[\text{sinc}\,\frac{(\nu-\nu')t_c}{2}\Big]\, e^{i(\nu-\nu')\tau}.
\end{aligned} \tag{5.1.10}$$

In this case, $\langle I(r,t)\rangle$ may not be treated as a constant, but a function of $(\nu-\nu')$ and τ. Note, although the integral over t_m is limited within a time window of t_c, the ensemble average

---

[2]Mathematically, integrating over parameter t_m is equivalent to integrating over time t. Physically, the two types of summation correspond to the following two pictures: (1) a measurement at time t deals with a large number of randomly distributed overlapped−partially-overlapped wavepackets, and (2) a large number of accumulated measurements happen randomly at different time.

has taken into account all possible random phases of these overlapped-partially-overlapped wavepackets.

In fact, under certain experimental condition, a measurement may not be able to "take into account all possible random phases". For instance, if only a few subfields participated the measurement, the intensity fluctuation $\Delta I(r,t)$ cannot be approximated to zero, in this case, the measured total field is the result of an incoherent superposition among limited number of wavepackets

$$E(r,t) = \sum_m e^{-i\omega_0(\tau-t_m)} \mathcal{F}_{(\tau-t_m)}\{a_m(\nu)\}, \qquad (5.1.11)$$

for example, the superposition of two wavepackets gives

$$E(r,t) = e^{-i\omega_0(\tau-t_1)} \mathcal{F}_{(\tau-t_1)}\{a_1(\nu)\} + e^{-i\omega_0(\tau-t_2)} \mathcal{F}_{(\tau-t_2)}\{a_2(\nu)\}, \qquad (5.1.12)$$

where the jth $(j = 1,2)$ wavepacket is centered at time $t = r/c + t_j$, i.e., the jth group of Fourier-modes have a common phase at $r = 0$, $t = t_j$. The instantaneous intensity is thus:

$$I(r,t) = |\mathcal{F}_{(\tau-t_1)}\{a_1(\nu)\}|^2 + |\mathcal{F}_{(\tau-t_2)}\{a_2(\nu)\}|^2$$
$$+ 2Re\left[e^{-i\omega_0(t_1-t_2)} \mathcal{F}^*_{(\tau-t_1)}\{a_1(\nu)\} \cdot \mathcal{F}_{(\tau-t_2)}\{a_2(\nu)\}\right]. \qquad (5.1.13)$$

The first term and second term of Eq. (5.1.13) are the results of the self-interference of wavepackt one and wavepacket 2, respectively; and the third term of Eq. (5.1.13) is the interference term between wavepacket one and wavepacket two. The first two terms are deterministic functions of space-time belong to $\langle I(r,t)\rangle$. The contribution of the third term, determined by the relative phase of the wavepacket one and two and their creation times t_1 and t_2, is indeterministic. Its value varies from pair to pair, from time to time and from measurement to measurement. We name it intensity fluctuation.

For any number of overlapped-partially-overlapped wavepackets, if the measurement cannot take into account all possible random phases, the indeterministic intensity fluctuation $\Delta I(r,t)$ will be observable from the output current of the photodetector:

$$I(r,t) = \sum_m |\mathcal{F}_{(\tau-t_m)}\{a_m(\nu)\}|^2$$
$$+ \sum_{m\neq n} \left[e^{i\omega_0(t_n-t_m)} \mathcal{F}^*_{(\tau-t_m)}\{a_m(\nu)\} \cdot \mathcal{F}_{(\tau-t_n)}\{a_n(\nu)\}\right]. \qquad (5.1.14)$$

The second term contributes a random value from time to time or from measurement to measurement in the neighborhood of $\langle I(r,t)\rangle$ as an intensity fluctuation

$$\Delta I(r,t) = \sum_{m\neq n} \left[e^{i\omega_0(t_n-t_m)} \mathcal{F}^*_{(\tau-t_m)}\{a(\nu)\} \cdot \mathcal{F}_{(\tau-t_n)}\{a(\nu)\}\right] \neq 0.$$

Figure 5.1.2 illustrates an experimentally observed intensity, $I(t)$, of an artificial pseudo-thermal light in the photon counting regime by a fast point-like photodetector fixed at a chosen coordinate. The figure reports the counting number, which is proportional to the intensity, at time $t \pm \Delta t/2$ with Δt a chosen short time window in the order of nanosecond. Both $\langle I(t)\rangle$ and $\Delta I(t)$ are observable in this measurement. It is easy to conclude from this measurement, in general, $\langle I(\mathbf{r},t)\rangle$= constant, and $\Delta I(\mathbf{r},t)$ takes a random value from time to time indeterministically in the neighborhood of $\langle I(\mathbf{r},t)\rangle$. In a nanosecond time window, the "weak-light" condition guaranties that only a limited number of subfields contribute to the measurement of $I(\mathbf{r},t)$. It is unlikely that a limited number of subfields could take

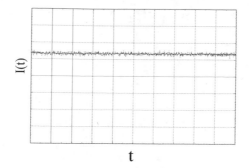

FIGURE 5.1.2 A typical measured intensity of pseudo-thermal light by a fast point-like photodetector in nanosecond scale. $I(t)$ fluctuated randomly in the neighborhood of a constant value. The pseudo-thermal light source contains a single-line continuous wave (CW) laser beam and a fast rotating defusing ground glass. The transversely expanded laser beam is scattered by millions of tiny scattering elements of the fast rotating ground glass. These tiny scattering elements play the role of point-like sub-sources. It is reasonable to model this pseudo-thermal light with incoherent subfields and coherent Fourier-modes.

all possible complex amplitudes. $\Delta I(\mathbf{r}, t)$ is therefore observable under certain experimental condition, such as measurement within a short time window.

(IV) Coherent subfields and incoherent Fourier-modes.

We shall now consider the model of the fourth type of radiation where the sub-sources radiate coherently. The Fourier-modes, however, are created randomly with random relative phases. Continuous wave (CW) laser beam with multi-cavity-modes is a closer example, except that the cavity-modes of the laser are usually discrete rather than continuous. The mechanism of stimulated emission forces all the sub-sources to radiate coherently, however, it may still leave the cavity modes with random relative phases. Since the Fourier-modes are assumed incoherent, the only surviving terms in Eq. (1.4.4) are those with $\omega = \omega'$. The expectation value of the intensity is thus

$$\langle I(r,t) \rangle = \int d\omega \sum_{m,n} a_m(\omega)\, a_n(\omega) \tag{5.1.15}$$

which includes to the first and second sums of Eq. (5.1.3). Eq. (5.1.15) reflects the incoherent nature of the Fourier-modes and the coherent nature of the sub-sources: the measured intensity is the sum of the sub-intensities of the Fourier modes; however, the amplitudes corresponding to the sub-sources add coherently with all cross-terms of the sub-sources. Eq. (5.1.15) can be further evaluated by assuming $a(\nu)$ to be a well-defined function for all sub-sources

$$\langle\, I(r,t)\,\rangle \cong N^2 \int d\omega\, a^2(\omega) = \int d\nu\, A^2(\omega), \tag{5.1.16}$$

where $A(\omega) = \sum_m a_m(\omega) \simeq N a(\omega)$.

Comparing with Eq. (5.1.4), we see that $\langle\, I(r,t)\,\rangle$ is N times greater. We thus have a "super radiator", a result of coherent superposition or constructive interference in terms of the subfields.

5.2 COHERENT AND INCOHERENT RADIATION—QUANTUM MECHANI-CAL PICTURE

Einstein's picture is consistent with quantum picture regarding to the coherence property of radiation. In this section, we show

(I) Coherent and incoherent radiation in single-photon state representation.

Assume a natural weak point-like light source and a far-field measurement of a point-like photodetector. Recall in section 4.6, we have approximated the state of such radiation field

$$|\Psi\rangle = \prod_m \left\{ |0\rangle + \epsilon\, c_m\, |\Psi_m\rangle \right\}$$

$$\simeq |0\rangle + \epsilon\Big[\sum_m c_m\, |\Psi_m\rangle\Big] + \epsilon^2\Big[\sum_{m<n} c_m\, c_n |\Psi_m\rangle|\Psi_n\rangle\Big] + \cdots$$

where, we assumed a common complex amplitude for each state $|\Psi_m\rangle$ created from the mth atomic transition, which may take arbitrary phases from transition to transition. Since $|\epsilon| \ll 1$, we listed the first-order and the second-order approximations. Following Glauber's theory, $G^{(1)}(\mathbf{r}, t)$ is calculated as follows

$$\begin{aligned}
&G^{(1)}(\mathbf{r},t)\\
&= \langle\langle \Psi | E^{(-)}(\mathbf{r},t) E^{(+)}(\mathbf{r},t) | \Psi\rangle\rangle_{\mathrm{En}}\\
&= \langle\langle \Psi |\Big[\sum_p \int d\omega\, \hat{a}_p^\dagger(\omega)\, g_p^*(\omega;\mathbf{r},t)\Big]\Big[\sum_q \int d\omega'\, \hat{a}_q(\omega') g_q(\omega';\mathbf{r},t)\Big] |\Psi\rangle\rangle_{\mathrm{En}}\\
&= \langle\langle \Psi |\Big[\sum_p \int d\omega\, \hat{a}_p^\dagger(\omega)\, g_p^*(\omega;\mathbf{r},t)\Big]|0\rangle\langle 0|\Big[\sum_q \int d\omega'\, \hat{a}_q(\omega') g_q(\omega';\mathbf{r},t)\Big] |\Psi\rangle\rangle_{\mathrm{En}}\\
&= \langle\Big[\sum_m \psi_m^*(\mathbf{r},t)\Big]\Big[\sum_n \psi_n(\mathbf{r},t)\Big]\rangle_{\mathrm{En}}
\end{aligned} \tag{5.2.1}$$

where we have defined the effective wavefunction of the mth photon,

$$\psi_m(\mathbf{r},t) \equiv \langle 0| E^{(+)}(\mathbf{r},t)|\Psi_m\rangle = \int d\omega\, a_m(\omega) e^{i\varphi_m}\, g_m(\omega;\mathbf{r},t)$$

by applying the completeness relation

$$\sum_n |n\rangle\langle n| = 1.$$

In the above calculation, we have assumed a common phase φ_m for the mth atomic transition, which may take arbitrary phases from transition to transition. Now, we further assume φ_m may vary from Fourier mode to Fourier mode, $\psi_m(\mathbf{r}, t)$ can be written as

$$\psi_m(\mathbf{r},t) = \int d\omega\, a_m(\omega) e^{i\varphi_m(\omega)}\, g_m(\omega;\mathbf{r},t). \tag{5.2.2}$$

In the far-field 1-D approximation, we thus have

$$\langle I(r,t)\rangle \propto \langle n(r,t)\rangle \propto \Big\langle \sum_m \int d\omega\, a_m(\omega) e^{-i\varphi_m(\omega)}\, e^{i\omega(\tau-t_m)}$$

$$\times \sum_n \int d\omega'\, a_n(\omega') e^{i\varphi_n(\omega')}\, e^{-i\omega(\tau-t_n)} \Big\rangle \tag{5.2.3}$$

which can be written in the following form

$$\langle I(r,t) \rangle$$

$$\propto \Big\langle \sum_{m=n} \int_{\omega=\omega'} d\omega\, a_m^2(\omega) \Big\rangle$$

$$+ \Big\langle \sum_{m \neq n} \int_{\omega=\omega'} d\omega \Big\{ a_m(\omega) a_n(\omega) e^{-i[\varphi_m(\omega) - \varphi_n(\omega)]} e^{-i\omega(t_m - t_n)} \Big\} \Big\rangle$$

$$+ \Big\langle \sum_{m=n} \int_{\omega \neq \omega'} d\omega d\omega' \Big\{ a_m(\omega) a_m(\omega') e^{-i[\varphi_m(\omega) - \varphi_m(\omega')]} \Big\} e^{i[\omega(\tau - t_m) - \omega'(\tau - t_n)]} \Big\rangle$$

$$+ \Big\langle \sum_{m \neq n} \int_{\omega \neq \omega'} d\omega d\omega' \Big\{ a_m(\omega) a_n(\omega') e^{-i[\varphi_m(\omega) - \varphi_n(\omega')]} \Big\} e^{i[\omega(\tau - t_m) - \omega'(\tau - t_n)]} \Big\rangle$$

which is the same as Eq.(5.1.3) obtained in Einstein's picture.

(II) Coherent and incoherent radiation in coherent state representation.

Assume the same simple measurement described above, except an arbitrary brightness light source, and the quantum coherent state representation is a better choice to describe the state of the measured radiation:

$$|\Psi\rangle = \prod_m |\{\alpha_m\}\rangle = \prod_{m,\,\omega} |\alpha_m(\omega)\rangle. \tag{5.2.4}$$

which is the same as Eq.(4.6.7), except we do not limit our discussion to the thermal state.

Following Glauber's theory, the probability of producing a photodection event at space-time coordinate $(\mathbf{r}, t)$ is proportional to $G^{(1)}(\mathbf{r}, t) = \langle \langle E^{(-)}(\mathbf{r}, t) E^{(+)}(\mathbf{r}, t) \rangle_{\mathrm{QM}} \rangle_{\mathrm{En}}$. Substitute the quantum state and the field operators into $G^{(1)}(\mathbf{r}, t)$, we have

$$G^{(1)}(\mathbf{r}, t) = \langle \langle \Psi | E^{(-)}(\mathbf{r}, t) E^{(+)}(\mathbf{r}, t) | \Psi \rangle \rangle_{\mathrm{En}}$$

$$= \Big\langle \langle \Psi | \Big[\sum_p \int d\omega\, a_p^*(\omega)\, g_p^*(\omega; \mathbf{r}, t) \Big]$$

$$\times \Big[\sum_q \int d\omega'\, a_q(\omega') g_q(\omega'; \mathbf{r}, t) \Big] | \Psi \rangle \Big\rangle_{\mathrm{En}}$$

$$= \Big\langle \Big[\sum_m \psi_m^*(\mathbf{r}, t) \Big] \Big[\sum_n \psi_n(\mathbf{r}, t) \Big] \Big\rangle_{\mathrm{En}} \tag{5.2.5}$$

where $\psi_m(\mathbf{r}, t)$ is the effective wavefunction of the mth subfield or the mth group of identical photons:

$$\psi_m(\mathbf{r}, t) \equiv \langle \Psi_m | E^{(+)}(\mathbf{r}, t) | \Psi_m \rangle = \int d\omega\, a_m(\omega) e^{i\varphi_m(\omega)}\, g_m(\omega; \mathbf{r}, t), \tag{5.2.6}$$

where we have given an additional freedom to the complex amplitudes, $\alpha_m(\omega) = a_m(\omega) e^{i\varphi_m(\omega)}$: its phase $\varphi_m(\omega)$ may vary from Fourier mode to Fourier mode. In the far-field 1-D approximation, we have

$$\langle I(r,t) \rangle \propto \Big\langle \sum_m \int d\omega\, a_m(\omega) e^{-i\varphi_m(\omega)}\, e^{i\omega(\tau - t_m)}$$

$$\times \sum_n \int d\omega'\, a_n(\omega') e^{i\varphi_n(\omega')}\, e^{-i\omega(\tau - t_n)} \Big\rangle \tag{5.2.7}$$

which can be written in the following form

$$
\langle I(r,t) \rangle
$$

$$
\propto \Big\langle \sum_{m=n} \int_{\omega=\omega'} d\omega\, a_m^2(\omega) \Big\rangle
$$

$$
+ \Big\langle \sum_{m \neq n} \int_{\omega=\omega'} d\omega \Big\{ a_m(\omega) a_n(\omega) e^{-i[\varphi_m(\omega)-\varphi_n(\omega)]} e^{-i\omega(t_m-t_n)} \Big\} \Big\rangle
$$

$$
+ \Big\langle \sum_{m=n} \int_{\omega \neq \omega'} d\omega d\omega' \Big\{ a_m(\omega) a_m(\omega') e^{-i[\varphi_m(\omega)-\varphi_m(\omega')]} \Big\} e^{i[\omega(\tau-t_m)-\omega'(\tau-t_n)]} \Big\rangle
$$

$$
+ \Big\langle \sum_{m \neq n} \int_{\omega \neq \omega'} d\omega d\omega' \Big\{ a_m(\omega) a_n(\omega') e^{-i[\varphi_m(\omega)-\varphi_n(\omega')]} \Big\} e^{i[\omega(\tau-t_m)-\omega'(\tau-t_n)]} \Big\rangle
$$

which is the same as Eq.(5.1.3) obtained in Einstein's picture.

5.3 TEMPORAL COHERENCE AND SPATIAL COHERENCE

The following discussions are in Einstein's picture, which is consistent with quantum picture in terms of coherence of radiation.

(I) Temporal coherence

The concept of temporal coherence of radiation implies two aspects of physics: (1) the coherent and incoherent superposition of radiations that are excited from temporally separated radiation events; (2) the superposition or interference between temporally delayed fields. Temporal coherence is an intrinsic property of the radiation. In this section we will focus our discussion on the physics of (1) and leave (2) to Chapter 7.

Regarding to temporal coherence, we may simplify our discussion to point radiation sources. A point source may consist of a large number of sub-sources such as trillions of atomic transitions that radiate a large number of spherical harmonic waves, labeled by different modes of frequency and different sub-sources within the point source. In the Maxwell electromagnetic wave theory, each spherical harmonic wave is a solution of the EM wave equation, and thus the superposition of all or part of these spherical harmonic subfields at a space-time point of observation is also a solution of the EM wave equation

$$
\mathbf{E}(\mathbf{r},t) = \sum_{\omega} \sum_{m=1}^{M} \hat{\mathbf{e}}\, \frac{E_m(\omega)}{r}\, e^{-i[\omega(t-t_m)-k(\omega)r]}, \tag{5.3.1}
$$

where $E_m(\omega) = a_m(\omega) e^{i\varphi_m(\omega)}$ is the complex amplitude for the mode ω excited by the mth sub-source or the mth radiation event, at time t_m. The amplitude distribution of the field is represented by $a_m(\omega)$, while the phase variation, among the subfields and the harmonic-modes is represented by $\varphi_m(\omega)$. The polarization is indicated by an unit vector $\hat{\mathbf{e}}$. To simplify the notation we select one polarization of the field, and the vector notation will be dropped in the following discussions. These harmonic waves, or subfields, are thus superposed at a space-time point either incoherently or coherently depending on their initial phases $e^{i\varphi_m(\omega)}$.

Eq. (5.3.1) shows that each subfield has spherical symmetry which is reasonable for a point source and a spherical boundary condition of infinity. The superposition of these spherical harmonic subfields, either incoherent, partial coherent, or coherent, has a null effect on the transverse spatial distribution of the field, but may lead to different temporal behavior of the radiation. For example, the coherent constructive-destructive interference among a large number of Fourier-modes of different frequencies may result in a wavepacket. A large number of wavepackets created at different times may also superposed coherently or incoherently. If these wavepackets created at different times are able to superposed coherently, we consider the radiation temporally coherent. Otherwise, the radiation is temporally incoherent.

In the previous section we have suggested four simple models to classify the radiation field:

(I) Incoherent subfields   &   Incoherent Fourier-modes;
(II) Coherent subfields   &   Coherent Fourier-mode;
(III) Incoherent subfields &   Coherent Fourier-mode;
(IV) Coherent subfields   &   Incoherent Fourier-mode.

In terms of the concept of coherence, radiation (I) is definitely temporally incoherent; radiation (II) is definitely temporally coherent. Radiation (III) is interesting since in the weak light condition the measurement at a space-time point may reveal that the radiation behaves like a wavepacket from time to time, caused by the coherent constructive-destructive interference among many Fourier-modes excited by each individual sub-source. In this situation we may classify radiation (III) as temporally coherent. When the light intensity gets stronger, however, the incoherent superposition of a large number of overlapped or partially overlapped wavepackets turns the radiation temporally incoherent. Radiation (IV) is very special. On one hand, the coherent superposition among subfields involves constructive interference of harmonic waves which builds up an enhanced Fourier-mode; on the other hand, the incoherent superposition of Fourier-modes prohibits the formation of a wavepacket. If the radiation source excites a single frequency mode, or if a single frequency is isolated from the multi-modes for observation, we may classify radiation (IV) as temporally coherent. However, if multi-modes cannot be avoided in the measurement, we may have to take into account the incoherent nature of the radiation (IV).

For a point source and free propagation, the subfields excited by all sub-sources take the same optical path when reaching a space-time point of observation. It is the constructive-destructive interference among different frequency modes that results in a wavepacket (field) or a pulse (intensity). For example, the coherent superposition of radiation (II) in which all subfields created at time t_m are in-phase, may excites a spherical wavepacket

$$E(r,t) = \frac{e^{-i(\omega_0 t - k_0 r + \varphi_0)}}{r} \int d\nu \left[\sum_m a_m(\nu) \right] e^{-i\nu\tau}$$

$$= \mathcal{F}_\tau \left\{ \sum_m a_m(\nu) \right\} \frac{e^{-i(\omega_0 t - k_0 r + \varphi_0)}}{r}, \qquad (5.3.2)$$

representing a spherical harmonic wave (carrier) propagating together with an envelope. The envelope is the Fourier transform of the spectral distribution function of the field, and propagates with the same speed as that of the phase of the spherical harmonic wave in free space. It is interesting to see that constructive-destructive interference turns continuous waves (or modes) into a pulse and enhances the intensity significantly within the pulse. Of course, the energy of the radiation must be conserved. The enhancement of the energy within the pulse (constructive interference) is at the price of losing energy outside the pulse (destructive interference).

In far-field observations the above spherical wavepacket can be approximated as a plane-wave wavepacket:

$$E(\mathbf{r},t) \simeq \mathcal{F}_\tau \left\{ \sum_m a_m(\nu) \right\} e^{-i(\omega_0 t - \mathbf{k}_0 \cdot \mathbf{r} + \varphi_0)}, \qquad (5.3.3)$$

where we have written the field as a polarized vector field and treated $a_m(\nu) = a_m/r \simeq$ constant.

(II) Spatial coherence

Similar to that of temporal coherence, the concept of spatial coherence involves two aspects of physics: (1) the coherent or incoherent superposition of subfields that are excited from spatially separated sub-sources; (2) the superposition or interference between spatially separated fields. Spatial coherence is also an intrinsic property of the radiation. In this section we will focus our discussion on the physics of (1) and leave (2) to Chapter 7.

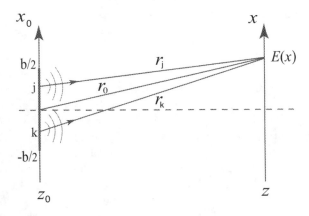

FIGURE 5.3.1 A 1-D radiation source of finite dimension consists of a large number of point-like sub-sources randomly distributed along the x_0-axis.

Instead of point source we now consider radiation source with finite dimension, for example, the 1-D source shown in Fig. 5.3.1, which consists of a large number of point-like sub-sources randomly distributed along the x_0-axis. The physics we consider now is the coherent or incoherent superposition among these subfields that are created from these spatially separated sub-sources. If these subfields are able to superpose coherently we consider the radiation spatially coherent; otherwise, the radiation is spatially incoherent. For instance, a TEM_{00} mode laser beam is considered spatially coherent; however, sunlight is considered spatially incoherent[3]. Compared with a point source, the geometry of spatially extended sources complicates the physics and mathematics. For a point radiation source and free propagation (direct point-to-point propagation), the subfields originated from all sub-sources take the same optical path when reaching a space-time point of observation. For a radiation source of finite dimension, however, the superposed subfields may not experience the same optical path length to reach the observer. The superposition is more complicated in this case.

To make the physics and mathematics of spatial coherence easily understandable we divide our discussion into three steps: (1) the study of the spatial coherent property of the source, which involves the superposition of subfields excited from a large number of coherent or incoherent sub-sources (in this section); (2) the study of diffraction of a spatial mode when passing through an aperture, which involves the superposition of a large number of secondary wavelets originating from each point of the primary wavefront of the mode (in chapter 5); (3) the propagation of a known field from one plane to another plane, which combines the physics and mathematics of both (1) and (2) (in chapter 5).

---

[3]Regarding to the radiation source, we name the laser as a spatial coherent light source and the sun as a spatial incoherent light source.

Assume the 1-D radiation source of Fig.5.3.1 consists of a large number of randomly distributed sub-sources along the x_0-axis from $x_0 = -b/2$ to $x_0 = b/2$. To simplify the mathematics, we assume single frequency radiation ($\omega = $ constant) and limits our discussion to one polarization[4]. The source plane is assigned as $z_0 = 0$. The observation plane of $z = $ constant is parallel to the source plane and located at a distance z. We are interested in knowing where and how the light is arriving on the plane of $z = $ constant. Basically, we need to calculate the intensity distribution, or the expectation value $\langle I(x,t) \rangle$ along the x-axis. Different from a point source, here each subfield excited by a sub-source corresponding to one point of the x_0-axis takes a different optical path to reach a point on the x-axis. The field $E(x,t)$ is the result of the superposition of these subfields

$$E(x,t) = \int_{-\frac{b}{2}}^{\frac{b}{2}} dx_0 \frac{E(x_0)}{r} e^{-i(\omega t - kr)}, \qquad (5.3.4)$$

where $E(x_0) = a(x_0)e^{i\varphi(x_0)}$ is the complex amplitude of the field excited at the sub-source of x_0. In Eq. (5.3.4) we have treated each subfield spherical wave centered at each sub-source using

$$r = \sqrt{z^2 + (x - x_0)^2}.$$

The expectation value of the intensity $\langle I(x,t) \rangle$ is thus written as

$$\langle I(x,t) \rangle = \langle \int_{-\frac{b}{2}}^{\frac{b}{2}} dx_0 \, dx_0' \frac{a(x_0)}{r} \frac{a(x_0')}{r'} e^{-i[\varphi(x_0)-\varphi(x_0')]} e^{-ik[r-r']} \rangle \qquad (5.3.5)$$

where $r = \sqrt{z^2 + (x - x_0)^2}$ and $r' = \sqrt{z^2 + (x - x_0')^2}$.

We will analyze two extreme cases: incoherent sub-sources and coherent sub-sources.

Case I: Incoherent sub-sources.

In the case of incoherent sub-sources the only surviving terms under the integral of Eq. (5.3.5) are those terms with $x_0 = x_0'$ when taking into account all possible realizations of the field with random values of $\varphi(x_0) - \varphi(x_0')$. The expectation value of the intensity at x, $\langle I(x,t) \rangle$, turns out to be a trivial sum of the sub-intensities from each sub-source at x_0:

$$\langle I(x,t) \rangle = \int_{-\frac{b}{2}}^{\frac{b}{2}} dx_0 \frac{a^2(x_0)}{r^2} \simeq \int_{-\frac{b}{2}}^{\frac{b}{2}} dx_0 \, I_{x_0}(x). \qquad (5.3.6)$$

Therefore, in the case of incoherent sub-sources, we expect to observe smoothly distributed light along the entire x-axis, i.e., $\langle I(x,t) \rangle \sim$ constant (in space and in time) when x is not too far from the optical axis.

In certain cases Eq. (5.3.6) is also written in terms of the angular diameter of the radiation source which is defined from the view point of the observer. For example, the angular size of the 1-D aperture in Fig. 5.3.1, $\Delta\theta$, is defined as the angle between two lines which connect the observation point of x with the source points of $x_0 = b/2$ and $x_0 = -b/2$. Eq. (5.3.6) may be written in terms of the angular variable θ as

$$\langle I(x,t) \rangle = \int_{-\frac{\Delta\theta}{2}}^{\frac{\Delta\theta}{2}} d\theta \, I(\theta) \sim I_0, \qquad (5.3.7)$$

where I_0 is a constant, if x is not too far from the optical axis, or if $I(\theta) \sim$ constant.

---

[4] We will keep this assumption in the next several sections unless otherwise specified.

In the discussion of the temporal behavior of radiation we found that incoherent sub-sources produce randomly distributed radiations in time. Now we have discovered similar behavior for incoherent sub-sources with random transverse spatial distribution. The physics is very simple: for a incoherent source, each sub-source radiates *independently and randmoly*. The sub-intensities, instead of the subfields, excited from these incoherent sub-sources are then simply added at any space-time point of observation.

Case II: Coherent sub-sources.

In the case of coherent sub-sources we analyze the following two special situations:

(1) $\varphi(x_0) = \varphi_0 = $ constant

While $\varphi(x_0) = $ constant for all subfields distributed along the x_0-axis, the field at space point x is the result of a coherent superposition of a large number of spherical harmonic waves each centered at x_0. Eq. (5.3.5) becomes

$$\langle I(x,t)\rangle = \left| \int_{-\frac{b}{2}}^{\frac{b}{2}} dx_0\, \frac{a(x_0)}{r}\, e^{-i(\omega t - kr)} \right|^2, \tag{5.3.8}$$

where $r = \sqrt{z^2 + (x - x_0)^2}$ is a function of x and x_0 for a chosen z. The coherent superposition indicated in Eq. (5.3.8) results in a diffraction pattern on the observation plane, implying a constructive-destructive interference. The diffraction pattern is easy to calculate numerically for any observation plane, either far-field or near-field. To have an analytical solution, however, is never easy for an arbitrary plane. We now consider far-field observation by applying the far-field Fraunhofer approximation

$$r = \sqrt{z^2 + (x - x_0)^2} \simeq r_0 - \vartheta\, x_0, \tag{5.3.9}$$

where angle ϑ is defined in Fig. (5.3.1). Substituting Eq. (5.3.9) into Eq. (5.3.8), the diffraction pattern is approximated as

$$\langle I(x,t)\rangle \simeq \left| \frac{e^{-i[\omega t - kr_0(x)]}}{r_0} \int_{-\frac{b}{2}}^{\frac{b}{2}} dx_0\, a_0\, e^{-ik\vartheta x_0} \right|^2 = I_0 \operatorname{sinc}^2 \frac{k\vartheta b}{2}, \tag{5.3.10}$$

where b is the width of the 1-D source, $\theta \simeq x/(z - z_0)$, and we have treated $a(x_0) \simeq a_0$ as a constant, which is reasonable for a randomly distribution.

Eq. (5.3.10) indicates a standard Fraunhofer diffraction pattern of a 1-D aperture, which is the result of coherent superposition of a large number of subfields each excited by a point sub-source distributed along the x_0-axis. The subfields are superposed *constructively* within the pattern and *destructively* outside the pattern. It is interesting to see that although each sub-source radiates spherical waves in all directions, when the sub-sources radiate coherently, we can only observe light within a certain limited angular region, $\Delta\vartheta \sim \lambda/b$. For visible light, a coherent source of a few millimeters in transverse dimension only radiates nearly collimated light with a diverging angle on the order of $\Delta\vartheta \sim 10^{-3}$ radians, which can be effectively treated as collimated radiation.

(2) $\varphi(x_0) = k_{x_0} x_0$

In certain experimental conditions, the complex amplitude of the subfield may have a phase factor of $e^{ik_{x_0} x_0}$, where k_{x_0} is a constant. This phase factor implies that any sub-source located at an arbitrary coordinate x_0 radiates with a constant relative phase, $\Delta\varphi = k_{x_0} \Delta x_0$, with respect to the sub-source at $x_0 + \Delta x_0$. Eq. (5.3.10) becomes

$$\langle\, I(x,t)\,\rangle \simeq \left| \frac{e^{-i[\omega t - kr_0(x)]}}{r_0} \int_{-\frac{b}{2}}^{\frac{b}{2}} dx_0\, a_0\, e^{ik_{x_0} x_0}\, e^{-ik\vartheta x_0} \right|^2$$

$$= I_0 \operatorname{sinc}^2 \frac{(k_{x_0} - k\vartheta)b}{2}. \tag{5.3.11}$$

Eq. (5.3.11) implies a similar far-field Fraunhofer diffraction pattern as shown in Eq. (5.3.10), except with a constant angular shift in the propagation direction $\theta_0 = k_{x_0}/k$.

Eqs. (5.3.10) and (5.3.11) have defined a wavepacket in the transverse dimension, which is the Fourier transform of the aperture function

$$E(x,t) \simeq \frac{e^{-i[\omega t - kr_0(x)]}}{r_0(x)} \int_{-\infty}^{\infty} dx_0\, A(x_0)\, e^{ik_x\, x_0}$$

$$= \mathcal{F}_{k_x}\bigl\{ A(x_0) \bigr\} \frac{e^{-i[\omega t - kr_0(x)]}}{r_0(x)}, \tag{5.3.12}$$

where $A(x_0)$ is named the "aperture function" consisting with a real function $|A(x_0)|$ and a phase $\varphi(x_0)$, and $k_x \sim k\vartheta$ is the transverse wavevector along the x direction also known as the "spatial frequency". The complex aperture function $A(x_0)$ of the 1-D source in Fig. 5.3.1 is usually written as

$$A(x_0) = \begin{cases} A_0\, e^{i\varphi(x_0)} & -b/2 \leq x_0 \leq b/2 \\[2mm] 0 & \text{otherwise} \end{cases}$$

where $\varphi(x_0) = \varphi_0 = $ constant in Eq. (5.3.10) and $\varphi(x_0) = k_{x_0} x_0$ in Eq. (5.3.11). The real-positive amplitude of the field along the 1-D aperture is described by $A(x_0)$, while the phase variation along the 1-D aperture is represented by $e^{i\varphi(x_0)}$. The wavepacket consists of a "carrier" spherical wave and a 1-D "envelope" $\mathcal{F}_{k_x}\bigl\{ A(x_0) \bigr\}$ in the transverse dimension. The envelope restricts the values of k_x within a certain limit, which implies a restricted propagation direction. The formation of the wavepacket is the result of a constructive-destructive interference among a large number of coherent subfields excited by the spatially coherent sub-sources.

In Eq. (5.3.12), the transverse coordinate x_0 and the transverse wavevector k_x are Fourier conjugate variables, and obviously, the far-field observation plane is effectively the Fourier transform plane of the aperture function. Based on the Fourier transform we may introduce a classical "uncertainty relation" between spatial variables x_0 and k_x (or p_x)

$$\Delta x_0\, \Delta k_x \geq 2\pi \quad \text{or} \quad \Delta x_0\, \Delta p_x \geq h, \tag{5.3.13}$$

where p_x is the transverse momentum and h is the Planck constant. Eq. (5.3.13) defines a "diffraction limit" for a spatially coherent radiation source. Laser beams with a TEM$_{00}$ mode are typically spatially coherent. An idealized laser beam propagates under its "diffraction limit": the greater the size of the laser beam the smaller the diverging angle.

It is interesting to see the similarities between the temporal wavepacket in the longitudinal dimension and the spatial wavepacket in the transverse dimension. The temporal coherence of the radiation produces temporal wavepackets, implying temporal constructive-destructive interference; and the spatial coherence of the radiation results in spatial wavepackets, implying spatial constructive-destructive interference. In both cases, the energy of the radiation is enhanced significantly in the region of constructive interference, at the price of losing energy in the region of destructive interference.

SUMMARY

In this chapter, we classified optical coherence. In fact, there have been two different types of coherence in optics: (1) an optical property based on which a radiation or a radiation source is named coherent or incoherent; and (2) the observable or unobservable interference between temporal delayed or spatially separated electromagnetic fields. In this chapter, we restricte all discussions on (1).

To specify the coherence property, or the state of a radiation field, we start from Einstein's granularity picture of light to introduce two category of superpositions in terms of Einstein's subfields and Maxwells Fourier-modes. We then show this simple model is consistent with that of quantum theory of light. The concepts of temporal coherence and spatial coherence are introduced and discussed in the end of this chapter.

Based on Einstein's picture of quantized electromagnetic field, assuming coherently or incoherently radiated Fourier-modes created from a point like source which contains a large number sub-sources, we classified the coherence property of light into four extreme cases:

(I) Incoherent subfields   &   Incoherent Fourier-modes;
(II) Coherent subfields   &   Coherent Fourier-mode;
(III) Incoherent subfields   &   Coherent Fourier-mode;
(IV) Coherent subfields   &   Incoherent Fourier-mode.

In the discussion of case (III), we introduced the the concepts of single-wavepacket, a few wavepackets, and a large number of overlapped and partially-overlaped wavepackets, each created from a randomly radiated sub-source. Although we did not introduce the concept of photon and formulate each subfield with a specific atomic transition, the purpose is obviously.

The optical coherence is then generalized to non-point-like sources. The concepts of spatial coherence and spatial wavepacket are introduced. The concepts of temporal coherence and spacial coherence are distinguished.

REFERENCES AND SUGGESTIONS FOR READING

[1] J.W. Goodman, *Introduction to Fourier Optics*, Roberts & Company, Englewood, 2005.

CHAPTER 6

Diffraction and Imaging

Imaging is an optical phenomenon that everyone is familiar with. However, not everyone knows that imaging is an interference phenomenon, and perhaps, even fewer people understand that a camera produced image under sunlight is the result of a single-photon interference: a photon interfering with the photon itself. To prepare the background knowledge for imaging, we start from diffraction and field propagation. In quantum optics, diffraction and imaging are regarded as the transverse behavior of a photon, or a subfield. Precisely, we are interested in determining the probability amplitude distribution of observing a photon on a transverse plane of $z = $ constant from a known probability amplitude distribution on another plane, such as $z_0 = 0$. In classical optics, diffraction and imaging are viewed as the transverse behavior of classical electromagnetic field. Precisely, classical optics calculate $E(\mathbf{r}, t)$ on a transverse plane of $z = $ constant from a known distribution of the field $E(\mathbf{r}_0, t_0)$ on another plane, such as $z_0 = 0$. Strictly speaking, all the results of this chapter can be obtained from classical electromagnetic theory, Maxwell's continuum picture of radiation gives the same diffraction pattern and image of an object as that of Einstein's granularity picture and the quantum theory of light. The quantum interference picture described in this chapter is a necessary preparatory knowledge for later discussions, such as that of "ghost imaging". Nevertheless, the quantum picture of classic imaging is helpful, in general, to learn and to understand the principles of quantum optics.

6.1 DIFFRACTION

To observe diffraction of minimum-width, a plane-wave is necessary. In modern optics laboratories, a large diameter laser beam with TEM$_{00}$ mode is popularly used for a good simulation. Before the invention of the laser, distant point-like radiation sources and optical collimators were widely used to provide such a plane-wave solution. A collimator consists of a pinhole, or a line aperture (1-D), and a lens system, as shown in Fig. (6.1.1). The pinhole is placed at the principal focus of the lens system and is illuminated by a discharge-tube. The use of the pinhole is to simulate a point radiation source. The lens system introduces appropriate phase delays to the spherical wavefront originated from the point source and turns it into a "flat" wavefront. The spherical radiation then becomes a well collimated plane-wave like light beam. Far-field observation of a point-like source is also a good approximation to simulate plane-wave.

Fig. 6.1.1 is a schematic setup for diffraction measurement. The setup consists of a plane-wave source of radiation and a 1-D aperture with finite dimension of b. The 1-D aperture is placed normally to the plane-wave along the x_0 axis as shown in Fig. 6.1.1(a). The new boundary condition will modify the plane-wave solution, slightly or significantly depending on the size of the aperture. This phenomenon has been known as "diffraction". After passing the aperture, the radiation at each point of x_0 may propagate to the observation plane which is placed at distance $(z - z_0)$ from the aperture. The radiation passing a coordinate x_0 takes a slightly different optical path to reach a point on the observation plane and to superpose with all other radiations coming from other points of the x_0-axis. The superposition of these radiations with slight relative phase delays produces a diffraction pattern by means of the intensity distribution along the x-axis (1-D).

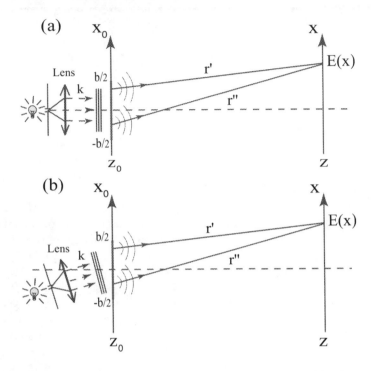

FIGURE 6.1.1 Schematic experimental setup of single-slit diffraction. (a) The normally incident light passes the 1-D aperture, propagates to and arrives at an arbitrary observation plane, either far-field or near-field. (b) The incident radiation illuminates the 1-D aperture from a non-normal incident angle of $\theta_0 \neq 0$. In both setups the constructive-destructive interferences result in a diffraction pattern. The far-field diffraction is named Fraunhofer while the near-field diffraction is called Fresnel.

$$
\begin{aligned}
&\langle I(x, z, t)\rangle \\
&= \Big\langle \sum_m E_m^*(x_m, z_m, t_m)\, g_m^*(x, z, t) \sum_n E_n(x_n, z_n, t_n)\, g_n(x, z, t) \Big\rangle \\
&= \Big\langle \sum_m \big| E_m(x_m, z_m, t_m)\, g_m(x, z, t) \big|^2 \Big\rangle \\
&\quad + \Big\langle \sum_{m \neq n} E_m^*(x_m, z_m, t_m)\, g_m^*(x, z, t) E_n(x_m, z_m, t_m)\, g_m(x, z, t) \Big\rangle
\end{aligned}
\tag{6.1.1}
$$

where $g_m(x, z, t)$ is the Green's function that propagates the mth subfield from its creation coordinate (x_m, z_m, t_m) at the radiation source to the observation coordinate (x, z, t). For incoherent radiation in thermal state, such as the sunlight or discharge-tube radiation, it is easy to understand that the observed diffraction pattern is the result of the mth subfield interfering with the mth subfield itself, no matter how complicated $g_m(x, z, t)$ is.

In Fig. 6.1.1(a), we have assumed a subfield incident normally to the diffraction aperture in the form of plane-wave. The subfield at each point along the x_0-axis is known:

$$E_m(x_0) = \begin{cases} |E_m|e^{i\varphi_m} & -b/2 \leq x_0 \leq b/2 \\ 0 & \text{otherwise,} \end{cases}$$

where φ_m is the phase of the mth plane-wave at the aperture. In Einstein's picture, the mth subfield or wavepacket propagates to the diffraction aperture with a "flat" wavefront. The electromagnetic field has a constant amplitude $|E_m|$ and a constant phase φ_m at each point along the x_0-axis at time t_0.

In Fig. 6.1.1(b), we have assumed a subfield incident with angle $\theta_0 \neq 0$ to the diffraction aperture in the form of plane-wave. The subfield at each point along the x_0-axis is known:

$$E_m(x_0) = \begin{cases} |E_m|e^{i(\varphi_m + \vec{k}_0 \cdot \vec{x}_0)} & -b/2 \leq x_0 \leq b/2 \\ 0 & \text{otherwise,} \end{cases}$$

where $\vec{k}_0$ is the 2-D wavevector of the mth incident plane-wave, $(\vec{k}_0 \cdot \vec{x}_0)$ is the relative phase delay of the mth plane-wave at each coordinate x_0 of the aperture. In Einstein's picture, the mth subfield or wavepacket propagates to the diffraction aperture with a "tilted flat" wavefront. The electromagnetic field has a constant complex amplitude E_m and a phase $(\varphi_m + \vec{k}_0 \cdot \vec{x}_0)$ at each point along the x_0-axis at time t_0.

According to Huygens's principle, each point on primary wavefront serves as the source of spherical secondary wavelets. The secondary wavelets at (x_0, z_0) is then propagating to the observation coordinate (x, z) at a later time t:

$$E_m(x_0)g_{x_0}(x, z, t) = \frac{E_m(x_0)}{r} e^{-i[\omega(t-t_0)-kr]},$$

where $r = \sqrt{(z-z_0)^2 + (x-x_0)^2}$. The field $E_m(x, z, t)$ in the observation plane is the superposition of all these secondary wavelets:

$$E_m(x, z, t) = \int_{-\frac{b}{2}}^{\frac{b}{2}} dx_0\, E_m(x_0)g_{x_0}(x, z, t)$$

$$= \int_{-\frac{b}{2}}^{\frac{b}{2}} dx_0 \frac{E_m(x_0)}{r} e^{-i[\omega(t-t_0)-kr]}. \tag{6.1.2}$$

The integral is easily completed under the far-field approximation for Fig. 6.1.1(a)

$$E_m(x, z, t) \simeq \frac{e^{-i[\omega t - kr_0(x) - \varphi_m]}}{r_0(x)} \int_{-\frac{b}{2}}^{\frac{b}{2}} dx_0\, |E_m|e^{-ik_x x_0}$$

$$= \text{sinc} \frac{k_x b}{2} \frac{e^{-i[\omega t - kr_0(x) - \varphi_m]}}{r_0(x)}. \tag{6.1.3}$$

where $k_x \simeq kx/(z-z_0) \simeq (2\pi/\lambda)\theta$. The integral results in a slightly different sinc-function for Fig. 6.1.1(b)

$$E_m(x, z, t) \simeq \frac{e^{-i[\omega t - kr_0(x) - \varphi_m]}}{r_0(x)} \int_{-\frac{b}{2}}^{\frac{b}{2}} dx_0\, |E_m|e^{i\vec{k}_0 \cdot \vec{x}_0} e^{-ik_x x_0}$$

$$= \text{sinc} \frac{(k_{0x} - k_x)b}{2} \frac{e^{-i[\omega t - kr_0(x) - \varphi_m]}}{r_0(x)}. \tag{6.1.4}$$

where k_{0x} represents the x-component of the mth incident plane-wave wavevector.

In both cases, we find the aperture of width b turns the input mth subfield of plane-wave into a spherical-like wave with an envelope of sinc-function. Mathematically, we may rewrite Eq.(6.1.3) and Eq.(6.1.4) into a slightly different form by introducing the complex aperture function $A_m(x_0)$,

$$A_m(x_0) = \begin{cases} |E_m|e^{ik_{0x}x_0} & -b/2 \leq x_0 \leq b/2 \\ 0 & \text{otherwise.} \end{cases}$$

For the plane-wave diffraction measurements shown in Fig. 6.1.1(a) and Fig. 6.1.1(b), $|E_m|$ is a non-zero constant within the diffraction aperture and zero elsewhere along the x_0-axis. The complex aperture function $A_m(x_0)$ is composed of a real aperture function $|A_m(x_0)|$ and a phase factor $e^{ik_{0x}x_0}$, $A(x_0) = |A(x_0)|e^{i\varphi(x_0)}$. The mth subfield in Eq.(6.1.3) and Eq.(6.1.4) for the setups of Fig. 6.1.1(a) and Fig. 6.1.1(b) along the x-axis is thus the Fourier transform of the complex aperture function

$$E_m(x,z,t) \simeq \frac{e^{-i[\omega t - kr_0(x) - \varphi_m]}}{r_0(x)} \int dx_0 \, A_m(x_0) \, e^{-ik_x x_0}$$

$$= \mathcal{F}_{k_x}\{A_m(x_0)\} \frac{e^{-i[\omega t - kr_0(x) - \varphi_m]}}{r_0(x)}. \tag{6.1.5}$$

It is interesting to find that the Fourier transform is completed by the mth subfield itself. In other words, the observed diffraction pattern is the result of the mth subfield interfering with the mth subfield itself. The observed diffraction sinc-functions for Fig. 6.1.1(a) and Fig. 6.1.1(b) indicate that the mth photon interfering with itself constructively when $\theta_0 - \theta \leq \lambda/b$ and destructively when $\theta_0 - \theta > \lambda/b$. The mth subfield of plane-wave is thus diffracted by the aperture into a spherical-like wave with an envelope of the Fourier transform of the complex aperture function.

In the above discussions, we have been focused on the transverse behavior of a subfield, or a photon. We have calculated the transverse distribution of the mth subfield, or the probability amplitude distribution of the mth photon, in the far-field plane as the Fourier transform of the known complex aperture function $A_m(x_0)$, or the probability amplitude distribution of the mth photon, in the plane of z_0. The subfield $E_m(x,z,t)$ in Einstein's picture plays the same role as the effective wave function $\varphi_m(x,z,t)$ of the mth photon which represents the probability amplitude to observe the mth photon in space-time point (x,z,t). In the beginning of this chapter, we have stated that we are interested in determining the probability amplitude distribution of observing a photon on a transverse plane of $z =$constant from a known probability amplitude distribution on another plane, such as z_0. At this point, we may conclude that we have achieved our goal, at least, for a far-field transverse plane.

Next, we calculate the expected diffraction pattern in the observation plane produced by a large number of incoherent subfields. The diffraction pattern can be estimated numerically from Eq. (6.1.2) for an arbitrary observation plane. In the following, we continue our far-field approximation in order to obtain an easy analytical solution. The far-field diffraction is named Fraunhofer diffraction to distinguish it from the near-field Fresnel diffraction. Under the Fraunhofer far-field approximation, adding all identical sub-diffraction patterns of the incoherent subfields, we obtain the Fraunhofer diffraction pattern for Fig. 6.1.1(a):

$$\langle I(x,t) \rangle \simeq \sum_m \left| \frac{e^{-i[\omega t - kr_0(x) - \varphi_m]}}{r_0(x)} \int_{-\frac{b}{2}}^{\frac{b}{2}} dx_0 \, |E_m| e^{-ik_x x_0} \right|^2$$

$$= I_0 \, \text{sinc}^2 \frac{k_x b}{2} = I_0 \, \text{sinc}^2 \frac{\pi b}{\lambda}\theta \tag{6.1.6}$$

indicating that each and all mth photon interfering with the mth photon itself constructively when $\theta \leq \lambda/b$ and destructively when $\theta > \lambda/b$. The diffraction pattern for Fig. 6.1.1(b) is slightly different due to the non-zero incident angle of the plane-wave:

$$
\langle I(x,t) \rangle \simeq \sum_m \left| \frac{e^{-i[\omega t - kr_0(x) - \varphi_m]}}{r_0(x)} \int_{-\frac{b}{2}}^{\frac{b}{2}} dx_0 \, |E_m| e^{ik_{0x}x_0} \, e^{-ik_x x_0} \right|^2
$$

$$
= I_0 \, \text{sinc}^2 \frac{(k_{0x} - k_x)b}{2} = I_0 \, \text{sinc}^2 \frac{\pi b}{\lambda}(\theta_0 - \theta). \tag{6.1.7}
$$

In terms of the complex aperture function, the observed diffraction patterns for the setups of Fig. 6.1.1(a) and Fig. 6.1.1(b) in a far-field plane is the Fourier transform of the complex aperture function

$$
\langle I(x,t) \rangle \simeq \sum_m \left| \frac{e^{-i[\omega t - kr_0(x) - \varphi_m]}}{r_0(x)} \int dx_0 \, A_m(x_0) \, e^{-ik_x x_0} \right|^2
$$

$$
\propto \sum_m \left| \mathcal{F}_{k_x} \{ A_m(x_0) \} \right|^2. \tag{6.1.8}
$$

We may conclude from Eq.(6.1.8), again, that the observed diffraction pattern is the result of the mth subfield interfering with the mth subfield itself. The measured intensity in the observation plane, $\langle I(x,t) \rangle$, is the sum of a large number of sub-intensities contributed by each subfield.

We have successfully calculated the Fraunhofer diffraction pattern from Einstein's granularity picture of light. The same diffraction pattern can also be obtained in Maxwell's continuum picture. The philosophy of the classical approach is very simple: regardless of the granularity of the field, only consider the final observed field, which is the result of superposition among a large number of subfields, at each point of the diffraction aperture, i.e., at each point along the x_0-axis, before the Fourier transform.

$$
E(x_0) = \sum_m E_m(x_0).
$$

The field in the observation plane is thus

$$
E(x,z,t) = \sum_m E_m(x_0) g_{x_0}(x,z,t) = \sum_m \frac{E_m(x_0)}{r} \, e^{-i[\omega(t-t_0) - kr]}.
$$

The expectation value of the intensity, $\langle I(x,t) \rangle$, measured in the observation plane is therefore

$$
\langle I(x,t) \rangle \simeq \int dx_0 \left[\sum_m E_m^*(x_0) \right] e^{ik_x x_0} \int dx_0' \left[\sum_m E_m(x_0') \right] e^{-ik_x' x_0}
$$

$$
= \int dx_0 \left[\sum_m |E_m(x_0)| \right] e^{ik_x x_0} \int dx_0' \left[\sum_m |E_m(x_0')| \right] e^{-ik_x' x_0}
$$

$$
\propto \left| \mathcal{F}_{k_x} \{ A(x_0) \} \right|^2. \tag{6.1.9}
$$

where

$$
A(x_0) \equiv \sum_m E_m(x_0) \tag{6.1.10}
$$

is defined as the aperture function.

By examining Eq.(6.1.8) and Eq.(6.1.9), we find φ_m does not have any contribution to the diffraction pattern observed from $\langle I(x,t) \rangle$ for incoherent radiation in thermal state, such as natural light. This is because in the measurement of thermal states, what we observe is the result of the mth subfield or photon, interfering with the mth subfield or photon itself. In fact, the random relative phases between subfields have played their important role in the calculation of the ensemble average or intensity expectations: all the $m \neq n$ terms vanish. In certain measurements, such as in the measurement of intensity fluctuation correlation $\langle \Delta I_1 \Delta I_2 \rangle$ of incoherent radiation in thermal state, the $m \neq n$ terms may not vanish, therefore, the relative phases between different subfields have to be taken into account. We will have detailed discussions later.

Nevertheless, in classical optics, especially when dealing with coherent radiation sources, such as a TEM$_{00}$ laser beam in which $\varphi_m = \varphi_0$, we can always treat $A(x_0)$ the complex aperture function of the **k**-th mode of the radiation along the x_0-axis. The diffraction of laser beam can be always treated classically. In this case, we usually treat the diffraction aperture as a coherent radiation source by considering the electromagnetic field distributed along the observation x-axis the superposition of the "secondary wavelets" of each plane-wavefront, according to Huygens's principle. In Fig. 6.1.1(a) and Fig. 6.1.1(b), the classical field $E(x,t)$ in the observation plane is thus

$$E(x,t) = \int_{-\frac{b}{2}}^{\frac{b}{2}} dx_0 \, \frac{A(x_0)}{r} \, e^{-i(\omega t - kr)}. \tag{6.1.11}$$

The constructive-destructive interference of the secondary wavelets results in a diffraction pattern on the observation plane. The diffraction can be easily estimated in the far-field approximation. For both cases of Fig. 5.3.1(a) and Fig. 5.3.1(b), we obtain the Fraunhofer diffraction pattern as the Fourier transform of the aperture function

$$\langle I(x,t) \rangle \simeq \left| \int dx_0 \, A(x_0) \, e^{-ik_x x_0} \right|^2 \propto \left| \mathcal{F}_{k_x}\{A(x_0)\} \right|^2. \tag{6.1.12}$$

where, again, $A(x_0) = |A(x_0)|e^{ik_{0x} x_0}$ is the classical complex aperture function of the **k**-th mode radiation along the 1-D x_0-axis.

It might be necessary to emphasize that although the diffraction patterns observed from the setups of Fig. 6.1.1(a) and Fig. 6.1.1(b) take the same Fourier transform of the aperture function, the diffraction patterns are relatively shifted at the 1-D observation plane along the x-axis, due to their different complex aperture functions. How much is the shift? It depends on the relative phase delay along the x_0-axis introduced by the incident plane-waves at angle $\theta_0 \neq 0$. Therefore, incoherent radiations created from thermal light sources with finite transverse dimension, such as the sunlight, cannot achieve a diffraction of minimum-width. Assuming a disk-like source with randomly distributed point-like sub-sources, the mth subfield created from the mth sub-source produces its own sub-diffraction pattern on the observation plane,

$$I_m(x,t) = I_{0m} \, \mathrm{sinc}^2 \frac{(k_{0x_m} - k_x)b}{2} = I_{0m} \, \mathrm{sinc}^2 \frac{\pi b}{\lambda}(\theta_{0m} - \theta) \tag{6.1.13}$$

where k_{0x_m} is the x_0-component of the wavevector of the mth subfield and θ_{0m} is the incident angle of the mth subfield. The final diffraction pattern observed along the x-axis is the sum of I_m

$$\langle I(x,t) \rangle = \sum_m I_{0m} \, \mathrm{sinc}^2 \frac{\pi b}{\lambda}(\theta_{0m} - \theta), \tag{6.1.14}$$

resulting in a broadened diffraction pattern. The broadening is determined by the angular diameter of the source $\Delta\theta_s$. If we consider each subfield produces a minimum diffraction satisfying $\Delta k_x \Delta x = 2\pi$, the broadened diffraction shall satisfy $\Delta k_x \Delta x > 2\pi$.

From the above analysis we conclude that the Fraunhofer diffraction of a plane-wave is the Fourier transform of the aperture function under the far-field approximation. We may turn this approximation into an exact Fourier transform with the help of an optical converging lens. Fig. 6.1.2 schematically illustrates such a setup. In the setup, an aperture is placed

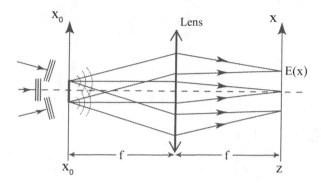

FIGURE 6.1.2 Schematic setup of a "Fourier transformer". A well collimated light passes through an aperture that is placed at the front focal plane of a lens. The diffraction pattern observed at the back focal plane of the lens, which is called the Fourier transform plane, represents the Fourier transform of the aperture function.

at the front focal plane of a lens. A plane-wave of incident angle θ_0 has a corresponding aperture function of $A(\vec{\rho}_0) = |A(\vec{\rho}_0)|e^{i\vec{k}_0 \cdot \vec{\rho}_0}$, where $\vec{\rho}_0$ is the 2-D transverse coordinate on the aperture plane. The plane-wave with incident angle θ_0 produces a Fourier transform of its complex aperture function $A(\vec{\rho}_0)$ on the observation plane that is located on the back focal point of the lens,

$$E(\vec{\rho}) \propto \mathcal{F}_{k_{\vec{\rho}}}\{A(\vec{\rho}_0)\} \tag{6.1.15}$$

where $k_{\vec{\rho}} \simeq k\vec{\rho}/f = (2\pi/\lambda)(\vec{\rho}/f)$ and $\vec{\rho}_0$ are the conjugate variables of the Fourier transform.

6.2   FIELD PROPAGATION

In this section we continue our discussion on the transverse behavior of radiation regarding to its propagation. Precisely, we are interested in determining $E_m(\mathbf{r}, t)$ as well as $E(\mathbf{r}, t) = \sum_m E_m(\mathbf{r}, t)$ on a transverse plane of z = constant from a known distribution of $E_m(\mathbf{r}_0, t_0)$ on a plane of $z_0 = 0$. We assume the field $E_m(\mathbf{r}_0, t_0)$ is generated by the mth sub-source of a radiation source that can be either point-like or spatially extended. The observation plane at z = constant is located at an arbitrary distance from plane $z_0 = 0$ that can be treated as either far-field or near-field. The goal is to find a general solution $E(\mathbf{r}, t)$, and thus $\langle I(\mathbf{r}, t)\rangle$, on the observation plane, based on the knowledge of $E_m(\mathbf{r}_0, t_0)$ according to the Maxwell electromagnetic wave theory. The use of Green's function makes this goal formally achievable.

In fact, we have discussed a special propagation of plane-waves from a known distribution of $A(\rho_0)$ on the plane of $z_0 = 0$ to a far-field observation plane of z = constant in previous section, namely "diffraction". In this section we first generalize the concept of propagation by introducing the Green's function, or propagator, which propagates each Fourier mode of the field from one space-time point to another space-time point, and then obtain the

Green's function or propagator to propagate the field from one plane of $z_0 = 0$ (plane σ_0) to another plane $z = $ constant (plane σ_0) in Fresnel near-field.

Unless $E_m(\mathbf{r}_0, t_0)$ is a non-analytic function in the space-time region of interests, there must exist a Fourier integral representation for the subfield $E_m(\mathbf{r}_0, t_0)$

$$E_m(\mathbf{r}_0, t_0) = \int d\mathbf{k}\, E_m(\mathbf{k})\, v_\mathbf{k}(\mathbf{r}_0, t_0)\, e^{-i\omega t_0}, \tag{6.2.1}$$

where $v_\mathbf{k}(\mathbf{r}_0, t_0)$ is a solution of the Helmholtz wave equation under appropriate boundary condition. The solution of the EM wave equation $v_\mathbf{k}(\mathbf{r}_0, t_0)\, e^{-i\omega t_0}$, namely the Fourier-mode, can be chosen as a set of plane-waves or spherical-waves, for example, depending on the boundary conditions. In Eq. (6.2.1), $E_m(\mathbf{k}) = a_m(\mathbf{k})e^{i\varphi_m(\mathbf{k})}$ is the complex amplitude of the Fourier-mode $\mathbf{k}$ of the mth subfield. In principle, we should be able to find an appropriate propagator or Green's function corresponding to the propagation of each mode under the Fourier integral from space-time point $(\mathbf{r}_0, t_0)$ to space-time point $\mathbf{r}, t$,

$$\begin{aligned} E_m(\mathbf{r}, t) &= \int d\mathbf{k}\, E_m(\mathbf{k})\, v_\mathbf{k}(\mathbf{r}_0, t_0)\, e^{-i\omega t_0}\, g(\mathbf{k}; \mathbf{r} - \mathbf{r}_0, t - t_0) \\ &= \int d\mathbf{k}\, E_m(\mathbf{k}; \mathbf{r}_0, t_0)\, g(\mathbf{k}; \mathbf{r} - \mathbf{r}_0, t - t_0), \end{aligned} \tag{6.2.2}$$

where $E_m(\mathbf{k}; \mathbf{r}_0, t_0) = E_m(\mathbf{k})\, v_\mathbf{k}(\mathbf{r}_0, t_0)\, e^{-i\omega t_0}$. The observed field $E_m(\mathbf{r}, t)$ is the result of the superposition of all possible Fourier modes, which are propagated from space-time point $(\mathbf{r}_0, t_0)$ to space-time point $\mathbf{r}, t$. In classical optics, when a Fourier mode propagates from one point to another point, the propagation gains a phase. In quantum optics, Fourier mode is an eigenfunction of the free-space Hamiltonian of the EM field. The time evolution of an eigenfunction of Hamiltonian gains a simple phase. The Green's function is easy to obtain classically and quantum mechanically.

The total field $E(\mathbf{r}, t)$ is the superposition of $\sum_m E_m(\mathbf{r}, t)$:

$$\begin{aligned} E(\mathbf{r}, t) &= \sum_m E_m(\mathbf{r}, t) \\ &= \sum_m \int d\mathbf{k}\, E_m(\mathbf{k})\, g(\mathbf{k}; \mathbf{r} - \mathbf{r}_0, t - t_0)\, v_\mathbf{k}(\mathbf{r}_0, t_0)\, e^{-i\omega t_0} \\ &= \sum_m \int d\mathbf{k}\, E_m(\mathbf{k}; \mathbf{r}_0, t_0)\, g(\mathbf{k}; \mathbf{r} - \mathbf{r}_0, t - t_0) \end{aligned} \tag{6.2.3}$$

where $E_m(\mathbf{k}, \mathbf{r}_0, t_0) = E_m(\mathbf{k})v_\mathbf{k}(\mathbf{r}_0, t_0)e^{-i\omega t_0}$. To simplify the mathematics and notation, we may drop the index m when dealing with the propagation of each Fourier-mode $\mathbf{k}$ from one plane to another plane. Physically, this is equivalent to a superposition, either coherent or incoherent, of all the subfields in the plane of z_0 and treated the individual propagation of the subfields classically

$$\begin{aligned} E(\mathbf{r}, t) &= \sum_m E_m(\mathbf{r}, t) \\ &= \int d\mathbf{k}\, \Big[\sum_m E_m(\mathbf{k}) \Big]\, g(\mathbf{k}; \mathbf{r} - \mathbf{r}_0, t - t_0)\, v_\mathbf{k}(\mathbf{r}_0, t_0)\, e^{-i\omega t_0} \\ &= \int d\mathbf{k}\, E(\mathbf{k}; \mathbf{r}_0, t_0)\, g(\mathbf{k}; \mathbf{r} - \mathbf{r}_0, t - t_0) \end{aligned} \tag{6.2.4}$$

where $E(\mathbf{k}; \mathbf{r}_0, t_0) = \sum_m E_m(\mathbf{k})v_\mathbf{k}(\mathbf{r}_0, t_0)e^{-i\omega t_0}$, which may involve a large number of subfields propagated from different space-time coordinates. In this case, the complicated spatial and temporal distribution of the subfields may complicate the superposition. When dealing with a propagation from a plane, such as σ_0 to another plane, such as σ, we define $E(\mathbf{k}; \mathbf{r}_0, t_0) \equiv A(\mathbf{k}; \vec{\rho}_0, z_0, t_0)$ aperture function of the σ_0 plane. Simplifying the mathematics, we usually set $z_0 = 0$ and $t_0 = 0$ and write $A(\mathbf{k}; \vec{\rho}_0, z_0, t_0)$ as $A(\mathbf{k}; \vec{\rho}_0)$ by ignoring the zeros. Although we are able to treat the field propagation classically, we cannot ignore the fact that the amplitude of the mode $\mathbf{k}$ may contain a large number of subfields and the subfields may have different phases at z_0 plane. If the mth subfield has a random phase, In certain cases, the random phases may not have any contribution and may not be problematic. In some cases, however, the random relative phases between subfields must be taken into account, for example, in the measurement of intensity fluctuation correlation $\langle \Delta I_1 \Delta I_2 \rangle$ of incoherent radiation in thermal state. We will have detailed discussions later.

For certain experimental setups, the field propagation from one space-time point $(\mathbf{r}_0, t_0)$ to another space-time point $(\mathbf{r}, t)$ may have to be broken into a few steps, $g = g_1 \times g_2 ... \times g_N$, in these cases

$$E(\mathbf{r}, t) = \int d\mathbf{k}\, E(\mathbf{k}; \mathbf{r}_0, t_0)\, g_1(\mathbf{k}; \mathbf{r}_1 - \mathbf{r}_0, t_1 - t_0) \qquad (6.2.5)$$
$$\times\, g_2(\mathbf{k}; \mathbf{r}_2 - \mathbf{r}_1, t_2 - t_1) ... \times g_N(\mathbf{k}; \mathbf{r} - \mathbf{r}_{N-1}, t - t_{N-1}),$$

where N represents the number of steps. The final $g(\mathbf{k}; \mathbf{r} - \mathbf{r}_0, t - t_0)$ can be quite different for different setups. No matter how complicated, the Green's function plays the same role in the propagation of each Fourier-mode from space-time point $(\mathbf{r}_0, t_0)$ to $(\mathbf{r}, t)$.

Fig. 6.2.1 illustrates an experimental setup in which the field travels freely from a finite size aperture A on the plane σ_0 to the Fresnel near-field observation plane σ. Based on Fig. 6.2.1, we evaluate $g_{\vec{\rho}_0, z_0, t_0}(\vec{\rho}, z, t)$, namely the Green's function for free-space Fresnel propagation.

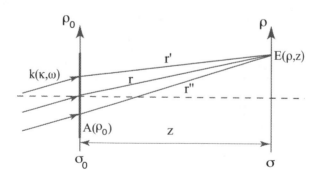

FIGURE 6.2.1   Schematic of free-space Fresnel propagation. The complex amplitude $A(\omega; \vec{\rho}_0, z_0) \equiv \sum_m E_m(\omega; \vec{\rho}_0, z_0, t_0)$, which is in the σ_0 plane, is the resulting field of a superposition among all possible input subfields at space-time coordinate $\vec{\rho}_0, z_0, t_0$, which may come from different spatial coordinates, either transversely distributed or longitudinally distributes, and may come from different early times. The field at a space-time point $(\vec{\rho}, z, t)$, which is in the σ plane, is the result of a superposition among all possible subfields coming from plane σ_0.

The field at a space-time point $(\vec{\rho}, z, t)$, which is in the observation plane σ, is the result of a superposition among all possible subfields coming from each point $\vec{\rho}_0, z_0, t_0$ in the plane

σ_0, see Fig. 6.2.1,

$$
\begin{aligned}
E(\vec{\rho}, z, t) &= \int d\mathbf{k} \int_{\sigma_0} d\vec{\rho}_0 \, \frac{A(\mathbf{k}; \vec{\rho}_0, z_0, t_0)}{|\mathbf{r} - \mathbf{r}_0|} \, e^{-i[\omega(t - t_0) - \mathbf{k} \cdot (\mathbf{r} - \mathbf{r}_0)]} \\
&= \int d\omega \int_{\sigma_0} d\vec{\rho}_0 \, \frac{A(\omega; \vec{\rho}_0, z_0, t_0)}{r} \, e^{-i[\omega(t - t_0) - kr]}
\end{aligned}
\tag{6.2.6}
$$

where $A(\omega; \vec{\rho}_0, z_0, t_0)$ is the complex aperture function of the mode ω at each transverse coordinate $\vec{\rho}_0$ of the σ_0 plane, which is also considered the known field distribution on the σ_0 plane; $\mathbf{r}_0$ and $\mathbf{r}$ are the coordinates in the σ_0 plane and σ plane, respectively, $r = |\mathbf{r} - \mathbf{r}_0|$ is the optical path propagates field $A(\omega; \vec{\rho}_0, z_0)$ from coordinate $(\vec{\rho}_0, z_0, t_0)$ to coordinate $(\vec{\rho}, z, t)$. To simplify the mathematics, we set $z_0 = 0$ and $t_0 = 0$, and thus $r = \sqrt{z^2 + |\vec{\rho} - \vec{\rho}_0|^2}$. In Eq. (6.2.6),

$$
g_{\vec{\rho}_0}(\omega; \vec{\rho}, z) = \frac{e^{-i(\omega t - kr)}}{r}
\tag{6.2.7}
$$

is the Green's function that propagates the mode ω from point $(\vec{\rho}_0, z_0, t_0)$ in the σ_0 plane to point $(\vec{\rho}, z, t)$ in the σ plane. It should be emphasized that in certain cases in which the propagation is not from point-to-point, for instance from plane-to-point or from plane-to-plane, the transverse component of the wavevector, $\vec{\kappa}$, may need to be specified. In this case, the Green's function is in the form of $g_{\sigma_0}(\vec{\kappa}, \omega; \vec{\rho}, z)$.

In the Fresnel near-field paraxial approximation, when $|\vec{\rho} - \vec{\rho}_0|^2 \ll z^2$ we take the first-order expansion of r in terms of z and $\vec{\rho}$

$$
r = \sqrt{z^2 + |\vec{\rho} - \vec{\rho}_0|^2} \simeq z \left(1 + \frac{|\vec{\rho} - \vec{\rho}_0|^2}{2z^2} \right).
$$

Thus $E(\vec{\rho}, z, t)$ can be approximated as

$$
E(\vec{\rho}, z, t) \simeq \int d\omega \int_{\sigma_0} d\vec{\rho}_0 \, \frac{A(\omega; \vec{\rho}_0)}{z} \, e^{i\frac{\omega}{c}z} \, e^{i\frac{\omega}{2cz}|\vec{\rho} - \vec{\rho}_0|^2} e^{-i\omega t},
\tag{6.2.8}
$$

where $e^{i\frac{\omega}{2cz}|\vec{\rho} - \vec{\rho}_0|^2}$ is known as the Fresnel phase factor. In Eq. (6.2.8), we have shortened $A(\omega; \vec{\rho}_0, z_0, t_0)$ to $A(\omega; \vec{\rho}_0)$ by ignoring the zeros.

We can then write $E(\vec{\rho}, z, t)$ in the following form

$$
\begin{aligned}
&E(\vec{\rho}, z, t) \\
&= \int d\omega \, e^{-i\omega t} \, \frac{e^{i\frac{\omega}{c}z}}{z} \int_{\sigma_0} d\vec{\rho}_0 \, A(\omega; \vec{\rho}_0) \, e^{i\frac{\omega}{2cz}|\vec{\rho} - \vec{\rho}_0|^2} \\
&= \int d\omega \, e^{-i\omega t} \, \frac{e^{i\frac{\omega}{c}z}}{z} \int_{\sigma_0} d\vec{\rho}_0 \, A(\omega; \vec{\rho}_0) \, G(|\vec{\rho} - \vec{\rho}_0|, \frac{\omega}{cz}).
\end{aligned}
\tag{6.2.9}
$$

where $G(|\vec{\alpha}|, \beta) = e^{i(\beta/2)|\alpha|^2}$, namely the Fresnel phase factor. It is straightforward to find that the Gaussian function $G(|\vec{\alpha}|, \beta)$ has the following properties:

$$
\begin{aligned}
G^*(|\vec{\alpha}|, \beta) &= G(|\vec{\alpha}|, -\beta), \\
G(|\vec{\alpha}|, \beta_1 + \beta_2) &= G(|\vec{\alpha}|, \beta_1) \, G(|\vec{\alpha}|, \beta_2), \\
G(|\vec{\alpha}_1 + \vec{\alpha}_2|, \beta) &= G(|\vec{\alpha}_1|, \beta) \, G(|\vec{\alpha}_2|, \beta) \, e^{i\beta \vec{\alpha}_1 \cdot \vec{\alpha}_2}, \\
\int d\vec{\alpha} \, G(|\vec{\alpha}|, \beta) \, e^{i\vec{\gamma} \cdot \vec{\alpha}} &= i\frac{2\pi}{\beta} \, G(|\vec{\gamma}|, -\frac{1}{\beta}).
\end{aligned}
\tag{6.2.10}
$$

Notice that the last equation in Eq. (6.2.10) is the Fourier transform of the $G(|\vec{\alpha}|, \beta)$ function. As we shall see in the following, these properties are very useful in simplifying the calculations of the Green's functions $g_{\vec{\rho}_0}(\omega; \vec{\rho}, z)$.

The Green's function that propagates a Fourier mode from point $\vec{\rho}_0$ in the σ_0 plane of $z_0 = 0$ to point $\vec{\rho}_0$ in the Fresnel near-field σ plane of $z = $ constant is thus approximated to be

$$g_{\vec{\rho}_0}(\omega; \vec{\rho}, z) = \frac{e^{i\frac{\omega}{c}z}}{z} G(|\vec{\rho} - \vec{\rho}_0|, \frac{\omega}{cz}). \qquad (6.2.11)$$

Now, we imagine to inserting a plane σ', which has an transverse dimension of infinity, between σ_0 and σ. This is equivalent having two consecutive Fresnel propagations over a distance of d_1 and d_2. Thus, the calculation of these consecutive Fresnel near-field propagations should yield the same Green's function as that of the above direct Fresnel near-field propagation shown in Eq. (6.2.11):

$$E(\omega, \vec{\kappa}; \vec{\rho}, z)$$
$$= C^2 \frac{e^{i\frac{\omega}{c}(d_1+d_2)}}{d_1 d_2} \int_{\sigma'} d\vec{\rho'} \int_{\sigma_0} d\vec{\rho}_0 \, A(\vec{\rho}_0) G(|\vec{\rho'} - \vec{\rho}_0|, \frac{\omega}{cd_1}) G(|\vec{\rho} - \vec{\rho'}|, \frac{\omega}{cd_2})$$
$$= C \frac{e^{i\frac{\omega}{c}z}}{z} \int_{\sigma_0} d\vec{\rho}_0 \, A(\vec{\rho}_0) \, G(|\vec{\rho} - \vec{\rho}_0|, \frac{\omega}{cz}) \qquad (6.2.12)$$

where C is a necessary normalization constant. The double integral of $d\vec{\rho}_0$ and $d\vec{\rho'}$ in Eq. (6.2.12) can be evaluated as

$$\int_{\sigma'} d\vec{\rho'} \int_{\sigma_0} d\vec{\rho}_0 \, A(\vec{\rho}_0) \, G(|\vec{\rho'} - \vec{\rho}_0|, \frac{\omega}{cd_1}) \, G(|\vec{\rho} - \vec{\rho'}|, \frac{\omega}{cd_2})$$
$$= \int_{\sigma_0} d\vec{\rho}_0 \, A(\vec{\rho}_0) \, G(\vec{\rho}_0, \frac{\omega}{cd_1}) \, G(\vec{\rho}, \frac{\omega}{cd_2})$$
$$\times \int_{\sigma'} d\vec{\rho'} \, G(\vec{\rho'}, \frac{\omega}{c}(\frac{1}{d_1} + \frac{1}{d_2})) \, e^{-i\frac{\omega}{c}(\frac{\vec{\rho}_0}{d_1} + \frac{\vec{\rho}}{d_2}) \cdot \vec{\rho'}}$$
$$= \frac{i2\pi c}{\omega} \frac{d_1 d_2}{d_1 + d_2} \int_{\sigma_0} d\vec{\rho}_0 \, A(\vec{\rho}_0) \, G(\vec{\rho}_0, \frac{\omega}{cd_1}) \, G(\vec{\rho}, \frac{\omega}{cd_2})$$
$$\times \, G(|\frac{\vec{\rho}_0}{d_1} + \frac{\vec{\rho}}{d_2}|, \frac{\omega}{c}(\frac{d_1 d_2}{d_1 + d_2}))$$
$$= \frac{i2\pi c}{\omega} \frac{d_1 d_2}{d_1 + d_2} \int_{\sigma_0} d\vec{\rho}_0 \, A(\vec{\rho}_0) \, G(|\vec{\rho} - \vec{\rho}_0|, \frac{\omega}{c(d_1 + d_2)})$$

where we have applied Eq. (6.2.10), and the integral of $d\vec{\rho'}$ has been taken to infinity. Substituting this result into Eq. (6.2.12), we thus have

$$C^2 \frac{i2\pi c}{\omega} \frac{e^{i\frac{\omega}{c}(d_1+d_2)}}{d_1 + d_2} \int_{\sigma_0} d\vec{\rho}_0 \, A(\vec{\rho}_0) \, G(|\vec{\rho} - \vec{\rho}_0|, \frac{\omega}{c(d_1 + d_2)})$$
$$= C \frac{e^{i\frac{\omega}{c}z}}{z} \int_{\sigma_0} d\vec{\rho}_0 \, A(\vec{\rho}_0) \, G(|\vec{\rho} - \vec{\rho}_0|, \frac{\omega}{cz}).$$

Therefore, the constant C must take the value of $C = -i\omega/2\pi c$. The Green's function that propagates a Fourier mode from point $\vec{\rho}_0$ in the σ_0 plane to point $\vec{\rho}$ in the σ plane of Fresnel near-field, namely the Fresnel propagator, is thus approximated to be

$$g_{\vec{\rho}_0}(\omega; \vec{\rho}, z) = \frac{-i\omega}{2\pi c} \frac{e^{i\frac{\omega}{c}z}}{z} G(|\vec{\rho} - \vec{\rho}_0|, \frac{\omega}{cz}). \qquad (6.2.13)$$

6.3 OPTICAL IMAGING

The concept of classic optical imaging was well developed in optics prior to the electromagnetic wave theory of light. The early theories of geometric optics provided quite a few phenomenological theories to explain the point-to-point relationship between an object plane and its image plane. In these theories radiation is treated as "ray of light"[1] and the image is explained as the result of the peculiar way of their propagation. A later theory of classical imaging, namely the theory of physical optics, is based on the concept of waves. Light is treated as waves that propagate to and interfere at a space-time point. The image is considered the result of constructive-destructive interference among these waves.

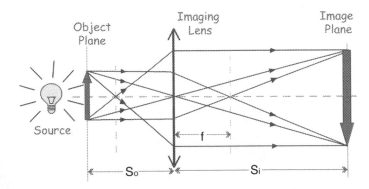

FIGURE 6.3.1 Imaging: a lens produces an *image* of an object in its image plane which is defined by the Gaussian thin-lens equation $1/s_i + 1/s_o = 1/f$. The object is illuminated by an incoherent light source. The concept of optical image is based on the existence of a point-to-point relationship between the radiation field in the object plane and the radiation field in the image plane: any radiation starting from a point on the object plane must collapse or stop at a unique point on the image plane.

Fig. 6.3.1 schematically illustrates a classic imaging setup. An object that is illuminated by an incoherent light source. Spatially incoherent radiations are reflected or scattered from the object plane. An imaging lens is used to image the randomly radiated or scattered radiations from the object onto an image plane which is defined by the "Gaussian thin lens equation"

$$\frac{1}{s_i} + \frac{1}{s_o} = \frac{1}{f}, \tag{6.3.1}$$

where s_o is the distance between the object and the imaging lens, s_i the distance between the imaging lens and the image plane, and f the focal length of the imaging lens. The Gaussian thin-lens equation defines an image plane which has a point-to-point relationship with the object plane: any radiation starting from a point on the object plane must collapse or stop at a unique point on the image plane. We will prove in this section that this point-to-point relationship is the result of *constructive-destructive interference*. Any radiation starting from a point on the object plane, either emitted from or scattered from, may propagate to any directions and take different optical paths to reach the image plane. However, only these propagated with equal optical paths are able to superpose constructively at one unique point on the image plane, while these that experience unequal path propagations will superpose destructively at all other points on the image plane. The use of the imaging lens makes this constructive-destructive interference possible.

---

[1]This "ray of light" is not Einstein's "bundle of rays".

A perfect point-to-point correspondence between the object plane and the image plane produces a perfect image. Illuminated by a radiation, the image produced by an imaging system is a reproduction, either magnified or demagnified, of the illuminated object, mathematically corresponding to a convolution between the intensity distribution function of the object plane, $|A(\vec{\rho}_o)|^2$, and a δ-function like image-forming function which characterizes the point-to-point relationship between the object plane and the image plane:

$$
\begin{aligned}
\langle I(\vec{\rho}_i)\rangle &= \langle \Big| \int_{obj} d\vec{\rho}_o \, A(\vec{\rho}_o) \, \delta(\vec{\rho}_o + \frac{\vec{\rho}_i}{\mu}) \Big|^2 \rangle \\
&= \int_{obj} d\vec{\rho}_o \, |A(\vec{\rho}_o)|^2 \, \delta(\vec{\rho}_o + \frac{\vec{\rho}_i}{\mu}) \\
&= |A(\vec{\rho}_o)|^2 \otimes \delta(\vec{\rho}_o + \frac{\vec{\rho}_i}{\mu}) \\
&\simeq |A(\vec{\rho}_i/\mu)|^2
\end{aligned}
\tag{6.3.2}
$$

where $\langle I(\vec{\rho}_i)\rangle = |A(\vec{\rho}_i/\mu)|^2$ is the expected intensity distribution function, i.e., the reproduction of the aperture function in the image plane, $A(\vec{\rho}_o)$ is the complex aperture function of the object plane, $\vec{\rho}_o$ and $\vec{\rho}_i$ are 2-D vectors of the transverse coordinates in the object plane and the image plane, respectively, and $\mu = s_i/s_o$ is the image magnification factor (μ may take a positive value or a negative value), the mathematical symbol $\otimes$ represents convolution. In Eq. (6.3.2), we have assumed an incoherent illumination, namely an incoherent imaging process in which $A(\vec{\rho}_o)$ contains a random phase.

In reality, limited by the finite size of the imaging system, we may never obtain a perfect point-to-point correspondence. The incomplete constructive-destructive interference turns the point-to-point correspondence into a point-to-"spot" relationship. The δ-function in the convolution of Eq. (6.3.2) has to be replaced by a point-spread function:

$$
\begin{aligned}
\langle I(\vec{\rho}_i)\rangle &= \langle \Big| \int_{obj} d\vec{\rho}_o \, A(\vec{\rho}_o) \, \mathrm{somb}\big[\frac{R}{s_o}\frac{\omega}{c}|\vec{\rho}_o + \frac{\vec{\rho}_i}{\mu}|\big] \Big|^2 \rangle \\
&= \langle \Big| A(\vec{\rho}_o) \otimes \mathrm{somb}\big[\frac{R}{s_o}\frac{\omega}{c}|\vec{\rho}_o + \frac{\vec{\rho}_i}{\mu}|\big] \Big|^2 \rangle \\
&= |A(\vec{\rho}_o)|^2 \otimes \mathrm{somb}^2\big[\frac{R}{s_o}\frac{\omega}{c}|\vec{\rho}_o + \frac{\vec{\rho}_i}{\mu}|\big]
\end{aligned}
\tag{6.3.3}
$$

where the sombrero-like function, or the Airy disk, is defined as

$$
\mathrm{somb}(x) = \frac{2J_1(x)}{x},
$$

and $J_1(x)$ is the first-order Bessel function, and R the radius of the imaging lens, and R/s_o is known as the numerical aperture of the imaging system. The sombrero-like point-spread function, or the Airy disk, defines the observed spot size on the image plane that is produced by the radiation field coming from point $\vec{\rho}_o$. It is clear from Eq. (6.3.3) that a larger imaging lens and shorter wavelength will result in a narrower point-spread function, and thus a higher spatial resolution of the image. The finite size of the observed spot on the image plane determines the spatial resolution of the imaging system.

It should be emphasized that we must not confuse a trivial "projection shadow" with an image. Similar to an x-ray photograph, projection makes a shadow of an object, instead of an image of the object. Fig. 6.3.2 distinguishes a projection shadow from an image. The object-shadow correspondence is essentially defined by the propagation direction of the light rays, and there is no unique imaging plane. The shadow can be found in any plane behind

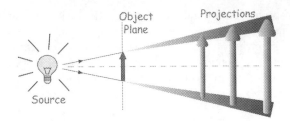

FIGURE 6.3.2 Projection: a light source illuminates an object and no image forming system is present, no image plane is defined, and only projections, or shadows, of the object can be observed.

the object. A projection shadow is the result of the simple "blocked-unblocked" effect of light, which is very different from an imaged image, both from a fundamental and from a practical viewpoint. There is no spatial resolution defined in terms of the Rayleigh's criterion for a projection shadow.

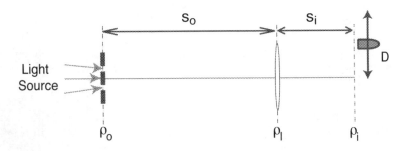

FIGURE 6.3.3 A typical imaging system. A lens of finite size is used to produce a magnified or demagnified image of an object with limited spatial resolution. To simplify the notation, we set $z_o = 0$, $t_o = 0$.

Figure 6.3.3 illustrates a simplified schematic experimental setup of an imaging system. Based on this simplified setup, we calculate and analyze the imaging process. The following calculation is divided into two parts: (1) we calculate the point-to-point "propagator" from the object plane to the image plane: the radiation started (emitted, scattered, reflected, or transimited) from a point in the object plane must collapse or stop at an unique point in the image plane; (2) we calculate the classic image as a convolution between the aperture function and the point-spread image-forming function. To simplify the mathematics, we assume the object is illuminated by a spatially incoherent radiation of single frequency. By assuming $\omega =$ constant, the following analysis and calculation will focus on the spatial behavior of the field and ignore the temporal coherence discussion, unless certain circumstances are specified.

(I) The point-to-point "propagator" from the object plane to the image plane.

Selecting a radiation field $E(\vec{\rho}_o, z_o = 0, t_o = 0)$ from an arbitrary point of the object plane, propagating $E(\vec{\rho}_o, 0, 0)$ to the image plane by means of an appropriate propagator $g_{\vec{\rho}_o}(\vec{\rho}_i, z_i, t_i)$,

$$
\begin{aligned}
E(\vec{\rho}_i, s_o + s_i, t_i) \\
= E(\vec{\rho}_o)\, g_{\vec{\rho}_o}(\vec{\rho}_i, z_i, t_i) \\
= \int_{lens} d\vec{\rho}_l\, E(\vec{\rho}_o)\, \left[g_{\vec{\rho}_o}(\vec{\rho}_l, z_l, t_l) \right] \left[g_{lens} \right] \left[g_{\vec{\rho}_l}(\vec{\rho}_i, z_i, t_i) \right]
\end{aligned}
\tag{6.3.4}
$$

where we have taken $(z_o = 0, t_o = 0)$ and ignored the zeros from $E(\vec{\rho}_o, z_o, t_o)$; $g_{\vec{\rho}_o}(\vec{\rho}_l, z_l, t_l)$ and $g_{\vec{\rho}_l}(\vec{\rho}_i, z_i, t_i)$ are the Green's functions that propagate the field $E(\vec{\rho}_o)$ from the object plane to the imaging lens plane, and from the imaging lens plane to the image plane, respectively; g_{Lens} is the Gaussian function introduced by the imaging lens. Due to the single frequency, $\omega = $ constant, assumption, we have dropped ω from the Green's functions, or propagators.

Assuming Fresnel near-field propagations from the object plane to the imaging lens plane, and from the imaging lens plane to the image plan, the following propagator $g_{\vec{\rho}_o}(\vec{\rho}_i, z_i, t_i)$ shall propagate field $E(\vec{\rho}_o)$ from the object plane to the image plane

$$
g_{\vec{\rho}_o}(\vec{\rho}_i, z_i, t_i)
$$

$$
= \int_{lens} d\vec{\rho}_l \left\{ \frac{-i\omega}{2\pi c} \frac{e^{i\frac{\omega}{c}s_o}}{s_o} G(|\vec{\rho}_l - \vec{\rho}_o|, \frac{\omega}{cs_o}) \right\}
$$

$$
\times \left\{ G(|\vec{\rho}_l|, -\frac{\omega}{cf}) \right\} \left\{ \frac{-i\omega}{2\pi c} \frac{e^{i\frac{\omega}{c}s_i}}{s_i} G(|\vec{\rho}_i - \vec{\rho}_l|, \frac{\omega}{cs_i}) \right\} \tag{6.3.5}
$$

where $\vec{\rho}_o$, $\vec{\rho}_l$, and $\vec{\rho}_i$ are two-dimensional vectors defined, respectively, in the object plane, imaging lens plane, and image plane. The first curly bracket is the free-space Fresnel propagator that propagates the field from the source/object plane to the imaging lens plane; the second curly bracket is the Gaussian function introduced by the imaging lens; the third curly brackets is the free-space Fresnel propagators that propagates the field from the imaging lens plane to the observation plane, i.e., the image plane. The Fresnel propagator includes a spherical wave function $e^{i\frac{\omega}{c}(z_j - z_k)}/(z_j - z_k)$ and a Fresnel phase factor $G(|\vec{\alpha}|, \beta) = e^{i(\beta/2)|\vec{\alpha}|^2} = e^{i\omega|\vec{\rho}_j - \vec{\rho}_k|^2/2c(z_j - z_k)}$. The third curly bracket adds a phase factor, $G(|\vec{\rho}_l|, -\frac{\omega}{cf}) = e^{-i\frac{\omega}{2cf}|\vec{\rho}_l|^2}$, which is introduced by the imaging lens.

Applying the properties of the Gaussian function, Eq. (6.3.5) can be simplified into the following form

$$
g_{\vec{\rho}_o}(\vec{\rho}_i, z_i, t_i)
$$

$$
= \frac{-\omega^2}{(2\pi c)^2 s_o s_i} e^{i\frac{\omega}{c}(s_o + s_i)} G(|\vec{\rho}_i|, \frac{\omega}{cs_i}) G(|\vec{\rho}_o|, \frac{\omega}{cs_o})
$$

$$
\times \int_{lens} d\vec{\rho}_l \, G(|\vec{\rho}_l|, \frac{\omega}{c}[\frac{1}{s_o} + \frac{1}{s_i} - \frac{1}{f}]) \, e^{-i\frac{\omega}{c}(\frac{\vec{\rho}_o}{s_o} + \frac{\vec{\rho}_i}{s_i}) \cdot \vec{\rho}_l}. \tag{6.3.6}
$$

The image plane is defined by the Gaussian thin-lens equation of Eq. (6.3.1), hence, the integral in Eq. (6.3.6) simplifies and gives, for a finite sized lens of radius R, the so called point-spread function of the imaging system:

$$
\text{somb}(x) = \frac{2J_1(x)}{x} \quad \text{with} \quad x = [\frac{R}{s_o} \frac{\omega}{c} |\vec{\rho}_o + \frac{\rho_i}{\mu}|]. \tag{6.3.7}
$$

Where, again, $J_1(x)$ is the first-order Bessel function and $\mu = s_i/s_o$ is the magnification of the imaging system. Replacing the somb-function with the integral, we have

$$
g_{\vec{\rho}_o}(\vec{\rho}_i, z_i, t_i)
$$

$$
= \frac{-\omega^2}{(2\pi c)^2 s_o s_i} e^{i\frac{\omega}{c}(s_o + s_i)} G(|\vec{\rho}_i|, \frac{\omega}{cs_i}) G(|\vec{\rho}_o|, \frac{\omega}{cs_o})
$$

$$
\times \text{somb}[\frac{R}{s_o} \frac{\omega}{c} |\vec{\rho}_o + \frac{\vec{\rho}_i}{\mu}|]. \tag{6.3.8}
$$

For a large sized imaging lens, the somb-function can be approximated as $\delta(|\vec{\rho}_o + \vec{\rho}_i/\mu|)$,

$$g_{\vec{\rho}_o}(\vec{\rho}_i, z_i, t_i) = \frac{-\omega^2 e^{i\varphi_{s_i}}}{(2\pi c)^2 s_o s_i} \delta(|\vec{\rho}_o + \frac{\vec{\rho}_i}{\mu}|) \tag{6.3.9}$$

indicating a point-to-point propagator: the field $E(\vec{\rho}_o, z_o, t_o)$ that is emitted from a point $(\vec{\rho}_o, z_o)$ in the object plane must collapse or stop at an unique point in the image plane $(\vec{\rho}_i = \mu\vec{\rho}_o, z_i = z_o + s_o + s_i)$. In Eq. (6.3.9), we have absorbed all phases into φ_{s_i}.

After the above point-to-spot propagation of Eq. (6.3.8) or the point-to-point propagation of Eq. (6.3.9), the intensity distribution of the radiation field $E(\vec{\rho}_i, s_o + s_i, t_i) = E(\vec{\rho}_o) g_{\vec{\rho}_o}(\vec{\rho}_i, z_i, t_i)$ in the image plane becomes

$$\langle |E(\vec{\rho}_o) g_{\vec{\rho}_o}(\vec{\rho}_i, z_i, t_i)|^2 \rangle \propto \mathrm{somb}^2[\frac{R}{s_o} \frac{\omega}{c} |\vec{\rho}_o + \frac{\vec{\rho}_i}{\mu}|] \tag{6.3.10}$$

for an imaging lens with finite diameter D, and

$$\langle |E(\vec{\rho}_o) g_{\vec{\rho}_o}(\vec{\rho}_i, z_i, t_i)|^2 \rangle \propto \delta(|\vec{\rho}_o + \frac{\vec{\rho}_i}{\mu}|) \tag{6.3.11}$$

for a large sized imaging lens as an approximation. It is interesting to find that although the field $E(\vec{\rho}_o)$ may propagate to all possible directions and may take all possible optical paths with different optical delays to arrive at the image plane, the point-to-spot propagator forces it to collapse or stop at an unique spot around point $\vec{\rho}_i$ of the image plane, and the point-to-point propagator forces it to collapse or stop at an unique point $\vec{\rho}_i$ of the image plane. This point-to-spot propagator and the point-to-point propagator are the results of a superposition that sums all subfields emitted from $\vec{\rho}_o$ and passed through each and all points, $\vec{\rho}_l$, of the imaging lens along all possible optical paths with different optical delays. These subfields interfere constructively at an unique spot (approximated as a point) of the image plane while their optical path lengths are equal, and interfere destructively at all other points of the image plane while their optical path lengths are unequal. The imaging lens makes this constructive-destructive interference possible.

(II) The classic image: a convolution between the aperture function and the point-spread image-forming function.

Examine the simplified experimental setup of Fig. 6.3.3, again, the measured intensity $\langle I(\vec{\rho}_i, z_i, t_i) \rangle$ in the image plane is calculated as follows

$$\langle I(\vec{\rho}_i, z_i, t_i) \rangle = \langle |E(\vec{\rho}_i, s_o + s_i)|^2 \rangle$$

$$= \langle |\int_{obj} d\vec{\rho}_o \, E(\vec{\rho}_o) \, g_{\vec{\rho}_o}(\vec{\rho}_i, z_i, t_i)|^2 \rangle$$

$$= \langle |\int_{obj} d\vec{\rho}_o \, A(\vec{\rho}_o) \, g_{\vec{\rho}_o}(\vec{\rho}_l, z_l, t_l) \, g_{lens} \int_{lens} d\vec{\rho}_l \, g_{\vec{\rho}_l}(\vec{\rho}_i, z_i, t_i)|^2 \rangle$$

$$= \langle |\int_{obj} d\vec{\rho}_o \, A(\vec{\rho}_o) \int_{lens} d\vec{\rho}_l \, g_{\vec{\rho}_o}(\vec{\rho}_l, z_l, t_l) \, g_{lens} \, g_{\vec{\rho}_l}(\vec{\rho}_i, z_i, t_i)|^2 \rangle. \tag{6.3.12}$$

Assuming Fresnel near-field propagation, the following Gaussian propagators shall be able to propagate each and all fields $E(\vec{\rho}_o)$ from the object plane to the image plane, and superpose at each point, $(\vec{\rho}_i, s_o + s_i)$, of the image plane,

$$E(\vec{\rho}_i, s_o + s_i)$$
$$= \int_{obj} d\vec{\rho}_o \int_{lens} d\vec{\rho}_l \left\{ A(\vec{\rho}_o) \right\} \left\{ \frac{-i\omega}{2\pi c} \frac{e^{i\frac{\omega}{c}s_o}}{s_o} G(|\vec{\rho}_l - \vec{\rho}_o|, \frac{\omega}{cs_o}) \right\}$$
$$\times \left\{ G(|\vec{\rho}_l|, -\frac{\omega}{cf}) \right\} \left\{ \frac{-i\omega}{2\pi c} \frac{e^{i\frac{\omega}{c}s_i}}{s_i} G(|\vec{\rho}_i - \vec{\rho}_l|, \frac{\omega}{cs_i}) \right\} \tag{6.3.13}$$

which is the same as Eq. (6.3.5), except the integral of $\vec{\rho}_o$.

Applying the properties of the Gaussian function, Eq. (6.3.13) can be simplified into the following form

$$E(\vec{\rho}_i, z = s_o + s_i)$$
$$= \frac{-\omega^2}{(2\pi c)^2 s_o s_i} e^{i\frac{\omega}{c}(s_o + s_i)} G(|\vec{\rho}_i|, \frac{\omega}{cs_i}) \int_{obj} d\vec{\rho}_o A(\vec{\rho}_o) G(|\vec{\rho}_o|, \frac{\omega}{cs_o})$$
$$\times \int_{lens} d\vec{\rho}_l G(|\vec{\rho}_l|, \frac{\omega}{c}[\frac{1}{s_o} + \frac{1}{s_i} - \frac{1}{f}]) e^{-i\frac{\omega}{c}(\frac{\vec{\rho}_o}{s_o} + \frac{\vec{\rho}_i}{s_i}) \cdot \vec{\rho}_l}. \tag{6.3.14}$$

The image plane is defined by the Gaussian thin-lens equation of Eq. (6.3.1), hence, the second integral simplifies and gives, for a finite sized lens of radius R, the so called point-spread function of the imaging system:

$$E(\vec{\rho}_i, z = s_o + s_i)$$
$$= \frac{-\omega^2}{(2\pi c)^2 s_o s_i} e^{i\frac{\omega}{c}(s_o + s_i)} G(|\vec{\rho}_i|, \frac{\omega}{cs_i}) \int_{obj} d\vec{\rho}_o \left[A(\vec{\rho}_o) G(|\vec{\rho}_o|, \frac{\omega}{cs_o}) \right]$$
$$\times somb[\frac{R}{s_o} \frac{\omega}{c} |\vec{\rho}_o + \frac{\vec{\rho}_i}{\mu}|] \tag{6.3.15}$$

where, again, $somb(x) = 2J_1(x)/x$, and $J_1(x)$ is the first-order Bessel function. The somb-function is the result of an integral that sums all subfields emitted from $\vec{\rho}_o$ and passed through each and all points, $\vec{\rho}_l$, of the lens along all possible optical paths with different optical delays. These subfields interfere constructively when their optical path lengths are equal and interfere destructively when their optical path lengths are unequal. The image observed from the measurement of $\langle I(\vec{\rho}_i, z_i) \rangle$ is thus

$$\langle I(\vec{\rho}_i) \rangle \propto \left\langle \left| \int_{obj} d\vec{\rho}_o A(\vec{\rho}_o) somb[\frac{R}{s_o} \frac{\omega}{c} |\vec{\rho}_o + \frac{\vec{\rho}_i}{\mu}|] \right|^2 \right\rangle$$
$$= \int_{obj} d\vec{\rho}_o |A(\vec{\rho}_o)|^2 somb^2[\frac{R}{s_o} \frac{\omega}{c} |\vec{\rho}_o + \frac{\vec{\rho}_i}{\mu}|]$$
$$\simeq |A(\vec{\rho}_o)|^2 \otimes \delta(|\vec{\rho}_o + \vec{\rho}_i/\mu|)$$
$$= |A(\vec{\rho}_i/\mu)|^2 \tag{6.3.16}$$

indicating a reproduction of the aperture function, i.e., an magnified or demagnified image of the object. An image, magnified by a factor of μ, is thus reproduced by the convolution between the squared moduli of the aperture function of the object and the point-spread function from an incoherent imaging process. In Eq. (6.3.16), we have assumed a large sized imaging lens and approximated the point-spread function a δ-function, $\delta(|\vec{\rho}_o + \vec{\rho}_i/\mu|)$; we have also absorbed the Fresnel phase $G(|\vec{\rho}_o|, \frac{\omega}{cs_o})$ into the complex aperture function, which is reasonable for incoherent imaging. Note, the Fresnel phase $G(|\vec{\rho}_o|, \frac{\omega}{cs_o})$ may have to be taken into account for coherent imaging.

Realistically and obviously, the finite spot size of point-to-spot propagator, which is defined by the point-spread function, will determine the *spatial resolution* of an imaging setup and thus limits the ability of an optical imaging system. Mathematically, the spatial resolution of a classic imaging setup can be estimated from the following convolution

$$\langle I(\vec{\rho}_i) \rangle \propto \int_{obj} d\vec{\rho}_o \, |A(\vec{\rho}_o)|^2 \, \mathrm{somb}^2 \big[\frac{R}{s_o} \frac{\omega}{c} |\vec{\rho}_o + \frac{\vec{\rho}_i}{\mu}| \big]$$

$$= |A(\vec{\rho}_o)|^2 \otimes \mathrm{somb}^2 \big[\frac{R}{s_o} \frac{\omega}{c} |\vec{\rho}_o + \frac{\vec{\rho}_i}{\mu}| \big]. \tag{6.3.17}$$

The most popular definition of the spatial resolution of imaging is perhaps the Rayleigh's criterion: the images of two nearby point objects are said to be *un-resolvable* when the center of one point-spread function falls on the first minimum of the point-spread function of the other. Fig. 6.3.4 qualitatively depicts this situation, in which the two point-spread functions have just become un-resolvable. To quantify this situation we consider a point

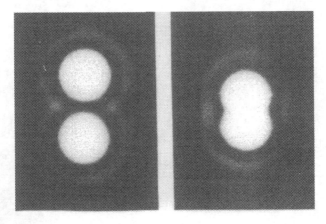

FIGURE 6.3.4 Left: non-overlapped images. The images of two nearby point objects are spatially resolvable. Right: overlapped images. The center of one point-spread function falls on the first minimum of the point-spread function of the other. This situation is defined as *un-resolvable* by Rayleigh's criterion.

object located at $\vec{\rho}_o = 0$ of the object plane. For an idealized imaging system, this point would have a unique corresponding image point on the image plane at $\vec{\rho}_i = 0$, which means a point-to-point relationship. Realistically, however, we have to take into account the point-to-spot relationship that is determined by the size of the point-spread function. The Rayleigh's criterion defines the size of the point-spread function by taking its first minimum, i.e.,

$$\frac{R\,\omega}{c\,s_i} \Delta|\vec{\rho}_i| \simeq 3.83 \quad \text{or} \quad \Delta|\vec{\rho}_i| \simeq 1.22 \, \frac{\pi\,c\,s_i}{R\,\omega} \tag{6.3.18}$$

Now, we consider another nearby point on the object plane $\vec{\rho}_o = \vec{a}$. In order to distinguish the image of $\vec{\rho}_o = \vec{a}$ from that of $\vec{\rho}_o = 0$, the value of $m|\vec{a}|$ cannot be smaller then $\Delta|\vec{\rho}_i|$. Therefore we must have

$$m|\vec{a}| \geq \frac{\pi\,c\,s_i}{R\,\omega},$$

where again $m = |\vec{\rho}_i|/|\vec{\rho}_o| = s_i/s_o$ is the magnification factor of the image. We thus have a minimum resolvable angular separation, or angular limit of resolution,

$$\Delta\theta_{min} \simeq \frac{|\vec{a}|}{s_o} \simeq 1.22 \, \frac{\pi\,c}{R\,\omega} = 1.22 \, \frac{\lambda}{D}, \tag{6.3.19}$$

where D is the diameter of the lens system. It is clear from Eqs. (6.3.3) and (6.3.19) that the use of large size imaging lenses and shorter wavelength radiation sources will result in narrower point-spread functions and smaller minimum resolvable angles, i.e., higher spatial resolution. To improve the spatial resolution, one of the efforts in the lithography industry is the use of shorter wavelengths. This effort is, however, limited to a certain level because of the inability of lenses to effectively work beyond a certain "cutoff" wavelength.

Eqs. (6.3.19) impose a diffraction-limited spatial resolution on an imaging system when the diameter of the imaging system and the wavelength of the light source are both fixed. This limit is fundamental in both classical optics and in quantum optics. Any violation would be considered a violation of the uncertainty principle.

6.4   FOURIER TRANSFORM VIA A LENS

We continue a similar calculation for the setup of Fig. 6.1.2. The object-aperture is located in the front focal plane of the converging lens, and is illuminated by a well collimated radiation. The following calculation will show the diffraction pattern observed in the back focal plane of the lens is the Fourier transform of the complex aperture function $A(\vec{\rho}_o)$. Comparing the setup of Fig. 6.1.2 with that of Fig. 6.3.1, what we need is to assign $s_o = f$ and $s_i = f$, and to complete the integrals of Eq. (6.3.14). We will first evaluate the integral over the lens. To simplify the mathematics we approximate the integral to infinity. Differing from the calculation for imaging resolution, the purpose of this evaluation is to find the Fourier transform. Thus, the approximation of an infinite lens is appropriate. By applying the properties of the Gaussian function listed in Eq. (6.2.10), the integral over the lens contributes the following function of $\vec{\rho}_o$ to the integral of $d\vec{\rho}_o$ in Eq. (6.3.14):

$$\int_{lens} d\vec{\rho}_l\, G(|\vec{\rho}_l|, \frac{\omega}{cf})\, e^{-i\frac{\omega}{c}(\frac{\vec{\rho}_o}{f} + \frac{\vec{\rho}}{f})\cdot\vec{\rho}_l} \propto C\, G(|\vec{\rho}_o|, -\frac{\omega}{cf})\, e^{-i\frac{\omega}{cf}\vec{\rho}_o\cdot\vec{\rho}},$$

where C absorbs all constants including a phase factor $G(|\vec{\rho}|, -\frac{\omega}{cf})$.

Replacing this result with the integral of $d\vec{\rho}_l$ in Eq. (6.3.14), under the condition of a well collimated light illumination of $\vec{\kappa} = 0$, we obtain:

$$E(\vec{\rho}) \propto \int_{obj} d\vec{\rho}_o\, a(\vec{\rho}_o)\, e^{i\varphi_0}\, e^{-i\frac{\omega}{cf}\vec{\rho}\cdot\vec{\rho}_o} = \mathcal{F}_{(\frac{\omega}{c}\frac{\vec{\rho}}{f})}\{a(\vec{\rho}_o)\}, \qquad (6.4.1)$$

which is the Fourier transform of the object-aperture function $A(\vec{\rho}_o)$ with conjugate variable $(\omega/c)(\vec{\rho}/f)$. In fact, $(\omega/c)(\vec{\rho}/f)$ is the transverse wavevector on the back focal plane of the lens. For a well collimated light illumination of $\vec{\kappa} \neq 0$, the far-field diffraction pattern will be shifted:

$$E(\vec{\rho}) \propto \int_{obj} d\vec{\rho}_o\, a(\vec{\rho}_o)e^{i\vec{\kappa}\cdot\vec{\rho}_o}\, e^{-i\frac{\omega}{cf}\vec{\rho}\cdot\vec{\rho}_o} = \mathcal{F}_{(\frac{\omega}{c}\frac{\vec{\rho}}{f} - \vec{\kappa})}\{a(\vec{\rho}_o)\}. \qquad (6.4.2)$$

SUMMARY

In this chapter, we studied the individual transverse behavior of a photon or a subfield and the collective transverse behavior of electromagnetic field, including their diffraction, propagation, and imaging. This chapter is another good excise to practice the superposition "principle" of electromagnetic field. Superposing each and all subfields that emitted and propagated from a plane with known distribution of radiation field, now, we are able to calculate the transverse distribution of radiation field or intensity in another arbitrary parallel plane, such as a Fraunhofer far-field plane or a Fresnel near-field plane, especially an image plane or a Fourier transform plane.

Diffraction is a special interference phenomenon. We have given a simple example in which the secondary wavelets of Huygens along a 1-D aperture propagate to and superpose at an observation plane

$$E(x,t) = \int_{-\frac{b}{2}}^{\frac{b}{2}} dx_0 \, \frac{A(x_0)}{r} \, e^{-i(\omega t - kr)}.$$

The constructive-destructive interference of the secondary wavelets results in a diffraction pattern which can be easily estimated in the far-field approximation:

$$\langle I(x,t) \rangle \simeq \left| \int dx_0 \, A(x_0) \, e^{-ik_x x_0} \right|^2 \propto \left| \mathcal{F}_{k_x} \{ A(x_0) \} \right|^2.$$

Mathematically, the observed Fraunhofer far-field diffraction is the Fourier transform of the complex aperture function.

A Fresnel near-field propagator which propagates a Fourier mode from point $\vec{\rho}_0$ in the σ_0 plane of $z_0 = 0$ to point $\vec{\rho}_0$ in the Fresnel near-field σ plane of $z = $ constant is derived in the section of field propagation

$$g_{\vec{\rho}_0}(\omega; \vec{\rho}, z) = \frac{-i\omega}{2\pi c} \, \frac{e^{i\frac{\omega}{c}z}}{z} \, G(|\vec{\rho} - \vec{\rho}_0|, \frac{\omega}{cz}).$$

This Fresnel near-field propagator will be used multiple times in later chapters to propagate radiation fields from one coordinate to another. In this chapter, it helps for the calculation of classic imaging.

Optical imaging implies a unique point-to-point relationship between the radiation field in the object plane and the radiation field in the image plane. In a classic imaging system, it is interesting to find that although an arbitrarily selected radiation field $E(\vec{\rho}_o)$ from the object plane may propagate to all possible directions and may take all possible optical paths with different optical delays to arrive at the image plane, the point-to-spot (point-to-point) propagator

$$g_{\vec{\rho}_o}(\vec{\rho}_i, z_i, t_i) = \frac{-\omega^2 e^{i\varphi_{s_i}}}{(2\pi c)^2 s_o s_i} \, \mathrm{somb}[\frac{R}{s_o} \frac{\omega}{c} |\vec{\rho}_o + \frac{\vec{\rho}_i}{\mu}|]$$
$$\simeq \frac{-\omega^2 e^{i\varphi_{s_i}}}{(2\pi c)^2 s_o s_i} \, \delta(|\vec{\rho}_o + \frac{\vec{\rho}_i}{\mu}|)$$

forces it to collapse or stop at an unique spot around point $\vec{\rho}_i$ (an unique point $\vec{\rho}_i$) of the image plane. This point-to-spot (point-to-point) propagator is the result of a superposition that sums all subfields emitted from $\vec{\rho}_o$ and passed through each and all points, $\vec{\rho}_l$, of the imaging lens along all possible optical paths with different optical delays. These subfields interfere constructively at an unique spot (approximated as a point) of the image plane while their optical path lengths are equal, and interfere destructively at all other points of the image plane while their optical path lengths are unequal. The imaging lens makes this constructive-destructive interference possible. The most important results derived from the imaging excise are:

$$\frac{1}{s_i} + \frac{1}{s_o} = \frac{1}{f},$$

which defines an image plane of the imaging system; and

$$\langle I(\vec{\rho}_i) \rangle \propto \langle \left| \int_{obj} d\vec{\rho}_o \, A(\vec{\rho}_o) \operatorname{somb}\left[\frac{R}{s_o} \frac{\omega}{c} |\vec{\rho}_o + \frac{\vec{\rho}_i}{\mu}|\right] \right|^2 \rangle$$

$$= \langle \left| A(\vec{\rho}_o) \otimes \operatorname{somb}\left[\frac{R}{s_o} \frac{\omega}{c} |\vec{\rho}_o + \frac{\vec{\rho}_i}{\mu}|\right] \right|^2 \rangle,$$

which yields an image of the object. In the case of incoherent imaging with a large sized imaging lens,

$$\langle I(\vec{\rho}_i) \rangle \simeq |A(\vec{\rho}_o)|^2 \otimes \delta(|\vec{\rho}_o + \vec{\rho}_i/\mu|) = |A(\vec{\rho}_i/\mu)|^2$$

indicating a reproduction of a magnified or demagnified real image of the object.

REFERENCES AND SUGGESTIONS FOR READING

[1] J.W. Goodman, *Introduction to Fourier Optics*, Roberts & Company, Englewood, 2005.

[2] E. Hecht, *Optics*, Addison-Wesley, 2001.

CHAPTER 7

First-Order Coherence of Light

The concept of first-order coherence was introduced in classical theory of light much earlier than that of the quantum theory. The degree of first-order coherence is defined to quantify the interference between temporally delayed or spatially separated electromagnetic waves. The superposed radiations are defined as first-order coherent if the interference fringes exhibit 100% modulation, or first-order incoherent if no interference fringes are observable. The radiation fields are considered as partial coherent if the modulation is less than 100%, however, greater than zero. The higher the degree of first-order coherence, the higher interference visibility we could observe. Although it is named "coherence" and is an intrinsic property of the radiation itself, either temporal or spatial, the concept is very different from the coherence property of light that we have discussed in Chapter 4. For a certain spectral bandwidth $\Delta\omega$ and spatial frequency $\Delta\vec{\kappa}$ or Δk_x (Δk_y), the interference observed in an interferometer is the same for laser light and thermal light. The measurement cannot distinguish a laser beam from thermal radiation by means of the degree of first-order coherence. Although laser radiation is named coherent, this does not mean we can observe interference for a temporal delay beyond certain limit. Similarly, thermal light, such as the sunlight, is considered incoherent radiation, but this does not prevent us to observe interference fringes under the "white-light" condition, i.e., in the neighborhood of equal optical paths of an interferometer. One should pay special attention to this.

In this chapter we start from defining the first-order coherence of light in the framework of classical optics. The goal of this chapter is to probe the quantum nature of optical coherence, which has been especially successful in dealing with the coherent phenomena of light at quantum level. In fact, we have implicitly introduced the quantum concepts of radiation by means of the Einstein's picture of light from Chapter 1, although no field quantization was involved. The large number of sub-sources, either radiate independently or coherently, are connected with a large number of atomic transitions. The very basic contribution to the radiation field from each sub-source is therefore the creation of a photon. In the bright light condition, a measurement may involve the creation and annihilation of a large number of photons. The individual behavior of a photon may not be that significant in the final observation. However, it does not prevent us to ask very simple questions: What would happen if a photon and only one photon presents in the observation? Can a photon interfere with itself? In a "single-photon interferometer", does a photon passes "both paths" or "which path"? In fact, even in the "bright light" condition, we may find that the interference is the result of a subfield interfering with the subfield itself. Connecting Einstein's bundle of ray, or subfield, with the concept of photon, we would have the same questions. Quantum theory of coherence deals with this kind of physics and gives its quantitative predictions and qualitative interpretations according to the principles of quantum mechanics. In the light of new technology, we are now able to study these problems experimentally. In certain aspects,

FIGURE 7.0.1   On one hand, in quantum theory of light a photon can never be divided into parts; on the other hand, we have never lost interference at single-photon's level. In fact, according to quantum theory, interference is a single-photon phenomenon. In Dirac's language: "... photon ... only interferes with itself".

we may say that quantum theory of coherence studies the interference of a photon with itself, or a group of photons interference with the group itself. In this regard, the first-order quantum coherence studies and quantifies the self-interference of a photon: a single-photon interfering with itself. The quantum degree of first-order coherence is introduced in terms of the superposition of single-photon amplitudes, either in the form of self-superposition of Einstein's subfield or in the form of self-superposition of effective wavefunction of a photon.

In this chapter, we first define the classical mutual-coherence function $\Gamma^{(1)}(\mathbf{r}_1, t_1; \mathbf{r}_2, t_2)$ and the normalized complex degree of first-order classical coherence $\gamma^{(1)}(\mathbf{r}_1, t_1; \mathbf{r}_2, t_2)$. We then focus on the measurement of first-order temporal coherence for a few simplified models of radiation, and introduce the concept of first-order quantum coherence $G^{(1)}(\mathbf{r}_1, t_1; \mathbf{r}_2, t_2)$ and the normalized degree of first-order quantum coherence $g^{(1)}(\mathbf{r}_1, t_1; \mathbf{r}_2, t_2)$. In the end of this chapter, we introduce the concepts of classical and quantum first-order spatial coherence.

7.1   FIRST-ORDER COHERENCE OF LIGHT—EM THEORY

To introduce the concepts of first-order coherence of light we analyze a typical Young's double-pinhole interferometer shown in Fig. 7.1.1. The upper and the lower pinholes P_1 and P_2 are located at coordinates $\mathbf{r}_1$ and $\mathbf{r}_2$, respectively. The observation is made by scanning a point-like photodetector on the far-field observation plane Σ, or by a photodetector array such as a CCD or a CMOS. The light source, a bright distant star that is treated as either a point-like source or an extended source with a certain angular size, the pinholes, and the observation plan are arranged symmetrically with respect to the optical axis as shown in Fig. 7.1.1.

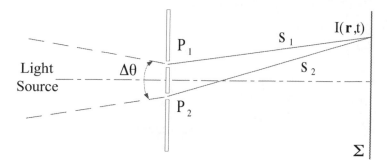

FIGURE 7.1.1 Schematic of the Young's double-pinhole interference experiment. The interference pattern is observed by a photodetector array such as a CCD or a CMOS, or by scanning a point-like photodetector on the observation plane Σ.

The radiation field at space-time coordinates $P_1(\mathbf{r}_1, t_1)$ and $P_2(\mathbf{r}_2, t_2)$ are superposed in a later time at each point on the transverse plane Σ for the observation of interference. Either observable or unobservable, the interference is determined by the intrinsic property of the radiation: (1) the maximum allowable optical delay for observing interference is determined by the spectral bandwidth of the field, the greater bandwidth $\Delta\omega$ the shorter allowable temporal delay between the superposed fields; (2) the maximum allowable transverse spatial separation between the two superposed fields is determined by the bandwidth of the spatial frequency $|\Delta\vec{\kappa}|$, the greater bandwidth of $|\Delta\vec{\kappa}|$ the smaller allowable separation between the two fields. The observations (1) and (2) are considered as the temporal coherence and spatial coherence of the radiation field, respectively, by definition. Although thermal light and laser radiation are distinguished as incoherent light and coherent radiation, both the thermal field and the laser beam can be superposed at a space-time point to cause interference. Whether observable or unobservable, the concept of first-order temporal coherence and spatial coherence apply to both.

Introducing first-order temporal and spatial coherence, we start with a simplified model of a disk-like radiation source at plane $z = 0$ with angular diameter $\Delta\theta$, as shown in Fig. 7.1.1. The expectation value of the intensity, $\langle I(\mathbf{r}, t)\rangle$, measured at space-time point $(\mathbf{r}, t)$ is:

$$
\begin{aligned}
\langle I(\mathbf{r}, t)\rangle &= \langle|E(\mathbf{r}, t)|^2\rangle \\
&= \langle|E(\mathbf{r}_1, t_1)\, g_1(\omega; \mathbf{r}, t) + E(\mathbf{r}_2, t_2)\, g_2(\omega; \mathbf{r}, t)|^2\rangle \\
&= \langle|E(\mathbf{r}_1, t - \tfrac{s_1}{c}) + E(\mathbf{r}_2, t - \tfrac{s_2}{c})|^2\rangle \\
&= \langle|E(\mathbf{r}_1, t_1) + E(\mathbf{r}_2, t_2)|^2\rangle
\end{aligned}
\tag{7.1.1}
$$

where $E(\mathbf{r}_1, t_1)$ and $E(\mathbf{r}_2, t_2)$ are the electromagnetic fields passed through P_1 and P_2 at times t_1 and t_2; $t_1 \equiv t - s_1/c$ and $t_2 \equiv t - s_2/c$ are the "earlier" times relative to the measurement time t; $g_j(\omega; \mathbf{r}, t)$, $j = 1, 2$, is the Green's function that propagates the field from the jth pinhole to $(\mathbf{r}, t)$ along the jth optical path. The early times t_1 and t_2 are defined by the observation time t and the optical paths s_1 and s_2: $(t - t_1 = s_1/c)$ and $(t - t_2 = s_2/c)$, or $[(t - t_1) - s_1/c = 0]$ and $[(t - t_2) - s_2/c = 0]$. Since the observed field at space-time coordinate $(\mathbf{r}, t)$, which has propagated through the jth pinhole along the jth optical path is

$$E_j(\mathbf{r}, t) = \sum_m \int d\omega \, E_m(\omega) \, g_m(\omega; \mathbf{r}_j, t_j) \, g_j(\omega; \mathbf{r}, t)$$

$$\simeq \sum_m \int d\omega \, E_m(\omega) \, g_m(\omega; \mathbf{r}_j, t_j) \, e^{-i\omega[(t-t_j)-s_j/c]}$$

$$= \sum_m \int d\omega \, E_m(\omega) \, g_m(\omega; \mathbf{r}_j, t - s_j/c)$$

$$= E(\mathbf{r}_j, t - s_j/c), \tag{7.1.2}$$

we conclude the observed field at $(\mathbf{r}, t)$ that has propagated along the jth optical path is the field that passed the jth pinhole at an early time $t_j = t - s_j/c$. In Eq. (7.1.2), m labels the mth point-like sub-source in the light source plane. For each subfield, on one hand, its creation time and annihilation time are defined by the physical processes of its creation and its annihilation; on the other hand, its propagation paths are also defined with measurable physical lengths. At a defined observation time t, the observer measures a different phase from its jth propagation, which is exactly the same phases when the subfield passed the jth pinhole at early time $t_j = t - s_j/c$.

In the above discussion, we have defined the "early times", $t_1 \equiv t - s_1/c$ and $t_2 \equiv t - s_2/c$, in terms of the observation time t by propagating the field backward in time, similar to the Heisenberg picture in quantum mechanics. We may reexamine the physical picture of Eq. (7.1.2) from a different direction:

$$E(\mathbf{r}_j, t_j) = \sum_m \int d\omega \, E_m(\omega) \, e^{-i[\omega(t_j - t_m) - \mathbf{k}_m \cdot (\mathbf{r}_j - \mathbf{r}_m)]}$$

$$= \sum_m \int d\omega \, E_m(\omega) \, e^{-i[\omega(t - s_j/c - t_m) - k r_{mj})]}$$

$$= \sum_m \int d\omega \, E_m(\omega) \, e^{-i[\omega(t - t_m) - k(s_j + r_{mj})]}$$

$$= E_j(\mathbf{r}, t) \tag{7.1.3}$$

where t_m is the creation time of the mth subfield, r_{mj} is the optical path from the mth point-like sub-source to the jth pinhole, $E_j(\mathbf{r}, t)$ is the measured field at space-time coordinate $(\mathbf{r}, t)$ that propagated along the jth optical path. Here, we propagate the field from space-time coordinate $(\mathbf{r}_m, t_m)$ of its creation, along the jth optical path, to space-time coordinate $(\mathbf{r}, t)$ of its observation, similar to the Schrödinger picture of quantum mechanics. However, the early times of t_1 and t_2 are still defined from the observation time t: $t_j = t - s_j/c$. Whether start from Heisenberg picture or from Schrödinger picture, the space-time coordinates of the creation-annihilation events and the optical path associated with the jth pinhole are defined physically. The physical phase delay between $E(\mathbf{r}_1, t - s_1/c) = E_1(\mathbf{r}, t)$ and $E(\mathbf{r}_2, t - s_2/c) = E_2(\mathbf{r}, t)$ determines the result of their superposition and thus the observed interference.

The expectation value of the intensity, $\langle I(\mathbf{r}, t) \rangle$, at space-time point $(\mathbf{r}, t)$ is thus:

$$\langle I(\mathbf{r}, t) \rangle = \langle |E(\mathbf{r}, t)|^2 \rangle$$

$$= \langle \, | \, E(\mathbf{r}_1, t - \frac{s_1}{c}) + E(\mathbf{r}_2, t - \frac{s_2}{c}) \, |^2 \, \rangle$$

$$= \langle E^*(\mathbf{r}_1, t_1) E(\mathbf{r}_1, t_1) \rangle + \langle E^*(\mathbf{r}_2, t_2) E(\mathbf{r}_2, t_2) \rangle$$

$$+ \langle E^*(\mathbf{r}_1, t_1) E(\mathbf{r}_2, t_2) \rangle + \langle E(\mathbf{r}_1, t_1) E^*(\mathbf{r}_2, t_2) \rangle \tag{7.1.4}$$

where, again, $t_1 = t - s_1/c$, $t_2 = t - s_2/c$, referring to the earlier times of the fields at the upper and the lower pinholes P_1 and P_2, respectively. It is recognized that the first two terms in Eq.(7.1.4) correspond to the light intensity passing through the upper and lower pinholes at space-time points $(\mathbf{r}_1, t_1)$ and $(\mathbf{r}_2, t_2)$, respectively. The third and fourth terms, which involve the cross product of the field at space-time point $(\mathbf{r}_1, t_1)$ [or $(\mathbf{r}_2, t_2)$] with its conjugate at space-time point $(\mathbf{r}_2, t_2)$ [or $(\mathbf{r}_1, t_1)$], give rise to an interference pattern at the observation plane and results in a sinusoidal modulation of the photocurrent as a function of the position of the scanning photodetector or the coordinate of the pixels of the CCD.

We break $\langle I(\mathbf{r}, t) \rangle$ of Eq. (7.1.4) into two groups:

(1) $\langle E^*(\mathbf{r}_1, t_1) E(\mathbf{r}_1, t_1) \rangle$ and $\langle E^*(\mathbf{r}_2, t_2) E(\mathbf{r}_2, t_2) \rangle$;
(2) $\langle E^*(\mathbf{r}_1, t_1) E(\mathbf{r}_2, t_2) \rangle$ and $\langle E(\mathbf{r}_1, t_1) E^*(\mathbf{r}_2, t_2) \rangle$.

We define the first group as the self-coherence function:

$$\Gamma^{(1)}(\mathbf{r}_1, t_1; \mathbf{r}_1, t_1) \equiv \langle E^*(\mathbf{r}_1, t_1) E(\mathbf{r}_1, t_1) \rangle \tag{7.1.5}$$
$$\Gamma^{(1)}(\mathbf{r}_2, t_2; \mathbf{r}_2, t_2) \equiv \langle E^*(\mathbf{r}_2, t_2) E(\mathbf{r}_2, t_2) \rangle.$$

and the second group as the mutual-coherence function:

$$\Gamma^{(1)}(\mathbf{r}_1, t_1; \mathbf{r}_2, t_2) \equiv \langle E^*(\mathbf{r}_1, t_1) E(\mathbf{r}_2, t_2) \rangle \tag{7.1.6}$$
$$\Gamma^{(1)}(\mathbf{r}_2, t_2; \mathbf{r}_1, t_1) \equiv \langle E(\mathbf{r}_1, t_1) E^*(\mathbf{r}_2, t_2) \rangle.$$

It is obvious that

$$\Gamma^{(1)}(\mathbf{r}_1, t_1; \mathbf{r}_2, t_2) = \Gamma^{(1)*}(\mathbf{r}_2, t_2; \mathbf{r}_1, t_1). \tag{7.1.7}$$

In connection with the concepts of classical statistics, the self-coherence function of Eq. (7.1.5) and the mutual-coherence function of Eq. (7.1.6) are recognized as the self-correlation function and the cross-correlation function, respectively, of the fields. Physically, $\Gamma(\mathbf{r}_1, t_1; \mathbf{r}_2, t_2)$ determines the visibility of the interference, which will be quantified in the following. The self-coherence function $\Gamma(\mathbf{r}_j, t_j; \mathbf{r}_j, t_j)$, $j = 1, 2$, defined in Eq.(7.1.5) represents the expectation value, or expectation function of intensity which has been discussed in section 1.4. Applying the mutual-coherence function and the self-coherence function, the expectation value of $I(\mathbf{r}, t)$ in Eq. (7.1.4) is written as

$$\langle I(\mathbf{r}, t) \rangle = \Gamma^{(1)}_{11} + \Gamma^{(1)}_{22} + \Gamma^{(1)}_{12} + \Gamma^{(1)}_{21}. \tag{7.1.8}$$

where we have used the shortened notation

$$\Gamma^{(1)}_{11} = \Gamma^{(1)}(\mathbf{r}_1, t_1; \mathbf{r}_1, t_1), \quad \Gamma^{(1)}_{22} = \Gamma^{(1)}(\mathbf{r}_2, t_2; \mathbf{r}_2, t_2),$$
$$\Gamma^{(1)}_{12} = \Gamma^{(1)}(\mathbf{r}_1, t_1; \mathbf{r}_2, t_2), \quad \Gamma^{(1)}_{21} = \Gamma^{(1)}(\mathbf{r}_2, t_2; \mathbf{r}_1, t_1).$$

Now, we introduce the normalized complex degree of first-order coherence by writing Eq. 7.1.8 in the following form

$$\langle I(\mathbf{r}, t) \rangle = \Gamma^{(1)}_{11} + \Gamma^{(1)}_{22} + 2\sqrt{\Gamma^{(1)}_{11} \Gamma^{(1)}_{22}} \, Re \, \gamma^{(1)}_{12}$$
$$= \langle I_1 \rangle + \langle I_2 \rangle + 2\sqrt{\langle I_1 \rangle \langle I_2 \rangle} \, |\gamma^{(1)}_{12}| \cos(\omega\tau) \tag{7.1.9}$$

where $\tau = \tau_1 - \tau_2 = (t_1 - t_2) - (r_1 - r_2)/c = (s_2 - s_1)/c + (r_2 - r_1)/c$, $(r_1 - r_2)/c$ is the relative time delay between the optical paths from the light source to pinholes P_1 and P_2.

In Eq. (7.1.9). we have defined the normalized complex degree of first-order coherence

$$\gamma^{(1)}(\mathbf{r}_1, t_1; \mathbf{r}_2, t_2) \equiv \frac{\Gamma^{(1)}(\mathbf{r}_1, t_1; \mathbf{r}_2, t_2)}{\left[\Gamma^{(1)}(\mathbf{r}_1, t_1; \mathbf{r}_1, t_1)\right]^{\frac{1}{2}} \left[\Gamma^{(1)}(\mathbf{r}_2, t_2; \mathbf{r}_2, t_2)\right]^{\frac{1}{2}}} \tag{7.1.10}$$

$$= \frac{\langle E^*(\mathbf{r}_1, t_1) E(\mathbf{r}_2, t_2) \rangle}{\left[\langle |E(\mathbf{r}_1, t_1)|^2 \rangle\right]^{\frac{1}{2}} \left[\langle |E(\mathbf{r}_2, t_2)|^2 \rangle\right]^{\frac{1}{2}}},$$

for $\Gamma(\mathbf{r}_1, t_1; \mathbf{r}_1, t_1) \neq 0$ and $\Gamma(\mathbf{r}_2, t_2; \mathbf{r}_2, t_2) \neq 0$. It is easy to find that

$$0 \leq |\gamma_{12}^{(1)}| \leq 1. \tag{7.1.11}$$

$|\gamma_{12}^{(1)}|$ is thus related to the visibility of the interference fringe modulation

$$V \equiv \frac{I_{MAX} - I_{MIN}}{I_{MAX} + I_{MIN}} = \frac{2\sqrt{\langle I_1 \rangle \langle I_2 \rangle}}{\langle I_1 \rangle + \langle I_2 \rangle} |\gamma_{12}^{(1)}|. \tag{7.1.12}$$

If $\langle I_1 \rangle = \langle I_2 \rangle = I_0/2$, which is perhaps the most common arrangement for an optimized interferometer, $|\gamma_{12}|$ is identical to the visibility of the interference modulation:

$$V = |\gamma_{12}^{(1)}|.$$

In this case, the expectation function of the intensity on the observation plane is thus:

$$\langle I(\mathbf{r}, t) \rangle = I_0 \left[1 + V \cos(\omega\tau)\right]. \tag{7.1.13}$$

The radiation fields at space-time points $(\mathbf{r}_1, t_1)$ and $(\mathbf{r}_2, t_2)$ are named first-order coherent, partially coherent and incoherent in terms of the value of $|\gamma_{12}^{(1)}|$:

1st-order Coherent	*if*	$|\gamma^{(1)}(\mathbf{r}_1, t_1; \mathbf{r}_2, t_2)| = 1$
1st-order Partially Coherent	*if*	$0 < |\gamma^{(1)}(\mathbf{r}_1, t_1; \mathbf{r}_2, t_2)| < 1$
1st-order Incoherent	*if*	$|\gamma^{(1)}(\mathbf{r}_1, t_1; \mathbf{r}_2, t_2)| = 0.$

7.2 FIRST-ORDER TEMPORAL COHERENCE—EINSTEIN'S PICTURE

In this section we focus on the first-order temporal coherence of light. To simplify the mathematics and the discussion, we assume a far-field point-like radiation source, $\Delta\theta = 0$, for the Young's double-slit experimental setup of Fig. 7.2.1.

(I) $\Gamma^{(1)}(\mathbf{r}_1, t_1; \mathbf{r}_2, t_2)$: incoherent subfields & incoherent Fourier modes

Assume a point-like light source which contains a large number of independent and randomly radiating point-like sub-sources, each sub-source creates a large set of incoherent Fourier-modes. The mutual-coherence function of $E(\mathbf{r}_1, t_1)$ and $E(\mathbf{r}_2, t_2)$ is expected to be

$$\Gamma^{(1)}(\mathbf{r}_1, t_1; \mathbf{r}_2, t_2)$$

$$= \langle \sum_m \int d\omega \, a_m(\omega) e^{-i\varphi_m(\omega)} \, e^{i[\omega t_1 - k(\omega)z_1]}$$

$$\times \sum_n \int d\omega' \, a_n(\omega') e^{i\varphi_n(\omega')} \, e^{-i[(\omega' t_2 - k'(\omega')z_2)]} \rangle$$

$$\simeq \sum_m \int d\omega \, a_m^2(\omega) e^{i[\omega(t_1 - t_2) - k(\omega)(z_1 - z_2)]}$$

$$\simeq \mathcal{F}_\tau \left\{ \sum_m a_m^2(\nu) \right\} e^{i\omega_0 \tau} \tag{7.2.1}$$

where the expectation operation or ensemble average has taken into account all possible realizations of the random phases in the sum of the subfields and the Fourier-modes. In

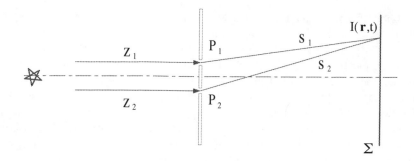

FIGURE 7.2.1 Schematic of the Young's double-slit interference experiment which measures the degree of first-order coherence of incoherent or coherent light. To simplify the discussion, we focus on the measurement of first-order temporal coherence by assuming a far-field point-like radiation source with a zero opening angle, $\Delta\theta = 0$, we have $r_1 \simeq z_1$, $r_2 \simeq z_2$ and $z_1 = z_2$.

Eq. (7.2.1), again, we have defined $\tau_1 = t_1 - z_1/c$, $\tau_2 = t_2 - z_2/c$, and $\tau = \tau_1 - \tau_2 = (s_2 - s_1)/c + (z_2 - z_1)/c$. In the case of $z_1 = z_2$, see Fig. 7.2.1, we have $\tau = (s_2 - s_1)/c$. In the simplified situation in which all the subfields hold identical spectrum distribution $a^2(\omega)$, it is easy to find that the optical delay of τ and the spectral bandwidth $\Delta\omega$ are restricted by the relation of $\Delta\omega\,\tau = 2\pi$ for a nonzero mutual-coherence function of $\Gamma_{12}^{(1)}$, and consequently for an observable interference.

The mutual-coherence function $\Gamma^{(1)}(\mathbf{r}_1, t_1; \mathbf{r}_2, t_2)$ of Eq. (7.1.6) can be calculated directly from the formally integrated Fourier transform of the fields for $E^*(\mathbf{r}_1, t_1)$ and $E(\mathbf{r}_2, t_2)$,

$$\Gamma^{(1)}(\mathbf{r}_1, t_1; \mathbf{r}_2, t_2)$$
$$= \Big\langle \mathcal{F}_{\tau_1}^* \Big\{ \sum_m a_m(\nu) e^{-i\varphi_m(\nu)} \Big\} e^{i(\omega_0 t_1 - k_0 z_1)}$$
$$\times \mathcal{F}_{\tau_2} \Big\{ \sum_n a_n(\nu') e^{i\varphi_n(\nu')} \Big\} e^{-i(\omega_0 t_2 - k_0 z_2)} \Big\rangle$$
$$= \mathcal{F}_\tau \Big\{ \sum_m a_m^2(\nu) \Big\} e^{i\omega_0 \tau} \tag{7.2.2}$$

where, again, the expectation operation is partially completed by taking into account all possible values of the random relative phases $\varphi_m(\nu) - \varphi_n(\nu')$. The only surviving terms in the integral are the diagonal terms of $\nu = \nu'$ and $m = n$. Again, we find that the optical delay of τ and the spectral bandwidth $\Delta\omega$ are restricted by $\Delta\omega\,\tau = 2\pi$ for a nonzero mutual-coherence function of $\Gamma_{12}^{(1)}$, and consequently an observable sinusoidal interference modulation.

Taking the result of Eq. (7.2.1) or Eq. (7.2.2), keeping only the nonzero contributions, the expectation value of $\langle I(\mathbf{r}, t)\rangle$ can be written into the following form,

$$\langle I(\mathbf{r}, t)\rangle = \Big\langle \sum_m \int d\omega\, E_m^*(\omega; \mathbf{r}, t) \sum_n \int d\omega'\, E_n(\omega'; \mathbf{r}, t) \Big\rangle$$
$$= \sum_m \int d\omega\, I_m(\omega; \mathbf{r}, t)$$
$$= \sum_m \int d\omega\, \Big| \frac{1}{\sqrt{2}} \big[E_m(\omega; z_1, t_1) + E_m(\omega; z_2, t_2) \big] \Big|^2$$
$$= \sum_m \int d\omega\, I_{m0}(\omega) \big[1 + \cos(\omega\,\tau) \big] \tag{7.2.3}$$

where we explicitly write $1/\sqrt{2}$ for later discussions. Eq. (7.2.3) indicates that the observed interference pattern is the sum of a large number of individual sub-interference patterns, $I_m(\omega; \mathbf{r}, t)$, each is produced by a Fourier-mode of ω associated with a sub-source of mth. In the neighborhood of $\tau = 0$, these individual sub-patterns coincide so that the intensity modulation can be easily observed ($V(0) \cong 100\%$). When the detector moves away from $\tau = 0$, i.e., the value of $|\tau|$ increases, the relative phase shifts between the sub-patterns of different Fourier-mode of ω also increases. The spread of these sub-patterns smoothes the light intensity distribution on the observation plane, and the interference visibility is then reduced from 100% to 0, which means a constant distribution. The interference visibility, $V(\tau)$, is determined by the maximum relative phase shift $\Delta\omega\tau$, where $\Delta\omega = \omega_{max} - \omega_{min}$ is the bandwidth of the field. If $\Delta\omega\tau \ll 2\pi$, i.e., the relative phase shifts are not large enough to produce noticeable separation between the sub-patterns of different ω for the value of τ, then the interference modulation will be observable. When the relative phase shifts increase, however, the sub-patterns of these Fourier-modes are significantly separated at the value of τ. The interference modulation visibility is then reduced to zero. In other words, for a certain spectral bandwidth of $\Delta\omega$, if the time delay $\tau \ll 2\pi/\Delta\omega$, the fields at t and $t + \tau$ are considered temporally coherent by definition. Otherwise, the fields are considered temporally incoherent by definition, which means no interference can be observed.

From Eq. (7.2.3) we may conclude, again, that each Fourier-mode associated with a sub-source only interferes with itself. However, we should not ignore that Eq. (7.2.3) is the result of an ensemble average on a large number of interference patterns produced by the superpositions between all possible different Fourier-modes associated with all possible different sub-sources. It is the averaging effect among these interference patterns, namely a destructive interference cancelation, between different modes associated with different sub-sources leads to Eq. (7.2.1), or Eq. (7.2.2), and consequently makes "each Fourier-mode associated with a sub-source only interferes with itself". We should always keep this in mind in the following discussions.

A complete destructive interference cancelation happens only when "taking into account all possible realizations of the field". In one measurement, the cancelation may not be complete, therefore, the interference pattern may randomly fluctuate in the neighborhood of its expectation of Eq. (7.2.3) in a nondeterministic manner. In a perfect interferometer this fluctuation will be the major contribution to the "noise" of the observation.

In summary, for a far-field point thermal source, $\Gamma^{(1)}(\mathbf{r}_1, t_1; \mathbf{r}_2, t_2)$ is a function of the temporal delay $\tau = \tau_1 - \tau_2 = (s_2 - s_1)/c + (z_2 - z_1)/c$, which implies that the temporal correlation function of a thermal field is invariant under the displacements of temporal variables, i.e., invariant for any time t. This is the characteristic of stationary fields. For stationary fields the temporal mutual-correlation function $\Gamma^{(1)}(\mathbf{r}_1, t_1; \mathbf{r}_2, t_2)$ is usually written as $\Gamma^{(1)}_{12}(\tau)$. Compared with the expectation value of the intensity of thermal radiation in Eq. (5.1.4), we find $\langle I(\mathbf{r}_1, t_1) \rangle = \Gamma^{(1)}_{11}(0)$ and $\langle I(\mathbf{r}_2, t_2) \rangle = \Gamma^{(1)}_{22}(0)$. The expectation function of the intensity on the observation plane of the Young's double-slit experiment illustrated in Figure 7.1.1 is thus

$$\langle I(\mathbf{r}, t) \rangle = \Gamma^{(1)}_{11}(0) + \Gamma^{(1)}_{22}(0) + 2 \, Re \, \Gamma^{(1)}_{12}(\tau) \qquad (7.2.4)$$

$$= \Gamma^{(1)}_{11}(0) + \Gamma^{(1)}_{22}(0) + 2\sqrt{\Gamma^{(1)}_{11}(0) \, \Gamma^{(1)}_{22}(0)} \, Re \, \gamma^{(1)}_{12}(\tau),$$

where $\Gamma^{(1)}_{11}(0) \, [\Gamma^{(1)}_{22}(0)]$ is calculated in Eq. (5.1.4), and $\Gamma^{(1)}_{12}(\tau)$ is calculated in Eq. (7.2.2). It can be seen from Eqs. (7.2.4) and (7.2.2) that the maximum interference occurs in the neighborhood of $\tau = 0$. The interference starts to be invisible at $\tau = 2\pi/\Delta\omega$. In Eq. (7.2.4) the normalized complex degree of first-order temporal coherence is defined by

$$\gamma_{12}^{(1)}(\tau) \equiv \frac{\Gamma_{12}^{(1)}(\tau)}{[\,\Gamma_{11}^{(1)}(0)\,]^{\frac{1}{2}}\,[\,\Gamma_{22}^{(1)}(0)\,]^{\frac{1}{2}}}.$$

The time delay

$$\tau = \frac{2\pi}{\Delta\omega} \equiv \tau_c \qquad (7.2.5)$$

is defined as the coherence time of the field and consequently $c\tau_c$ is defined as the coherence length of the field (longitudinal). The fields with separation τ are named temporally coherent when $|\gamma^{(1)}(\tau)| = 1$, corresponding to $\tau \ll \tau_c$; temporally partially coherent when $0 < |\gamma(\tau)| < 1$, corresponding to $0 < \tau < \tau_c$; and temporally incoherent when $|\gamma^{(1)}(\tau)| = 0$, corresponding to $\tau \geq \tau_c$. Therefore, when we say a thermal radiation source has a coherence time τ_c or a coherence length $c\tau_c$ we mean that first-order interference is observable between any fields with a temporal separation of $\tau < \tau_c$.

(II) $\Gamma^{(1)}(\mathbf{r}_1, t_1; \mathbf{r}_2, t_2)$: natural light, consisting of a large number of overlapped and partially overlapped wavepackets.

In Einstein's picture, the radiation field $E(\mathbf{r}, t)$ at space-time $(\mathbf{r}, t)$ is the result of a superposition among a large number of randomly distributed subfields, or wavepackets, randomly created from a large number of independent point-like sub-sources, such as billions of independent atomic transitions. We label the subfield or wavepacket associated with the mth sub-source with parameter t_m, indicating the initial creation time of the wavepacket at the mth sub-source. Each photodetection event at space-time point $(\mathbf{r}, t)$ involves a large number of randomly created and randomly distributed overlapped−partially-overlapped wavepackets. The referred earlier fields $E(\mathbf{r}_1, t_1)$ and $E(\mathbf{r}_2, t_2)$ have the same nature in terms of the superposition of the subfields. Substituting the formally integrated wavepackets of $E^*(\mathbf{r}_1, t_1)$ and $E(\mathbf{r}_2, t_2)$ into $\langle E^*(\mathbf{r}_1, t_1)E(\mathbf{r}_2, t_2)\rangle$, the mutual-coherence function $\Gamma^{(1)}(\mathbf{r}_1, t_1; \mathbf{r}_2, t_2)$ is culated as

$$\Gamma^{(1)}(\mathbf{r}_1, t_1; \mathbf{r}_2, t_2)$$
$$= \langle \sum_{m=1}^{M} e^{i\omega_0 \tau_{1m}} \mathcal{F}_{\tau_{1m}}^* \{a(\nu)\} \sum_{n=1}^{N} e^{-i\omega_0 \tau_{2n}} \mathcal{F}_{\tau_{2n}} \{a(\nu)\} \rangle$$
$$= \langle \sum_{m=n} e^{i\omega_0 \tau} \mathcal{F}_{\tau_{1m}}^* \{a(\nu)\} \mathcal{F}_{\tau_{2m}} \{a(\nu)\} \rangle$$
$$+ \langle \sum_{m \neq n} e^{i\omega_0 [(t_{2n} - t_{1m}) + \tau]} \mathcal{F}_{\tau_{1m}}^* \{a(\nu)\} \mathcal{F}_{\tau_{2n}} \{a(\nu)\} \rangle \qquad (7.2.6)$$

where m and n label the mth and nth wavepacket. We have also defined $\tau_{1m} = (t - t_m) - s_1/c - z_1/c = \tau_1 - t_m$ and $\tau_{2n} = (t - t_n) - s_2/c - z_2/c = \tau_2 - t_n$, $\tau = \tau_1 - \tau_2 = (s_2 - s_1)/c + (z_2 - z_1)/c$. It is convenient to break up the sum into two groups. If we consider a random distribution of the subfield or wavepackets, i.e., stochastically radiated atomic transitions with random relative phases, or at arbitrary initial creation times of t_m and t_n, the second group ($m \neq n$) of the sum vanishes when taking into account all possible relative phases or all possible values of $t_m - t_n$ in the superposition. This result indicates that *the only observable interference is the interference in which the mth subfield or wavepacket interfering with the mth subfield itself.* Interference between two different subfields or wavepackets becomes unobservable after the ensemble average.

Accordingly, $\langle I(\mathbf{r}, t) \rangle$, the expectation function of the intensity in the observation plane is the sum of a large number of individual sub-interferences, each is the result of an individual wavepacket interfering with the wavepacket itself

$$\langle I(\mathbf{r}, t) \rangle \simeq \sum_m \Big| \frac{1}{\sqrt{2}} \big[\mathcal{F}_{\tau_{1m}}\{a(\nu)\} e^{-i\omega_0 \tau_{1m}} + \mathcal{F}_{\tau_{2m}}\{a(\nu)\} e^{-i\omega_0 \tau_{2m}} \big] \Big|^2. \tag{7.2.7}$$

Due to the common relative delay $\tau = \tau_1 - \tau_2$, all the sub-interference-patterns comprise the same sinusoidal modulation. The identical sub-interference-patterns add at the observation plane.

Now we approximate the sum into an integral of t_m, similar to what we have done earlier. The mutual-coherence function $\Gamma^{(1)}(\mathbf{r}_1, t_1; \mathbf{r}_2, t_2)$ is calculated as

$$\Gamma^{(1)}(\mathbf{r}_1, t_1; \mathbf{r}_2, t_2)$$
$$= \langle e^{i\omega_0 \tau} \sum_m \mathcal{F}^*_{\tau_{1m}}\{a(\nu)\} \, \mathcal{F}_{\tau_{2m}}\{a(\nu)\} \rangle$$
$$\simeq e^{i\omega_0 \tau} \int_\infty dt_m \, \mathcal{F}^*_{(\tau_1 - t_m)}\{a(\nu)\} \, \mathcal{F}_{(\tau_2 - t_m)}\{a(\nu)\}$$
$$= \mathcal{F}_\tau\{A^2(\nu)\} \, e^{i\omega_0 \tau} \tag{7.2.8}$$

where, again, $\tau = \tau_1 - \tau_2 = (s_2 - s_1)/c + (z_2 - z_1)/c$. The mutual-coherence function, and consequently the interference visibility, is quantitatively determined by how much the two wavepackets, $\mathcal{F}_{\tau_{1m}}\{a(\nu)\}$ and its delayed conjugate $\mathcal{F}^*_{\tau_{2m}}\{a(\nu)\}$, are overlapped in space-time. The integral has a maximum value when $\tau = \tau_1 - \tau_2 = 0$. The radiation fields become first-order incoherent when $\tau \geq 2\pi/\Delta\omega$. Again,

$$\tau_c = 2\pi/\Delta\omega$$

is called the coherence time of the field, which is nothing but the temporal width of the wavepackets.

The integral over t_m in Eq. (7.2.8) is mathematically equivalent to a time integral and, consequently, an autocorrelation or a self-convolution of the wavepacket:

$$\Gamma^{(1)}(\mathbf{r}_1, t_1; \mathbf{r}_2, t_2) = e^{-i\omega_0 \tau} \int_\infty dt \, \mathcal{F}^*_{\tau_1}\{E(\nu)\} \, \mathcal{F}_{\tau_2}\{E(\nu)\}$$
$$= \int_\infty dt' \, E^*(t') \, E(t' - \tau). \tag{7.2.9}$$

Applying the Wiener-Khintchine theorem, we have

$$\Gamma^{(1)}(\mathbf{r}_1, t_1; \mathbf{r}_2, t_2) = \int_\infty dt \, E^*(t) \, E(t - \tau) = \int_\infty d\nu \, |E(\nu)|^2 \, e^{-i\nu\tau}.$$

(III) $\Gamma^{(1)}(\mathbf{r}_1, t_1; \mathbf{r}_2, t_2)$: a single-photon wavepacket.

Now, we turn our discussion into an interesting and controversial topic of quantum optics: interference of a single-photon wavepacket. We assume the measurement involves only a single subfield or wavepacket at a time period of $\Delta t > \tau_c$. This experimental condition is achievable by using a weak light source at single-photon's level. The question is: do we observe interference? In classical optics, even if a wavepacket or subfield only carries energy of $\hbar\omega$, this wavepacket can still be divided into two[1] by taking path s_1 (passing through

---

[1]We may name it partial-wavepacket or partial-subfield to distinguish it from Einstein's quantized sub-field that carries quantized energy of $\hbar\omega$.

pinhole P_1) and/or path s_2 (passing through pinhole P_2), and then to be superposed at coordinate $(\mathbf{r}, t)$. The instantaneous intensity $I(\mathbf{r}, t)$ is the result of a superposition of two partial-wavepackets, one passing through pinhole P_1, another passing through pinhole P_2:

$$
I(\mathbf{r}, t) = \Big| \frac{1}{\sqrt{2}} \big[\mathcal{F}_{(\tau_1 - t_0)}\{a(\nu)\} e^{-i\omega_0(\tau_1 - t_0)} + \mathcal{F}_{(\tau_2 - t_0)}\{a(\nu)\} e^{-i\omega_0(\tau_2 - t_0)} \big] \Big|^2
$$

$$
= \frac{1}{2} \Big\{ \big| \mathcal{F}_{(\tau_1 - t_0)}\{a(\nu)\} \big|^2 + \big| \mathcal{F}_{(\tau_2 - t_0)}\{a(\nu)\} \big|^2
$$

$$
+ 2\, Re \big[\mathcal{F}^*_{(\tau_1 - t_0)}\{a(\nu)\} \mathcal{F}_{(\tau_2 - t_0)}\{a(\nu)\} e^{i\omega_0 \tau} \big] \Big\}, \tag{7.2.10}
$$

where, again, $a(\nu)$ is the amplitude for the Fourier-mode of frequency ω of the wavepacket, $\tau_1 = t_1 - z_1/c$, $\tau_2 = t_2 - z_2/c$, t_0 is the initial creation time of the wavepacket. It is easy to see from Eq. (7.2.10) that an interference pattern is potentially observable when the two partial-wavepackets overlap at the observation points $(\mathbf{r}, t)$. The maximum interference occurs in the neighborhood of $\tau = \tau_1 - \tau_2 \sim 0$ when the two partial-wavepackets completely overlap. For $\tau > 2\pi/\Delta\omega$, the two Fourier transforms cannot take nonzero values simultaneously, the third term of Eq. (7.2.10) vanishes, and consequently no interference is observable when the temporal delay is greater than the temporal width of the wavepacket. Obviously, the coherence time of a single-photon wavepacket is $\tau_c = 2\pi/\Delta\omega$, or concisely: the coherence time of a single-photon wavepacket equals the temporal width of the wavepacket.

On one hand, it is interesting to find that *one subfield or photon produces an interference pattern* of $I(\mathbf{r}, t)$ based on classical superposition. On the other hand, we are facing a serious question: how to measure the interference pattern? Suppose we place a 2-D photodetector, CCD or CMOS, in the observation plane, a photon can only produce one photoelectron from a pixel of the CCD or CMOS. The registration of one pixel cannot provide any meaningful information about the interference. A timely accumulative measurement of photon-counting is always necessary in this case. After a timely accumulative counting on a large number of subfields or wavepackets, one by one, we find the accumulated charges in each pixel of the CCD or CMOS, which is proportional to the number of photons registered at coordinate $\mathbf{r}$, within the accumulative time period of ΔT

$$
Q(\mathbf{r}) \propto \int_{\Delta T} dt\, I(\mathbf{r}, t) \propto n(\mathbf{r}) \propto \langle I(\mathbf{r}, t) \rangle_{\Delta T}. \tag{7.2.11}
$$

The interference pattern is observable from $Q(\mathbf{r})$ or $n(\mathbf{r})$ after a timely accumulative photon counting.

Another serious question then naturally arises: what do we mean *each wavepacket produces an interference-pattern*? In classical optics, this means the energy of the subfield $\hbar\omega$ is distributed on the entire interference pattern and the expected amount of energy crossing a unit area per unit time, precisely the Poynting vector or the intensity, of the electromagnetic field, is modulated sinusoidally as a function of the space-time coordinate $(\mathbf{r}, t)$. There is no problem in classical theory to have an interference pattern produced by the mth subfield in terms of an intensity distribution $I_m(\mathbf{r}, t)$ in space-time, even if the mth wavepacket only carries the energy $\hbar\omega$ of a single-photon. However, definitely, this is not the picture corresponding to the above photon-counting measurement. The energy $\hbar\omega$ of a subfield must be localized within one pixel of the photodetector array. If the energy $\hbar\omega$ of a subfield is distributed on the entire interference pattern, when a photoelectron is excited from a pixel of the photodetector array at $(\mathbf{r}, t)$ by taking an amount of energy $\hbar\omega$ from the electromagnetic field, how much time does it take for the energy from other coordinates of the interference pattern to arrive at that pixel? Remember, Einstein had asked a similar question to his students based on his two light-year big wavepacket[2]. Although the distance

---

[2]Einstein's story was given at the end of section 2.3.

between two points of an interference pattern may not be as big as a light-year, we must answer the same question.

From the measurement point of view, the interference pattern produced by a subfield means nothing but a probability distribution function $P(\mathbf{r}, t)$: the probability for a subfield, or a photon, to produce a photodetection event at space-time coordinate $(\mathbf{r}, t)$. In an accumulative measurement involving a large number of M wavepackets, there will be $MP(\mathbf{r})$ photodetection events occurring at coordinate $\mathbf{r}$, corresponding to the time averaged intensity $\langle I(\mathbf{r}, t) \rangle_{\Delta T}$.

(IV) $\Gamma(\mathbf{r}_1, t_1; \mathbf{r}_2, t_2)$: two independent single-photon wavepackets.

Assuming two subfields, mth and nth, created from two independent and randomly radiated atomic transitions, are superposed through an Young's double-pinhole interferometer for interference observation. The mth wavepacket passes the upper pinhole and propagates along path s_1, and the nth wavepacket passes the lower pinhole and propagates along path s_2. The instantaneous intensity $I(\mathbf{r}, t)$ in the observation plane is calculated as follows

$$
\begin{aligned}
I(&\mathbf{r}, t) \\
&= \left| \mathcal{F}_{\tau_{1m}} \{a_m(\nu)\} e^{-i\omega_0 \tau_{1m}} + \mathcal{F}_{\tau_{2n}} \{a_n(\nu)\} e^{-i\omega_0 \tau_{2n}} \right|^2 \\
&= \left| \mathcal{F}_{\tau_{1m}} \{a_m(\nu)\} \right|^2 + \left| \mathcal{F}_{\tau_{2n}} \{a_n(\nu)\} \right|^2 \\
&\quad + 2\, Re \left[\mathcal{F}_{\tau_{1m}}^* \{a_m(\nu)\} \mathcal{F}_{\tau_{2n}} \{a_2(\nu)\} e^{i\omega_0 [(t_n - t_m) + \tau]} \right],
\end{aligned} \tag{7.2.12}
$$

where $\tau_{1m} = \tau_1 - t_m$ and $\tau_{2n} = \tau_2 - t_n$; t_m and t_n are the initial creation times of the two wavepackets at the mth and the nth independent sub-sources. It is easy to see from Eq. (7.2.12) that an instantaneous interference pattern is potentially observable when the two wavepackets are overlapped at $(\mathbf{r}, t)$.

Again, we are facing a serous problem: how to measure the two wavepacket produced interference pattern? Suppose we place a 2-D photodetector, CCD or CMOS, in the observation plane, two photons can only produce two photoelectrons from two pixels of the CCD or CMOS. The registration of two pixels cannot provide any meaningful information about the interference. Similar to the single-photon case, a timely accumulative measurement of photon-counting is always necessary. In a timely accumulative measurement, each pair of wavepackets may produce two photoelectrons at a time t to charge two electronic integrators. From time to time, or from wavepacket-pair to wavepacket-pair, the photocurrents keep charging the integrators until achieving an observable level. The first-order temporal mutual-coherence function of two independent single-photon wavepackets is expected to be

$$
\begin{aligned}
\Gamma^{(1)}&(\mathbf{r}_1, \mathbf{r}_2) \\
&= \left\langle \sum_{m,n} \mathcal{F}_{\tau_{1m}}^* \{a(\nu)\} \mathcal{F}_{\tau_{2n}} \{a(\nu)\} e^{i\omega_0 [(t_n - t_m) + \tau]} \right\rangle \\
&= \left\langle \sum_{m,n} e^{i\omega_0 (t_n - t_m)} \left[\mathcal{F}_{\tau_{1m}}^* \{a(\nu)\} \mathcal{F}_{\tau_{2n}} \{a(\nu)\} e^{i\omega_0 \tau} \right] \right\rangle \\
&\simeq 0,
\end{aligned} \tag{7.2.13}
$$

where timely accumulated measurement is considered by taking into account $t_n - t_m =$ random number from time to time or from wavepacket-pair to wavepacket-pair. In Eq.(7.2.13), $\tau_{1m} = \tau_1 - t_m = (t - t_m) - (s_1 + z_1)/c$, $\tau_{2n} = \tau_2 - t_n = (t - t_n) - (s_2 + z_2)/c$, $\tau = (s_2 - s_1)/c + (z_2 - z_1)/c$. Note, the above calculation only calculate the mutual-coherence function produced by two different subfields: the mth subfield that is created from the mth atomic transition and the nth subfield that is created from the nth atomic transition.

$\Gamma^{(1)}(\mathbf{r}_1, \mathbf{r}_2) \simeq 0$ comes from the averaging of $(t_n - t_m)$: each m-nth pair of wavepackets produces an "instantaneous" interference pattern with a special phase of $\omega_0[(t_n - t_m) + \tau]$; due to the randomness of t_m and t_n from wavepacket-pair to wavepacket-pair, the averaged sinusoidal function of $\cos\{\omega_0[(t_n - t_n) + \tau]\}$ results in a value of zero. Based on Eq. (7.2.13) we may conclude that the interference between independent wavepackets at the single-photon's level is practically non-observable.

However, if we can force the two "independent" sub-sources radiating in phase, i.e., achieving $(t_n - t_m)$ =constant, these identical "instantaneous" interference patterns would add together accumulatively produce an observable interference pattern from the CCD or CMOS. The degree of the first-order coherence of the two in-phased single-photon wavepackets is determined by the value of $(t_n - t_m) = $ constant. When $(t_n - t_m) = 0$, $\Gamma^{(1)}(\mathbf{r}_1, t_1; \mathbf{r}_2, t_2)$ achieves its maximum value, we consider the two independent single-photon wavepackets first-order coherent; however, when $(t_n - t_m) > 2\pi/\Delta\omega$ the two wavepackets cannot have nonzero values simultaneously, $\Gamma^{(1)}(\mathbf{r}_1, t_1; \mathbf{r}_2, t_2) = 0$, we consider the two independent wavepackets first-order incoherent.

(V) $\Gamma^{(1)}(\mathbf{r}_1, t_1; \mathbf{r}_2, t_2)$: radiation of laser.

Case (1): A laser pulse

A laser pulse with enough power is able to produce an observable interference pattern on the observation plane of an Young's double-pinhole interferometer. A modern 2-D photodetector array, CCD or CMOS, is able to monitor the "one-shot" interference pattern $I(\mathbf{r}, t)$ of a laser pulse.

We have shown in our early discussions that the mutual-coherence function $\Gamma(\mathbf{r}_1, t_1; \mathbf{r}_2, t_2)$ can be calculated from the classical fields $E(\mathbf{r}_1, t_1)$ and $E(\mathbf{r}_2, t_2)$ passing through the upper pinhole P_1 and the lower pinhole P_2, respectively, at early times $t_1 = t - s_1/c$ and $t_2 = t - s_2/c$, where t refers the measurement time of the photodetector. The mutual-coherence function $\langle E^*(\mathbf{r}_1, t_1) E(\mathbf{r}_2, t_2) \rangle$ is found to be:

$$\Gamma^{(1)}(\mathbf{r}_1, t_1; \mathbf{r}_2, t_2) = \langle E^*(\mathbf{r}_1, t_1) E(\mathbf{r}_2, t_2) \rangle$$
$$= \mathcal{F}^*_{\tau_1}\{E(\nu)\} \mathcal{F}_{\tau_2}\{E(\nu)\} e^{i\omega_0 \tau} \qquad (7.2.14)$$

by *taking into account all possible realizations of the fields* $E(\mathbf{r}_1, t_1)$ and $E(\mathbf{r}_2, t_2)$.

In Einstein's picture of light, *taking into account all possible realizations of the fields* means *taking into account all possible random phases of the subfields*. A laser pulse is the result of coherent superposition among a large number of subfields and Fourier modes.

$$\Gamma^{(1)}_{12}(\mathbf{r}_1, t_1; \mathbf{r}_2, t_2) = \left\langle \sum_m E^*_m(\mathbf{r}_1, t_1) \sum_n E_n(\mathbf{r}_2, t_2) \right\rangle$$
$$= \sum_m \mathcal{F}^*_{\tau_1}\{a_m(\nu)\} \sum_n \mathcal{F}_{\tau_2}\{a_n(\nu)\} e^{i\omega_0 \tau}$$
$$= \mathcal{F}^*_{\tau_1}\{A(\nu)\} \mathcal{F}_{\tau_2}\{A(\nu)\} e^{i\omega_0 \tau} \qquad (7.2.15)$$

where $A(\nu) = \sum_m a(\nu)$. The expectation function is calculated by taking into account a constant relative phase for all subfields created from a large number of coherently radiated sub-sources. Similar to the wavepacket at single-photon's level, the degree of first-order coherence is determined by the overlapping-nonoverlapping of the wavepackets. At the neighborhood of $\tau \sim 0$, $\Gamma^{(1)}(\mathbf{r}_1, t_1; \mathbf{r}_2, t_2)$ achieves its maximum value, we consider the fields $E(\mathbf{r}_1, t_1)$ and $E(\mathbf{r}_2, t_2)$ first-order coherent; however, when $\tau > 2\pi/\Delta\omega$ the two wavepackets cannot have nonzero values simultaneously, $\Gamma^{(1)}(\mathbf{r}_1, t_1; \mathbf{r}_2, t_2) = 0$, we define the fields of $E(\mathbf{r}_1, t_1)$ and $E(\mathbf{r}_2, t_2)$ first-order incoherent, although the laser pulse itself is considered as coherent radiation.

From Eq.(7.2.15) we find that $\Gamma^{(1)}(\mathbf{r}_1, t_1; \mathbf{r}_2, t_2)$ indicates a non-stationary field, which is consistent with the nature of a laser pulse. Due to the non-stationary nature, we need to pay attention to the normalized degree of first-order coherence function $\gamma_{12}(\tau_1, \tau_2)$,

$$\gamma_{12}^{(1)}(\tau_1, \tau_2) = \frac{\mathcal{F}_{\tau_1}^* \{A(\nu)\} \mathcal{F}_{\tau_2} \{A(\nu)\} e^{i\omega_0 \tau}}{\sqrt{\left|\mathcal{F}_{\tau_1} \{A(\nu)\}\right|^2 \left|\mathcal{F}_{\tau_2} \{A(\nu)\}\right|^2}}, \tag{7.2.16}$$

under the condition of

$$\left|\mathcal{F}_{\tau_1} \{A(\nu)\}\right|^2 \left|\mathcal{F}_{\tau_2} \{A(\nu)\}\right|^2 \neq 0.$$

One may find that Eq. (7.2.16) leads to $|\gamma_{12}^{(1)}(\tau_1, \tau_2)| = 1$ for real functions, even if the two wavepackets only slightly overlap. This is because the product of the self-coherence functions $\Gamma_{11}^{(1)}(\tau_1)$ and $\Gamma_{22}^{(1)}(\tau_2)$, which is used for normalization, behaves the same as that of the mutual-coherence function $\Gamma_{12}^{(1)}(\tau_1, \tau_2)$. Examine Eq. (7.1.12), we find the visibility of the interference fringe modulation is very different from the degree of fist-order coherence $|\gamma_{12}^{(1)}(\tau_1, \tau_2)|$. In this case, the interference visibility has to be estimated from its definition of Eq. (7.1.12).

Case (2): CW laser radiation.

Continuous wave (CW) laser radiation may contain a single cavity mode, a few cavity modes, or a large number of cavity modes, each centered at frequency ω_{0j} with finite bandwidth $\Delta\omega_j$, where j labels the jth cavity mode. Each cavity mode can be treated as either a coherent wavepacket of temporal width $2\pi/\Delta\omega_j$ or a single frequency mode of $\Delta\omega_j \sim 0$. In principle, our earlier treatments of $\Gamma^{(1)}(\mathbf{r}_1, t_1; \mathbf{r}_2, t_2)$ may apply to CW laser radiation, except each wavepacket has a different center frequency ω_{0j}.

In the following, we estimate the first-order coherence function of CW laser radiation by approximating the fields $E(\mathbf{r}_1, t_1)$ and $E(\mathbf{r}_2, t_2)$ as a set of overlapped-partially overlapped wavepackets each centered at ω_{0j} with spectral bandwidth $\Delta\omega_j$,

$$\begin{aligned}
&\Gamma^{(1)}(\mathbf{r}_1, t_1; \mathbf{r}_2, t_2) \\
&= \Big\langle \sum_j \mathcal{F}_{\tau_{1j}}^* \{A_{1j}(\nu)\} e^{i\omega_{0j}(\tau_{1j})} \sum_k \mathcal{F}_{\tau_{2k}} \{A_{2k}(\nu)\} e^{-i\omega_{0k}(\tau_{2k})} \Big\rangle \\
&= \Big\langle \sum_{j=k} \mathcal{F}_{\tau_{1j}}^* \{A_{1j}(\nu)\} \mathcal{F}_{\tau_{2j}} \{A_{2j}(\nu)\} e^{i\omega_{0j}\tau} \Big\rangle \\
&\quad + \Big\langle \sum_{j \neq k} \mathcal{F}_{\tau_{1j}}^* \{A_{1j}(\nu)\} \mathcal{F}_{\tau_{2k}} \{A_{2k}(\nu)\} e^{i[\omega_{0j}(\tau_{1j}) - \omega_{0k}(\tau_{2k})]} \Big\rangle,
\end{aligned} \tag{7.2.17}$$

where $A_{1j}(\nu)$ and $A_{2k}(\nu)$ are the amplitude distribution function of the jth cavity mode along pass one and pass two, respectively, with $\nu = \omega_j - \omega_{0j}$.

The result of Eq. (7.2.17) depends on the coherent or incoherent relation between the cavity modes. For incoherent cavity modes, the second term in the summation vanishes, the observable sinusoidal modulation comes from the self-interferences of each cavity mode with itself:

$$\Gamma^{(1)}(\mathbf{r}_1, t_1; \mathbf{r}_2, t_2) = \sum_j \mathcal{F}_{\tau_{1j}}^* \{A_{1j}(\nu)\} \mathcal{F}_{\tau_{2j}} \{A_2(\nu)\} e^{i\omega_{0j}\tau}. \tag{7.2.18}$$

For coherent cavity modes, either mode-locked or other coherent relation, both first term and the second term of the summation have to be taken into account. In both cases, the beating frequencies between cavity modes may be observable.

Case (3): two independent laser pulses

In Einstein's picture, each laser pulse, or laser wavepacket may consist of a large number of identical and coherently superposed sub-wavepackets. Ensemble average is physically meaningful for the measurement of the interference between a pair of laser wavepackets. Taking into account all possible realizations of the fields $E(\mathbf{r}_1, t_1)$ and $E(\mathbf{r}_2, t_2)$ from the j-kth pair of independent laser wavepackets, the mutual-coherence function $\langle E^*(\mathbf{r}_1, t_1)E(\mathbf{r}_2, t_2)\rangle$ is found to be:

$$\Gamma^{(1)}(\mathbf{r}_1, t_1; \mathbf{r}_2, t_2) = \mathcal{F}^*_{\tau_{1j}}\{A_j(\nu)\}\,\mathcal{F}_{\tau_{2k}}\{A_k(\nu)\}\,e^{i\omega_0[(t_k - t_j) + \tau]}, \tag{7.2.19}$$

where, we have assumed the two independent laser pulses have the same central frequency ω_0; again, $A_j(\nu)$ is the amplitude for the mode ω of the jth laser wavepacket, t_j and t_k are the initial creation times of the two individual laser pulses. Note, here we have used j and k to label the jth and the kth laser pulses to distinguish the mth and the nth quantized subfield in Einstein's picture. It is easy to see that within a selected pair of laser wavepackets, $(t_k - t_j)$ holds a well-defined value. If the measurement is completed within a pair of laser pulses, the interference will be observable. In fact, the interference between two independent lasers was experimentally demonstrated by Mandel during the years of 1960's to 1970's after the invention of the laser.[3]

What will happen if the measurement involves a large number of individual pairs of wavepackets accumulatively? Can we still observe interference? As we have discussed earlier, in this kind of measurement, the finally measured interference pattern on the observation plane is the sum of a large number of sub-interference-patterns, each is produced by a pair of wavepackets

$$\langle I(\mathbf{r}, t)\rangle$$
$$\simeq \sum_{j,k}\left\{\left|\mathcal{F}_{\tau_{1j}}\{A_j(\nu)\}\right|^2 + \left|\mathcal{F}_{\tau_{2k}}\{A_k(\nu)\}\right|^2\right.$$
$$\left. + 2\,Re\,\mathcal{F}^*_{\tau_{1j}}\{A_1(\nu)\}\mathcal{F}_{\tau_{2k}}\{A_2(\nu)\}\,e^{i\omega_0[(t_k - t_j) + \tau]}\right\}, \tag{7.2.20}$$

where index j and k label the j-kth laser wavepacket-pair. If the laser wavepacket-pairs are "phase-locked", where "phase-lock" means forcing the two lasers to generate their wavepackets at $t_k - t_j = $ constant for all j-k pairs to have the relative phase $\omega_0[(t_k - t_j) + \tau] = \omega_0[\text{constant} + \tau]$, which is independent of j-k, the sub-interference-patterns would be identical from pulse-pair to pulse-pair, and consequently, the timely accumulative observation would be the sum of these identical sub-interference-patterns. In this case the timely accumulative interference, which involves a large number of phase-locked laser pulses, is observable.

7.3 FIRST-ORDER SPATIAL COHERENCE—EINSTEIN'S PICTURE

The finite bandwidth of the temporal spectrum $\Delta\omega$ is not the only factor determining the degree of first-order coherence and the visibility of first-order interference; the finite bandwidth of the spatial frequency $\Delta\vec{\kappa}$ is another important factor we have to take into

---

[3]These experiments stimulated a great deal of attention on a fundamental issue: can interference take place between two different photons? The debate was partially provoked from a statement of Dirac: "...photon... only interferes with itself. Interference between two different photons never occurs".

account regarding the visibility of interference observed from an Young's double-pinhole interferometer. This concern leads to the concepts of spatial coherence and the degree of first-order spatial coherence. The spatial coherence of light is directly related to the transverse dimension, or the angular size of the light source.

(I) Spatial coherence: thermal radiation source

In the following discussion, we assume a double-pinhole interferometer facing a distant star of finite angular size. The distant star, which consists of a large number of randomly radiated and randomly distributed independent point-like sub-sources of radiation, has an angular diameter of $\Delta\theta$ relative to the interferometer. To simplify the mathematics and to focus on the physics of spatial coherence, we assume each of the mth subfield radiated from each of the mth point-like sub-sources of the distant star, $E_m(\mathbf{r}, t)$, monochromatic. From the view point of the Young's double-pinhole interferometer, which is schematically illustrated in Fig. 7.3.1, each point-like sub-source is identified by an angular coordinate θ (1-D), and each subfield radiated from a sub-source has a different transverse wavevector along the x-axis, $k_x \sim k\theta$ (1-D). The fields $E(\mathbf{r}_1, t_1)$ at pinhole P_1 and $E(\mathbf{r}_2, t_2)$ at pinhole P_2 are treated as a superposition of all the randomly radiated and randomly distributed subfields from the entire surface of the distant star.

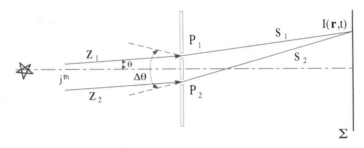

FIGURE 7.3.1 Schematic of the Young's double-slit interference experiment. The source is a distant star of finite angular size ($\Delta\theta \neq 0$) consisting of a large number of randomly radiated and randomly distributed point-like sub-sources, or atomic transitions. To simplify the mathematics, we restrict our calculations in 1-D.

The mutual-coherence function $\Gamma^{(1)}(\mathbf{r}_1, t_1; \mathbf{r}_2, t_2)$ is then written as

$$\Gamma^{(1)}(\mathbf{r}_1, t_1; \mathbf{r}_2, t_2)$$
$$= \langle \sum_{m=1}^{M} E_m^*(\mathbf{r}_1, t_1) \sum_{n=1}^{N} E_n(\mathbf{r}_2, t_2) \rangle \qquad (7.3.1)$$
$$= \langle \sum_{m=n} E_m^*(\mathbf{r}_1, t_1) E_m(\mathbf{r}_2, t_2) \rangle + \langle \sum_{m \neq n} E_m^*(\mathbf{r}_1, t_1) E_n(\mathbf{r}_2, t_2) \rangle$$

where m and n label each of the independent contributions of a point-like sub-source. It is convenient to break up the sum into two groups. The second group ($m \neq n$) of the sum vanishes in the expectation calculation when taking into account all possible values of the random relative phase differences $\varphi_m - \varphi_n$. Thus $\Gamma^{(1)}(\mathbf{r}_1, t_1; \mathbf{r}_2, t_2)$ becomes

$$\Gamma^{(1)}(\mathbf{r}_1, t_1; \mathbf{r}_2, t_2) = \sum_{m=1}^{M} E_m^*(\mathbf{r}_1, t_1) \, E_m(\mathbf{r}_2, t_2). \qquad (7.3.2)$$

Eq. (7.3.2) indicates a summation of a large number of individual interferences each is the result of a self-interference: a subfield interfering with the subfield itself.

We can transfer Eq.(7.3.2) into a simple integral by assuming a random distribution of the sub-sources on the distant star

$$\Gamma^{(1)}(\mathbf{r}_1, t_1; \mathbf{r}_2, t_2)$$

$$= \sum_{m=1}^{M} E_{1m}^* \, e^{i[\omega(t - \frac{s_1}{c}) - k(z_{1m})]} \, E_{2m} e^{-i[\omega(t - \frac{s_2}{c}) - k(z_{2m})]}$$

$$= e^{i\omega\tau_s} \sum_{m=1}^{M} a_{1m} a_{2m} \, e^{ik(z_{2m} - z_{1m})}$$

$$\cong e^{i\omega\tau_s} \int_{-\frac{\Delta\theta}{2}}^{\frac{\Delta\theta}{2}} (\frac{I_0}{\Delta\theta}) \, d\theta \, e^{ik\theta b}$$

$$= I_0 \operatorname{sinc}(\frac{\pi b \Delta\theta}{\lambda}) \, e^{i\omega\tau_s}, \tag{7.3.3}$$

where $\tau_s = (s_2 - s_1)/c$ and $I_0 \sim \sum_{m=1}^{M} a_{1m} a_{2m} \sim$ constant. To simply the mathematics, we have restricted our calculation in 1-D, and the integral is taken over the entire angular diameter of the star along the x direction from $-\Delta\theta/2$ to $\Delta\theta/2$.

The normalized degree of first-order spatial coherence of the two fields at the upper and the lower pinholes is thus:

$$\gamma^{(1)}(\mathbf{r}_1, t_1; \mathbf{r}_2, t_2) = \operatorname{sinc}(\frac{\pi b \Delta\theta}{\lambda}) \, e^{i\omega\tau_s} = \operatorname{sinc}(\Delta k_x b) \, e^{i\omega\tau_s}. \tag{7.3.4}$$

Here $\gamma^{(1)}(\mathbf{r}_1, t_1; \mathbf{r}_2, t_2)$ is a function of the separation b between the upper and the lower pinholes and the angular size $\Delta\theta$ of the distant star. The two fields at the upper and lower pinholes are said to be spatially coherent when $b \ll \lambda/\Delta\theta$, or $b \ll 2\pi/\Delta k_x$, ($|\gamma^{(1)}(\mathbf{r}_1, t_1; \mathbf{r}_2, t_2)| \cong 1$), and the two fields are said to be spatially incoherent when $b \geq \lambda/\Delta\theta$, or $b \geq 2\pi/\Delta k_x$, ($|\gamma^{(1)}(\mathbf{r}_1, t_1; \mathbf{r}_2, t_2)| \cong 0$). For a point source, $\Delta\theta \sim 0$, and consequently $|\gamma_{12}| \cong 1$ for any value of b. This means the radiation fields excited by a point radiation source are spatially coherent despite the spatial separation between the fields.

The interference pattern on the observation plane can be written as:

$$\langle I(\mathbf{r}, t) \rangle = I_0 [1 + \operatorname{sinc}(\frac{\pi b \Delta\theta}{\lambda}) \cos(\omega\tau_s)], \tag{7.3.5}$$

with an interference visibility

$$V = |\gamma_{12}^{(1)}| = \operatorname{sinc}(\frac{\pi b \Delta\theta}{\lambda}) = \operatorname{sinc}(\Delta k_x b). \tag{7.3.6}$$

Notice we have simplified the calculation of the $\gamma^{(1)}$ function by assuming a monochromatic plane-wave of single wavelength λ, so that $V = |\gamma_{12}^{(1)}|$ is independent of τ_s.

We may find the following physical picture useful for understanding the concept of spatial coherence. In Fig. 7.3.1, we have assumed a large number, $M \sim \infty$, of randomly radiated and randomly distributed point-like sub-sources. The subfields created from these sub-sources are all emitted with random relative phases. The only observable interferences are the self-interferences of these subfields. The cross-interference between subfields excited from different sub-sources cancels completely while taking into account all possible values of the relative phases between subfields. The measured intensity on the observation plane is thus

$$\langle I(\mathbf{r}, t) \rangle = \sum_{m=1}^{M} I_m(\mathbf{r}, t)$$

$$= \sum_{m=1}^{M} \left| \frac{1}{\sqrt{2}} \left[E_m(\mathbf{r}_1, t_1) + E_m(\mathbf{r}_2, t_2) \right] \right|^2$$

$$= \sum_{m=1}^{M} I_{m0} \left\{ 1 + \cos[\omega \tau_s + k(z_{2m} - z_{1m})] \right\}$$

$$\simeq \sum_{m=1}^{M} I_{m0} \left[1 + \cos(\omega \tau_s + k\theta_m b) \right] \tag{7.3.7}$$

We may consider that each point sub-source on the distant star, identified by angle θ_m, produces a Young's double-slit sinusoidal interference pattern on the observation plane with a relative phase shift $k(z_{2m} - z_{1m}) \sim k\theta_m b \sim k_x b$. The maximum relative phase separation between these individual patterns is $k(\Delta\theta b)$, or $\Delta k_x b$, corresponding to the phase shift between the interference patterns excited by the sub-sources of $\Delta\theta$ and $-\Delta\theta$. Therefore, when $2\pi\Delta\theta b/\lambda \ll 2\pi$, i.e., $b \ll \lambda/\Delta\theta$, or $b \ll 2\pi/\Delta k_x$, the relative phase shifts are not large enough to produce noticeable separation between the sub-interference-patterns and so the interference modulation is observable. However, when the relative phase shifts increase to a certain value of $2\pi\Delta\theta b/\lambda \sim 2\pi$, i.e., $b \sim \lambda/\Delta\theta$ or $b \sim 2\pi/\Delta k_x$, the individual patterns become significantly separated. The spread of the patterns smoothes the light intensity distribution on the observation plane so the interference pattern can no longer be identified, and the interference modulation visibility is reduced from 100% to 0%. Based on the above observation we introduce the concept of spatial coherence of the field. For a given light source with angular size $\Delta\theta$, the fields $E(\mathbf{r}_1, t)$ and $E(\mathbf{r}_2, t)$ are considered as spatially coherent if their transverse spatial separation is less than

$$l_c = \lambda/\Delta\theta = 2\pi/\Delta k_x, \tag{7.3.8}$$

where l_c is defined as the transverse coherence length of the radiation. Any fields with spatial separation beyond l_c are considered as spatially incoherent, which implies no observable first-order interference from a double-slit interferometer. The degree of first-order spatial coherence, defined in Eq. (7.1.10) and calculated in Eq. (7.3.4), is a quantitative measure of the Young's double-slit interference for a radiation source of finite angular diameter $\Delta\theta$.

Taking up an early suggestion by Fizeau, Michelson designed a stellar interferometer based on the mechanism of Young's double-pinhole interference. One important application of the Michelson stellar interferometer is the measurement of the angular size of a distant star or the angular separation between distant double stars. The principle and operation of this stellar interferometer is quite simple. What one needs to do is to manipulate the separation between the two pinholes from $b = 0$, point by point, to a critical value $b = l_c$. If one can make an accurate judgement at the critical value of $b = l_c$ at which the interference pattern becomes invisible, this value of l_c can be used to estimate of the angular size of the distant star. According to Eq. (7.3.5), the double-slit interferometer starts to lose its interference at $l_c = \lambda/\Delta\theta$. The angular size of the distant star, $\Delta\theta = \lambda/l_c$, is thus measured with certain accuracy. Of course, making an accurate judgement is never easy, as there are too many physical parameters that contribute to the instability of an interference pattern.

(II) Coherent radiation source

For coherent radiation sources the condition of having an observable Young's double-pinhole interference pattern is obvious: the two pinholes must be illuminated by the radiation simultaneously. We have discussed the propagation of spatially coherent radiation, such as a TEM_{00} laser beam, in sections 2.3 and 3.2. The coherent radiation propagates in a collimated manner with diffraction limited diverging angle $\Delta\vartheta = \lambda/b$ in 1-D, or $\Delta\vartheta = 1.22\lambda/D$ in 2-D, where D is the diameter of the source. If the spatially coherent radiation is regarded as a wavepacket in the transverse dimension, the above condition indicates that the distance between the two pinholes must be less than the transverse width of the spatial wavepacket. In other words, the spatial separation between the fields $E(\mathbf{r}_1, t_1)$ and $E(\mathbf{r}_2, t_2)$ must be within the spatial coherence of the field, which is the transverse width of the spatial wavepacket in this case, in order to have observable interference.

(III) Spatial coherence: a single-photon wavepacket.

Now, again, we turn our discussion into an interesting and controversial topic of quantum optics: interference of a single-photon wavepacket. We assume the light source is so weak that only a single subfield or wavepacket of $\hbar\omega$ is received at the double-pinhole interferometer within a time period of the observation. The observed single-photon wavepacket may be created from any point-like sub-sources that are randomly distributed on the surface of the distant star. The questions are: (1) Do we observe interference from the double-slit interferometer? (2) If we do, how to measure it?

Based on our previous discussions in this section, the answer to question (1) is positive:

$$I(\mathbf{r}, t) = \left| \frac{1}{\sqrt{2}} \left[E_m(\mathbf{r}_1, t_1) + E_m(\mathbf{r}_2, t_2) \right] \right|^2$$
$$= I_{m0} \left\{ 1 + \cos[\omega\tau_s + k(z_{2m} - z_{1m})] \right\}$$
$$\simeq I_{m0} \left[1 + \cos(\omega\tau_s + k\theta b) \right]. \tag{7.3.9}$$

Again, in classical theory, although the mth subfield only carries energy of $\hbar\omega$, this wavepacket can be divided into two by taking path s_1 (passing through pinhole P_1) and/or path s_2 (passing through pinhole P_2), and then to be superposed at coordinate $(\mathbf{r}, t)$. The classical intensity distribution on the observation plane is the result of a superposition of two partial-wavepackets, one passing through pinhole P_1, another passing through pinhole P_2.

The answer to question (2) is the same as our previous answer in this section: photon counting. Placing a modern photon-counting photodetector array in the observation plane, one wavepacket or subfield can only excite one photoelectron from one element of the photodetector array, which may not give us any meaningful information about the interference. However, after a timely accumulative counting on a large number of subfields or wavepackets, one by one, we find the accumulated charges in each photoelement, which is proportional to the number of photons registered at coordinate $(\mathbf{r})$, within the accumulative time period of ΔT

$$Q(\mathbf{r}) \propto \int_{\Delta T} dt\, I(\mathbf{r}, t) \propto n(\mathbf{r}) \propto \langle I(\mathbf{r}, t) \rangle_{\Delta T}$$
$$= I_0 \left[1 + \text{sinc}\left(\frac{\pi b \Delta\theta}{\lambda} \right) \cos(\omega\tau_s) \right]. \tag{7.3.10}$$

The interference pattern is observable from $Q(\mathbf{r})$ or $n(\mathbf{r})$ after a timely accumulative photon counting.

We are then facing the same serious question: how long does it take for the energy from other part of the interference pattern to arrive at the excited single CCD pixal? In the classical picture of a subfield, its energy of $\hbar\omega$ should be uniformly distributed on the entire interference pattern. If we insist the classical picture of a photon, we would be easily tripped to the above question as Einstein did to himself and to his students.

From the view point of quantum theory, the entire interference pattern "produced" by a subfield of $\hbar\omega$ means nothing but a probability distribution function $P(\mathbf{r}, t)$: the probability for a subfield, or a photon, to produce a photodetection event at space-time coordinate $(\mathbf{r}, t)$. In an accumulative measurement involving a large number of M photons, there will be $MP(\mathbf{r})$ photodetection events occurring at coordinate $\mathbf{r}$, corresponding to the time averaged intensity $\langle I(\mathbf{r}, t) \rangle_{\Delta T}$.

We are ready to introduce the concept of quantum first-order coherence of light.

7.4 FIRST-ORDER COHERENCE OF LIGHT—QM THEORY

To be consistent with the analysis of first-order classical coherence, our introduction of first-order quantum coherence starting from the same Young's double-pinhole interferometer of Figure 7.1.1. The concept of quantum degree of first-order coherence will be introduced during the analysis of the experiment. To emphasize the quantum nature of the concept, in this section, we will focus on the following photon counting type measurements: (1) the measurement of a photon; (2) the accumulative measurement of a large number of photons.

The schematic setup of the Young's double-pinhole interference experiment is the same as that shown in 7.1.1, except that the measurement device is a single-photon counting detector array, either a CCD or a CMOS working in photon counting mode, or a point-like photon counting detector scannable on the observation plane. Following the photodetection theory introduced in Chapter 3, the probability of observing a photoelectron event from a point-like photodetector by annihilating a photon at space-time point $(\mathbf{r}, t)$ is:

$$P(\mathbf{r}, t) \propto \langle \hat{E}^{(-)}(\mathbf{r}, t) \hat{E}^{(+)}(\mathbf{r}, t) \rangle$$
$$= \left\langle \langle \hat{E}^{(-)}(\mathbf{r}, t) \, \hat{E}^{(+)}(\mathbf{r}, t) \rangle_{\mathrm{QM}} \right\rangle_{\mathrm{Ensemble}} \tag{7.4.1}$$

where $\hat{E}^{(-)}$ and $\hat{E}^{(+)}$ are the negative frequency and the positive frequency field operators, respectively, the expectation value is calculated by averaging the field operators over the quantum state which may be a pure state or a mixed state, depending on the light source. For the simply experimental setup of Fig. 7.1.1, the field operators can be approximated as follows:

$$\hat{E}^{(+)}(\mathbf{r}, t) \simeq \int d\omega \, \hat{a}(\omega) \frac{1}{\sqrt{2}} \left[g_0(\omega; \mathbf{r}_1, t_1) \, g_1(\omega; \mathbf{r}, t) + g_0(\omega; \mathbf{r}_2, t_2) \, g_2(\omega; \mathbf{r}, t) \right]$$

$$\simeq \int d\omega \, \hat{a}(\omega) \frac{1}{\sqrt{2}} \left[g_0(\omega; \mathbf{r}_1, t_1) \, e^{-i[\omega(t - t_1) - \omega(s_1)/c]} \right.$$
$$\left. + g_0(\omega; \mathbf{r}_2, t_2) \, e^{-i[\omega(t - t_2) - \omega(s_2)/c]} \right]$$

$$= \hat{E}^{(+)}(\mathbf{r}_1, t_1) + \hat{E}^{(+)}(\mathbf{r}_2, t_2)$$

$$\hat{E}^{(-)}(\mathbf{r}, t) \simeq \int d\omega \, \hat{a}^\dagger(\omega) \frac{1}{\sqrt{2}} \left[g_0^*(\omega; \mathbf{r}_1, t_1) \, g_1^*(\omega; \mathbf{r}, t) + g_0^*(\omega; \mathbf{r}_2, t_2) \, g_2^*(\omega; \mathbf{r}, t) \right]$$

$$\simeq \int d\omega \, \hat{a}^\dagger(\omega) \frac{1}{\sqrt{2}} \left[g_0^*(\omega; \mathbf{r}_1, t_1) \, e^{i[\omega(t - t_1) - \omega(s_1)/c]} \right.$$
$$\left. + g_0^*(\omega; \mathbf{r}_2, t_2) \, e^{i[\omega(t - t_2) - \omega(s_2)/c]} \right]$$

$$= \hat{E}^{(-)}(\mathbf{r}_1, t_1) + \hat{E}^{(-)}(\mathbf{r}_2, t_2), \tag{7.4.2}$$

by defining $t_j \equiv t - s_j/c$ (free-space), $j = 1, 2$, s_j denotes the optical path from the jth pinhole to the point-like photon counting detector. In Eq. (7.4.2), $g_0(\omega; \mathbf{r}_j, t_j)$ is the Green's function, or the field propagator, propagating the ω mode of the field from the point-like source to the jth pinhole of $(\mathbf{r}_j, t_j)$; $g_j(\omega; \mathbf{r}, t)$ is the Green's function, or the field operator, propagating the ω mode of the field from the jth pinhole to the observation coordinate $(\mathbf{r}, t)$. To be consistent with the concept of probability and probability amplitude, all the above Green's functions, or the field propagators, are normalized. In Eq. (7.4.2), again, we have absorbed constant $1/\sqrt{2}$ into the field operators.

We conclude that the field $E(\mathbf{r}, t)$ at space-time point $(\mathbf{r}, t)$ is the result of a superposition of the fields $E(\mathbf{r}_1, t_1)$ and $E(\mathbf{r}_2, t_2)$ which passed the upper pinhole P_1 and the lower pinhole P_2, respectively, at earlier times t_1 and t_2,

$$
\begin{aligned}
&\langle \hat{E}^{(-)}(\mathbf{r}, t)\, \hat{E}^{(+)}(\mathbf{r}, t) \rangle \\
&= \langle\, [\hat{E}^{(-)}(\mathbf{r}_1, t_1) + \hat{E}^{(-)}(\mathbf{r}_2, t_2)][\hat{E}^{(+)}(\mathbf{r}_1, t_1) + \hat{E}^{(+)}(\mathbf{r}_2, t_2)] \rangle \\
&= \langle\, \hat{E}^{(-)}(\mathbf{r}_1, t_1)\hat{E}^{(+)}(\mathbf{r}_1, t_1) \rangle + \langle\, \hat{E}^{(-)}(\mathbf{r}_2, t_2)\hat{E}^{(+)}(\mathbf{r}_2, t_2) \rangle \\
&\quad + \langle\, \hat{E}^{(-)}(\mathbf{r}_1, t_1)\hat{E}^{(+)}(\mathbf{r}_2, t_2) \rangle + \langle\, \hat{E}^{(-)}(\mathbf{r}_2, t_2)\hat{E}^{(+)}(\mathbf{r}_1, t_1) \rangle.
\end{aligned}
\tag{7.4.3}
$$

We define the first-order self-coherence function or self-correlation function between field $E(\mathbf{r}_j, t_j)$, $j = 1, 2$, and the field itself as:

$$
\begin{aligned}
G^{(1)}(\mathbf{r}_1, t_1; \mathbf{r}_1, t_1) &\equiv \langle\, \hat{E}^{(-)}(\mathbf{r}_1, t_1)\, \hat{E}^{(+)}(\mathbf{r}_1, t_1) \rangle = G^{(1)}_{11}, \\
G^{(1)}(\mathbf{r}_2, t_2; \mathbf{r}_2, t_2) &\equiv \langle\, \hat{E}^{(-)}(\mathbf{r}_2, t_2)\, \hat{E}^{(+)}(\mathbf{r}_2, t_2) \rangle = G^{(1)}_{22},
\end{aligned}
\tag{7.4.4}
$$

and the first-order mutual-coherence function or cross-correlation function between fields $E(\mathbf{r}_1, t_1)$ and $E(\mathbf{r}_2, t_2)$ as:

$$
\begin{aligned}
G^{(1)}(\mathbf{r}_1, t_1; \mathbf{r}_2, t_2) &\equiv \langle\, \hat{E}^{(-)}(\mathbf{r}_1, t_1)\, \hat{E}^{(+)}(\mathbf{r}_2, t_2) \rangle = G^{(1)}_{12}, \\
G^{(1)}(\mathbf{r}_2, t_2; \mathbf{r}_1, t_1) &\equiv \langle\, \hat{E}^{(-)}(\mathbf{r}_2, t_2)\, \hat{E}^{(+)}(\mathbf{r}_1, t_1) \rangle = G^{(1)}_{21}.
\end{aligned}
\tag{7.4.5}
$$

Obviously,

$$
G^{(1)}(\mathbf{r}_1, t_1; \mathbf{r}_2, t_2) = G^{(1)*}(\mathbf{r}_2, t_2; \mathbf{r}_1, t_1).
$$

The probability of producing a photoelectron by annihilating a photon at space-time point $(\mathbf{r}, t)$ is thus written in terms of the first-order coherence functions of the fields $E(\mathbf{r}_1, t_1)$ and $E(\mathbf{r}_2, t_2)$:

$$
\begin{aligned}
P(\mathbf{r}, t) \propto\ & G^{(1)}(\mathbf{r}_1, t_1; \mathbf{r}_1, t_1) + G^{(1)}(\mathbf{r}_2, t_2; \mathbf{r}_2, t_2) \\
& + G^{(1)}(\mathbf{r}_1, t_1; \mathbf{r}_2, t_2) + G^{(1)}(\mathbf{r}_2, t_2; \mathbf{r}_1, t_1) \\
= & G^{(1)}_{11} + G^{(1)}_{22} + G^{(1)}_{12} + G^{(1)}_{21}.
\end{aligned}
\tag{7.4.6}
$$

The first two terms are the self-coherence or self-correlation functions and the last two terms are the mutual-coherence functions or cross-correlation functions.

In the above discussion we have used "coherence function" and "correlation function" to define the same mathematical expression. The name of "coherence" emphasizes the interference nature of the phenomenon. The terminology of "correlation" emphasizes the statistical process of the measurement. In quantum mechanics, the superposition of quantum amplitudes quantitively specifies the interference behavior of a single-photon. However, one measurement of a photon cannot provide us any meaningful information about the physics except a record of a photoelectron event. To learn meaningful physics from an experimental observation, a large number of single-photon based measurements are always necessary. Quantum theory of light predicts the probability for a photodetection event to occur at space-time coordinate $(\mathbf{r}, t)$ and specifies the cause of the nontrivial statistical distribution function as the result of interference. We will use either "coherence function" or "correlation function" in the following discussion. Keep in mind that both names represent the same physics from the same measurement.

In the view of quantum theory of light, after a photon is created at the light source, it has certain probabilities to pass the upper pinhole P_1 and the lower pinhole P_2. The Young's double-pinhole interferometer provides two different yet indistinguishable alternative ways for a photon to produce a photodetection event at $(\mathbf{r}, t)$. The superposition of the quantum amplitudes determines a nontrivial probability distribution function for observing a photon at $(\mathbf{r}, t)$. If one insists to give each term in Eq. (7.4.6) a physical meaning, the self-correlation function corresponds to the probability of having the photon passing through either the upper or the lower pinholes (which-path); the cross-correlation functions (third and the fourth terms), which determine the interference, correspond to the probabilities of having the photon passing through both upper and lower pinholes (both-path) simultaneously; however, one has to deal with "negative probability" problem.[4]

The degree of first-order coherence, namely the normalized first-order coherence function or correlation function, between radiation fields $E(\mathbf{r}_1, t_1)$ at space-time $(\mathbf{r}_1, t_1)$ and $E(\mathbf{r}_2, t_2)$ at space-time $(\mathbf{r}_2, t_2)$ is defined as

$$
\begin{aligned}
&g^{(1)}(\mathbf{r}_1, t_1; \mathbf{r}_2, t_2) \\
&\equiv \frac{\langle \hat{E}^{(-)}(\mathbf{r}_1, t_1) \hat{E}^{(+)}(\mathbf{r}_2, t_2) \rangle}{\left[\langle \hat{E}^{(-)}(\mathbf{r}_1, t_1) \hat{E}^{(+)}(\mathbf{r}_1, t_1) \rangle \right]^{\frac{1}{2}} \left[\langle \hat{E}^{(-)}(\mathbf{r}_2, t_2) \hat{E}^{(+)}(\mathbf{r}_2, t_2) \rangle \right]^{\frac{1}{2}}} \\
&= \frac{G^{(1)}(\mathbf{r}_1, t_1; \mathbf{r}_2, t_2)}{\left[G^{(1)}(\mathbf{r}_1, t_1; \mathbf{r}_1, t_1)\right]^{\frac{1}{2}} \left[G^{(1)}(\mathbf{r}_2, t_2; \mathbf{r}_2, t_2)\right]^{\frac{1}{2}}}.
\end{aligned}
\tag{7.4.7}
$$

$|g^{(1)}(\mathbf{r}_1, t_1; \mathbf{r}_2, t_2)|$ takes a value between 1 and 0. Similar to that of the classical theory, we name the quantized fields as first-order coherent fields, first-order partial coherent fields and first-order incoherent fields accordingly:

1st-order Coherent	if	$\|g^{(1)}(\mathbf{r}_1, t_1; \mathbf{r}_2, t_2)\| = 1$
1st-order Partial Coherent	if	$0 < \|g^{(1)}(\mathbf{r}_1, t_1; \mathbf{r}_2, t_2)\| < 1$
1st-order Incoherent	if	$\|g^{(1)}(\mathbf{r}_1, t_1; \mathbf{r}_2, t_2)\| = 0$.

In the Young's double-pinhole interferometer, the probability of having a "click" event at space-time point $(\mathbf{r}, t)$ is then written in terms of the degree of first-order coherence function $g^{(1)}(\mathbf{r}_1, t_1; \mathbf{r}_2, t_2)$:

$$
P(\mathbf{r}, t) \propto 1 + Re\, g^{(1)}(\mathbf{r}_1, t_1; \mathbf{r}_2, t_2)
\tag{7.4.8}
$$

where, to simplify the physical picture, we have assumed $G^{(1)}(\mathbf{r}_1, t_1; \mathbf{r}_1, t_1) = G^{(1)}(\mathbf{r}_2, t_2; \mathbf{r}_2, t_2)$. The value of $g^{(1)}(\mathbf{r}_1, t_1; \mathbf{r}_2, t_2)$ is a measure of the interference visibility. The relationship between $g^{(1)}(\mathbf{r}_1, t_1; \mathbf{r}_2, t_2)$ and the interference visibility is similar to that of the classical theory.

From Eqs. 7.4.4 and 7.4.7, the first-order coherence function and the normalized first-order coherence function between $E(\mathbf{r}_1, t_1)$ and $E(\mathbf{r}_2, t_2)$ obviously have the following property:

$$
G^{(1)}(\mathbf{r}_1, t_1; \mathbf{r}_2, t_2) = G^{(1)*}(\mathbf{r}_2, t_2; \mathbf{r}_1, t_1)
\tag{7.4.9}
$$

$$
g^{(1)}(\mathbf{r}_1, t_1; \mathbf{r}_2, t_2) = g^{(1)*}(\mathbf{r}_2, t_2; \mathbf{r}_1, t_1)
$$

---

[4]The cross-correlation function in the double-pinhole experiment is a sinusoidal function with both positive (constructive interference) and negative (destructive interference) values. If one insists to interpret the cross-correlation as the probability of having a photon passing through both upper and lower pinholes (both-path) simultaneously, one cannot avoid facing "negative probability" problem in this simple classic interference measurement.

Quantum theory has provided us with a set of tools, formulated in Heisenberg representation, for the evaluation of first-order optical coherence and correlation. Light fields at two space-time points are said to be coherent if interference is observable in principle. The higher the visibility of the interference, the higher the degree of the coherence. Different quantum states may produce different coherence functions for a certain measurement.

The calculation of the first-order coherence functions $G^{(1)}(\mathbf{r}_j, t_j; \mathbf{r}_k, t_k)$, $j = 1, 2$, $k = 1, 2$, for a pure state is straightforward

$$G^{(1)}(\mathbf{r}_j, t_j; \mathbf{r}_k, t_k) = \langle \Psi | \hat{E}^{(-)}(\mathbf{r}_j, t_j) \hat{E}^{(+)}(\mathbf{r}_k, t_k) | \Psi \rangle, \tag{7.4.10}$$

where $|\Psi\rangle$ is the state of the field. Mixed state is usually characterized by a density operator or matrix,

$$\hat{\rho} = \sum_n P_n |\Psi_n\rangle\langle\Psi_n|,$$

where P_n is the probability of finding the field in a given set of pure state $|\Psi_n\rangle$. The first-order coherence functions are formally calculated from

$$
\begin{aligned}
G^{(1)}&(\mathbf{r}_j, t_j; \mathbf{r}_k, t_k) \\
&= \left\langle \langle \hat{E}^{(-)}(\mathbf{r}_j, t_j) \hat{E}^{(+)}(\mathbf{r}_k, t_k) \rangle_{\mathrm{QM}} \right\rangle_{\mathrm{Ensemble}} \\
&= tr\left[\hat{\rho}\, \hat{E}^{(-)}(\mathbf{r}_j, t_j)\, \hat{E}^{(+)}(\mathbf{r}_k, t_k) \right] \\
&= \sum_n P_n \langle \Psi_n | \hat{E}^{(-)}(\mathbf{r}_j, t_j)\, \hat{E}^{(+)}(\mathbf{r}_k, t_k) | \Psi_n \rangle \\
&= \sum_n P_n\, G_n^{(1)}(\mathbf{r}_j, t_j; \mathbf{r}_k, t_k).
\end{aligned}
\tag{7.4.11}
$$

Eq. 7.4.11 is a statistical weighted sum over all individual contributions of the $G_n^{(1)}$s.

7.5 FIRST-ORDER TEMPORAL QUANTUM COHERENCE OF LIGHT

In this section, we discuss and calculate the first-order temporal quantum coherence for different quantum states. We will find the concept of effective wavefunction for a photon and for a system of quanta is helpful in these excises. Similar to the classical treatment of

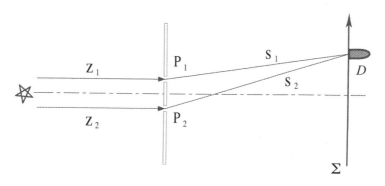

FIGURE 7.5.1 Schematic of the same Young's double-slit interferometer, which measures the temporal coherence of light produced from a far-field weak point-like source containing a large number of atoms that are ready for two-level atomic transitions.

first-order temporal coherence in section 6.2, we start from the following analysis based on a similar Young's double-slit experimental setup of Fig. 7.5.1 with a distant point-like light source at $\mathbf{r} = 0$ which contains of a large number of incoherent or coherent sub-sources, except a point-like photon counting detector is used to replace the analog photodetector.

(I) A single-photon wavepacket.

Considering a weak distant point-like source at single-photon's level from which only one photon is observable at the interferometer within the responds time of the photon counting detector. What are the self-coherence function and mutual-coherence function for a single-photon wavepacket? This excise serves two main purposes: (1) learning the basic skills for calculating first-order quantum coherence; (2) introducing the concept of effective wavefunction of photon.

The concept of single-photon wavepacket has been introduced in Chapter 1 and 2. Following our early discussions, we take the following approximation to characterize the state of the measured single-photon field:

$$|\Psi\rangle = \int d\omega \, f(\omega) \, \hat{a}^\dagger(\omega)|0\rangle, \tag{7.5.1}$$

indicating a single-photon state in the form of coherent superposition of Fock states $\hat{a}^\dagger(\omega)|0\rangle = |...0, 1_\omega, ..., 0, ...\rangle$, and $f(\omega)$ is the normalized complex probability amplitude of the Fock state, which may contain a phase factor $e^{i\varphi(\omega)}$ for the mode ω, as we have discussed in Chapter 2. We will write this phase factor explicitly into the form of $f(\omega)e^{i\omega t_0}$ in most of the following calculations. In the notation of $f(\omega)e^{i\omega t_0}$, $f(\omega)$ is the real and positive amplitude.

For the simply experimental setup of Fig. 7.5.1 , the field operators can be approximated as:

$$\hat{E}^{(+)}(z_j, t_j) \simeq \int d\omega \, \hat{a}(\omega) \, g_0(\omega; \mathbf{r}_j, t_j)]$$
$$= \int d\omega \, \hat{a}(\omega) \, e^{-i(\omega t_j - k z_j)} = \int d\omega \, \hat{a}(\omega) \, e^{-i\omega \tau_j} \tag{7.5.2}$$

where $\tau_j = t_j - z_j/c$ (free-space, and $t_0 = 0$, $z_0 = 0$). If $t_0 \neq 0$, $\tau_j = (t_j - t_0) - z_j/c$.

The first-order quantum mutual-coherence function $G^{(1)}(z_1, t_1; z_2, t_2)$ is thus calculated as follows:

$$
\begin{aligned}
&G^{(1)}(z_1, t_1; z_2, t_2) \\
&= \langle\Psi| \hat{E}^{(-)}(z_1, t_1) \, \hat{E}^{(+)}(z_2, t_2) \, |\Psi\rangle \\
&= \sum_n \langle\Psi|\hat{E}^{(-)}(z_1, t_1)|n\rangle\langle n|\hat{E}^{(+)}(z_2, t_2)|\Psi\rangle \\
&= \langle\Psi|\hat{E}^{(-)}(z_1, t_1)|0\rangle\langle 0|\hat{E}^{(+)}(z_2, t_2)|\Psi\rangle \\
&= \Big[\int d\omega \, f^*(\omega) \, \langle 0|\hat{a}(\omega)\Big] \Big[\int d\omega'' \, \hat{a}^\dagger(\omega'') \, e^{i(\omega'' t_1 - k'' z_1)}\Big]|0\rangle \\
&\quad \times \langle 0|\Big[\int d\omega''' \, \hat{a}(\omega''') \, e^{-i(\omega''' t_2 - k''' z_2)}\Big] \Big[\int d\omega' f(\omega') \, \hat{a}^\dagger(\omega')|0\rangle \Big] \\
&= \Big\{ \int d\omega \, f^*(\omega) \, e^{i(\omega t_1 - k z_1)}\Big\}\Big\{ \int d\omega' f(\omega') \, e^{-i(\omega' t_2 - k' z_2)}\Big\} \\
&= \Big[e^{i\omega_0(\tau_1 - t_0)} \mathcal{F}^*_{\tau_1 - t_0}\big\{ f(\nu)\big\} \Big]\Big[e^{-i\omega_0(\tau_2 - t_0)} \mathcal{F}_{\tau_2 - t_0}\big\{ f(\nu)\big\} \Big] \\
&= \Psi^*(z_1, t_1) \, \Psi(z_2, t_2) \tag{7.5.3}
\end{aligned}
$$

where we have used notation $f(\omega)e^{i\omega t_0}$ in the Fourier transform. In Eq. 7.5.3, we have also defined $\nu = \omega - \omega_0$, $t_j = t - s_j/c$ (free-space), $j = 1, 2$, and $\tau_j = t_j - z_j/c$ (free-space) as usual. To derive the above result, we have applied the completeness relation

$$\sum_n |n\rangle\langle n| = 1.$$

For a pure state of single-photon wavepacket, $G^{(1)}(z_1, t_1; z_2, t_2)$ cannot be written as a stationary function of $t_1 - t_2$ in general; however, it can be easily written as a product of two time dependent wavepackets, or effective wavefunctions, which has been defined in Chapter 3, $\Psi^*(z_1, t_1)$ and $\Psi(z_2, t_2)$ with

$$\Psi(z_j, t_j) = \langle 0|\hat{E}^{(+)}(z_j, t_j)|\Psi\rangle$$
$$= e^{-i\omega_0(\tau_j - t_0)}\mathcal{F}_{\tau_j - t_0}\{f(\nu)\}. \tag{7.5.4}$$

Here we have introduced t_0 to indicate the creation time of the wavepacket from the atomic transition. Sometimes, we write the wavepackets in the form

$$\Psi(z_j, t_j) = e^{-i\omega_0(\tau_j - t_0)}\mathcal{F}_{\tau_j}\{f(\nu)\},$$

in which the phase factor is embedded into the complex amplitude $f(\nu)$. The Fourier transform of the complex function $f(\nu)$, which contains a phase factor $e^{i\omega t_0}$, will shift the wavepacket from τ_j to $\tau_j - t_0$. It is clear that the first-order temporal coherence is determined by the overlapping and non-overlapping of the effective wavefunction with its conjugate. The first-order quantum coherence time is obviously the temporal width of the single-photon wavepacket $\tau_c = 2\pi/\Delta\omega$, agreeing with the classical case.

Similar to the mutual-coherence function $G^{(1)}(z_1, t_1; z_2, t_2)$, the self-correlation functions $G^{(1)}(z_1, t_1; z_1, t_1)$ and $G^{(1)}(z_2, t_2; z_2, t_2)$ can be expressed as the products of the time dependent wavepackets $\Psi(z_1, t_1)$ and $\Psi(z_2, t_2)$ with their conjugates:

$$G^{(1)}(z_1, t_1; z_1, t_1) = \left|\mathcal{F}_{\tau_1 - t_0}\{f(\nu)\}\right|^2 = \left|\Psi(z_1, t_1)\right|^2$$
$$G^{(1)}(z_2, t_2; z_2, t_2) = \left|\mathcal{F}_{\tau_2 - t_0}\{f(\nu)\}\right|^2 = \left|\Psi(z_2, t_2)\right|^2. \tag{7.5.5}$$

The probability of having a photodetection event (a "click") at space-time point $(\mathbf{r}, t)$ is thus written in terms of the superposition of the wavepackets $\Psi(z_1, t_1)$ and $\Psi(z_2, t_2)$:

$$P(\mathbf{r}, t) \propto \left|\Psi(\mathbf{r}, t)\right|^2$$
$$= \left|\frac{1}{\sqrt{2}}[\Psi(z_1, t_1) + \Psi(z_2, t_2)]\right|^2$$
$$= \frac{1}{2}\left[|\Psi(z_1, t_1)|^2 + |\Psi(z_2, t_2)|^2 + \Psi^*(z_1, t_1)\Psi(z_2, t_2) + \Psi(z_1, t_1)\Psi^*(z_2, t_2)\right]. \tag{7.5.6}$$

Although no wavefunction is defined for a photon in quantum mechanics, it is interesting to see that the effective wavefunction $\Psi(\mathbf{r}, t)$ plays the role of wavefunction in Eq. (7.5.6). The effective wavefunction represents the probability amplitude of observing a photon at space-time point $(\mathbf{r}, t)$, or the probability amplitude of having a photodetection event at space-time point $(\mathbf{r}, t)$. In the above experiment, the effective wavefunction $\Psi(\mathbf{r}, t)$ has two different yet indistinguishable[5] amplitudes $\Psi(z_1, t_1)$ and $\Psi(z_2, t_2)$, namely the probability amplitude of observing a photon at (z_1, t_1) and the probability amplitude of observing a photon at (z_2, t_2), respectively. Each amplitude is in the form of a wavepacket in space-time. The total effective wavefunction $\Psi(\mathbf{r}, t)$ is the result of the superposition of two effective wavefunctions or two

---

[5]Note, here "indistinguishable" refers to a photodetection event at space-time point $(\mathbf{r}, t)$.

dynamical wavepackets in space-time. The effective wavefunction reflects the wave-particle nature of a photon. The normalized degree of first-order temporal coherence function is thus written in terms of the effective wavefunction of a photon:

$$g^{(1)}(z_1, t_1; z_2, t_2) = \frac{\Psi^*(z_1, t_1)\, \Psi(z_2, t_2)}{\left[\,|\,\Psi(z_1, t_1)\,|^2\,\right]^{\frac{1}{2}} \left[\,|\,\Psi(z_2, t_2)\,|^2\,\right]^{\frac{1}{2}}}. \tag{7.5.7}$$

(II) A large number of randomly distributed and randomly created wavepackets—thermal state.

For the same Young's double-pinhole interferometer of Fig. 7.5.1, we assume a far-field weak point-like light source containing a large number of independent and randomly radiated atoms that are ready for two-level atomic transitions. Taking the lowest-order contribution to a photodetection event of D, see Chapter 2, the first-order temporal mutual-coherence function and self-coherence function can be calculated from

$$G^{(1)}(z_j, t_j; z_k, t_k) = \left\langle \langle \Psi | \hat{E}^{(-)}(z_j, t_j) \hat{E}^{(+)}(z_k, t_k) | \Psi \rangle \right\rangle_{\text{En}}, \tag{7.5.8}$$

where $j = 1, 2$, $k = 1, 2$, and

$$|\Psi\rangle = \sum_m c_m \int d\omega\, f_m(\omega)\, \hat{a}_m^\dagger(\omega)\,|\,0\,\rangle = \sum_m c_m |\Psi_m\rangle,$$

representing the first-order approximated state of thermal field, namely the incoherently superposed single-photon states. $G^{(1)}(z_j, t_j; z_k, t_k)$ can be also calculated from the density operator,

$$G^{(1)}(z_j, t_j; z_k, t_k) = tr\, \hat{\rho}^{(1)} \hat{E}^{(-)}(z_j, t_j) \hat{E}^{(+)}(z_k, t_k) \tag{7.5.9}$$

with

$$\hat{\rho}^{(1)} = \sum_m P_m\, |\Psi_m\rangle\langle\Psi_m|.$$

The only difference between Eq. 7.5.8 and Eq. 7.5.9 is that Eq. (7.5.9) considers ensemble average before quantum expectation. In Eq. (7.5.9) ensemble average has been completed in the process of calculating the density operator. To have a complete view on the quantum mechanical treatment, the following excises is based on Eq. 7.5.8, which will complete the ensemble average after the quantum expectation.

Similar to Eq. (7.5.2), the field operators $\hat{E}^{(-)}(z_j, t_j)$ and $\hat{E}^{(+)}(z_k, t_k)$ can be approximated as

$$\hat{E}^{(-)}(z_j, t_j) = \sum_m \int d\omega\, \hat{a}_m^\dagger(\omega)\, g_m(\omega; z_j, t_j)$$

$$= \sum_m \int d\omega\, \hat{a}_m^\dagger(\omega)\, e^{i\omega\tau_j}$$

$$\hat{E}^{(+)}(z_k, t_k) = \sum_n \int d\omega'\, \hat{a}_n(\omega')\, g_n^*(\omega'; z_k, t_k)$$

$$= \sum_n \int d\omega'\, \hat{a}_n(\omega')\, e^{-i\omega'\tau_k}. \tag{7.5.10}$$

where $g_m(\omega; z_j, t_j)$ is the Green's function that propagates the ω mode of the mth subfield from (z_m, t_m) to z_j, t_j, $j = 1, 2$.

Following Eq. (7.5.8), the mutual-coherence function $G^{(1)}(z_1, t_1; z_2, t_2)$ is calculated as

$$
\begin{aligned}
G^{(1)}&(z_1, t_1; z_2, t_2) \\
&= \Big\langle \sum_m c_m^* \langle \Psi_m | \hat{E}^{(-)}(z_1, t_1)\, \hat{E}^{(+)}(z_2, t_2) \sum_n c_n |\Psi_n\rangle \Big\rangle_{\mathrm{En}} \\
&= \Big\langle \sum_l \sum_m c_m^* \langle \Psi_m | \hat{E}^{(-)}(z_1, t_1)|l\rangle\langle l|\hat{E}^{(+)}(z_2, t_2) \sum_n c_n |\Psi_n\rangle \Big\rangle_{\mathrm{En}} \\
&= \Big\langle \sum_m c_m^* \langle \Psi_m | \hat{E}^{(-)}(z_1, t_1)|0\rangle\langle 0|\hat{E}^{(+)}(z_2, t_2) \sum_n c_n |\Psi_n\rangle \Big\rangle_{\mathrm{En}} \\
&= \Big\langle \sum_m c_m^* \psi_m^*(z_1, t_1) \sum_n c_n \psi_n(z_2, t_2) \Big\rangle_{\mathrm{En}} \\
&= \big\langle \Psi^*(z_1, t_1)\Psi(z_2, t_2) \big\rangle_{\mathrm{En}}
\end{aligned}
\tag{7.5.11}
$$

where we have defined an effective wavefunction for the nth photon:

$$
\psi_n(z_j, t_j) \equiv \langle 0|\hat{E}^{(+)}(z_j, t_j)|\Psi_n\rangle.
\tag{7.5.12}
$$

and an effective wavefunction for the ensemble of the photon system:

$$
\Psi(z_j, t_j) \equiv \langle 0|\hat{E}^{(+)}(z_j, t_j)|\Psi\rangle.
\tag{7.5.13}
$$

Obviously, the effective wavefunction of the photon system is the result of a superposition of the single-photon effective wavefunctions

$$
\Psi(z_j, t_j) = \sum_n c_n\, \psi_n(z_j, t_j).
\tag{7.5.14}
$$

The above superposition is consistent with Einstein's model of thermal field. The effective wavefunction of a single-photon plays the same role as that of Einstein's subfield. In fact, we could simply replace Einstein's subfield with the effective wave function of single-photon in all early analysis for thermal radiation and obtain the same results.

Substituting the effective wavefunctions into Eq. 7.5.11, the quantum mutual-coherence function $G^{(1)}(z_1, t_1; z_2, t_2)$ turns to be

$$
\begin{aligned}
G^{(1)}&(z_1, t_1; z_2, t_2) \\
&\propto \Big\langle \sum_{m,n} c_m^* c_n e^{-i\omega_0(t_{0m} - t_{0n})} e^{i\omega_0 \tau} \mathcal{F}_{\tau_1 - t_{0m}}^* \{f(\nu)\} \mathcal{F}_{\tau_2 - t_{0n}} \{f(\nu)\} \Big\rangle_{\mathrm{En}} \\
&\simeq \sum_n |c_n|^2 e^{i\omega_0 \tau} \mathcal{F}_{\tau_1 - t_{0n}}^* \{f(\nu)\} \mathcal{F}_{\tau_2 - t_{0n}} \{f(\nu)\} \\
&= \sum_n |c_n|^2 \psi_n^*(z_1, t_1)\psi_n(z_2, t_2).
\end{aligned}
\tag{7.5.15}
$$

Eq. (7.5.15) indicates that the first-order temporal coherence is determined by the overlapping and non-overlapping of the effective wavefunction of each single-photon and its conjugate. The physical picture behind the first-order quantum coherence of thermal field, which consists a large number of randomly distributed and randomly created wavepackets, is similar to that of single-photon wavepacket. This suggests that the first-order coherence time of thermal field should also be determined by the temporal width of each single-photon wavepackets $2\pi/\Delta\omega_m$, where $\Delta\omega_m$ is the spectrum of the mth single-photon wavepacket. Assuming identical spectrum for all measured randomly radiated and randomly distributed photons, the sum of m can be approximated as the following convolution

$$G^{(1)}(z_1, t_1; z_2, t_2)$$

$$\simeq \int dt_n\, e^{i\omega_0\tau} \mathcal{F}^*_{\tau_1 - t_{0n}}\{f(\nu)\} \mathcal{F}_{\tau_2 - t_{0n}}\{f(\nu)\}$$

$$= \mathcal{F}_\tau\{a^2(\nu)\}\, e^{i\omega_0\tau} \qquad (7.5.16)$$

which agrees with Eq. (7.2.8). i.e., the coherence time of thermal field calculated from the quantum theory of first-order coherence is the same as that calculated from Einstein's granularity picture.

The calculation of the self-coherence function $G^{(1)}(z_j, t_j; z_j, t_j)$, $j = 1, 2$ is straightforward

$$G^{(1)}(z_j, t_j; z_j, t_j)$$

$$= \Big\langle \sum_m c_m^* \langle \Psi_m | \hat{E}^{(-)}(z_j, t_j)\, \hat{E}^{(+)}(z_j, t_j) \sum_n c_n |\Psi_n\rangle \Big\rangle_{\text{En}}$$

$$= \Big\langle \sum_l \langle \sum_m c_m^* \langle \Psi_m | \hat{E}^{(-)}(z_j, t_j) | l \rangle \langle l | \hat{E}^{(+)}(z_j, t_j) \sum_n c_n |\Psi_n\rangle \Big\rangle_{\text{En}}$$

$$= \Big\langle \big| \langle 0 | \hat{E}^{(+)}(z_j, t_j) \sum_n c_n |\Psi_n\rangle \big|^2 \Big\rangle_{\text{En}}$$

$$= \Big\langle \big| \Psi(z_j, t_j) \big|^2 \Big\rangle_{\text{En}} \qquad (7.5.17)$$

Substituting the effective wavefunction of Eq. 7.5.4 into Eq. 7.5.17, the self-coherence function $G^{(1)}(z_j, t_j; z_j, t_j)$ is evaluated from the following ensemble average

$$G^{(1)}(z_j, t_j; z_j, t_j)$$

$$= \Big\langle \sum_{m,n} c_m^* c_n e^{-i\omega_0(t_{0m} - t_{0n})} \mathcal{F}^*_{\tau_j - t_{0m}}\{f(\nu)\} \mathcal{F}_{\tau_j - t_{0n}}\{f(\nu)\} \Big\rangle_{\text{En}}$$

$$\simeq \sum_n |c_n|^2 \big| \mathcal{F}_{\tau_j - t_{0n}}\{f(\nu)\} \big|^2$$

$$= \sum_n |c_n|^2 \big| \psi_n(z_j, t_j) \big|^2 \qquad (7.5.18)$$

where, again, we have assumed a large number of atomic transitions with the same spectrum distribution and the same central frequency ω_0, however random relative phases. The physics behind Eq. (7.5.18), especially the cancelation of the cross terms between the mth and the nth effective wavefunction, is similar to that of the Einstein's picture of thermal light with a large number of wavepackets produced from a large number of independent and randomly radiated sub-sources, except the concept of Einstein's subfield is replaced with the concept of effective wavefunction of a photon.

The probability of observing a photoelectron event from the photon counting detector D at space-time coordinate $(\mathbf{r}, t)$ is therefore:

$$P(\mathbf{r}, t) \propto G_{11}^{(1)} + G_{22}^{(1)} + G_{12}^{(1)} + G_{21}^{(1)}$$

$$= \sum_n |c_n|^2 \Big| \frac{1}{\sqrt{2}} \big[\psi_n(z_1, t_1) + \psi_n(z_2, t_2) \big] \Big|^2. \qquad (7.5.19)$$

Eq. (7.5.19) indicates that for thermal light, or a radiation system that consists of a large number of radomly distributed and randomly created photons, "...photon... only interferes with itself. Interference between two different photons never occurs". The physical picture is very clear: each independently and randomly created photon may produce an interference pattern by means of a probability distribution function of that photon in space-time. The sum of these individual single-photon interferences yields the final observed interference. Replacing the effective wavefunctions $\Psi_n(z_j, t_j)$ in Eq. (7.5.19) with their Fourier transform representation as wavepackets, we obtain

$$
\begin{aligned}
P(\mathbf{r}, t) &\propto \sum_n \frac{|c_n|^2}{2} \left[\left| \mathcal{F}_{\tau_1 - t_{0n}} \{f(\nu)\} \right|^2 + \left| \mathcal{F}_{\tau_2 - t_{0n}} \{f(\nu)\} \right|^2 \right. \\
&\quad \left. + 2Re\, e^{i\omega_0 \tau} \mathcal{F}^*_{\tau_1 - t_{0n}} \{f(\nu)\} \mathcal{F}_{\tau_2 - t_{0n}} \{f_n(\nu)\} \right] \\
&\simeq \int dt_{0n} \frac{1}{2} \left[\left| \mathcal{F}_{\tau_1 - t_{0n}} \{f(\nu)\} \right|^2 + \left| \mathcal{F}_{\tau_2 - t_{0n}} \{f(\nu)\} \right|^2 \right. \\
&\quad \left. + 2Re\, e^{i\omega_0 \tau} \mathcal{F}^*_{\tau_1 - t_{0n}} \{f(\nu)\} \mathcal{F}_{\tau_2 - t_{0n}} \{f_n(\nu)\} \right] \\
&\propto 1 + \mathcal{F}_\tau \{|f(\nu)|^2\} \cos \omega_0 \tau, \quad\quad\quad (7.5.20)
\end{aligned}
$$

where we have assumed a random and continuous distribution of t_{0n} with $|c_n|^2 =$ constant. Again, $\tau = \tau_1 - \tau_2 = (s_2 - s_1)/c$ for the experimental setup of Fig. 7.5.1. The integral over t_0 is mathematically equivalent to a time integral. The results of the integrals are easily obtained by applying the Parseval theorem:

$$
G^{(1)}(z_j, t_j; z_j, t_j) \simeq \int dt_0 \left| \mathcal{F}_{\tau_j - t_0} \{f(\nu)\} \right|^2 \simeq 1,
$$

and the Wiener-Khintchine theorem:

$$
\begin{aligned}
G^{(1)}(z_1, t_1; z_2, t_2) &\simeq e^{i\omega_0 \tau} \int dt_0\, \mathcal{F}^*_{\tau_1 - t_0} \{f(\nu)\} \mathcal{F}_{\tau_2 - t_0} \{f(\nu)\} \\
&\simeq e^{i\omega_0 \tau} \mathcal{F}_\tau \{|f(\nu)|^2\}.
\end{aligned}
$$

The calculation is similar to what we have done earlier for the Einstein's subfield, except the spectrum distribution function $f(\nu)$ is normalized for the quantum effective wavefunction:

$$
\int d\nu \, |f(\nu)|^2 = 1.
$$

(III) Coherent state.

Considering the same Young's double-pinhole interferometer, which provides two different optical paths for propagating its input radiation to a photodetector for photodetection. In this exercise, we model the input radiation as a multi-mode coherent state with one polarization, which is written as a product state of a set of single-mode coherent states

$$
|\{\alpha\}\rangle = \prod_\omega |\alpha_\omega\rangle,
$$

where we have simplify the problem to 1-D. In terms of the concept of photon, $|\{\alpha\}\rangle$ represents the state of a group of identical photons with $|\alpha|^2$ the mean number of identical indistinguishable photons.

The following excise is for the calculation of 1-D temporal mutual-coherence function:

$$
\begin{aligned}
G^{(1)}&(z_1, t_1; z_2, t_2) \\
&= \langle \{\alpha\}| \, \hat{E}^{(-)}(z_1, t_1)\hat{E}^{(+)}(z_2, t_2) \, |\{\alpha\}\rangle \\
&= \langle \{\alpha\}| \int d\omega \, \hat{a}^\dagger(\omega) \, e^{i\omega\tau_1} \int d\omega' \, e^{-i\omega'\tau_2} \, \hat{a}(\omega') \, |\{\alpha\}\rangle \\
&= \int d\omega \, a(\omega)e^{-i\varphi(\omega)} \, e^{i\omega\tau_1} \int d\omega' \, a(\omega')e^{i\varphi(\omega')} \, e^{-i\omega'\tau_2} \\
&= \Psi^*(z_1, t_1) \, \Psi(z_2, t_2),
\end{aligned}
\tag{7.5.21}
$$

where

$$
\Psi(z_j, t_j) \equiv \int d\omega \, a(\omega)e^{i\varphi(\omega)} \, e^{-i\omega\tau_j}
\tag{7.5.22}
$$

is defined as the effective wavefunction of a group of identical photons, with $\alpha(\omega) = |\alpha(\omega)|e^{i\varphi(\omega)}$ the eigenvalue of the annihilation operator $\hat{a}(\omega)$, which is also the complex amplitude of the Fourier mode ω of the wavepacket. Obviously, the first-order coherence time of coherent state is the temporal width of the effective wavefunction $\psi(z_j, t_j)$, or the coherent field $E(z_j, t_j)$, $\tau_c = 2\pi/\Delta\omega$.

In fact, the effective wavefunction of a group identical photons defined in Eq. (7.5.22) has the same mathematical form and plays the same role as that of the classical field $E(z_j, t_j)$. It is interesting to see that the first-order temporal mutual-coherence function $G^{(1)}(z_1, t_1; z_2, t_2)$ of coherent state is formally the same as that of the classical fields $E(z_1, t_1)$ and $E(z_2, t_2)$. The rest part of calculations are similar to that of the classical treatment.

(IV) Thermal field in coherent state representation

Considering the coherent state representation of thermal field that has been introduced in Eq.(4.6.7),

$$
|\Psi\rangle = \prod_m |\Psi_m\rangle = \prod_m \prod_\omega |\alpha_m(\omega)\rangle,
$$

where we have explicitly separated index m which identifies the mth photon or the mth group of identical photons, from ω which indicates the ωth mode of 1-D.

Following Eqs. (7.5.10) and (7.5.21), the first-order temporal mutual-coherence function of the measured thermal field is thus

$$
\begin{aligned}
G^{(1)}&(z_1, t_1; z_2, t_2) \\
&= \Big\langle \prod_m \langle\Psi_m| \, \hat{E}^{(-)}(z_1, t_1)\hat{E}^{(+)}(z_2, t_2) \prod_n |\Psi_n\rangle \Big\rangle_{\text{En}} \\
&= \Big\langle \sum_m \int d\omega \, a_m(\omega)e^{-i\varphi_m(\omega)} \, e^{i\omega\tau_1} \sum_n \int d\omega' \, a_n(\omega')e^{i\varphi_n(\omega')} \, e^{-i\omega'\tau_2} \Big\rangle_{\text{En}} \\
&= \Big\langle \sum_m \psi_m^*(z_1, t_1) \sum_n \psi_n(z_2, t_2) \Big\rangle_{\text{En}} \\
&= \sum_m \psi_m^*(z_1, t_1) \, \psi_m(z_2, t_2),
\end{aligned}
\tag{7.5.23}
$$

where

$$
\psi_m(z_j, t_j) = \int d\omega \, a_m(\omega)e^{i\varphi_m(\omega)} e^{-i\omega\tau_j}
$$

is defined as the effective wavefunction of the mth group identical photons, agreeing with our early result that was derived from Einstein's granularity model of thermal field, except replacing the subfields with the effective wavefunction. Obviously, the first-order coherence time of thermal field calculated from coherence state representation is $\tau_c = 2\pi/\Delta\omega$, agreeing with our early results.

Following a similar calculation, the first-order self-coherence function of thermal field in coherent state representation is

$$G^{(1)}(z_j, t_j; z_j, t_j) = \sum_m \left| \psi_m(z_j, t_j) \right|^2 \tag{7.5.24}$$

where, again, $\psi_m(z_j, t_j)$ is the effective wavefunction of the mth group identical photons.

7.6   FIRST-ORDER SPATIAL QUANTUM COHERENCE OF LIGHT

Similar to the discussion of first-order spatial classical coherence, we assume a double-pinhole interferometer facing a weak distant star of finite angular size. The distant star, which consists of a large number of randomly radiated and randomly distributed independent atomic transitions, has an angular diameter of $\Delta\theta$ relative to the interferometer, as shown in Fig. 7.3.1. The observed radiation field is in thermal state at single-photon's level. A photon counting CCD or CMOS or a scannable point-like photon counting detector is placed on the observation plane of the interferometer to measure the interference. In this section, we introduce the concept of first-order spatial quantum coherence through a simple calculation based on the single-photon state representation.

The thermal state in single-photon state representation can be written as:

$$|\Psi\rangle \simeq \sum_m c_m |\Psi_m\rangle = \sum_m \int d\mathbf{k}\, f_m(\mathbf{k})\, \hat{a}_m^\dagger(\mathbf{k})|0\rangle \tag{7.6.1}$$

where c_m is absorbed into $f_m(\omega)$.

The field operators are similar to that of Eq. (7.5.10), except approximated into a more general form of 3-D:

$$\hat{E}^{(+)}(\mathbf{r}, t) = \sum_m \int d\mathbf{k}\, \hat{a}_m(\mathbf{k})\, g_m(\mathbf{k}; \mathbf{r}, t)$$

$$\hat{E}^{(-)}(\mathbf{r}, t) = \sum_m \int d\mathbf{k}\, \hat{a}_m^\dagger(\mathbf{k})\, g_m^*(\mathbf{k}; \mathbf{r}, t) \tag{7.6.2}$$

where $g_m(\mathbf{k}; \mathbf{r}, t)$ is the Green's function that propagates the $\mathbf{k}$-mode of the mth subfield from the mth atomic transition to space-time coordinate $(\mathbf{r}, t)$.

The first-order quantum mutual-coherence function and self-coherence functions, respectively, are calculated in the following,

$$G^{(1)}(\mathbf{r}_1, t_1; \mathbf{r}_2, t_2)$$

$$= \Big\langle \sum_m c_m \langle \Psi_m | \hat{E}^{(-)}(\mathbf{r}_1, t_1) \hat{E}^{(+)}(\mathbf{r}_2, t_2) \sum_n c_n |\Psi_n\rangle \Big\rangle_{En}$$

$$= \Big\langle \sum_m \psi_m^*(\mathbf{r}_1, t_1) \sum_n \psi_n(\mathbf{r}_2, t_2) \Big\rangle$$

$$= \sum_m \psi_m^*(\mathbf{r}_1, t_1) \psi_m(\mathbf{r}_2, t_2) \tag{7.6.3}$$

and

$$
\begin{aligned}
G^{(1)}(\mathbf{r}_j, t_j; \mathbf{r}_j, t_j) & \\
&= \Big\langle \sum_m c_m \langle \Psi_m | \hat{E}^{(-)}(\mathbf{r}_j, t_j) \hat{E}^{(+)}(\mathbf{r}_j, t_j) \sum_n c_n | \Psi_n \rangle \Big\rangle_{En} \\
&= \Big\langle \sum_m \psi_m^*(\mathbf{r}_j, t_j) \sum_n \psi_n(\mathbf{r}_j, t_j) \Big\rangle \\
&= \sum_m |\psi_n(\mathbf{r}_j, t_j)|^2
\end{aligned}
\tag{7.6.4}
$$

where $\psi_m(\mathbf{r}_j, t_j)$, $j = 1, 2$, is defined as the effective wavefunction of the mth photon,

$$
\psi_m(\mathbf{r}_j, t_j) = \int d\mathbf{k}\, f_m(\mathbf{k})\, g_m(\mathbf{k}; r_j, t_j).
\tag{7.6.5}
$$

Replacing the effective wavefunctions with Einstein's subfields, we find $G^{(1)}(\mathbf{r}_1, t_1; \mathbf{r}_2, t_2)$ is the same as $\Gamma^{(1)}(\mathbf{r}_1, t_1; \mathbf{r}_2, t_2)$ that has been calculated in section 6.3. We thus have

$$
G^{(1)}(\mathbf{r}_1, t_1; \mathbf{r}_2, t_2) = G_0 \operatorname{sinc}\left(\frac{\pi b \Delta\theta}{\lambda}\right) e^{i\omega\tau_s}.
\tag{7.6.6}
$$

where $\tau_s = (s_2 - s_1)/c$; and the normalized quantum degree of first-order spatial coherence is:

$$
g^{(1)}(\mathbf{r}_1, t_1; \mathbf{r}_2, t_2) = \operatorname{sinc}\left(\frac{\pi b \Delta\theta}{\lambda}\right) e^{i\omega\tau_s} = \operatorname{sinc}\left(\Delta k_x\, b\right) e^{i\omega\tau_s}.
\tag{7.6.7}
$$

The probability for a single-photon to produce a photoelectron event at space-time coordinate $(\mathbf{r}, t)$ is thus a sinusoidal function of $\tau_s = (s_2 - s_1)/c$,

$$
P(\mathbf{r}, t) \propto 1 + \operatorname{sinc}\left(\frac{\pi b \Delta\theta}{\lambda}\right) \cos\left(\omega\tau_s\right),
\tag{7.6.8}
$$

with modulation visibility

$$
V = |g_{12}^{(1)}| = \operatorname{sinc}\left(\frac{\pi b \Delta\theta}{\lambda}\right) = \operatorname{sinc}\left(\Delta k_x\, b\right),
\tag{7.6.9}
$$

which is determined mainly by the angular diameter $\Delta\theta$ of the light source. Similar to the classical first-order spatial coherence, we define $\Delta l_c \equiv \lambda / \Delta\theta$ the first-order spatial quantum coherence length of the thermal field. Obviously, the interference is unobservable when the spatial separation of the two pinholes is greater than the first-order spatial quantum coherence length $b > \Delta l_c$.

Note, similar to that of the first-order spatial classical coherence in section 6.3, to simplify the mathematics, we have approximated a monochromatic thermal field.

After a timely accumulative counting on a large number of photons, one by one, we find the number of photons registered in each photoelement of the CCD or CMOS, or registered in the scannable point-like photodetector on the observation plane, is modulated sinusoidally as a function of $\tau_s = (s_2 - s_1)/c$ with modulation visibility $\operatorname{sinc}(\pi b \Delta\theta / \lambda)$:

$$
n(\mathbf{r}) \propto n_0 [1 + \operatorname{sinc}\left(\frac{\pi b \Delta\theta}{\lambda}\right) \cos\left(\omega\tau_s\right)]
\tag{7.6.10}
$$

where n_0 is the mean number of photons accumulatively received by the Young's double-pinhole interferometer. The interference pattern on the observation plane is observable from $n(\mathbf{r})$ after a timely accumulative photon counting. Again, the visibility of the interference modulation is determined by the angular diameter $\Delta\theta$ of the light source. The interference achieves maximum visibility for a point-like source when $\Delta\theta \sim 0$, and vanishes when the separation of the two pinholes becomes greater then the first-order spatial quantum coherence length $b > \Delta l_c$.

7.7   PHOTON AND EFFECTIVE WAVEFUNCTION

If the measured electromagnetic field can be characterized by single-photon state representation, it is found that

$$
\begin{aligned}
G^{(1)}&(\mathbf{r}_i, t_i; \mathbf{r}_j, t_j) \\
&= \langle \Psi | \hat{E}^{(-)}(\mathbf{r}_i, t_i)\, \hat{E}^{(+)}(\mathbf{r}_j, t_j) | \Psi \rangle \\
&= \sum_n \langle \Psi | \hat{E}^{(-)}(\mathbf{r}_i, t_i) | n \rangle \langle n | \hat{E}^{(+)}(\mathbf{r}_j, t_j) | \Psi \rangle \\
&= \langle \Psi | \hat{E}^{(-)}(\mathbf{r}_i, t_i) | 0 \rangle \langle 0 | \hat{E}^{(+)}(\mathbf{r}_j, t_j) | \Psi \rangle \\
&= \Psi^*(\mathbf{r}_i, t_i)\, \Psi(\mathbf{r}_j, t_j)
\end{aligned}
\tag{7.7.1}
$$

where, in the third line, we have used the completeness relation

$$
\sum_n |n\rangle\langle n| = 1.
\tag{7.7.2}
$$

The effective wavefunction is formulated in Eq. 7.7.1 as

$$
\Psi(\mathbf{r}_j, t_j) = \langle 0 | \hat{E}^{(+)}(\mathbf{r}_j, t_j) | \Psi \rangle
\tag{7.7.3}
$$

where $j = 1, 2$.

It is easy to find the following linear superposition relationship between the effective wavefunctions:

$$
\begin{aligned}
\Psi(\mathbf{r}, t) &= \langle 0 | \hat{E}^{(+)}(\mathbf{r}, t) | \Psi \rangle \\
&= \langle 0 | \frac{1}{\sqrt{2}} \big[\hat{E}^{(+)}(\mathbf{r}_1, t_1) + \hat{E}^{(+)}(\mathbf{r}_2, t_2) \big] | \Psi \rangle \\
&= \frac{1}{\sqrt{2}} \big[\Psi(\mathbf{r}_1, t_1) + \Psi(\mathbf{r}_2, t_2) \big]
\end{aligned}
\tag{7.7.4}
$$

associated with the linear superposition of the fields

$$
\hat{E}^{(+)}(\mathbf{r}, t) = \frac{1}{\sqrt{2}} \big[\hat{E}^{(+)}(\mathbf{r}_1, t_1) + \hat{E}^{(+)}(\mathbf{r}_2, t_2) \big].
$$

(I) Heisenberg picture and Schrödinger picture.

$G^{(1)}(\mathbf{r}_1, t_1; \mathbf{r}_2, t_2)$ is referred to as the coherence function or correlation function between fields at space-time points $(\mathbf{r}_1, t_1)$ and $(\mathbf{r}_2, t_2)$. We use $(\mathbf{r}_1, t_1)$ and $(\mathbf{r}_2, t_2)$ to mark the space-time dependence of the field operators. As we have emphasized earlier, however, $G^{(1)}(\mathbf{r}_1, t_1; \mathbf{r}_2, t_2)$ is not directly measured at these two points. The photodetection event happens at space-time point $(\mathbf{r}, t)$. $t_1 = t - s_1/c$, $t_2 = t - s_2/c$, referring to the earlier times of the fields at the upper and the lower pinholes P_1 and P_2, respectively. In the Heisenberg picture, the states (or wavefunctions) are independent of time. The time dependence is carried by the operator.

On the other hand, the concept of wavefunction belongs to the Schrödinger picture. In the Schrödinger representation, the time dependence is carried with the wavefunction. The counting rate of a photodetector is proportional to the probability of producing a photo-electron by annihilating a photon at space-time point $(\mathbf{r}, t)$, which equals the normal square of the probability amplitude, or the effective wavefunction of a photon, $\Psi(\mathbf{r}, t)$. The total effective wavefunction (probability amplitude) $\Psi(\mathbf{r}, t)$ is the result of the superposition of two earlier effective wavefunctions (probability amplitudes): $\Psi(\mathbf{r}_1, t_1)$ and $\Psi(\mathbf{r}_2, t_2)$. Although having different optical paths, $\Psi(\mathbf{r}_1, t_1)$ and $\Psi(\mathbf{r}_2, t_2)$ contribute to the same photoelectron event at space-time $(\mathbf{r}, t)$ by means of quantum mechanical superposition. The physical picture associated with the concept of the effective wavefunction of a photon is the same as the wavefunction of the Schrödinger picture for a mass particle.

The effective wavefunction is a very useful concept. It is calculated from the Heisenberg representation but interpreted in the Schrödinger picture.

Although the effective wavefunction is formally defined in the Heisenberg representation, it should be consistent with that of the Schrödinger representation,

$$\Psi(r, t) = \langle 0 \, | \, \hat{E}^{(+)}(r, t) \, | \, \Psi(0) \rangle \tag{7.7.5}$$

$$= \langle 0| \int d\omega \, g(\omega; r, t) \, \hat{a}(\omega) \int d\omega' \, f(\omega') \, \hat{a}^\dagger(\omega)|0\rangle$$

$$= \langle 0| \int d\omega \, \hat{a}(\omega) \int d\omega' \, f(\omega') \, g(\omega'; r, t) \, \hat{a}^\dagger(\omega')|0\rangle$$

$$= \langle 0| \int d\omega \, \hat{a}(\omega) \, | \, \Psi(t) \rangle$$

$$= \langle 0| \, \hat{E}^{(+)}(0) \, | \, \Psi(t) \rangle,$$

where we have simplified the discussion in 1-dimension and assumed $\mathbf{r}_0 = 0$ and $t_0 = 0$ as usual. In Eq. (7.7.5) we have successfully evolved the initial state $| \, \Psi(0) \rangle$ to the finally measured state of $| \, \Psi(t) \rangle$ with the help of the Green's function $g(\omega; r, t)$.

(II) The physics.

Substituting the single-photon state into Eq. 7.7.3, the temporal effective wavefunction of the photon is calculated to be

$$\Psi(z_j, t_j) = \mathcal{F}_{\tau_j}\{f(\nu)\} \, e^{-i\omega_0 \tau_j} \tag{7.7.6}$$

where $\tau_j = t_j - z_j/c$ (free-space). We will assume all the effective wavefunction calculations in the rest of this chapter in free-space unless otherwise specified. Eq. (7.7.6) indicates a typical wavepacket in space-time.

The schematic picture of a Gaussian wavepacket is the same as shown in Fig. 1.2.1. There should be no surprise that Eq. (7.7.6) is no different from the classical wavepacket of Eq. (1.2.3). Eq. (7.7.6) and Eq. (1.2.3) describe the same dynamic properties of light, except that the effective wavefunction is associated with the quantized field and the concept of photon, or the dynamic picture of a photon in space-time.

(III) Photon interferes with itself.

In terms of the effective wavefunction and the concept of photon, we may now view the Young's double-pinhole experiment from a new perspective: a photon is emitted from the light source at space-time $(\mathbf{r} = 0, t = 0)$ (event 1) and later excites an atom at space-time coordinate $(\mathbf{r}, t)$ to produce a photoelectron (event 2). The photon has two possible paths between the source and the photon counting detector. The two possible paths quantum mechanically correspond to two probability amplitudes that contribute to the superposition of the effective wavefunction. The probability to produce a photoelectron at space-time point $(\mathbf{r}, t)$ is determined by the relative "phase" delay and "group" delay between the two amplitudes, i.e., the delay between the two wavepackets $\Psi(\mathbf{r}_1, t_1)$ and $\Psi(\mathbf{r}_2, t_2)$. If $\Psi(\mathbf{r}_1, t_1)$ and $\Psi(\mathbf{r}_2, t_2)$ are well separated at space-time point $(\mathbf{r}, t)$, there would be no interference. If $\Psi(\mathbf{r}_1, t_2)$ and $\Psi(\mathbf{r}_2, t_2)$ are indistinguishable or partially indistinguishable at space-time point $(\mathbf{r}, t)$, interference is observable. The greater the overlapping is, the greater the interference visibility and the higher the degree of coherence will be. Figure 7.7.1

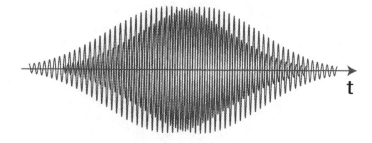

FIGURE 7.7.1 Schematic of two partially overlapped Gaussian wavepackets at the photon counting detector. Interference is expected with fringe visibility of less than 100%.

schematically shows partially overlapped Gaussian wavepackets. Interference is expected with fringe visibility of less than 100%.

Based on the above physical picture, it seems that we have enough reason to conclude that "a photon only interferes with itself". We may never know through which slit (or both slits) a single photon has been passing in the Young's double-pinhole experiment. We can, however, accurately predict the counting rate of the photodetector as a function of the relative delay between two effective wavefunctions of the photon.

7.8 MEASUREMENT OF THE FIRST-ORDER COHERENCE OF LIGHT

We have emphasized earlier that in the quantum theory of light, a measurement of a photon or a wavepacket cannot provide us any meaningful knowledge of physics, except a record of an photoelectron event. A time accumulative measurement, which involves the measurement of a large number of photons, are always necessary. This type of measurement is named photon counting. An idealized point-like photon counting detector and associated counting circuit, namely a photon counting device, counts how many photoelectron event occur within a chosen time interval at a chosen spatial coordinate of the photodetector. Basically, it provides a time averaged counting rate of photoelectrons

$$R_d = \int_T dt \, \langle \, \hat{E}^{(-)}(\mathbf{r}, t) \hat{E}^{(+)}(\mathbf{r}, t) \, \rangle, \tag{7.8.1}$$

where T is the accumulative time interval of the photon counting. In the Young's double-pinhole interferometer, a photon counting device measures the time averaged first-order coherence function by means of an interference pattern

$$R_d = \int_T dt \, \big[G^{(1)}(\mathbf{r}_1, t_1; \mathbf{r}_1, t_1) + G^{(1)}(\mathbf{r}_2, t_2; \mathbf{r}_2, t_2)$$
$$+ \, G^{(1)}(\mathbf{r}_1, t_1; \mathbf{r}_2, t_2) + G^{(1)}(\mathbf{r}_2, t_2; \mathbf{r}_1, t_1) \big]. \tag{7.8.2}$$

As we have learned that time average and ensemble average are equivalent for stationary field. Time averaging may not have any effect on the first-order coherence function of a stationary field; however, a time average may not be trivial for time dependent effective wavefunction, or the first-order coherence function. What do we learn about the first-order coherence of a single-photon state after the time average?

Example (I): Single-photon pure state.

The following discussion will focus on the first-order mutual-coherence function or cross-correlation function $G^{(1)}(\mathbf{r}_1, t_1; \mathbf{r}_2, t_2)$. In the case of single-photon pure state, the first-order mutual-coherence function is a product of two wavepackets in space-time: $G^{(1)}(\mathbf{r}_1, t_1; \mathbf{r}_2, t_2) = \Psi^*(\mathbf{r}_1, t_1)\,\Psi(\mathbf{r}_2, t_2)$. In general, it cannot be written as a time independent function of $t_1 - t_2$.

The time averaged $G^{(1)}(\mathbf{r}_1, t_1; \mathbf{r}_2, t_2)$ of the single-photon pure state, which is measurable in an Young's double-pinhole interferometer, can be formally calculated as

$$\int_T dt\ \Psi^*(z_1, t_1)\,\Psi(z_2, t_2)$$
$$\simeq \int_{T \sim \infty} dt\, e^{i(\omega_0 t_1 - k_0 z_1)}\,\mathcal{F}^*_{\tau_1}\{f(\nu)\}\,e^{-i(\omega_0 t_2 - k_0 z_2)}\,\mathcal{F}_{\tau_2}\{f(\nu)\}.$$
$$= \mathcal{F}_\tau\{|f(\nu)|^2\}\,e^{-i\omega_0 \tau} \tag{7.8.3}$$

where, again, $\nu = \omega - \omega_0$, ω_0 is the center frequency of the spectrum, and $\tau = [(z_2 + s_2) - (z_1 + s_1)]/c$ for arbitrary z_1 and z_2.

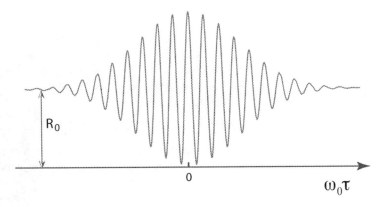

FIGURE 7.8.1 Counting rate of the photon counting detector as a function of $\omega_0 \tau$ in the Young's double-pinhole experiment. The observed interference measures the time averaged first-order coherence. In this observation, a Gaussian spectrum is assumed.

Substituting Eq. 7.8.3 into Eq. 7.8.2 and assuming equal chance for a photon to pass the upper and the lower pinholes, the time averaged counting rate of the photon counting detector at a chosen coordinate is thus a function of τ:

$$R_d(\tau) = R_0\left[1 + \mathcal{F}_\tau\{|f(\nu)|^2\}\,\cos \omega_0 \tau\right] \tag{7.8.4}$$

where R_0 is a constant that is calculated from the time integral of the self-correlation function. Figure 7.8.1 shows a measured result of a photon counting detector. Its time averaged counting rate is a function of τ. This is the only and all information we can directly observe from the measurement.

Example (II): Single-photon mixed state.

For a single-photon mixed state, the calculation of the time averaged first-order mutul-coherence function is straightforward:

$$\int_T dt\, G^{(1)}(z_1, t_1; z_2, t_2) = \int_T dt\, tr\left[\hat{\rho}\hat{E}^{(-)}(z_1, t_1)\hat{\rho}\hat{E}^{(+)}(z_2, t_2)\right]$$
$$= \int_T dt \sum_n P_n\, G_n^{(1)}(z_1, t_1; z_2, t_2)$$
$$\propto \mathcal{F}_\tau\{|f(\nu)|^2\}\,\cos \omega_0 \tau. \tag{7.8.5}$$

It is clearly shown that the time averaging is equivalent to the statistical ensemble averaging in this case. Although after ensemble averaging, time averaging becomes a trivial calculation in this example, a time accumulative integral is always necessary for a photon counting measurement. The observed interference pattern is thus the same as that of Example (I):

$$R_d(\tau) = R_0 \left[1 + \mathcal{F}_\tau \{ |f(\nu)|^2 \} \cos \omega_0 \tau \right].$$

A time averaged measurement of first-order coherence cannot distinguish a pure state from mixed state. This point has been emphasized in early chapters from the classical point of view.

SUMMARY

In this chapter, we introduced the concept of first-order coherence of light, based on the electromagnetic theory of light and the quantum theory of light, to qualify and quantify the interference ability between temporally delayed or spatially separated radiation fields. We defined the concepts of first-order self-coherence function, mutual-coherence function, and the degree of first-order coherence in terms of the electromagnetic theory of light, including Einstein's granularity picture, and the quantum theory of light. We also introduced the concepts of coherence time and coherence length for different radiations.

In the electromagnetic theory of light, the first-order self-coherence function and mutual-coherence function are defined as

$$\Gamma^{(1)}(\mathbf{r}_1, t_1; \mathbf{r}_1, t_1) \equiv \langle E^*(\mathbf{r}_1, t_1) \, E(\mathbf{r}_1, t_1) \rangle$$

$$\Gamma^{(1)}(\mathbf{r}_2, t_2; \mathbf{r}_2, t_2) \equiv \langle E^*(\mathbf{r}_2, t_2) \, E(\mathbf{r}_2, t_2) \rangle$$

$$\Gamma^{(1)}(\mathbf{r}_1, t_1; \mathbf{r}_2, t_2) \equiv \langle E^*(\mathbf{r}_1, t_1) \, E(\mathbf{r}_2, t_2) \rangle$$

$$\Gamma^{(1)}(\mathbf{r}_2, t_2; \mathbf{r}_1, t_1) \equiv \langle E(\mathbf{r}_1, t_1) \, E^*(\mathbf{r}_2, t_2) \rangle.$$

The normalized degree of first-order coherence is defined and calculated as

$$\gamma(\mathbf{r}_1, t_1; \mathbf{r}_2, t_2) \equiv \frac{\Gamma(\mathbf{r}_1, t_1; \mathbf{r}_2, t_2)}{\left[\Gamma(\mathbf{r}_1, t_1; \mathbf{r}_1, t_1) \right]^{\frac{1}{2}} \left[\Gamma(\mathbf{r}_2, t_2; \mathbf{r}_2, t_2) \right]^{\frac{1}{2}}}$$

$$= \frac{\langle E^*(\mathbf{r}_1, t_1) \, E(\mathbf{r}_2, t_2) \rangle}{\left[\langle | E(\mathbf{r}_1, t_1) |^2 \rangle^{\frac{1}{2}} \langle | E(\mathbf{r}_2, t_2) |^2 \rangle \right]^{\frac{1}{2}}}.$$

The degree of first-order coherence is usually measured through an interferometer which produces either a temporal delay or a spatial separation between the superposed radiation fields at space-time point $(\mathbf{r}, t)$. It is interesting to find that in Einstein's granularity picture of light, the first-order self-coherence function and mutual-coherence function are the results of a subfield interfering with the subfield itself. Consequently, the observed first-order interference is the result of a subfield interfering with the subfield itself.

In the quantum theory of light, the first-order self-coherence function and mutual-coherence function are defined as

$$G^{(1)}(\mathbf{r}_1, t_1; \mathbf{r}_1, t_1) \equiv \langle \hat{E}^{(-)}(\mathbf{r}_1, t_1) \, \hat{E}^{(+)}(\mathbf{r}_1, t_1) \rangle$$

$$G^{(1)}(\mathbf{r}_2, t_2; \mathbf{r}_2, t_2) \equiv \langle \hat{E}^{(-)}(\mathbf{r}_2, t_2) \, \hat{E}^{(+)}(\mathbf{r}_2, t_2) \rangle$$

$$G^{(1)}(\mathbf{r}_1, t_1; \mathbf{r}_2, t_2) \equiv \langle \hat{E}^{(-)}(\mathbf{r}_1, t_1) \, \hat{E}^{(+)}(\mathbf{r}_2, t_2) \rangle$$

$$G^{(1)}(\mathbf{r}_2, t_2; \mathbf{r}_1, t_1) \equiv \langle \hat{E}^{(-)}(\mathbf{r}_2, t_2) \, \hat{E}^{(+)}(\mathbf{r}_1, t_1) \rangle.$$

The normalized quantum degree of first-order coherence is defined and calculated as

$$g^{(1)}(\mathbf{r}_1, t_1; \mathbf{r}_2, t_2)$$

$$\equiv \frac{\langle \hat{E}^{(-)}(\mathbf{r}_1, t_1)\hat{E}^{(+)}(\mathbf{r}_2, t_2)\rangle}{\left[\langle \hat{E}^{(-)}(\mathbf{r}_1, t_1)\hat{E}^{(+)}(\mathbf{r}_1, t_1)\rangle\right]^{\frac{1}{2}}\left[\langle \hat{E}^{(-)}(\mathbf{r}_2, t_2)\hat{E}^{(+)}(\mathbf{r}_2, t_2)\rangle\right]^{\frac{1}{2}}}$$

$$= \frac{G^{(1)}(\mathbf{r}_1, t_1; \mathbf{r}_2, t_2)}{\left[G^{(1)}(\mathbf{r}_1, t_1; \mathbf{r}_1, t_1)\right]^{\frac{1}{2}}\left[G^{(1)}(\mathbf{r}_2, t_2; \mathbf{r}_2, t_2)\right]^{\frac{1}{2}}}.$$

In the process of calculating $G^{(1)}(\mathbf{r}_j, t_j; \mathbf{r}_k, t_k)$s, $j = 1, 2; k = 1, 2$, we further developed and emphasized the concepts of effective wavefunction of a photon and the effective wavefunction of an ensemble of photons in the single-photon state representation

$$\psi_m(\mathbf{r}, t) \equiv \langle 0|\hat{E}^{(+)}(\mathbf{r}, t)|\Psi_m\rangle$$

$$\Psi(\mathbf{r}, t) \equiv \langle 0|\hat{E}^{(+)}(\mathbf{r}, t)|\Psi\rangle.$$

In the coherent state representation, the mth group of identical photons and the ensemble of entire groups

$$\psi_m(\mathbf{r}, t) \equiv \langle \Psi_n|\hat{E}^{(+)}(\mathbf{r}, t)|\Psi_n\rangle$$

$$\Psi(\mathbf{r}, t) \equiv \langle \Psi|\hat{E}^{(+)}(\mathbf{r}, t)|\Psi\rangle.$$

with the relationship of superposition

$$\Psi(\mathbf{r}, t) = \sum_m c_m \, \psi_m(\mathbf{r}, t).$$

It is interesting to find that in quantum theory, the first-order self-coherence function and mutual-coherence function are the results of a photon interfering with the photon itself. Consequently, the observed first-order interference is the result of a photon interfering with the photon itself

In this chapter, we give a few detailed analysis on temporal and spatial coherence. These excises are helpful in understanding the physics of first-order interference phenomena.

REFERENCES AND SUGGESTIONS FOR READING

[1] R.J. Glauber, Phys. Rev. **130**, 2529 (1963); Phys. Rev. **131**, 2766 (1963).

[2] M. Born and E. Wolf, *Principle of Optics*, Cambridge, 2002.

[3] R. Loudon, *The Quantum Theory of Light*, Oxford Science, Oxford, 2000.

[4] M.O. Scully and M.S. Zubairy, *Quantum Optics*, Cambridge, 1997.

CHAPTER 8

Second-Order Coherence of Light

In the history of optical science, we were satisfied with the measurement of mean intensity $\langle I(\mathbf{r}, t)$ of light, namely the measurement of first-order coherence of light, for optical studies and observations, until the middle of 20th century when two astrophysicists Hanbury Brown and Twiss (HBT) introduced their "intensity interferometer". Different from all other optical interferometers, HBT used two independent photodetectors to measure the intensity correlation of light, $\langle I(\mathbf{r}_1, t_1) I(\mathbf{r}_2, t_2) \rangle$, at two separated space-time coordinates, now known as the second-order coherence of light. Whether it is named "second-order correlation" or "second-order coherence" refers to the same physical phenomenon. "Correlation" emphasizes its statistical nature, while "coherence" emphasizes its interference nature. Figure 8.0.1 is a schematic illustration of a HBT interferometer. Comparing with Fig. 7.3.1, we can easily

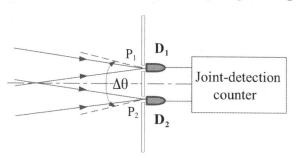

FIGURE 8.0.1 Schematic setup of a HBT interferometer. The interferometer is similar to the classic Young's double-pinhole interferometer and the Michelson stellar interferometer, except two photodetectors, either in analog mode or in photon counting mode, are placed behind the pinholes for joint-detection of the radiations at space-time coordinates $(\mathbf{r}_1, t_1)$ (P_1) and $(\mathbf{r}_2, t_2)$ (P_2). The joint-detection integrator or counter can be either a current-current correlator (analog mode) or a photon counting coincidence circuit (photon counting mode).

find the differences between a classic Young's double-slit interferometer and a HBT interferometer. The classic Young's double-slit interferometer measures the first-order coherence of fields $E(\mathbf{r}_1, t_1)$ and $E(\mathbf{r}_2, t_2)$, namely the first-order coherence function $\Gamma^{(1)}(\mathbf{r}_1, t_1; \mathbf{r}_2, t_2)$ or $G^{(1)}(\mathbf{r}_1, t_1; \mathbf{r}_2, t_2)$ as well as the normalized degree of first-order coherence $\gamma^{(1)}(\mathbf{r}_1, t_1; \mathbf{r}_2, t_2)$ or $g^{(1)}(\mathbf{r}_1, t_1; \mathbf{r}_2, t_2)$ by means of the measurement of mean intensity (analog) or mean photon number (photon-counting) at space-time coordinate $(\mathbf{r}, t)$. In a classic Young's double-slit interferometer, the fields $E(\mathbf{r}_1, t_1)$ and $E(\mathbf{r}_2, t_2)$ are superposed and measured at a chosen space coordinate $\mathbf{r}$ at a later time t, $t = t_1 + s_1/c = t_2 + s_2/c$, by a photodetector. Neither $\Gamma^{(1)}(\mathbf{r}_1, t_1; \mathbf{r}_2, t_2)$ $[\gamma^{(1)}(\mathbf{r}_1, t_1; \mathbf{r}_2, t_2)]$ nor $G^{(1)}(\mathbf{r}_1, t_1; \mathbf{r}_2, t_2)$ $[g^{(1)}(\mathbf{r}_1, t_1; \mathbf{r}_2, t_2)]$ are directly measured at space-time points $(\mathbf{r}_1, t_1)$ and $(\mathbf{r}_2, t_2)$. A photodetection event can never happen at two different space-time coordinates. However, the second-order coherence function $\Gamma^{(2)}(\mathbf{r}_1, t_1; \mathbf{r}_2, t_2)$ that is defined and calculated based on the electromagnetic theory of light, and $G^{(2)}(\mathbf{r}_1, t_1; \mathbf{r}_2, t_2)$ that is defined and calculated based on the quantum theory of light, as well as the normalized degree of second-order coherence $\gamma^{(2)}(\mathbf{r}_1, t_1; \mathbf{r}_2, t_2)$ and $g^{(2)}(\mathbf{r}_1, t_1; \mathbf{r}_2, t_2)$, are measured directly by two independent photodetectors, either in the analog mode or in the photon counting mode, at space-time points $(\mathbf{r}_1, t_1)$ and $(\mathbf{r}_2, t_2)$, respectively and jointly.

In this chapter, we first define the second-order coherence function $\Gamma^{(2)}(\mathbf{r}_1, t_1; \mathbf{r}_2, t_2)$, which is formulated from the electromagnetic theory of light, including Einstein's quantized granularity picture of light, and $G^{(2)}(\mathbf{r}_1, t_1; \mathbf{r}_2, t_2)$, which is formulated from the quantum theory of light, in the process of introducing the Hanbury Brown and Twiss interferometer (HBT interferometer). We then give a few simple calculations and discussions on the degree of second-order coherence for incoherent and coherent light as well as for entangled photon pairs. In the process of calculating $G^{(2)}(\mathbf{r}_1, t_1; \mathbf{r}_2, t_2)$, we introduce the concept of two-photon effective wavefunction $\Psi(\mathbf{r}_1, t_1; \mathbf{r}_2, t_2)$ of a pair of jointly detected photons or two groups of jointly detected identical photons and emphasize the differences between the two-photon effective wavefunctions of product state and entangled state. In the end of this chapter, we will extend the concept of second-order coherence to Nth-order coherence $(N > 2)$ and analyze a few simple measurements on the third-order temporal and spatial coherence of thermal field.

8.1 SECOND-ORDER COHERENCE—FORMULATED FROM MAXWELL'S CONTINUUM EM THEORY

Assume that two analog mode photodetectors, D_1 and D_2 in Fig. 8.1.1, are used to measure the instantaneous intensities of light at $(\mathbf{r}_1, t_1)$ and $(\mathbf{r}_2, t_2)$, respectively. The output currents of the two photodetectors are amplified by two independent Radio-Frequency (RF) amplifiers. An electronic linear multiplier, or RF mixer, is used to multiply the two currents $i_1(t)$ and $i_2(t)$ at time t. The output voltage of the linear multiplier is proportional to $i_1(t) \times i_2(t)$, which is proportional to the jointly measured instantaneous intensities, $I(\mathbf{r}_1, t_1) \times I(\mathbf{r}_1, t_1)$, by the two analog photodetectors:

$$
\begin{aligned}
V_{12}(t) &\propto i_1(t) \times i_2(t) \\
&\propto I(\mathbf{r}_1, t_1)\, I(\mathbf{r}_2, t_2) \\
&= E^*(\mathbf{r}_1, t_1)\, E(\mathbf{r}_1, t_1)\, E^*(\mathbf{r}_2, t_2)\, E(\mathbf{r}_2, t_2)
\end{aligned}
\tag{8.1.1}
$$

where $V_{12}(t)$ is the output voltage of the linear multiplier, $i_1(t)$ and $i_2(t)$ are the electronically amplified photocurrent of D_1 and D_2, respectively, at time t of the multiplication; $t_1 = t - \tau_1^e$, $t_2 = t - \tau_2^e$ are the early times that are defined by the electronic time delays τ_1^e and τ_2^e, including the delays of the detectors, the amplifiers and the adjustable delay-line cables. The above statements have assumed an idealized correlation measurement by neglecting "unavoidable" time averages on t_1, t_2, and t: implying *instantaneous* responses of the photodetectors and the associated electronics, including the linear multiplier. The instantaneous intensities $I(\mathbf{r}_1, t_1)$ and $I(\mathbf{r}_2, t_2)$ are identified by the electronic delays. Notice that t_1 and t_2 of the two photodetection events as well as the relative time delay $t_1 - t_2 = \tau_1^e - \tau_2^e$ are all defined by the electronics in this setup.

The second-order coherence function $\Gamma^{(2)}(\mathbf{r}_1, t_1; \mathbf{r}_2, t_2)$, in terms of either temporal delay or spatial separation between the measured intensities of $I(\mathbf{r}_1, t_1)$ and $I(\mathbf{r}_2, t_2)$, formulated

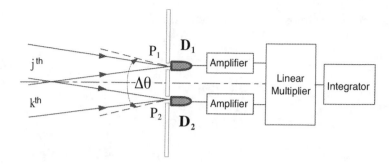

FIGURE 8.1.1 Schematic setup for the measurement of second-order coherence function of $\Gamma^{(2)}(\mathbf{r}_1, t_1; \mathbf{r}_2, t_2)$, or second-order correlation function of $\langle I(\mathbf{r}_1, t_1) I(\mathbf{r}_2, t_2)\rangle$ in terms of either temporal delay or spatial separation between the measured intensities of $I(\mathbf{r}_1, t_1)$ and $I(\mathbf{r}_2, t_2)$. Two analog photodetectors, D_1 and D_2, are placed behind the pinholes for joint-detection of the radiations. An electronic linear multiplier, or RF mixer, is followed to multiply the two currents $i_1(t)$ and $i_2(t)$ at time t for joint measurement of D_1 and D_2.

from the electromagnetic field theory of light, is thus

$$\Gamma^{(2)}(\mathbf{r}_1, t_1; \mathbf{r}_2, t_2) = \langle I(\mathbf{r}_1, t_1) I(\mathbf{r}_2, t_2)\rangle$$
$$= \langle E^*(\mathbf{r}_1, t_1)E(\mathbf{r}_1, t_1)E^*(\mathbf{r}_2, t_2)E(\mathbf{r}_2, t_2)\rangle \tag{8.1.2}$$

and the normalized degree of second-order coherence is defined as

$$\gamma^{(2)}(\mathbf{r}_1, t_1; \mathbf{r}_2, t_2) = \frac{\langle E^*(\mathbf{r}_1, t_1)E(\mathbf{r}_1, t_1) E^*(\mathbf{r}_2, t_2)E(\mathbf{r}_2, t_2)\rangle}{\langle E^*(\mathbf{r}_1, t_1)E(\mathbf{r}_1, t_1)\rangle\langle E^*(\mathbf{r}_2, t_2)E(\mathbf{r}_2, t_2)\rangle} \tag{8.1.3}$$

where the expectation or ensemble average, $\langle ... \rangle$ denotes, again, *taking into account all possible realizations of the field*.

Perhaps, the easiest expectation is a factorizable $\Gamma^{(2)}(\mathbf{r}_1, t_1; \mathbf{r}_2, t_2)$

$$\Gamma^{(2)}(\mathbf{r}_1, t_1; \mathbf{r}_2, t_2) = \langle I(\mathbf{r}_1, t_1)\rangle\langle I(\mathbf{r}_2, t_2)\rangle$$
$$= \Gamma^{(1)}(\mathbf{r}_1, t_1; \mathbf{r}_1, t_1)\Gamma^{(1)}(\mathbf{r}_2, t_2; \mathbf{r}_2, t_2), \tag{8.1.4}$$

with the corresponding degree of second-order coherence

$$\gamma^{(2)}(\mathbf{r}_1, t_1; \mathbf{r}_2, t_2) = 1. \tag{8.1.5}$$

The physics behind a factorizable (non-factorizable) $\Gamma^{(2)}(\mathbf{r}_1, t_1; \mathbf{r}_2, t_2)$ with $\gamma^{(2)} = 1$ ($\gamma^{(2)} \neq 1$) will be given in later discussions. Statistically, a factorizable $\Gamma^{(2)}(\mathbf{r}_1, t_1; \mathbf{r}_2, t_2)$ with $\gamma^{(2)} = 1$ means the two measured intensities $I(\mathbf{r}_1, t_1)$ and $I(\mathbf{r}_2, t_2)$ are independent with no correlation. The nontrivial second-order correlation and anti-correlation, respectively, are defined with

$$\gamma^{(2)}(\mathbf{r}_1, t_1; \mathbf{r}_2, t_2) > 1 \tag{8.1.6}$$

and

$$\gamma^{(2)}(\mathbf{r}_1, t_1; \mathbf{r}_2, t_2) < 1. \tag{8.1.7}$$

The following is a classic calculation of $\Gamma^{(2)}(\mathbf{r}_1, t_1; \mathbf{r}_2, t_2)$ for thermal radiation in Maxwell's continuous picture of light. To simplify the mathematics, we consider a 1-D joint measurement of two point-like photodetectors D_1 and D_2, and a point-like thermal source,

$$\Gamma^{(2)}(r_1, t_1; r_2, t_2)$$

$$= \Big\langle \int d\omega \, E_1^*(\omega) \, e^{i\omega\tau_1} \int d\omega' \, E_1(\omega') \, e^{-i\omega'\tau_1}$$

$$\times \int d\omega'' \, E_2^*(\omega'') \, e^{i\omega''\tau_2} \int d\omega''' \, E_2(\omega''') \, e^{-i\omega'''\tau_2} \Big\rangle$$

$$= \Big\langle \int d\omega \, d\omega' \, d\omega'' \, d\omega''' \, E_1^*(\omega) E_1(\omega') E_2^*(\omega'') E_2(\omega''') e^{i(\omega-\omega')\tau_1} \, e^{i(\omega''-\omega''')\tau_2} \Big\rangle$$

$$= \int_{\omega=\omega'} d\omega \, |E_1(\omega)|^2 \int_{\omega''=\omega'''} d\omega'' \, |E_2(\omega'')|^2$$

$$+ \int_{\omega=\omega'''} d\omega \, E_1^*(\omega) E_2(\omega) \, e^{i\omega(\tau_1-\tau_2)} \int_{\omega'=\omega''} d\omega' \, E_2^*(\omega') E_1(\omega') \, e^{-i\omega'(\tau_1-\tau_2)}$$

$$= \Gamma_{11}^{(1)} \Gamma_{22}^{(1)} + \Gamma_{12}^{(1)} \Gamma_{21}^{(1)} \tag{8.1.8}$$

where the ensemble average has taken account all possible realizations of the fields, i.e., all possible random phases of the Fourier modes. The survival terms from the ensemble average can be written in terms of the first-order coherence functions of the field

$$\Gamma_{11}^{(1)} \equiv \int_{\omega=\omega'} d\omega \, |E_1(\omega)|^2 = \langle I_1 \rangle = \text{constant}$$

$$\Gamma_{22}^{(1)} \equiv \int_{\omega''=\omega'''} d\omega'' \, |E_2(\omega'')|^2 = \langle I_2 \rangle = \text{constant}$$

$$\Gamma_{12}^{(1)} \equiv \int_{\omega=\omega'''} d\omega \, E_1^*(\omega) E_2(\omega) \, e^{i\omega(\tau_1-\tau_2)}$$

$$\simeq \mathcal{F}_\tau \{a^2(\nu)\} e^{i\omega_0\tau}$$

$$\Gamma_{21}^{(1)} \equiv \int_{\omega'=\omega''} d\omega' \, E_2^*(\omega') E_1(\omega') \, e^{-i\omega'(\tau_1-\tau_2)}$$

$$\simeq \mathcal{F}_\tau^* \{a^2(\nu)\} e^{-i\omega_0\tau}. \tag{8.1.9}$$

where $\tau \equiv \tau_1 - \tau_2$. In Eq.(8.1.9) we have approximated equal amplitudes $|E_1(\nu)| = |E_2(\nu)| = a(\nu)$, which is reasonable for our simply 1-D joint measurement. It is easy to find the nontrivial correlation, if there is any, must be in the second term of Eq.(8.1.8): $\Gamma_{12}^{(1)} \Gamma_{21}^{(1)} = |\Gamma_{12}^{(1)}|^2$. The temporal second-order coherence function of thermal field is thus approximated as follows

$$\Gamma^{(2)}(r_1, t_1; r_2, t_2) \simeq \Gamma_0 \big[1 + \big| \mathcal{F}_\tau \{a^2(\nu)\} \big|^2 \big]. \tag{8.1.10}$$

The result of Eq.(8.1.10) is under idealized experimental condition: two perfect photodetectors and a perfect correlation measurement circuit for instantaneous joint photodetection. The "unavoidable" time averages on t_1, t_2, and t are all ignored. In a realistic measurement, however, the finite response times of the photodectors and the correlation measurement circuit must be taken into account. Although the finite response times of D_1 and D_2 as well as the correlation measurement circuit have no contribution to the constant $\Gamma_{11}^{(1)} \Gamma_{22}^{(1)}$, they may affect the nontrivial correlation of $\Gamma_{12}^{(1)} \Gamma_{21}^{(1)}$ significantly. We will discuss this effect later.

8.2 SECOND-ORDER COHERENCE IN EINSTEIN'S GRANULARITY PICTURE OF LIGHT

$\Gamma^{(2)}(\mathbf{r}_1, t_1; \mathbf{r}_2, t_2)$ is usually named "classical second-order coherence function", although it does not restrict itself within the classical Maxwell's continuum picture of radiation. Einstein's granularity picture of quantized light has been widely used for calculating $\Gamma^{(2)}(\mathbf{r}_1, t_1; \mathbf{r}_2, t_2)$ of thermal field. Naming it "classical" may not be approximate in Einstein's picture; however, as Einstein realized that his concept of "bundle of ray" is still under the frame work of electromagnetic theory, we consider $\Gamma^{(2)}(\mathbf{r}_1, t_1; \mathbf{r}_2, t_2)$ the second-order coherence function formulated from electromagnetic theory.

In the following we evaluate the second-order coherence function $\Gamma^{(2)}(\mathbf{r}_1, t_1; \mathbf{r}_2, t_2)$ of thermal radiation in Einstein's granularity picture by assuming a light source of finite size that contains a large number of independent and randomly radiating point-like sub-sources (such as spontaneous atomic transitions). The radiation fields $E(\mathbf{r}_1, t_1)$ and $E(\mathbf{r}_2, t_2)$ measured at photodetectors D_1 and D_2, respectively, are the results of superposition among a large number of subfields, labeled by $E_m(\mathbf{r}_m, t_m)$, originated from each of these independent sub-sources. The second-order coherence function $\Gamma^{(2)}(\mathbf{r}_1, t_1; \mathbf{r}_2, t_2)$ is therefore

$$
\begin{aligned}
&\Gamma^{(2)}(\mathbf{r}_1, t_1; \mathbf{r}_2, t_2) \\
&= \left\langle E^*(\mathbf{r}_1, t_1) E(\mathbf{r}_1, t_1) E^*(\mathbf{r}_2, t_2) E(\mathbf{r}_2, t_2) \right\rangle \\
&= \left\langle \sum_{m,n,p,q} E_m^*(\mathbf{r}_1, t_1)\, E_n(\mathbf{r}_1, t_1) E_p^*(\mathbf{r}_2, t_2)\, E_q(\mathbf{r}_2, t_2) \right\rangle \\
&= \sum_m E_m^*(\mathbf{r}_1, t_1)\, E_m(\mathbf{r}_1, t_1) \sum_n E_n^*(\mathbf{r}_2, t_2)\, E_n(\mathbf{r}_2, t_2) \\
&\quad + \sum_{m \neq n} E_m^*(\mathbf{r}_1, t_1)\, E_n(\mathbf{r}_1, t_1)\, E_n^*(\mathbf{r}_2, t_2)\, E_m(\mathbf{r}_2, t_2) \\
&= \sum_{m,n} \left| \frac{1}{\sqrt{2}} \big[E_m(\mathbf{r}_1, t_1) E_n(\mathbf{r}, t_2) + E_n(\mathbf{r}_1, t_1) E_m(\mathbf{r}_2, t_2) \big] \right|^2
\end{aligned}
\tag{8.2.1}
$$

where m, n, p, q label the randomly created and randomly distributed subfields and their corresponding point-like sub-sources. Considering the random phases of the subfields, as the result of an interference cancelation when taking into account all possible random phases of the subfields, the only surviving terms from the ensemble sum are the following: (1) $m = q, n = p$, (2) $m = p, n = q$. It is interesting to find that in Einstein's granularity picture of quantized thermal field, $\Gamma^{(2)}(\mathbf{r}_1, t_1; \mathbf{r}_2, t_2)$ is the sum of a set of the following superposition or interference

$$
\begin{aligned}
&\Gamma_{mn}^{(2)}(\mathbf{r}_1, t_1; \mathbf{r}_2, t_2) \\
&\equiv \left| \frac{1}{\sqrt{2}} \big[E_m(\mathbf{r}_1, t_1) E_n(\mathbf{r}, t_2) + E_n(\mathbf{r}_1, t_1) E_m(\mathbf{r}_2, t_2) \big] \right|^2
\end{aligned}
\tag{8.2.2}
$$

corresponding to two different yet indistinguishable alternatives for a pair of randomly created and randomly paired subfields, or photons, to produce a joint photodetection event: (1) the mth subfield, or photon, produces a photodetection event at D_1 and the nth subfield, or photon, produces a photodetection event at D_2; (2) the nth subfield, or photon, produces a photodetection event at D_1 and the mth subfield, or photon, produces a photodetection event at D_2, namely, a two-photon interference: a randomly created and randomly paired subfields (photons) interfering with the pair itself.

On the other hand, we find

$$\langle I(\mathbf{r}_1, t_1)\rangle = \sum_m E_m^*(\mathbf{r}_1, t_1)\, E_m(\mathbf{r}_1, t_1)$$

$$\langle I(\mathbf{r}_2, t_2)\rangle = \sum_n E_n^*(\mathbf{r}_2, t_2)\, E_n(\mathbf{r}_2, t_2) \tag{8.2.3}$$

and

$$\langle \Delta I(\mathbf{r}_1, t_1)\Delta I(\mathbf{r}_2, t_2)\rangle$$
$$= \langle \sum_{m\neq n} E_m^*(\mathbf{r}_1, t_1)\, E_n(\mathbf{r}_1, t_1) \sum_{p\neq q} E_p^*(\mathbf{r}_2, t_2)\, E_q(\mathbf{r}_2, t_2)\rangle$$
$$= \sum_{m\neq n} E_m^*(\mathbf{r}_1, t_1)\, E_n(\mathbf{r}_1, t_1)\, E_n^*(\mathbf{r}_2, t_2)\, E_m(\mathbf{r}_2, t_2) \tag{8.2.4}$$

We thus write $\Gamma^{(2)}(\mathbf{r}_1, t_1; \mathbf{r}_2, t_2)$ as the follows:

$$\Gamma^{(2)}(\mathbf{r}_1, t_1; \mathbf{r}_2, t_2)$$
$$= \langle I(\mathbf{r}_1, t_1)\rangle\langle I(\mathbf{r}_2, t_2)\rangle + \langle \Delta I(\mathbf{r}_1, t_1)\Delta I(\mathbf{r}_2, t_2)\rangle. \tag{8.2.5}$$

The first terms of Eq. (8.2.5) is a product of two mean intensities measured by D_1 and D_2, respectively. The second term of Eq. (8.2.5) is the intensity fluctuation correlation measured by D_1 and D_2, jointly. For thermal radiations, the first term usually contributes a constant to $\Gamma^{(2)}(\mathbf{r}_1, t_1; \mathbf{r}_2, t_2)$, and the second term contributes a nontrivial function of space-time $(\mathbf{r}_1, t_1)$ and $(\mathbf{r}_2, t_2)$ to $\Gamma^{(2)}(\mathbf{r}_1, t_1; \mathbf{r}_2, t_2)$, in terms of either temporal delay or spatial separation between the two measurements. The nontrivial second-order coherence function of thermal field is observable from the correlation measurement of its intensity fluctuations. It is clear, in Einstein's picture, the intensity fluctuation correlation is the result of two-photon interference, corresponding to the cross interference term of the two-photon superposition in Eq.(8.2.1) and Eq.(8.2.2). In principle, it is a coherent phenomenon. We thus name it two-photon interference induced intensity fluctuation correlation. In terms of the measurement, the correlation is observed from $\Delta I(\mathbf{r}_1, t_1) = \sum_{m\neq n} E_m^*(\mathbf{r}_1, t_1)E_n(\mathbf{r}_1, t_1)$, which is measured by D_1 and $\Delta I(\mathbf{r}_2, t_2) = \sum_{p\neq q} E_p^*(\mathbf{r}_2, t_2)E_q(\mathbf{r}_2, t_2)$, which is measured by D_2.

In an idealized measurement by means of two idealized photodetectors and an idealized correlation measurement circuit, in which all intensity fluctuations measured by D_1 and D_2 contribute to the nontrivial correlation of $\Gamma^{(2)}(\mathbf{r}_1, t_1; \mathbf{r}_2, t_2)$, $\langle \Delta I(\mathbf{r}_1, t_1)\Delta I(\mathbf{r}_2, t_2)\rangle$ of thermal field is usually written in terms of the first-order coherence functions

$$\langle \Delta I(\mathbf{r}_1, t_1)\Delta I(\mathbf{r}_2, t_2)\rangle$$
$$\simeq \sum_m E_m^*(\mathbf{r}_1, t_1)E_m(\mathbf{r}_2, t_2) \sum_n E_n^*(\mathbf{r}, t_2)E_n(\mathbf{r}_1, t_1)$$
$$= \Gamma^{(1)}(\mathbf{r}_1, t_1; \mathbf{r}_2, t_2)\Gamma^{(1)}(\mathbf{r}_2, t_2; \mathbf{r}_1, t_1). \tag{8.2.6}$$

$\Gamma^{(2)}(\mathbf{r}_1, t_1; \mathbf{r}_2, t_2)$ of thermal field is also formally written in terms of the first-order coherence functions:

$$\Gamma^{(2)}(\mathbf{r}_1, t_1; \mathbf{r}_2, t_2) = \Gamma_{11}^{(1)}\Gamma_{22}^{(1)} + \Gamma_{12}^{(1)}\Gamma_{21}^{(1)} = \Gamma_{11}^{(1)}\Gamma_{22}^{(1)} + |\Gamma_{12}^{(1)}|^2. \tag{8.2.7}$$

where we have written $\Gamma^{(1)}(\mathbf{r}_j, t_j; \mathbf{r}_k, t_k)$, $j = 1, 2; k = 1, 2$, with the short-hand notations

$\Gamma_{jk}^{(1)}$,

$$\Gamma_{11}^{(1)} = \sum_m E_m^*(\mathbf{r}_1, t_1) E_m(\mathbf{r}_1, t_1) \quad \Gamma_{22}^{(1)} = \sum_n E_n^*(\mathbf{r}_2, t_2) E_n(\mathbf{r}_2, t_2)$$

$$\Gamma_{12}^{(1)} = \sum_m E_m^*(\mathbf{r}_1, t_1) E_m(\mathbf{r}_2, t_2) \quad \Gamma_{21}^{(1)} = \sum_n E_n^*(\mathbf{r}_2, t_2) E_n(\mathbf{r}_1, t_1). \tag{8.2.8}$$

Adapting our previous results of Chapter 7, we recognize the last term of Eq. (8.2.7), $\Gamma_{12}^{(1)}\Gamma_{21}^{(1)}$, contributes a nontrivial function of coordinates $(\mathbf{r}_1, t_1)$ and $(\mathbf{r}_2, t_2)$ to $\Gamma^{(2)}(\mathbf{r}_1, t_1; \mathbf{r}_2, t_2)$, while the first term of Eq. (8.2.7), $\Gamma_{11}^{(1)}\Gamma_{22}^{(1)}$, contributes a constant to $\Gamma^{(2)}(\mathbf{r}_1, t_1; \mathbf{r}_2, t_2)$, despite any temporal delay or spatial separation of the two measurements. Under the condition of perfect correlation measurement, mathematically, the second-order coherence function of thermal field can be calculated from the first-order coherence function.

In last section, we derived the second-order coherence function of Eq. (8.1.8) (in 1-D) from Maxwell's continuum theory of light for thermal field. In fact, Eq. (8.1.8) also has a corresponding physical meaning in Einstein's granularity picture of light. To understand this, we consider a thermal radiation with broad spectrum, such as sunlight or blackbody radiation. Comparing with the broad spectrum of the total field, each quantized bundle of ray, or subfield, can be treated as a single mode light quantum with energy $\hbar\omega$, as Einstein did in 1905. The indexes of m and n can be replaced by ω. Similar to Eq. (8.2.1), Eq. (8.1.8) can be written as the following superposition:

$$\Gamma^{(2)}(r_1, t_1; r_2, t_2)$$
$$= \langle \sum_\omega E^*(\omega) e^{i\omega\tau_1} \sum_{\omega'} E(\omega') e^{-i\omega'\tau_1} \sum_{\omega''} E^*(\omega'') e^{i\omega''\tau_2} \sum_{\omega'''} E(\omega''') e^{-i\omega'''\tau_2} \rangle$$
$$= \sum_{\omega=\omega'} |E(\omega)|^2 \sum_{\omega''=\omega'''} |E(\omega'')|^2$$
$$+ \sum_{\omega=\omega'''} E^*(\omega) E(\omega) e^{i\omega(\tau_1-\tau_2)} \sum_{\omega'=\omega''} E^*(\omega') E(\omega') e^{-i\omega'(\tau_1-\tau_2)}$$
$$= \langle | \sum_{\omega,\omega'} \frac{1}{\sqrt{2}} [E(\omega) e^{i\omega\tau_1} E(\omega') e^{i\omega'\tau_2} + E(\omega) e^{i\omega\tau_2} E(\omega') e^{i\omega'\tau_1}] |^2 \rangle \tag{8.2.9}$$

corresponding to two different yet indistinguishable alternatives for a pair of randomly created and randomly paired subfields, or photons with energy $\hbar\omega$ and $\hbar\omega'$ to produce a joint photodetection event: (1) the light quantum of $\hbar\omega$ produces a photodetection event at D_1 while the light quantum of $\hbar\omega'$ produces a photodetection event at D_2; (2) the light quantum of $\hbar\omega'$ produces a photodetection event at D_1 while the light quantum of $\hbar\omega$ produces a photodetection event at D_2, namely, two-photon interference: a randomly created and randomly paired light quanta (photons) interfering with the pair itself.

Consequently, the calculation of $\Gamma_{12}^{(1)}$ and $\Gamma_{21}^{(1)}$ becomes much easier when dealing with continuous and broad spectrum thermal radiation in which each subfield can be approximated as a single mode light quantum with energy $\hbar\omega$,

$$\Gamma_{12}^{(1)} \simeq \int_\omega d\omega |E^*(\omega)|^2 (\omega) e^{i\omega(\tau_1-\tau_2)} = \mathcal{F}_\tau\{a^2(\nu)\} e^{i\omega_0\tau}$$

$$\Gamma_{21}^{(1)} \simeq \int_{\omega'} d\omega' |E^*(\omega')|^2 e^{-i\omega'(\tau_1-\tau_2)} = \mathcal{F}_\tau^*\{a^2(\nu)\} e^{-i\omega_0\tau} \tag{8.2.10}$$

where ω_0 is the central frequency of the thermal field.

It should be emphasized that, here, $\Gamma_{12}^{(1)}$ and $\Gamma_{21}^{(1)}$ are measured by two independent photodetectors D_1 and D_2 at different space-time coordinates $(\mathbf{r}_1, t_1)$ and $(\mathbf{r}_2, t_2)$. Although

Eq. (8.2.7) is helpful for mathematical calculations, we should keep in mind that $\Gamma_{12}^{(1)}$ and $\Gamma_{21}^{(1)}$ in Eq. (8.2.7) are measured differently in the first-order coherence function that was defined in Chapter 7, especially when taking into account non-idealized correlation measurement devices due to the "slow" responses of the electronics. We will emphasize this point again in the following discussions.

We now calculate the second-order temporal coherence function and the second-order spatial coherence function of thermal field by assuming idealized photodetectors and correlation measurement electronics.

(I) Second-order temporal coherence of thermal field

To simplify the mathematics, we assume a point-like thermal radiation source at $\mathbf{r} = 0$ and a far-field measurement. The second-order temporal coherence calculation is straightforward by applying the wavepacket model of Einstein's subfield, where each sub-source emits a wavepacket at time t_{0m}. Eq. (8.2.1) is thus formally written as

$$\Gamma^{(2)}(z_1, t_1; z_2, t_2)$$
$$= \sum_{m,n} \frac{1}{2} \Big| e^{-i\omega_0 \tau_{1m}} \mathcal{F}_{\tau_{1m}}\{a_m(\nu)\} e^{-i\omega_0 \tau_{2n}} \mathcal{F}_{\tau_{2n}}\{a_n(\nu)\}$$
$$+ e^{-i\omega_0 \tau_{1n}} \mathcal{F}_{\tau_{1n}}\{a_n(\nu)\} e^{-i\omega_0 \tau_{2m}} \mathcal{F}_{\tau_{2m}}\{a_m(\nu)\} \Big|^2, \qquad (8.2.11)$$

where, again, $\mathcal{F}_{\tau_{1m}}\{a_m(\nu)\}$ and $\mathcal{F}_{\tau_{2m}}\{a_m(\nu)\}$ are the Fourier transforms of the measured subfields with $\tau_{1m} = (t_1 - t_{0m}) - z_1/c = \tau_1 - t_{0m}$ and $\tau_{2m} = (t_2 - t_{0m}) - z_2/c = \tau_2 - t_{0m}$, ω_{0m} and ω_{0n} are the carrier frequencies of the mth and nth wavepackets, respectively. Eq. (8.2.11) indicates a two-photon interference: a randomly created and randomly paired wavepackets, excited from the mth and the nth sub-sources of a point-like light source, propagate with different temporal delays, produce a joint photodetection event of D_1 and D_2 with two different yet indistinguishable alternatives: (1) the mth and the nth wavepackets are detected by D_1 and D_2, respectively; and (2) the mth and the nth wavepackets are detected by D_2 and D_1, respectively[1]. The superposition of the above two alternatives of each mth-nth pair implies a two-photon interference phenomenon concealed in the joint measurement of D_1 and D_2 at (z_1, t_1) and (z_2, t_2), and the sum of a large set of two-photon interferences results in the second-order coherence function $\Gamma^{(2)}(z_1, t_1; z_2, t_2)$.

Assuming a perfect correlation measurement, $\Gamma^{(2)}(z_1, t_1; z_2, t_2)$ is calculated as follows

$$\Gamma^{(2)}(z_1, t_1; z_2, t_2)$$
$$= \sum_m \big| \mathcal{F}_{\tau_{1m}}\{a_m(\nu)\} \big|^2 \sum_n \big| \mathcal{F}_{\tau_{2n}}\{a_n(\nu)\} \big|^2$$
$$+ \Big| \sum_m e^{i\omega_{0m}(\tau_1 - \tau_2)} \mathcal{F}_{\tau_{1m}}^*\{a_m(\nu)\} \mathcal{F}_{\tau_{2m}}\{a_m(\nu)\} \Big|^2$$
$$= \int dt_m \, \big| \mathcal{F}_{\tau_1 - t_m}\{a(\nu)\} \big|^2 \int dt_n \, \big| \mathcal{F}_{\tau_2 - t_n}\{a(\nu)\} \big|^2$$
$$+ \Big| \int d\omega_{0m} e^{i\omega_{0m}(\tau_1 - \tau_2)} \mathcal{F}_{\tau_1 - t_2}\{a^2(\nu)\} \Big|^2$$
$$= \Gamma_{11}^{(1)} \Gamma_{22}^{(1)} + \big| \Gamma_{12}^{(1)} \big|^2 \qquad (8.2.12)$$

---

[1]In the language of quantum mechanics, this superposition corresponds to a two-photon interference phenomenon, which involves a superposition between two different yet indistinguishable two-photon probability amplitudes: (1) photon m and photon n are annihilated at D_1 and D_2, respectively; and (2) photon m and photon n are annihilated at D_2 and D_1, respectively.

where $|\mathcal{F}_{\tau_1-\tau_2}\{a^2(\nu)\}|^2$ is the result of a convolution, i.e., the result of the Wiener-Khintchine theorem. $\Gamma_{11}^{(1)}$ and $\Gamma_{22}^{(1)}$ are easy to estimate by applying the Parseval theorem, we discuss $\Gamma_{12}^{(1)}$ in the following two different cases:

Case (1): $\omega_{0m} = \omega_{0n} = \omega_0$

In this case, all the measured wavepackets have the same carrier frequency ω_0. The normalized degree of second-order coherence function $\gamma^{(2)}(z_1, t_1; z_2, t_2)$ is therefore

$$\gamma^{(2)}(z_1, t_1; z_2, t_2) \cong 1 + |\mathcal{F}_\tau\{a^2(\nu)\}|^2. \tag{8.2.13}$$

where $\tau \equiv \tau_1 - \tau_2$.

Case (2): $\omega_{0m} \neq \omega_{0n}$

We further assume broad spectrum thermal field in which the bandwidth of the total field is much greater than that of the individual wavepacket, such as sunlight or blackbody radiation. In this case, an additional sum or integral on ω_{0m} is necessary.

$$\begin{aligned}
\Gamma_{12}^{(1)} &= \int d\omega_{0m} e^{i\omega_{0m}(\tau_1-\tau_2)} \mathcal{F}_{\tau_1-\tau_2}\{a^2(\nu)\} \\
&\simeq \int d\omega_{0m} a^2(\omega_{0m}) e^{i\omega_{0m}(\tau_1-\tau_2)} \\
&= \mathcal{F}_{\tau_1-\tau_2}\{a^2(\nu')\} e^{i\omega_0(\tau_1-\tau_2)}
\end{aligned} \tag{8.2.14}$$

where ω_0 is the central frequency of the broad spectrum. In Eq. (8.2.14) we have treated $\mathcal{F}_{\tau_1-\tau_2}\{a^2(\nu)\}$ as the mean amplitude, $a^2(\omega_{0m})$, of the wavepacket of carrier frequency ω_{0m}. In case (2), $\gamma^{(2)}(z_1, t_1; z_2, t_2)$ is thus approximated as

$$\gamma^{(2)}(z_1, t_1; z_2, t_2) \cong 1 + |\mathcal{F}_\tau\{a^2(\nu)\}|^2.$$

which is formally the same as Eq. (8.2.13), except a broader spectrum of the total field and a different central frequency. We will restrict most of our discussions on second-order temporal coherence of thermal field for the case of $\omega_{0m} = \omega_{0n} = \omega_0$, except otherwise specified.

It is obvious that the second-order temporal coherence as well as the degree of second-order temporal coherence of thermal field depend on the relative delay

$$\tau = \tau_1 - \tau_2 = [(t - \tau_1^e) - z_1/c] - [(t - \tau_2^e) - z_2)/c] = (z_2 - z_1)/c + (\tau_2^e - \tau_1^e)$$

and it is symmetric with respect to τ

$$\gamma^{(2)}(\tau) = \gamma^{(2)}(-\tau). \tag{8.2.15}$$

It should be emphasized that t_1 and t_2 are defined from the joint-detection time t, but not from the source, consistent with the Heisenberg picture. Basically, τ is determined by the relative optical and electrical path delays. Experimentally, τ can be "scanned" by either hardware or software in either analog measurement or in photon counting measurement.

Apparently, the nontrivial second-order temporal correlation explored a paradoxical behavior of thermal radiation: although D_1 and D_2 are randomly excited by the subfields from time to time at any time t_1 and t_2, if one of them is excited at a certain time the other one has a twice greater chance of being excited simultaneously at $t_1 = t_2$. It seems that the randomly radiated and randomly distributed subfields in thermal state must be non-randomly "bunched" temporally in joint measurement between two distant photodetectors.

(II) Second-order far-field spatial coherence of thermal field

Assuming a perfect degree of second-order temporal coherence by means of $\gamma^{(2)}(\tau) \simeq 2$, we evaluate the second-order far-field spatial coherence function of thermal field in terms of the transverse spatial separation of two point-like photodetectors D_1 and D_2. This calculation is under Einstein's granularity picture by assuming a disk-like radiation source with finite angular diameter, $\Delta\theta$, or transverse size of D that contains a large number of independent and randomly distribute point-like sub-sources (spontaneous atomic transitions) on the entire disk, such as the mth and nth sub-sources. To simplify the mathematics, we restrict our calculation to 1-D. A simplified schematic experimental setup for measuring the second-order spatial coherence or correlation function is shematically shown in Fig. 8.2.1. The angular size of the radiation source, defined as the angle subtended by the source at the detector, is $\Delta\theta$ ($\Delta\theta \sim D/z$, with $z_1 = z_2 = z$, D is the transverse size of the source).

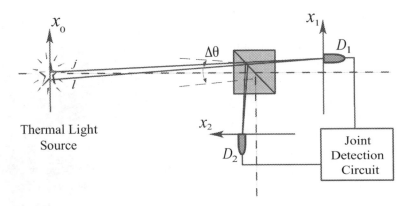

FIGURE 8.2.1 Schematic experimental setup for measuring the second-order spatial coherence function $\Gamma^{(2)}(x_1, t_1; x_2, t_2)$ of thermal field. Imagine the measurement is in the far-field zone of a distant star.

We start the calculation of the second-order coherence function of thermal field from

$$\Gamma^{(2)}(x_1, t_1; x_2, t_2)$$
$$= \sum_{m,n} \left| \frac{1}{\sqrt{2}} \left[E_m(x_1, t_1) E_n(x_2, t_2) + E_n(x_1, t_1) E_m(x_2, t_2) \right] \right|^2 \tag{8.2.16}$$

where we have simplified the problem in 1-D, and assume all the mth and the mth sub-sources are randomly distributed along the x_0 axis and spatially distinguishable. The mth and the nth sub-sources are identified by their transverse coordinates on the x_0 axis. Eq. (8.2.16) indicates a two-photon interference: two independent subfields, created by the mth sub-source at x_{0m} and the nth sub-sources at x_{0n}, excited a joint photodetection event between D_1 and D_2 with two different yet indistinguishable alternatives: (1) the mth and the nth wavepackets are detected by D_1 and D_2, respectively; and (2) the mth and the nth wavepackets are detected by D_2 and D_1, respectively.[2]

Further simplifying the mathematics, we assume monochromatic radiation. Substituting the plane-wave approximation into Eq. (8.2.16), and assuming $t_1 \simeq t_2$ by arranging a

---

[2]In the language of quantum mechanics, again, this superposition corresponds to a two-photon interference phenomenon, which involves a superposition between two different yet indistinguishable two-photon probability amplitudes: (1) photon m and photon n are annihilated at D_1 and D_2, respectively; and (2) photon m and photon n are annihilated at D_2 and D_1, respectively.

symmetrical optical and electronic experimental setup for the photodetection events of D_1 and D_2, $\Gamma^{(2)}(x_1, x_2)$ is approximately

$$\Gamma^{(2)}(x_1, x_2)$$

$$\simeq \int dx_0 \int dx_0' \Big| \frac{1}{\sqrt{2}} \big[a(x_0) e^{i\varphi(x_0)} e^{-ikr(x_0,x_1)} a(x_0') e^{i\varphi(x_0')} e^{-ikr(x_0',x_2)}$$

$$+ a(x_0') e^{i\varphi(x_0')} e^{-ikr(x_0',x_1)} a(x_0) e^{i\varphi(x_0)} e^{-ikr(x_0,x_2)} \big] \Big|^2$$

$$\simeq \int dx_0\, a^2(x_0) \int dx_0'\, a^2(x_0') + \Big| \int dx_0\, a^2(x_0)\, e^{-ikx_0(x_1-x_2)/z} \Big|^2$$

$$\simeq I_0^2 \, [1 + \operatorname{sinc}^2 (\frac{\pi \Delta\theta(x_1 - x_2)}{\lambda})], \tag{8.2.17}$$

where we have applied the far-field approximation and treated $a(x_0) \simeq a$ as a constant. Comparing with our early calculation of the first-order spatial coherence function for the distant star, we see that $x_1 - x_2$ is equivalent to the spatial separation b between the two pinholes. Notice that, due to the chosen positive directions of x_1 and x_2, we have $x_1 - x_2 = b$.

The degree of second-order spatial coherence $\gamma^{(2)}$ is thus

$$\gamma^{(2)}(x_1, t_1; x_2, t_2) = 1 + \operatorname{sinc}^2 \big[\frac{\pi \Delta\theta(x_1 - x_2)}{\lambda} \big]. \tag{8.2.18}$$

If the angular size $\Delta\theta$ of the thermal source is not too small, for short wavelength radiation the sinc-function in Eqs. (8.2.17) and (8.2.18) quickly drops from its maximum to minimum when $x_1 - x_2$ goes from zero to a value such that $\Delta\theta(x_1 - x_2)/\lambda = 1$. In this case we effectively have a "point"-to-"point" relationship between the x_1 plane and the x_2 plane. Notice, Eqs. (8.2.17) and (8.2.18) are functions of $x_1 - x_2$, which is independent of the absolute values of either x_1 or x_2. This is a very important and useful property of thermal field. It signifies whatever transverse coordinate x_1 we choose for D_1, there is a unique position x_2 for D_2 where the maximum joint-detection between D_1 and D_2 is expected, i.e., maximum constructive interference between $E_m(x_1, t_1)E_n(x_2, t_2)$ and $E_n(x_1, t_1)E_m(x_2, t_2)$ in Eq. (8.2.16) is observable at that unique position.

(III) Second-order near-field spatial coherence of thermal field

In the following, we attempt a calculation starting from Eq. (8.2.1) for the Fresnel near-field second-order spatial coherence function of thermal field $\Gamma^{(2)}(\vec{\rho}_1, z_1; \vec{\rho}_2, z_2)$,

$$\Gamma^{(2)}(\vec{\rho}_1, z_1; \vec{\rho}_2, z_2)$$

$$= \sum_{m,n} \Big| \frac{1}{\sqrt{2}} \big[E_m(\vec{\rho}_1, z_1) E_n(\vec{\rho}_2, z_2) + E_n(\vec{\rho}_1, z_1) E_m(\vec{\rho}_2, z_2) \big] \Big|^2$$

$$= \sum_{m,n} \Gamma_{mn}^{(2)}(\vec{\rho}_1, z_1; \vec{\rho}_2, z_2), \tag{8.2.19}$$

where we have used the transverse and longitudinal coordinates of $\vec{\rho}$ and z to specify the spatial coordinates of the photodetectors. In Eq. (8.2.19) we have ignored the temporal variables by assuming a perfect degree of second-order temporal coherence as usual. In the near-field we apply the Fresnel approximation to propagate the field from each sub-source to the photodetectors. Eq. (8.2.19) can be written in terms of the Green's functions:

$$\Gamma_{mn}^{(2)}(\vec{\rho}_1, z_1; \vec{\rho}_2, z_2)$$

$$= I_{mn}^2 \Big| \frac{1}{\sqrt{2}} \big[g_m(\vec{\rho}_1, z_1)\, g_n(\vec{\rho}_2, z_2) + g_m(\vec{\rho}_2, z_2)\, g_n(\vec{\rho}_1, z_1) \big] \Big|^2 \tag{8.2.20}$$

with

$$g_m(\vec{\rho}_j, z_j) = \frac{c_0}{z_j} e^{i\frac{\omega z_j}{c}} e^{i\frac{\omega}{2cz_j}|\vec{\rho}_j - \vec{\rho}_m|^2}.$$

where c_0 is a normalization consistent. In Eq. (8.2.20), we have also assumed $|E_m|^2 = |E_n|^2 = I_{mn}$. Next, we approximate the sum of $\Gamma_{mn}^{(2)}(\vec{\rho}_1, z_1; \vec{\rho}_2, z_2)$ in terms of the mth and nth sub-source into the integrals of $\vec{\rho}_0$ and $\vec{\rho}_0'$ on the entire source plane:

$$\Gamma^{(2)}(\vec{\rho}_1, z_1; \vec{\rho}_2, z_2)$$
$$= I_0^2 \Big[\Big| \int d\vec{\rho}_0\, g_{\vec{\rho}_0}(\vec{\rho}_1, z_1) \int d\vec{\rho}_0'\, g_{\vec{\rho}_0'}(\vec{\rho}_2, z_2) \Big|^2$$
$$+ \int d\vec{\rho}_0\, g_{\vec{\rho}_0}^*(\vec{\rho}_1, z_1) g_{\vec{\rho}_0}(\vec{\rho}_2, z_2) \int d\vec{\rho}_0'\, g_{\vec{\rho}_0'}^*(\vec{\rho}_2, z_2) g_{\vec{\rho}_0'}(\vec{\rho}_1, z_1) \Big]$$
$$= \Gamma_{11}^{(1)}\Gamma_{22}^{(1)} + \Gamma_{12}^{(1)}\Gamma_{21}^{(1)} \tag{8.2.21}$$

Substitute the Green's functions into Eq. (8.2.21), we obtain

$$\Gamma_{11}^{(1)}\Gamma_{22}^{(1)} \sim \text{constant},$$

and

$$\Gamma_{12}^{(1)}(\vec{\rho}_1, z_1; \vec{\rho}_2, z_2)$$
$$\propto \frac{1}{z_1 z_2} \int d\vec{\rho}_0\, a^2(\vec{\rho}_0)\, e^{-i\frac{\omega}{c}z_1}\, e^{-i\frac{\omega}{2cz_1}|\vec{\rho}_1 - \vec{\rho}_0|^2}\, e^{i\frac{\omega}{c}z_2}\, e^{i\frac{\omega}{2cz_2}|\vec{\rho}_2 - \vec{\rho}_0|^2},$$

Assuming $a^2(\vec{\rho}_0) \sim$ constant, and taking $z_1 = z_2 = d$, we obtain

$$\Gamma_{12}^{(1)}(\vec{\rho}_1; \vec{\rho}_2) \propto \int d\vec{\rho}_0\, e^{-i\frac{\omega}{2cd}|\vec{\rho}_1 - \vec{\rho}_0|^2}\, e^{i\frac{\omega}{2cd}|\vec{\rho}_2 - \vec{\rho}_0|^2}$$
$$\propto e^{-i\frac{\omega}{2cd}(|\vec{\rho}_1|^2 - |\vec{\rho}_2|^2)} \int d\vec{\rho}_0\, e^{i\frac{\omega}{cd}(\vec{\rho}_1 - \vec{\rho}_2)\cdot\vec{\rho}_0}$$
$$\propto e^{-i\frac{\omega}{2cd}(|\vec{\rho}_1|^2 - |\vec{\rho}_2|^2)} \text{somb}\Big[\frac{R}{d}\frac{\omega}{c}|\vec{\rho}_1 - \vec{\rho}_2|\Big], \tag{8.2.22}$$

where we have assumed a disk-like light source with a finite radius of R, and again, $\text{somb}(x) = J_1(x)/x$, $J_1(x)$ is the first-order Bessel Function. In Eq. (8.2.22) we have absorbed all constants into the proportionality constant. The second-order spatial correlation function $\Gamma^{(2)}(\vec{\rho}_1; \vec{\rho}_2)$ is thus

$$\Gamma^{(2)}(\vec{\rho}_1, \vec{\rho}_2) = I_0^2 \Big[1 + \text{somb}^2\big(\frac{R}{d}\frac{\omega}{c}|\vec{\rho}_1 - \vec{\rho}_2|\big)\Big]. \tag{8.2.23}$$

Consequently, the degree of second-order spatial coherence is

$$\gamma^{(2)}(\vec{\rho}_1, \vec{\rho}_2) = 1 + \text{somb}^2\big(\frac{R}{d}\frac{\omega}{c}|\vec{\rho}_1 - \vec{\rho}_2|\big). \tag{8.2.24}$$

For a large value of $2R/d \sim \Delta\theta$, where $\Delta\theta$ is the angular size of the radiation source viewed at the photodetectors, the point-to-"spot" sombrero-like function can be approximated as a δ-function of $|\vec{\rho}_1 - \vec{\rho}_2|$. We thus effectively have a "point"-to-"point" correlation between the transverse planes of $z_1 = d$ and $z_2 = d$.

The above discussions on the second-order spatial coherence or correlation function of thermal field, either far-field or near-field, ended with an interesting result. Analogous to EPR's language, the photodetectors D_1 and D_2 have equal chance to be triggered at any position on the transverse planes of $z_1 = z_2$, however, if D_1 is triggered at a certain transverse position of $\vec{\rho}_1$, D_2 has twice chance to be trigged at a transverse position $\vec{\rho}_2 = \vec{\rho}_1$.

8.3 SECOND-ORDER COHERENCE—FORMULATED FROM QUANTUM THEORY OF LIGHT

Now we introduce the second-order quantum coherence function, $G^{(2)}(\mathbf{r}_1, t_1; \mathbf{r}_2, t_2)$, which is formulated from the quantum theory of light. One may ask immediately: why do we need quantum theory if Einstein's granularity picture or the concept of quantized subfield gives a successful explanation to the nontrivial second-order coherence function of thermal field? (1) Einstein's granularity picture does not work for a certain quantum states, such as entangled states; (2) Although Einstein's picture gives a reasonable solution to the nontrivial second-order coherence function of thermal field, the concept of subfield itself is still under the framework of classical electromagnetic field theory. Recall our earlier story about the size of a single photon after one-year propagation in which Einstein trapped himself and his students into a self-contradictory situation. We have a much serious problem when dealing with second-order correlation of thermal field. In addition to the collapse of one single-wavepacket we are facing the collapse of two wavepackets, and the two collapses are correlated in space-time! Either in their temporal correlation or spatial correlation, the randomly created and randomly paired wavepackets, respectively, may collapse to any localized space-time points, however, if one wavepacket collapses to a certain space-time point, the other one has twice the chance to collapse at a unique localized space-time area. Einstein realized that his concept of quantized realistic electromagnetic subfield may imply an action-at-a-distance. Einstein may accept the two-photon interference picture of the second-order coherence of thermal field, however, Einstein may ask another serious question: how long does it take for the nonlocal superposition of the two subfields to complete? We may have to accept the quantum mechanical concepts of probability and probability amplitude as Bohr suggested from the beginning of quantum theory.

Figure 8.3.1 illustrates a schematic setup of the measurement for second-order coherence function $G^{(2)}(\mathbf{r}_1, t_1; \mathbf{r}_2, t_2)$. We assume a weak light source at single-photon's level. The measured radiation is in a certain quantum state, for instance, in thermal state, in coherent state, or in entangled state. To simplify the discussion, we assume idealized point-like photodetectors, D_1 and D_2, operated in photon counting mode. D_1 and D_2 are both scannable along the longitudinal axis and on the transverse plan as shown in the figure for joint-detection measurement. The registration times of the photodetection events of D_1 and D_2 are recorded along two synchronized time axis t_1 and t_2, each is divided into a large number of "short" time windows Δt. The width of the time window is adjustable either

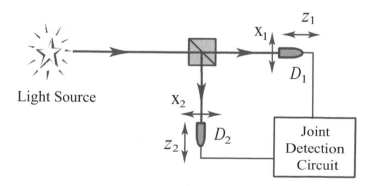

FIGURE 8.3.1 A schematic measurement of second-order coherence function $G^{(2)}(\mathbf{r}_1, t_1; \mathbf{r}_2, t_2)$ for quantum states created from different radiation sources. D_1 and D_2 are idealized point-like photodetectors, and are both scannable along their longitudinal axes and on their transverse planes. For most of the following discussions, we assume D_1 and D_2 both operated at photon counting mode. D_1 and D_2 together with a joint-detection circuit plays the role of a coincidence counter.

by hardware or software of the circuit. If a photon excites a photoelectron from D_1 within the jth time window, the counting circuit registers an "one" in the jth time window. If no photoelectron is excited within the kth time window, the circuit records a "zero" in the kth time window. After a certain time accumulation, the statistical correlation of $n_1(\mathbf{r}_1, t_1)$ and $n_2(\mathbf{r}_2, t_2)$ is then calculated from:

$$\langle n_1(\mathbf{r}_1, t_1)\, n_2(\mathbf{r}_2, t_2)\rangle = \frac{1}{N} \sum_{j=1}^{N} n_1(\mathbf{r}_1, t_{1j}) n_2(\mathbf{r}_2, t_{2j}) \tag{8.3.1}$$

where j labels the jth synchronized time window, t_1 and t_2 are defined by the electronic delays, $t_1 = t_j - \tau_1^e$ and $t_2 = t_j - \tau_2^e$. To simplify the discussion, we usually, but not necessarily, assume equal electronic delays $\tau_1^e = \tau_2^e$ and name the jointly observed photodetection events "coincidences".

Following Glauber's theory, the probability of observing a pair of photons, either randomly paired photons by chance or coherently prepared photon pairs from a certain physical process, to excite a joint-detection event at $(\mathbf{r}_1, t_1)$ and $(\mathbf{r}_2, t_2)$ is

$$G^{(2)}(\mathbf{r}_1, t_1; \mathbf{r}_2, t_2)$$
$$\equiv \langle\, E^{(-)}(\mathbf{r}_1, t_1) E^{(-)}(\mathbf{r}_2, t_2) E^{(+)}(\mathbf{r}_2, t_2) E^{(+)}(\mathbf{r}_1, t_1)\,\rangle$$
$$= \langle\langle\, E^{(-)}(\mathbf{r}_1, t_1) E^{(-)}(\mathbf{r}_2, t_2) E^{(+)}(\mathbf{r}_2, t_2) E^{(+)}(\mathbf{r}_1, t_1)\,\rangle_{\text{QM}}\rangle_{\text{En}} \tag{8.3.2}$$

which is recognized as the expectation value of the normal-ordered field operators at space-time coordinates $(\mathbf{r}_1, t_1)$ and $(\mathbf{r}_2, t_2)$. In the view of quantum theory of light, $G^{(2)}(\mathbf{r}_1, t_1; \mathbf{r}_2, t_2)$ is thus a measure of the probability for jointly observing a pair of photons at $(\mathbf{r}_1, t_1)$ and $(\mathbf{r}_2, t_2)$ from a radiation field. Differing from the first-order coherence function, the second-order coherence function is directly measurable by two photodetectors at space-time coordinates $(\mathbf{r}_1, t_1)$ and $(\mathbf{r}_2, t_2)$. It is very important to keep in mind, here, t_1 and t_2 are the physical times of the two photodetection events, or the registration times of a pair of photons. In the first-order coherence function $G^{(1)}(\mathbf{r}_1, t_1; \mathbf{r}_2, t_2)$, however, t_1 and t_2 are the "referred early times" of a photodetection event at $(\mathbf{r}, t)$.

To be consistent with the theory of first-order coherence, a normalized second-order quantum coherence function, namely the degree of second-order quantum coherence is defined accordingly:

$$g^{(2)}(\mathbf{r}_1, t_1; \mathbf{r}_2, t_2)$$
$$\equiv \frac{\langle\, E^{(-)}(\mathbf{r}_1, t_1) E^{(-)}(\mathbf{r}_2, t_2) E^{(+)}(\mathbf{r}_2, t_2) E^{(+)}(\mathbf{r}_1, t_1)\,\rangle}{\langle\, E^{(-)}(\mathbf{r}_1, t_1) E^{(+)}(\mathbf{r}_1, t_1)\,\rangle \langle\, E^{(-)}(\mathbf{r}_2, t_2) E^{(+)}(\mathbf{r}_2, t_2)\,\rangle}. \tag{8.3.3}$$

The second-order quantum coherence function $G^{(2)}(\mathbf{r}_1, t_1; \mathbf{r}_2, t_2)$ as well as the degree of second-order quantum coherence $g^{(2)}(\mathbf{r}_1, t_1; \mathbf{r}_2, t_2)$ of a radiation is obviously determined by the state of the jointly measured photon pair. There is no surprise that different two-photon state results in different second-order coherence or correlation. In the following, we calculate the degree of second-order quantum coherence $g^{(2)}(\mathbf{r}_1, t_1; \mathbf{r}_2, t_2)$ of thermal field in the single-photon state representation and in the coherent state representation.

8.4   SECOND-ORDER QM COHERENCE OF THERMAL STATE

In this section, we evaluate the second-order quantum coherence function for thermal field.

(I) $G^{(2)}(\mathbf{r}_1, t_1; \mathbf{r}_2, t_2)$ of thermal field in single-photon state representation

Suppose the light source is weak enough that no more than one pair of photons are observed within the chosen coincidence time window Δt. Applying the same model for thermal field that has been discussed in Chapter 3, we may approximate the state of the thermal field as

$$|\Psi\rangle = \prod_m \left\{ |0\rangle + \epsilon\, c_m \, |\Psi_m\rangle \right\}$$

$$\simeq |0\rangle + \epsilon \Big[\sum_m c_m \, |\Psi_m\rangle \Big] + \epsilon^2 \Big[\sum_{m<n} c_m \, c_n |\Psi_m\rangle |\Psi_n\rangle \Big] + \; ...$$

where each c_m and c_n takes a random phase from transition to transition. Since $|\epsilon| \ll 1$, we only listed the first-order and the second-order approximations on ϵ. The necessary lowest-order approximation that contributes to the second-order correlation measurement is,

$$|\Psi\rangle = \sum_{m<n} c_m \, c_n |\Psi_m\rangle |\Psi_n\rangle \tag{8.4.1}$$

with

$$|\Psi_m\rangle = \int d\mathbf{k}\, f_m(\mathbf{k})\, \hat{a}_m^\dagger(\mathbf{k})|0\rangle.$$

The state of Eq. (8.4.1) is a mixed two-photon state which characterizes the state of a large number of jointly measured photon pairs, each is created from a randomly radiated and randomly paired atomic transitions, as an inhomogeneous ensemble. We usually write the state of Eq. (8.4.1) in the form of mixed ensemble,

$$\hat{\rho} = \sum_{m,n} P_{mn}|\Psi_m\rangle |\Psi_n\rangle \langle \Psi_n| \langle \Psi_m| \tag{8.4.2}$$

where P_{mn} is the probability for observing the m-nth pair of subfields, which is obviously a constant for thermal state.

The field operators are similar to that of Eq. (7.5.10), except approximated into a more general form:

$$\hat{E}^{(+)}(\mathbf{r},t) = \sum_m \int d\mathbf{k}\, \hat{a}_m(\mathbf{k})\, g_m(\mathbf{k};\mathbf{r},t)$$

$$\hat{E}^{(-)}(\mathbf{r},t) = \sum_m \int d\mathbf{k}\, \hat{a}_m^\dagger(\mathbf{k})\, g_m^*(\mathbf{k};\mathbf{r},t) \tag{8.4.3}$$

where $g_m(\mathbf{k};\mathbf{r},t)$ is the Green's function that propagates the $\mathbf{k}$-mode of the mth subfield from the mth sub-source to space-time coordinate $(\mathbf{r},t)$. We have also ignored the constants. We thus have

$$G^{(2)}(\mathbf{r}_1,t_1;\mathbf{r}_2,t_2)$$

$$= \left\langle \langle \Psi| \hat{E}^{(-)}(\mathbf{r}_1,t_1)\hat{E}^{(-)}(\mathbf{r}_2,t_2)|0\rangle \langle 0| \hat{E}^{(+)}(\mathbf{r}_2,t_2)\hat{E}^{(+)}(\mathbf{r}_1,t_1)|\Psi\rangle \right\rangle_{En}$$

$$= \left\langle \sum_m c_m^* \psi_m^*(\mathbf{r}_1,t_1) \sum_n c_n^* \psi_n^*(\mathbf{r}_2,t_2) \sum_q c_q \psi_q(\mathbf{r}_2,t_2) \sum_p c_p \psi_p(\mathbf{r}_1,t_1) \right\rangle_{En}$$

$$= \sum_m \psi_m^*(\mathbf{r}_1,t_1)\psi_m(\mathbf{r}_1,t_1) \sum_n \psi_n^*(\mathbf{r}_2,t_2)\psi_n(\mathbf{r}_2,t_2)$$

$$+ \sum_{m\neq n} \psi_m^*(\mathbf{r}_1,t_1)\psi_n(\mathbf{r}_1,t_1)\psi_n^*(\mathbf{r}_2,t_2)\psi_m(\mathbf{r}_2,t_2)$$

$$\propto \sum_{m,n} \left| \frac{1}{\sqrt{2}}[\psi_m(\mathbf{r}_1,t_1)\psi_n(\mathbf{r}_2,t_2) + \psi_m(\mathbf{r}_2,t_2)\psi_n(\mathbf{r}_1,t_1)] \right|^2 \tag{8.4.4}$$

where $\psi_m(\mathbf{r}_j, t_j)$, $j = 1, 2$, is defined as the effective wavefunction of the mth photon at space-time coordinate $(\mathbf{r}_j, t_j)$,

$$\psi_m(\mathbf{r}_j, t_j) \equiv \langle 0|\hat{E}^{(+)}(\mathbf{r}_2, t_2)|\Psi_m\rangle = \int d\mathbf{k}\, f_m(\mathbf{k})\, g_m(\mathbf{k}; r_j, t_j). \tag{8.4.5}$$

In Eq. (8.4.4) we have approximated $|c_m| = $ constant, which is a reasonable approximation for thermal state. We may find immediately that the second-order coherence function $G^{(2)}(\mathbf{r}_1, t_1; \mathbf{r}_2, t_2)$ of thermal state, which is formulated from quantum theory of light in single-photon state representation, is the same as the second-order coherence function $\Gamma^{(2)}(\mathbf{r}_1, t_1; \mathbf{r}_2, t_2)$ that we have calculated previously in Einstein's picture, except the sub-field is replaced by the effective wavefunction of a photon. We also find the second-order coherence function $G^{(2)}(\mathbf{r}_1, t_1; \mathbf{r}_2, t_2)$ of thermal state is the result of two-photon interference, which involves the superposition of two different yet indistinguishable two-photon amplitudes: (1) the mth photon is annihilated at $(\mathbf{r}_1, t_1)$ and the nth photon is annihilated at $(\mathbf{r}_2, t_2)$; and (2) the mth photon is annihilated at $(\mathbf{r}_2, t_2)$ and the nth photon is annihilated at $(\mathbf{r}_1, t_1)$. We name this superposition two-photon interference: a pair of randomly created and randomly distributed photons in thermal state interfering with the pair itself.

Similar to Eq.(8.2.3) in Einstein's picture, we find the constant part of $G^{(2)}(\mathbf{r}_1, t_1; \mathbf{r}_2, t_2)$ is a product of the following two mean number of photons measured by D_1 and D_2, respectively,

$$\langle n(\mathbf{r}_1, t_1)\rangle = \sum_m \psi_m^*(\mathbf{r}_1, t_1)\, \psi_m(\mathbf{r}_1, t_1)$$

$$\langle n(\mathbf{r}_2, t_2)\rangle = \sum_n \psi_n^*(\mathbf{r}_2, t_2)\, \psi_n(\mathbf{r}_2, t_2), \tag{8.4.6}$$

except the subfield is replaced by the effective wavefunction of a photon. The cross-interference term of Eq.(8.4.4), which is the nontrivial term of the second-order coherence function $G^{(2)}(\mathbf{r}_1, t_1; \mathbf{r}_2, t_2)$, corresponding to the photon number fluctuation correlation,

$$\langle \Delta n(\mathbf{r}_1, t_1)\Delta n(\mathbf{r}_2, t_2)\rangle$$
$$= \sum_{m \neq n} \psi_m^*(\mathbf{r}_1, t_1)\psi_n(\mathbf{r}_1, t_1)\psi_n^*(\mathbf{r}_2, t_2)\psi_m(\mathbf{r}_2, t_2) \tag{8.4.7}$$

is also the same as the intensity fluctuation correlation we have calculated previously in Einstein's picture, except the subfield is replaced by the effective wavefunction of a photon.

In the case of perfect measurement system, including idealized photodetectors and idealized correlation measurement electronics, the second-order coherence function, $G^{(2)}(\mathbf{r}_1, t_1; \mathbf{r}_2, t_2)$, of thermal state can be formally written in terms of its first-order quantum coherence functions:

$$G^{(2)}(\mathbf{r}_1, t_1; \mathbf{r}_2, t_2) = G_{11}^{(1)} G_{22}^{(1)} + G_{12}^{(1)} G_{21}^{(1)} = G_{11}^{(1)} G_{22}^{(1)} + |G_{12}^{(1)}|^2. \tag{8.4.8}$$

where

$$G_{11}^{(1)} = \sum_m \psi_m^*(\mathbf{r}_1, t_1)\psi_m(\mathbf{r}_1, t_1) \quad G_{22}^{(1)} = \sum_n \psi_n^*(\mathbf{r}_2, t_2)\psi_n(\mathbf{r}_2, t_2)$$

$$G_{12}^{(1)} = \sum_m \psi_m^*(\mathbf{r}_1, t_1)\psi_m(\mathbf{r}_2, t_2) \quad G_{21}^{(1)} = \sum_n \psi_n^*(\mathbf{r}_2, t_2)\psi_n(\mathbf{r}_1, t_1). \tag{8.4.9}$$

It should be emphasized that, here, $G_{12}^{(1)}$ and $G_{21}^{(1)}$ are physically measured by two independent photodetectors D_1 and D_2 at different space-time coordinates $(\mathbf{r}_1, t_1)$ and $(\mathbf{r}_2, t_2)$, which is very different from the measurement of the first-order coherence function. Although

Eq. (8.4.8) is helpful for mathematical calculations, we must always keep this in mind, especially when applying non-idealized measurement system. The realistic finite response time of the photodetectors and the correlation measurement circuit will complicate the calculation, especially when dealing with thermal field of broadband spectrum.

(II) Two-photon effective wavefunction of thermal state in single-photon state representation

In fact, the effective wavefunction of the m-nth photon pair in thermals state can be defined directly from a product state

$$|\Psi_{mn}\rangle = |\Psi_m\rangle|\Psi_n\rangle, \tag{8.4.10}$$

with

$$|\Psi_m\rangle = \int d\omega\, f_m(\omega)\, \hat{a}^\dagger(\omega)|\,0\,\rangle, \quad |\Psi_n\rangle = \int d\omega\, f_n(\omega)\, \hat{b}^\dagger(\omega)|\,0\,\rangle,$$

where the complex amplitudes $f_m(\omega)$ and $f_n(\omega)$ will be treated with identical spectrum distribution, but random phases that are determined by the initial condition of the mth and nth atomic transitions. The second-order coherence function of the m-nth photon pair is thus

$$
\begin{aligned}
G^{(2)}_{mn}&(\mathbf{r}_1, t_1; \mathbf{r}_2, t_2)\\
&= \langle\Psi_{mn}|\, E^{(-)}(\mathbf{r}_1, t_1)E^{(-)}(\mathbf{r}_2, t_2)E^{(+)}(\mathbf{r}_2, t_2)E^{(+)}(\mathbf{r}_1, t_1)\,|\Psi_{mn}\rangle\\
&= \sum_l \langle\Psi_{mn}|\, E^{(-)}(\mathbf{r}_1, t_1)E^{(-)}(\mathbf{r}_2, t_2)\,|l\rangle \langle l|\, E^{(+)}(\mathbf{r}_2, t_2)E^{(+)}(\mathbf{r}_1, t_1)\,|\Psi_{mn}\rangle\\
&= \langle\Psi_{mn}|E^{(-)}(\mathbf{r}_1, t_1)E^{(-)}(\mathbf{r}_2, t_2)|0\rangle\langle 0|E^{(+)}(\mathbf{r}_2, t_2)E^{(+)}(\mathbf{r}_1, t_1)|\Psi_{mn}\rangle\\
&= \big|\psi_{mn}(\mathbf{r}_1, t_1; \mathbf{r}_2, t_2)\big|^2,
\end{aligned}
\tag{8.4.11}
$$

where, in the third line, we have used the completeness relation

$$\sum_l |l\rangle\langle l| = 1.$$

The effective wavefunction of the m-nth photon pair,

$$\psi_{mn}(\mathbf{r}_1, t_1; \mathbf{r}_2, t_2) \equiv \langle 0|E^{(+)}(\mathbf{r}_2, t_2)E^{(+)}(\mathbf{r}_1, t_1)|\Psi_{mn}\rangle. \tag{8.4.12}$$

The second-order quantum coherence function $G^{(2)}(\mathbf{r}_1, t_1; \mathbf{r}_2, t_2)$ of thermal field can be also calculated from the density operator of Eq. (8.4.2), which characterizes an ensemble of mixed two-photon product states,

$$
\begin{aligned}
G^{(2)}&(\mathbf{r}_1, t_1; \mathbf{r}_2, t_2)\\
&= \big\langle\big\langle\, E^{(-)}(\mathbf{r}_1, t_1)E^{(-)}(\mathbf{r}_2, t_2)E^{(+)}(\mathbf{r}_2, t_2)E^{(+)}(\mathbf{r}_1, t_1)\,\big\rangle_{\text{QM}}\big\rangle_{\text{En}}\\
&= tr\,\big[\hat{\rho}\, E^{(-)}(\mathbf{r}_1, t_1)E^{(-)}(\mathbf{r}_2, t_2)\, E^{(+)}(\mathbf{r}_2, t_2)E^{(+)}(\mathbf{r}_1, t_1)\big]\\
&= \sum_{m,n} P_{mn}\, \langle\Psi_{mn}|\, E^{(-)}(\mathbf{r}_1, t_1)E^{(-)}(\mathbf{r}_2, t_2)E^{(+)}(\mathbf{r}_2, t_2)E^{(+)}(\mathbf{r}_1, t_1)\,|\Psi_{mn}\rangle\\
&= \sum_{m,n} P_{mn}\, G^{(2)}_{mn}(\mathbf{r}_1, t_1; \mathbf{r}_2, t_2) = \sum_{m,n} P_{mn}\, |\psi_{mn}(\mathbf{r}_1, t_1; \mathbf{r}_2, t_2)|^2,
\end{aligned}
\tag{8.4.13}
$$

where P_{mn} is the probability to find the radiation in the two-photon product state $|\Psi_{mn}\rangle$. Obviously, $P_{mn} = $ constant for thermal field, and thus we have

$$G^{(2)}(\mathbf{r}_1, t_1; \mathbf{r}_2, t_2) = \sum_{m,n} |\psi_{mn}(\mathbf{r}_1, t_1; \mathbf{r}_2, t_2)|^2. \tag{8.4.14}$$

(III) $G^{(2)}(\mathbf{r}_1, t_1; \mathbf{r}_2, t_2)$ of thermal state in coherent state representation

Considering the thermal state in coherent state representation that has been introduced in Eq.(4.6.7),

$$|\Psi\rangle = \prod_m |\Psi_m\rangle = \prod_{m,\mathbf{k}} |\alpha_m(\mathbf{k})\rangle,$$

where we have explicitly separated index m, indicating the mth photon or the mth group of identical photons, from $\mathbf{k}$, indicating the $\mathbf{k}$th mode. Coherent state may represent a group of identical photons with $\bar{n} = |\alpha| \gg 1$. Coherent state representation is not only simplified the calculation of second-order coherence function significantly, but also made it possible to calculate the coherence function of a group identical photons. We will restrict our discussion at single-photon's level, $\bar{n} = |\alpha| \ll 1$. In this case, we may approximate it the state of a single-photon as we have discussed in Chapter 4.

Under this restriction, the state of the mth photon in the coherent state representation, $|\Psi_m\rangle$ is treated as pure state; the ensemble of the randomly created and randomly distributed photons are in mixed state with random relative phases. Since $|\alpha_m(\mathbf{k})\rangle$ is an eigenstate of the annihilation operator with an eigenvalue $\alpha_m(\mathbf{k})$,

$$\hat{a}_m(\mathbf{k})|\alpha_m(\mathbf{k})\rangle = \alpha_m(\mathbf{k})|\alpha_m(\mathbf{k})\rangle,$$

therefore

$$\hat{a}_m(\mathbf{k})|\Psi\rangle = \alpha_m(\mathbf{k})|\Psi\rangle.$$

The field operators are the same as shown in Eq. (8.4.3). We thus have

$$\langle\Psi|\hat{E}^{(-)}(\mathbf{r}, t)|\Psi\rangle = \langle\Psi|\Big[\sum_m \int d\mathbf{k}\, \hat{a}_m^\dagger(\mathbf{k})\, g_m^*(\mathbf{k}; \mathbf{r}, t)\Big]|\Psi\rangle$$

$$= \langle\Psi|\Big[\sum_m \int d\mathbf{k}\, \alpha_m^*(\mathbf{k})\, g_m^*(\mathbf{k}; \mathbf{r}, t)\Big]|\Psi\rangle$$

$$= \sum_m \psi_m^*(\mathbf{r}, t)$$

$$\langle\Psi|\hat{E}^{(+)}(\mathbf{r}, t)|\Psi\rangle = \langle\Psi|\Big[\sum_n \int d\mathbf{k}\, \hat{a}_n(\mathbf{k})\, g_n(\mathbf{k}; \mathbf{r}, t)\Big]|\Psi\rangle$$

$$= \langle\Psi|\Big[\sum_n \int d\mathbf{k}\, \alpha_n(\mathbf{k})\, g_n(\mathbf{k}; \mathbf{r}, t)\Big]|\Psi\rangle$$

$$= \sum_n \psi_n(\mathbf{r}, t) \tag{8.4.15}$$

Therefore, we have

$$G^{(2)}(\mathbf{r}_1, t_1; \mathbf{r}_2, t_2)$$
$$= \Big\langle \sum_m \psi_m^*(\mathbf{r}_1, t_1) \sum_n \psi_n^*(\mathbf{r}_2, t_2) \sum_q \psi_q(\mathbf{r}_2, t_2) \sum_p \psi_p(\mathbf{r}_1, t_1) \Big\rangle_{\text{En}}$$
$$= \sum_m \psi_m^*(\mathbf{r}_1, t_1)\psi_m(\mathbf{r}_1, t_1) \sum_n \psi_n^*(\mathbf{r}_2, t_2)\psi_n(\mathbf{r}_2, t_2)$$
$$+ \sum_{m \neq n} \psi_m^*(\mathbf{r}_1, t_1)\psi_n(\mathbf{r}_1, t_1)\psi_n^*(\mathbf{r}_2, t_2)\psi_m(\mathbf{r}_2, t_2)$$
$$= \sum_{m,n} \Big| \frac{1}{\sqrt{2}} [\psi_m(\mathbf{r}_1, t_1)\psi_n(\mathbf{r}_2, t_2) + \psi_n(\mathbf{r}_1, t_1)\psi_m(\mathbf{r}_2, t_2)] \Big|^2 \tag{8.4.16}$$

where $\psi_m(\mathbf{r}_j, t_j)$, $j = 1, 2$, is defined as the effective wavefunction of the mth photon at space-time coordinate $(\mathbf{r}_j, t_j)$ of the jth photodetector D_j,

$$\psi_m(\mathbf{r}_j, t_j) = \int d\mathbf{k}\, \alpha_m(\mathbf{k})\, g_m(\mathbf{k}; \mathbf{r}_j, t_j).$$

The second-order coherence function $G^{(2)}(\mathbf{r}_1, t_1; \mathbf{r}_2, t_2)$ of thermal field, which is formulated from quantum theory of light in coherent state representation, is the result of two-photon interference which involves the superposition of two different yet indistinguishable two-photon amplitudes: (1) the mth photon, or the mth group identical photons, is observed at $(\mathbf{r}_1, t_1)$ and the nth photon, or the nth group identical photons, is observed at $(\mathbf{r}_2, t_2)$; and (2) the mth photon, or the mth group identical photons, is observed at $(\mathbf{r}_2, t_2)$ and the nth photon, or the nth group identical photons, is observed at $(\mathbf{r}_1, t_1)$. We name this superposition two-photon interference: a pair of randomly created and randomly distributed photons in thermal state interfering with the pair itself.

Similar to Eq.(8.4.6), we find the constant part of $G^{(2)}(\mathbf{r}_1, t_1; \mathbf{r}_2, t_2)$ is a product of the following two mean number of photons measured by D_1 and D_2, respectively,

$$\langle n(\mathbf{r}_1, t_1) \rangle = \sum_m \psi_m^*(\mathbf{r}_1, t_1)\, \psi_m(\mathbf{r}_1, t_1)$$

$$\langle n(\mathbf{r}_2, t_2) \rangle = \sum_n \psi_n^*(\mathbf{r}_2, t_2)\, \psi_n(\mathbf{r}_2, t_2),$$

while the cross-interference term, which is the nontrivial term of the second-order coherence function $G^{(2)}(\mathbf{r}_1, t_1; \mathbf{r}_2, t_2)$, corresponding to the photon number fluctuation correlation measured by D_1 and D_2, jointly,

$$\langle \Delta n(\mathbf{r}_1, t_1) \Delta n(\mathbf{r}_2, t_2) \rangle = \sum_{m \neq n} \psi_m^*(\mathbf{r}_1, t_1)\psi_n(\mathbf{r}_1, t_1)\psi_n^*(\mathbf{r}_2, t_2)\psi_m(\mathbf{r}_2, t_2)$$

is also the same as the photon number fluctuation correlation we have calculated in the single-photon state representation.

We find, again, the second-order quantum coherence function $G^{(2)}(\mathbf{r}_1, t_1; \mathbf{r}_2, t_2)$ calculated from the quantum coherent state representation is the same as the second-order coherence function $\Gamma^{(2)}(\mathbf{r}_1, t_1; \mathbf{r}_2, t_2)$ that we have calculated previously in Einstein's picture, except the subfield is replaced by the effective wavefunction of a photon or a group of identical photons.

In the case of perfect measurement system, including idealized photodetectors and idealized correlation measurement electronics, $G^{(2)}(\mathbf{r}_1, t_1; \mathbf{r}_2, t_2)$ can be formally written in terms of the first-order quantum coherence functions:

$$G^{(2)}(\mathbf{r}_1, t_1; \mathbf{r}_2, t_2) = G^{(1)}_{11} G^{(1)}_{22} + G^{(1)}_{12} G^{(1)}_{21} = G^{(1)}_{11} G^{(1)}_{22} + |G^{(1)}_{12}|^2. \tag{8.4.17}$$

It should be emphasized that, again, here, $G^{(1)}_{12}$ is measured by two independent photodetectors D_1 and D_2 at different space-time coordinates $(\mathbf{r}_1, t_1)$ and $(\mathbf{r}_2, t_2)$, which is very different from the measurement of the first-order coherence function. The realistic finite response time of the photodetectors and correlation measurement circuit will complicate the calculation, especially when dealing with thermal state of broadband spectrum.

We have mentioned that coherent state may represent the state of a group of identical photons when $\bar{n} = |\alpha| \gg 1$. The effective wavefunction $\psi_m(\mathbf{r}, t)$ represents the space-time behavior of the mth group of identical photons. In this case, we need to modify the statement about two-photon interference slightly: $G^{(2)}(\mathbf{r}_1, t_1; \mathbf{r}_2, t_2)$ of thermal field, which is formulated from quantum theory of light in coherent state representation, is the result of two-photon interference: a pair of randomly created and randomly distributed groups of identical photons interfering with the pair itself.

(IV) Two-photon effective wavefunction of thermal state in coherent state representation

Similar to the single-photon state representation, we may define a two-photon effective wavefunction for the m-nth pair of photons, or the m-nth pair of groups of identical photons, in the coherent state representation

$$\psi_{mn}(\mathbf{r}_1, t_1; \mathbf{r}_2, t_2) = \psi_m(\mathbf{r}_1, t_1)\psi_n(\mathbf{r}_2, t_2).$$

The quantum second-order coherence function $G^{(2)}(\mathbf{r}_1, t_1; \mathbf{r}_2, t_2)$ is thus written as the superperposition between two-photon effective wavefunctions

$$G^{(2)}(\mathbf{r}_1, t_1; \mathbf{r}_2, t_2)$$
$$= \sum_{m,n} \left| \frac{1}{\sqrt{2}} [\psi_{mn}(\mathbf{r}_1, t_1; \mathbf{r}_2, t_2) + \psi_{nm}(\mathbf{r}_1, t_1; \mathbf{r}_2, t_2)] \right|^2.$$

In fact, the effective wavefunction of the m-nth photon pair, or the m-nth pair of groups of identical photons, can be defined directly from a product state

$$|\Psi_{mn}\rangle = |\Psi_m\rangle|\Psi_n\rangle,$$

with

$$|\Psi_m\rangle = \prod_{\mathbf{k}} |\alpha_m(\mathbf{k})\rangle, \quad |\Psi_n\rangle = \prod_{\mathbf{k}'} |\alpha_n(\mathbf{k}')\rangle,$$

The second-order coherence function of the m-nth photon pair is thus

$$G_{mn}^{(2)}(\mathbf{r}_1, t_1; \mathbf{r}_2, t_2)$$
$$= \langle\Psi_{mn}| E^{(-)}(\mathbf{r}_1, t_1)E^{(-)}(\mathbf{r}_2, t_2)E^{(+)}(\mathbf{r}_2, t_2)E^{(+)}(\mathbf{r}_1, t_1) |\Psi_{mn}\rangle$$
$$= \langle\Psi_{mn}| E^{(-)}(\mathbf{r}_1, t_1)E^{(-)}(\mathbf{r}_2, t_2) |\Psi_{mn}\rangle \langle\Psi_{mn}| E^{(+)}(\mathbf{r}_2, t_2)E^{(+)}(\mathbf{r}_1, t_1) |\Psi_{mn}\rangle$$
$$= \left| \psi_{mn}(\mathbf{r}_1, t_1; \mathbf{r}_2, t_2) \right|^2,$$

The effective wavefunction of the m-nth photon pair, which is characterized by a product state, is defined as

$$\psi_{mn}(\mathbf{r}_1, t_1; \mathbf{r}_2, t_2) = \langle\Psi_{mn}|E^{(+)}(\mathbf{r}_2, t_2)E^{(+)}(\mathbf{r}_1, t_1)|\Psi_{mn}\rangle. \tag{8.4.18}$$

Thermal state concerns an ensemble of mixed two-photon product states, the second-order quantum coherence function can be calculated from the density operator:

$$G^{(2)}(\mathbf{r}_1, t_1; \mathbf{r}_2, t_2)$$
$$= \left\langle \left\langle E^{(-)}(\mathbf{r}_1, t_1)E^{(-)}(\mathbf{r}_2, t_2)E^{(+)}(\mathbf{r}_2, t_2)E^{(+)}(\mathbf{r}_1, t_1) \right\rangle_{\mathrm{QM}} \right\rangle_{\mathrm{En}}$$
$$= tr\left[\hat{\rho}\, E^{(-)}(\mathbf{r}_1, t_1)E^{(-)}(\mathbf{r}_2, t_2)\, E^{(+)}(\mathbf{r}_2, t_2)E^{(+)}(\mathbf{r}_1, t_1) \right]$$
$$= \sum_{m,n} \langle\Psi_{mn}| E^{(-)}(\mathbf{r}_1, t_1)E^{(-)}(\mathbf{r}_2, t_2)E^{(+)}(\mathbf{r}_2, t_2)E^{(+)}(\mathbf{r}_1, t_1) |\Psi_{mn}\rangle$$
$$= \sum_{m,n} G_{mn}^{(2)}(\mathbf{r}_1, t_1; \mathbf{r}_2, t_2) = \sum_{m,n} |\psi_{mn}(\mathbf{r}_1, t_1; \mathbf{r}_2, t_2)|^2,$$

where we have considered P_{mn} =constant, again, for thermal state.

(V) Second-order quantum temporal coherence function of thermal state

Assuming a far-field joint photon counting measurement similar to that of Fig. 8.3.1, the mth and the nth photon both have 50%:50% chances to be detected by D_1 and D_2 in the following two different yet indistinguishable alternative ways: (1) the mth photon is annihilated at D_1 and the nth photon is annihilated at D_2; (2) the mth photon is annihilated at D_2 and the nth photon is annihilated at D_1. The 1-D effective two-photon wavefunction of the mnth photon pair is calculated from the field operators and the product state

$$\psi_{mn}(z_1, t_1; z_2, t_2) = \frac{1}{\sqrt{2}}[\psi_{mn}(z_1, t_1; z_2, t_2) + \psi_{nm}(z_1, t_1; z_2, t_2)]$$

where

$$
\begin{aligned}
\psi_{mn}(z_1, t_1; z_2, t_2) &= \langle \Psi_{mn} | \hat{E}_n^{(+)}(z_2, t_2) \hat{E}_m^{(+)}(z_1, t_1) | \Psi_{mn} \rangle \\
&= \langle \Psi_m | \hat{E}_m^{(+)}(z_1, t_1) | \Psi_m \rangle \langle \Psi_n | \hat{E}_n^{(+)}(z_2, t_2) | \Psi_n \rangle \\
&= \psi_m(z_1, t_1) \psi_n(z_2, t_2) \\
\psi_{nm}(z_1, t_1; z_2, t_2) &= \langle \Psi_{mn} | \hat{E}_n^{(+)}(z_1, t_1) \hat{E}_m^{(+)}(z_2, t_2) | \Psi_{mn} \rangle \\
&= \langle \Psi_n | \hat{E}_n^{(+)}(z_1, t_1) | \Psi_n \rangle \langle \psi_m | \hat{E}_m^{(+)}(z_2, t_2) | \Psi_m \rangle \\
&= \psi_n(z_1, t_1) \psi_m(z_2, t_2).
\end{aligned}
$$

The two effective wavefunctions in the above equation represent two different yet indistinguishable two-photon amplitudes: (1) the mth and nth photon or group identical photons are measured at D_1 and D_2, respectively; (2) the mth and nth photon or group identical photons are measured at D_2 and D_1, respectively. The superposition of the above two-photon amplitudes yields a nontrivial second-order correlation:

$$
\begin{aligned}
G^{(2)}&(z_1, t_1; z_2, t_2) \\
&\propto \sum_{m,n} \frac{1}{2} \left| \psi_{mn}(z_1, t_1; z_2, t_2) + \psi_{nm}(z_1, t_1; z_2, t_2) \right|^2 \\
&= G_{11}^{(1)} G_{22}^{(1)} + G_{12}^{(1)} G_{21}^{(1)},
\end{aligned}
$$

with, again,

$$
\begin{aligned}
G_{11}^{(1)} G_{22}^{(1)} &= \sum_m \left| \psi_m(z_1, t_1) \right|^2 \sum_n \left| \psi_n(z_2, t_2) \right|^2 \\
&= \sum_m \left| \mathcal{F}_{\tau_1 - t_{0m}}\{f(\nu)\} \right|^2 \sum_n \left| \mathcal{F}_{\tau_2 - t_{0n}}\{f(\nu)\} \right|^2 \\
G_{12}^{(1)} G_{21}^{(1)} &= \sum_{m,n} \psi_m^*(z_1, t_1) \psi_m(z_2, t_2) \psi_n^*(z_2, t_2) \psi_n(z_1, t_1) \\
&= \sum_m \left[\mathcal{F}_{\tau_1 - t_{0m}}^*\{f(\nu)\} \mathcal{F}_{\tau_2 - t_{0m}}\{f(\nu)\} \right] \\
&\quad \times \sum_n \left[\mathcal{F}_{\tau_2 - t_{0n}}^*\{f(\nu)\} \mathcal{F}_{\tau_1 - t_{0n}}\{f(\nu)\} \right].
\end{aligned}
$$

where we have assumed perfect joint measurement system, including the photon counting detectors and the joint measurement circuit, and have taken the wavepackets in the form of their Fourier transforms. It is clear shown in the above excise this peculiar behavior is determined by the product nature of the two-photon state.

Based on the above analysis, either in terms of the single-photon state representation or in terms of the coherent state representation, we may conclude that the calculation of the second-order quantum coherence function $G^{(2)}(\mathbf{r}_1, t_1; \mathbf{r}_2, t_2)$ of thermal field, either temporal or spatial, is the same as that of our previously calculated second-order coherence $\Gamma^{(2)}(\mathbf{r}_1, t_1; \mathbf{r}_2, t_2)$ by simply replacing Einstein's subfield with the effective wavefunction of a photon or a group of identical photons. Consequently, repeating the previous calculations in Einstein's picture, we may conclude the same second-order temporal-spatial coherence function as well as the degree of second-order temporal-spatial coherence of thermal field. Physically, we have not been able to find any differences between the second-order temporal and spatial coherence functions measured in joint photon-counting mode and measured in continuous analog mode.

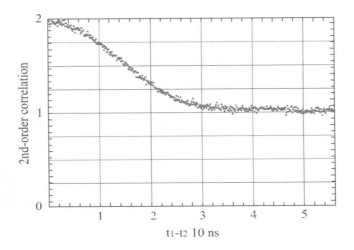

FIGURE 8.4.1 Experimental data for the measurement of second-order temporal coherence function of a pseudo-thermal field by joint photon-counting. The horizontal axis $t_1 - t_2$, in the unit of 10^{-8} second, labels the registration time differences of the photon pair measured by photon-counting detectors D_1 and D_2 at $z_1 = z_2$.

Figure 8.4.1 is a typical measurement on the degree of second-order temporal coherence of a pseudo-thermal field, $g^{(2)}(t_1 - t_2)$, in photon-counting regime. The experimental results clearly show that the photons have twice greater chance of been jointly detected at $t_1 = t_2$ by photodetectors D_1 and D_2 located at $z_1 = z_2$. These experimental effects were considered "photon bunching" in the history. The observation of "photon bunching" is quite a surprise, which apparently contradict with the random nature of thermal light. Remember, the two photons in thermal state are created randomly with equal probabilities at any time and any sub-source, t_{0m} and t_{0n}. There is no physical mechanism to create photons in "bunching" from stochastic radiation processes. In the view of quantum optics, "photon bunching" is nothing more than a *two-photon interference* phenomenon: although the photons are all created independently and randomly from the thermal light source, the two-photon amplitudes are superposed constructively when $\tau_1 - \tau_2 = 0$. It should be emphasized that *two-photon interference* is not the interference between two photons. The interference takes place between two different yet indistinguishable alternatives, namely the two-photon amplitudes that contributed to a joint photodetection event.

Although the quantum interference picture for the phenomenon is quite straightforward, one may feel uncomfortable, perhaps, due to the following reasons: (1) The superposition is "nonlocal" which involves two separated space-time coordinates through the measurement of two independent photodetectors. (2) The superposition of two-photon amplitudes has no counterpart in the Maxwell electromagnetic wave theory of light, especially, in the case of thermal light. Thermal light is considered "classical" radiation in some theories. However, we should not forget quantum theory begins with the study of thermal light: blackbody is a typical thermal source for generating thermal radiation, and the Planck theory which leads to the quantization of electromagnetic fields, is all about blackbody radiation.

(VI) Second-order quantum near-field spatial coherence function of thermal state

We have concluded that the quantum second-order coherence function $G^{(2)}(\mathbf{r}_1, t_1; \mathbf{r}_2, t_2)$ can be calculated from Einstein's picture by simply replacing the quantized subfields with the effective wavefunctions. The following is an excise to derive the second-order quantum near-field spatial coherence function $G^{(2)}(\vec{\rho}_1, z_1; \vec{\rho}_2, z_2)$ from our previously calculated second-order near-field coherence function $\Gamma^{(2)}(\vec{\rho}_1, z_1; \vec{\rho}_2, z_2)$

$$
\begin{aligned}
& G^{(2)}(\vec{\rho}_1, z_1; \vec{\rho}_2, z_2) \\
&= \sum_{m,n} \Big| \frac{1}{\sqrt{2}} \big[\psi_{mn}(\vec{\rho}_1, z_1; \vec{\rho}_2, z_2) + \psi_{nm}(\vec{\rho}_1, z_1; \vec{\rho}_2, z_2) \big] \Big|^2 \\
&= \sum_{m,n} \Big| \frac{1}{\sqrt{2}} \big[\psi_m(\vec{\rho}_1, z_1) \psi_n(\vec{\rho}_2, z_2) + \psi_n(\vec{\rho}_1, z_1) \psi_m(\vec{\rho}_2, z_2) \big] \Big|^2,
\end{aligned} \tag{8.4.19}
$$

by replacing Einstein's subfield with quantum mechanical (QM) effective function. In the above calculation, gain, we have ignored the temporal variables by assuming a perfect second-order temporal correlation similar to our early calculations for second-order spatial correlation. Substitute the Green's function or the near-field Fresnel propagator into the effective wavefunctions, either derived from single-photon state representation or from coherence state representation, we have

$$
\begin{aligned}
& G^{(2)}_{mn}(\vec{\rho}_1, z_1; \vec{\rho}_2, z_2) \\
&= \Big| |\alpha_{mn}|^2 \frac{1}{\sqrt{2}} \big[g_m(\vec{\rho}_1, z_1) g_n(\vec{\rho}_2, z_2) + g_n(\vec{\rho}_1, z_1) g_m(\vec{\rho}_2, z_2) \big] \Big|^2
\end{aligned} \tag{8.4.20}
$$

with

$$
g_m(\vec{\rho}_j, z_j) = \frac{c_o}{z_j} e^{i\frac{\omega z_j}{c}} e^{i\frac{\omega}{2c z_j}|\vec{\rho}_j - \vec{\rho}_m|^2}.
$$

where c_0 is a normalization consistent. Now, consider the sum of $G^{(2)}_{mn}(\vec{\rho}_1, z_1; \vec{\rho}_2, z_2)$ by turning the sum of m and n into the integrals of $\vec{\rho}_0$ and $\vec{\rho}_0'$ on the entire source plane:

$$
\begin{aligned}
& G^{(2)}(\vec{\rho}_1, z_1; \vec{\rho}_2, z_2) \\
&= G_0^2 \Big[\Big| \int d\vec{\rho}_0 g_{\vec{\rho}_0}(\vec{\rho}_1, z_1) \int d\vec{\rho}_0' g_{\vec{\rho}_0'}(\vec{\rho}_2, z_2) \Big|^2 \\
&\quad + \int d\vec{\rho}_0 g^*_{\vec{\rho}_0}(\vec{\rho}_1, z_1) g_{\vec{\rho}_0}(\vec{\rho}_2, z_2) \int d\vec{\rho}_0' g^*_{\vec{\rho}_0'}(\vec{\rho}_2, z_2) g_{\vec{\rho}_0'}(\vec{\rho}_1, z_1) \Big] \\
&= G_{11}^{(1)} G_{22}^{(1)} + G_{12}^{(1)} G_{21}^{(1)}
\end{aligned} \tag{8.4.21}
$$

Mathematically, Eq. (8.4.21) is the same as Eq. (8.2.21), except a constant G_0^2. We then follow the same mathematical procedure as that in the calculation of $\Gamma^{(2)}(\vec{\rho}_1, \vec{\rho}_2)$, the second-order near-field spatial quantum correlation function is calculated to be

$$G^{(2)}(\vec{\rho}_1, \vec{\rho}_2) = G_0^2 \left[1 + \text{somb}^2 \left(\frac{R}{d} \frac{\omega}{c} |\vec{\rho}_1 - \vec{\rho}_2| \right) \right], \tag{8.4.22}$$

which is the same as $\Gamma^{(2)}(\vec{\rho}_1, \vec{\rho}_2)$. Consequently, the degree of second-order spatial coherence is

$$g^{(2)}(\vec{\rho}_1, \vec{\rho}_2) = 1 + \text{somb}^2 \left(\frac{R}{d} \frac{\omega}{c} |\vec{\rho}_1 - \vec{\rho}_2| \right), \tag{8.4.23}$$

which is also the same as $\gamma^{(2)}(\vec{\rho}_1, \vec{\rho}_2)$ calculated from Einstein's picture. This result has been experimentally confirmed.

8.5   SECOND-ORDER COHERENCE OF COHERENT LIGHT

Figure 8.5.1 is a schematic experimental setup for measuring the second-order temporal and spatial coherence of coherent light. A well-collimated laser beam is divided by a 50%:50% beamsplitter. Photodetectors D_1 and D_2 are scannable longitudinally and transversely in the far-field zones of the transmitted and the reflected arms for the joint-detection of the radiation. The joint-detection circuit contains a current-current linear multiplier which has been described earlier. The output of the current-current linear multiplier, which is proportional to the temporal coherence function $\Gamma^{(2)}(z_1, t_1; z_2, t_2)$, or the spatial coherence function $\Gamma^{(2)}(x_1, t_1; x_2, t_2)$ is recorded as a function of $\tau_1 = t_1 - z_1/c$ and $\tau_2 = t_2 - z_2/c$ either by scanning the electronic delays or the longitudinal optical delays; or as a function of x_1 and x_2 by scanning D_1 and D_2 transversely, as shown in the figure.

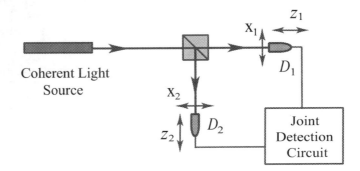

FIGURE 8.5.1 A schematic measurement of the second-order temporal coherence function $\Gamma^{(2)}(z_1, t_1; z_2, t_2)$ or the spatial coherence function $\Gamma^{(2)}(x_1, t_1; x_2, t_2)$ of a coherent radiation source.

(I) Second-order coherence of coherent light—formulated from electromagnetic field theory

We have classified light into coherent radiation and incoherent radiation in Chapter 4 under the framework of classical Maxwell electromagnetic theory of light. In Einstein's picture, coherent light refers coherently superposed electromagnetic fields in terms of sub-sources and Fourier modes. In general, the second-order coherence function $\Gamma^{(2)}(\mathbf{r}_1, t_1; \mathbf{r}_2, t_2)$ of coherent radiation is simply a factorizeable function of two expected intensities measured by D_1 and D_2 at coordinates $(\mathbf{r}_1, t_1)$ and $(\mathbf{r}_2, t_2)$, respectively,

$$\Gamma^{(2)}(\mathbf{r}_1, t_1; \mathbf{r}_2, t_2) = \langle E^*(\mathbf{r}_1, t_1) E(\mathbf{r}_1, t_1) E^*(\mathbf{r}_2, t_2) E(\mathbf{r}_2, t_2) \rangle$$
$$= \Gamma^{(1)}(\mathbf{r}_1, t_1; \mathbf{r}_1, t_1) \, \Gamma^{(1)}(\mathbf{r}_2, t_2; \mathbf{r}_2, t_2), \tag{8.5.1}$$

and consequently the normalized degree of second-order coherence of coherent radiation turns to be

$$\gamma^{(2)}(\mathbf{r}_1, t_1; \mathbf{r}_2, t_2) = 1. \tag{8.5.2}$$

To simplify the physical picture, we discuss the following two extreme cases:

(1) Second-order temporal correlation of coherent light

In Chapter 4, we have given four simplified models of radiation to clarify four different types of superposition among a large number of sub-fields and Fourier-modes. A Q-switched laser pulse or a single-mode laser beam can be treated as a radiation source that generates a large number of coherently radiated sub-sources and coherently excited Fourier-modes. Similar to our early analysis, we agin assume a Gaussian spectrum in each of the sub-radiation. The coherent superposition of the identical subfields and the Fourier-modes of Gaussian distribution together produce a Gaussian wavepackets as given in Eq. (1.2.9),

$$E(z, t) = E_0 \, e^{-\tau^2/4\sigma^2} e^{-i[\omega_0 t - k(\omega_0)z - \varphi_0]}.$$

It is easy to find that the second-order temporal coherence function $\Gamma^{(2)}(z_1, t_1; z_2, t_2)$ becomes a factorizeable function of $\Gamma^{(1)}(z_1, t_1; z_1, t_1)$ and $\Gamma^{(1)}(z_2, t_2; z_2, t_2)$

$$\begin{aligned}
\Gamma^{(2)}(z_1, t_1; z_2, t_2) &= \langle E^*(z_1, t_1)E(z_1, t_1)E^*(z_2, t_2)E(z_2, t_2) \rangle \\
&= I_0 e^{-\tau_1^2/2\sigma^2} \left| e^{-i[\omega_0 t_1 - k(\omega_0)z_1 - \varphi_0]} \right|^2 \\
&\quad \times I_0 e^{-\tau_2^2/2\sigma^2} \left| e^{-i[\omega_0 t_2 - k(\omega_0)z_2 - \varphi_0]} \right|^2 \\
&= \Gamma^{(1)}(z_1, t_1; z_1, t_1) \, \Gamma^{(1)}(z_2, t_2; z_2, t_2)
\end{aligned} \tag{8.5.3}$$

which is the product of two mean intensities or two first-order self-coherence functions, and consequently,

$$\gamma^{(2)}(z_1, t_1; z_2, t_2) = 1.$$

This type second-order correlation is named "trivial" correlation. Statistically, this trivial correlation means no correlation between the two temporally separated radiations measured at (z_1, t_1) and (z_2, t_2). Although $\gamma^{(2)}(z_1, t_1; z_2, t_2) = 1$, it does not prevent observable second-order interference in the joint measurement of D_1 and D_2, if first-order interferences are observable in their individual measurements of mean intensities, or first-order self-coherence functions, respectively.

In idealized correlation measurements, the second-order temporal coherence function can be rewritten as a product of first-order mutual-coherence functions $\Gamma^{(1)}(z_1, t_1; z_2, t_2)$ and $\Gamma^{(1)}(z_2, t_2; z_1, t_1)$

$$\begin{aligned}
\Gamma^{(2)}(z_1, t_1; z_2, t_2) &= \langle E^*(z_1, t_1)E(z_1, t_1)E^*(z_2, t_2)E(z_2, t_2) \rangle \\
&= I_0 e^{-(\tau_1^2 + \tau_2^2)/4\sigma^2} e^{i[\omega_0(t_1 - t_2) - k(\omega_0)(z_1 - z_2) - \varphi_0]} \\
&\quad \times I_0 e^{-(\tau_2^2 + \tau_1^2)/4\sigma^2} e^{-i[\omega_0(t_1 - t_2) - k(\omega_0)(z_1 - z_2) - \varphi_0]} \\
&= \Gamma^{(1)}(z_1, t_1; z_2, t_2) \, \Gamma^{(1)}(z_2, t_2; z_1, t_1)
\end{aligned} \tag{8.5.4}$$

We may find four straightforward consequences from the above discussion:

(a) The factorizeable $\Gamma^{(2)}$ is a product of two first-order self-coherence functions $\Gamma_{11}^{(1)}$ and $\Gamma_{22}^{(1)}$ or a product of two first-order mutual-coherence functions $\Gamma_{12}^{(1)}$ and $\Gamma_{21}^{(1)}$.

(b) The factorizeable $\Gamma^{(2)}$ means no statistical correlation between the two temporally separated radiations measured at (z_1, t_1) and (z_2, t_2), although the radiation may hold a well-defined temporal profile in a deterministic manner.

(c) Although $\Gamma^{(2)}$ can be a function of time and can be a function of the temporal separation between the two measurement events, $\gamma^{(2)}(z_1, t_1; z_2, t_2) = 1$ is always true. Moreover, $\gamma^{(2)}$ is defined only when $\Gamma^{(1)}(z_1, t_1; z_1, t_1) \neq 0$ and $\Gamma^{(1)}(z_2, t_2; z_2, t_2) \neq 0$.

(2) Second-order spatial correlation of coherent light

Next, we consider the measurement of the second-order spatial (transverse) coherence of a coherent radiation. The schematic experimental setup is the same as Fig. 8.3.1. In this measurement, D_1 and D_2 are scanning in the transversely, instead of longitudinally. To simplify the mathematics, we calculate the second-order coherence in far-field and in 1-D. Taking the results of chapter 6, we find

$$
\begin{aligned}
\Gamma^{(2)}(x_1, x_2) &= \langle\, E^*(x_1)E(x_1)E^*(x_2)E(x_2)\,\rangle \\
&= I_{01}\,\mathrm{sinc}^2\frac{\pi D x_1}{\lambda z_1}\, I_{02}\,\mathrm{sinc}^2\frac{\pi D x_2}{\lambda z_2} \\
&= \Gamma^{(1)}(x_1, x_1)\,\Gamma^{(1)}(x_2, x_2)
\end{aligned}
\tag{8.5.5}
$$

where D is the diameter of the laser beam (transverse size of the source) and λ is the wavelength of the radiation, and consequently,

$$
\gamma^{(2)}(x_1, x_2) = 1.
$$

The physics of Eq. (8.5.5) is very clear: the transmitted and the reflected laser beam each produces a Fraunhofer diffraction pattern on the observation planes, individually, resulting in a factorizeable second-order coherence function $\Gamma^{(2)}(x_1, x_2) = \Gamma^{(1)}(x_1, x_1)\Gamma^{(1)}(x_2, x_2)$.

It is not difficult to generalize the 1-D solution $\Gamma^{(2)}(x_1, x_2)$ to 2-D $\Gamma^{(2)}(\vec{\rho}_1, \vec{\rho}_2)$

$$
\begin{aligned}
\Gamma^{(2)}(\vec{\rho}_1, \vec{\rho}_2) &= \langle\, E^*(\vec{\rho}_1)E(\vec{\rho}_1)E^*(\vec{\rho}_2)E(\vec{\rho}_2)\,\rangle \\
&= I_{01}\left[\mathrm{somb}^2\frac{D\omega}{z_1 c}|\vec{\rho}_1|\right] I_{02}\left[\mathrm{somb}^2\frac{D\omega}{z_2 c}|\vec{\rho}_2|\right] \\
&= \Gamma^{(1)}(\vec{\rho}_1, \vec{\rho}_1)\,\Gamma^{(1)}(\vec{\rho}_2, \vec{\rho}_2).
\end{aligned}
\tag{8.5.6}
$$

and consequently,

$$
\gamma^{(2)}(\vec{\rho}_1, \vec{\rho}_2) = 1.
$$

(II) Second-order quantum coherence of coherent state.

Now, we calculated the second-order coherence function, or correlation function, of coherent state from the quantum formula. Coherent state is well-defined in quantum mechanics. We assume a far-field measurement of two photon counting photodetectors on a multi-mode coherent state,

$$
|\Psi\rangle = |\{\alpha\}\rangle = \prod_{\mathbf{k}} |\alpha(\mathbf{k})\rangle.
$$

In addition, we assume a pure state in which all modes are in phase. For a far-field measurement, the field operators at $(\mathbf{r}_1, t_1)$ and $(\mathbf{r}_2, t_2)$ are:

$$
\begin{aligned}
\hat{E}^{(-)}(\mathbf{r}_1, t_1) &= \int d\mathbf{k}\, \hat{a}^\dagger(\mathbf{k})\, e^{i(\omega t_1 - \mathbf{k}\cdot\mathbf{r}_1)}, \\
\hat{E}^{(+)}(\mathbf{r}_2, t_2) &= \int d\mathbf{k}'\, \hat{a}(\mathbf{k}')\, e^{-i(\omega' t_2 - \mathbf{k}'\cdot\mathbf{r}_2)}.
\end{aligned}
\tag{8.5.7}
$$

Since coherent state is an eigen state of the annihilation operator, the calculation is straightforward,

$$G^{(2)}(\mathbf{r}_1, t_1; \mathbf{r}_2, t_2)$$
$$= \prod_{\mathbf{k}} \langle \alpha(\mathbf{k}) | \hat{E}^{(-)}(\mathbf{r}_1, t_1) \hat{E}^{(-)}(\mathbf{r}_2, t_2) \hat{E}^{(+)}(\mathbf{r}_2, t_2) \hat{E}^{(+)}(\mathbf{r}_1, t_1) \prod_{\mathbf{k}'} |\alpha(\mathbf{k}') \rangle$$
$$= \int d\mathbf{k} \, \alpha^*(\mathbf{k}) \, e^{i(\omega t_1 - \mathbf{k} \cdot \mathbf{r}_1)} \int d\mathbf{k}'' \, \alpha^*(\mathbf{k}'') \, e^{i(\omega'' t_2 - \mathbf{k}'' \cdot \mathbf{r}_2)}$$
$$\times \int d\mathbf{k}''' \, \alpha(\mathbf{k}''') \, e^{-i(\omega''' t_2 - \mathbf{k}''' \cdot \mathbf{r}_2)} \int d\mathbf{k}' \, \alpha(\mathbf{k}') \, e^{-i(\omega' t_1 - \mathbf{k}' \cdot \mathbf{r}_1)}$$
$$= \Psi^*(\mathbf{r}_1, t_1) \Psi^*(\mathbf{r}_2, t_2) \Psi(\mathbf{r}_2, t_2) \Psi(\mathbf{r}_1, t_1)$$
$$= G^{(1)}(\mathbf{r}_1, t_1; \mathbf{r}_1, t_1) \, G^{(1)}(\mathbf{r}_2, t_2; \mathbf{r}_2, t_2) \tag{8.5.8}$$

where $\Psi(\mathbf{r}_j, t_j)$, $j = 1, 2$, is the effective wavefunction defined in Chapter 6. In idealized correlation measurements, we also have

$$G^{(2)}(\mathbf{r}_1, t_1; \mathbf{r}_2, t_2) = \Psi^*(\mathbf{r}_1, t_1) \Psi^*(\mathbf{r}_2, t_2) \Psi(\mathbf{r}_2, t_2) \Psi(\mathbf{r}_1, t_1)$$
$$= G^{(1)}(\mathbf{r}_1, t_1; \mathbf{r}_2, t_2) \, G^{(1)}(\mathbf{r}_2, t_2; \mathbf{r}_1, t_1). \tag{8.5.9}$$

It is easy to find from Eq. (8.5.8) the degree of second-order coherence of coherent radiation is a trivial constant one:

$$g^{(2)}(\mathbf{r}_1, t_1; \mathbf{r}_2, t_2) = 1. \tag{8.5.10}$$

To simplify the physical picture, again, we discuss the following two extreme cases:

(1) Second-order temporal coherence measurement, assuming $\Delta \vec{\kappa} = 0$ and $\Delta \omega \neq 0$.

The calculation gives the same result as that of Eq.(8.5.3), if we assume a Gaussian spectrum distribution centered at ω_0,

$$G^{(2)}(z_1, t_1; z_2, t_2) = G_0 e^{-\tau_1^2 / 2\sigma^2} G_0 e^{-\tau_2^2 / 2\sigma^2}$$
$$= G^{(1)}(z_1, t_1; z_1, t_1) \, G^{(1)}(z_2, t_2; z_2, t_2) \tag{8.5.11}$$

and consequently,

$$g^{(2)}(z_1, t_1; z_2, t_2) = 1.$$

(2) Second-order spatial coherence measurement, assuming $\Delta \omega = 0$ and $\Delta \vec{\kappa} \neq 0$.

The calculation gives the same result as that of Eq.(8.5.5), if we assume a similar rectangular spatial spectrum distribution centered at $\vec{\kappa} = 0$

$$G^{(2)}(x_1, x_2) = G_{01} \text{sinc}^2 \frac{\pi D x_1}{\lambda z_1} G_{02} \text{sinc}^2 \frac{\pi D x_2}{\lambda z_2}$$
$$= G^{(1)}(x_1, x_1) \, G^{(1)}(x_2, x_2) \tag{8.5.12}$$

where D is the diameter of the laser beam (transverse size of the source) and λ is the wavelength of the radiation. Again, we have simplified the calculation in 1-D. We thus have

$$g^{(2)}(x_1, x_2) = 1.$$

The above results are consistent with that of the electromagnetic field theory. In the electromagnetic field theory of light and the quantized quantum theory of light, we both conclude the second-order coherence function of coherent light is factorizeable into a product of the first-order self-coherence functions, and the degree of second-order coherence of coherent light is a constant of one. This kind of correlation is usually called trivial correlation.

In the quantum theory of light, the above result is easy to understand. Coherent state is a pure state corresponding to a vector in Hilbert space, and represents the state of a group of identical photons. The coherent property of a group of identical photons is restricted within the group, namely, the only interference effect we can observe is the interference of the group with the group itself, unless the group is able to be distinguished into two or more sub-groups of identical photons. A pair of distinguishable groups of identical photons, either random pair or entangled pair, is able to interfere with the pair itself, and thus produce the nontrivial second-order coherence function.

A factorizeable $\Gamma^{(2)}(\mathbf{r}_1, t_1; \mathbf{r}_2, t_2)$ and $G^{(2)}(\mathbf{r}_1, t_1; \mathbf{r}_2, t_2)$ as well as a constant $\gamma^{(2)}(\mathbf{r}_1, t_1; \mathbf{r}_2, t_2) = 1$ and $g^{(2)}(\mathbf{r}_1, t_1; \mathbf{r}_2, t_2) = 1$ means the jointly measured two temporally and/or spatially separated radiations do not have statistical correlation. However, a trivial second-order correlation does not prevent observable interference in the joint measurement of D_1 and D_2, if first-order interferences are observable in the mean intensity, or mean photon number, measurement of D_1 and D_1. The observed interference in the joint measurement of D_1 and D_2 is simply a product of the first-order interferences observed respectively, in the mean intensity, or mean photon number, measurements of D_1 and D_2.

8.6   SECOND-ORDER COHERENCE OF ENTANGLED STATE AND NUMBER STATE

Quantum entanglement will be discussed in Chapter 10. In this section, we simply introduce an entangled state and a number state of $n = 2$ to replace the incoherent or coherent radiation sources in the experimental setup of Fig. 8.3.1, and calculate their second-order quantum coherence function $G^{(2)}(\mathbf{r}_1, t_1; \mathbf{r}_2, t_2)$.

(I) Entangled biphoton state of SPDC.

In the following, an entangled two-photon state generated from a nonlinear optical process of SPDC will be applied for the calculation of the second-order coherence function $G^{(2)}(\mathbf{r}_1, t_1; \mathbf{r}_2, t_2)$. Roughly speaking, an entangled two-photon state is a non-factorizeable state of two-photons: the state cannot be written as a product of that of two photons. In other words, the two photons cannot be considered as independent. The two-photon state of SPDC is a very special entangled two-particle state. The two subsystems are "perfectly correlated", i.e., by measuring the state of one, the state of the other one is determined with certainty, although none of the subsystems is in a defined state before the measurement. The entangled signal-idler photon pair of SPDC was named biphoton by Klyshko. To simplify the mathematics, the following calculation is approximated in 1-D by considering the following non-factorizeable biphoton state with total photon number $n = 2$

$$|\Psi\rangle = \int d\omega_s d\omega_i \, \delta(\omega_s + \omega_i - \omega_p) \, f(\omega_s, \omega_i) \, \hat{a}^\dagger(\omega_s) \hat{a}^\dagger(\omega_i) |0\rangle$$

$$\simeq \int_{-\infty}^{\infty} d\nu \, f(\nu) \, a^\dagger(\omega_s^0 + \nu) \, a^\dagger(\omega_i^0 - \nu)|0\rangle \tag{8.6.1}$$

with

$$\omega_s = \omega_s^0 + \nu, \quad \omega_i = \omega_i^0 - \nu, \quad \omega_s + \omega_i = \omega_p = \text{constant}$$

where $\delta(\omega_p - \omega_s - \omega_i)$ has been applied, ω_s^0 and ω_i^0 are the center frequencies for the signal mode ω_s and the idler mode ω_i, respectively. We have generalized the integral to infinity with the help of the spectral density function $f(\nu)$. The biphoton state of Equation 8.6.1 is a pure state by means: (1) the state of a signal-idler pair, which produced a joint photodetection event, is a state vector in the form of a coherent superposition of all possible Fock states $|..., 0, 1_{\omega_s}, ..., 1_{\omega_i}, 0, ...\rangle$; (2) a large number of jointly measured signal-idler photon pairs are in the same state, which can be treated as a homogenous ensemble.

Assuming a simple far-field 1-D measurement in which a point-like photon counting detector D_1 registers a photodetection event at (z_1, t_1) and another point-like photon counting detector D_2 observes another photodetection event at (z_2, t_2), the second-order temporal coherence function is thus calculated as

$$
\begin{aligned}
G^{(2)}&(z_1, t_1; z_2, t_2) \\
&= \langle \Psi | \hat{E}^{(-)}(z_1, t_1) \hat{E}^{(-)}(z_2, t_2) \hat{E}^{(+)}(z_2, t_2) \hat{E}^{(+)}(z_1, t_1) | \Psi \rangle \\
&= \sum_n \langle \Psi | \hat{E}^{(-)}(z_1, t_1) \hat{E}^{(-)}(z_2, t_2) | n \rangle \langle n | \hat{E}^{(+)}(z_2, t_2) \hat{E}^{(+)}(z_1, t_1) | \Psi \rangle \\
&= | \langle 0 | \hat{E}^{(+)}(z_2, t_2) \, \hat{E}^{(+)}(z_1, t_1) | \Psi \rangle |^2 \\
&= | \Psi(z_1, t_1; z_2, t_2) |^2
\end{aligned}
\tag{8.6.2}
$$

where, in the third line, we have used the completeness relation

$$
\sum_n |n\rangle\langle n| = 1.
$$

The effective wavefunction of a pair of entangled signal-idler photons, characterized by the state of Eq. (8.6.1), is thus defined in Eq. (8.6.2)

$$
\Psi(z_1, t_1; z_2, t_2) = \langle 0 | \hat{E}^{(+)}(z_2, t_2) \, \hat{E}^{(+)}(z_1, t_1) | \Psi \rangle.
\tag{8.6.3}
$$

The two-photon effective wavefunction is then calculated as

$$
\begin{aligned}
\Psi&(z_1, t_1; z_2, t_2) \\
&= \langle 0 | \left[\int d\omega \, \hat{a}(\omega) \, e^{-i(\omega t_2 - k z_2)} \right] \left[\int d\omega' \, \hat{a}(\omega') \, e^{-i(\omega' t_1 - k' z_1)} \right] \\
&\quad \times \left[\int_{-\infty}^{\infty} d\nu \, f(\nu) \, a^{\dagger}(\omega_s^0 + \nu) \, a^{\dagger}(\omega_i^0 - \nu) | 0 \rangle \right] \\
&\cong \int d\omega \int d\omega' \int_{-\infty}^{\infty} d\nu \, f(\nu) \, e^{-i\omega' \tau_2} \, e^{-i\omega \tau_1} \langle 0 | \hat{a}(\omega) \, \hat{a}(\omega') \, a^{\dagger}(\omega_s^0 + \nu) \, a^{\dagger}(\omega_i^0 - \nu) | 0 \rangle.
\end{aligned}
\tag{8.6.4}
$$

(1) Non-Degenerate, $\omega_s^0 \neq \omega_i^0$, and the photodetectors D_1 that is located at z_1 only receives the ω_s modes and D_2 that is located at z_2 only receives the ω_i modes. The non-degenerate modes can be distinguished experimentally either by spectral filters or by polarization analyzers if the two modes are orthogonal polarized. In this case, the only surviving term in Equation (8.6.4) is,

$$
\begin{aligned}
\Psi(z_1, t_1; z_2, t_2) &\cong \int_{-\infty}^{\infty} d\nu \, f(\nu) \, e^{-i(\omega_i^0 - \nu)\tau_2} \, e^{-i(\omega_s^0 + \nu)\tau_1} \\
&= e^{-i(\omega_s^0 \tau_1 + \omega_i^0 \tau_2)} \, \mathcal{F}_{\tau_2 - \tau_1} \{ f(\nu) \}.
\end{aligned}
\tag{8.6.5}
$$

It is clear that the two-photon effective wavefunction of the biphoton system cannot be written as a product of two effective wavefunctions of signal photon and idler photon. Different from the randomly created and randomly paired photons in thermal state, the entangled signal-idler photon pair is characterized by a nonlocal wavepacket instead of two wavepackets. We will discuss this in detail in Chapter 9. The second-order coherence function is thus

$$G^{(2)}(z_1, t_1; z_2, t_2) = \left| \mathcal{F}_{\tau_2 - \tau_1} \{ f(\nu) \} \right|^2. \tag{8.6.6}$$

(2) Degenerate, $\omega_s^0 = \omega_i^0$, the photodetectors D_1 that is located at z_1 and D_2 that is located at z_2 cannot distinguish the ω_s and the ω_i modes. In this case, the surviving terms are these when $\omega = \omega_s^0 + \nu$ with $\omega' = \omega_i^0 - \nu$, and $\omega = \omega_i^0 - \nu$ with $\omega' = \omega_s^0 + \nu$, thus

$$
\begin{aligned}
\Psi(&z_1, t_1; z_2, t_2) \\
&\cong \int_{-\infty}^{\infty} d\nu \, f(\nu) \left[e^{-i(\omega_i^0 - \nu)\tau_2} e^{-i(\omega_s^0 + \nu)\tau_1} + e^{-i(\omega_s^0 + \nu)\tau_2} e^{-i(\omega_i^0 - \nu)\tau_1} \right] \\
&= e^{-i\omega_p(\tau_2 + \tau_1)} \left[\int_{-\infty}^{\infty} d\nu \, f(\nu) \, e^{-i\nu(\tau_1 - \tau_2)} + \int_{-\infty}^{\infty} d\nu \, f(\nu) \, e^{-i\nu(\tau_2 - \tau_1)} \right] \\
&= \Psi(\tau_1 + \tau_2) \left[\Psi(\tau_1 - \tau_2) + \Psi(\tau_2 - \tau_1) \right] \\
&= \Psi_{si}(z_1, t_1; z_2, t_2) + \Psi_{is}(z_1, t_1; z_2, t_2)
\end{aligned}
\tag{8.6.7}
$$

where $\Psi(\tau_1 - \tau_2) = \Psi^*(\tau_2 - \tau_1)$. In the degenerate case, the two-photon effective wavefunction is the result of a superposition of two wavepackets $\Psi_{si}(z_1, t_1; z_2, t_2)$ and $\Psi_{is}(z_1, t_1; z_2, t_2)$ corresponding to two different yet indistinguishable alternative ways for the entangled two photons to produce a joint-detection event: (1) the signal produces a photodetection event at $D - 1$ and the idler produces a photodetection event at $D - 2$; (2) the idler produces a photodetection event at $D - 1$ and the signal produces a photodetection event at $D - 2$ Again, $\Psi(z_1, t_1; z_2, t_2)$ cannot be written as the product of two effective wavefunctions. The entangled signal-idler photons are characterized by a non-factorizeable 2-D function, which is a nonlocal wavepacket in space-time.

In either cases the second-order coherence or correlation function is written as the normal square of the non-factorizable effective two-photon wavefunction

$$G^{(2)}(z_1, t_1; z_2, t_2) = \left| \Psi(z_1, t_1; z_2, t_2) \right|^2. \tag{8.6.8}$$

The second-order coherence function of entangles biphoton state is very different from that of thermal radiation. For entangled states, it is impossible to write $G^{(2)}(z_1, t_1; z_2, t_2)$ in terms of the product of $G^{(1)}$s.

It is interesting to see that $G^{(2)}(z_1, t_1; z_2, t_2)$ will be a delta-like function of $\tau_2 - \tau_1$ if $\Delta\nu \sim \infty$. This means that if D_1 registers a photon at t_1 then there is a unique precise time t_2 for D_2 to register its twin, even though none of the photons is localized in time and the two photodetectors may be separated at very large distance. This kind of correlation has been historically named Einstein-Podolsky-Rosen (EPR) correlation.

It is also interesting to see that an anti-correlation is observable if the biphoton superposition achieves its destructive interference condition

$$
\begin{aligned}
\Psi(&z_1, t_1; z_2, t_2) \\
&\cong \int_{-\infty}^{\infty} d\nu \, f(\nu) \left[e^{-i(\omega_i^0 - \nu)\tau_2} e^{-i(\omega_s^0 + \nu)\tau_1} - e^{-i(\omega_s^0 + \nu)\tau_2} e^{-i(\omega_i^0 - \nu)\tau_1} \right] \\
&= \Psi_{si}(z_1, t_1; z_2, t_2) - \Psi_{is}(z_1, t_1; z_2, t_2).
\end{aligned}
\tag{8.6.9}
$$

The degree of second-order coherence achieves its minimum value when $\Psi_{si}(z_1, t_1; z_2, t_2)$ and $\Psi_{is}(z_1, t_1; z_2, t_2)$ overlap completely in space-time. The joint-detection counting rate will be

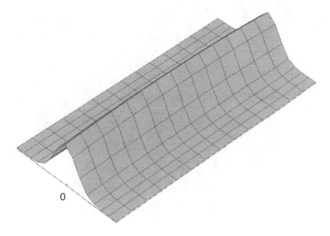

FIGURE 8.6.1 A schematic envelope of a biphoton wavepacket with a Gaussian shape along $\tau_1 - \tau_2$. The wavepacket is uniformly distributed along $\tau_1 + \tau_2$ due to the assumption of $\omega_p = $ constant.

zero when $\tau_1 = \tau_2$. This means that if D_1 registers a photon at t_1 then D_2 is forbidden to register its twin at $t_1 = t_2$, if $z_1 = z_2$ is chosen.

A schematic picture of a biphoton wavepacket described by Eq. (8.6.5), is illustrated in Figure 8.6.1. It shows clearly that $\Psi(z_1, t_1; z_2, t_2)$ of the entangled two-photon system is represented by a non-factorizeable 2-D wavepacket, instead of a product of two wavepackets. The non-factorizeable wavepacket, again, from a different point of view, illustrates the entangled nature of the two-photon state. One should not be surprised by the fact that the system does not have two individual wavepackets associated with the signal photon and the idler photon. The biphoton state is formed from a coherent superposition of Fock states of $n = 2$, instead of a product of two individual single-photon wavepackets. The coherent superposition of these Fock states results in the non-factorizeable effective wavefunction or the biphoton wavepacket. This is very different from the classical case. It may be easier to see the connection between the picture of a single-photon wavepacket and the concept of a photon. It is definitely not easy to accept the fact that the connection between the biphoton wavepacket and the concept of biphoton is the same as that of the single-photon. The biphoton wavepacket describes the dynamic property of the entangled photon pair as one system in space-time. We will return to the discussion of these important physics again later.

Phenomenologically, unlike the single-photon wavepacket, (1) the biphoton wavepacket is dynamically propagating along the axis z_1 and z_2 in different directions; however, (2) the propagation is never independent. The joint photoelectron events happen in such a way that whenever a photoelectron event occurs at space-time $(\mathbf{r}_1, t_1)$, the other one can happen only in the neighborhood of space-time point $(\mathbf{r}_2, t_2)$, which is related to $(\mathbf{r}_1, t_1)$ by $(t_1 - t_2) - (z_1 - z_2)/c = 0$. The statement of point (1) is formulated by $e^{-i(\omega_0 t_2 - k_0 z_2)} e^{-i(\omega_0 t_1 - k_0 z_1)}$ or $e^{-i\omega_0(\tau_1 + \tau_2)}$ along the $\tau_1 + \tau_2$ axis. The statement of point (2) is formulated by $\mathcal{F}_{\tau_1 - \tau_2}\{f(\Omega)\}$ along the $\tau_1 - \tau_2$ axis. This phenomenon is interesting. On one hand, the signal photon and the idler photon are both in mixed single-photon state with randomly distributed probability of producing a photoelectron event along both z_1 and z_2 axes in space and along t_1 and t_2 in time. It means that neither the signal photon nor the idler photon is localized in space-time. On the other hand, the space-time correlation between the joint photoelectron events, $(t_1 - t_2) - (z_1 - z_2)/c = 0$, determines the location and time of finding the idler (signal) photon after the annihilation of the signal (idler) photon. How could the detection of the signal (idler) photon localize the idler (signal) photon? The problem gets more interesting when the two photodetection events become space-like separated events: the two detectors could be separated in light-years but the determination, however, is made instantaneously. It seems that we may have to give up the concept of two photons when thinking about an EPR two-photon state. The signal-idler system might be better treated as a whole, as described by the non-factorizeable biphoton wavepacket. If one insists that the signal photon and the idler photon are two individual systems, the EPR paradox may be unavoidable. This point will be emphasized again in Chapter 9.

(II): Number state of $n_\omega = 2$

In Chapter 2 we have given an example of number state of $n_\omega = 2$

$$|\Psi\rangle = \sum_\omega f(\omega)[\hat{a}^\dagger(\omega)]^2|0\rangle \simeq \int_{-\infty}^\infty d\nu f(\nu)[a^\dagger(\omega_0 + \nu)]^2|0\rangle \qquad (8.6.10)$$

The two-photon effective wavefunction is calculated as

$$
\begin{aligned}
\Psi(z_1, &t_1; z_2, t_2) \\
&= \langle 0|\left[\int d\omega\, a(\omega)\, e^{-i(\omega t_2 - k z_2)}\right]\left[\int d\omega'\, a(\omega')\, e^{-i(\omega' t_1 - k' z_1)}\right] \\
&\quad \times \left[\int_{-\infty}^\infty d\nu\, f(\nu)\, a^\dagger(\omega_0 + \nu)\, a^\dagger(\omega_0 + \nu)|0\rangle\right] \\
&\cong \int d\omega \int d\omega' \int_{-\infty}^\infty d\nu\; f(\nu)\, e^{-i\omega' \tau_2}\, e^{-i\omega \tau_1} \\
&\quad \times \langle 0|\, a(\omega)\, a(\omega')\, a^\dagger(\omega_0 + \nu)\, a^\dagger(\omega_0 + \nu)|0\rangle.
\end{aligned}
\qquad (8.6.11)
$$

The photodetectors D_1 that is located at z_1 and D_2 that is located at z_2 cannot distinguish the two modes of $\omega = \omega_0 + \nu$. In this case, the surviving terms are these when $\omega = \omega' = \omega_0 + \nu$, thus

$$
\begin{aligned}
\Psi(z_1, t_1; z_2, t_2) &\cong \int_{-\infty}^\infty d\nu\, f(\nu)\left[e^{-i(\omega_0 + \nu)(\tau_2 + \tau_1)}\right] \\
&= e^{-i\omega_0(\tau_2 + \tau_1)}\, \mathcal{F}_{\tau_2 + \tau_1}\{f(\nu)\}.
\end{aligned}
\qquad (8.6.12)
$$

Again, $\Psi(z_1, t_1; z_2, t_2)$ cannot be written as the product of $\Psi(z_1, t_1)$ and $\Psi(z_2, t_2)$. The number state of $n_\omega = 2$ is also characterized by a non-factorizable 2-dimensional function. However, this 2-D function is very different from either Equation (8.6.5) or Equation (8.6.7). The second-order coherence is thus

$$G^{(2)}(z_1, t_1; z_2, t_2) = \left|\mathcal{F}_{\tau_2 + \tau_1}\{f(\nu)\}\right|^2. \qquad (8.6.13)$$

It is interesting to see that $G^{(2)}(z_1, t_1; z_2, t_2)$ will be a delta-like function of $\tau_2 + \tau_1$ if $\Delta\nu \sim \infty$. In this case, $(t_1 + t_2) \simeq (z_1 + z_2)/c$. The physics behind is very different from that of the entangled states too.

8.7 MEASUREMENT OF SECOND-ORDER COHERENCE

Similar to the first-order coherence measurement, the second-order coherence is also based on the measurement of a large number of joint photodetection events. Unlike the first-order coherence function $G^{(1)}(\mathbf{r}_1, t_1; \mathbf{r}_2, t_2)$, the second-order coherence function $G^{(2)}(\mathbf{r}_1, t_1; \mathbf{r}_2, t_2)$ is a directly measurable function of space-time variables. Therefore, $G^{(2)}(\mathbf{r}_1, t_1; \mathbf{r}_2, t_2)$ can be measured by means of time averaged joint photodetection counting rate as a function of chosen space variables or by examining the statistical histogram of a large number of joint photodetection events as a function of $t_1 - t_2$. The photon registration events of the photodetectors D_1 and D_2 can be recorded in the accuracy of 10^{-12} second in modern technologies. $G^{(2)}$ thus can be measured as a function of $t_1 - t_2$ in picosecond resolution by recording the registration time for each individual photon pairs. In the following, we show two examples of $G^{(2)}$ measurements in photon counting regime.

(I) Measurement of $G^{(2)}(t_1 - t_2)$ and $g^{(2)}(t_1 - t_2)$.

In these type of measurements, the spatial coordinates of the photodetectors, $\mathbf{r}_1$ and $\mathbf{r}_2$, are chosen to be constants during the measurement. The measurement electronics is able to read and record the registration times t_1 and t_2 or the registration time difference $t_1 - t_2$ for each of the measured photon pairs. A statistical histogram, showing the number of joint photodetection events against $t_1 - t_2$, is made after counting a large number of joint photodetection events. This distribution corresponds to the second-order correlation function $G^{(2)}(t_1 - t_2)$. In previous section, we have shown a typical measured degree of second-order coherence function $g^{(2)}(t_1 - t_2)$, i.e., the normalized second-order coherence function, for thermal (pseudo-thermal) field with narrow spectrum in Fig. 8.4.1.

Figure 8.7.1 is an measured $G^{(2)}(t_1 - t_2)$ function for entangled photon pairs of SPDC. The horizontal axis $t_1 - t_2$ labels the registration time differences of the photon pairs, corresponding to $\tau_1 - \tau_2$ with constant spatial coordinates $\mathbf{r}_1$ and $\mathbf{r}_2$. The vertical axis indicates the number of photon pairs that are observed with the value of $t_1 - t_2$. We may

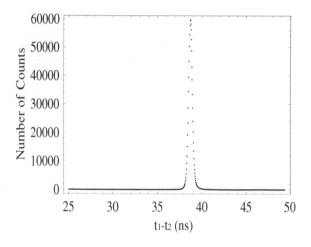

FIGURE 8.7.1 Second-order correlation function for entangled photon pairs. The horizontal axis $t_1 - t_2$, which labels the measured registration time difference of the photon pair, corresponds to $\tau_1 - \tau_2$ with fixed values of z_1 and z_2. The calculated temporal with of $G^{(2)}(t_1 - t_2)$ is a few femtoseconds; the measured width, however, is about 700 picoseconds, due to the relatively slow photodetectors.

draw two conclusions from Fig. 8.7.1 directly: (1) the entangled photon pair is likely to be generated simultaneously; and (2) the measured $t_1 - t_2$ are independent of the creation time of the pair.

The above two examples are typical measurements of $G^{(2)}(t_1 - t_2)$ and $g^{(2)}(t_1 - t_2)$ in photon counting. The measurement electronics records the joint photodetection time difference $t_1 - t_2$ for each photon pairs. The observed function is based on the statistics of a large number of photon pairs. The distribution function or the histogram that illustrates number of counts against $t_1 - t_2$, statistically corresponds to the second-order correlation function $G^{(2)}(t_1 - t_2)$ that is calculated based on the knowledge of the state of the radiation.

Due to the limited ability of the photodetectors in determining the registration time of a photoelectron, obviously, the responds time of the photodetectors and the associated electronics will affect the measurement of $G^{(2)}(t_1 - t_2)$. For instance, the calculated width of $G^{(2)}(t_1 - t_2)$ for entangled signal-idler pairs of SPDC is typically in the order of a few femtosecond to a few hundred femtosecond. The observed width in Fig. 8.7.1, however, is about 700 ps. The broadening is due to the slow response of the photodetectors. In the observation of a photoelectron event, accompanied with the annihilation of a photon at time t the photodetector exports an pulse of electric current to an electronic circuit which is able to analyze the pulse and to determine the electronic registration time $\tilde{t}$ within a certain uncertainty. The jitter of the leading edge of the pulse as well as the fluctuations of the pulse height, all contribute to the uncertainty of the electronic registration times, or the measured times, of $\tilde{t}_1$ and $\tilde{t}_2$. In Chapter 1, we have characterized this uncertainty as a response function of the photodetector, $D(\tilde{t} - t)$, where t is the photon annihilation time and $\tilde{t}$ is the electronic registration time. In section 8.5, we have treated the joint photodetection measurement of D_1 and D_2 as a convolution between the response functions and the second-order coherence function $G^{(2)}(t_1 - t_2)$:

$$G^{(2)}(\tilde{t}_1 - \tilde{t}_2) = \frac{1}{t_c^2} \int dt_1 \int dt_2 \, G^{(2)}(t_1 - t_2) \, D(\tilde{t}_1 - t_1) \, D(\tilde{t}_2 - t_2) \tag{8.7.1}$$

where we have chosen:

$$\int dt \, D(\tilde{t} - t) = t_c, \tag{8.7.2}$$

where t_c is defined as the response time of the photodetector. When the width of the response functions are much narrower then that of $G^{(2)}(t_1 - t_2)$, the response function can be treated as delta functions, $D(\tilde{t} - t) \sim t_c \delta(\tilde{t} - t)$. In this extreme case, the measured second-order correlation function will reveal the theoretical expectation of $G^{(2)}(t_1 - t_2)$:

$$G^{(2)}(\tilde{t}_1 - \tilde{t}_2) = \int dt_1 \int dt_2 \, G^{(2)}(t_1 - t_2) \, \delta(\tilde{t}_1 - t_1) \, \delta(\tilde{t}_2 - t_2)$$
$$= G^{(2)}(t_1 - t_2). \tag{8.7.3}$$

However, when the response time is in the same order or even greater then the width of the correlation function, the situation is completely different. When the width of the response functions are much wider then that of $G^{(2)}(t_1 - t_2)$, the second-order correlation function itself can then be treated as a delta function. In this extreme case, Eq. (8.7.1) turns into the following convolution between the response functions of the two photon counting detectors:

$$G^{(2)}(\tilde{t}_1 - \tilde{t}_2) = G^{(2)}(0) \left(\frac{\tau_0}{t_c^2} \right) \int d\tau \, D(\tilde{t}_1 - \tilde{t}_2 - \tau) \, D(\tau). \tag{8.7.4}$$

Where we have chosen:

$$\int dt \, G^{(2)}(t) = G^{(2)}(0)\tau_0 \simeq \int dt \, G^{(2)}(0)\tau_0 \, \delta(t), \tag{8.7.5}$$

where $\tau_0 \sim \tau_c$ is a constant and is named as the second-order correlation time. Eq. (8.7.4) indicates: (1) The width of the observed $G^{(2)}(\tilde{t}_1 - \tilde{t}_2)$ is now determined by the responds function of the photodetectors, which could be significantly broadened. This is what has happened in the measurement of Fig. 8.7.1. (2) The relative slow response time of the photodetectors may reduce the magnitude of the measured second-order correlation. For Gaussian type responds functions the convolution yields a Gaussian function of $G^{(2)}(\tilde{t}_1 - \tilde{t}_2)$ with a reduced central value of $G^{(2)}(0)$. The reduction factor is roughly τ_0/t_c.

The above reduction may not be a problem in the measurement of entangled states, however, it may affect the measurement of $G^{(2)}$ function for thermal or pseudo-thermal light significantly. The $G^{(2)}$ function of thermal or pseudo-thermal field has two terms corresponding to (1) the product of two measured mean photon numbers and (2) the photon number fluctuation correlation,

$$G^{(2)}(\tilde{t}_1 - \tilde{t}_2) \propto \frac{1}{t_c^2} \int dt_1 \int dt_2 \, \langle n_1 n_2 \rangle D(\tilde{t}_1 - t_1) \, D(\tilde{t}_2 - t_2)$$

$$= \frac{1}{t_c^2} \int dt_1 \int dt_2 \left[\langle n_1 \rangle \langle n_2 \rangle + \langle \Delta n_1 \Delta n_2 \rangle \right] D(\tilde{t}_1 - t_1) \, D(\tilde{t}_2 - t_2). \tag{8.7.6}$$

The time average over t_c may reduce the magnitude of the second term many orders smaller than that of the first term. For example, the time integral does not change the value of the first term of Eq. (8.7.6) for a CW thermal or pseudo-thermal field with broad spectrum

$$\frac{1}{t_c^2} \int dt_1 \int dt_2 \left[\langle n_1 \rangle \langle n_2 \rangle D(\tilde{t}_1 - t_1) \, D(\tilde{t}_2 - t_2) \right.$$

$$= \frac{1}{t_c^2} \left\{ \left[\bar{n}_1 \bar{n}_2 \int dt_1 \, D(\tilde{t}_1 - t_1) \int dt_2 \, D(\tilde{t}_2 - t_2) \right] \right.$$

$$= \bar{n}_1 \bar{n}_2. \tag{8.7.7}$$

However, it reduces the value of the second term significantly when $\tau_0 \ll t_c$

$$\frac{1}{t_c^2} \int dt_1 \int dt_2 \, \langle \Delta n_1 \Delta n_2 \rangle D(\tilde{t}_1 - t_1) D(\tilde{t}_2 - t_2)$$

$$\simeq \frac{\bar{n}_1 \bar{n}_2}{t_c^2} \int dt_1 \int dt_2 \, \tau_0 \delta(t_1 - t_2) D(\tilde{t}_1 - t_1) D(\tilde{t}_2 - t_2)$$

$$\simeq \frac{\bar{n}_1 \bar{n}_2}{t_c^2} \tau_0 \int d\tau \, D(\tilde{t}_1 - \tilde{t}_2 - \tau) \, D(\tau) \tag{8.7.8}$$

For Gaussian type responds functions, the magnitude of the second term is thus roughly τ_0/t_c times smaller than that of the first term. The coherence time of sunlight is in the order of 10^{-15} second. A nanosecond photodetector yields $\tau_0/t_c \sim 10^{-6}$. It is definitely uneasy to distinguish the nontrivial photon number fluctuation correlation, which is one part of 10^6 relative to the trivial constant $\bar{n}_1 \bar{n}_2$, from the measurement of $G^{(2)}(\tilde{t}_1 - \tilde{t}_2)$. This has been the major difficulty for HBT measurement of sunlight and other type natural radiations.

How to identify the nontrivial second-order coherence of thermal field with short coherence time τ_0? The obvious approach is to use fast photodetectors with $t_c \sim \tau_0$. The state-of-the-art technology, however, has not been able to achieve that fast yet. One realistic approach, perhaps, is to use short pulsed thermal or pseudo-thermal light. If the temporal width of $\langle n_j \rangle$, $j = 1, 2$, is in the same order of τ_0, the time integral in Eq. 8.7.6 shall have similar effects on the trivial correlation part of $G^{(2)}$

$$\frac{1}{t_c^2} \int dt_1 \int dt_2 \langle n(t_1) \rangle \langle n(t_2) \rangle D(\tilde{t}_1 - t_1) D(\tilde{t}_2 - t_2)$$

$$\simeq \frac{\bar{n}_1 \bar{n}_2}{t_c^2} \int dt_1 \int dt_2 \, \tau_0 \delta(t_1 - t_2) D(\tilde{t}_1 - t_1) D(\tilde{t}_2 - t_2)$$

$$\simeq \frac{\bar{n}_1 \bar{n}_2}{t_c^2} \tau_0 \int d\tau \, D(\tilde{t}_1 - \tilde{t}_2 - \tau) \, D(\tau) \tag{8.7.9}$$

which is the same as Eq. (8.7.8). In Eq. (8.7.9), we have approximated the short pulsed $\langle n(t_1) \rangle \langle n(t_2) \rangle$ a δ-function, $\langle n(t_1) \rangle \langle n(t_2) \rangle \sim \bar{n}_1 \bar{n}_2 \tau_0 \delta(t_1 - t_2)$. This mechanism may be suitable for short pulsed HBT type applications, such as short puled X-ray ghost microscope, although the spectrum of an X-ray source is much wider than that of sunlight. Regarding sunlight HBT measurements, of course, it is impossible to have a short pulsed sun, however, we can definitely "gate" our photodetectors to "expose" D_1 and D_2 within a short time window. In addition to a specially designed measurement circuit to separate $\langle \Delta n_1 \Delta n_2 \rangle$ from $\langle n_1 \rangle \langle n_2 \rangle$, an observable nontrivial correlation of $\langle \Delta n_1 \Delta n_2 \rangle$ is achievable for sunlight HBT measurements.

(II) About "coincidence" measurement.

A coincidence measurement does not measure $G^{(2)}(t_1 - t_2)$, but rather cumulatively counts the joint detection events that fall into a certain coincidence time window Δt_c around a chosen value of $t_1 - t_2 = \tau$, where τ is a time constant which is determined by a particular experimental arrangement. Mathematically, this is equivalent to have a time integral on the second-order coherence function $G^{(2)}(t_1 - t_2)$:

$$R_c = \int_{\Delta T} dt_1\, dt_2\, G^{(2)}(t_1 - t_2)\, S(t_1 - t_2), \tag{8.7.10}$$

where R_c is the coincidence counting rate and $S(t_1 - t_2)$ represents the time window of the coincidence circuit

$$S(t_1 - t_2) = \begin{cases} 1 & \tau - \Delta t_c/2 \leq t_1 - t_2 \leq \tau + \Delta t_c/2 \\ 0 & \text{otherwise.} \end{cases}$$

(III) Measurement of $G^{(2)}(z_1 - z_2)$ or $G^{(2)}(\vec{\rho}_1 - \vec{\rho}_2)$.

The longitudinal and transverse spatial correlation measurements have been discussed earlier. A coincidence counter or a linear multiplier and associated electronics can be used to measure the correlation by scanning the point-like photodetector longitudinally or transversely. In the photon counting measurements, the coincidence counter counts and records all the the joint-detection events fall into the coincidence time window of $\tau - \Delta t_c/2 \leq t_1 - t_2 \leq \tau + \Delta t_c/2$ as a function of $z_1 - z_2$ or $\vec{\rho}_1 - \vec{\rho}_2$. In the current-current correlation measurements, the linear multiplier and associated electronics integrates and records the output reading of the linear multiplier that is proportional to $i_1(t)i_2(t) \propto I(t_1)I(t_2)$. The delays of the electronic cables between the linear multiplier and the photodetectors D_1 and D_2 define the early times $t_1 = t - \tau_1^e$, $t_2 = t - \tau_2^e$ of the measured photodetection events,[3] as well as $t_1 - t_2 = \tau_1^e - \tau_2^e$. $G^{(2)}(z_1 - z_2)$ is measured by means of reading and recording the coincidence counting rate or the integrated output of the linear multiplier as a function of $z_1 - z_2$ or $\vec{\rho}_1 - \vec{\rho}_2$. The variation of $z_1 - z_2$ or $\vec{\rho}_1 - \vec{\rho}_2$ can be achieved by scanning one of the photodetectors along its longitudinal axis or its transverse plane.[4]

It should be emphasized that the temporal and the transverse correlation may not be treated as independent. For instance, in the measurement of the second-order transverse spatial coherence function $G^{(2)}(\vec{\rho}_1 - \vec{\rho}_2)$, the value of $\tau_1 - \tau_2$ must be carefully chosen to achieve a non-zero constant value (usually the maximum value) of $G^{(2)}(z_1, t_1; z_2, t_2)$ during the scanning of the transverse coordinates of the point-like photodetectors.

---

[3]The linear multiplier measures the current-current correlation of the photodetectors D_1 and D_2 at time t: $V_{12}(t) \propto i_1(t)i_2(t) \propto I(t_1)I(t_2)$, where $V_{12}(t)$ is the output reading of the linear multiplier at time t, and $i_1(t)$ and $i_2(t)$ are the output currents of the photodetectors at time t. The delays of the electronic cables between the linear multiplier and the photodetectors D_1 and D_2 define the early times $t_1 = t - \tau_1^e$, $t_2 = t - \tau_2^e$ of the measured photodetection events, as well as $t_1 - t_2 = \tau_1^e - \tau_2^e$.

[4]The linear multiplier can also be used for the measurement of $G^{(2)}(t_1 - t_2)$ of stationary light by varying the cable length of the delay-line or by other means of the electronic delays with fixed values of $\mathbf{r}_1$ and $\mathbf{r}_2$.

8.8 THE HANBURY BROWN AND TWISS INTERFEROMETER

In 1956, Hanbury Brown and Twiss published two interferometers that measure the second-order temporal coherence function and the second-order spatial coherence function of thermal field, namely, the so called temporal intensity interferometer and spatial intensity interferometer.

(I) Temporal HBT interferometer

The temporal intensity interferometer of HBT is schematically shown in Fig. 8.8.1. In

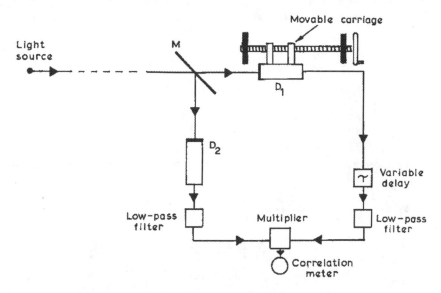

FIGURE 8.8.1 Schematic of the historical Hanbury Brown and Twiss temporal intensity interferometer. This HBT interferometer measures the second-order temporal coherence of thermal radiation in the joint-detection of two spatially separated photodetectors D_1 and D_2. In the time of HBT, the response time of the best correlation multiplier may achieve 10^{-6} second. The coherent time of a natural white light is in the order of 10^{-14} second. HBT coupled a "low-pass filter" with each of their photomultipliers to increase the contrast of the measured $\Gamma^{(2)}(z_1, t_1; z_2, t_2)$ to an observable level.

this interferometer, radiation emitted from a far-field thermal radiation source (such as a distant star) is divided by a beamsplitter into two beams of equal intensities. The two independent light beams are then measured by two photodetectors D_1 and D_2, one of which can be scanned longitudinally along the optical path. The output photocurrents of D_1 and D_2 are recorded, respectively, and multiplied electronically by a linear multiplier at a later time, $t = t_1 + \tau_1^e = t_2 + t_1^e$. HBT found that although both $\langle I_1 \rangle$ and $\langle I_2 \rangle$ kept constant values during the scanning, the statistical correlation of $\langle I_1 I_2 \rangle$, however, turned to be a nontrivial function of $\tau = \tau_1 - \tau_2 = (z_2 - z_1)/c + (\tau_2^e - \tau_1^e)$,

$$\langle I(\tau_1) I(\tau_2) \rangle \propto 1 + \alpha \left| \mathcal{F}_\tau \{ a^2(\nu) \} \right|^2 \tag{8.8.1}$$

where, again, $\tau_j = (t - \tau_j^e) - z_j/c = t_j - z_j/c$, $j = 1, 2$, $0 < \alpha < 1$ is a positive real number that determines the contrast of the measured second-order temporal coherence function $\gamma^{(2)}(z_1, t_1; z_2, t_2)$. When $\alpha = 1$, Eq. (8.8.1) is the same as our theoretical expectation of Eq. (8.2.11). The experimentally observed contrasts of $\gamma^{(2)}(z_1, t_1; z_2, t_2)$ from historical HBT interferometers, are quite low with $\alpha \ll 1$.

As we have discussed in previous section, the low contrast $\alpha \ll 1$ was the result of (1) relatively broadband spectrum of their measured thermal field; (2) relatively slower photodetectors and associated correlation measurement electronics used in the historical HBT interferometer. Recall that Eq. (8.2.11) was derived by assuming a "perfect" correlation measurement with "perfect" photodetectors and "perfect" electronics. Due to the limited ability of the photodetectors in determining the registration time of a photoelectron, the relative slow responds time of the photodetectors and the associated electronics will affect the measurement of $G^{(2)}(t_1 - t_2)$.

The width of the response functions of the photodetectors used in the historical HBT interferometers are much wider than that of the $G^{(2)}(t_1 - t_2)$ of natural light in their astrophysics observations. The second-order correlation function can be treated as a delta function in the convolution between the response functions of the two photon counting detectors:

$$G^{(2)}(\tilde{t}_1 - \tilde{t}_2)$$
$$= \frac{1}{t_c^2} \int dt_1 \int dt_2 \, G^{(2)}(t_1 - t_2) \, D(\tilde{t}_1 - t_1) \, D(\tilde{t}_2 - t_2)$$
$$\propto \frac{1}{t_c^2} \int dt_1 \int dt_2 \left[\langle n_1 \rangle \langle n_2 \rangle + \langle \Delta n_1 \Delta n_2 \rangle \right] D(\tilde{t}_1 - t_1) \, D(\tilde{t}_2 - t_2). \qquad (8.8.2)$$

The time integral does not change the value of the first term of Eq. (8.8.2) simply because it is a constant

$$\frac{1}{t_c^2} \int dt_1 \int dt_2 \left[\langle n_1 \rangle \langle n_2 \rangle D(\tilde{t}_1 - t_1) \, D(\tilde{t}_2 - t_2) \right]$$
$$= \frac{1}{t_c^2} \left\{ \left[\bar{n}_1 \bar{n}_2 \int dt_1 \, D(\tilde{t}_1 - t_1) \int dt_2 \, D(\tilde{t}_2 - t_2) \right] \right.$$
$$= \bar{n}_1 \bar{n}_2.$$

However, it reduces the value of the second term significantly when $\tau_0 \ll t_c$

$$\frac{1}{t_c^2} \int dt_1 \int dt_2 \, \langle \Delta n_1 \Delta n_2 \rangle D(\tilde{t}_1 - t_1) D(\tilde{t}_2 - t_2)$$
$$\simeq \frac{\bar{n}_1 \bar{n}_2}{t_c^2} \int dt_1 \int dt_2 \, \tau_0 \delta(t_1 - t_2) D(\tilde{t}_1 - t_1) D(\tilde{t}_2 - t_2)$$
$$\simeq \frac{\bar{n}_1 \bar{n}_2}{t_c^2} \tau_0 \int d\tau \, D(\tilde{t}_1 - \tilde{t}_2 - \tau) \, D(\tau)$$

For Gaussian type responds functions, the magnitude of the second term is thus roughly τ_0 / t_c times smaller than that of the first term. The coherence time of sunlight is in the order of 10^{-15} second. A nanosecond photodetector yields $\tau_0 / t_c \sim 10^{-6}$. It is definitely uneasy to distinguish the nontrivial photon number fluctuation correlation, which is one part of 10^6 relative to the trivial constant $\bar{n}_1 \bar{n}_2$, from the measurement of $G^{(2)}(\tilde{t}_1 - \tilde{t}_2)$. This has been the major difficulty for HBT measurement of sunlight and other type natural radiations.

In the beginning of 1970s, right after the invention of laser, the concept of pseudo-thermal light was introduced into the measurement of the second-order coherence of thermal field to achieve $\alpha \sim 1$. A pseudo-thermal light source usually contains a laser beam and a fast rotating defusing ground glass. The transversely expanded laser beam is scattered by millions of tiny scattering elements of the fast rotating ground glass. These tiny scattering elements play the role of point-like sub-sources. Millions of randomly scattered subfields, each may contain a number of identical-photons, are scattered from each tiny scattering elements of the ground glass with random relative phases. The spectral bandwidth of a laser beam, especially operated in single-mode or single-frequency, is usually narrow enough to

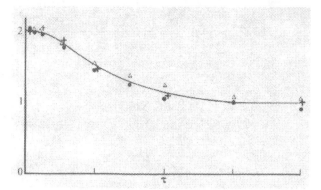

FIGURE 8.8.2 Historical measurements of the second-order temporal coherence function of thermal field. Thermal radiation has almost twice the chance of been captured by two individual photodetectors at $\tau = \tau_1 - \tau_2 \sim 0$, although the thermal radiation is incoherently, randomly, and stochastically radiated from the source. Note, most of these measurements use "pseudo-thermal" light sources with relatively narrower spectrum.

satisfy $\tau_0 \sim t_c$. Figure 8.8.2 is a collection of historically measured temporal degree of second-order coherence function of thermal or pseudo-thermal field by using similar experimental setups of the HBT interferometer. In these measurements, the two photodetectors D_1 and D_2 have almost twice chance to be excited at $\tau = \tau_1 - \tau_2 \sim 0$ then that of $\tau = \tau_1 - \tau_2 > \tau_c$, where τ_c is the coherence time of the thermal field.

(II) Far-field spatial HBT interferometer.

Hanbury Brown and Twiss published their spatial intensity interferometer intensity interferometer in the same year as that of their temporal intensity interferometer. In fact, in 1956, Hanbury Brown and Twiss had successfully utilized the second-order far-field spatial coherence of thermal field

$$\gamma^{(2)}(x_1, t_1; x_2, t_2) = 1 + \text{sinc}^2 \frac{\pi \Delta \theta}{\lambda}(x_1 - x_2) \tag{8.8.3}$$

in astrophysics applications for measuring the angular diameter of distant stars or the angular separation of distant double-stars. A typical spatial HBT spatial intensity interferometer for such applications is schematically shown in Fig. 8.8.3. The spatial HBT interferometer is similar to the Michelson stellar interferometer, except that the observation is the second-order spatial correlation measured by two independent photodetectors, instead of the visibility of the first-order interference pattern measured by one photodetector. The long-base-line HBT intensity interferometer has been widely used in modern astronomical observations. The HBT interferometer is especially useful for measuring celestial bodies of smaller angular size and their angular separation. When the saparation between D_1 and D_2 is "scanned" from $x_1 - x_2 = 0$ to $x_1 - x_2 = b$, see Fig. 8.8.3, the measured intensity fluctuation correlation drops from its maximum value of "one" to its minimum value of "zero". The greater the value of b, the smaller the $\Delta \theta$ that is measurable. In modern applications, b, which is usually called "base line", can be as long as kilometers or more.

(III) Near-field spatial HBT interferometer.

In the early days, spatial HBT intensity interferometer was considered working in far-field only. Until the advent of the HBT interferometer for half a century, we proved that the spatial HBT interferometer can also work in the Fresnel near-field too. The second-order

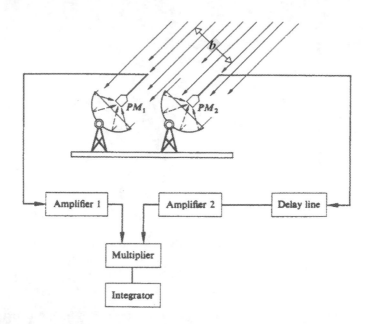

FIGURE 8.8.3 Schematic of a spatial HBT interferometer for astrophysics applications. The interferometer measures the angular size of a distant thermal radiation source by means of its second-order spatial coherence, or spatial correlation. When the separation between D_1 and D_2 is "scanned" from $x_1 - x_2 = 0$ to $x_1 - x_2 = b$, the measured intensity fluctuation correlation drops from its maximum value of "one" to its minimum value of "zero". The greater the value of b, the smaller the $\Delta\theta$ that is measurable. In modern applications, b, which is usually called the "base line", can be as long as kilometers or more.

spatial coherence function of thermal field $G^{(2)}(\mathbf{r}_1, t_1; \mathbf{r}_2, t_2)$ in near-field was experimentally demonstrated by Scarcelli et al. in the years 2005 to 2006. Their experimental setup, schematically illustrated in Fig. 8.3.1, is similar to that of the historical HBT experiment, except the far-field distant star is replaced by a Fresnel near-field pseudo-thermal radiation source. An unfolded version of the schematic is shown in Fig. 8.8.4, which might be easier for analyzing the physics. Scarcelli et al. found that although the single detector counting rates or the output currents of D_1 and D_2 were monitored, respectively, to be constants during the measurement, nontrivial second-order correlations were observable with almost 50% contrast while the two point-like-photodetectors, D_1 and D_2, are aligned symmetrically on the transverse planes of x_1 and x_2, as indicated in the upper and the lower cases of Fig. 8.8.4. In the upper measurement, D_1 and D_2 are aligned symmetrically on the optical axis. A sinc-like function of $\langle \Delta I_1 \Delta I_2 \rangle$ is observed by scanning either D_1 or D_2 transversely in the neighborhood of the optical axis. When D_1 is moved a few millimeters up (or down) from its symmetrical position, as shown in the middle, the nontrivial correlation disappears with $\langle \Delta I_1 \Delta I_2 \rangle \sim 0$ when scanning either D_1 or D_2 in the neighborhood of that unsymmetrical position. In the lower measurement, D_2 is moved up (or down) to a symmetrical position, again with respect to D_1. A similar sinc-like function of $\langle \Delta I_1 \Delta I_2 \rangle$ is observed by scanning either D_1 or D_2 in the neighborhood of their new symmetrical position. Notice, *equal distance* between the photodetectors and the light source is required for the observation of the sinc-function like correlation. For a large source of transverse dimension, a few millimeter difference may cause a complete disappearing of the nontrivial correlation.

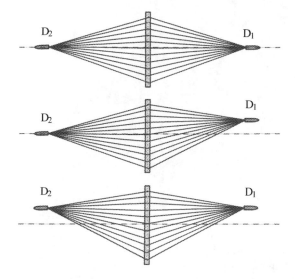

FIGURE 8.8.4 Second-order near-field spatial correlation measurement by Scarcelli *et al.* Upper: D_1 and D_2 are placed at equal distances from the source and aligned symmetrically on the optical axis, a nontrivial $g^{(2)}(x_1 - x_2)$ is observed with maximum value of ~ 2; Middle: D_1 is moved up to a non-symmetrical position, $g^{(2)}(x_1 - x_2)$ becomes a constant of 1; Lower: D_2 is moved up to the symmetrical position with D_1, the nontrivial $g^{(2)}(x_1 - x_2)$ is observable with a maximum value of ~ 2, again.

What is the reason for $\langle \Delta I_1 \Delta I_2 \rangle \sim 0$ when D_1 is moved a few millimeters up (or down) from its symmetrical position? Is it possible the spatial move caused a temporal delay beyond the temporal coherence of the thermal field? A second-order temporal coherence of the pseudo-thermal field used in the near-field HBT experiment was measured in the same experimental setup and under the same experimental condition. Scarcelli *et al* found the coherent time of their pseudo-thermal field is in the order of μs, corresponding to an optical delay of ~ 300 meters. A few millimeters up (or down) of the photodetectors in Fresnel near-field can never result in such a delay.

The near-field second-order coherence function has been calculated in last section

$$g^{(2)}(\vec{\rho}_1, \vec{\rho}_2) = 1 + \text{somb}^2 \left(\frac{\pi \Delta \theta}{\lambda} |\vec{\rho}_1 - \vec{\rho}_2| \right). \tag{8.8.4}$$

The source-angular-size dependent of the point-to-spot correlation has been experimentally confirmed in 1-D measurements. Fig. 8.8.6 shows the experimental results of Zhou *et al* with different angular sized sources. The fitting curves are calculated from Eq. (8.8.4). The calculated $\gamma^{(2)}(x_1 - x_2)$ functions agree with the experimental results within the experimental error. It is interesting to find that a point-to-"point" HBT correlation is achievable with a large angular sized thermal field source.

The above experimental observations are quite surprise. First, it tells us that our 50 years belief about the far-field condition of HBT correlation is never true; the nontrivial second-order spatial correlation of thermal field is observable in the Fresnel near-field. Second, apparently, we have observed an EPR type nonlocal phenomenon: although D_1 and D_2 both show constant counting rates when scanning their transverse positions, if D_1 observes a photon or wavepacket at a transverse position x_1, D_2 has twice chance to observe its randomly paired photon or wavepacket at a certain transverse coordinate x_2. This nontrivial second-order near-field spatial correlation is not only interesting in fundamental concerns

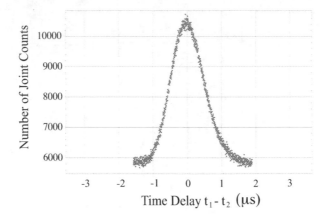

FIGURE 8.8.5 Second-order temporal coherence of the pseudo-thermal radiation used in the near-field HBT experiment. The coherent time is found in the order of μs, corresponding to a longitudinal optical delay of $\sim$300 meters.

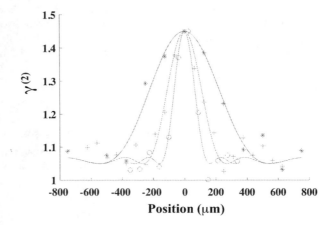

FIGURE 8.8.6 Measured point-to-"spot" spatial correlation of a pseudo-thermal radiation with different source sizes (diamaters) 1mm (blue star), 2mm (green cross) and 4mm (red circle), respectively. The fitting curves are calculated from Eq. (8.8.4).

but also useful in practical applications. The point-to-point correlation between two near-field planes has turned into a useful new imaging technology: lensless ghost imaging, which is especially useful for these radiations for which no effective imaging lenses can be used to produce the classic point-to-point image forming function. We will have a detailed discussion on the lensless ghost imaging in Chapter 10.

(III) The physical cause of the HBT correlation

What is the physical cause of the HBT correlation? What is the reason for two randomly created and randomly distributed photons to have twice chance of being jointly captured within a temporal delay that equals the coherence time of the thermal field and within a transverse area that equals the coherence area of the thermal radiation? From the view point of quantum optics, either in terms of Einstein's granularity picture of light or in terms of the concepts of quantum mechanics, HBT correlation is a two-photon interference phenomenon: a pair of randomly created and randomly paired subfields or photons interfering with the pair itself. A large number of constructively-destructively added two-photon interferences produced the nontrivial temporal and spatial intensity fluctuation correlation of HBT.

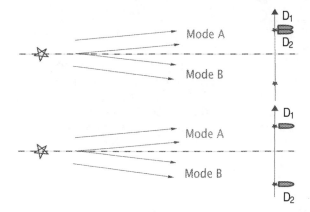

FIGURE 8.8.7 A phenomenological interpretation of the historical HBT experiment. Upper: the two photodetectors receive identical modes of the far-field radiation and thus experience identical intensity fluctuations. The joint measurement of D_1 and D_2 gives a maximum value of $\langle \Delta I_1 \Delta I_2 \rangle$. Lower: the two photodetectors receive different modes of the far-field radiation. In this case the joint measurement gives $\langle \Delta I_1 \Delta I_2 \rangle = 0$.

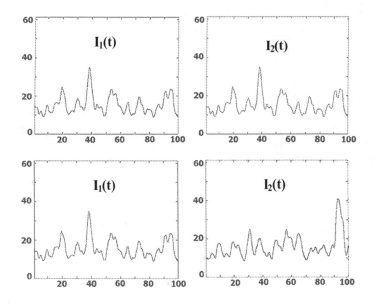

FIGURE 8.8.8 The two upper (lower) curves of $I(t)$ corresponds to the upper (lower) configuration in Fig 8.8.7.

Interestingly, the physical cause of HBT correlation remains yet the subject of today's debate. A naive argument of rejecting two-photon interference considers thermal light "classical": "How could quantum theory apply to classical light?" Perhaps this argument forgot that blackbody radiation is a typical thermal radiation and Planck's quantization is indeed for thermal field. We cannot forget that the starting point of quantum theory is the classically incomprehensible behavior of the thermal field! A profound reservation about two-photon interference might be its non-local nature. The concept of nonlocal interference is indeed uneasy to accept.

Historically, a widely accepted theory suggests that the temporal and spatial HBT correlations of thermal field are pre-prepared in the radiation source as a statistical property of thermal light. Thermal light is named "correlated light" accordingly. It is true, statistically, the intensity of light, or the number of photons, created from a thermal source fluctuates temporally and spatially. However, we cannot find any physical mechanism to "correlate" these pre-prepared fluctuations. For example, the HBT interferometers on earth receive radiations from all possible locations of the sun, including its upper edge and lower edge. The intensity of light, or the number of photons, created from the upper edge of the sun and from the lower edge of the sun may both fluctuate statistically. Do we have any physical mechanism that is able to prepare radiations from the upper edge of the sun and from the lower edge of the sun with identical intensity fluctuations or photon number fluctuations? If there is any, this mechanism is indeed nonlocal. The creation process of sunlight is stochastic. Either in the view of classical electromagnetic theory of light or in the view of quantum theory of light, sunlight should be created randomly and distributed randomly.

It was HBT suggested a reasonable interpretation to the HBT correlation. In a HBT intensity interferometer, see Fig. 8.8.3, the measurement is in the far-field of the thermal radiation source, which is equivalent to the Fourier transform plane. When D_1 and D_2 are moved side by side, the two detectors measure the same spatial mode of the radiation field. HBT believe that the measured intensities should have the same fluctuations while the two photodetectors receive the same spatial mode and thus yield a maximum value of $\langle \Delta I(\mathbf{k}) \Delta I(\mathbf{k}) \rangle$ and gives $\gamma^{(2)} \sim 2$. When the two photodetectors move apart to a certain distance, D_1 and D_2 start to measure different spatial modes of the radiation field. In this case the measured intensities have different fluctuations. The measurement yields $\langle \Delta I(\mathbf{k}) \Delta I(\mathbf{k}') \rangle = 0$ and gives $\gamma^{(2)} \sim 1$. Figures 8.8.7 and 8.8.8 illustrate the above two different situations.

For fifty years, this theory has convinced us to believe that the observation of the nontrivial second-order correlation of thermal field only occurs in the Fraunhofer far-field. What will happen if we move the two HBT photodetectors to the Fresnel "near-field"[5] as shown in the unfolded schematic of Fig. 8.8.9? Is the nontrivial second-order correlation still observable in the Fresnel near-field? In fact, the second-order correlation of thermal field in the Fresnel near-field has been observed by Scarcelli *et al.* during the years 2005 to 2006.

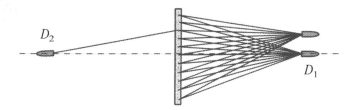

FIGURE 8.8.9 (1) each photodetector is able to receive radiations from a large number of sub-sources or spatial modes; (2) in a joint-detection between D_1 and D_2, the two photodetectors have more chances to be triggered by radiations coming from different sub-sources and different spatial modes. The ratio between the joint-detections triggered by radiations coming from the same sub-source (mode) and these triggered by different sub-sources (modes) is roughly $N/N^2 = 1/N$; and (3) despite the scanning position of D_1 (or D_2) the ratio of $1/N$ does not change.

Examining Fig. 8.8.9, it is easy to find that in the Fresnel near-field: (1) each photodetector, D_1 and D_2, is able to receive radiations from a large number of sub-sources, which are spatially distributed on the entire source, and spatial modes, which are angularly distributed

---

[5]The concept of "near-field" was defined by Fresnel to be distinct from the Fraunhofer far-field. The Fresnel near-field is defined for a light source with angular size satisfying $\Delta\theta > \lambda/D$, where D is the diameter of the source. The Fresnel near-field is different from the "near-surface-field". The "near-surface-field" considers a distance of a few wavelengths from a surface.

within the entire opening angle of the source; (2) in a joint-detection between D_1 and D_2, the two photodetectors have more chances to be triggered by radiations coming from different sub-sources and different spatial modes. The ratio between the joint-detections triggered by radiations coming from the same sub-source (mode) and these triggered by different sub-sources (modes) is roughly $N/N^2 = 1/N$; and (3) despite the scanning position of D_1 (or D_2) the ratio of $1/N$ does not change. If we believe the HBT correlation of thermal field comes from the measurement of identical spatial mode $\langle \Delta I(\mathbf{k}) \Delta I(\mathbf{k}) \rangle$ and the measurement of different spatial modes results in $\langle \Delta I(\mathbf{k}) \Delta I(\mathbf{k}') \rangle = 0$, we would easily conclude a constant second-order spatial coherence in the Fresnel near-field. i.e., $\langle I_1 I_2 \rangle \sim 1$ despite the scanning position of D_1 (or D_2), which is in false.

We know that the observed intensity fluctuation of thermal light may have different causes. Statistically, intensity of light, or number of photons, created from a thermal source may fluctuate. The index variations of the medium of propagation may also induce additional intensity fluctuations. However, these random fluctuations can never produce a statistical correlation of $\langle \Delta I_1 \Delta I_2 \rangle$, especially, a nontrivial $\langle \Delta I_1(\mathbf{r}_1, t_1) \Delta I_2(\mathbf{r}_2, t_2) \rangle$ that depends on the temporal delay and spatial separation of the photodetection events of D_1 and D_2. We may say: the HBT correlation is *observed* from the intensity fluctuations, or the photon number fluctuations, of the thermal field, however, it is neither *caused* by the statistical fluctuations originated from the stochastic photon creation processes, nor *caused* by the additional fluctuations induced by its propagation medium.

8.9 NTH-ORDER COHERENCE OF LIGHT

In this section we introduce the concept of higher-order coherence or correlation of light. Formulated from electromagnetic field, the Nth-order ($N > 2$) coherence of light is defined as

$$
\begin{aligned}
&\Gamma^{(N)}(\mathbf{r}_1, t_1; \mathbf{r}_2, t_2; \ldots; \mathbf{r}_N, t_N) \\
&= \big\langle E^*(\mathbf{r}_1, t_1) E(\mathbf{r}_1, t_1) \, E^*(\mathbf{r}_2, t_2) E(\mathbf{r}_2, t_2) \ldots E^*(\mathbf{r}_N, t_N) E(\mathbf{r}_N, t_N) \big\rangle
\end{aligned} \tag{8.9.1}
$$

and the degree of Nth-order coherence is defined as

$$
\begin{aligned}
&\gamma^{(N)}(\mathbf{r}_1, t_1; \mathbf{r}_2, t_2; \ldots; \mathbf{r}_N, t_N) \\
&= \frac{\Gamma^{(N)}(\mathbf{r}_1, t_1; \mathbf{r}_2, t_2; \ldots; \mathbf{r}_N, t_N)}{\Gamma^{(1)}(\mathbf{r}_1, t_1; \mathbf{r}_1, t_1) \Gamma^{(1)}(\mathbf{r}_2, t_2; \mathbf{r}_2, t_2) \ldots \Gamma^{(1)}(\mathbf{r}_N, t_N; \mathbf{r}_N, t_N)}
\end{aligned} \tag{8.9.2}
$$

where the ensemble average, $\langle \ldots \rangle$ denotes, again, *taking into account all possible realizations of the field*.

Formulated from quantum theory of light, the Nth-order ($N > 2$) coherence of light is defined as

$$
\begin{aligned}
&G^{(N)}(\mathbf{r}_1, t_1; \mathbf{r}_2, t_2; \ldots; \mathbf{r}_N, t_N) \\
&= \big\langle \langle \hat{E}^{(-)}(\mathbf{r}_1, t_1) \hat{E}^{(-)}(\mathbf{r}_2, t_2) \ldots \hat{E}^{(-)}(\mathbf{r}_N, t_N) \\
&\quad \times \hat{E}^{(+)}(\mathbf{r}_N, t_N) \ldots \hat{E}^{(+)}(\mathbf{r}_2, t_2) \hat{E}^{(+)}(\mathbf{r}_1, t_1) \rangle_{\text{QM}} \big\rangle_{\text{En}}
\end{aligned} \tag{8.9.3}
$$

and the degree of Nth-order coherence is defined as

$$
\begin{aligned}
&g^{(N)}(\mathbf{r}_1, t_1; \mathbf{r}_2, t_2; \ldots; \mathbf{r}_N, t_N) \\
&= \frac{G^{(N)}(\mathbf{r}_1, t_1; \mathbf{r}_2, t_2; \ldots; \mathbf{r}_N, t_N)}{G^{(1)}(\mathbf{r}_1, t_1; \mathbf{r}_1, t_1) G^{(1)}(\mathbf{r}_2, t_2; \mathbf{r}_2, t_2) \ldots G^{(1)}(\mathbf{r}_N, t_N; \mathbf{r}_N, t_N)}
\end{aligned} \tag{8.9.4}
$$

where $\langle \ldots \rangle_{\text{En}}$ denotes, again, an ensemble average.

There should be no surprise that the Nth-order coherence function of thermal field is a nontrivial function of space-time coordinates $(\mathbf{r}_1, t_1; \mathbf{r}_2, t_2 \ldots; \mathbf{r}_N, t_N)$ of the N-fold joint photodetection events, caused by N-photon interference: the randomly created and randomly grouped N photons in thermal state interfering with the group itself. To show this, we start from the discussion of third-order coherence of thermal light in terms of Einstein's granularity picture and the quantum theory of light, we than generalize it to the Nth-order coherence function of thermal radiation.

(I) Third-order coherence of thermal light

We first calculate the third-order coherence of thermal field in Einstein's picture. Assuming a simple experimental setup, in which a thermal light source is facing three photodetectors, D_1, D_2, and D_3, either in analog modes or in photon counting modes. D_1, D_2, and D_3 as well as the correlation circuit are prepared for a three-folding joint measurement or a three-folding joint photon counting. In Einstein's picture, the thermal radiation source constants of a large number of randomly distributed and randomly radiating independent point-like sub-sources, such as trillions of independent and randomly radiating atomic transitions, contribute to the measurement of each of the three detectors. Each point-like sub-source contributes an independent wavepacket as a subfield of complex amplitude $E_m = a_m e^{i\varphi_m}$, where a_m is the real and positive amplitude of the mth subfield and φ_m is a *random* phase associated with the mth sub-field. Basically, we have the following pictures for the source: (I) a large number of independent point-like sub-sources distributed randomly in space (counted spatially); (II) each point-source contains a large number of independently and randomly radiating atoms (counted temporally); and (III) a large number of sub-sources, either counted spatially or temporally, may contribute to each of the independent radiation mode $(\vec{\kappa}, \omega)$ at each of the individual point photodetectors (counted by mode),

It is easy to find that the expectation value of the first-order jth self-correlation function, or the jth intensity measured by the jth photodetector, D_j, $j = 1, 2, 3$, is a constant,

$$\Gamma^{(1)}(\mathbf{r}_j, t_j; \mathbf{r}_j, t_j) = \langle E^*(\mathbf{r}_j, t_j) E(\mathbf{r}_j, t_j) \rangle = \text{constant},$$

where the expectation operation has taken into account all possible values of the phases of the subfields. Although each and all the measured intensities are constants, it does not prevent a nontrivial third-order coherence in the joint measurement of three independent photodetectors.

$$\begin{aligned}
\Gamma^{(3)}&(\mathbf{r_1}, t_1; \mathbf{r_2}, t_2; \mathbf{r_3}, t_3) \\
&= \langle E^*(\mathbf{r_1}, t_1) E(\mathbf{r_1}, t_1) E^*(\mathbf{r_2}, t_2) E(\mathbf{r_2}, t_2) E^*(\mathbf{r_3}, t_3) E(\mathbf{r_3}, t_3) \rangle \\
&= \langle \sum_a E_{a1}^* \sum_b E_{b1} \sum_c E_{c2}^* \sum_d E_{d2} \sum_e E_{e3}^* \sum_f E_{f3} \rangle \\
&= \sum_{a,b,c} \Big| \frac{1}{\sqrt{6}} \big[E_{a1} E_{b2} E_{c3} + E_{a1} E_{b3} E_{c2} + E_{a2} E_{b1} E_{c3} \\
&\qquad\qquad + E_{a2} E_{b3} E_{c1} + E_{a3} E_{b1} E_{c2} + E_{a3} E_{b2} E_{c1} \big] \Big|^2,
\end{aligned} \tag{8.9.5}$$

where E_{aj}, $a = 1, 2, \ldots, \infty$, $j = 1, 2, 3$, is short for $E_a(\mathbf{r}_j, t_j)$, indicating the field at coordinate $(\mathbf{r}_j, t_j)$ that is originated from the ath sub-source. Similar to the calculation of the second-order coherence function of thermal light, a partial ensemble average has been taken in Eq. (8.9.5) by means of *taking into account all possible random phase values of the subfields*.

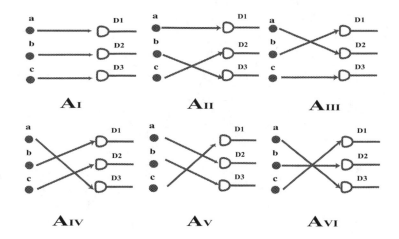

FIGURE 8.9.1 Three independent quantized subfields, or photons, a, b, c have six alternatives of triggering a joint-detection event between D_1, D_2, and D_3. The sum of a large number of these interferences yield a nontrivial third-order coherence function of thermal field $\Gamma^{(3)}(\mathbf{r_1}, t_1; \mathbf{r_2}, t_2; \mathbf{r_3}, t_3)$.

Eq.(8.9.5) indicates that $\Gamma^{(3)}(\mathbf{r_1}, t_1; \mathbf{r_2}, t_2; \mathbf{r_3}, t_3)$ is the sum of a large number of interferences, and each interference corresponds to the following superposition

$$\Gamma_{abc}^{(3)} = \left| \frac{1}{\sqrt{6}} \left[E_{a1} E_{b2} E_{c3} + E_{a1} E_{b3} E_{c2} + E_{a2} E_{b1} E_{c3} \right. \right.$$
$$\left. \left. + E_{a2} E_{b3} E_{c1} + E_{a3} E_{b1} E_{c2} + E_{a3} E_{b2} E_{c1} \right] \right|^2. \tag{8.9.6}$$

The superposed six terms in Eq. (8.9.6) indicate six different yet indistinguishable alternatives for three independent quantized subfields, or photons, to trigger a three-fold joint-detection event of D_1, D_2, and D_3. These six alternatives are schematically illustrated in Fig.8.9.1.

Similar to the second-order coherence function of thermal light, in an idealized correlation measurement by assuming three perfect photodetectors and a perfect three-fold correlation circuit, in general, we may write $\Gamma^{(3)}(\mathbf{r_1}, t_1; \mathbf{r_2}, t_2; \mathbf{r_3}, t_3)$ in terms of the first-order coherence functions

$$\Gamma^{(3)}(\mathbf{r_1}, t_1; \mathbf{r_2}, t_2; \mathbf{r_3}, t_3)$$
$$= \Gamma_{11}^{(1)} \Gamma_{22}^{(1)} \Gamma_{33}^{(1)} + |\Gamma_{12}^{(1)}|^2 \Gamma_{33}^{(1)} + |\Gamma_{23}^{(1)}|^2 \Gamma_{11}^{(1)}$$
$$+ |\Gamma_{31}^{(1)}|^2 \Gamma_{22}^{(1)} + \Gamma_{12}^{(1)} \Gamma_{23}^{(1)} \Gamma_{31}^{(1)} + \Gamma_{21}^{(1)} \Gamma_{32}^{(1)} \Gamma_{13}^{(1)} \tag{8.9.7}$$

Equation (8.9.7) can be normalized as

$$\gamma^{(3)}(\mathbf{r_1}, t_1; \mathbf{r_2}, t_2; \mathbf{r_3}, t_3)$$
$$= 1 + |\gamma_{12}^{(1)}|^2 + |\gamma_{13}^{(1)}|^2 + |\gamma_{23}^{(1)}|^2 + \gamma_{12}^{(1)} \gamma_{23}^{(1)} \gamma_{31}^{(1)} + \gamma_{21}^{(1)} \gamma_{32}^{(1)} \gamma_{13}^{(1)}. \tag{8.9.8}$$

The third-order coherence formulated from quantum theory can be easily calculated. Similar to the first-order and second-order coherence of thermal field, we may model thermal radiation either in single-photon state representation or in coherent state representation . In either cases, the second-order quantum coherence of thermal field $G^{(N)}(\mathbf{r_1}, t_1; \mathbf{r_2}, t_2; \mathbf{r_3}, t_3)$

can be easily calculated

$$G^{(3)}(\mathbf{r_1}, t_1; \mathbf{r_2}, t_2; \mathbf{r_3}, t_3)$$

$$= \langle \langle \hat{E}^{(-)}(\mathbf{r_1}, t_1) \hat{E}^{(-)}(\mathbf{r_2}, t_2) \hat{E}^{(-)}(\mathbf{r_3}, t_3) \hat{E}^{(+)}(\mathbf{r_3}, t_3) \hat{E}^{(+)}(\mathbf{r_2}, t_2) \hat{E}^{(+)}(\mathbf{r_1}, t_1) \rangle_{\text{QM}} \rangle_{\text{En}}$$

$$= \sum_{a,b,c} \left| \frac{1}{\sqrt{6}} [\psi_{a1}\psi_{b2}\psi_{c3} + \psi_{a1}\psi_{b3}\psi_{c2} + \psi_{a2}\psi_{b1}\psi_{c3} \right.$$

$$\left. + \psi_{a2}\psi_{b3}\psi_{c1} + \psi_{a3}\psi_{b1}\psi_{c2} + \psi_{a3}\psi_{b2}\psi_{c1}] \right|^2, \tag{8.9.9}$$

where ψ_{aj} (ψ_{bj}) (ψ_{cj}), $j = 1, 2, 3$, is the effective wavefunction of the ath (bth) (cth) photon that is created from the ath (bth) (cth) sub-source and is annihalized at the jth photodetector D_j. Eq.(8.9.9) indicates that $G^{(3)}(\mathbf{r_1}, t_1; \mathbf{r_2}, t_2; \mathbf{r_3}, t_3)$ is the sum of a large number of interferences, and each interference corresponds to the following superposition of three-photon amplitudes

$$G^{(3)}_{abc} = \left| \frac{1}{\sqrt{6}} [\psi_{a1}\psi_{b2}\psi_{c3} + \psi_{a1}\psi_{b3}\psi_{c2} + \psi_{a2}\psi_{b1}\psi_{c3} \right.$$

$$\left. + \psi_{a2}\psi_{b3}\psi_{c1} + \psi_{a3}\psi_{b1}\psi_{c2} + \psi_{a3}\psi_{b2}\psi_{c1}] \right|^2. \tag{8.9.10}$$

The superposed six terms in Eq. (8.9.10) indicate six different yet indistinguishable alternatives for three randomly created photons to trigger a three-fold joint-detection event of D_1, D_2, and D_3. These six three-photon amplitudes are also schematically illustrated in Fig.8.9.1.

In an idealized correlation measurement by assuming three perfect photodetectors and a perfect three-fold correlation measurement circuit, we may write $G^{(3)}(\mathbf{r_1}, t_1; \mathbf{r_2}, t_2; \mathbf{r_3}, t_3)$ in terms of the first-order quantum coherence functions

$$G^{(3)}(\mathbf{r_1}, t_1; \mathbf{r_2}, t_2; \mathbf{r_3}, t_3)$$

$$= G^{(1)}_{11} G^{(1)}_{22} G^{(1)}_{33} + |G^{(1)}_{12}|^2 G^{(1)}_{33} + |G^{(1)}_{23}|^2 G^{(1)}_{11}$$

$$+ |G^{(1)}_{31}|^2 G^{(1)}_{22} + G^{(1)}_{12} G^{(1)}_{23} G^{(1)}_{31} + G^{(1)}_{21} G^{(1)}_{32} G^{(1)}_{13} \tag{8.9.11}$$

Equation (8.9.11) can be normalized as

$$g^{(3)}(\mathbf{r_1}, t_1; \mathbf{r_2}, t_2; \mathbf{r_3}, t_3)$$

$$= 1 + |g^{(1)}_{12}|^2 + |g^{(1)}_{13}|^2 + |g^{(1)}_{23}|^2 + g^{(1)}_{12} g^{(1)}_{23} g^{(1)}_{31} + g^{(1)}_{21} g^{(1)}_{32} g^{(1)}_{13}. \tag{8.9.12}$$

No surprise, we found that $G^{(3)}(\mathbf{r_1}, t_1; \mathbf{r_2}, t_2; \mathbf{r_3}, t_3)$ has the same form as $\Gamma^{(3)}(\mathbf{r_1}, t_1; \mathbf{r_2}, t_2; \mathbf{r_3}, t_3)$, except replacing Einstein's sub-fields with the effective wavefunctions.

The third-order temporal and spatial coherence function of pseudo-thermal field have been experimentally measured by Zhou et al. around the year of 2010. The experimental setup of Zhou et al. is similar to that of the HBT interferometer, except the use of three photodetectors and a 3-fold joint-detection coincidence counter.

(1) Third-order temporal coherence of thermal light.

In the third-order temporal correlation measurement of Zhou et al., their three point-like photodetectors were set with equal distances, $z_1 = z_2 = z_3$, from their point-like pseudo-thermal source (simulated by a pinhole).

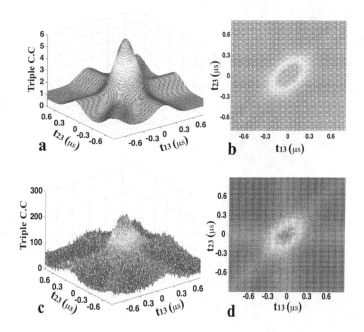

FIGURE 8.9.2 Calculated (Top, a and b) and measured (bottom, c and d) third-order temporal correlation of thermal light. The 3-D three-photon joint detection histogram is plotted as a function of $t_{13} \equiv t_1 - t_3$ and $t_{23} \equiv t_2 - t_3$. The simulation function is calculated from Eq. (8.9.13). In addition, the single detector counting rates for D_1, D_2, and D_3 are all monitored to be constants. The experimental data and the simulation are in agreement within the experimental error.

Based on the experimentally setup of Zhou *et al.*, the calculation of the third-order temporal coherence of $\gamma^{(3)}(t_1, t_2, t_3)$ or $g^{(3)}(t_1, t_2, t_3)$ is straightforward:

$$g^{(3)}(t_1, t_2, t_3) = 1 + \operatorname{sinc}^2\left[\frac{\Delta\omega(t_1 - t_2)}{2\pi}\right]$$
$$+ \operatorname{sinc}^2\left[\frac{\Delta\omega(t_2 - t_3)}{2\pi}\right] + \operatorname{sinc}^2\left[\frac{\Delta\omega(t_3 - t_1)}{2\pi}\right]$$
$$+ 2\operatorname{sinc}\left[\frac{\Delta\omega(t_1 - t_2)}{2\pi}\right]\operatorname{sinc}\left[\frac{\Delta\omega(t_2 - t_3)}{2\pi}\right]\operatorname{sinc}\left[\frac{\Delta\omega(t_3 - t_1)}{2\pi}\right]. \quad (8.9.13)$$

In Eq. (8.9.13), we have assumed a constant spectrum distribution $f(\omega)$ within the bandwidth of the radiation field $\Delta\omega$ to simplify the mathematics. It is easy to see that when $t_1 = t_2 = t_3$, $g^{(3)}(t_1, t_2, t_3) = 6$. The third-order correlation function of thermal field achieves a maximum contrast of 6 to 1 ($\sim$71% visibility).

The measured $g^{(3)}(t_1, t_2, t_3)$ is shown in Fig. 8.9.2. The experimental observation of Zhou *et al.* and the simulation are in agreement within the experimental error.

(2) Third-order near-field spatial coherence of thermal light.

In the following, we briefly discuss the third-order near-field spatial coherence of thermal light. Now, considering a measurement similar to that of the second-order near-field spatial coherence measurement, except the scanning of three point-like photodetectors in the transverse planes of $\vec{\rho}_1$, $\vec{\rho}_2$, and $\vec{\rho}_3$, respectively. We assume a perfect joint-measurement system, including three perfect photodetectors and a perfect 3-fold joint-measurement coincidence circuit, and a maximum temporal correlation at single wavelength thermal field to simplify the mathematics.

Similar to the calculation of near-field second-order spatial coherence, we applying the near-field Fresnel Green's function to propagate each subfield from each sub-source to the photodetectors,

$$
g_a(\vec{\rho}_j, z_j) = \frac{c_0}{z_j} e^{i\frac{\omega z_j}{c}} e^{i\frac{\omega}{2cz_j}|\vec{\rho}_j - \vec{\rho}_a|^2}.
$$

where c_0 is a normalization consistent. The third-order spatial coherence functions $\Gamma^{(3)}_{abc}(\vec{\rho}_1, z_1; \vec{\rho}_2, z_2; \vec{\rho}_3, z_3)$ and $G^{(3)}_{abc}(\vec{\rho}_1, z_1; \vec{\rho}_2, z_3; \vec{\rho}_3, z_3)$, respectively, can be written in terms of the Green's functions:

$$
\Gamma^{(3)}_{abc}(\vec{\rho}_1, z_1; \vec{\rho}_2, z_2; \vec{\rho}_3, z_3) = \sum_{a,b,c} |E_a|^2 |E_b|^2 |E_c|^2
$$

$$
\left| \frac{1}{\sqrt{6}} \left[g_a(\vec{\rho}_1, z_1) g_b(\vec{\rho}_2, z_2) g_c(\vec{\rho}_3, z_3) + g_a(\vec{\rho}_1, z_1) g_b(\vec{\rho}_3, z_3) g_c(\vec{\rho}_2, z_2) \right. \right.
$$

$$
+ g_a(\vec{\rho}_2, z_2) g_b(\vec{\rho}_1, z_1) g_c(\vec{\rho}_3, z_3) + g_a(\vec{\rho}_2, z_2) g_b(\vec{\rho}_3, z_3) g_c(\vec{\rho}_1, z_1)
$$

$$
\left. \left. + g_a(\vec{\rho}_3, z_2) g_b(\vec{\rho}_1, z_1) g_c(\vec{\rho}_2, z_2) + g_a(\vec{\rho}_3, z_3) g_b(\vec{\rho}_2, z_2) g_c(\vec{\rho}_1, z_1) \right] \right|^2
$$

and

$$
G^{(3)}_{abc}(\vec{\rho}_1, z_1; \vec{\rho}_2, z_2; \vec{\rho}_3, z_3) = \sum_{a,b,c} |\alpha_a|^2 |\alpha_b|^2 |\alpha_c|^2
$$

$$
\left| \frac{1}{\sqrt{6}} \left[g_a(\vec{\rho}_1, z_1) g_b(\vec{\rho}_2, z_2) g_c(\vec{\rho}_3, z_3) + g_a(\vec{\rho}_1, z_1) g_b(\vec{\rho}_3, z_3) g_c(\vec{\rho}_2, z_2) \right. \right.
$$

$$
+ g_a(\vec{\rho}_2, z_2) g_b(\vec{\rho}_1, z_1) g_c(\vec{\rho}_3, z_3) + g_a(\vec{\rho}_2, z_2) g_b(\vec{\rho}_3, z_3) g_c(\vec{\rho}_1, z_1)
$$

$$
\left. \left. + g_a(\vec{\rho}_3, z_2) g_b(\vec{\rho}_1, z_1) g_c(\vec{\rho}_2, z_2) + g_a(\vec{\rho}_3, z_3) g_b(\vec{\rho}_2, z_2) g_c(\vec{\rho}_1, z_1) \right] \right|^2
$$

Now, consider the sum of $\Gamma^{(3)}_{abc}(\vec{\rho}_1, z_1; \vec{\rho}_2, z_2; \vec{\rho}_3, z_3)$ and $G^{(3)}_{abd}(\vec{\rho}_1, z_1; \vec{\rho}_2, z_2; \vec{\rho}_3, z_3)$ by turning the sum of a, b, and c into the integrals of $\vec{\rho}_0$, $\vec{\rho}_0'$, and $\vec{\rho}_0''$ on the entire source plane:

$$
\Gamma^{(3)}(\vec{\rho}_1, \vec{\rho}_2, \vec{\rho}_3) = \sum_{a,b,c} \Gamma^{(3)}_{abc}(\vec{\rho}_1, \vec{\rho}_2, \vec{\rho}_3)
$$

$$
= \Gamma^{(1)}(\vec{\rho}_1, \vec{\rho}_1) \Gamma^{(1)}(\vec{\rho}_2, \vec{\rho}_2) \Gamma^{(1)}(\vec{\rho}_3, \vec{\rho}_3)
$$

$$
+ \left| \Gamma^{(1)}(\vec{\rho}_1, \vec{\rho}_2) \right|^2 \Gamma^{(1)}(\vec{\rho}_3, \vec{\rho}_3)
$$

$$
+ \left| \Gamma^{(1)}(\vec{\rho}_2, \vec{\rho}_3) \right|^2 \Gamma^{(1)}(\vec{\rho}_1, \vec{\rho}_1)
$$

$$
+ \left| \Gamma^{(1)}(\vec{\rho}_3, \vec{\rho}_1) \right|^2 \Gamma^{(1)}(\vec{\rho}_2, \vec{\rho}_2)
$$

$$
+ \Gamma^{(1)}(\vec{\rho}_1, \vec{\rho}_2) \Gamma^{(1)}(\vec{\rho}_2, \vec{\rho}_3) \Gamma^{(1)}(\vec{\rho}_3, \vec{\rho}_1)
$$

$$
+ \Gamma^{(1)}(\vec{\rho}_2, \vec{\rho}_1) \Gamma^{(1)}(\vec{\rho}_3, \vec{\rho}_2) \Gamma^{(1)}(\vec{\rho}_1, \vec{\rho}_3) \tag{8.9.14}
$$

where

$$
\Gamma^{(1)}_{jk} = I_0 \int d\vec{\rho}_0 \, g^*_{\vec{\rho}_0}(\vec{\rho}_j, z_j) g_{\vec{\rho}_0}(\vec{\rho}_k, z_k)]
$$

where we have absorbed all constants into constant I_0.

$$G^{(3)}(\vec{\rho}_1, \vec{\rho}_2, \vec{\rho}_3) = \sum_{a,b,c} G^{(3)}_{abc}(\vec{\rho}_1, \vec{\rho}_2, \vec{\rho}_3)$$

$$= G^{(1)}(\vec{\rho}_1, \vec{\rho}_1) G^{(1)}(\vec{\rho}_2, \vec{\rho}_2) G^{(1)}(\vec{\rho}_3, \vec{\rho}_3)$$
$$+ \left| G^{(1)}(\vec{\rho}_1, \vec{\rho}_2) \right|^2 G^{(1)}(\vec{\rho}_3, \vec{\rho}_3)$$
$$+ \left| G^{(1)}(\vec{\rho}_2, \vec{\rho}_3) \right|^2 G^{(1)}(\vec{\rho}_1, \vec{\rho}_1)$$
$$+ \left| G^{(1)}(\vec{\rho}_3, \vec{\rho}_1) \right|^2 G^{(1)}(\vec{\rho}_2, \vec{\rho}_2)$$
$$+ G^{(1)}(\vec{\rho}_1, \vec{\rho}_2) G^{(1)}(\vec{\rho}_2, \vec{\rho}_3) G^{(1)}(\vec{\rho}_3, \vec{\rho}_1)$$
$$+ G^{(1)}(\vec{\rho}_2, \vec{\rho}_1) G^{(1)}(\vec{\rho}_3, \vec{\rho}_2) G^{(1)}(\vec{\rho}_1, \vec{\rho}_3) \qquad (8.9.15)$$

where

$$G^{(1)}_{jk} = |\alpha_0|^2 \int d\vec{\rho}_0 \, g^*_{\vec{\rho}_0}(\vec{\rho}_j, z_j) g_{\vec{\rho}_0}(\vec{\rho}_k, z_k)]$$

where we have absorbed all constants into constant $|\alpha_0|^2$.

Completing the integral on $\vec{\rho}_0$ for each $\Gamma^{(1)}_{jk}$ and $G^{(1)}_{jk}$ and taking $z_j = z_k = d$, similar to the near-field second-order spatial coherence, the normalized third-order coherence function $\gamma^{(3)}(\vec{\rho}_1, \vec{\rho}_2, \vec{\rho}_3)$ and $g^{(3)}(\vec{\rho}_1, \vec{\rho}_2, \vec{\rho}_3)$, respectively, is then approximated as

$$g^{(3)}(\vec{\rho}_1, \vec{\rho}_2, \vec{\rho}_3)$$
$$\propto 1 + \mathrm{somb}^2 \left[\frac{D}{d} \frac{\pi}{\lambda} |\vec{\rho}_1 - \vec{\rho}_2| \right]$$
$$+ \mathrm{somb}^2 \left[\frac{D}{d} \frac{\pi}{\lambda} |\vec{\rho}_2 - \vec{\rho}_3| \right] + \mathrm{somb}^2 \left[\frac{D}{d} \frac{\pi}{\lambda} |\vec{\rho}_3 - \vec{\rho}_1| \right]$$
$$+ 2 \, \mathrm{somb} \left[\frac{D}{d} \frac{\pi}{\lambda} |\vec{\rho}_1 - \vec{\rho}_2| \right] \mathrm{somb} \left[\frac{D}{d} \frac{\pi}{\lambda} |\vec{\rho}_2 - \vec{\rho}_3| \right] \mathrm{somb} \left[\frac{D}{d} \frac{\pi}{\lambda} |\vec{\rho}_3 - \vec{\rho}_1| \right], \qquad (8.9.16)$$

where $D/d = \Delta\theta$ is the angular size of the source from the view points of the photodetectors.

In 1-D, the normalized third-order coherence function $\gamma^{(3)}(x_1, x_2, x_3)$ or $g^{(3)}(x_1, x_2, x_3)$ is approximated to be

$$g^{(3)}(x_1, x_2, x_3)$$
$$\propto 1 + \mathrm{sinc}^2 \left[\frac{D}{d} \frac{\pi}{\lambda} (x_1 - x_2) \right]$$
$$+ \mathrm{sinc}^2 \left[\frac{D}{d} \frac{\pi}{\lambda} (x_2 - x_3) \right] + \mathrm{somb}^2 \left[\frac{D}{d} \frac{\pi}{\lambda} (x_3 - x_1) \right]$$
$$+ 2 \, \mathrm{sinc} \left[\frac{D}{d} \frac{\pi}{\lambda} (x_1 - x_2) \right] \mathrm{sinc} \left[\frac{D}{d} \frac{\pi}{\lambda} (x_2 - x_3) \right] \mathrm{sinc} \left[\frac{D}{d} \frac{\pi}{\lambda} (x_3 - x_1) \right]. \qquad (8.9.17)$$

For a large angular sized thermal source, the sinc-like functions effectively turn into δ-functions of $(x_j - x_k)$, $j = 1, 2, 3$, $k = 1, 2, 3$. The point-to-point-to-point nontrivial correlation between three transverse planes encourages three-photon lensless ghost imaging of thermal light with enhanced spatial resolution. The three-photon ghost imaging of thermal light exhibits a number of unusual interesting features that may be useful for certain applications.

(II) Nth-order coherence of thermal light.

The concept and calculation on thermal light third-order near-field spatial coherence are easily generalizable to higher-orders. The physics and mathematics may all start from Eq. (8.9.18) or Eq. (8.9.19), which is considered the key equation to understand the N-photon interference nature of the nontrivial Nth-order coherence of thermal field.

The Nth-order ($N \geq 2$) coherence of thermal light can be extended from $\Gamma^{(3)}(\mathbf{r}_1, t_1; \mathbf{r}_2, t_2; \mathbf{r}_3, t_3)$ to $\Gamma^{(N)}(\mathbf{r}_1, t_1; \dots \mathbf{r}_N, t_N)$,

$$
\begin{aligned}
\Gamma^{(N)} & (\mathbf{r}_1, t_1; \dots; \mathbf{r}_N, t_N) \\
& \equiv \langle E^*(\mathbf{r}_1, t_1) E(\mathbf{r}_1, t_1) \dots E^*(\mathbf{r}_N, t_N) E(\mathbf{r}_N, t_N) \rangle \\
& \simeq \sum_{a,b,c\dots} \left| \frac{1}{\sqrt{N!}} \left[\sum_{N!} E_{a1} E_{b2} E_{c3} \dots \right] \right|^2,
\end{aligned}
\tag{8.9.18}
$$

where $\sum_{N!} E_{a1} E_{b2} E_{c3} \dots$ means the sum of $N!$ terms of N-fold subfields, permuting in the orders of $1, 2, 3 \dots, N, 1, 3, 2 \dots, N, 2, 1, 3 \dots, N$, etc., corresponding to $N!$ alternative ways for the N independent subfields (photons), created from a large number of independent subsources that are labeled by $a, b, c \dots$, to produce a N-fold joint photodetection event between N independent photodetectors.

Replacing the subfields by the effective wavefunctions, the quantum Nth-order ($N \geq 2$) coherence of thermal light can be easily extend from $G^{(3)}(\mathbf{r}_1, t_1; \mathbf{r}_2, t_2; \mathbf{r}_3, t_3)$ to $G^{(N)}(\mathbf{r}_1, t_1; \dots \mathbf{r}_N, t_N)$,

$$
\begin{aligned}
G^{(N)} & (\mathbf{r}_1, t_1; \dots; \mathbf{r}_N, t_N) \\
& \equiv \langle \langle \hat{E}^{(-)}(\mathbf{r}_1, t_1) \dots \hat{E}^{(-)}(\mathbf{r}_N, t_N) \hat{E}^{(+)}(\mathbf{r}_N, t_N) \dots \hat{E}^{(+)}(\mathbf{r}_1, t_1) \rangle_{\text{QM}} \rangle_{\text{En}} \\
& \simeq \sum_{a,b,c\dots} \left| \frac{1}{\sqrt{N!}} \left[\sum_{N!} \psi_{a1} \psi_{b2} \psi_{c3} \dots \right] \right|^2,
\end{aligned}
\tag{8.9.19}
$$

(III) Nth-order coherence of coherent light.

Similar to the second-order coherence of coherent state, there should be no surprise that the degree of the Nth-order coherence light in coherent state is one. In fact, it is easy to show that the Nth-order coherence function of coherent light is factorizable into N independent first-order self-correlation functions (mean intensities), and consequently the normalized degree of Nth-order coherence of coherent radiation is

$$
\begin{aligned}
\gamma_{\text{coh}}^{(N)}(\mathbf{r}_1, t_1; \mathbf{r}_2, t_2 \dots; \mathbf{r}_N, t_N) &= 1 \\
g_{\text{coh}}^{(N)}(\mathbf{r}_1, t_1; \mathbf{r}_2, t_2 \dots; \mathbf{r}_N, t_N) &= 1.
\end{aligned}
\tag{8.9.20}
$$

SUMMARY

In this chapter, we introduced the concept of second-order coherence of light by introducing the Hanbury Brown and Twiss interferometer.

In the electromagnetic theory of light, second-order coherence of light is measured by two photodetectors D_1 and D_2 at a space-time coordinates $(\mathbf{r}_1, t_1)$ and $(\mathbf{r}_2, t_2)$ by means of the measurement of intensity correlation:

$$
\begin{aligned}
\langle I(\mathbf{r}_1, t_1) \, I(\mathbf{r}_2, t_2) \rangle &= \langle E^*(\mathbf{r}_1, t_1) E(\mathbf{r}_1, t_1) E^*(\mathbf{r}_2, t_2) E(\mathbf{r}_2, t_2) \rangle \\
&\equiv \Gamma^{(2)}(\mathbf{r}_1, t_1; \mathbf{r}_2, t_2)
\end{aligned}
$$

and the degree of second-order coherence is defined and calculated as

$$\gamma^{(2)}(\mathbf{r}_1, t_1; \mathbf{r}_2, t_2) = \frac{\langle I(\mathbf{r}_1, t_1) I(\mathbf{r}_2, t_2) \rangle}{\langle I(\mathbf{r}_1, t_1) \rangle \langle I(\mathbf{r}_2, t_2) \rangle}$$

$$= \frac{\langle E^*(\mathbf{r}_1, t_1) E(\mathbf{r}_1, t_1) E^*(\mathbf{r}_2, t_2) E(\mathbf{r}_2, t_2) \rangle}{\langle E^*(\mathbf{r}_1, t_1) E(\mathbf{r}_1, t_1) \rangle \langle E^*(\mathbf{r}_2, t_2) E(\mathbf{r}_2, t_2) \rangle}.$$

Very different from the first-order coherence, the second-order coherence is measured by two photodetectors, jointly and respectively, at space-time coordinates $(\mathbf{r}_1, t_1)$ and $(\mathbf{r}_2, t_2)$.

In the quantum theory of light, second-order coherence measures the probability of observing a joint photodetection event at space-time coordinates $(\mathbf{r}_1, t_1)$ and $(\mathbf{r}_2, t_2)$ of two photodetectors D_1 and D_2:

$$\langle n(\mathbf{r}_1, t_1) n(\mathbf{r}_2, t_2) \rangle \propto P(\mathbf{r}_1, t_1; \mathbf{r}_2, t_2)$$

$$\propto \left\langle \langle \hat{E}^{(-)}(\mathbf{r}_1, t_1) \hat{E}^{(-)}(\mathbf{r}_2, t_2) \hat{E}^{(+)}(\mathbf{r}_2, t_2) \hat{E}^{(+)}(\mathbf{r}_1, t_1) \rangle_{\text{QM}} \right\rangle_{\text{En}}$$

$$\equiv G^{(2)}((\mathbf{r}_1, t_1; \mathbf{r}_2, t_2))$$

which is recognized as the expectation of field operators in normal-order. Again, very different from first-order coherence, the second-order coherence is measured by two photodetectors, jointly and respectively, at space-time coordinates $(\mathbf{r}_1, t_1)$ and $(\mathbf{r}_2, t_2)$. The degree of second-order quantum coherence is defined and calculated as

$$g^{(2)}(\mathbf{r}_1, t_1; \mathbf{r}_2, t_2) = \frac{\langle n(\mathbf{r}_1, t_1) n(\mathbf{r}_2, t_2) \rangle}{\langle n(\mathbf{r}_1, t_1) \rangle \langle n(\mathbf{r}_2, t_2) \rangle}$$

$$= \frac{\langle \hat{E}^{(-)}(\mathbf{r}_1, t_1) \hat{E}^{(-)}(\mathbf{r}_2, t_2) \hat{E}^{(+)}(\mathbf{r}_2, t_2) \hat{E}^{(+)}(\mathbf{r}_1, t_1) \rangle}{\langle \hat{E}^{(-)}(\mathbf{r}_1, t_1) \hat{E}^{(+)}(\mathbf{r}_1, t_1) \rangle \langle \hat{E}^{(-)}(\mathbf{r}_2, t_2) \hat{E}^{(+)}(\mathbf{r}_2, t_2) \rangle}.$$

In the process of calculating $G^{(2)}(\mathbf{r}_1, t_1; \mathbf{r}_2, t_2)$ we introduced the concepts of effective wavefunction of a photon pair in product state, which is factorizable,

$$\Psi_{mn}(\mathbf{r}_1, t_1; \mathbf{r}_2, t_2) = \Psi_m(\mathbf{r}_1, t_1) \Psi_n(\mathbf{r}_2, t_2)$$

and in entangled state, which is non-factorizable, as well as in number state, which is non-factorizable as well

$$\Psi_{mn}(\mathbf{r}_1, t_1; \mathbf{r}_2, t_2) \neq \Psi_m(\mathbf{r}_1, t_1) \Psi_n(\mathbf{r}_2, t_2).$$

It is interesting to find that in the view of quantum theory of light, including Einstein's granularity theory of light, the second-order coherence or correlation, in general, is the results of two-photon interference: a pair of photons interfering with the pair itself. For example, the HBT correlation is the result of a randomly created and randomly paired photons interfering with the pair itself:

$$G^{(2)}(\mathbf{r}_1, t_1; \mathbf{r}_2, t_2) = \sum_{m,n} \left| \Psi_{mn}(\mathbf{r}_1, t_1; \mathbf{r}_2, t_2) + \Psi_{nm}(\mathbf{r}_1, t_1; \mathbf{r}_2, t_2) \right|^2.$$

This chapter put a considerable effort to distinguish the two-photon interference nature of the nontrivial second-order correlation of thermal field from that of the thermal source produced intensity fluctuations. In this chapter, a few detailed analysis on temporal and spatial second-order coherence are given, including thermal state, coherence state, entangled state, and number state. These exercises are helpful in understanding the physics of the second-order interference phenomena. The nontrivial second-order correlation of light is the result of a pair of photons, or a paired groups of identical photons, interfering with the pair itself.

In the last part of this chapter, we generalized the second-order coherence to Nth-order

$$\Gamma^{(N)}(\mathbf{r_1}, t_1; \ldots; \mathbf{r_N}, t_N)$$
$$\equiv \langle E^*(\mathbf{r_1}, t_1) E(\mathbf{r_1}, t_1) \ldots E^*(\mathbf{r_N}, t_N) E(\mathbf{r_N}, t_N) \rangle,$$

and

$$G^{(N)}(\mathbf{r_1}, t_1; \ldots; \mathbf{r_N}, t_N)$$
$$\equiv \langle \hat{E}^{(-)}(\mathbf{r_1}, t_1) \ldots \hat{E}^{(-)}(\mathbf{r_N}, t_N) \hat{E}^{(+)}(\mathbf{r_N}, t_N) \ldots \hat{E}^{(+)}(\mathbf{r_1}, t_1) \rangle,$$

and extended the interference picture from a pair of photons (subfields)in thermal state to N-photons (subfields) in thermal state

$$\Gamma^{(N)}(\mathbf{r_1}, t_1; \ldots; \mathbf{r_N}, t_N) \simeq \sum_{j,k,l\ldots} \left| \frac{1}{\sqrt{N!}} \left[\sum_{N!} E_{j1} E_{k2} E_{l3} \ldots \right] \right|^2,$$

and

$$G^{(N)}(\mathbf{r_1}, t_1; \ldots; \mathbf{r_N}, t_N) \simeq \sum_{j,k,l\ldots} \left| \frac{1}{\sqrt{N!}} \left[\sum_{N!} \Psi_{j1} \Psi_{k2} \Psi_{l3} \ldots \right] \right|^2,$$

where $\sum_{N!} E_{j1} E_{k2} E_{l3} \ldots$ and $\sum_{N!} \Psi_{j1} \Psi_{k2} \Psi_{l3} \ldots$ mean the sum of $N!$ terms of N-fold subfields and effective wavefunctions, permuted in the orders of $1, 2, 3 \ldots, N$, $1, 3, 2 \ldots, N$, $2, 1, 3 \ldots, N$, etc., corresponding to $N!$ alternative ways for N independent subfields or photons to produce a N-fold joint photodetection event.

REFERENCES AND SUGGESTIONS FOR READING

[1] R. Hanbury-Brown, and R.Q. Twiss, Nature, **177**, 27 (1956); **178**, 1046, (1956); **178**, 1447 (1956).

[2] R. Hanbury-Brown, *Intensity Interferometer*, Taylor and Francis Ltd, London, 1974.

[3] U. Fano, Am J. Phys., **29**, 539 (1961).

[5] R.J. Glauber, *Quantum Optics*, Eds. Kay and A. Maitland, Academic Press, New York, 1970.

[6] W. Martienssen and E. Spiller, Am. J. Phys. **32**, 919 (1964).

[7] A. Valencia, G. Scarcelli, M. D'Angelo, and Y.H. Shih, Phys. Rev. Lett. **94**, 063601 (2005).

[8] G. Scarcelli, V. Berardi, and Y.H. Shih, Phys. Rev. Lett. **96**, 063602 (2006).

[9] J. Liu and Y.H. Shih, Phys. Rev. A **79**, 203819 (2009).

[10] Y. Zhou, J. Simon, J. Liu, and Y.H. Shih, Phys. Rev. A, (2010).

[11] R. Loudon, *The Quantum Theory of Light*, Oxford Science, 2000.

[12] M.O. Scully and M.S. Zubairy, *Quantum Optics*, Cambridge, 1997.

CHAPTER 9

Quantum Entanglement

In quantum theory, *a particle* is allowed to be in a state of coherent superposition among a set of orthogonal states. A vivid picture of this concept is the "Schrödinger cat": *a cat* is alive and dead simultaneously. This would be no surprise if one means a large number of an ensemble of cats with 50% alive and 50% dead. But, "Schrödinger cat" is *a cat*. We are talking about *a cat* being alive and dead simultaneously. In mathematics, the best word to characterize the relationship between states "alive" and "dead" is perhaps "orthogonal". In quantum mechanics, the superposition of orthogonal states is used to describe the state of a quantum object, or a particle. The superposition principle is indeed a mystery compared to our everyday classical life experience.

In this chapter, we turn to another surprising consequence of quantum mechanics, namely, quantum entanglement. Quantum entanglement involves a multi-particle system and coherent superposition of orthogonal multi-particle states. The Schrödinger cat is perhaps still the best example to picture quantum entanglement. Now, we are talking about a pair of Schrödinger cats "propagating" to distant locations. The two cats are nonclassical by means of the following two features: (1) each of the cats is in the state of alive and dead simultaneously; (2) the two has to be observed both alive or both dead whenever we look at them, despite the distance between the two. There would be probably no surprise if our observation is based on a large number of twin cats that are prepared to be half alive-alive and another half dead-dead. In this case, obviously, we have 50% chance to observe an alive-alive pair and 50% chance to observe a dead-dead pair for each joint-observation. However, we are talking about *a pair* of cats, i.e., each pair of cats, to be in the state of alive-alive and dead-dead simultaneously, and, in addition, each of the cats in the pair must be alive and dead simultaneously. It seems impossible to have such cats in classical reality.

Beyond the superposition of orthogonal single-particle states, the superposition of orthogonal multi-particle states represents an even more troubling concept in classical theory. It is not only because the superposition of multi-particle states has no counterpart in classical theory, but also because it represents nonlocal behavior of particles which may never be explained classically.

9.1 EPR EXPERIMENT AND EPR STATE

The concept of quantum entanglement started in 1935. Einstein, Podolsky, and Rosen (EPR) suggested a *gedankenexperiment* and introduced an entangled two-particle system based on the superposition of two-particle wavefunction. The EPR system composes two distant interaction-free particles which are characterized by the following wavefunction:

$$\Psi(x_1, x_2) = \frac{1}{2\pi\hbar} \int dp_1 dp_2 \, \delta(p_1 + p_2) \, e^{ip_1(x_1 - x_0)/\hbar} e^{ip_2 x_2/\hbar}$$
$$= \delta(x_1 - x_2 - x_0) \tag{9.1.1}$$

where $e^{ip_1(x_1 - x_0)/\hbar}$ and $e^{ip_2 x_2/\hbar}$ are the eigenfunctions, with eigenvalues $p_1 = p$ and $p_2 = -p$, respectively, of the momentum operators $\hat{p}_1$ and $\hat{p}_2$ associated with particles 1 and 2; x_1 and x_2 are the coordinate variables to describe the positions of particles 1 and 2, respectively; and x_0 is a constant. The EPR state is very peculiar. Although there is no interaction between the two distant particles, the two-particle superposition cannot be factorized into a product of two individual superposition of two particles. Quantum theory does not prevent such states.

What can we learn from the EPR state of Eq. (9.1.1)?

(1) In coordinate representation, the wavefunction is a delta function: $\delta(x_1 - x_2 - x_0)$. The two particles are always separated in space with a constant value of $x_1 - x_2 = x_0$, although the coordinates x_1 and x_2 of the two particles are both unspecified.

(2) The delta wavefunction $\delta(x_1 - x_2 - x_0)$ is the result of the superposition of the plane wavefunctions of free particle one, $e^{ip_1(x_1 - x_0)/\hbar}$, and free particle two, $e^{ip_2 x_2/\hbar}$, with a particular distribution $\delta(p_1 + p_2)$. It is $\delta(p_1 + p_2)$ that made the superposition special: although the momentum of particle one and particle two may take on any values, the delta function restricts the superposition with only these terms in which the total momentum of the system takes a constant value of zero.

Now, we transfer the wavefunction from coordinate representation to momentum representation:

$$\Psi(p_1, p_2) = \frac{1}{2\pi\hbar} \int dx_1 dx_2 \, \delta(x_1 - x_2 - x_0) \, e^{-ip_1(x_1 - x_0)/\hbar} e^{-ip_2 x_2/\hbar}$$
$$= \delta(p_1 + p_2). \tag{9.1.2}$$

What can we learn from the EPR state of Eq. (9.1.2)?

(1) In momentum representation, the wavefunction is a delta function: $\delta(p_1 + p_2)$. The total momentum of the two-particle system takes a constant value of $p_1 + p_2 = 0$, although the momenta p_1 and p_2 are both unspecified.

(2) The delta wavefunction $\delta(p_1 + p_2)$ is the result of the superposition of the plane wavefunctions of free particle one, $e^{-ip_1(x_1 - x_0)/\hbar}$, and free particle two, $e^{-ip_2 x_2/\hbar}$, with a particular distribution $\delta(x_1 - x_2 - x_0)$. It is $\delta(x_1 - x_2 - x_0)$ that made the superposition special: although the coordinates of particle one and particle two may take on any values, the delta function restricts the superposition with only these terms in which $x_1 - x_2$ is a constant value of x_0.

In an EPR system, the value of the momentum (position) for neither single subsystem is determined. However, if one of the subsystems is measured to be at a certain momentum (position), the other one is determined with a unique corresponding value, despite the distance between them. An idealized EPR state of a two-particle system is therefore characterized by $\Delta(p_1 + p_2) = 0$ and $\Delta(x_1 - x_2) = 0$ simultaneously, even if the momentum and position of each individual free particle are completely undefined, i.e., $\Delta p_j \sim \infty$ and $\Delta x_j \sim \infty$, $j = 1, 2$. In other words, each of the subsystems may have completely random values or all possible values of momentum and position in the course of their motion, but the correlations of the two subsystems are determined with certainty whenever a joint measurement is performed.

The EPR states of Eq. (9.1.1) and Eq. (9.1.2) are simply the results of the quantum mechanical *superposition of two-particle states*. The physics behind EPR states is far beyond the acceptable limit of Einstein.

Does a free particle have a defined momentum and position in the state of Eq. (9.1.1) and Eq. (9.1.2), regardless of whether we measure it or not? On one hand, the momentum and position of neither independent particle is specified and the superposition is taken over all possible values of the momentum and position. We may have to believe that the particles do not have any defined momentum and position, or have all possible values of momentum and position within the superposition, during the course of their motion. On the other hand, if the measured momentum or position of one particle uniquely determines the momentum or position of the other distant particle, it would be hard for anyone who believes no action-at-a-distance to imagine that the momentum and position of the two particles are not predetermined with defined values before the measurement. EPR thus put us into a paradoxical situation. It seems reasonable for us to ask the same question that EPR had asked in 1935: "Can quantum-mechanical description of physical reality be considered complete?"

In their 1935 article, Einstein, Podolsky, and Rosen argued that the existence of the entangled two-particle state of Eq. (9.1.1) and Eq. (9.1.2), a straightforward quantum mechanical superposition of two-particle state, led to the violation of the uncertainty principle of quantum theory. To draw their conclusion, EPR started from the following criteria:

Locality: there is no action-at-a-distance;

Reality: "if, without in any way disturbing a system, we can predict with certainty the value of a physical quantity, then there exist an element of physical reality corresponding to this quantity". According to the delta wavefunctions, we can predict with certainty the outcome result of measuring the momentum (position) of particle 1 by measuring the momentum (position) of particle 2, and the measurement of particle 2 cannot cause any disturbance to particle 1, if the measurements are space-like separated events. Thus, the momentum and position of particle 1 must both be elements of physical reality regardless of whether we measure it or not. This, however, is not allowed by quantum theory. Now consider:

Completeness: "every element of the physical reality must have a counterpart in the complete theory". This led to the question as the title of their 1935 article: "Can Quantum-Mechanical Description of Physical Reality Be Considered Complete?"

EPR's arguments were never appreciated by Copenhagen. One objection was regarding EPR's criterion of physical reality: "it is too narrow", was a criticism from Bohr. It is perhaps not easy to find a wider criterion. A memorable quote from Wheeler, "No elementary quantum phenomenon is a phenomenon until it is a recorded phenomenon", summarizes what Copenhagen has been trying to tell us. By 1927, most physicists accepted the Copenhagen interpretation as the standard view of quantum formalism. Einstein, however, refused to compromise. As Pais recalled in his book, during a walk around 1950, Einstein suddenly stopped and "asked me if I really believed that the moon (pion) exists only if I look at it".

There has been arguments considering $\Delta(p_1 + p_2)\Delta(x_1 - x_2) = 0$ a violation of the uncertainty principle. This argument is false. It is easy to find that $p_1 + p_2$ and $x_1 - x_2$ are not conjugate variables. As we know, non-conjugate variables correspond to commuting operators in quantum mechanics, if the corresponding operators exist.[1] To have $\Delta(p_1+p_2) = 0$ and $\Delta(x_1 - x_2) = 0$ simultaneously, or to have $\Delta(p_1+p_2)\Delta(x_1-x_2) = 0$, is not a violation of the uncertainty principle. This point can be easily seen from the following two dimensional

---

[1] It is possible that a measurable physical variable has no quantum mechanical operator associated with, such as time t. From this perspective, an uncertainty relation based on variables rather than operators is more general. We should also remember that there is no position operator defined for a photon, however, this does not prevent us to discuss the uncertainty of observing a corresponding photodetection event in terms of its spatial coordinate or position Δx, and the uncertain relation of $\Delta x \Delta p \geq h$.

Fourier transform:

$$\Psi(x_1, x_2) = \frac{1}{2\pi\hbar} \int dp_1\, dp_2\, \delta(p_1 + p_2)\, e^{ip_1(x_1-x_0)/\hbar}\, e^{ip_2 x_2/\hbar}$$

$$= \frac{1}{2\pi\hbar} \int d(p_1 + p_2)\, \delta(p_1 + p_2)\, e^{i(p_1+p_2)(x_1'+x_2)/2\hbar}$$

$$\times \int d(p_1 - p_2)/2\, e^{i(p_1-p_2)(x_1'-x_2)/2\hbar}$$

$$= 1 \times \delta(x_1 - x_2 - x_0)$$

where $x' = x_1 - x_0$;

$$\Psi(p_1, p_2) = \frac{1}{2\pi\hbar} \int dx_1\, dx_2\, \delta(x_1 - x_2 - x_0)\, e^{-ip_1(x_1-x_0)/\hbar}\, e^{-ip_2 x_2/\hbar}$$

$$= \frac{1}{2\pi\hbar} \int d(x_1' + x_2)\, e^{-i(p_1+p_2)(x_1'+x_2)/2\hbar}$$

$$\times \int d(x_1' - x_2)/2\, \delta(x_1' - x_2)\, e^{-i(p_1-p_2)(x_1'-x_2)/2\hbar}$$

$$= \delta(p_1 + p_2) \times 1.$$

The Fourier conjugate variables are $(x_1 + x_2) \Leftrightarrow (p_1 + p_2)$ and $(x_1 - x_2) \Leftrightarrow (p_1 - p_2)$. Although it is possible to have $\Delta(x_1 - x_2) \sim 0$ and $\Delta(p_1 + p_2) \sim 0$ simultaneously, the uncertainty relations must hold for the Fourier conjugates $\Delta(x_1 + x_2)\Delta(p_1 + p_2) \geq \hbar$, and $\Delta(x_1 - x_2)\Delta(p_1 - p_2) \geq \hbar$; with $\Delta(p_1 - p_2) \sim \infty$ and $\Delta(x_1 + x_2) \sim \infty$.

In fact, in their 1935 paper, Einstein-Podolsky-Rosen never questioned $\Delta(x_1-x_2)\Delta(p_1+p_2) = 0$ as a violation of the uncertainty principle. The violation of the uncertainty principle was probably not Einstein's concern at all, although their 1935 paradox was based on the argument of the uncertainty principle. What really bothered Einstein so much? In all his life, Einstein, a true believer of realism, never accepted that a particle does not have a defined momentum and position during its motion, but rather is specified by probability amplitude of certain momentum and position. "God does not play dice" was the most vivid criticism from Einstein to refuse the Schrödinger cat. The entangled two-particle system was used as an example to clarify and to reinforce Einstein's realistic opinion. To Einstein, the acceptance of Schrödinger cat probably means action-at-a-distance or an inconsistency between quantum mechanics and the theory of relativity, when dealing with the entangled EPR two-particle system. Let us follow Copenhagen to consider that *each particle* in an EPR pair has no defined momentum and position, or has all possible momentum and position within the superposition, i.e., assuming $\Delta p_j \neq 0$, $\Delta x_j \neq 0$, $j = 1, 2$, for *each single-particle* until the measurement. Assume the measurement devices are particle counting devices for the measurement of position. For each registration of a particle, the measurement device records a value of its position. No one can predict what value is registered for each measurement; the best knowledge we may have is the probability to register that value. If we further assume no physical interaction between the two distant particles and believe no action-at-a-distance in nature, we would also believe that no matter how the two particles are created, the two registered values must be independent of each other. Thus, the value of $x_1 - x_2$ is unpredictable within the uncertainties of Δx_1 and Δx_2. The above statement is also valid for the momentum measurement. Therefore, after a set of measurements on a large number of particle pairs, the statistical uncertainty of the measurement on $p_1 + p_2$ and $x_1 - x_2$ must obey the following inequalities:

$$\Delta(p_1 + p_2) = \sqrt{(\Delta p_1)^2 + (\Delta p_2)^2} > Max(\Delta p_1, \Delta p_2)$$

$$\Delta(x_1 - x_2) = \sqrt{(\Delta x_1)^2 + (\Delta x_2)^2} > Max(\Delta x_1, \Delta x_2). \tag{9.1.3}$$

Eq. (9.1.3) is obviously true in statistics, especially when we are sure that no disturbance is possible between the two independent local measurements. The condition of "no disturbance is possible between the two independent local measurements " can be easily realized by making the two measurement events space-like separated events. The classical inequality of Eq. (9.1.3) would not allow $\Delta(p_1 + p_2) = 0$ and $\Delta(x_1 - x_2) = 0$ as required in the EPR state, unless $\Delta p_1 = 0$, $\Delta p_2 = 0$, $\Delta x_1 = 0$ and $\Delta x_2 = 0$, simultaneously. Unfortunately, the assumption of $\Delta p_1 = 0$, $\Delta p_2 = 0$, $\Delta x_1 = 0$, $\Delta x_2 = 0$ cannot be true because it violates the uncertainty relations $\Delta p_1 \Delta x_1 \geq \hbar$ and $\Delta p_2 \Delta x_2 \geq \hbar$.

In a non-perfect entangled system, the uncertainties of $p_1 + p_2$ and $x_1 - x_2$ may differ from zero. Nevertheless, the measurements may still satisfy the EPR inequalities:

$$\Delta(p_1 + p_2) < min(\Delta p_1, \Delta p_2)$$
$$\Delta(x_1 - x_2) < min(\Delta x_1, \Delta x_2). \tag{9.1.4}$$

The apparent contradiction between the classical inequality Eq. (9.1.3) and the EPR inequality Eq. (9.1.4) deeply troubled Einstein. While one sees the measurements of $p_1 + p_2$ and $x_1 - x_2$ of the two distant individual free particles satisfying Eq. (9.1.4), but believing Eq. (9.1.3), one might easily be trapped into concluding either there is a violation of the uncertainty principle or there exists action-at-a-distance.

Is it possible to have a "better" theory which provides correct predictions of the behavior of a particle similar to quantum theory and, at the same time, respects its description of physical reality by EPR as "complete"? The followers of Einstein's realism have tried in many different ways to formulate a realistic theory of quantum mechanics. It was Bohm who first attempted a version of a so called "hidden variable theory", which seemed to satisfy these requirements. The hidden variable theory was successfully applied to many different quantum phenomena until 1964, when Bell proved a theorem to show that an inequality, which is violated by certain quantum mechanical statistical predictions, can be used to distinguish local hidden variable theory from quantum mechanics. Since then, the testing of Bell's inequalities became a standard instrument for the study of fundamental problems of quantum theory. The experimental testing of Bell inequality started from the early 1970's. Most of the historical experiments concluded the violation of Bell inequality and thus disapproved the local hidden variable theory. A detailed discussion of Bell theorem and experimental testing of Bell inequality will be given later.

In the following discussion, we examine a set of simple yet popular classical models that attempted to simulate the behavior of the entangled EPR system. These models are based on the statistical behavior of an ensemble of particles instead of the superposition state of a particle. In other words, it is not based on the concept of Schrödinger cat (a cat is alive and dead simultaneously, or a pair of cats that is in the state of alive-alive and dead-dead simultaneously), but rather, a large number of twin-cats in which 50% are alive twins and 50% are dead twins.

We discuss three models in the following:

(1) In model one, each single pair of particles holds defined momenta $p_1 = $ constant and $p_2 = $ constant with $p_1 + p_2 = 0$. From pair to pair, the values of p_1 and p_2 may vary significantly. The sum of p_1 and p_2, however, keeps constant of zero. Thus, each joint detection of the two distant particles measures precisely the constant values of p_1 and p_2 and measures $p_1 + p_2 = 0$. The uncertainties of Δp_1 and Δp_2 only have statistical meaning in terms of the measurements of an ensemble. This model successfully simulated $\Delta p_1 \sim \infty$, $\Delta p_2 \sim \infty$, and $\Delta(p_1 + p_2) = 0$ based on the measurement of a large number of classically correlated particle pairs. This is, however, only half of the EPR story. Can we have $\Delta x_1 \sim \infty$, $\Delta x_2 \sim \infty$, and $\Delta(x_1 - x_2) = 0$ in this model? We do have $\Delta x_1 \sim \infty$ and $\Delta x_2 \sim \infty$, otherwise the uncertainty principle will be violated; the position correlation, however, can never achieve $\Delta(x_1 - x_2) = 0$ by any means.

(2) In model two, each single pair of particles holds defined position $x_1 = $ constant and $x_2 = $ constant with $x_1 - x_2 = x_0$. From pair to pair, the values of x_1 and x_2 may vary significantly. The difference of x_1 and x_2, however, keeps constant of x_0. Thus, each joint detection of the two distant particles measures precisely the constant values of x_1 and x_2 and measures $x_1 - x_2 = x_0$. The uncertainties of Δx_1 and Δx_2 only have statistical meaning in terms of the measurements of an ensemble. This model successfully simulated $\Delta x_1 \sim \infty$, $\Delta x_2 \sim \infty$, and $\Delta(x_1 - x_2) = 0$ based on the measurement of a large number of classically correlated particle pairs. This is, however, only half of the EPR story. Can we have $\Delta p_1 \sim \infty$, $\Delta p_2 \sim \infty$, and $\Delta(p_1 + p_2) = 0$ in this model? We do have $\Delta p_1 \sim \infty$ and $\Delta p_2 \sim \infty$, otherwise the uncertainty principle will be violated; the momentum correlation, however, can never achieve $\Delta(p_1 + p_2) = 0$ by any means.

The above two models of classically correlated particle pairs can never achieve both $\Delta(p_1 + p_2) = 0$ and $\Delta(x_1 - x_2) = 0$. What would happen if we combine the two parts together? This leads to the third model of classical simulation.

(3) In model three, among a large number of classically correlated particle pairs, we assume 50% in model one and the other 50% in model two. The $p_1 + p_2$ measurements would have 50% chance with $p_1 + p_2 = 0$ and 50% chance with $p_1 + p_2 = $ random value. On the other hand, the $x_1 - x_2$ measurements would have 50% chance with $x_1 - x_2 = x_0$ and 50% chance with $x_1 - x_2 = $ random value. What are the statistical uncertainties on the measurements of $(p_1 + p_2)$ and $(x_1 - x_2)$ in this case? If we focus on only these events of model one, the statistical uncertainty on the measurement of $(p_1 + p_2)$ is $\Delta(p_1 + p_2) = 0$, and if we focus on these events of model two, the statistical uncertainty on the measurement of $(x_1 - x_2)$ is $\Delta(x_1 - x_2) = 0$; however, if we consider all the measurements together, the statistical uncertainties on the measurements of $(p_1 + p_2)$ and $(x_1 - x_2)$, are both infinity: $\Delta(p_1 + p_2) = \infty$ and $\Delta(x_1 - x_2) = \infty$.

In conclusion, classically correlated particle pairs may partially simulate EPR correlation with three type of optimized observations:

(1) $\Delta(p_1 + p_2) = 0$ (100%) & $\Delta(x_1 - x_2) = \infty$ (100%);
$\Delta p_1 = 0, \Delta p_2 = 0; \Delta x_1 = \infty, \Delta x_2 = \infty$.

(2) $\Delta(x_1 - x_2) = 0$ (100%) & $\Delta(p_1 + p_2) = \infty$ (100%);
$\Delta x_1 = 0, \Delta x_2 = 0; \Delta p_1 = \infty, \Delta p_2 = \infty$.

(3) $\Delta(p_1 + p_2) = 0$ (50%) & $\Delta(x_1 - x_2) = 0$ (50%);
$\Delta p_1 = \infty, \Delta p_2 = \infty; \Delta x_1 = \infty, \Delta x_2 = \infty$.

Within one setup of experimental measurements, only the entangled EPR states result in the simultaneous observation of

$$\Delta(p_1 + p_2) = 0 \ (100\%) \ \& \ \Delta(x_1 - x_2) = 0 \ (100\%)$$
$$\Delta p_1 \sim \infty, \quad \Delta p_2 \sim \infty, \quad \Delta x_1 \sim \infty, \quad \Delta x_2 \sim \infty.$$

We thus have another tool, besides the testing of Bell inequality, to distinguish quantum entangled states from classically correlated particle pairs.

9.2 ENTANGLED STATE VS PRODUCT STATE AND CLASSICALLY COR-RELATED STATE

(1) Product state:

In general, a product state describes the behavior of a system composed of two or more independent sub-systems. Usually, the measurement of a product state involves the jointly-detection of two or more independent particles, such as the coincidence measures between two or more particle counting detectors within a certain time window. The experimental setup is similar to that of the EPR correlation measurement, however, the measurement only deals with independent particles. The independent particle system can be characterized as a product state, i.e., the state of the system is factorizable into a product of states of two or more sub-systems. For example

$$\hat{\rho} = \hat{\rho}_1 \times \hat{\rho}_2. \tag{9.2.1}$$

The density matrix $\hat{\rho}$ in Eq. (9.2.1) characterizes a quantum system composed of two independent sub-systems of $\hat{\rho}_1$ and $\hat{\rho}_1$. In Eq. (9.2.1), each independent sub-system can be in any state. The simplest case is a product of two pure states:

$$|\Psi\rangle = \sum_{a,b} f(a)g(b)\,|a\rangle\,|b\rangle = \sum_a f(a)\,|a\rangle \cdot \sum_b g(b)\,|b\rangle\,, \tag{9.2.2}$$

where $\{|a\rangle\}$ and $\{|b\rangle\}$ are two sets of orthogonal vectors for subsystems 1 and 2, respectively, and

$$\hat{\rho} = |\Psi\rangle\langle\Psi|, \quad \hat{\rho}_1 = \sum_{a,a'} f(a)f^*(a')|a\rangle\langle a'|, \quad \hat{\rho}_2 = \sum_{b,b'} f(b)f^*(b')|b\rangle\langle b'|.$$

(2) Entangled state:

Differing from product states, entangled states describe the behavior of entangled quantum systems. The entangled two-particle state was mathematically formulated by Schrödinger. Consider a pure state for a system composed of two spatially separated sub-systems,

$$\hat{\rho} = |\Psi\rangle\,\langle\Psi|\,, \quad |\Psi\rangle = \sum_{a,b} c(a,b)\,|a\rangle\,|b\rangle \tag{9.2.3}$$

where, again, $\{|a\rangle\}$ and $\{|b\rangle\}$ are two sets of orthogonal vectors for subsystems 1 and 2, respectively, and $\hat{\rho}$ is the density matrix. If $c(a,b)$ does not factor into a product of the form $f(a) \times g(b)$, then it follows that the state does not factor into a product state for subsystems 1 and 2:

$$\hat{\rho} \neq \hat{\rho}_1 \times \hat{\rho}_2 \tag{9.2.4}$$

the state was defined by Schrödinger as the entangled state.

Following this notation, the first classic example of a two-particle entangled state, the EPR state of Eq. (9.1.1) and Eq. (9.1.2), is thus written as:

$$|\Psi\rangle_{\text{EPR}} = \sum_{p_1,p_2} \delta(p_1 + p_2)\,|p_1\rangle|p_2\rangle$$

$$= \sum_{x_1,x_2} \delta(x_1 - x_2 + x_0)\,|x_1\rangle|x_2\rangle, \tag{9.2.5}$$

where we have described the EPR entangled system as the coherent superposition of the momentum eigenstates as well as the coherent superposition of the position eigenstates of each particle. One clear property of the EPR state is its independence of vector bases. The two δ-functions in Eq. (9.2.5) represent, respectively and simultaneously, the perfect position-position correlation and momentum-momentum correlation. Although the two distant particles are interaction-free, the superposition selects only the eigenstates which are specified by the δ-function. We may use the following statement to summarize the surprising feature of the EPR state: *the values of the momentum and the position for neither interaction-free single subsystem is determinate. However, if one of the subsystems is measured to be at a certain value of momentum and/or position, the momentum and/or position of the other one is 100% determined, despite the distance between them.*

It is necessary to emphasize that Eq. (9.2.5) is true, 'simultaneously, in the conjugate momentum and position space.

(3) Classically correlated state:

There exist many different types of classically correlated states. The following are two typical classically correlated states that have been used to simulate the EPR state of Eq. (9.2.5):

$$\hat{\rho} = \sum_{p_1,p_2} \delta(p_1 + p_2)\, |\, p_1\, \rangle |\, p_2\, \rangle \langle\, p_2\, |\langle\, p_1\, |, \tag{9.2.6}$$

or

$$\hat{\rho} = \sum_{x_1,x_2} \delta(x_1 - x_2 + x_0)\, |\, x_1\, \rangle |\, x_2\, \rangle \langle\, x_2\, |\langle\, x_1\, |. \tag{9.2.7}$$

Differing from entangled states, Eq. (9.2.6) and Eq. (9.2.7) cannot be simultaneously true for one system as we have discussed earlier.

(4) Entangled states in spin variables:

A different classic example of entangled two-particle system was suggested by Bohm. Instead of using continuous space-time variables, Bohm simplified the entangled two-particle state to discrete spin variables. EPR-Bohm state is a singlet state of two spin 1/2 particles:

$$|\Psi\rangle = \frac{1}{\sqrt{2}}\left[\, |\uparrow\rangle_1 |\downarrow\rangle_2 - |\downarrow\rangle_1 |\uparrow\rangle_2\, \right] \tag{9.2.8}$$

where the kets $|\uparrow\rangle$ and $|\downarrow\rangle$ represent states of spin "up" and spin "down", respectively, along an *arbitrary* direction. Again, for this state, *the spin of neither particle is determined; however, if one particle is measured to be spin up along a certain direction, the other one must be spin down along that direction, despite the distance between the two spin 1/2 particles.* Similar to the original EPR state, Eq. (9.2.8) is independent of the choice of the spin directions and the eigenstates of the associated non-commuting spin operators. It is easy to show that Eq. (9.2.8) is true, simultaneously, in the three orthogonal spin representations:

$$\begin{aligned}
|\Psi\rangle &= \frac{1}{\sqrt{2}}\left[\, |\, \hat{x}^+\, \rangle_1 |\, \hat{x}^-\, \rangle_2 - |\, \hat{x}^-\, \rangle_1 |\, \hat{x}^+\, \rangle_2\, \right] \\
&= \frac{1}{\sqrt{2}}\left[\, |\, \hat{y}^+\, \rangle_1 |\, \hat{y}^-\, \rangle_2 - |\, \hat{y}^-\, \rangle_1 |\, \hat{y}^+\, \rangle_2\, \right] \\
&= \frac{1}{\sqrt{2}}\left[\, |\, \hat{z}^+\, \rangle_1 |\, \hat{z}^-\, \rangle_2 - |\, \hat{z}^-\, \rangle_1 |\, \hat{z}^+\, \rangle_2\, \right]. \tag{9.2.9}
\end{aligned}$$

The above two equations are very different from the classically correlated state. It is easy to show that classically correlated states are coordinate dependent. The state does not hold the same form if choosing the other orthogonal spin directions.

The most widely used entangled two-particle states might have been the "Bell states" (or EPR-Bohm-Bell states). Bell states are a set of polarization states for a pair of entangled photons. The four Bell states which form a complete orthonormal basis of two-photon state are usually represented as

$$|\Phi_{12}^{(\pm)}\rangle = \frac{1}{\sqrt{2}}\big[\,|\,0_1\,0_2\,\rangle \pm |\,1_1\,1_2\,\rangle\big],$$

$$|\Psi_{12}^{(\pm)}\rangle = \frac{1}{\sqrt{2}}\big[\,|\,0_1\,1_2\,\rangle \pm |\,1_1\,0_2\,\rangle\big] \tag{9.2.10}$$

where $|0\rangle$ and $|1\rangle$ represent two arbitrary orthogonal polarization bases, for example, $|\,0\,\rangle = |H\rangle$ (horizontal) and $|\,1\,\rangle = |V\rangle$ (vertical) linear polarization, respectively. We will have a detailed discussion on Bell states later.

9.3 ENTANGLED BIPHOTON STATE

The state of a signal-idler photon pair created in the nonlinear optical process of SPDC is a typical EPR state. Roughly speaking, the process of SPDC involves sending a pump laser beam into a nonlinear material, such as a non-centrosymmetric crystal. Occasionally, the nonlinear interaction leads to the annihilation of a high frequency pump photon and the simultaneous creation of a pair of lower frequency signal-idler photons into an entangled two-photon state:

$$|\Psi\rangle = \Psi_0 \sum_{s,i} \delta\,(\omega_s + \omega_i - \omega_p)\,\delta\,(\mathbf{k}_s + \mathbf{k}_i - \mathbf{k}_p)\,a_s^\dagger(\mathbf{k}_s)\,a_i^\dagger(\mathbf{k}_i)\,|\,0\rangle \tag{9.3.1}$$

where ω_j, $\mathbf{k}_j$ (j = s, i, p) are the frequency and wavevector of the signal (s), idler (i), and pump (p), $a_s^\dagger$ and $a_i^\dagger$ are creation operators for the signal and the idler photon, respectively, and Ψ_0 is a normalization constant. We have assumed a CW monochromatic laser pump, i.e., ω_p and $\mathbf{k}_p$ are considered as constants. The two delta functions in Eq. (9.3.1) are technically named as phase matching condition:

$$\omega_p = \omega_s + \omega_i, \qquad \mathbf{k}_p = \mathbf{k}_s + \mathbf{k}_i. \tag{9.3.2}$$

The names *signal* and *idler* are historical leftovers. The names probably came about due to the fact that in the early days of SPDC, most of the experiments were done with non-degenerate processes. One radiation was in the visible range (and thus easily detected, the signal), and the other was in IR range (usually not detected, the idler). We will see in the following discussions that the role of the idler is not less than that of the signal. The SPDC process is referred to as type-I if the signal and idler photons have identical polarizations, and type-II if they have orthogonal polarizations. The process is said to be *degenerate* if the SPDC photon pair have the same free space wavelength (e.g. $\lambda_i = \lambda_s = 2\lambda_p$), and *nondegenerate* otherwise. In general, the pair exit the crystal *non-collinearly*, that is, propagate to different directions defined by the second equation in Eq. (9.3.2) and the Snell's law. Of course, the pair may also exit *collinearly*, in the same direction, together with the pump.

The state of the signal-idler pair can be derived, quantum mechanically, by the first order perturbation theory with the help of the nonlinear interaction Hamiltonian. The SPDC interaction arises in a nonlinear crystal driven by a pump laser beam. The polarization, i.e., the dipole moment per unit volume, is given by

$$P_i = \chi_{i,j}^{(1)} E_j + \chi_{i,j,k}^{(2)} E_j E_k + \chi_{i,j,k,l}^{(3)} E_j E_k E_l + ... \tag{9.3.3}$$

where $\chi^{(m)}$ is the *mth* order electrical susceptibility tensor. In SPDC, it is the second order nonlinear susceptibility $\chi^{(2)}$ that plays the role. The second order nonlinear interaction Hamiltonian can be written as

$$H = \epsilon_0 \int_V d\mathbf{r} \, \chi^{(2)}_{ijk} \, E_i E_j E_k \tag{9.3.4}$$

where the integral is taken over the interaction volume V.

It is convenient to use the Fourier representation for the electrical fields in Eq. (9.3.4):

$$\mathbf{E}(\mathbf{r}, t) = \int d\mathbf{k} \, [\, \mathbf{E}^{(-)}(\mathbf{k})e^{-i(\omega(\mathbf{k})t - \mathbf{k}\cdot\mathbf{r})} + \mathbf{E}^{(+)}(\mathbf{k})e^{i(\omega(\mathbf{k})t - \mathbf{k}\cdot\mathbf{r})} \,]. \tag{9.3.5}$$

Substituting Eq. (9.3.5) into Eq. (9.3.4) and keeping only the terms of interest, we obtain the SPDC Hamiltonian in the interaction representation:

$$\hat{H}_{\text{int}}(t) = \epsilon_0 \int_V d\mathbf{r} \int d\mathbf{k}_s \, d\mathbf{k}_i \, \chi^{(2)}_{lmn} \hat{E}^{(+)}_{p\,l} e^{i(\omega_p t - \mathbf{k}_p\cdot\mathbf{r})}$$
$$\times \hat{E}^{(-)}_{s\,m} e^{-i(\omega_s(\mathbf{k}_s)t - \mathbf{k}_s\cdot\mathbf{r})} \hat{E}^{(-)}_{i\,n} e^{-i(\omega_i(\mathbf{k}_i)t - \mathbf{k}_i\cdot\mathbf{r})} + h.c., \tag{9.3.6}$$

where $h.c.$ stands for Hermitian conjugate. To simplify the calculation, we have also assumed the pump field to be plane and monochromatic with wave vector $\mathbf{k}_p$ and frequency ω_p.

It is easily noticeable that in Eq. (9.3.6), the volume integration can be done for some simplified cases. At this point, we assume that V is infinitely large. Later, we will see that the finite size of V in longitudinal and/or transversal directions may have to be taken into account. For an infinite volume V, the interaction Hamiltonian Eq. (9.3.6) is written as

$$\hat{H}_{\text{int}}(t) = \epsilon_0 \int d\mathbf{k}_s \, d\mathbf{k}_i \, \chi^{(2)}_{lmn} \, \hat{E}^{(+)}_{p\,l} \hat{E}^{(-)}_{s\,m} \hat{E}^{(-)}_{i\,n}$$
$$\times \delta(\mathbf{k}_p - \mathbf{k}_s - \mathbf{k}_i)e^{i(\omega_p - \omega_s(\mathbf{k}_s) - \omega_i(\mathbf{k}_i))t} + h.c. \tag{9.3.7}$$

It is reasonable to consider the pump field classical, which is usually a laser beam, and quantize the signal and idler fields, which are both in single-photon level:

$$\hat{E}^{(-)}(\mathbf{k}) = -i\sqrt{\frac{2\pi\hbar\omega}{V}}a^\dagger(\mathbf{k}),$$
$$\hat{E}^{(+)}(\mathbf{k}) = i\sqrt{\frac{2\pi\hbar\omega}{V}}a(\mathbf{k}), \tag{9.3.8}$$

where $a^\dagger(\mathbf{k})$ and $a(\mathbf{k})$ are photon creation and annihilation operators, respectively. The state of the emitted photon pair can be calculated by applying the first order perturbation

$$|\Psi\rangle = -\frac{i}{\hbar} \int dt \, H_{int}(t) \, |0\rangle. \tag{9.3.9}$$

By using vacuum $|0\rangle$ for the initial state in Eq. (9.3.9), we assume that there is no input radiation in any signal and idler modes, that is, we have a SPDC process.

Further assuming an infinite interaction time, evaluating the time integral in Eq. (9.3.9) and omitting altogether the constants and slow (square root) functions of ω, we obtain the *entangled* two-photon state of Eq. (9.3.1) in the form of integral:

$$|\Psi\rangle = \Psi_0 \int d\mathbf{k}_s d\mathbf{k}_i \, \delta[\omega_p - \omega_s(\mathbf{k}_s) - \omega_i(\mathbf{k}_i)]\delta(\mathbf{k}_p - \mathbf{k}_s - \mathbf{k}_i)a^\dagger_s(\mathbf{k}_s)a^\dagger_i(\mathbf{k}_i)|0\rangle \tag{9.3.10}$$

where Ψ_0 is a normalization constant which has absorbed all omitted constants. Eq. (9.3.10) has been used in chapter 3 for the calculation of second-order correlation function.

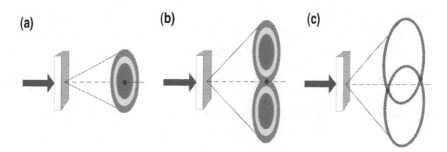

FIGURE 9.3.1 Three widely used SPDC setups. (a) Type-I SPDC. (b) Collinear degenerate type-II SPDC. Two rings overlap at one region. (c) Non-collinear degenerate type-II SPDC. For clarity, only two degenerate rings, one for e-polarization and the other for o-polarization, are shown.

The way of achieving phase matching, i.e., the way of achieving the delta functions in Eq. (9.3.10) basically determines how the signal-idler pair "looks". For example, in a negative uniaxial crystal, one can use a linearly polarized pump laser beam as an extraordinary ray of the crystal to generate a signal-idler pair both polarized as the ordinary rays of the crystal, which is defined as type-I phase matching. One can alternatively generate a signal-idler pair with one ordinary polarized and another extraordinary polarized, which is defined as type II phase matching. Fig. 9.3.1 shows three examples of SPDC two-photon source. All three schemes have been widely used for different experimental purposes. Technical details can be found from text books and research references in nonlinear optics.

The two-photon state in the forms of Eq. (9.3.1) or Eq. (9.3.10) is a pure state, which describes the behavior of a signal-idler photon pair mathematically. Does the signal or the idler photon in the EPR state of Eq. (9.3.1) or Eq. (9.3.10) have a defined energy and momentum regardless of whether we measure it or not? Quantum mechanics answers: No! However, if one of the subsystems is measured with a certain energy and momentum, the other one is determined with certainty, despite the distance between them.

In the above calculation of the two-photon state we have approximated an infinite large volume of nonlinear interaction. For a finite volume of nonlinear interaction, we may write the state of the signal-idler photon pair in a more general form:

$$|\Psi\rangle = \int d\mathbf{k_s}\, d\mathbf{k_i}\, F(\mathbf{k}_s, \mathbf{k}_i)\, a_i^\dagger(\mathbf{k}_s)\, a_s^\dagger(\mathbf{k}_i)|0\rangle \qquad (9.3.11)$$

where

$$F(\mathbf{k}_s, \mathbf{k}_i) = \epsilon\, \delta(\omega_p - \omega_s - \omega_i)\, f(\Delta_z L)\, h_{tr}(\vec{\kappa}_1 + \vec{\kappa}_2)$$

$$f(\Delta_z L) = \int_L dz\, e^{-i(k_p - k_{sz} - k_{iz})z}$$

$$h_{tr}(\vec{\kappa}_1 + \vec{\kappa}_2) = \int_A d\vec{\rho}\, h_{tr}(\vec{\rho})\, e^{-i(\vec{\kappa}_s + \vec{\kappa}_i)\cdot\vec{\rho}}$$

$$\Delta_z = k_p - k_{sz} - k_{iz} \qquad (9.3.12)$$

where ϵ is named as parametric gain, ϵ is proportional to the second order electric susceptibility $\chi^{(2)}$ and is usually treated as a constant; L is the length of the nonlinear interaction; the integral in $\vec{\kappa}$ is evaluated over the cross section A of the nonlinear material illuminated by the pump, $\vec{\rho}$ is the transverse coordinate vector, $\vec{\kappa}_j$ (with $j = s, i$) is the transverse wavevector of the signal and idler, and $f(|\vec{\rho}|)$ is the transverse profile of the pump, which can be treated as a Gaussion in most of the experimental conditions. The functions $f(\Delta_z L)$ and $h_{tr}(\vec{\kappa}_1 + \vec{\kappa}_2)$ can be approximated as δ-functions for an infinitely long ($L \sim \infty$) and wide ($A \sim \infty$) nonlinear interaction region. The reason we have chosen the form of Eq. (9.3.12) is to separate the "longitudinal" and the "transverse" correlations. We will show that $\delta(\omega_p - \omega_s - \omega_i)$ and $f(\Delta_z L)$ together can be rewritten as a function of $\omega_s - \omega_i$. To simplify the mathematics, we assume near co-linearly SPDC. In this situation, $|\vec{\kappa}_{s,i}| \ll |\mathbf{k}_{s,i}|$.

Basically, function $f(\Delta_z L)$ determines the "longitudinal" space-time correlation. Finding the solution of the integral is straightforward:

$$f(\Delta_z L) = \int_0^L dz\, e^{-i(k_p - k_{sz} - k_{iz})z} \tag{9.3.13}$$
$$= e^{-i\Delta_z L/2}\operatorname{sinc}(\Delta_z L/2).$$

Now, we consider $f(\Delta_z L)$ with $\delta(\omega_p - \omega_s - \omega_i)$ together, and taking advantage of the δ-function in frequencies by introducing a detuning frequency ν to evaluate function $f(\Delta_z L)$:

$$\omega_s = \omega_s^0 + \nu$$
$$\omega_i = \omega_i^0 - \nu$$
$$\omega_p = \omega_s + \omega_i = \omega_s^0 + \omega_i^0. \tag{9.3.14}$$

The dispersion relation $k(\omega)$ allows us to express the wave numbers through the detuning frequency ν:

$$k_s \approx k(\omega_s^0) + \nu \frac{dk}{d\omega}\Big|_{\omega_s^0} = k(\omega_s^0) + \frac{\nu}{u_s},$$
$$k_i \approx k(\omega_i^0) - \nu \frac{dk}{d\omega}\Big|_{\omega_i^0} = k(\omega_i^0) - \frac{\nu}{u_i} \tag{9.3.15}$$

where u_s and u_i are group velocities for the signal and the idler, respectively. Now, we connect Δ_z with the detuning frequency ν:

$$\Delta_z = k_p - k_{sz} - k_{iz}$$
$$= k_p - \sqrt{(k_s)^2 - (\vec{\kappa}_s)^2} - \sqrt{(k_i)^2 - (\vec{\kappa}_i)^2}$$
$$\cong k_p - k_s - k_i + \frac{(\vec{\kappa}_s)^2}{2k_s} + \frac{(\vec{\kappa}_i)^2}{2k_i}$$
$$\cong k_p - k(\omega_s^0) - k(\omega_i^0) + \frac{\nu}{u_s} - \frac{\nu}{u_i} + \frac{(\vec{\kappa}_s)^2}{2k_s} + \frac{(\vec{\kappa}_i)^2}{2k_i}$$
$$\cong D\nu \tag{9.3.16}$$

where $D \equiv 1/u_s - 1/u_i$. We have also applied $k_p - k(\omega_s^0) - k(\omega_i^0) = 0$ and $|\vec{\kappa}_{s,i}| \ll |\mathbf{k}_{s,i}|$. The "longitudinal" wavevector correlation function is rewritten as a function of the detuning frequency ν: $f(\Delta_z L) \cong f(\nu DL)$. In addition to the above approximations, we have inexplicitly assumed the angular independence of the wavevector $k = n(\theta)\omega/c$. For type II SPDC, the refraction index of the extraordinary-ray depends on the angle between the wavevector and the optical axis and an additional term appears in the expansion. Making the approximation valid, we have restricted our calculation to near-collinear process. Thus, for a good approximation, in the near-collinear experimental setup:

$$\Delta_z L \cong \nu DL = (\omega_s - \omega_i)DL/2. \tag{9.3.17}$$

Type-I degenerate SPDC is a special case. Due to the fact that $u_s = u_i$, and hence,

$D = 0$, the expansion of $k(\omega)$ should be carried out up to the second order. Instead of (9.3.17), we have

$$\Delta_z L \cong -\nu^2 D' L = -(\omega_s - \omega_i)^2 D' L / 4 \tag{9.3.18}$$

where

$$D' \equiv \frac{d}{d\omega}\left(\frac{1}{u}\right)\Big|_{\omega^0}.$$

The two-photon state of the signal-idler pair is then approximated as

$$|\Psi\rangle = \int d\nu \, d\vec{\kappa}_s \, d\vec{\kappa}_i \, f(\nu) \, h_{tr}(\vec{\kappa}_s + \vec{\kappa}_i) \, a_s^\dagger(\omega_s^0 + \nu, \vec{\kappa}_s) \, a_i^\dagger(\omega_i^0 - \nu, \vec{\kappa}_i)|0\rangle \tag{9.3.19}$$

where the normalization constant has been absorbed into $f(\nu)$.

9.4   EPR CORRELATION OF ENTANGLED BIPHOTON SYSTEM

EPR state is pure state that characterize the behavior of an entangled biphoton system. The state is in the form of coherent superposition among a set of specially selected two-photon states. The EPR correlation, or the nontrivial second-order coherence $G^{(2)}(\mathbf{r}_1, t_1; \mathbf{r}_2, t_2)$, is the result of a nonlocal interference among this peculiar set of biphoton amplitudes, which specifies the probability for jointly detecting a photon pair at coordinates $(\mathbf{r}_1, t_1)$ and $(\mathbf{r}_2, t_2)$. In principle, one joint-detection event of an EPR pair involves the superposition of all biphoton amplitudes that are specified by the EPR state. One measurement of an EPR pair is able to provide all information about the interference. A question naturally comes in this regard: Can we then observe the EPR correlation from the measurement of one EPR pair? Generally speaking, in quantum mechanics we may never learn any meaningful physics from the measurement of one pair of particles. On one hand, in photon counting, the outcome of a measurement is just a *yes* (a count or a "click") or *no* (no count). In a joint measurement of two photon counting detectors, the outcome of *yes* means a *yes-yes* or a "click-click" joint-detection event. A joint-detection event of *yes* may occur at any pair of space-time coordinates $(\mathbf{r}_1, t_1; \mathbf{r}_2, t_2)$ for a measurement, but definitely cannot occur at all points of an interference pattern. On the other hand, quantum mechanics does not predict a precise coordinates $(\mathbf{r}_1, t_1)$ and $(\mathbf{r}_2, t_2)$ for a photon pair to appear. Rather, quantum mechanics predict the probability for a joint-detection event to occur at a space-time point $(\mathbf{r}_1, t_1$ and $\mathbf{r}_2, t_2)$. To learn the EPR correlation, an ensemble measurement on a large number of *identical* pairs are necessary, where *identical* means that all pairs which are involved in the measurement must be prepared in the same EPR state. The measurement of quantum coherence or correlation is typically statistical. Statistically, if the outcome of a joint measurement has its maximum probability for *yes* to occur at a certain set of values of physical observable or under a certain relationship between physical variables and 100% *no* otherwise, the measured quantum system is considered to have an EPR correlation on that observable. As a good example, EPR's *gedankenexperiment* suggested to us a system of particle pairs with perfect correlation $\delta(x_1 - x_2 + x_0)$ in position. To confirm the EPR correlation, we need to observe *yes* only when the positions of the two distant detectors satisfy $x_1 - x_2 = x_0$, and 100% *no* otherwise when $x_1 - x_2 \neq x_0$. To observe this peculiar EPR correlation, a realistic approach is to measure the joint-detection counting rate at each pair of chosen coordinates x_1 and x_2 by scanning all possible values of $x_1 - x_2$. In quantum optics, this means the measurement of the second-order correlation function of $G^{(2)}(\tau_1 - \tau_2)$ (longitudinal), and/or $G^{(2)}(\vec{\rho}_1 - \vec{\rho}_2)$ (transverse), where $\tau_j = t_j - z_j/c$, $j = 1, 2$, and $\vec{\rho}_j$ is the transverse coordinate of the *jth* point-like photon counting detector.

Now, we exam the second-order coherence function of the entangled signal-idler photon pair of SPDC

$$G^{(2)}(\mathbf{r}_1, t_1; \mathbf{r}_2, t_2) = \langle \hat{E}^{(-)}(\mathbf{r}_1, t_1)\hat{E}^{(-)}(\mathbf{r}_2, t_2)\hat{E}^{(+)}(\mathbf{r}_2, t_2)\hat{E}^{(+)}(\mathbf{r}_1, t_1)\rangle$$

where $\hat{E}^{(-)}$ and $\hat{E}^{(+)}$ are the negative-frequency and the positive-frequency field operators of the detection events at space-time points $(\mathbf{r}_1, t_1)$ and $(\mathbf{r}_2, t_2)$. For the entangled biphoton state of SPDC,

$$G^{(2)}(\mathbf{r}_1, t_1; \mathbf{r}_2, t_2) = |\langle 0 | \hat{E}^{(+)}(\mathbf{r}_2, t_2) \hat{E}^{(+)}(\mathbf{r}_1, t_1) | \Psi \rangle |^2$$
$$= |\Psi(\mathbf{r}_1, t_1; \mathbf{r}_2, t_2)|^2$$

where $|\Psi\rangle$ is the biphoton state, and $\Psi(\mathbf{r}_1, t_1; \mathbf{r}_2, t_2)$ is the biphoton effective wavefunction that we have discussed in Chapter 7. To evaluate $G^{(2)}(\mathbf{r}_1, t_1; \mathbf{r}_2, t_2)$ and $\Psi(\mathbf{r}_1, t_1; \mathbf{r}_2, t_2)$, we need to propagate the field operators from the source to the space-time coordinates $(\mathbf{r}_1, t_1)$ and $(\mathbf{r}_2, t_2)$.

In general, the field operator $\hat{E}^{(+)}(\mathbf{r}, t)$ at space-time point $(\mathbf{r}, t)$ can be written in terms of the Green's function which propagates each Fourier mode from space-time point $(\mathbf{r}_0, t_0)$ to $(\mathbf{r}, t)$:

$$\hat{E}^{(+)}(\mathbf{r}, t) = \sum_{\mathbf{k}} \hat{E}^{(+)}(\mathbf{k}, \mathbf{r}_0, t_0)\, g(\mathbf{k}, \mathbf{r} - \mathbf{r}_0, t - t_0). \tag{9.4.1}$$

To simplify the notation, we have assumed one polarization in Eq. (9.4.1). The Green's function $g(\mathbf{k}, \mathbf{r} - \mathbf{r}_0, t - t_0)$ is also named optical transfer function. For different experimental setup, $g(\mathbf{k}, \mathbf{r} - \mathbf{r}_0, t - t_0)$ can be quite different.

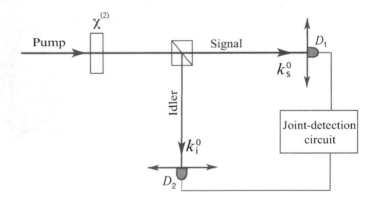

FIGURE 9.4.1 Simplified measurement of second-order temporal and spatial coherence or correlation for entangled biphoton state: the signal-idler photon pair are received by two point-like photodetectors D_1 (signal) and D_2 (idler), respectively and jointly. Assuming paraxial propagation of the signal-idler, the z_1 and z_2 are chosen along the central wavevector $\mathbf{k}_s^0$ and $\mathbf{k}_i^0$.

Considering an idealized simple experimental setup shown in Fig. 9.4.1, in which the signal-idler pair is received by two point-like photon counting detectors D_1 (signal) and D_2 (idler), respectively and jointly, for the measurements of the second-order temporal (longitudinal) coherence function $G^{(2)}(\tau_1 - \tau_2)$ and spatial (transverse) coherence function $G^{(2)}(\vec{\rho}_1 - \vec{\rho}_2)$. To simplify the mathematics, we further assume paraxial experimental condition. In the discussion of longitudinal and transverse correlation measurements, it is convenient to write the field $\hat{E}^{(+)}(\mathbf{r}_j, t_j)$ in terms of its longitudinal and transverse space-time variables under the Fresnel paraxial approximation:

$$\hat{E}^{(+)}(\vec{\rho}_j, z_j, t_j) \cong \int d\omega \, d\vec{\kappa} \; g(\vec{\kappa}, \omega; \vec{\rho}_j, z_j) \, e^{-i\omega t_j} \hat{a}(\omega, \vec{\kappa})$$

$$\cong \int d\omega \, d\vec{\kappa} \, \gamma(\vec{\kappa}, \omega; \vec{\rho}_j, z_j) \, e^{-i\omega \tau_j} \hat{a}(\omega, \vec{\kappa}) \qquad (9.4.2)$$

where $g(\vec{\kappa}, \omega; \vec{\rho}_j, z_j) = \gamma(\vec{\kappa}, \omega; \vec{\rho}_j, z_j) \, e^{i\omega z_j/c}$ is the spatial part of the Green's function, $\vec{\rho}_j$ and z_j are the transverse and longitudinal coordinates of the jth photodetector and $\vec{\kappa}$ is the transverse wavevector. We have chosen $z_0 = 0$ and $t_0 = 0$ at the output plane of the SPDC. For convenience, all constants associated with the field are absorbed into $g(\vec{\kappa}, \omega; \vec{\rho}_j, z_j)$.

The two-photon effective wavefunction $\Psi(\vec{\rho}_1, z_1, t_1; \vec{\rho}_2, z_2, t_2)$ is thus calculated as

$$\Psi(\vec{\rho}_1, z_1, t_1; \vec{\rho}_2, z_2, t_2)$$

$$= \langle 0| \int d\omega' \, d\vec{\kappa}' \; g(\vec{\kappa}', \omega'; \vec{\rho}_2, z_2) \, e^{-i\omega' t_2} \hat{a}(\omega', \vec{\kappa}')$$

$$\times \int d\omega'' \, d\vec{\kappa}'' \; g(\vec{\kappa}'', \omega''; \vec{\rho}_1, z_1) \, e^{-i\omega'' t_1} \hat{a}(\omega'', \vec{\kappa}'')$$

$$\times \int d\nu \, d\vec{\kappa}_s \, d\vec{\kappa}_i \, f(\nu) \, h_{tr}(\vec{\kappa}_s + \vec{\kappa}_i) \, a_s^\dagger(\omega_s^0 + \nu, \vec{\kappa}_s) \, a_i^\dagger(\omega_i^0 - \nu, \vec{\kappa}_i)|0\rangle$$

$$= e^{-i(\omega_s^0 \tau_1 + \omega_i^0 \tau_2)} \int d\nu \, d\vec{\kappa}_s \, d\vec{\kappa}_i \, f(\nu) \, h_{tr}(\vec{\kappa}_s + \vec{\kappa}_i)$$

$$\times \; e^{-i\nu(\tau_1 - \tau_2)} \, \gamma(\vec{\kappa}_s, \nu; \vec{\rho}_1, z_1) \, \gamma(\vec{\kappa}_i, -\nu; \vec{\rho}_2, z_2). \qquad (9.4.3)$$

Although Eq. (9.4.3) cannot be factorized into a product of longitudinal and transverse integrals, it is not difficult to calculate the temporal correlation and the transverse spatial correlation separately by choosing suitable experimental conditions.

Typical experiment may be designed for measuring either temporal (longitudinal) or spatial (transverse) correlation only. Thus, based on different experimental setups, we may simplify the calculation to either temporal (longitudinal) correlation:

$$\Psi(\tau_1; \tau_2) = \Psi_0 \, e^{-i(\omega_s^0 \tau_1 + \omega_i^0 \tau_2)} \int d\nu \, f(\nu) \, e^{-i\nu(\tau_1 - \tau_2)} \qquad (9.4.4)$$

or spatial (transverse) correlation:

$$\Psi(\vec{\rho}_1, z_1; \vec{\rho}_2, z_2)$$

$$= \Psi_0 \int d\vec{\kappa}_s \, d\vec{\kappa}_i \, h_{tr}(\vec{\kappa}_s + \vec{\kappa}_i) \, g(\vec{\kappa}_s, \omega_s; \vec{\rho}_1, z_1) \, g(\vec{\kappa}_i, \omega_i; \vec{\rho}_2, z_2). \qquad (9.4.5)$$

In certain experimental conditions, we may approximate $h_{tr}(\vec{\kappa}_s + \vec{\kappa}_i) \sim \delta(\vec{\kappa}_s + \vec{\kappa}_i)$.

(I): Biphoton temporal correlation.

To measure the biphoton temporal correlation of SPDC, we select a pair of transverse wavevector $\vec{\kappa}_s = -\vec{\kappa}_i$ in Eq. (9.4.3) by using appropriate optical apertures. The effective two-photon wavefunction is thus simplified to that of Eq. (9.4.4)

$$\Psi(\tau_1; \tau_2) \cong \Psi_0 \, e^{-i(\omega_s^0 \tau_1 + \omega_i^0 \tau_2)} \int d\nu \, f(\nu) \, e^{-i\nu(\tau_1 - \tau_2)}$$

$$= \left[\Psi_0 \, e^{-\frac{i}{2}(\omega_s^0 + \omega_i^0)(\tau_1 + \tau_2)} \right] \left[\mathcal{F}_{\tau_1 - \tau_2} \{f(\nu)\} \, e^{-\frac{i}{2}(\omega_s^0 - \omega_i^0)(\tau_1 - \tau_2)} \right] \qquad (9.4.6)$$

where, again, $\mathcal{F}_{\tau_1 - \tau_2} \{f(\nu)\}$ is the Fourier transform of the spectrum amplitude function

$f(\nu)$. Eq. (9.4.6) indicates a 2-D wavepacket: a δ-function like narrow envelope along axis $\tau_1 - \tau_2$ with constant amplitude in axis $\tau_1 + \tau_2$. For fixed positions of the photodetectors D_1 and D_2, the 2-D wavepacket means the following: the signal-idler pair may be jointly detected at any time; however, if the signal is registered at a certain time t_1, the idler must be registered at a unique time of $t_2 \sim t_1 - (z_1 - z_2)/c$. In other words, although the joint detection of the pair may happen at any times of t_1 and t_2 with equal probability ($\Delta(t_1+t_2) \sim \infty$), the registration time difference of the pair must happen within $\Delta(t_1-t_2) \sim 0$. A schematic of the wavepacket has been shown in Chapter 7.

As we have discussed in Chapter 8, the longitudinal correlation function $G^{(2)}(\tau_1 - \tau_2)$ is thus

$$G^{(2)}(\tau_1 - \tau_2) \propto |\mathcal{F}_{\tau_1-\tau_2}\{f(\nu)\}|^2,$$

which is a δ-function-like function in the case of SPDC. Thus, we have shown the entangled signal-idler photon pair of SPDC hold a typical EPR correlation in energy and time:

$$\Delta(\omega_s + \omega_i) \sim 0 \ \& \ \Delta(t_1 - t_2) \sim 0$$

with
$$\Delta\omega_s \sim \infty, \ \Delta\omega_i \sim \infty, \ \Delta t_1 \sim \infty, \ \Delta t_2 \sim \infty.$$

(II): Biphoton spatial correlation.

In the following, we study the spatial correlation $G^{(2)}(\vec{\rho}_1, z_1; \vec{\rho}_2, z_2)$ of entangled biphoton state in three different measurements: (1) in the output plane of the source; (2) in the Fraunhofer far-field; and (3) in the Fresnel near-field. Similar to that of the biphoton temporal correlation, we will calculate the effective biphoton wavefunction of the signal-idler pair under the above three conditions. To simplify the mathematics we will concentrate our calculation on the spatial function only by choosing a pair of monochromatic conjugate frequencies ω_s and ω_i. In this case, the effective two-photon wavefunction of Eq. (9.4.3) is simplified to that of Eq. (9.4.5):

$$\Psi(\vec{\rho}_1, z_1; \vec{\rho}_2, z_2) = \Psi_0 \int d\vec{\kappa}_s \, d\vec{\kappa}_i \, \delta(\vec{\kappa}_s + \vec{\kappa}_i) \, g(\vec{\kappa}_s, \omega_s, \vec{\rho}_1, z_1) \, g(\vec{\kappa}_i, \omega_i, \vec{\rho}_2, z_2)$$

where we have assumed $h_{tr}(\vec{\kappa}_s + \vec{\kappa}_i) \sim \delta(\vec{\kappa}_s + \vec{\kappa}_i)$, which is reasonable for a large transverse sized SPDC.

(1): $\Psi(\vec{\rho}_1, z_1; \vec{\rho}_2, z_2)$ in the output plane of SPDC.

The integral of $d\vec{\kappa}_s$ and $d\vec{\kappa}_i$ can be evaluated easily with the help of the EPR type transverse wavevector correlation $\delta(\vec{\kappa}_s + \vec{\kappa}_i)$:

$$\int d\vec{\kappa}_s \, d\vec{\kappa}_i \, \delta(\vec{\kappa}_s + \vec{\kappa}_i) \, e^{-i\vec{\kappa}_s \cdot \vec{\rho}_s} \, e^{-i\vec{\kappa}_i \cdot \vec{\rho}_i}$$

$$= \int d(\vec{\kappa}_s + \vec{\kappa}_i) \, \delta(\vec{\kappa}_s + \vec{\kappa}_i) \, e^{-i(\vec{\kappa}_s + \vec{\kappa}_i) \cdot (\vec{\rho}_s + \vec{\rho}_i)/2}$$

$$\times \int d(\vec{\kappa}_s - \vec{\kappa}_i) \, e^{-i(\vec{\kappa}_s - \vec{\kappa}_i) \cdot (\vec{\rho}_s - \vec{\rho}_i)/2}$$

$$\simeq 1 \times \delta(\vec{\rho}_s - \vec{\rho}_i), \tag{9.4.7}$$

where we have assumed a uniformly distributed signal-idler fields with incident (or output) angle $\theta_s \sim |\vec{\kappa}_s|/|\mathbf{k}_s|$ and $\theta_i \sim |\vec{\kappa}_i|/|\mathbf{k}_i|$ relative to the output plane: $A(\vec{\rho}_s) \sim e^{-i\vec{\kappa}_s \cdot \vec{\rho}_s}$ and $A(\vec{\rho}_i) \sim e^{-i\vec{\kappa}_i \cdot \vec{\rho}_i}$. Eq. (9.4.7) indicates an idealized 2-D wavepacket in terms of its transverse spatial coordinates: a normalized constant distribution along the axis of $(\vec{\rho}_s + \vec{\rho}_i)$, which is the Fourier transform of $\delta(\vec{\kappa}_s + \vec{\kappa}_i)$; and a δ-function along the axis of $(\vec{\rho}_s - \vec{\rho}_i)$, which is the Fourier transform of a normalized constant.

Thus, we have shown that the entangled signal-idler photon pair of SPDC holds a typical EPR correlation in transverse momentum and position, which is very close to the original proposal of EPR:

$$\Delta(\vec{\kappa}_s + \vec{\kappa}_i) \sim 0 \quad \& \quad \Delta(\vec{\rho}_s - \vec{\rho}_i) \sim 0$$
$$\text{with} \quad \Delta\vec{\kappa}_s \sim \infty, \quad \Delta\vec{\kappa}_i \sim \infty, \quad \Delta\vec{\rho}_s \sim \infty, \quad \Delta\vec{\rho}_i \sim \infty.$$

In EPR's language, we may never know where to locate the signal photon and the idler photon on the output plane of the source, however, if the signal (idler) is found at a certain position, the idler (signal) must be observed at a corresponding unique position. Simultaneously, the signal photon and the idler may have any transverse momentum, however, if the signal (idler) is measured with a transverse momentum, the transverse momentum of the idler (signal) is uniquely determined with certainty. In the *collinear* SPDC: the signal-idler pair is always emitted from the same point in the output plane of the two-photon source, $\vec{\rho}_s = \vec{\rho}_i$, and if one of them propagates slightly off from the collinear axes, the other one must propagate to the opposite direction with $\vec{\kappa}_s = -\vec{\kappa}_i$. This interesting behavior has been experimentally utilized in quantum imaging.

One may not have too much problem to face the above statement. The interaction of spontaneous parametric down-conversion is nevertheless a local phenomenon occurs inside the nonlinear material. The nonlinear interaction coherently creates mode in pairs that satisfy the phase match conditions of Eq. (9.3.2) which is also named as the energy conservation and momentum conservation. The signal-idler photon pair can be excited to any of these coupled modes or in all of these coupled modes simultaneously, resulting in a particular two-photon superposition. It is this "selected" two-photon superposition made the signal-idler pair comes out from the same point of the source and propagates to opposite directions with $\vec{\kappa}_s = -\vec{\kappa}_i$.[2]

(2): $\Psi(\vec{\rho}_1, z_1; \vec{\rho}_2, z_2)$ in the Fraunhofer far-field.

The two-photon superposition is getting more interesting when the signal-idler is propagated to a large distance, either by free propagation or guided by optical components such as lenses. In classical opinion, the signal photon and the idler photon are considered independent whenever they are released from the source because there is no interaction anymore between the pair in free space. Therefore, the signal photon and the idler photon should have independent and random distributions in terms of their transverse position $\vec{\rho}_1$ and $\vec{\rho}_2$. This classical picture, however, is incorrect. It is found that the signal-idler biphoton system would not lose its entangled nature in transverse position. This interesting behavior has been experimentally observed in quantum imaging, indicating an EPR type correlation of $\delta(\vec{\rho}_1 - \vec{\rho}_2)$. The sub-diffraction limit spatial resolution observed in the "quantum lithography" experiment and the nonlocal correlation observed in the "ghost imaging" experiment are both the results of two-photon superposition. Two-photon superposition does happen to a distant joint detection event of a signal-idler photon pair. There is no surprise one would have difficulties to face this effect. The two-photon superposition is a nonlocal concept in this case. The biphoton amplitudes in the two-photon superposition correspond to different

---

[2]Mathematically, we have approved $\vec{\kappa}_s \simeq -\vec{\kappa}_i$ and $\vec{\rho}_s \simeq \vec{\rho}_i$ in SPDC. Physically, this peculiar behavior distinguishes the entangled biphoton radiation from classical light. Applying $\vec{\kappa}_s = -\vec{\kappa}_i$ directly to the calculation, the integral of Eq. (9.4.7) can be simplified as

$$\int d\vec{\kappa}_s \, d\vec{\kappa}_i \, \delta(\vec{\kappa}_s + \vec{\kappa}_i) \, e^{-i\vec{\kappa}_s \cdot \vec{\rho}_s} \, e^{-i\vec{\kappa}_i \cdot \vec{\rho}_i} \simeq \int d\vec{\kappa}_s \, e^{-i\vec{\kappa}_s \cdot (\vec{\rho}_s - \vec{\rho}_i)} \simeq \delta(\vec{\rho}_s - \vec{\rho}_i),$$

and effectively gives $\vec{\rho}_s \simeq \vec{\rho}_i$. Physically and mathematically, based on $\vec{\kappa}_s \simeq -\vec{\kappa}_i$ and $\vec{\rho}_s \simeq \vec{\rho}_i$ we may find a better way to view the physics behind the observation and to simplify the calculation significantly.

yet indistinguishable alternative ways of triggering a joint photodetection event at distance. There is no counterpart of such concept in classical theory and may never be understood classically.

Now we consider propagating the signal-idler pair away from the source to the far-field observation points $(\vec{\rho}_1, z_1)$ and $(\vec{\rho}_2, z_2)$, respectively. To simplify the discussion, we place D_1 and D_2 on the plane of $z_1 = z_2$. In the far-field, either achieved by moving D_1 and D_2 to distances or by the use of a Fourier transform lens, we take a first-order approximation in the phase delay, i.e., $r \simeq r_0 - \vec{\kappa} \cdot \vec{\rho}_0$, where r is the distance from point $\vec{\rho}_0$ to D_1 or D_2, r_0 is the distance from the origin point of the coordinate system which is defined as the center point in the output plane of the SPDC, to D_1 or D_2. In this case, the effective biphoton wavefunction is approximated as:

$$\Psi(\vec{\rho}_1, z_1; \vec{\rho}_2, z_2) \simeq \langle 0 | \frac{e^{-i(\omega t_1 - k_1 r_{01})}}{r_{01}} \int d\vec{\rho}_0 \, A(\vec{\rho}_0) \, e^{-i\vec{\kappa}_1 \cdot \vec{\rho}_0} \, \hat{a}(\vec{\kappa}_1)$$

$$\times \frac{e^{-i(\omega t_2 - k_2 r_{02})}}{r_{02}} \int d\vec{\rho}'_0 \, A(\vec{\rho}'_0) \, e^{-i\vec{\kappa}_2 \cdot \vec{\rho}'_0} \, \hat{a}(\vec{\kappa}_2)$$

$$\times \int d\vec{\kappa}_s \, d\vec{\kappa}_i \, \delta(\vec{\kappa}_s + \vec{\kappa}_i) \, \hat{a}^\dagger(\vec{\kappa}_s) \, \hat{a}^\dagger(\vec{\kappa}_i) \, | 0 \rangle$$

$$\propto \int d\vec{\rho}_0 \, d\vec{\rho}'_0 \, \delta(\vec{\rho}_0 - \vec{\rho}'_0) e^{-i\vec{\kappa}_1 \cdot \vec{\rho}_0} \, e^{-\vec{\kappa}_2 \cdot \vec{\rho}'_0}$$

$$\propto 1 \times \delta(\vec{\rho}_1 + \vec{\rho}_2). \tag{9.4.8}$$

The δ-function of $\delta(\vec{\rho}_0 - \vec{\rho}'_0)$ in Eq. (9.4.8) is obtained from the double integral of $d\vec{\kappa}_s$ and $d\vec{\kappa}_i$ with $A(\vec{\rho}_0) \sim A_0 e^{-i\vec{\kappa}_s \cdot \vec{\rho}_0}$ and $A(\vec{\rho}'_0) \sim A_0 e^{-i\vec{\kappa}_i \cdot \vec{\rho}'_0}$ where A_0 is a constant. The physics is very clear in the above calculation. (1) The far-field plane is the Fourier transform plane of the SPDC. Each point on the Fourier transform plane plane corresponds to a $\vec{\kappa}_s$ or $\vec{\kappa}_i$. The δ-function $\delta(\vec{\rho}_1 + \vec{\rho}_2)$ on the Fourier transform plane confirms $\delta(\vec{\kappa}_s + \vec{\kappa}_i)$. (2) The 2-D wavepacket in transverse spatial coordinates is the result of a superposition among a large number of biphoton amplitudes. Each biphoton amplitude starts from a point on the output plane of SPDC (at $\vec{\rho}_0 = \vec{\rho}'_0$) and ends at point photodetectors D_1 and D_2. The 2-D wavepacket indicates a typical EPR correlation: we may observe the signal photon and the idler photon at any point on a far-field plane of the entangled biphoton source, however, if one of them is observed at a certain point the other one can only be observed at a unique point.

In the the above calculation, we have treated all integrals to infinity. This approximation yields δ-function for constant distributions. In reality, the finite size of the biphoton source or the applied lenses in the experimental setup may have to be taken into account. Let us assume a finite integral on a circular area from $|\vec{\rho}_0| = 0$ to $|\vec{\rho}_0| = R$ for degenerate SPDC ($\omega_s = \omega_i$), the integral of Eq. (9.4.8) turns

$$\Psi(\vec{\rho}_1, z_1; \vec{\rho}_2, z_2) \propto \int_A d\vec{\rho}_0 \, e^{-i(\vec{\kappa}_1 + \vec{\kappa}_2) \cdot \vec{\rho}_0} \simeq \frac{2 J_1 \big[R \, |\vec{\kappa}_1 + \vec{\kappa}_2| \big]}{R \, |\vec{\kappa}_1 + \vec{\kappa}_2|}$$

$$= \text{somb} \Big[\frac{R}{z} \frac{\omega}{c} (\vec{\rho}_1 + \vec{\rho}_2) \Big], \tag{9.4.9}$$

indicating a Fraunhofer diffraction pattern in the joint-detection of D_1 and D_2, which is a somb-function of $\vec{\rho}_1 + \vec{\rho}_2$. For large sized SPDC, nevertheless, δ-functions are good approximations.

(3): $\Psi(\vec{\rho}_1, z_1; \vec{\rho}_2, z_2)$ in the Fresnel near-field.

We now consider the Fresnel near-field measurement by moving D_1 and D_2 to the near-field of the biphoton source. We further assume the two-photon source has a finite but large transverse dimension. Under this simple experimental setup, the Green's function, or the optical transfer function describing arm-j, $j = 1, 2$, in which the signal and the idler freely propagate to photodetector D_1 and D_2, respectively. Substitute the $g_j(\omega, \vec{\kappa}; z_j, \vec{\rho}_j)$ of free propagation, $j = 1, 2$, into Eq. (9.4.5), the effective wavefunction is thus

$$
\begin{aligned}
\Psi(\vec{\rho}_1, &z_1; \vec{\rho}_2, z_2) \\
&= \Psi_0 \int d\vec{\kappa}_s \, d\vec{\kappa}_i \, \delta(\vec{\kappa}_s + \vec{\kappa}_i) \, e^{-i\vec{\kappa}_s \cdot \vec{\rho}_0} \, e^{-i\vec{\kappa}_i \cdot \vec{\rho}'_0} \\
&\quad \times \int_A d\vec{\rho}_0 \, \frac{-i\omega_s}{2\pi c z_1} \, e^{i\frac{\omega_s}{c} z_1} \, G(|\vec{\rho}_1 - \vec{\rho}_s|, \frac{\omega_s}{c z_1}) \\
&\quad \times \int_A d\vec{\rho}'_0 \, \frac{-i\omega_i}{2\pi c z_2} \, e^{i\frac{\omega_i}{c} z_2} \, G(|\vec{\rho}_2 - \vec{\rho}'_0|, \frac{\omega_i}{c z_2}) \\
&\simeq \frac{-\omega_s \omega_i}{2\pi^2 c^2 z_1 z_2} \, e^{i(\frac{\omega_s}{c} z_1 + \frac{\omega_i}{c} z_2)} \int_A d\vec{\rho}_0 \, e^{i[\frac{\omega_s}{2c z_1}|\vec{\rho}_1 - \vec{\rho}_0|^2 + \frac{\omega_i}{c z_2}|\vec{\rho}_2 - \vec{\rho}_0|^2]}
\end{aligned}
\tag{9.4.10}
$$

where $\vec{\rho}_0$ ($\vec{\kappa}_s$) and $\vec{\rho}'_0$ ($\vec{\kappa}_i$) are the transverse coordinates (wavevectors) for the signal and the idler fields, respectively, defined on the output plane of the biphoton source. The superposition of the above large number of biphoton amplitudes will produce a biphoton Fresnel diffraction pattern as a function of $\vec{\rho}_1$ and $\vec{\rho}_2$. Mathematically, it may not be easy to find an analytical solution for arbitrary $\vec{\rho}_1$ and $\vec{\rho}_2$, numerical solutions are always helpful for comparing with the experimental observation.

9.5 SUBSYSTEM IN AN ENTANGLED TWO-PHOTON STATE

The entangled EPR two-particle state is a pure state. The precise correlation of the subsystems is completely described by the state. The measurement, however, is not necessarily always for the two-particle system. It is an experimental choice to study only a subsystem and to ignore the other. What can we learn about a subsystem from these kinds of measurements? Mathematically, it is easy to show that by taking a partial trace of a two-particle pure state, the state of the subsystems are both in mixed states with entropy greater than zero. One can only learn statistical properties of the subsystems in this kind of measurement. In the following, again, we use the signal-idler pair of SPDC as an example to explore the physics.

(I) The state of the signal (or idler):

The biphoton state of SPDC is a pure state that satisfies

$$
\hat{\rho}^2 = \hat{\rho}, \quad \hat{\rho} \equiv |\Psi\rangle \langle \Psi|
\tag{9.5.1}
$$

where $\hat{\rho}$ is the density matrix operator corresponding to the biphoton state of SPDC. The single photon state of the signal and idler,

$$
\hat{\rho}_s = tr_i |\Psi\rangle \langle \Psi|, \quad \hat{\rho}_i = tr_s |\Psi\rangle \langle \Psi|,
\tag{9.5.2}
$$

are not. To calculate the signal (idler) state from the biphoton state, we have to take a partial trace, as usual, summing over the idler (signal) modes.

We assume a type II SPDC. The orthogonally polarized signal and idler are degenerate in frequency around $\omega^0 \cong \omega_p/2$. To simplify the discussion, by assuming appropriate experimental conditions, we trivialize the transverse part of the state and write the biphoton state in the following simplified form:

$$|\Psi\rangle = \Psi_0 \int d\nu \, \Phi(\text{DL}\nu) \, a_s^\dagger(\omega^0 + \nu) \, a_i^\dagger(\omega^0 - \nu) \, |0\rangle$$

where $\Phi(\text{DL}\nu)$ is a *sinc*-like function:

$$\Phi(\text{DL}\nu) = \frac{1 - e^{-i\text{DL}\nu}}{i\text{DL}\nu}$$

which is a function of the crystal length L, and the difference of inverse group velocities of the signal (ordinary) and the idler (extraordinary), $D \equiv 1/u_o - 1/u_e$. The constant Ψ_0 is calculated from the normalization $tr\,\hat{\rho} = \langle \Psi \mid \Psi \rangle = 1$. It is easy to calculate and to find $\hat{\rho}^2 = \hat{\rho}$ for the biphoton state of the signal-idler pair.

Summing over the idler modes, the density matrix of signal is given by

$$\hat{\rho}_s = \Psi_0^2 \int d\nu \, |\Phi(\nu)|^2 \, a_s^\dagger(\omega^0 + \nu) \, |0\rangle \, \langle 0| \, a_s(\omega^0 + \nu) \tag{9.5.3}$$

with

$$|\Phi(\nu)|^2 = \text{sinc}^2 \frac{\text{DL}\nu}{2} \tag{9.5.4}$$

where all constants coming from the integral have been absorbed into Ψ_0. First, we find immediately that $\hat{\rho}_s^2 \neq \hat{\rho}_s$. It means the state of the signal is a mixed state (as is the idler). Second, it is very interesting to find that the spectrum of the signal dependents on the group velocity of the idler. This, however, should not come as a surprise, because the state of the signal photon is calculated from the biphoton state by summing over the idler modes.

The spectrum of the signal and idler has been experimentally verified by Strekalov *et al* by using a Michelson interferometer in a standard Fourier spectroscopy type measurement. The measured interference pattern is shown in Fig. 9.5.1. The envelope of the sinusoidal modulations (in segments) is fitted very well by two "notch" functions (upper and lower part of the envelope). The experimental data agrees with the theoretical analysis of the experiment.

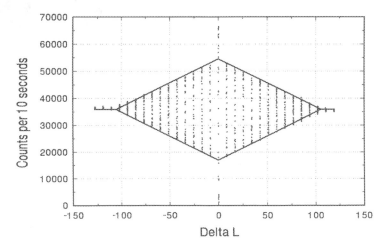

FIGURE 9.5.1 Experimental data indicated a "double notch" envelope. Each of the doted single vertical line contines many cycles of sinusoidal modulation.

The following is a simple calculation to explain the observed "notch" function. We first define the field operators:

$$\hat{E}^{(+)}(t, z_d) = \hat{E}_1^{(+)}\left(t - \frac{z_1}{c}, z_0\right) + \hat{E}_2^{(+)}\left(t - \frac{z_2}{c}, z_0\right)$$

where z_d is the position of the photodetector, z_0 is the input point of the interferometer, $t_1 = t - \frac{z_1}{c}$ and $t_2 = t - \frac{z_2}{c}$, respectively, are the early times before propagated to the photodetector at time t with time delays of z_1/c and z_2/c, where z_1 and z_2 are the optical paths in arm 1 and arm 2 of the interferometer. We have defined a very general field operator which is the superposition of two early fields propagated individually through arm 1 and arm 2 of any type of interferometer. The counting rate of the photon counting detector is thus

$$\begin{aligned}
R_d &= tr\left[\hat{\rho}_s \hat{E}^{(-)}(t, z_d)\hat{E}^{(+)}(t, z_d)\right] \\
&= \Psi_0^2 \int d\nu \, |\Phi(\nu)|^2 \, \Big|\langle 0| \Big[\hat{E}_1^{(+)}\left(t - \frac{z_1}{c}, z_0\right) \\
&\quad + \hat{E}_2^{(+)}\left(t - \frac{z_1}{c}, z_0\right)\Big] a_s^\dagger(\omega^0 + \nu) |0\rangle \Big|^2 \\
&\propto 1 + Re\left\{e^{-i\omega^0\tau} \int d\nu \, \mathrm{sinc}^2 \frac{\mathrm{DL}\nu}{2} \, e^{-i\nu\tau}\right\}
\end{aligned} \tag{9.5.5}$$

where $\tau = (z_1 - z_2)/c$. The Fourier transform of $\mathrm{sinc}^2(\mathrm{DL}\nu/2)$ has a "notch" shape. It is noticed that the base of the "notch" function is determined by parameter DL of the SPDC, which is easily confirmed from the experiment.

(II) The entropy problem:

Now we turn to another interesting aspect of physics, namely the physics of entropy. Entropy is an important concept in the information theory. The concept, named as Von Neuman entropy, is given by the state

$$S = -tr\left(\hat{\rho}\log\hat{\rho}\right). \tag{9.5.6}$$

It is easy to find that the entropy of the entangled two-photon pure state is zero. The entropy of its subsystems, however, are both greater than zero. The value of the Von Neuman entropy can be numerically evaluated from the measured spectrum. Note that the density operator of the subsystem is diagonal. Taking its trace is simply performing an integral over the frequency spectrum with the measured spectrum function. It is straightforward to find the entropy of the subsystems $S_s > 0$. This is an expected result due to the nature of the mixed state of the subsystems. Considering that the entropy of the two-photon system is zero and the entropy of the subsystems are both greater than zero, does this mean that negative entropy is present somewhere in the entangled two-photon system? According to classical information theory, for the entangled two-photon system, $S_s + S_{s|i} = 0$, where $S_{s|i}$ is the conditional entropy. It is this conditional entropy that must be negative, which means that *given the result of a measurement over one particle, the result of measurement over the other must yield negative information.* This paradoxical statement is similar and, in fact, closely related to the EPR "paradox". It comes from the same philosophy as that of the EPR.

9.6   BIPHOTON IN DISPERSIVE MEDIA

In this section we study the propagation of entangled photons in optical dispersive medium. Dispersion contains rich physics in general. It is not our goal to discuss in detail about biphoton dispersion, here, we give a simple analysis on the propagation of the entangled singnal-idler pair of a CW pumped SPDC in group dispersive optical fibers. As we have learned, although a biphoton wavepacket is defined with the two-photon state of SPDC, there is no wavepacket defined with the subsystems of the signal and the idler. After propagating the signal and the idler a certain distance in optical fibers, do we expect a "broadened" signal or idler photon? It is interesting to find that there is no wavepacket either broadened or unbroadened associated with the propagation of either signal or idler individually. It is the biphoton wavefunction $\psi(t_1, r_1; t_2, r_2)$ or the second-order correlation function $G^{(2)}(t_1, r_1; t_2, r_2)$ that is broadened in the dispersive medium. The biphoton wavepacket in dispersive medium propagates like a classical optical pulse.

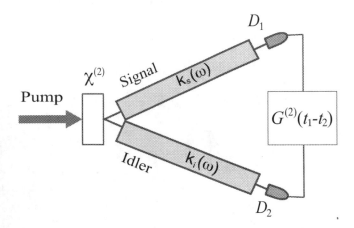

FIGURE 9.6.1 Broadening of $G^{(2)}(t_1 - t_2)$ due to the propagation of the signal and idler photon pair in group dispersive media.

Let us consider a simple experiment as shown in Fig. 9.6.1. We use a biphoton source of a CW laser pumped SPDC and two point-like photon counting detectors with a photon counting coincidence circuit to setup a standard second-order temporal correlation measurement. The photodetectors D_1 and D_2 are placed in the far-field zone of z_1 and z_2, respectively, for the detection of the signal and the idler photons. The photon-counting circuit records the registration time difference, $t_1 - t_2$, for each signal (registered at t_1)—idler (registered at t_2) pair. After measuring a large number of pairs, the circuit reports a histogram that summarizes how many total number of pairs (vertical axis) are measured with each value of $t_1 - t_2$ (horizontal axis). The measured histogram, number of joint detection counting rate against $t_1 - t_2$, corresponds to the seconder-order temporal correlation function $G^{(2)}(t_1 - t_2)$.

If we set up our experiment in the vacuum, or realistically in the air in which the dispersion can be ignored, and observed a certain function $G^{(2)}(t_1 - t_2)$ in the temporal correlation measurement. As we have learned earlier, the observed function $G^{(2)}(t_1 - t_2)$ is mainly determined by the phase matching condition of the SPDC. There is no contribution from the group broadening of either signal or idler propagation. Now, we put dispersive medium, such as optical fibers, between the two-photon source and the photodetectors D_1 and D_2. The dispersive medium has unignoreable first-order and second-order dispersion:

$$\frac{dk(\omega)}{d\omega}\Big|_{\omega_s^0} = k' \neq 0 \quad \text{and} \quad \frac{d^2k(\omega)}{d\omega^2}\Big|_{\omega_i^0} = k'' \neq 0$$

at the neighborhood of the wavelengths ω_s^0 and ω_i^0, where ω_s^0 and ω_i^0 are the central frequencies of the signal-idler radiation. The following calculation will answer the questions about what will happen to the $G^{(2)}(t_1 - t_2)$ function: Is the function $G^{(2)}(t_1 - t_2)$ broadened by the dispersive propagation of the signal and the idler?

The state of the signal-idler biphoton pair of SPDC, generated by a CW laser beam at frequency ω_p, can be written in the following simplified form as we have derived earlier:

$$|\Psi\rangle = \int_{-\infty}^{\infty} d\nu \, f(\nu) \, a_s^\dagger(\omega_s^0 + \nu) a_i^\dagger(\omega_i^0 - \nu) \, |0\rangle$$

where, again, ω_s^0 and ω_i^0 are the central frequencies of the signal-idler; ν is the returning frequency. The spectral amplitude, $f(\nu)$, provides all the information about the correlation properties of the two-photon light. $f(\nu)$ is basically determined by the longitudinal part of the phase matching in SPDC. Based on the simplified biphoton state and the field operators,

$$\hat{E}_1^{(+)} = \int d\omega_s \, a(\omega_s) \, e^{-i[\omega_s t_1 - k(\omega_s) z_1]}$$

$$\hat{E}_2^{(+)} = \int d\omega_i \, a(\omega_i) \, e^{-i[\omega_i t_2 - k(\omega_i) z_2]}$$

the biphoton wavepacket is calculated to be:

$$\psi(t_1, r_1; t_2, r_2) = \int d\omega_s f(\omega_s) e^{-i[\omega_s t_1 - k(\omega_s) z_1]} e^{-i[\omega_i t_2 - k(\omega_i) z_2]}. \tag{9.6.1}$$

Considering the first-order and the second-order dispersion of the propagation along z_1 and z_2, The biphoton wavepacket turns out to be

$$\psi(t_1, z_1; t_2, z_2) = \{ e^{-i\{(\omega_s^0 t_1 + \omega_i^0 t_2) - [k(\omega_s^0) z_1 + k(\omega_i^0) z_2]\}} \}$$

$$\times \int_{-\infty}^{\infty} d\nu \, f(\nu) \, e^{i(k_s'' z_1 + k_i'' z_2)(\nu^2/2)} \, e^{-i\nu[(t_1 - t_2) - (\frac{z_1}{u_s} - \frac{z_2}{u_i})]}$$

$$= \{ e^{-i\{(\omega_s^0 t_1 + \omega_i^0 t_2) - [k(\omega_s^0) z_1 + k(\omega_i^0) z_2]\}} \}$$

$$\times \mathcal{F}_\tau \{ f(\nu) e^{i(k_s'' z_1 + k_i'' z_2)(\nu^2/2)} \}, \tag{9.6.2}$$

where we have defined

$$\tau \equiv (t_1 - t_2) - [z_1/u_s - z_2/u_i]$$

$$= [t_1 - z_1/u_s] - [t_2 - z_2/u_i]$$

$$\equiv \tau_1 - \tau_2.$$

The second-order temporal coherence function $G^{(2)}(\tau)$ is thus

$$G^{(2)}(\tau) \propto \left| \mathcal{F}_\tau \{ f(\nu) e^{i(k_s'' z_1 + k_i'' z_2)(\nu^2/2)} \} \right|^2. \tag{9.6.3}$$

In Eq. (9.6.2) and Eq. (9.6.3), $\mathcal{F}_\tau \{ f(\nu) e^{i(k_s'' z_1 + k_i'' z_2)(\nu^2/2)} \}$ denotes as the Fourier transform of function $f(\nu) e^{i(k_s'' z_1 + k_i'' z_2)(\nu^2/2)}$. It is easy to find mathematically that the Gaussian function $e^{i(k_s'' z_1 + k_i'' z_2)\nu^2/2}$, corresponding to the second-order dispersion in both optical paths of the signal and the idler, contributes to the broadening of the biphoton wavepacket $\psi(t_1, z_1; t_2, z_2)$ as well as the second-order correlation function $G^{(2)}(\tau)$.

In summary, we have the following conclusions regarding to the propagation of biphoton wavepacket in dispersion media:

(1) The concept of broadening of a biphoton wavepacket is fundamentally different from the broadening of two classical wavepackets. The broadening of the second-order correlation

function in classical case, which can be measured by an optical auto-correlator, is due to the broadening of each measured individual pulses. In the measurement of entangled biphoton system, however, there is no wavepacket either broadened or unbroadened associated with the propagation of either signal or idler individually. It is the $G^{(2)}(\tau)$ function been broadened by the group dispersion media in 2-D space of $\tau = \tau_1 - \tau_2$. Comparing with a classical pulse propagating in group dispersion medium, we find that it is $G^{(2)}(\tau)$ in Eq. (9.6.3) plays the role of a pulse, except propagating in the 2-D space of $\tau = \tau_1 - \tau_2$. The second-order correlation function $G^{(2)}(t_1, 0; t_2, 0)$ observed at the output plane of SPDC has been calculated earlier

$$G^{(2)}(t_1 - t_2) \propto \left| \mathcal{F}_{t_1 - t_2} \{ f(\nu) \} \right|^2 \sim \delta(t_1 - t_2)$$

for a large sized $\chi^{(2)}$ material. The δ-function like second-order correlation function will be broadened by the dispersive media after a certain distance propagation, which is given in Eq. (9.6.3). The broadening process is similar to that of a laser pulse propagating in the same dispersive medium.[3]

(2) Nonlocal dispersion cancellation: one may have realized already an interesting aspect of the physics from Eq. (9.6.2). It is easy to see that any positive (negative) group-dispersion of the signal can be canceled nonlocally by applying a negative (positive) dispersive medium in the path of the idler by achieving

$$k_s'' z_1 + k_i'' z_2 = 0. \tag{9.6.4}$$

This interesting behavior has been named "nonlocal two-photon dispersion cancellation". No classical interpretation seems to be possible to explain this effect. Suppose we have two weak light pulses propagating through the positive and negative group-dispersion media channels, respectively, to reach D_1 and D_2 for second-order correlation measurement. In the view of classical physics, each pulse will be broadened locally and independently by the dispersion medium, and thus causing a statistically broadened temporal width of the second-order correlation function

$$\sigma_{12} \geq \sqrt{\sigma_1^2 + \sigma_2^2}, \tag{9.6.5}$$

where σ_{12}, σ_1, and σ_2 are the broadened widths of the correlation function and the two propagating pulses. Two-photon dispersion cancelation is useful in certain applications, such as nonlocal timing and positioning and distant clock synchronization, in which delta-function-like EPR correlation in $t_1 - t_2$ or in $z_1 - z_2$ is expected.

SUMMARY

In this chapter, we discussed the physics of quantum entanglement. Quantum entanglement has been considered as one of the most surprising consequences of quantum mechanics. We

---

[3]Behaving like classical pulse in optical fiber, when the length of a group-dispersive medium is greater than its "characteristic dispersion length", $z_{dis} = \Delta \tau_0^2 / 2\pi k''$, where $\Delta \tau_0$ is the unbroadened initial width of $G^{(2)}$, the second-order correlation function $G^{(2)}(\tau)$ is found to be

$$G^{(2)}(\tau) \sim \left| f\left(\frac{\tau}{k_s'' z_1 + k_i'' z_2} \right) \right|^2$$

i.e., achieving the same function as the spectrum of the SPDC. Furthermore, the width of $G^{(2)}(\tau)$, $\Delta\tau$, is proportional to the propagation distance along the dispersive medium

$$\Delta\tau = \Delta\tau_0 \frac{z}{z_{dis}}.$$

In the far-field-zone, i.e., $z > z_{dis}$, $\Psi(t_1, z_1; t_2, z_2)$ acquires a stable form which takes the same "shape" as the spectrum function, $f(\nu)$, and the width of $G^{(2)}(\tau)$ is linearly proportional to the length of the known dispersive medium. This effect has been observed experimentally by Valencia $et\ al.$.

have addressed three important issues in this chapter: (1) the surprising EPR correlation is the result of multi-particle (photon) interference; (2) the multi-particle (photon) superposition is nonlocal, which occurs at distant space-time coordinates and involves multi-particle (photon) detection events through the measurement of distant particle-detectors (photon-detectors); (3) the result of multi-particle interference of entangled EPR state is special and peculiar by means of its non-factorizable effective wavefunction, which means that the entangled particles may not be considered as independent, even if the particles are separated in large distances and are interaction-free.

In this chapter, we introduced the historical EPR *gedankenexperiment*, the EPR-Bohm spin 1/2 particle system

$$|\Psi\rangle = \frac{1}{\sqrt{2}} \left[\, |\uparrow\rangle_1 \, |\downarrow\rangle_2 - |\downarrow\rangle_1 \, |\uparrow\rangle_2 \, \right]$$

and the EPR-Bohm-Bell states

$$|\Phi_{12}^{(\pm)}\rangle = \frac{1}{\sqrt{2}} \left[\, | \, 0_1 \, 0_2 \, \rangle \pm | \, 1_1 \, 1_2 \, \rangle \right],$$

$$|\Psi_{12}^{(\pm)}\rangle = \frac{1}{\sqrt{2}} \left[\, | \, 0_1 \, 1_2 \, \rangle \pm | \, 1_1 \, 0_2 \, \rangle \right].$$

We analyzed a typical entangled biphoton state of SPDC

$$|\Psi\rangle = \Psi_0 \sum_{s,i} \delta \left(\omega_s + \omega_i - \omega_p \right) \delta \left(\mathbf{k}_s + \mathbf{k}_i - \mathbf{k}_p \right) a_s^\dagger(\mathbf{k}_s) \, a_i^\dagger(\mathbf{k}_i) \, | \, 0 \rangle.$$

We studied the generation of the biphoton state and calculated the second-order correlation and the effective wavefunction of biphoton in far-field and near-field. We have also discussed the behavior of the signal photon and the idler photon as sub-systems of the biphoton state. The biphoton state of SPDC is a simple yet a good example of entangled multi-particle system. These analysis are helpful in understanding the basic physics of quantum entanglement.

In this chapter, we addressed the fundamental concerns of Einstein-Podolsky-Rosen about quantum theory on *locality*, *reality*, and *completeness*.

In addition, we studied the subsystems of an entangled two-photon state with an interesting aspect of physics: "negative" entropy.

REFERENCES AND SUGGESTIONS FOR READING

[1] A. Einstein, B. Podolsky, and N. Rosen, Phys. Rev. **47**, 777 (1935).

[2] E.Schrödinger, Naturwissenschaften **23**, 807, 823, 844 (1935); translations appear in *Quantum Theory and Measurement*, ed. J.A. Wheeler and W.H. Zurek, Princeton University Press, New York, (1983).

[3] D. Bohm, *Quantum Theory*, Prentice Hall Inc., New York, (1951).

[4] Y.H. Shih, "Two-Photon Entanglement and Quantum Reality", *Advances in Atomic, Molecular, and Optical Physics*, ed., B. Bederson and H. Walther, Academic Press, Cambridge, 1997; "Entangled Photons", IEEE J. of Selected Topics in Quantum Electronics, **9**, No. 6, (2003).

[5] D.N. Klyshko, *Photon and Nonlinear Optics*, Gordon and Breach Science, New York, (1988); A. Yariv, *Quantum Electronics*, John Wiley and Sons, New York, (1989); R.W. Boyd, *Nonlinear Optics*, Academic Press, San Diego, 1992.

[6] M.H. Rubin, D.N. Klyshko, Y.H. Shih, and A.V. Sergienko, Phys. Rev. A, **50**, 5122 (1994).

[7] A.V. Sergienko, Y.H. Shih, and M.H. Rubin, JOSAB, **12**, 859 (1995).

[8] D.V. Strekalov, T.B. Pittman, and Y.H. Shih, Phys. Rev. A, **57**, 567 (1998).

[9] D.V. Strekalov, Y.H. Kim, and Y.H. Shih, Phys. Rev. A, Vol. 60, 2685 (1999).

[10] J.D. Franson, Phys. Rev. A, **45**, 3126 (1992); **80**, 032119 (2009); **81**, 023825 (2010).

[11] A. Valencia, M.V. Chekhova, A. Trifonov, and Y.H. Shih, Phys. Rev. Lett., **88**, 183601 (2002).

CHAPTER 10

Two-Photon Interferometry I: Biphoton Interference

Two-photon interferometry started to play an important role in quantum optics since 1980's. In that time, Dirac was criticized to be mistaken because he stated in his book, *The Principles of Quantum Mechanics*, that "...photon... only interferes with itself. Interference between two different photons never occurs". Two-photon interference was considered as the interference between two photons.

Is two-photon interference the interference of two photons? In this chapter, we will provide a negative answer to this question: No! Two-photon interference is not the interference between two photons. Two-photon interference is the result of the superposition of two-photon amplitudes, a nonclassical entity corresponding to different yet indistinguishable alternatives which lead to a joint photodetection event. In Dirac's language, "... a pair of photon only interferes with the pair itself ...".

In fact, neither classical theory nor quantum theory suggested the interference between different photons. Classical theory views optical interference the result of coherent superposition between electromagnetic waves. In quantum theory, the superposition occurs between quantum amplitudes, which corresponding to alternative ways of annihilating a photon in a photoelectron event, or annihilating a pair of photons in a joint-detection event between two individual photodetectors, or annihilating a group of N-photons in a N-fold joint-detection event between N individual photodetectors. Perhaps, the idea of interference between different photons came from the successful experimental observation of the first-order interference between radiations of independent light sources. The concept of interference between different photons was introduced in the middle of 1960's when Mandel demonstrated the interferences between two individual CW laser beams and between two laser pulses produced from two synchronized lasers.[1] Since then, interference between two photons became a hot and controversial topic in the study of optical interferometry. Before 1980's most experimental and theoretical studies focused on the first-order phenomena between coherent laser sources, which involve the measurement of a large number of identical photons. It is hard to distinguish the behavior of one photon, two photons, or a few photons from a bright laser beam or coherent electromagnet laser field, unless reducing the intensity to single-photon's level. However, at single-photon's level, on one hand, one "exposure" of two

---

[1](1) In the CW case, in order to obtain an observable interference pattern, the exposure time of the graphic film has to be less than the coherent time of the laser radiation. Although the observed interference patterns may differ from one exposure to another significantly, there is no doubt about interference in each individual exposure. (2) In the case of pulsed lasers, the two pulses have to be synchronized to overlap in time. The interference pattern may differ from pulse to pulse with random phases, the experiment of Mandel *et al.* observed clear interference pattern in each exposure of a pair of overlapped pulses.

individual photons cannot provide us an observable interference pattern either spatially or temporally. On the other hand, a statistical measurement based on a large number of photons averages out any possible observable interference pattern due to the random phases between "patterns" from one "exposure" to another "exposure". We may never be able to draw a definite conclusion to support or to reject the idea of interference of two photons.

The situation changed in the middle of 1980's since the observation of two-photon interference of an entangled photon pair of SPDC. In a two-photon interferometer, Alley and Shih successfully demonstrated the interference of an orthogonally polarized signal-idler photon pair and reported in 1986. In that experiment, the signal photon and the idler photon were prepared with well-defined orthogonal polarization, either in the linear $|X\rangle|Y\rangle$ base or in the circular $|R\rangle|L\rangle$ base. With the help of a beamsplitter the photon pair seemed lost their original polarization and exhibited a typical EPR-Bohm-Bell correlation with the violation of a Bell inequality. How cold a pair of well polarized photon turn into an entangled Bell state in which no polarization for either photon is specified? Alley and Shih provided an interpretation of the phenomenon as the result of a superposition between two different yet indistinguishable biphoton amplitudes.[2] In addition to the Bell-type polarization correlation, a two-photon correlation ($g^{(2)}(0) = 2$) and anti-correlation ($g^{(2)}(0) = 0$) were also observed with different chosen set of orientations of the polarization analyzers. The two-photon interference picture provided reasonable interpretation to that observation too: under the condition of complete overlapping, when the two biphoton amplitudes superpose in phase (with a "+" sign in between) the joint-detection of the two distant photodetectors reaches its maximum counting rate as the sign of "correlation", i.e., $g^{(2)}(0) = 2$; when the two biphoton amplitudes superpose out of phase (with a "−" sign in between) the joint-detection counting rate achieves its minimum value as the signature of "anti-correlation", i.e., $g^{(2)}(0) = 0$; when the two biphoton amplitudes superpose with other relative phases, a sinusoidal joint-detection counting rate comes out as the sign of two-photon interference. In 1987, Hong, Ou, Mandel, using a similar two-photon interferometer, demonstrated an anti-correlation "dip" by selecting one polarization. Hong, Ou, and Mandel provided a different interpretation of the phenomenon as the interference between two different photons: the observation of anti-correlation requires the signal photon and the idler photon to "meet" at the beamsplitter, indicating the interference between the signal photon and the idler photon. The debate took a decade to draw its conclusion. We will address this fundamentally important debate in the following section. The physics of two-photon interferometry will be discussed in the rest of this chapter in the process of analyzing a few historical two-photon interference experiments.

10.1 IS TWO-PHOTON INTERFERENCE THE INTERFERENCE OF TWO PHOTONS?

Figure 10.1.1 illustrates a typical historical two-photon interference experiment for the observation of biphoton anti-correlation[3]. The entangled signal-idler photon pair generated in SPDC is mixed by a 50%:50% near-normal incident beamsplitter,[4] BS, and detected by two detectors, D_1 and D_2, for coincidences. Balancing the signal and idler optical paths

---

[2]The terminology *biphoton* was introduced later by Klyshko.

[3]In fact, both biphoton anti-correlation and correlation are observable from the Alley-Shih interferometer.

[4]The first a few historical "dip" experiments adopted this near-normal incidence configuration from Alley and Shih's two-photon interferometer which used orthogonal polarization for observing both anti-correlation and correlation. Near-normal incidence is the only configuration to achieve 50%-50% transmission-reflection for both S and P polarization. In fact, it is unnecessary to choose near-normal incident beamsplitter for an interferometer which uses one polarization for "dip" measurement only. For one polarization, 50%-50% can be achieved at any incident angle.

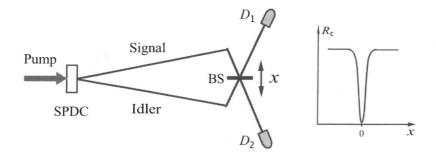

FIGURE 10.1.1 Schematic of a typical two-photon interferometer. A anti-correlation is observable in coincidences of D_1 and D_2 when scanning the beamsplitter around its balanced position $x = 0$. It is quite tempting to rely on a incorrect picture which somehow envisions the anti-correlation as arising from two individual photons of a given signal-idler pair meeting at the beamsplitter: interference between the signal photon and the idler photon.

by positioning the beamsplitter, one can observe a "null" in coincidences which indicates a complete destructive interference. When the optical path difference is increased from zero to unbalanced values, a coincidence curve of "dip" is observed. The width of the "dip" equals the coherence length of the signal and idler, which is mainly determined by the spectral bandwidth of the filters placed in front of D_1 and D_2.

It is quite tempting to rely on a picture which somehow envisions the interference as arising from two individual photons of a given signal-idler pair "meeting" at the beam-splitter. Loosely speaking, indistinguishability leads to interference, for one sees that when the condition for total destructive interference is held, the two optical paths of the inter-ferometer are of exactly the same length. It appears the signal and idler photons "meet" at the beamsplitter, and then it becomes impossible to distinguish which photon caused which single detector detection event. Destructive interference between the signal and idler photons occurs.

The picture of "interference between two photons" is further reinforced by the fact that changing the position of the beamsplitter from its balanced position, which begins to make these paths distinguishable, will bring about a degradation of interference. The coincidence counting rate seems to depend on the amount of overlap of the signal "wavepacket" and the idler "wavepacket" that is achieved. The shape of the "dip" is determined by the temporal convolution of the signal "wavepacket" and the idler "wavepacket", and therefore provides information about them. The "dip" was interpreted as the result of "photon bunching" or "anti-bunching" effects as well.

Is the above explanation correct? Does the observation of the anti-correlation "dip" mean the interference of the signal photon and the idler photon at the beamsplitter? Is the statement of Dirac failed in this experiment? We will provide an answer to these questions in the following way.

Let us examine a slightly modified experiment that is illustrated in Fig. 10.1.2. The experimental set up is similar to Fig. 10.1.1, except that the signal has two paths: one path length is L_l (longer path), the other is L_s (short path) with $L_l - L_s \equiv 2\Delta L$, such that $l_c^{s,i} \ll \Delta L \ll l_c^p$, where $l_c^{s,i}$ and l_c^p are the coherence length of the signal-idler field and the pump field, respectively. Experimentally, $l_c^{s,i}$ is usually determined by the chosen spectral filters for the photodetectors, l_c^p is determined by the spectral bandwidth of the pump laser. Due to the condition of $\Delta L > l_c^{s,i}$, although the setup provides two paths for the signal, there is no observable first-order interference of the signal photon itself. The counting rates of the single detectors, D_1 and D_2, respectively, remain constant. When the position of the

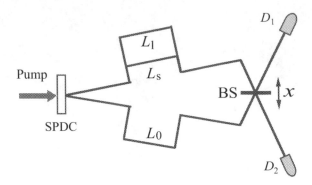

FIGURE 10.1.2 Schematic of a modified two-photon interferometer. In contrast with Fig. 10.1.1, there exists two optical paths, L_l (long) and L_s (short), for the signal while the idler takes optical paths $L_0 = (L_l + L_s)/2$ when BS is placed at its balanced position $x = 0$. The optical paths are chosen to satisfy $l_c^{s,i} \ll \Delta L \ll l_c^p$, where $\Delta L \equiv (L_l - L_s)/2$. In this setup, the signal photon and the idler photon never "meet" at the beamsplitter when taking $x = 0$. Do we expect an anti-correlation "dip" at $x = 0$?

beamsplitter BS is chosen to be $x = 0$, the path length of the idler arm takes its value of L_0 such that $L_0 = (L_l + L_s)/2$, i.e., L_0 takes the middle value of L_l and L_s. In this case, $L_l - L_0 = L_0 - L_s = \Delta L$, so that the signal photon and the idler photon are unable to "meet" at the beamsplitter due to the experimental condition of $\Delta L > l_c^{s,i}$.

Based on the idea of "distinguishability of two photons", the interference arising from the overlap of the signal and idler "wavepackets", "dips" are expected to appear for two positions of the beamsplitter only, i.e., $x = \pm\Delta L/2$. In these two cases the idler photon has a 50% chance of overlaping with the signal photon. This partial distinguishability results in that the contrast of these two dips should be at most 50%. When $x = 0$, however, the signal photon and the idler photon do not meet. There is no overlap of the signal and idler photon "wavepackets" because of $\Delta L > l_c^{s,i}$. Moreover, the detectors fire at random: in 50% of the joint detections D_1 fires ahead of D_2 by $\tau = \Delta L/c$; in the other 50% the opposite happens. So, no interference is expected.

Figure 10.1.3 shows the experimental result which tells quite a different story. We observe a high contrast interference "dip" in the middle ($x = 0$). In addition, the "dip" can turn into a "peak", or any Gaussian-like function between the "peak" and "dip", if the experimental conditions are slightly changed. The transition from "dip" to "peak" depends on $\phi = \omega_p \tau$, where $\tau = \Delta L/c$ is the time delay between the long path and the short path. Fixing $x = 0$ and varying ϕ, by slightly increasing or decreasing the value of ΔL, we observe a sinusoidal fringe in the joint detection counting rate, which is shown in Fig. 10.1.4, corresponding to a transition from "dip" to "peak" shown in Fig. 10.1.3. The experimental data indicates that it is not a necessary condition to have the signal photon and the idler photon "meeting" each other at the beamsplitter for observing two-photon correlation, anti-correlation, or two-photon interference. The idea of "destructive interference between signal and idler photons" has failed to give a correct prediction. Thus, the "dip" or "peak" may not be considered as the interference between the signal and idler photons. Indeed, two-photon interference is not the interference between two individual photons. Two-photon interference arises from the superposition of two-photon amplitudes, different yet indistinguishable alternatives that result in a click-click joint-detection event between two photodetectors. In this regard, Dirac is correct. His statement is still valid if we modify it slightly: "...*biphoton*... only interferes with itself. Interference between two different *biphotons* never occurs". Probably, Dirac's

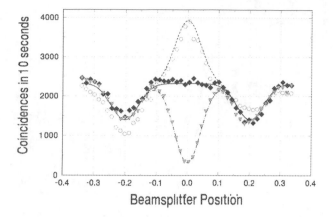

FIGURE 10.1.3 A high contrast anti-correlation "dip" is observed at $x = 0$, where the signal photon is unable to "meet" with the idler photon. In addition, the destructive "dip" can turn to a constructive "peak" when L_l-L_s is slightly changed.

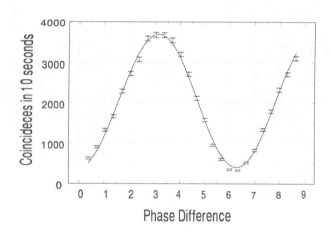

FIGURE 10.1.4 The dip-peak transition is measured as a sinusoidal function of ϕ, where $\phi = \omega_p \tau$ and $\tau \equiv \Delta L/c$.

statement "...photon... only interferes with itself" is confusing from the beginning, we may modify his statement as follows:

"Interference is the result of the superposition of quantum amplitudes, a nonclassical entity corresponding to different yet indistinguishable alternatives which lead to a photodetection event or a joint photodetection event. Interference between different photons or photon pairs never occurs".

(I): Analysis of the historical anti-correlation experiment of Fig. 10.1.1.

Let us first analyze the historical experiments of Fig. 10.1.1. As we have discussed earlier, the joint detection counting rate, R_c, of detectors D_1 and D_2 on the time interval T is given by the Glauber theory in a general form:

$$R_c \propto \int_T dt_1 dt_2 \, G^{(2)}(\mathbf{r}_1, t_1; \mathbf{r}_2, t_2)$$

$$= \int_T dt_1 dt_2 \, \langle \, \Psi \, | \, \hat{E}_1^{(-)} \hat{E}_2^{(-)} \hat{E}_2^{(+)} \hat{E}_1^{(+)} \, | \, \Psi \, \rangle$$

$$= \int_T dt_1 dt_2 \, \left| \langle \, 0 \, | \, \hat{E}_2^{(+)} \hat{E}_1^{(+)} \, | \, \Psi \, \rangle \right|^2 \tag{10.1.1}$$

where $\hat{E}_j^{(\pm)}$, $j = 1, 2$, are positive and negative-frequency components of the field at detectors D_1 and D_2, respectively, and $|\Psi\rangle$ is the state of the signal-idler photon pair:

$$|\Psi\rangle \simeq \int d\omega_p \, g(\omega_p) \int d\omega_s \, f(\omega_p, \omega_s) a_s^\dagger(\omega_s) \, a_i^\dagger(\omega_p - \omega_s) \, | 0 \rangle \tag{10.1.2}$$

where, again, we will concentrate to the temporal part of the state by selecting a pair of conjugate mode $\mathbf{k}_s$ and $\mathbf{k}_i$ for the measurement. Different from earlier discussions, here, $\omega_p = $ constant has been released. We assume a well collimated pump beam ($\Delta \vec{\kappa} \sim 0$) with longitudinal mode distribution function $g(\omega_p)$. In the study of two-photon interference, we need to deal with finite bandwidth of pump, especially in the case of pulse pumped SPDC.

The field $\hat{E}_1^{(+)}$ and $\hat{E}_2^{(+)}$ both have two contributions. Propagating the field operators from the source to the photodetector, and ignoring the transverse part of the Green's function:

$$\hat{E}_1^{(+)} = \frac{1}{\sqrt{2}} \left[i \int d\omega \, E_0(\omega) \, e^{-i\omega\tau_1^R} a_s(\omega) + \int d\omega \, E_0(\omega) \, e^{-i\omega\tau_1^T} a_i(\omega) \right]$$

$$\hat{E}_2^{(+)} = \frac{1}{\sqrt{2}} \left[\int d\omega \, E_0(\omega) \, e^{-i\omega\tau_2^T} a_s(\omega) + i \int d\omega \, E_0(\omega) \, e^{-i\omega\tau_2^R} a_i(\omega) \right] \tag{10.1.3}$$

where the superscript R and T stand for reflection and transmission, again, only one polarization is considered. $E_0(\omega) = \sqrt{\hbar\omega/2\epsilon_0 V}$, V is the quantization volume, $\tau \equiv t - z/c$, again, z is the longitudinal coordinate along the optical path. Similar to earlier calculations, we will treat $E_0(\omega)$ as a constant.

Applying the biphoton state of the signal-idler pair to Eq. (10.1.1), it is easy to find that the effective two-photon wavefunction has two amplitudes:

$$\Psi_{21} = \langle \, 0 \, | E_2^{(+)} E_1^{(+)} | \, \Psi \, \rangle = \Psi(\tau_2^T, \tau_1^T) - \Psi(\tau_2^R, \tau_1^R). \tag{10.1.4}$$

where $\Psi(\tau_2^T, \tau_1^T) = \langle 0 | E(\tau_2^T) E(\tau_1^T) | \Psi \rangle$ corresponds to the case when the signal and idler are both transmitted at the beamsplitter, while $\Psi(\tau_2^R, \tau_1^R) = \langle 0 | E(\tau_2^R) E(\tau_1^R) | \Psi \rangle$ corresponds to their reflection. The normalization constant has been absorbed into each of the amplitudes. The superposition of these two different yet indistinguishable two-photon amplitudes, or biphoton wavepackets, which contribute to a click-click joint-detection event between photodetectors D_1 and D_2, determine the probability of having a joint-detection at space-time $(\mathbf{r}_1, t_1; \mathbf{r}_2, t_2)$:

$$G^{(2)}(\mathbf{r}_1, t_1; \mathbf{r}_2, t_2)$$

$$= \left| \Psi(\tau_2^T, \tau_1^T) - \Psi(\tau_2^R, \tau_1^R) \right|^2 \tag{10.1.5}$$

$$= \left| \Psi(\tau_2^T, \tau_1^T) \right|^2 + \left| \Psi(\tau_2^R, \tau_1^R) \right|^2$$

$$- \Psi^*(\tau_2^T, \tau_1^T)\Psi(\tau_2^R, \tau_1^R) - \Psi(\tau_2^T, \tau_1^T)\Psi^*(\tau_2^R, \tau_1^R).$$

The biphoton interference is thus observable in the coincidence counting rate

$$R_c \propto \int_T dt_1 \, dt_2 \, \big| \, \Psi(\tau_2^T, \tau_1^T) - \Psi(\tau_2^R, \tau_1^R) \, \big|^2. \tag{10.1.6}$$

Examining Eqs. (10.1.5) and (10.1.6), when $\Psi(\tau_2^T, \tau_1^T)$ and $\Psi(\tau_2^R, \tau_1^R)$ are completely "overlap" in space-time, or are completely indistinguishable in the joint detection events, the coincidence counting rate is expected to be "null". Shifting the position of the beamsplitter from its balanced position, which begins to make these wavepackets non-overlapping, or distinguishable, will bring about a degradation of interference, i.e., observing a "dip" in coincidences. This is the result of the convolution of the biphoton wavepackets. In addition, if one could change the "-" to "+", for example, by playing with the polarization of the photon pair, one can make a "peak" instead of a "dip". Both "dip" and "peak" can be easily observed in a polarization two-photon interferometer, which will be discussed latter.

Further computing the interference as a function of the optical path difference of the two-photon interferometer, we need to calculate the biphoton wavepackets and their convolution. The biphoton wavepacket of SPDC has been calculated earlier in the case of monochromatic plane wave pump. Similar to the earlier calculation, except taking into account of the pump distribution function $g(\omega_p)$, the biphoton wavepacket is thus:

$$\Psi(\tau_2, \tau_1) = \Psi_0 \int d\omega_p \, g(\omega_p) \, e^{-i\frac{1}{2}\omega_p(\tau_2+\tau_1)}$$

$$\times \int d\omega_s \, f(\omega_p, \omega_s) \, e^{-i\frac{1}{2}(\omega_s-\omega_i)(\tau_2-\tau_1)} \tag{10.1.7}$$

where Ψ_0 absorbs all constants from the field and the state. To simplify the mathematics, we start with a factorizable integral by imposing the following approximations:

$$\omega_s = \omega_s^0 + \nu, \quad \omega_i = \omega_i^0 - \nu,$$
$$\omega_s^0 + \omega_i^0 \simeq \omega_p^0, \quad \omega_p = \omega_p^0 + \nu_p \tag{10.1.8}$$

where ω_s^0, ω_i^0, and ω_p^0 are the center frequencies for the signal, idler and pump, respectively. ω_s^0, ω_i^0, and ω_p^0 are considered as constants. Eq. (10.1.7) is then simplified to a product of two functions,

$$\Psi(\tau_2, \tau_1) = \Psi_0 \, v(\tau_2 + \tau_1) \, u(\tau_2 - \tau_1), \tag{10.1.9}$$

with

$$v(\tau_2 + \tau_1) = e^{-i\omega_p^0(\tau_2+\tau_1)/2} \int_{-\infty}^{\infty} d\nu_p \, g(\nu_p) \, e^{-i\nu_p(\tau_2+\tau_1)/2} \tag{10.1.10}$$

and

$$u(\tau_2 - \tau_1) = e^{-i\frac{1}{2}(\omega_s^0-\omega_i^0)(\tau_2-\tau_1)} \int_{-\infty}^{\infty} d\nu \, f(\nu) \, e^{-i\nu(\tau_2-\tau_1)}. \tag{10.1.11}$$

Basically, we have assumed the integral of $u(\tau_2 - \tau_1)$ independent of ω_p. This approximation is valid only for narrow bandwidth of $g(\nu_p)$ such as that of a CW laser pump. This approximation cannot be used for short pulse pump, especially in the case of femtosecond laser pumped SPDC. The discussion for ultra-short pulse pumped SPDC will be given later.

The functions $v(\tau_2 + \tau_1)$ and $u(\tau_2 - \tau_1)$ can be written in terms of the Fourier transforms of $g(\nu_p) \to \mathcal{F}_{\tau_+}\{g(\nu_p)\}$ and $f(\nu) \to \mathcal{F}_{\tau_-}\{f(\nu)\}$, where $\tau_+ \equiv (\tau_2 + \tau_1)/2$ and $\tau_- \equiv \tau_2 - \tau_1$. The effective two-photon wavefunction, or biphoton wavepacket, is given by

$$\Psi(\tau_2, \tau_1) = \Psi_0 \, \mathcal{F}_{\tau_+}\{g(\nu_p)\} \, e^{-i\omega_p^0\tau_+} \, \mathcal{F}_{\tau_-}\{f(\nu)\} \, e^{-i\omega_d^0\tau_-}$$

$$= \Psi_0 \, \mathcal{F}_{\tau_+}\{g(\nu_p)\} \, \mathcal{F}_{\tau_-}\{f(\nu)\} \, e^{-i\omega_s^0\tau_2} \, e^{-i\omega_i^0\tau_1} \tag{10.1.12}$$

where $\omega_d^0 \equiv \frac{1}{2}(\omega_s^0 - \omega_i^0)$.

The cross interference term in Eq. (10.1.6) is calculated as follows:

$$\int dt_1 dt_2 \, \Psi^*(\tau_2^T, \tau_1^T) \, \Psi(\tau_2^R, \tau_1^R)$$

$$= |\Psi_0|^2 \int dt_+ \, \mathcal{F}_{\tau_+^T}^* \{g(\nu_p)\} \, \mathcal{F}_{\tau_+^R} \{g(\nu_p)\} \, e^{i\omega_p^0(\tau_+^T - \tau_+^R)}$$

$$\times \int dt_- \, \mathcal{F}_{\tau_-^T}^* \{f(\nu)\} \, \mathcal{F}_{\tau_-^R} \{f(\nu)\} \, e^{i\omega_d^0(\tau_-^T - \tau_-^R)}$$

$$\simeq |\Psi_0|^2 \int dt_- \, \mathcal{F}_{t_-}^* \{f(\nu)\} \, \mathcal{F}_{t_- -\delta} \{f(\nu)\}$$

$$= |\Psi_0|^2 \, \mathcal{F}_{t_-}^* \{f(\nu)\} \otimes \mathcal{F}_{t_- -\delta} \{f(\nu)\} \qquad (10.1.13)$$

where $t_+ \equiv t_2 + t_1$, $t_- \equiv t_2 - t_1$, and $\delta = [(z_2^R - z_1^R) - (z_2^T - z_1^T)]/c$ is the optical path difference introduced by moving the beamsplitter upward from its balanced position in the two-photon interferometer of Fig. 10.1.1. We have assumed degenerate ($\omega_d^0 = 0$) type I SPDC in the above calculation. The coincidence counting rate R_c is therefore

$$R_c(\delta) = R_0 \left[1 - \mathcal{F}_{t_-}^* \{f(\nu)\} \otimes \mathcal{F}_{t_- -\delta} \{f(\nu)\}\right], \qquad (10.1.14)$$

where R_0 is a constant. Assuming Gaussian wavepackets, the convolution yields a Gaussian function $|\Psi_0|^2 e^{-\delta^2/\tau_c^2}$ with $\tau_c = l_c^{s,i}/c$ the coherence time of the signal and idler fields. The coincidence counting rate R_c is approximately

$$R_c(\delta) = R_0 \left[1 - e^{-\delta^2/\tau_c^2}\right].$$

It is now clear that the observed "dip" is a biphoton "destructive interference" phenomenon. Mathematically, the "dip" is the result of a convolution, or cross correlation, of the biphoton wavepackets, $\Psi(\tau_2^T, \tau_1^T)$ and $\Psi(\tau_2^R, \tau_1^R)$, along τ_- axis. Fig. 10.1.5 is a con-

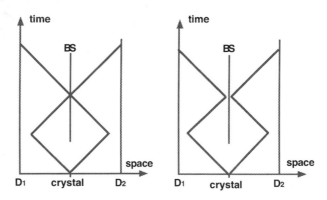

FIGURE 10.1.5 Conceptual Feynman diagrams. The beamsplitter is represented by the thin vertical lines. The biphoton apmlitudes, or biphoton wavepackets, are represented by "straight lines". Left: $\Psi(z_2^R, t_2; z_1^R, t_1)$ (reflect-reflect); Right: $\Psi(z_2^T, t_2; z_1^T, z_1)$ (transmit-transmit). The two Feyman alternatives both contribute to a "click-click" joint-detection event of D_1 and D_2.

ceptual Feynman diagram for the two-photon interference experiment of Fig. 10.1.1. While the beamsplitter is in its balanced position, the two Feynman alternatives (reflect-reflect vs transmit-transmit) are indistinguishable. Moving the beamsplitter away from its balanced position, the optical path difference of the two Feynman paths are no longer the same, corresponds to the moving away of the 2-D wavepacket along the τ_- axis. The superposition takes place between the transmit-transmit and the reflect-reflect biphoton amplitudes, instead of the signal photon and the idler photon. In Dirac's language: it is the interference of biphoton itself, but not the interference between the signal and the idler photons.

In general, the biphoton interference can occur in two different ways: (1) the convolution takes place along τ_- direction; (2) the convolution takes place along τ_+ direction. In the experiment shown in Fig. 10.1.1, the biphoton wavepackets, $\Psi(\tau_2^T, \tau_1^T)$ and $\Psi(\tau_2^R, \tau_1^R)$ completely "overlap" along τ_+, since $\tau_+^T = \tau_+^R$ in any position of the beamsplitter. $\Psi(\tau_2^R, \tau_1^R)$, however, moves away from $\Psi(\tau_2^T, \tau_1^T)$ along τ_-, when scanning the beamsplitter from its balanced position.

(II) Analysis of the modified "dip" experiment in Fig. 10.1.2.

In the view of biphoton interference, we now present an interpretation for the experiment of Fig. 10.1.2. Differing from that of the experiment of Fig. 10.1.1, the special experimental setup in Fig. 10.1.2 achieves four alternatives of producing a joint-detection event between D_1 and D_2. The effective biphoton wavefunction thus consists of four amplitudes:

$$\Psi_{21} = \Psi(\tau_2^{LT}, \tau_1^{0T}) - \Psi(\tau_2^{0R}, \tau_1^{LR}) + \Psi(\tau_2^{ST}, \tau_1^{0T}) - \Psi(\tau_2^{0R}, \tau_1^{SR})$$

where the superscript L, S, and 0 represent the long path, the short path, and the middle path of Fig. 10.1.2, respectively. Consequently, $G^{(2)}$ has sixteen terms contributing to the coincidence photon counting

$$G^{(2)} = \left| \Psi(\tau_2^{LT}, \tau_1^{0T}) - \Psi(\tau_2^{0R}, \tau_1^{LR}) + \Psi(\tau_2^{ST}, \tau_1^{0T}) - \Psi(\tau_2^{0R}, \tau_1^{SR}) \right|^2.$$

However, due to the experimental condition that we have chosen, $L_l - L_0 = L_0 - L_s \equiv \Delta L \gg l_c^{s,i}$, where, again, $l_c^{s,i}$ is the coherence length of the signal and idler, only four cross terms are non-zero. We have the following eight non-zero contributions to the coincidence counting rate of D_1 and D_2:

$$R_c \propto \int_T dt_1 dt_2 \Big\{ |\Psi(\tau_2^{LT}, \tau_1^{0T})|^2 + |\Psi(\tau_2^{0R}, \tau_1^{LR})|^2$$
$$+ |\Psi(\tau_2^{ST}, \tau_1^{0T})|^2 + |\Psi(\tau_2^{0R}, \tau_1^{SR})|^2$$
$$- \Psi^*(\tau_2^{LT}, \tau_1^{0T})\Psi(\tau_2^{0R}, \tau_1^{SR}) - \Psi(\tau_2^{LT}, \tau_1^{0T})\Psi^*(\tau_2^{0R}, \tau_1^{SR})$$
$$- \Psi^*(\tau_2^{ST}, \tau_1^{0T})\Psi(\tau_2^{0R}, \tau_1^{LR}) - \Psi(\tau_2^{ST}, \tau_1^{0T})\Psi^*(\tau_2^{0R}, \tau_1^{LR}) \Big\}. \qquad (10.1.15)$$

The interference cross terms are calculated as:

$$\Psi^*(\tau_2^{LT}, \tau_1^{0T})\Psi(\tau_2^{0R}, \tau_1^{SR})$$
$$= |\Psi_0|^2 e^{-i\omega_p^0 \frac{\Delta L}{c}} \mathcal{F}_{t_+}^* \{g(\nu_p)\} \mathcal{F}_{t_+ + \frac{\Delta L}{c}} \{g(\nu_p)\} \mathcal{F}_{t_-}^* \{f(\nu)\} \mathcal{F}_{t_- -\delta} \{f(\nu)\},$$
$$\Psi^*(\tau_2^{ST}, \tau_1^{0T}) \Psi(\tau_2^{0R}, \tau_1^{LR})$$
$$= |\Psi_0|^2 e^{i\omega_p^0 \frac{\Delta L}{c}} \mathcal{F}_{t_+}^* \{g(\nu_p)\} \mathcal{F}_{t_+ - \frac{\Delta L}{c}} \{g(\nu_p)\} \mathcal{F}_{t_-}^* \{f(\nu)\} \mathcal{F}_{t_- -\delta} \{f(\nu)\},$$

where $\delta = [(r_2^{0R} - r_1^{SR}) - (r_2^{LT} - r_1^{0T})]/c$ is the additional optical path difference introduced by moving the beamsplitter upward from its "balanced" position $x = 0$.

The Feynman paths for this experiment are illustrated in Fig. 10.1.6. The upper two correspond to $\Psi(\tau_2^{LT}, \tau_1^{0T})$ and $\Psi(\tau_2^{0R}, \tau_1^{SR})$; the lower two correspond to $\Psi(\tau_2^{ST}, \tau_1^{0T})$ and $\Psi(\tau_2^{0R}, \tau_1^{LR})$. Notice that if $\Delta L \ll l_c^p$, the upper two and the lower two Feynman paths, respectively, are indistinguishable by means of the click-click joint photodetection of D_1 and D_2.

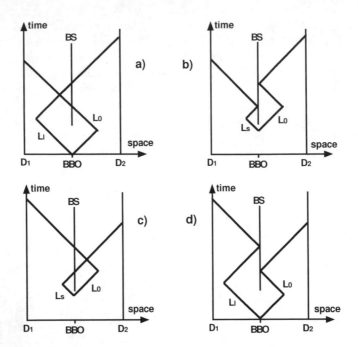

FIGURE 10.1.6 Conceptual Feynman diagrams. a) and b) are two amplitudes for a joint detection such that D_1 fires ahead of D_2; c) and d) are two amplitudes in the reversed order. If $\Delta L \ll l_p^{coh}$, the upper two and the lower two, respectively, are indistinguishable.

By increasing or decreasing ΔL or δ, we have two freedom to "shift" the 2-D biphoton wavepackets, independently, along τ_+ and τ_- axes:

$$\left[\mathcal{F}_{t_+}^* \{g(\nu_p)\} \otimes \mathcal{F}_{t_+ \pm \frac{\Delta L}{c}} \{g(\nu_p)\} \right] \left[\mathcal{F}_{t_-}^* \{f(\nu)\} \otimes \mathcal{F}_{t_- - \delta} \{f(\nu)\} \right].$$

For a chosen value of $\delta \ll l_c^p$, the upper pair and the lower pair of 2-D wavepackets illustrated in Fig. 10.1.6 are almost 100% overlapped along the τ_+ axis, respectively, yields $\mathcal{F}_{t_+}^* \{g(\nu_p)\} \otimes \mathcal{F}_{t_+ \pm \Delta L/c} \{g(\nu_p)\} \sim 1$. The joint detection counting rate in the neighborhood of $x = 0$ is thus:

$$R_c(\delta) = R_0 \left[1 - \cos \phi \, \mathcal{F}_{t_-}^* \{f(\nu)\} \otimes \mathcal{F}_{t_- - \delta} \{f(\nu)\} \right]. \tag{10.1.16}$$

For Gaussian wavepacket along the t_- axis, R_c is approximately

$$R_c(\delta) = R_0 \left[1 - \cos \phi \, e^{-\delta^2/\tau_c^2} \right]. \tag{10.1.17}$$

Eq. (10.1.16) indicates a $\sim 100\%$ interference modulation while scanning the beamsplitter in the neighborhood of $x = 0$. It is noticed that the phase factor $\phi = \omega_p^0(\Delta L/c)$ plays an important role in this measurement. Subsequently setting the phase ϕ to be equal to π, 0, or $\pi/2$ we observe respectively a peak, dip, or flat coincidence rate R_c distribution centered at the "balanced" position of the beamsplitter, agreeing with the experimental results shown in Fig. 10.1.3 and Fig. 10.1.4. The mechanism of manipulating phase ϕ along τ_+ axis is very useful for the preparation of Bell states. We will learn more about it in next section.

10.2 TWO-PHOTON INTERFERENCE WITH ORTHOGONAL POLARIZA-TION

In the history of two-photon interferometry, a great driving force was the experimental testing of Bell's inequality. In fact, the first historical two-photon interferometer was designed for that purpose. Before the discussions of Bell's states and Bell's inequality, we analyze a simple two-photon interferometer with a pair of orthogonal polarized photons and a pair of independent polarization analyzers that is schematically illustrated in Fig. 10.2.1.

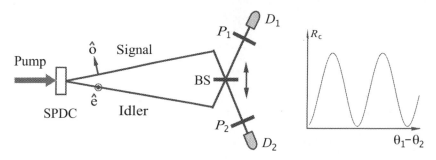

FIGURE 10.2.1 Schematic of an Alley-Shih two-photon interferometer. The type-II SPDC produces an orthogonally polarized signal-idler pair. BS is a 50%-50% beamsplitter for both $\hat{o}$ and $\hat{e}$ polarized signal and idler photons. The polarization analyzers P_1 and P_2 are oriented at any chosen angles θ_1 and θ_2 for polarization correlation measurement. Fixing BS at $x = 0$ by examining the correlation "peak" and the anti-correlation "dip", a sinusoidal polarization correlation of $\sin^2(\theta_1 - \theta_2)$ is observed in the coincidences of D_1 and D_2.

Assuming an idealized biphoton source generates an orthogonal polarized signal-idler pair in the following state:

$$|\Psi\rangle \simeq \int d\nu \, f(\nu) \, \hat{o} \, a_s^\dagger(\omega_s^0 + \nu) \, \hat{e} \, a_i^\dagger(\omega_i^0 - \nu) \, | \, 0\rangle \tag{10.2.1}$$

where $\hat{o}$ and $\hat{e}$ are unit vectors along the o-ray and the e-ray polarization direction of the SPDC crystal. In Eq. (10.2.1), we have assumed perfect phase matching $\omega_s + \omega_i - \omega_p = 0$ and $k_{s,o} + k_{i,e} - k_p = 0$. Suppose the polarizers of the detectors D_1 and D_2 are set at angles θ_1 and θ_2, relative to the polarization direction of the o-ray of the SPDC crystal, respectively, the field operators can be written as

$$E_1^{(+)} = \frac{1}{\sqrt{2}} \left[i \int d\omega \, E_0(\omega) \, e^{-i\omega\tau_1^R} \, \hat{\theta}_1 a_s(\omega) + \int d\omega \, E_0(\omega) \, e^{-i\omega\tau_1^T} \, \hat{\theta}_1 a_i(\omega) \right]$$

$$E_2^{(+)} = \frac{1}{\sqrt{2}} \left[\int d\omega \, E_0(\omega) \, e^{-i\omega\tau_2^T} \, \hat{\theta}_2 a_s(\omega) + i \int d\omega \, E_0(\omega) \, e^{-i\omega\tau_2^R} \, \hat{\theta}_2 a_i(\omega) \right]$$

where $\hat{\theta}_j$, $j = 1, 2$, is the unit vector along the jth analyzer direction. The effective wavefunction that contribute to the joint-detection events of D_1 and D_2 is calculated to be

$$\Psi_{21} = (\hat{\theta}_1 \cdot \hat{e})(\hat{\theta}_2 \cdot \hat{o}) \, \Psi(\tau_2^T, \tau_1^T) - (\hat{\theta}_1 \cdot \hat{o})(\hat{\theta}_2 \cdot \hat{e}) \, \Psi(\tau_2^R, \tau_1^R). \tag{10.2.2}$$

The joint detection counting rate of D_1 and D_2 is thus:

$$R_c \propto \int_T dt_1 dt_2 \left| (\hat{\theta}_1 \cdot \hat{\mathbf{e}})(\hat{\theta}_2 \cdot \hat{\mathbf{o}}) \, \Psi(\tau_2^T, \tau_1^T) - (\hat{\theta}_1 \cdot \hat{\mathbf{o}})(\hat{\theta}_2 \cdot \hat{\mathbf{e}}) \, \Psi(\tau_1^R, \tau_2^R) \right|^2$$

$$= R_0 \left\{ \sin^2\theta_1 \cos^2\theta_2 + \cos^2\theta_1 \sin^2\theta_2 - \sin\theta_1 \cos\theta_2 \cos\theta_1 \sin\theta_2 \right.$$

$$\left. \times \frac{1}{R_0} \int_T dt_1 dt_2 \left[\Psi^*(\tau_2^T, \tau_1^T) \Psi(\tau_2^R, \tau_2^R) + \Psi(\tau_2^T, \tau_1^T) \Psi^*(\tau_2^R, \tau_2^R) \right] \right\} \qquad (10.2.3)$$

where $\Psi(\tau_2^T, \tau_1^T)$ and $\Psi(\tau_2^R, \tau_1^R)$ are the transmitted-transmitted and reflected-reflected biphoton wavepackets with the following time averaging:

$$\int_T dt_1 dt_2 \, |\Psi(\tau_1, \tau_2)|^2 = R_0.$$

The third term of Eq. (10.2.3) determines the degree of two-photon coherence. Considering degenerate CW laser pumped SPDC, the biphoton wavepacket of Eq. (10.1.9) can be simplified as:

$$\Psi(\tau_2, \tau_1) = \Psi_0 \, \mathcal{F}_{\tau_-} \left\{ f(\nu) \right\},$$

where we have absorbed the phase factor $e^{-i\omega_p(\tau_1 + \tau_2)/2}$ into Ψ_0. The coefficient of $(\sin\theta_1 \cos\theta_2 \cos\theta_1 \sin\theta_2)$ in the third term of Eq. (10.2.3) is thus

$$\int_T dt_1 dt_2 \, \mathcal{F}_{\tau_2^T - \tau_1^T} \left\{ f(\nu) \right\} \mathcal{F}_{\tau_2^R - \tau_1^R} \left\{ f(\nu) \right\}$$

$$= \mathcal{F}_{t_-} \left\{ f(\nu) \right\} \otimes \mathcal{F}_{t_- - \delta} \left\{ f(\nu) \right\},$$

where, again, δ is the optical path difference introduced by moving the beamsplitter from its balanced position of $x = 0$.

In a polarization two-photon interferometer, we will be able to observe two biphoton interference effects:

(1) Anti-correlation "dip" and correlation "peak".

In this measurement, we fix θ_1 and θ_2, such that $\theta_1 = 45°$ with $\theta_2 = 45°$ or $\theta_1 = 45°$ with $\theta_2 = -45°$, an anti-correlation–"dip" or a correlation–"peak" as function of δ will be observed in the coincidence counting rate R_c when scanning δ in the neighborhood of $x = 0$,

$$R_c(\delta) = R_0 \left[1 \mp \mathcal{F}_{t_-}^* \{ f(\nu) \} \otimes \mathcal{F}_{t_- - \delta} \{ f(\nu) \} \right]. \qquad (10.2.4)$$

(2) Polarization correlation.

In this measurement, we make $\delta = 0$ to achieve complete overlaping between biphoton wavepackets $\Psi(\tau_2^T, \tau_1^T)$ and $\Psi(\tau_2^R, \tau_1^R)$. The coefficient of $(\sin\theta_1 \cos\theta_2 \cos\theta_1 \sin\theta_2)$ in the third term of Eq. (10.2.3) achieves its maximum value of 2. The coincidence counting rate will be a function of $\theta_1 - \theta_2$ when manipulating the relative angle of the two polarization analyzers,

$$R_c(\theta_1, \theta_2) = R_0 \sin^2(\theta_1 - \theta_2). \qquad (10.2.5)$$

This result is equivalent to the polarization correlation measurement for the Bell's state

$$|\Psi\rangle = \frac{1}{\sqrt{2}} \left[|X_1\rangle |Y_2\rangle - |Y_1\rangle |X_2\rangle \right],$$

where $|X_1\rangle$ and $|X_1\rangle$ are defined as the polarization state that are respectively coincide with the o-ray and e-ray polarization direction of SPDC. Bell's states and polarization correlation will be discussed in detail in Chapter 14.

Since Einstein-Podolsky-Rosen published their 1935 paper, the concept of "physical reality" bacame an important subject for physicists and philosophers to study. In the early 1950's, Bohm simplified the Einstein-Podolsky-Rosen state of 1935 to discrete spin variables by introducing the singlet state of two spin $1/2$ particles:

$$|\Psi\rangle = \frac{1}{\sqrt{2}}\left[|\uparrow\rangle_1|\downarrow\rangle_2 - |\downarrow\rangle_1|\uparrow\rangle_2\right] \qquad (10.2.6)$$

where the kets $|\uparrow\rangle$ and $|\downarrow\rangle$ represent states of spin "up" and spin "down", respectively, along an *arbitrary* direction. For the EPR-Bohm state, the spin of neither particle is determined; however, if one particle is measured to be spin up along a certain direction, the other one must be spin down along that direction, despite the distance between the two spin $1/2$ particles and the orientation of the Stern-Gerlach analyzers (SGA). The nonlocal behavior of this two-particle system leads to the questions of EPR-Bohm: Are the two spin $1/2$ particles prepared with defined spins at the source and in the course of their propagation? Is spin a physical reality of a particle independent of the observation?

In the Alley-Shih experiment, the same question was asked in a slightly different way: if two particles each is prepared with well-defined spin, can we expect similar nonlocal behavior? This question lead to their 1986 experiment. With the help of a two-photon interferometer, Alley and Shih discovered that a pair of photons with well-defined polarization can give similar EPR-Bohm type correlation. Since than, the complete set of Bell states have been experimentally observed

$$|\Psi^{(\pm)}\rangle = \frac{1}{\sqrt{2}}\left[|H_1\rangle|V_2\rangle \pm |V_1\rangle|H_2\rangle\right]$$

$$|\Phi^{(\pm)}\rangle = \frac{1}{\sqrt{2}}\left[|H_1\rangle|H_2\rangle \pm |V_1\rangle|V_2\rangle\right]. \qquad (10.2.7)$$

Where $|H\rangle$ and $|V\rangle$, respectively, indicates well-defined horizontal and vertical polarization. In fact, any set of orthogonal polarization can be used to construct Bell states. In general, we use polarization state vector $|X\rangle$ and $|Y\rangle$, which can be defined in any orthogonal orientation, to replace $|H\rangle$ and $|V\rangle$.

This observation has been puzzling us for two decades. (1) There seems nothing "hidden" in this experiment. The signal photon and the idler photon both have well-defined polarization before entering into the interferometer; (2) The signal field and the idler field are first-order incoherent, the incoherent superposition of the singal-idler fields cannot change the polarization of the signal and idler, either during the course of their propagation or in the process of their annihilation; (3) It seems neither "correlation nor "anti-correlation" is the intrinsic property of the photon source, or the state of the single photon and idler photon, since Alley-Shih observed both "correlation" and "anti-correlation" from one experiment utilizing the same single–idler photon source.

What is the cause of the nonlocal EPR-Bohm-Bell correlation for a pair of photons with well-defined polarization? We have attempted to introduce the concept of two-photon (two-particle) interference since 1986. In fact, this concept has been applied in the above analysis of the Alley-Shih experiment. In this regard, the nonlocal behavior of the EPR-Bohm spin $1/2$ particles is a two-particle interference phenomenon. The EPR-Bohm state specifies a coherent superposition of two-particle amplitudes, corresponding to two different yet indistinguishable alternative ways for the two spin $1/2$ particles to trigger a joint-detection event through the two distant SGAs. We will continue our discussion on the concept of physical reality and the physics behind this interesting observation.

10.3   FRANSON INTERFEROMETER

In 1989, Franson proposed an interferometer to explore the surprising behavior of entangled photon pairs. Figure 10.3.1 is a schematic setup of a Franson interferometer, which consists of an entangled biphoton source, and a pair of classic unbalanced interferometers with photon counting detectors coupled at their output ports. An pair of entangled photons, such as the signal photon and the idler photon of SPDC are sent into the unbalanced interferometers 1 and 2, respectively. The photon counting detectors are used for monitoring the single-detector counting rates, independently, and for observing the joint-detection counting rate, coincidently. The optical path differences of the two interferometers, ΔL_1 and ΔL_1 are both chosen to be much greater than the coherence length, $l_c^{s,i}$, of the signal-idler field, thus, there is no observable first-order interference in the single-detector counting rates of D_1 and D_2 when increasing or decreasing the values of ΔL_1 and ΔL_2 either individually or simultaneously. The joint-detection counting rate of D_1 and D_2, however, shows $\sim 100\%$ interference if the operation of the two interferometers satisfy the following conditions: (1) the photon pair only passes through the long-long and the short-short paths of the interferometers; (2) $|\Delta L_1 - \Delta L_2| \ll l_c^{s,i}$; (3) $\Delta L_1 + \Delta L_2 \ll l_c^p$ where l_c^p is the coherence length of the pump for SPDC. The surprising observation is the result of a biphoton interference phenomenon.

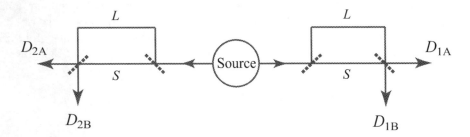

FIGURE 10.3.1   Schematic setup of a Franson interferometer.

In the following calculation we assume condition (1) is satisfied, i.e., there are only two alternatives, $\Psi(\tau_1^L, \tau_2^L)$ and $\Psi(\tau_1^S, \tau_2^S)$, contribute to a joint photodetection event of D_1 and D_2. The coincidence counting rate of D_1 and D_2 is thus

$$
\begin{aligned}
R_c \propto &\int_T dt_1\, dt_2\, \big|\, \Psi(\tau_1^L, \tau_2^L) + \Psi(\tau_1^S, \tau_2^S) \big|^2 \\
= &\int_T dt_1\, dt_2\, \big[\, \big|\, \Psi(\tau_1^L, \tau_2^L) \big|^2 + \big|\, \Psi(\tau_1^S, \tau_2^S) \big|^2 \\
&+ \Psi^*(\tau_1^L, \tau_2^L)\, \Psi(\tau_1^S, \tau_2^S) + \Psi(\tau_1^L, \tau_2^L)\, \Psi^*(\tau_1^S, \tau_2^S) \big]
\end{aligned}
\tag{10.3.1}
$$

where the subscripts L and S of τ label the long path and the short path of the *jth* classic interferometer, $j = 1, 2$. The cross term is the nontrivial term that determines the interference. Now, we further assume a biphoton wavepacket of SPDC, $\Psi(\tau_1, \tau_2) \sim \Psi_0 v(\tau_1 + \tau_2) u(\tau_1 - \tau_2)$, as shown in Eq. (10.1.9). The interference term can be written as:

$$
\begin{aligned}
&\int_T dt_1\, dt_2\, \Psi^*(\tau_1^L, \tau_2^L)\, \Psi(\tau_1^S, \tau_2^S) \\
&= e^{i\,\omega_p^0(\Delta L_1 + \Delta L_2)/2c}\, \big[\, \mathcal{F}^*_{\tau_1^L + \tau_2^L}\{g(\nu_p)\} \otimes \mathcal{F}_{\tau_1^S + \tau_2^S}\{g(\nu_p)\} \big] \\
&\quad \times\, e^{i(\omega_s^0 - \omega_i^0)(\Delta L_1 - \Delta L_2)/2c}\, \big[\, \mathcal{F}^*_{\tau_1^L - \tau_2^L}\{f(\nu)\} \otimes \mathcal{F}_{\tau_1^S - \tau_2^S}\{f(\nu)\} \big].
\end{aligned}
\tag{10.3.2}
$$

It is easy to see that the two convolutions in the brackets require the satisfaction of conditions (2) and (3) for observing interference from a Franson interferometer. The interference pattern has two parts of sinusoidal modulation: the sum frequency $\omega_s^0 + \omega_i^0 = \omega_p^0$ and the beating frequency $\omega_s^0 - \omega_i^0$. If degenerate SPDC is applied, and if one manipulates the the optical path difference of the interferometers simultaneously with $\Delta L_1 = \Delta L_2 = \Delta L$, the interference pattern keeps the sum frequency only as predicated by Franson in 1989:

$$R_c \propto 1 + V \cos\left(\omega_p \tau\right) \tag{10.3.3}$$

where $\tau \equiv \Delta L/c$ is the time delay between the long and short paths of the interferometer and V is the interference visibility, which is evaluated from the two convolutions in Eq. (10.3.2).

Franson interferometer has been studied intensively in the 1990s with the use of entangled two-photon source of SPDC. Most of the interesting physics associated with Franson interferometer have been experimentally observed.

The implementation of condition (1) is not that straightforward even if we are given an entangled biphoton source of SPDC, two standard Mach-Zehnder interferometers, and proper joint photodetection electronics. Naturally, there are four alternative ways in which the signal-idler photon pair may contribute to a joint photodetection event. Besides $\Psi(\tau_1^L, \tau_2^L)$ and $\Psi^S(\tau_1, \tau_2^S)$, the other two alternatives, or biphoton amplitudes, $\Psi(\tau_1^L, \tau_2^S)$ and $\Psi(\tau_1^S, \tau_2^L)$ do not despair automatically. The joint photodetection counting rate of D_1 and D_2 is the result of a superposition that contains four alternatives:

$$R_c \propto \int_T dt_1\, dt_2 \left| \Psi(\tau_1^L, \tau_2^L) + \Psi(\tau_1^S, \tau_2^S) + \Psi(\tau_1^L, \tau_2^S) + \Psi(\tau_1^S, \tau_2^L) \right|^2. \tag{10.3.4}$$

Due to the operation condition of the interferometer, $\Delta L_{1,2} > l_c^{s,i}$, however, only one cross term has non-zero contribution to the interference, which is the same as shown in Eq. (10.3.2). In this case, one would observe the same interference pattern as that of Eq. (10.3.1), except the maximum interference visibility is reduced from 100% to 50%:

$$R_c \propto 1 + \frac{1}{2} V \cos\left(\omega_p \tau\right). \tag{10.3.5}$$

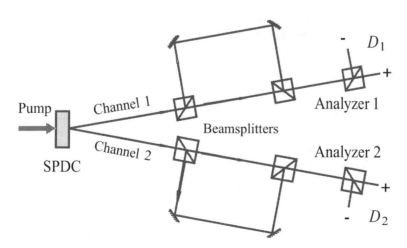

FIGURE 10.3.2 Scheme setup of a Franson interferometer. The entangled biphoton source is a type II noncollinear SPDC. The clever use of polarization guarantees the implementation of condition (1): only $\Psi(\tau_1^L, \tau_2^L)$ and $\Psi(\tau_1^S, \tau_2^S)$ contribute to a joint photodetection event.

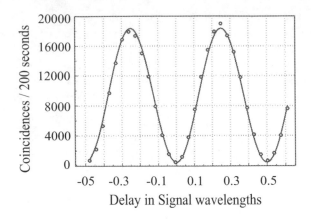

FIGURE 10.3.3   An earlier experimental data of Strekalov *et al.* reported $(95.0 \pm 1.4)\%$ interference visibility in joint detection counting rate. The single detector counting rates of D_1 and D_2, however, both kept constants while tuning the optical path differences of ΔL.

Figure 10.3.2 schematically illustrates a clever realization of Franson interference by Strekalov *et al.* from which $\sim 100\%$ interference visibility was observed. The entangled two-photon source is a non-collinear type II SPDC similar to that shown in Fig. 14.4.1. The unbalanced Mach-Zehnder interferometer in channel 1 and 2 is implemented by a long quartz rod followed with a Pockels cell. The quartz rods delay the slow polarization component relative to the fast one due to their birefringence. The Pockels cell, by applying an adjustable DC voltage, is for "fine-tuning" of the optical path difference, ΔL, of the interferometer. The birefringent delay of the interferometer is carefully chosen to be greater than the coherence time of the measured signal-idler field, which is mainly determined by the bandwidth of the spectral filters placed in front of D_1 and D_2. The fast-slow axes of the quartz rods as well as that of the Pockels cell are both oriented carefully to provide the $o_1 - e_2$ and $e_1 - o_2$ amplitudes with long-long and short-short optical paths, thus satisfy condition (1) of the Franson interferometer. Following the quartz rods and Pockels cells, in channel 1 and 2, are two polarization analyzers, A_1 and A_2. The axes of the analyzers are oriented at $45°$ relative to that of the quartz rod and the Pockels cell. The joint photodetection events are recorded as a function of the optical path difference ΔL with the help of a coincidence circuits in nanosecond time window. A $(95.0 \pm 1.4)\%$ visibility of interference pattern specified by Eq. (10.3.3) was reported in an earlier publication of Strekalov *et al.*, see Fig. (10.3.3). Recent measurements of Franson interferometer have observed $\sim 100\%$ interference visibility with statistical errors a few orders smaller.

The high degree two-photon coherence observed in Franson interferometer is considered as a demonstration of the nonlocal EPR inequality in energy. As we know, the loss of first-order interference in each of the Mach-Zehnder interferometer indicates a considerable large uncertainty in $\Delta \hbar \omega_{s,i}$, at least $\Delta \hbar \omega_{s,i} > 2\pi \hbar c / \Delta L$. In contrast, the two-photon interference pattern has shown quiet a high degree of visibility, which indicates

$$\Delta(\hbar \omega_s + \hbar \omega_i) \ll min(\Delta \hbar \omega_s, \Delta \hbar \omega_i).$$

In EPR's language, the energy of neither signal photon nor idler photon is defined in the course of their preparation and propagation; however, if one is measured with a certain value the other one must be measured with a unique value.

10.4 TWO-PHOTON GHOST INTERFERENCE

A two-photon interference experiment reported by Strekalov *et al.* in 1995 surprised the physics community. The experiment was named as "ghost" interference soon after the publication. The experiment itself is quite simple. The signal photon and the idler photon of SPDC are propagated to different directions to trigger two distant point-like photon counting detectors D_1 and D_2, respectively. On the way of its propagation, the signal passed a standard Young's double-slit, while the idler propagated freely to reach D_2. Due to the poor spatial coherence of the signal field, there is no observable standard first-order Young's interference when scanning D_1 transversely behind the double-slit. A high visibility second-order double-slit interference-diffraction pattern, however, was observed in the joint photodetection counting rate of D_1 and D_2 when D_2 was scanned across the "empty" idler beam while D_1 was placed in a fixed position behind the double-slit. The name of "ghost" was given because of the surprising nonlocal feature of the phenomenon.

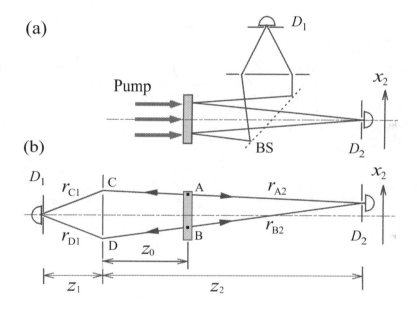

FIGURE 10.4.1 Simplified experimental scheme (a) and the "unfolded" version (b).

We now understood the observation is a two-photon interference phenomenon. The very special physics explored in the ghost interference experiment might have been its apparent nonlocal behavior: by scanning D_2 across the idler beam, how could one observe the interference pattern produced by the signal beam in distance?

The schematic experimental setup of the historical ghost interference experiment is illustrated in Fig. 10.4.1. A pair of orthogonal polarized signal-idler photon is prepared by a near-collinear degenerate type-II SPDC. The signal and the idler are separated by a polarization beamsplitter. The signal passes through a Young's double-silt (or single-slit) aperture and then travels about $1m$ to meet a point-like photon counting detector D_1. The idler travels to the far-field zone to feed into an optical fiber which is mated with a photon counting detector D_2. During the joint-detection measurement, D_1 is fixed at a point behind the double-slit while the horizontal transverse coordinate, x_2, of the fiber input tip, which is equivalent to that of D_2, is scanned by a step motor.

Figure 10.4.2 is the ghost interference-diffraction pattern published by Strekalov *et al.*. The coincidence counting rate is reported as a function of x_2, which is obtained by scanning D_2 (the fiber tip) across the idler beam, whereas the double-slit is in the signal beam. The

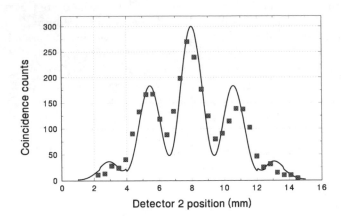

FIGURE 10.4.2 Typical observed interference-diffraction pattern. The solid curve is a theoretical fitting. The calculation has taken into account the finite size of D_1 and D_2, resulting in less than 100% interference visibility. In this measurement, D_1 was fixed in a symmetrical position behind the double-slit in the signal beam, while D_2 was scanned across the "empty" idler beam. If D_1 is moved to an asymmetrical point, which results in unequal distance to the two slits, the interference-diffraction pattern is observed to be simply shifted to one side.

Young's double-silt has a slit-width of $a = 0.15mm$ and slit-distance of $d = 0.47mm$. The interference period is measured to be $2.7 \pm 0.2mm$ and the half-width of the envelope is estimated to be about $8mm$. By curve fittings, it is easy to find that the observation is a standard Young's interference pattern, i.e., a sinusoidal oscillation with a sinc-function envelope:

$$R_c \propto \text{sinc}^2\left(\frac{\pi a x_2}{\lambda z_2}\right) \cos^2\left(\frac{\pi d x_2}{\lambda z_2}\right). \tag{10.4.1}$$

The interference pattern in Fig. 10.4.2 which is described by Eq. (10.4.1) was taken when D_1 was placed in a symmetrical point between the double-slit. If D_1 is moved to an asymmetrical point, which results in unequal distances to the two slits, the interference-diffraction pattern is shifted from the current symmetrical position to one side of x_2. Similar to the ghost imaging experiment, the remarkable feature here is that z_2 is the distance from the slits' plane, which is in the signal beam, back through BS to the SPDC crystal and then along the idler beam to the scanning fiber tip of detector D_2 (see Fig. 10.4.1). The calculated interference period and half-width of the sinc-function from Eq.(10.4.1) are $2.67mm$ and $8.4mm$, respectively.

Although the interference-diffraction pattern is observed in coincidences, *the single detector counting rates are both observed to be constant* when scanning detector D_1 and D_2. It seems reasonable not to have any interference modulation in the single counting rate of D_2, which is located in the "empty" idler beam. Of interest, however, is that the absence of the interference-diffraction structure in the single counting rate of D_1, which is behind the double-slit, is mainly due to the poor spatial coherence or the considerable large divergence of the signal beam, $\Delta\theta \gg \lambda/d$.

To explain the ghost interference, a simple model is presented in the following. The basic concept of the model is, again, two-photon interferometry of entangled state. In Chapter

9, we have introduced two EPR δ-functions $\delta(\vec{\rho}_s - \vec{\rho}_i)$ and $\delta(\vec{k}_s + \vec{k}_i)$ for near-collinear degenerate SPDC, which means: (1) the signal-idler pair may come out from any point on the output plane of the SPDC, however, if the signal is measured at a certain position the idler must be emitted from the same position; (2) the signal and idler may propagate to any directions around the pump beam, however, if the signal is observed in a certain direction the idler must be emitted to the opposite direction with equal angle relative to the pump. This peculiar entanglement nature of the signal-idler two-photon system determines the only two possible two-photon amplitudes in Fig. 10.4.1, when signal passes through the double-slit aperture while the idler triggers D_2. The coherent superposition is taken between these two-photon amplitudes.

Encouraged by the EPR δ-functions, Klyshko suggested an "Klyshko picture" to treat the SPDC crystal as a mirror in terms of "usual" geometrical optics in the following manner: we envision the output plane as a "hinge point" and "unfold" the schematic of Fig. 10.4.1a into that shown in Fig. 10.4.1b. Based on the unfolded Klyshko picture of Fig. 10.4.1b, we now give an quantitative calculation of the experiment.

The joint detection counting rate R_c is proportional to the probability of jointly detecting the signal-idler pair by detectors D_1 and D_2,

$$R_c \propto G^{(2)} = \langle \Psi | \hat{E}_1^{(-)} \hat{E}_2^{(-)} \hat{E}_2^{(+)} \hat{E}_1^{(+)} | \Psi \rangle = \left| \langle 0 | \hat{E}_2^{(+)} \hat{E}_1^{(+)} | \Psi \rangle \right|^2 \tag{10.4.2}$$

where $|\Psi\rangle$ is the two-photon state of SPDC. Let us simplify the mathematics by using the following "two-mode" expression for the state, bearing in mind that the EPR δ-functions have been taken into account based on the "straight line" picture of Fig. 10.4.1.

$$|\Psi\rangle = \epsilon \left[a_s^\dagger a_i^\dagger e^{i\varphi_A} + b_s^\dagger b_i^\dagger e^{i\varphi_B} \right] |0\rangle \tag{10.4.3}$$

where ϵ is a normalization constant that is proportional to the pump field (classical) and the nonlinearity of the crystal, φ_A and φ_B are the phases of the pump field at A and B, and $a_j^\dagger$ ($b_j^\dagger$) are the photon creation operators for the upper (lower) mode in Fig. 10.4.1 ($j = s, i$). In terms of the Copenhagen interpretation, one may say that the interference is due to the uncertainty in the birth-place (A or B in Fig. 10.4.1) of a signal-idler pair.

In Eq. (10.4.2) the fields at the detectors are given by:

$$\begin{aligned} \hat{E}_1^{(+)} &= \hat{a}_s \, exp(ik \, r_{A1}) + \hat{b}_s \exp(ik \, r_{B1}) \\ \hat{E}_2^{(+)} &= \hat{a}_i \, \exp(ik \, r_{A2}) + \hat{b}_i \exp(ik \, r_{B2}) \end{aligned} \tag{10.4.4}$$

where r_{Ai} (r_{Bi}) are the optical path lengths from region A (B) along the upper (lower) path to the ith detector. Substituting Eqs. (10.4.3) and (10.4.4) into Eq. (10.4.2),

$$G^{(2)} \propto \left| e^{i(kr_A + \varphi_A)} + e^{i(kr_B + \varphi_B)} \right|^2 \propto 1 + \cos[k \, (r_A - r_B)] \tag{10.4.5}$$

where we have assumed $\varphi_A = \varphi_B$ in the second line of Eq. (14.5.3). We have also defined the overall optical path lengths between the detectors D_1 and D_2 along the upper and lower paths (see Fig. 10.4.1): $r_A \equiv r_{A1} + r_{A2} = r_{C1} + r_{C2}$, $r_B \equiv r_{B1} + r_{B2} = r_{D1} + r_{D2}$, where r_{Ci} and r_{Di} are the respective path lengths from the slits C and D to the ith detector.

If the optical paths from the fixed detector D_1 to the two slits are equal, i.e., $r_{C1} = r_{D1}$, and if $z_2 \gg d^2/\lambda$ (far field), then $r_A - r_B = r_{C2} - r_{D2} \cong x_2 d/z_2$, and Eq. (10.4.5) can be written as:

$$R_c \propto \cos^2\left(\frac{\pi d x_2}{\lambda z_2}\right). \tag{10.4.6}$$

Equation (10.4.6) has the form of standard Young's double-slit interference pattern. Here, again, z_2 is the unusual distance from the slits plane, which is in the signal beam, back through BS to the crystal and then along the idler beam to the scanning fiber tip of detector D_2.

If the optical paths from the fixed detector D_1 to the two slits are unequal, i.e., $r_{C1} \neq r_{D1}$, the interference pattern will be shifted from the symmetrical position of Eq.(10.4.6) to an asymmetrical position of Eq. (10.4.5). This interesting phenomenon has been observed and discussed following the discussion of Fig. 10.4.2.

To calculate the "ghost" diffraction effect of a single-slit, we need an integral of the effective two-photon wavefunction over the slit width (the superposition of infinite number of probability amplitudes results in a click-click coincidence detection event):

$$R_c \propto \Big| \int_{-a/2}^{a/2} dx_0 \, \exp[-ik \, r(x_0, x_2)] \Big|^2 \cong \mathrm{sinc}^2 \Big(\frac{\pi a x_2}{\lambda z_2} \Big) \tag{10.4.7}$$

where $r(x_0, x_2)$ is the distance between points x_0 and x_2, x_0 belongs to the slit's plane, and the inequality $z_2 \gg a^2/\lambda$ is applied (far field approximation).

Repeating the above calculations, the combined interference-diffraction joint detection counting rate for the double-slit case is given by:

$$R_c \propto \mathrm{sinc}^2 \Big(\frac{\pi a x_2}{\lambda z_2} \Big) \cos^2 \Big(\frac{\pi d x_2}{\lambda z_2} \Big) \tag{10.4.8}$$

which is the same function as that of Eq. (10.4.1) obtained from experimental data fittings. If the finite size of the detectors and the divergence of the pump are taken into account by a convolution, the interference visibility will be reduced. These factors have been considered in the theoretical plots of Fig. 10.4.2.

Similar to Franson interferometer which demonstrated the nonlocal EPR inequality in energy, the ghost interference experiment has explored another nonlocal EPR inequality in momentum. As we know, the loss of first-order spatial coherence of the signal (idler) indicates a considerable large uncertainty in the transverse component of its momentum, $\Delta \vec{\kappa}_{s,i}$. In contrast, the two-photon interference pattern has shown quiet a high degree of two-photon spatial coherence, which indicates

$$\Delta(\vec{\kappa}_s + \vec{k}_i) \ll min(\Delta \vec{k}_s, \Delta \vec{\kappa}_i).$$

In EPR's language, the transverse momentum of neither signal photon nor idler photon is defined in the course of their preparation and propagation; however, if one is measured with a certain value the other one must be measured with a unique value despite the distance between the two measurements.

SUMMARY

In this chapter, we discussed the physics of biphoton interferometry. Two decades ago, based on the discovery of biphoton interference, Dirac was criticized to be mistaken because he stated that "... photon ... only interferes with itself". The debate about if biphoton interference is the interference between two photons began since that time. This chapter started from the question: Is two-photon interference the interference of two photons? Through the analysis of a few biphoton interference experiments, we concluded that two-photon interference is not the interference of two photons. Two-photon interference involves the superposition of two-photon amplitudes. Perhaps, Dirac's statement "...photon... only interferes with itself" is a bit confusing, we may modify his statement as follows:

"Interference is the result of the superposition of quantum amplitudes, a nonclassical entity corresponding to different yet indistinguishable alternatives which lead to a photodetection event or a joint photodetection event. Interference between different photons or photon pairs never occurs".

In this chapter, we analyzed several typical biphoton interference experiments, including biphoton correlation "peak" and anti-correlation "dip", Bell-type polarization correlation measurement of biphoton pairs, biphoton interference in Franson interferometer, and biphoton ghost interference. Through these analysis we demonstrated the details on the superposition of biphoton amplitudes, and the overlapping-convolution of the 2-D non-factorizable biphoton wavepackets along its $(\tau_1 - \tau_2)$ and/or $(\tau_1 + \tau_2)$ axes.

REFERENCES AND SUGGESTIONS FOR READING

[1] P.A. Dirac, *The principle of quantum mechanics*, Oxford University Press, 1982.

[2] G. Magyar and L. Mandel, Nature **198**, 255 (1963); R.L. Pfleegor and L. Mandel, Phys. Rev., **159** 1084 (1967).

[3] C.O. Alley and Y.H. Shih, *Foundations of Quantum Mechanics in the Light of New Technology*, Ed., M. Namiki, Physical Society of Japan, Tokyo, 47 (1986); Y.H. Shih and C.O. Alley, Phys. Rev. Lett., **61**, 2921, (1988).

[4] C.K. Hong, Z.Y. Ou and L. Mandel, Phys. Rev. Lett., **59**, 2044 (1987); Z.Y. Ou and L. Mandel, Phys. Rev. Lett., **62**, 50 (1988).

[5] D.V. Strekalov, T.B. Pittman, and Y.H. Shih, Phys. Rev. A, **57**, 567 (1998); T.B. Pittman *et al.*, Phys. Rev. Lett., **77**, 1917 (1996).

[6] Y.H. Shih and A.V. Sergienko, Phys. Rev. A, **50**, 2564 (1994); A.V. Sergienko, Y.H. Shih, and M.H. Rubin, JOSAB, **12**, 859 (1995).

[7] T.E. Kiess, Y.H. Shih, A.V. Sergienko, and C.O. Alley, Phys. Rev. Lett., **71**, 3893 (1993).

[8] P.G. Kwiat *et al.*, Phys. Rev. Lett., **75**, 4337 (1995).

[9] M.H. Rubin, D.N. Klyshko, Y.H. Shih, and A.V. Sergienko, Phys. Rev. A, **50**, 5122 (1994).

[10] Y.H. Shih, "Two-Photon Entanglement and Quantum Reality", *Advances in Atomic, Molecular, and Optical Physics*, ed., B. Bederson and H. Walther, Academic Press, Cambridge, 1997.

[11] J.D. Franson, Phys. Rev. Lett., **62**, 2205 (1989).

[12] P.G. Kwiat, A.M. Steinberg, and R.Y. Chiao, Phys. Rev. A, **47**, 2472 (1993).

[13] P.R. Tapster, J,G. Rarity, and P.C.M. Owens, Phys. Rev. Lett., **73**, 1923 (1994).

[14] D.V. Strekalov *et al.*, Phys. Rev. A **54**, R1 (1996).

[15] D.V. Strekalov, A.V. Sergienko, D.N. Klyshko, and Y.H. Shih, Phys. Rev. Lett., Vol. 74, 3600 (1995).

[16] P.G. Kwiat, A.M. Steinberg, and R.Y. Chiao, Phys. Rev. A, **45**, 7729 (1992).

CHAPTER 11

Two-Photon Interferometry II: Two-Photon Interference of Thermal Field

The study of entangled states greatly advanced our understanding about two-photon interferometry. Two-photon interference is not the interference between two photons, it is about a pair of photons interfering with the pair itself. In the language of quantum theory, two-photon interference is the result of superposition between different yet indistinguishable two-photon probability amplitudes. Is the concept of two-photon amplitude applicable only to the entangled states? Is two-photon interference occurring only with entangled photon pair? The answer is negative. In this chapter we study two-photon interference of "thermal field", or "thermal light". Thermal light is usually created from a natural light source, such as the sun, containing a large number of randomly radiated spontaneous atomic transitions. Thermal light is traditionally defined as "classical" light.[1] It is not the philosophy of this book to classify light into quantum and classical. Our interests are at (1) generalize the quantum theory of two-photon interferometry to thermal radiation: a randomly created and randomly paired photons interfering with the pair itself; (2) distinguish the quantum mechanical concept of nonlocal two-photon interference from the classical concept of local statistical correlation of light caused in the preparation process; (3) recognize the relationship between two-photon interference and intensity fluctuation correlation of thermal field: two-photon interference of thermal light is observed from the intensity fluctuation correlation, however, this type of intensity fluctuation correlation is not preprepared at the light source, instead this type of intensity fluctuation correlation is the result of two-photon interference. We may name it two-photon interference induced intensity fluctuation correlation. Thermal light may not be named as "correlated" radiation at all. There is no physical reason to assume the randomly created and randomly distributed photons in thermal state preprepared with "correlation", either "bunching" or "anti-bunching", at the radiation source.

Under certain conditions, the observed second-order interference can be factorized into a product of two individual classic first-order interferences. In this case, the interference is not only observable from the intensity fluctuation correlation measurement, or from the photon number fluctuation correlation measurement, of two photodetectors, but also observable in the counting rate of each individual photodetector. We consider this kind of two-photon

---

[1]There exist a number of definitions to classify "classical light" and "quantum light". One of the commonly used definitions considers thermal light classical because its positive P-function.

interference as trivial second-order phenomenon. We will avoid this type of second-order interference in this chapter by arranging the experimental condition in such a way under which no classic first-order interferences are observable.

It may not be uneasy to accept the quantum concept of two-photon interference for "classical" thermal radiation. However, this is not the first time in the history of physics we apply quantum mechanical concepts to "classical" light. We should not forget that it was Planck's theory of blackbody radiation originated the theory of quantum physics. Indeed, the radiation Planck dealt with was "classical" thermal light.

11.1 TWO-PHOTON INTERFERENCE BETWEEN SPATIALLY SEPARATED INCOHERENT THERMAL FIELDS

We start from a simple Young's double-slit, or double-pinhole, interferometer which is schematically illustrated in Figure 11.1.1. This interferometer is the same as the classic Young's double-slit, or double-pinhole, interferometer, except (1) the measurements are not only $\langle n(x_1) \rangle \propto \langle I(x_1) \rangle$ and $\langle n(x_2) \rangle \propto \langle I(x_2) \rangle$, but also $\langle \Delta n(x_1) \Delta n(x_2) \rangle \propto \langle \Delta I(x_1) \Delta I(x_2) \rangle$. (2) The separation between the upper slit-A and the lower slit-B is much greater than the spatial coherence length of the thermal field, $d \gg l_c$, i.e., the thermal fields at pinhole A and at pinhole B are first-order incoherent $G^{(1)}(\mathbf{r}_A, t_A; \mathbf{r}_B, t_B) = 0$ and thus $\langle \Delta n(\mathbf{r}_A) \Delta n(\mathbf{r}_B) \rangle \propto \langle \Delta I(\mathbf{r}_A) \Delta I(\mathbf{r}_B) \rangle = 0$. No first-order interferences are observable from $\langle n(x_1) \rangle \propto \langle I(x_1) \rangle$ and $\langle n(x_2) \rangle \propto \langle I(x_2) \rangle$. Do we expect observing interference in the photon number fluctuation correlation measurement $\langle \Delta n(x_1) \Delta n(x_2) \rangle \propto \langle \Delta I(x_1) \Delta I(x_2) \rangle$ when scanning D_1 and/or D_2 along their x-axis?

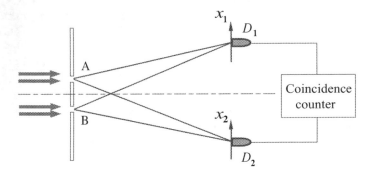

FIGURE 11.1.1 Schematic of a simple Young's double-slit, or double-pinhole, interference experiment. The separation between the upper slit-A and the lower slit-B is much greater than the spatial coherence length of the thermal field, $d \gg l_c$, i.e., the thermal fields at pinhole A and at pinhole B are first-order incoherent $G^{(1)}(\mathbf{r}_A, t_A; \mathbf{r}_B, t_B) = 0$. Two point-like photodetectors, D_1 and D_2, are scannable along their x-axises in the far-field of the interferometer. The single-detector counting rates of D_1 and D_2 are monitored respectively; and the coincidence counting rate of D_1 and D_2 is monitored jointly, during the scanning of D_1 and D_2. To simplify the mathematics, we assume line-like slits, or pint-like pinholes.

In the view of quantum theory of light, the single-detector counting rate monitors the first-order interference, which is the result of superposition between single-photon amplitudes. The joint photodetection counting rate observes the second-order interference, which is the result of superposition between two-photon amplitudes. In this experiment, fields A and B are first-order incoherent, i.e., the first-order mutual coherence function $G^{(1)}(\mathbf{r}_A, t_A; \mathbf{r}_B, t_B) = 0$ $G^{(1)}(\mathbf{r}_A, t_A; \mathbf{r}_B, t_B) = 0$ and thus $\langle \Delta n(\mathbf{r}_A) \Delta n(\mathbf{r}_B) \rangle \propto \langle \Delta I(\mathbf{r}_A) \Delta I(\mathbf{r}_B) \rangle = 0$. This condition is achieved by either using two independent light sources with random relative phase, or by separating the double slits, or pinholes, beyond the spatial coherence area of a thermal field. Under this experimental condition, obviously, there is no observable first-order interference. The counting rate of D_1 and D_2 are both constants during their scanning. Do we expect to observe interference modulation in the joint photodetection D_1 and D_2 when $G^{(1)}(\mathbf{r}_A, t_A; \mathbf{r}_B, t_B) = 0$?

(I) Two-photon Young's interference in Einstein's granularity picture of light

In the following, we calculate the two-photon Young's interference in the joint measurement of $\langle n_1 n_2 \rangle \propto \langle I_1 I_2 \rangle \propto G^{(2)}(x_1, t_1; x_2, t_2)$ in Einstein's granularity picture,

$$
\begin{aligned}
&G^{(2)}(x_1, t_1; x_2, t_2) \\
&= \Big\langle \sum_{m,n,p,q} E_m^*(x_1, t_1) E_n(x_1, t_1) E_p^*(x_2, t_2) E_q(x_2, t_2) \Big\rangle \\
&= \sum_{m=n} E_m^*(x_1, t_1) E_m(x_1, t_1) \sum_{m=n} E_n^*(x_2, t_2) E_n(x_2, t_2) \\
&\quad + \sum_{m \neq n} E_m^*(x_1, t_1) E_n(x_1, t_1) E_n^*(x_2, t_2) E_m(x_2, t_2) \\
&\simeq \sum_{m,n} \big| E_m(x_1, t_1) E_n(x_2, t_2) + E_n(x_1, t_1) E_m(x_2, t_2) \big|^2 \\
&= \sum_{m,n} \big| (E_{mA1} + E_{mB1})(E_{nA2} + E_{nB2}) + (E_{nA1} + E_{nB1})(E_{mA2} + E_{mB2}) \big|^2 \\
&= \sum_{m,n} \big| E_{mA1} E_{nA2} + E_{mA1} E_{nB2} + E_{mB1} E_{nA2} + E_{mB1} E_{nB2} \\
&\qquad + E_{nA1} E_{mA2} + E_{nA1} E_{mB2} + E_{nB1} E_{mA2} + E_{nB1} E_{mB2} \big|^2
\end{aligned}
\tag{11.1.1}
$$

where we have approximated the sum of $(m \neq n)$ to the sum of (m, n) by ignoring the $m = n$ term, which is reasonable for a large number of subfields. We have also used short-hand notation E_{mAj} and E_{mBj}, $j = 1, 2$, to specify the subfields at D_j that coming from pinhole-A and pinhole-B, respectively. Completing the sum for each term in the $|E_{mB1} E_{nA2} + \dots + E_{nA1} E_{mB2}|^2$, we find that a number of cross terms vanish due to the experimental condition $d > l_c$. We may group the surviving terms into the following four superpositions,

$$
\begin{aligned}
&G^{(2)}(x_1, t_1; x_2, t_2) \\
&= \sum_{m,n} \big| E_{mA1} E_{nA2} + E_{nA1} E_{mA2} \big|^2 + \sum_{m,n} \big| E_{mB1} E_{nB2} + E_{nB1} E_{mB2} \big|^2 \\
&\quad + \sum_{m,n} \big| E_{mA1} E_{nB2} + E_{nB1} E_{mA2} \big|^2 + \sum_{m,n} \big| E_{mB1} E_{nA2} + E_{nA1} E_{mB2} \big|^2 \\
&\equiv G_{AA}^{(2)}(x_1, t_1; x_2, t_2) + G_{BB}^{(2)}(x_1, t_1; x_2, t_2) \\
&\quad + G_{AB}^{(2)}(x_1, t_1; x_2, t_2) + G_{BA}^{(2)}(x_1, t_1; x_2, t_2).
\end{aligned}
\tag{11.1.2}
$$

We may calculate $G_{A1,A2}^{(2)}$, $G_{B1,B2}^{(2)}$, $G_{A1,B2}^{(2)}$, and $G_{B1,A2}^{(2)}$, separately, as four groups of distinguishable two-photon interferences. Focusing on the second-order spatial correlation and simplifying the the mathematics, in the following calculations, we assume perfect second-order temporal coherence with idealized photodetectors and idealized correlation circuit. We will write out the spatial variables only as usual.

$$G_{AA}^{(2)}(x_1; x_2)$$

$$= \sum_{m,n} |E_m g_{mA}(x_1) E_n g_{nA}(x_2) + E_n g_{nA}(x_1) E_m g_{mA}(x_2)|^2$$

$$\propto \sum_m |E_m g_{mA}(x_1)|^2 \sum_n |E_n g_{nA}(x_2)|^2$$

$$+ \sum_n |E_n g_{nA}(x_1)|^2 \sum_m |E_m g_{mA}(x_2)|^2$$

$$+ \sum_{m,n} E_m^* g_{mA}(x_1) E_n g_{nA}(x_1) E_n^* g_{nA}(x_2) E_m g_{mA}(x_2)$$

$$+ \sum_{m,n} E_n^* g_{nA}(x_1) E_m g_{mA}(x_1) E_m^* g_{mA}(x_2) E_n g_{nA}(x_2)$$

$$\propto \langle I(x_1) \rangle \langle I(x_2) \rangle + \langle \Delta I(x_1) \Delta I(x_2) \rangle$$

$$= G_0^{\langle I_1 \rangle \langle I_2 \rangle} + G_0^{\langle \Delta I_1 \Delta I_2 \rangle} \tag{11.1.3}$$

which is the result of a superposition between two-photon amplitudes: (1) the mth subfield passes pinhole-A then propagates to D_1 and the nth subfield passes pinhole-A then propagates to D_2; (2) the mth subfield passes pinhole-A then propagates to D_2 and the nth subfield passes pinhole-A then propagates to D_1. Note, we have especially labeled the two constants with upper indexes of $\langle n_1 \rangle \langle n_2 \rangle$ and $\langle \Delta n_1 \Delta n_2 \rangle$, although they have equal values.

$$G_{AB}^{(2)}(x_1; x_2)$$

$$= \sum_{m,n} |E_m g_{mA}(x_1) E_n g_{nB}(x_2) + E_n g_{nB}(x_1) E_m g_{mA}(x_2)|^2$$

$$\propto \sum_m |E_m g_{mA}(x_1)|^2 \sum_n |E_n g_{nB}(x_2)|^2$$

$$+ \sum_n |E_n g_{nA}(x_1)|^2 \sum_m |E_m g_{mB}(x_2)|^2$$

$$+ \sum_{m,n} E_m^* g_{mA}(x_1) E_n g_{nB}(x_1) E_n^* g_{nB}(x_2) E_m g_{mA}(x_2)$$

$$+ \sum_{m,n} E_n^* g_{nA}(x_1) E_m g_{mB}(x_1) E_m^* g_{mB}(x_2) E_n g_{nA}(x_2)$$

$$= \langle I(x_1) \rangle \langle I(x_2) \rangle + \langle \Delta I(x_1) \Delta I(x_2) \rangle$$

$$= G_0^{\langle I_1 \rangle \langle I_2 \rangle} + G_0^{\langle \Delta I_1 \Delta I_2 \rangle} \cos \frac{2\pi d}{\lambda z}(x_1 - x_2) \tag{11.1.4}$$

which is the result of a superposition between two-photon amplitudes; (3) the mth subfield passes pinhole-A then propagates to D_1 and the nth subfield passes pinhole-B then propagates to D_2; (4) the mth subfield passes pinhole-A then propagates to D_2 and the nth subfield passes pinhole-B then propagates to D_1.

In Eqs. (11.1.3) and (11.1.4), $g_m(x_j)$ is the Green's function that propagates the mth subfield from the source to D_j at (x_j), $j = 1, 2$,

$$g_{mA}(x_j) = g_m(\mathbf{r}_A) g_A(x_j)$$
$$g_{nB}(x_j) = g_n(\mathbf{r}_B) g_B(x_j)$$

with $g_m(\mathbf{r}_A)$ [$g_n(\mathbf{r}_B)$] representing the Green's function propagating the mth (nth) subfield from the mth (nth) sub-source to pinhole-A (pinhole-B), $g_A(x_j)$ and $g_B(x_j)$ the Green's functions propagating the subfields from pinhole-A and pinhole-B to D_j at (x_j), respectively, where A labels the upper path (passing pinhole-A), B labels the lower path (passing pinhole-B). Note, in the above calculations, we have assumed line-like slits, or point-like pinholes, i.e., the size of the slits or pinholes can be approximated to be infinitely small.

$G_{BB}^{(2)}(x_1; x_2)$ and $G_{BA}^{(2)}(x_1; x_2)$ can be calculated in a similar way,

$$
\begin{aligned}
&G_{BB}^{(2)}(x_1; x_2) \\
&= \sum_{m,n} |E_m g_{mB}(x_1) E_n g_{nB}(x_2) + E_n g_{nB}(x_1) E_m g_{mB}(x_2)|^2 \\
&= G_0^{\langle I_1 \rangle \langle I_2 \rangle} + G_0^{\langle \Delta I_1 \Delta I_2 \rangle}
\end{aligned}
\tag{11.1.5}
$$

which is the result of a superposition between two-photon amplitudes; (5) the mth subfield passes pinhole-B then propagates to D_1 and the nth subfield passes pinhole-B then propagates to D_2; (6) the mth subfield passes pinhole-B then propagates to D_2 and the nth subfield passes pinhole-B then propagates to D_1.

$$
\begin{aligned}
&G_{BA}^{(2)}(x_1; x_2) \\
&= \sum_{m,n} |E_m g_{mB}(x_1) E_n g_{nA}(x_2) + E_n g_{nA}(x_1) E_m g_{mB}(x_2)|^2 \\
&= G_0^{\langle I_1 \rangle \langle I_2 \rangle} + G_0^{\langle \Delta I_1 \Delta I_2 \rangle} \cos \frac{2\pi d}{\lambda z}(x_1 - x_2)
\end{aligned}
\tag{11.1.6}
$$

which is the result of a superposition between two-photon amplitudes; (7) the mth subfield passes pinhole-B then propagates to D_1 and the nth subfield passes pinhole-A then propagates to D_2; and (8) the mth subfield passes pinhole-B then propagates to D_2 and the nth subfield passes pinhole-A then propagates to D_1.

When scanning D_1 and D_2 along their x-axises, taking into account all possible joint photodetection events contributed from $G_{A1,A2}^{(2)}$, $G_{B1,B2}^{(2)}$, $G_{A1,B2}^{(2)}$, and $G_{B1,A2}^{(2)}$, the jointly measured photon number correlation, which is proportional to the intensity correlation, of D_1 and D_2 is thus

$$
\langle n_1(x_1) n_2(x_2) \rangle \propto 1 + \frac{1}{2} \cos \frac{2\pi d}{\lambda z}(x_1 - x_2).
\tag{11.1.7}
$$

A sinusoidal two-photon interference pattern with 50% contrast is observable from the joint measurement $\langle n_1 n_2 \rangle$, despite the experimental condition of $d > l_c$.

Adding only the photon number fluctuation correlations, or the intensity fluctuation correlations, calculated in $G_{A1,A2}^{(2)}$, $G_{B1,B2}^{(2)}$, $G_{A1,B2}^{(2)}$, and $G_{B1,A2}^{(2)}$, the jointly measured photon number fluctuation correlation, which is proportional to the intensity fluctuation correlation, of D_1 and D_2 is therefore

$$
\langle \Delta n_1(x_1) \Delta n_2(x_2) \rangle \propto 1 + \cos \frac{2\pi d}{\lambda z}(x_1 - x_2).
\tag{11.1.8}
$$

A sinusoidal two-photon interference pattern with 100% visibility is observable from the joint measurement of $\langle \Delta n_1 \Delta n_2 \rangle$, despite the experimental condition of $d > l_c$.

Differ from classic Young's double-pinhole interference, it is interesting to find that the Green's functions $g_m(\mathbf{r}_A, t_A)$ and $g_m(\mathbf{r}_B, t_B)$, which propagate the subfields from the source plane to the double-pinhole plane, have no effects on the two-photon interferences. What is the reason for $g_m(\mathbf{r}_A, t_A)$ and $g_m(\mathbf{r}_B, t_B)$ contribute to the first-order interference but not to the second-order interference? Let us first exam the classic Young's double-pinhole interferences which are, respectively, measured by D_1 and D_2,

$$\langle I(x_j) \rangle = \sum_m \left| E_m\, g_m(x_j) \right|^2$$

$$= \sum_m \left| E_m\, \frac{1}{\sqrt{2}} [g_{mA}(x_j) + g_{mB}(x_j)] \right|^2$$

$$= \sum_m |E_m|^2 |\frac{1}{\sqrt{2}} [g_m(\mathbf{r}_A)g_A(x_j) + g_m(\mathbf{r}_B, t_B)g_B(x_j)] |^2$$

$$\simeq I_0 \left[1 + \mathrm{sinc}\frac{\pi d \Delta\theta_s}{\lambda} \cos\frac{2\pi d}{\lambda z} x_j \right] \tag{11.1.9}$$

where $g_m(x_j)$ is the Green's function that propagates the mth subfield from the source to D_j at (x_j), $j = 1, 2$. The mth subfield, or photon, has two different yet indistinguishable alternatives to produce a photodetection event of D_j: (1) the mth subfield, or photon, passes upper-pinhole A and then is annihilated by D_j; (2) the mth subfield, or photon, passes lower-pinhole B and then is annihilated by D_j:

$$E(x_j) = \sum_m E_m(x_j) = \sum_m \frac{E_m}{\sqrt{2}} [g_{mA}(x_j) + g_{mB}(x_j)]] \tag{11.1.10}$$

In Eq.(11.1.9), the sinc-function is defined as $\mathrm{sinc}(x) \equiv \sin(x)/x$, which is the result of $\sum_m g_m^*(\mathbf{r}_A)g_m(\mathbf{r}_B)$ in the cross term of the above first-order superposition:

$$\sum_m g_m^*(\mathbf{r}_A)g_m(\mathbf{r}_B) \propto \int_{-\frac{\theta_s}{2}}^{\frac{\theta_s}{2}} d\theta_m\, e^{-i\frac{2\pi d}{\lambda}\theta_m} \simeq \mathrm{sinc}\frac{\pi d \Delta\theta_s}{\lambda} \tag{11.1.11}$$

where θ_m is the 1-D angular coordinate of the mth subfield and $\Delta\theta_s$ is the angular diameter of the light source relative to the double-pinhole interferometer. The sinc-function determines the visibility, or the contrast, of the classic Young's double-pinhole interference. The interference reaches its maximum visibility when the argument of the sinc-function is assigned with a value closer to zero. This can be achieved by applying a point-like source, $\Delta\theta_s \sim 0$, or other mechanisms to obtain $(d\Delta\theta_s)/\lambda \sim 0$. The interference vanishes when setting the argument of the sinc-function with a value of π or greater, for example achieving $d \gg \lambda/\Delta\theta_s$ as we have arranged in this experiment. Similar calculation of the above sinc-function has been given in Chapter 6.

On the other hand, in the joint measurement of $\langle n(x_1)n(x_2) \rangle$, it is the cross terms of two-photon superposition in $G_{A1,B2}^{(2)}$ and $G_{B1,A2}^{(2)}$ contribute sinusoidal modulations. For example the cross term of $G_{AB}^{(2)}(x_1, x_2)$:

$$\sum_{m,n} E_m^* g_m^*(x_A)g_A^*(x_1) E_n g_n(x_B)g_B(x_1)$$

$$\times E_n^* g_n^*(x_B)g_B^*(x_2) E_m g_m(x_A)g_A(x_2)$$

$$\propto \sum_{m,n} \left[g_m^*(x_A)g_m(x_A)\, g_n(x_B)g_n^*(x_B) \right]$$

$$\times g_A^*(x_1)g_B(x_1)g_A(x_2)g_B^*(x_2)$$

$$\propto e^{-i\frac{2\pi d}{\lambda z}(x_1-x_2)}, \tag{11.1.12}$$

where $g_m^*(x_A)g_m(x_A)$ and $g_n(x_B)g_n^*(x_B)$ are normalized to one as usual, which means the visibility of the two-photon interference does not depend on the angular size of the light source and thus the spatial coherence length l_c.

Next, we show that the Young's interference calculated from Einsteins' picture is consistent with that predicted of quantum theory of light. For comparison, two simple quantum mechanical analyses are given as follows.

(II) Two-photon Young's interference in single-photon state representation.

In the following, we calculate $\langle n(x_1, t_1) \, n(x_2, t_2) \rangle$ from the single-photon state representation:

$$|\Psi\rangle \simeq \prod_m \left(|0\rangle + \epsilon \, \hat{a}_m^\dagger |0\rangle \right)$$

$$\simeq |0\rangle + \epsilon \sum_m \hat{a}_m^\dagger |0\rangle + \epsilon^2 \sum_{m<n} \hat{a}_m^\dagger \hat{a}_n^\dagger |0\rangle + \dots \qquad (11.1.13)$$

Under weak light condition, we may keep the necessary lowest-order terms for calculating the counting rate of D_1 and D_2, respectively, and the coincidence counting rate of D_1 and D_2, jointly. The joint photodetection counting rate is proportional to the second-order coherence function $G^{(2)}(x_1, t_1; x_2, t_2)$,

$$G^{(2)}(x_1, t_1; x_2, t_2)$$
$$= \left\langle \langle \hat{E}^{(-)}(x_1, t_1) \hat{E}^{(-)}(x_2, t_2) \hat{E}^{(+)}(x_2, t_2) \hat{E}^{(+)}(x_1, t_1) \rangle_{\text{QM}} \right\rangle_{\text{En}}$$
$$\simeq \left\langle \left| \langle 0 | \sum_p [\hat{E}_{pA2}^{(+)} + \hat{E}_{pB2}^{(+)}] \sum_q [\hat{E}_{qA1}^{(+)} + \hat{E}_{qB1}^{(+)}] \sum_{m,n} \hat{a}_m^\dagger \hat{a}_n^\dagger |0\rangle \right|^2 \right\rangle_{\text{En}}$$
$$= \sum_{m,n} \left| \psi_{mA1} \psi_{nA2} + \psi_{mA1} \psi_{nB2} + \psi_{mB1} \psi_{nA2} + \psi_{mB1} \psi_{nB2} \right.$$
$$\left. + \psi_{nA1} \psi_{mA2} + \psi_{nA1} \psi_{mB2} + \psi_{nB1} \psi_{mA2} + \psi_{nB1} \psi_{mB2} \right|^2 \qquad (11.1.14)$$

with

$$\psi_{mAj} = \langle 0 | \hat{a}_m \hat{a}_m^\dagger |0\rangle e^{ik(r_{mA} + r_{Aj})}, \quad \psi_{nBj} = \langle 0 | \hat{a}_n \hat{a}_n^\dagger |0\rangle e^{ik(r_{nB} + r_{Bj})}$$

the effective wavefunctions of the mth photon and the nth photon measured by D_j, where the subindex A and B label slit-A and slit-B, respectively, r_{mA} (r_{nB}) is the optical path between the mth (nth) sub-source and slit-A (slit-B), r_{Aj} (r_{Bj}) is the optical path connecting slit-A (slit-B) and D_j. Again, to simplify the mathematics, perfect second-order temporal correlation, or monochromatic thermal field, is applied to the above second-order spatial coherence calculation, as usual.

Similar to the calculation in Einstein's granularity picture, completing the sum for each term in the $|\psi_{mB1}\psi_{nA2} + \dots + \psi_{nA1}\psi_{mB2}|^2$, we find that a number of cross terms vanish due to the experimental condition $d > l_c$. We may group the surviving terms into the following four superpositions,

$$G^{(2)}(x_1, t_1; x_2, t_2)$$

$$= \sum_{m,n} \left| \psi_{mA1} \psi_{nA2} + \psi_{nA1} \psi_{mA2} \right|^2 + \sum_{m,n} \left| \psi_{mB1} \psi_{nB2} + \psi_{nB1} \psi_{mB2} \right|^2$$

$$+ \sum_{m,n} \left| \psi_{mA1} \psi_{nB2} + \psi_{nB1} \psi_{mA2} \right|^2 + \sum_{m,n} \left| \psi_{mB1} \psi_{nA2} + \psi_{nA1} \psi_{mB2} \right|^2$$

$$\equiv G^{(2)}_{AA}(x_1, t_1; x_2, t_2) + G^{(2)}_{BB}(x_1, t_1; x_2, t_2)$$

$$+ G^{(2)}_{AB}(x_1, t_1; x_2, t_2) + G^{(2)}_{B}(x_1, t_1; x_2, t_2). \tag{11.1.15}$$

Under the assumption of perfect second-order temporal coherence, it is not difficult to see that the two-photon interferences in $G^{(2)}_{AA}(x_1, x_2)$ and $G^{(2)}_{BB}(x_1, x_2)$ contribute two constants to the joint photodetection of D_1 and D_2, for example:

$$G^{(2)}_{AA}(x_1, x_2)$$

$$= \sum_{m,n} \left| \psi_{mA1} \psi_{nA2} + \psi_{nA1} \psi_{mA2} \right|^2$$

$$\propto \sum_{m} \left| \psi_{mA}(x_1) \right|^2 \sum_{n} \left| \psi_{nA}(x_2) \right|^2 + \sum_{m,n} \psi^*_{mA}(x_1) \psi_{nA}(x_1) \psi^*_{nA}(x_2) \psi_{mA}(x_2)$$

$$+ \sum_{n} \left| \psi_{nA}(x_1) \right|^2 \sum_{m} \left| \psi_{mA}(x_2) \right|^2 + \sum_{m,n} \psi_{mA}(x_1) \psi^*_{nA}(x_1) \psi_{nA}(x_2) \psi^*_{mA}(x_2)$$

$$= G^{\langle n_1 \rangle \langle n_2 \rangle}_0 + G^{\langle \Delta n_1 \Delta n_2 \rangle}_0 \tag{11.1.16}$$

where, again, we have especially labeled the two constants with upper indexes of $\langle n_1 \rangle \langle n_2 \rangle$ and $\langle n_1 n_2 \rangle$, although they have equal values.

It is $G^{(2)}_{AB}(x_1, x_2)$ and $G^{(2)}_{BA}(x_1, x_2)$ contribute a two-photon interference to the joint photodetection of D_1 and D_2, for example:

$$G^{(2)}_{AB}(x_1, x_2)$$

$$= \sum_{m,n} \left| \psi_{mA1} \psi_{nB2} + \psi_{nB1} \psi_{mA2} \right|^2$$

$$= \sum_{m} \left| \psi_{mA}(x_1) \right|^2 \sum_{n} \left| \psi_{nB}(x_2) \right|^2 + \sum_{m,n} \psi^*_{mA}(x_1) \psi_{nB}(x_1) \psi^*_{nB}(x_2) \psi_{mA}(x_2)$$

$$+ \sum_{n} \left| \psi_{nB}(x_1) \right|^2 \sum_{m} \left| \psi_{mA}(x_2) \right|^2 + \sum_{m,n} \psi_{mA}(x_1) \psi^*_{nB}(x_1) \psi_{nB}(x_2) \psi^*_{mA}(x_2)$$

$$= G^{\langle n_1 \rangle \langle n_2 \rangle}_0 + G^{\langle \Delta n_1 \Delta n_2 \rangle}_0 \cos \frac{2\pi d}{\lambda z}(x_1 - x_2). \tag{11.1.17}$$

The photon number correlation which is proportional to the second-order spatial coherence function is therefore a two-photon interference pattern with 50% contrast (33% visibility):

$$\langle n_1(x_1) n_2(x_2) \rangle \propto 1 + \frac{1}{2} \cos \frac{2\pi d}{\lambda z}(x_1 - x_2). \tag{11.1.18}$$

If the correlation measurement is designed for the photon number fluctuations only, the four cross terms of the superposition in $G^{(2)}_{AA}(x_1, x_2)$, $G^{(2)}_{BB}(x_1, x_2)$, $G^{(2)}_{AB}(x_1, x_2)$ and $G^{(2)}_{BA}(x_1, x_2)$ contribute a two-photon interference pattern with 100% visibility (100% contrast):

$$\langle \Delta n_1(x_1) \, \Delta n_2(x_2) \rangle \propto 1 + \cos \frac{2\pi d}{\lambda z}(x_1 - x_2). \tag{11.1.19}$$

(III) Two-photon Young's interference in coherent state representation.

Assuming a large number of randomly created photons, such as a natural radiation; or a large number of randomly created groups of identical photons, such as a pseudo-thermal field scattered from laser beam by rotating ground glass, in thermal state:

$$|\Psi\rangle = \prod_m |\{\alpha_m\}\rangle \tag{11.1.20}$$

which is written in the quantum coherent state representation. Coherent state representation simplifies the calculation of second-order coherence function significantly, and also made it possible to calculate the interference of a pair of two groups of identical photons interfering with the pair itself. The coherent state may represent a group of identical photons with $\bar{n} = |\alpha| \gg 1$. To not bring additional confusion to the simple physics we intended to discuss here, we will restrict our discussion at single-photon level, $\bar{n} = |\alpha| \ll 1$. Under the condition of $\bar{n} = |\alpha| \ll 1$, we may consider the state a single-photon state in which m labels the mth quantized radiation or photon created from the mth atomic transition.

Following Glauber's theory, the probability to produce a joint photodetection event at D_1 and D_2 shown in Fig. 11.1.1 is proportional to the second-order coherence function $G^{(2)}(x_1, t_1; x_2, t_2)$,

$$
\begin{aligned}
&G^{(2)}(x_1, t_1; x_2, t_2) \\
&= \left\langle \langle \hat{E}^{(-)}(x_1, t_1)\hat{E}^{(-)}(x_2, t_2)\hat{E}^{(+)}(x_2, t_2)\hat{E}^{(+)}(x_1, t_1)\rangle_{\text{QM}} \right\rangle_{\text{En}} \\
&= \left\langle |\langle \Psi|\hat{E}^{(+)}(x_2, t_2)\hat{E}^{(+)}(x_1, t_1)|\Psi\rangle|^2 \right\rangle_{\text{En}} \\
&= \left\langle \Big| \prod_m \langle\{\alpha_m\}| \sum_p [\hat{E}_{pA}^{(+)}(x_2, t_2) + \hat{E}_{pB}^{(+)}(x_2, t_2)] \right. \\
&\qquad \left. \times \sum_q [\hat{E}_{qA}^{(+)}(x_1, t_1) + \hat{E}_{qB}^{(+)}(x_1, t_1)] \prod_n |\{\alpha_n\}\rangle \Big|^2 \right\rangle_{\text{En}} \tag{11.1.21} \\
&= \sum_{m,n} \big| \psi_{mA1}\psi_{nA2} + \psi_{mA1}\psi_{nB2} + \psi_{mB1}\psi_{nA2} + \psi_{mB1}\psi_{nB2} \\
&\qquad + \psi_{nA1}\psi_{mA2} + \psi_{nA1}\psi_{mB2} + \psi_{nB1}\psi_{mA2} + \psi_{nB1}\psi_{mB2} \big|^2
\end{aligned}
$$

with

$$\psi_{mAj} = \alpha_m e^{ik(r_{mA} + r_{Aj})}, \quad \psi_{nBj} = \alpha_n e^{ik(r_{nB} + r_{Bj})}$$

the effective wavefunctions of the mth photon and the nth photon measured by D_j, where the subindex A and B label slit-A and slit-B, respectively, r_{mA} (r_{nB}) is the optical path between the mth (nth) sub-source and slit-A (slit-B), r_{Aj} (r_{Bj}) is the optical path connecting slit-A (slit-B) and D_j. To simplify the mathematics, perfect second-order temporal correlation, or monochromatic thermal field is applied to the above second-order spatial coherence calculation, as usual.

Similar to the calculation in Einstein's granularity picture, completing the sum for each term in the $|\psi_{mB1}\psi_{nA2} + ... + \psi_{nA1}\psi_{mB2}|^2$, we find that a number of cross terms vanish due to the experimental condition $d > l_c$. We may group the surviving terms into the following four superpositions,

$$G^{(2)}(x_1, t_1; x_2, t_2)$$
$$= \sum_{m,n} \left| \psi_{mA1}\psi_{nA2} + \psi_{nA1}\psi_{mA2} \right|^2 + \sum_{m,n} \left| \psi_{mB1}\psi_{nB2} + \psi_{nB1}\psi_{mB2} \right|^2$$
$$+ \sum_{m,n} \left| \psi_{mA1}\psi_{nB2} + \psi_{nB1}\psi_{mA2} \right|^2 + \sum_{m,n} \left| \psi_{mB1}\psi_{nA2} + \psi_{nA1}\psi_{mB2} \right|^2$$
$$\equiv G_{AA}^{(2)}(x_1, t_1; x_2, t_2) + G_{BB}^{(2)}(x_1, t_1; x_2, t_2)$$
$$+ G_{AB}^{(2)}(x_1, t_1; x_2, t_2) + G_B^{(2)}(x_1, t_1; x_2, t_2). \tag{11.1.22}$$

The above nonlocal superposition represents a quantum phenomenon, namely "two-photon interference": two randomly created and randomly paired photons interfering with the pair itself. Different from Einstein's picture which is still under the framework of electromagnetic theory of light, effective wavefunction represents a concept of quantum mechanics: probability amplitude.

It is unnecessary to repeat the rest of calculations that we have completed in Einstein's granularity picture, replacing the quantized subfield with the effective wavefunction of photon, and assuming perfect second-order temporal coherence, we obtain the same results:

$$\langle n_1(x_1) n_2(x_2) \rangle \propto 1 + \frac{1}{2}\cos\frac{2\pi d}{\lambda z}(x_1 - x_2)$$
$$\langle \Delta n_1(x_1) \Delta n_2(x_2) \rangle \propto 1 + \cos\frac{2\pi d}{\lambda z}(x_1 - x_2). \tag{11.1.23}$$

(IV) Experimental demonstration of two-photon Young's interferometer

Smith and Shih demonstrated the above two-photon Young's double-slit interference phenomenon recently. Their experimental setup is schematically illustrated in Fig. 11.1.2.

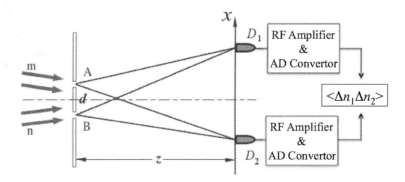

FIGURE 11.1.2 Two-photon Young's double-slit interference experiment. The interferometer is a standard Young's double-slit interferometer, except (1) the measurements are not only $\langle n(x_1) \rangle \propto \langle I(x_1) \rangle$ and $\langle n(x_2) \rangle \propto \langle I(x_2) \rangle$, but also $\langle \Delta n(x_1) \Delta n(x_2) \rangle \propto \langle \Delta I(x_1) \Delta I(x_2) \rangle$. (2) The separation between the upper slit-A and the lower slit-B is much greater than the spatial coherence length of the thermal field, $d \gg l_c$. Consequently, no first-order interferences are observable from $\langle n(x_1) \rangle \propto \langle I(x_1) \rangle$ and $\langle n(x_2) \rangle \propto \langle I(x_2) \rangle$. Furthermore, the experimental condition also means $\langle \Delta n(\mathbf{r}_A) \Delta n(\mathbf{r}_B) \rangle \propto \langle \Delta I(\mathbf{r}_A) \Delta I(\mathbf{r}_B) \rangle = 0$. Do we expect observing interference in the photon number fluctuation correlation measurement $\langle \Delta n(x_1, t_1) \Delta n(x_2, t_2) \rangle \propto \langle \Delta I(x_1, t_1) \Delta I(x_2, t_2) \rangle$ when scanning D_1 and/or D_2 along the x-axis?

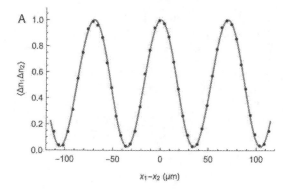

FIGURE 11.1.3 A typically two-photon Young's interference pattern is observed from $\langle \Delta n(x_1)\Delta n(x_2)\rangle$ under the experimental condition of $d \gg l_c$. During the scanning of D_1 and D_2, their single-detector counts are also monitored $\langle n(x_1)\rangle \sim$ constant and $\langle n(x_2)\rangle \sim$ constant. The absence of the first-order interferences guarantees $G_{AB}^{(1)} = 0$.

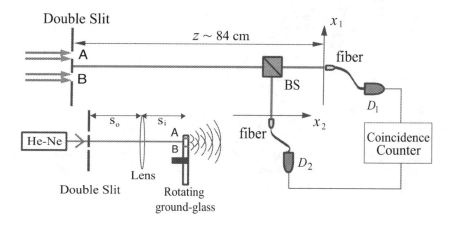

FIGURE 11.1.4 Schematic setup of a nontrivial two-photon interference experiment of Scarcelli *et al.* (2004). The experiment is prepared in such a way that slit-A and slit-B can be treated as two independent incoherent thermal sources with completely independent random fluctuations $\langle \Delta I_A \Delta I_B\rangle = 0$. The fields from A and B are mutually incoherent with $G_{AB}^{(1)} = 0$. No first-order interferences are observable from single-detector counting rates of D_1 and D_2, respectively. Do we expect observing interference in the coincidence counting rate of D_1 and D_2 when scanning D_1 and/or D_2 along their x-axis?

A typical measured two-photon interference pattern of Smith and Shih is reported in Fig. 11.1.3, which is in good agreement with the above predictions that are calculated either from Einstein's granularity picture of light, or from quantum theory of light in coherent state representation and in single-photon state representation.

In fact, the first two-photon Young's double-slit interference phenomenon with spatially incoherent thermal fields was experimental demonstration in 2004 by Scarcelli *et al.*. Their experimental setup is shown in Fig. 11.1.4. In this experiment, the measured pseudo-thermal fields, E_A and E_B, can be treated as independent incoherent thermal radiations coming from two independent thermal sources A and B. The two independent incoherent thermal sources are simulated by a He-Ne laser beam, a double-slit, and a fast rotating ground glass: a He-Ne laser beam is first impinging on slit-A and slit-B, a converging lens is followed to image slit-A and slit-B, respectively, onto the fast rotating ground glass. The two images of slit-A and slit-B can be treated as two independent incoherent thermal

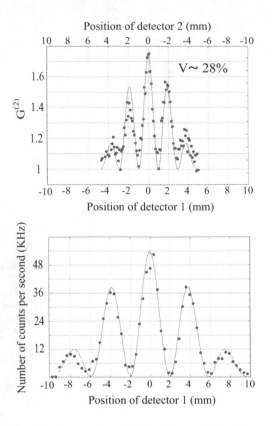

FIGURE 11.1.5 Upper: Normalized second-order interference-diffraction pattern vs position of the detectors when $G_{AB}^{(1)} = 0$. In this measurement, x_1 and x_2 were chosen with equal value but opposite direction. Lower: Equivalent first-order interference-diffraction pattern of a He-Ne laser beam satisfying $G_{AB}^{(1)} \simeq 1$.

light sources with completely independent random fluctuations. It is easy to see that their experimental setup has achieved $G_{AB}^{(1)} = 0$ and $\langle \Delta I_A \Delta I_B \rangle = 0$. Two transversely scannable optical fibers, each coupled with a photon counting detector, are facing the simulated two weak thermal sources A and B in the far-field. The single-detector counting rate and the joint-detection counting rate of D_1 and D_2 are monitored, respectively, during the scanning of the two fiber tips. In their measurement, x_1 and x_2 were scanned with equal value but moved to opposite directions, i.e., $x_1 = -x_2$.

The measured two-photon interference-diffraction pattern is shown in the upper plot of Fig. 11.1.5. It is interesting to see that the interference-diffraction pattern is twice as narrow as the standard interference-diffraction pattern of He-Ne light and with interference modulation twice faster as that of the standard pattern, as if it was produced by a source of light with half the wavelength of the He-Ne laser. For comparison, a first-order interference-diffraction pattern of a He-Ne laser beam is also shown in the lower plot of Fig 11.1.5. In Fig. 11.1.5 the solid line represents a fitting curve of standard interference-diffraction. The visibility of the pattern is about $\sim(28\pm1)\%$ (contrast $\sim(36\pm1)\%$). During the measurement, the single-detector counting rates of D_1 and D_2 were both observed as constants over the entire scanning range, which demonstrates the absence of any first order mutual-coherence between fields A and B.

It is worth mentioning that in the experiment of Scarcelli *et. al.*, their slit-A and slit-B cannot be treated as line-like slits. Taking into account of the finite width of the sources A and B, their 1-D interference-diffraction pattern is expected to be:

$$g^{(2)}(x_1, x_2) = 1 + \text{sinc}^2\left[\frac{\pi b(x_1 - x_2)}{\lambda(z - z_0)}\right]\cos^2\frac{\pi d}{\lambda z}(x_1 - x_2)] \tag{11.1.24}$$

where b is the 1-D width of the light sources A and B, d is the separation distance between the incoherent light sources A and B. The observed two-photon interference agrees with the theoretical expectation within their experimental error.

If Scarcelli's experiment was measuring photon number fluctuation correlation $\langle\Delta n(x_1)\Delta n(x_2)\rangle$, which is proportional to the intensity fluctuation correlation $\langle\Delta I(x_1)\Delta I(x_2)\rangle$, the two-photon interference visibility may achieve $\sim 100\%$ while no first-order interferences are observable in the measurements of D_1 and D_2, respectively.

11.2 TWO-PHOTON INTERFERENCE BETWEEN TEMPORALLY SEPA- RATED INCOHERENT THERMAL FIELDS

Based on the theoretical analysis and experimental demonstration of the above Young's double-slit interference phenomenon, we conclude that two-photon interference is observable from spatially separated incoherent thermal fields. Now we ask: do we expect to observe two-photon interference from temporally separated incoherent thermal fields? For instance, no first-order interferences are observable from a Mach-Zehnder interferometer when $(L - S)/c > \tau_c$, where τ_c is the coherence time of the thermal field. Is it possible to observe interference modulation from the joint measurement of $\langle n(z_1, t_1)n(z_2, t_2)\rangle$ or $\langle\Delta n(z_1, t_1)\Delta n(z_2, t_2)\rangle$ under the experimental condition of $(L - S)/c > \tau_c$?

In the view of quantum theory of light, we do expect to observe two-photon interference from temporally separated incoherent thermal fields in the measurement of photon number correlation $\langle n(z_1, t_1)n(z_2, t_2)\rangle$, which is proportional to intensity correlation $\langle I(z_1, t_1)I(z_2, t_2)\rangle)$, or photon number fluctuation correlation $\langle\Delta n(z_1, t_1)\Delta n(z_2, t_2)\rangle$, which is proportional to intensity fluctuation correlation $\langle\Delta I(z_1, t_1)\Delta I(z_2, t_2)\rangle$, under certain experiment conditions.

(I) Two-photon interference in Einstein's granularity picture

In the following, we calculate the second-order coherence $G^{(2)}(z_1, t_1; z_2, t_2)$, which is proportional to $\langle n(z_1, t_1)n(z_2, t_2)\rangle$, for the Mach-zehnder interferometer of Fig. 11.2.1 in Einstein's granularity picture.

$$G^{(2)}(z_1, t_1; z_2, t_2)$$
$$= \Big\langle \sum_{m,n,p,q} E_m^*(z_1, t_1)E_n(z_1, t_1)E_p^*(z_2, t_2)E_q(z_2, t_2)\Big\rangle$$
$$= \sum_{m=n} E_m^*(z_1, t_1)E_m(z_1, t_1) \sum_{m=n} E_n^*(z_2, t_2)E_n(z_2, t_2)$$
$$+ \sum_{m\neq n} E_m^*(z_1, t_1)E_n(z_1, t_1)E_n^*(z_2, t_2)E_m(z_2, t_2)$$
$$\simeq \sum_{m,n}\big|E_m(z_1, t_1)E_n(z_2, t_2) + E_n(z_1, t_1)E_m(z_2, t_2)\big|^2$$
$$= \sum_{m,n}\big|(E_{mL1} + E_{mS1})(E_{nL2} + E_{nS2}) + (E_{nL1} + E_{nS1})(E_{mL2} + E_{mS2})\big|^2$$
$$= \sum_{m,n}\big|E_{mL1}E_{nL2} + E_{mL1}E_{nS2} + E_{mS1}E_{nL2} + E_{mS1}E_{nS2}$$
$$+ E_{nL1}E_{mL2} + E_{nL1}E_{mS2} + E_{nS1}E_{mL2} + E_{nS1}E_{mS2}\big|^2 \tag{11.2.1}$$

where we have approximated the sum of $(m \neq n)$ to the sum of (m, n) by ignoring the

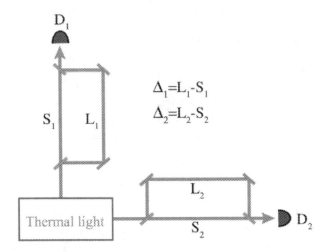

FIGURE 11.2.1 Schematic setup of a two-photon Mach-Zehnder interferometer. Under the experimental condition $(L - S)/c > \tau_c$, no first-order interferences are observable from the measurements of $\langle n_1 \rangle$ and $\langle n_2 \rangle$, respectively. Is it possible to observe interference from the measurement of $\langle n_1 n_2 \rangle$ or $\langle \Delta n_1 \Delta n_2 \rangle$?

$m = n$ term, which is reasonable for a large number of subfields; we have also used short-hand notation E_{mLj} and E_{mSj}, $j = 1, 2$, to specify the subfields at D_j that coming from the long-path and the short-path of the interferometer, respectively. Completing the sum for each term in the $|E_{mL1}E_{nL2} + ... + E_{nS1}E_{mS2}|^2$, we find that a number of cross terms vanish due to the experimental condition of $(L - S)/c > \tau_c$. We may group the surviving terms into the following four superpositions,

$$
\begin{aligned}
&G^{(2)}(z_1, t_1; z_2, t_2) \\
&= \sum_{m,n} \left| E_{mL1}E_{nL2} + E_{nL1}E_{mL2} \right|^2 + \sum_{m,n} \left| E_{mS1}E_{nS2} + E_{nS1}E_{mS2} \right|^2 \\
&\quad + \sum_{m,n} \left| E_{mL1}E_{nS2} + E_{nS1}E_{mL2} \right|^2 + \sum_{m,n} \left| E_{mS1}E_{nL2} + E_{nL1}E_{mS2} \right|^2 \\
&\equiv G^{(2)}_{LL}(z_1, t_1; z_2, t_2) + G^{(2)}_{SS}(z_1, t_1; z_2, t_2) \\
&\quad + G^{(2)}_{LS}(z_1, t_1; z_2, t_2) + G^{(2)}_{SL}(z_1, t_1; z_2, t_2).
\end{aligned}
\tag{11.2.2}
$$

$G^{(2)}_{LL}(z_1, t_1; z_2, t_2)$ is the result of a two-photon superposition between amplitudes (1) the mth wavepacket passes through the long-path of the interferometer and excites a photoelectron at D_1 while the nth wavepacket passes through the long-path of the interferometer and excites a photoelectron at D_2; (2) the nth wavepacket passes through the long-path of the interferometer and excites a photoelectron at D_1 while the mth wavepacket passes through the long-path of the interferometer and excites a photoelectron at D_2.

$G^{(2)}_{SS}(z_1, t_1; z_2, t_2)$ is the result of a two-photon superposition between amplitudes (3) the mth wavepacket passes through the short-path of the interferometer and excites a photoelectron at D_1 while the nth wavepacket passes through the short-path of the interferometer and excites a photoelectron at D_2; (4) the nth wavepacket passes through the short-path of the interferometer and excites a photoelectron at D_1 while the mth wavepacket passes through the short-path of the interferometer and excites a photoelectron at D_2.

$G_{LS}^{(2)}(z_1, t_1; z_2, t_2)$ is the result of a two-photon superposition between amplitudes (5) the mth wavepacket passes through the long-path of the interferometer and excites a photo-electron at D_1 while the nth subfield passes through the short-path of the interferometer and excites a photoelectron at D_2; (6) the nth wavepacket passes through the short-path of the interferometer and excites a photoelectron at D_1 while the mth wavepacket passes through the long-path of the interferometer and excites a photoelectron at D_2.

$G_{SL}^{(2)}(z_1, t_1; z_2, t_2)$ is the result of a two-photon superposition between amplitudes (7) the mth wavepacket passes through the short-path of the interferometer and excites a photo-electron at D_1 while the nth wavepacket passes through the long-path of the interferometer and excites a photoelectron at D_2; (8) the nth wavepacket passes through the long-path of the interferometer and excites a photoelectron at D_1 while the mth wavepacket passes through the short-path of the interferometer and excites a photoelectron at D_2.

We may calculate $G_{L_1, L_2}^{(2)}$, $G_{S_1, S_2}^{(2)}$, $G_{L_1, S_2}^{(2)}$, and $G_{S_1, L_2}^{(2)}$ separately, as four groups of distinguishable two-photon interferences.

It is straightforward to find that $G_{L_1, L_2}^{(2)}$ and $G_{S_1, S_2}^{(2)}$ contribute two constants to the measurement of $G^{(2)}(z_1, t_1; z_2, t_2)$:

$$
\begin{aligned}
G_{LL}^{(2)}&(z_1, t_1; z_2, t_2) \\
&= \sum_{m,n} \Big| \mathcal{F}_{\tau_{L1}}\{a_m(\nu)\} e^{-i\omega_0(\tau_{L1})} \mathcal{F}_{\tau_{L2}}\{a_n(\nu)\} e^{-i\omega_0(\tau_{L2})} \\
&\qquad + \mathcal{F}_{\tau_{L1}}\{a_n(\nu)\} e^{-i\omega_0(\tau_{L1})} \mathcal{F}_{\tau_{L2}}\{a_m(\nu)\} e^{-i\omega_0(\tau_{L2})} \Big|^2 \\
&= \sum_m \big| \mathcal{F}_{\tau_{L1}}\{a_m(\nu)\} \big|^2 \sum_n \big| \mathcal{F}_{\tau_{L2}}\{a_n(\nu)\} \big|^2 \\
&\quad + \sum_n \big| \mathcal{F}_{\tau_{L1}}\{a_n(\nu)\} \big|^2 \sum_m \big| \mathcal{F}_{\tau_{L2}}\{a_m(\nu)\} \big|^2 \\
&\quad + \sum_m \mathcal{F}_{\tau_{L1}}^*\{a_m(\nu)\} \mathcal{F}_{\tau_{L2}}\{a_m(\nu)\} \sum_n \mathcal{F}_{\tau_{L2}}^*\{a_n(\nu)\} \mathcal{F}_{\tau_{L1}}\{a_n(\nu)\} \\
&\quad + \sum_m \mathcal{F}_{\tau_{L1}}\{a_m(\nu)\} \mathcal{F}_{\tau_{L2}}^*\{a_m(\nu)\} \sum_n \mathcal{F}_{\tau_{L2}}\{a_n(\nu)\} \mathcal{F}_{\tau_{L1}}^*\{a_n(\nu)\} \\
&= G_0^{\langle I_1 \rangle \langle I_2 \rangle} + G_0^{\langle \Delta I_1 \Delta I_2 \rangle}.
\end{aligned} \tag{11.2.3}
$$

$G_{L_1, S_2}^{(2)}$ and $G_{S_1, L_2}^{(2)}$ may contribute constants or sinusoidal interferences to the measurement of $G^{(2)}(z_1, t_1; z_2, t_2)$, depending on the experimental setups and conditions. Observing two-photon interference, for example, the cross interference term of the superposition in

$$
\begin{aligned}
G_{LS}^{(2)}&(z_1, t_1; z_2, t_2) \\
&= \sum_{m,n} \Big| \mathcal{F}_{\tau_{L1}}\{a_m(\nu)\} e^{-i\omega_0(\tau_{L1})} \mathcal{F}_{\tau_{S2}}\{a_n(\nu)\} e^{-i\omega_0(\tau_{S2})} \\
&\qquad - \mathcal{F}_{\tau_{S1}}\{a_n(\nu)\} e^{-i\omega_0(\tau_{S1})} \mathcal{F}_{\tau_{L2}}\{a_m(\nu)\} e^{-i\omega_0(\tau_{L2})} \Big|^2
\end{aligned} \tag{11.2.4}
$$

must be able to take a non-zero value. Examining the cross term of Eq. (11.2.4)

$$
\begin{aligned}
&\sum_{m,n} \mathcal{F}_{(t_1 - t_m) - \frac{z_{L1}}{c}}^*\{a_m(\nu)\} e^{i\omega_0(\tau_{A1} - t_m)} \mathcal{F}_{(t_1 - t_n) - \frac{z_{S1}}{c}}\{a_n(\nu)\} e^{-i\omega_0(\tau_{S1} - t_n)} \\
&\quad \times \mathcal{F}_{(t_2 - t_n) - \frac{z_{S2}}{c}}^*\{a_n(\nu)\} e^{i\omega_0(\tau_{S2} - t_n)} \mathcal{F}_{(t_2 - t_m) - \frac{z_{L2}}{c}}\{a_m(\nu)\} e^{-i\omega_0(\tau_{L2} - t_m)},
\end{aligned}
$$

keeping this term non-zero, mathematically, we find the following conditions must be satisfied:

(A) the conjugated wavepackets $\mathcal{F}^*_{\tau_{L1}-t_m}\{a_m(\nu)\}$ and $\mathcal{F}_{\tau_{S1}-t_n}\{a_n(\nu)\}$ $[\mathcal{F}^*_{\tau_{S2}-t_n}\{a_n(\nu)\}$ and $\mathcal{F}_{\tau_{L2}-t_m}\{a_m(\nu)\}]$ must overlap at (z_1, t_1) $[(z_2, t_2)]$.

(B) the conjugated wavepackets $\mathcal{F}^*_{\tau_{L1}-t_m}\{a_m(\nu)\}$ and $\mathcal{F}_{\tau_{L2}-t_m}\{a_m(\nu)\}$ $[\mathcal{F}^*_{\tau_{S2}-t_n}\{a_n(\nu)\}$ and $\mathcal{F}_{\tau_{S1}-t_n}\{a_n(\nu)\}]$ must overlap within the coincidence time window.

If the above experimental conditions are physically satisfied, the cross term turns to be

$$\int dt_m\, \mathcal{F}^*_{(t_1-t_m)-\frac{z_{L1}}{c}}\{a_m(\nu)\} \mathcal{F}_{(t_2-t_m)-\frac{z_{L2}}{c}}\{a_m(\nu)\}$$

$$\times \int dt_n\, \mathcal{F}^*_{(t_2-t_n)-\frac{z_{S2}}{c}}\{a_n(\nu)\} \mathcal{F}_{(t_1-t_n)-\frac{z_{S1}}{c}}\{a_n(\nu)\}$$

$$\simeq 1.$$

It is not difficult to satisfy the above experimental conditions. For a CW thermal light source, either natural light source or pseudo-thermal light source, there are chances for a randomly created and randomly paired wavepackets to satisfy condition (A). It is relatively easier to satisfy condition (B) for pseudo-thermal fields. We can easily find a CW laser with narrow enough spectrum bandwidth. The selected laser should be able to produce pseudo-thermal wavepackets with wide enough temporally width for the joint measurement of D_1 and D_2. Using a natural thermal light with broad bandwidth of spectrum, such as the sunlight, although condition (B) may not be satisfied naturally while the coincidence time window cannot be set short enough to be comparable with the temporal width of the wavepackets due to the slow time response of the electronics; artificially, we can design an electronic integrator to make the temporal width of the output currents of D_1 and D_2 wider enough to have non-zero values, simultaneously, within the coincidence time window. Phyisically, the product of $\Delta i_{mn}(t_1) \times \Delta i_{nm}(t_2)$ gives a non-zero output, and thus keep the cross interference term a non-zero value for all m-n th pair of wavepackets.

In this case, the cross interference term turns to be

$$\sum_m \mathcal{F}^*_{\tau_{L1}}\{a_m(\nu)\}\mathcal{F}_{\tau_{L2}}\{a_m(\nu)\} \sum_n \mathcal{F}^*_{\tau_{S2}}\{a_n(\nu)\}\mathcal{F}_{\tau_{S1}}\{a_n(\nu)\}$$

$$\times e^{i\omega_0[(\tau_{L1}-\tau_{L2})-(\tau_{S1}-\tau_{S2})]}$$

$$+\sum_m \mathcal{F}_{\tau_{L1}}\{a_m(\nu)\}\mathcal{F}^*_{\tau_{L2}}\{a_m(\nu)\} \sum_n \mathcal{F}_{\tau_{S2}}\{a_n(\nu)\}\mathcal{F}^*_{\tau_{S1}}\{a_n(\nu)\}$$

$$\times e^{-i\omega_0[(\tau_{L1}-\tau_{L2})-(\tau_{S1}-\tau_{S2})]}$$

$$= G_0^{\langle\Delta I_1 \Delta I_2\rangle}\cos\{\omega_0[(\tau_{L1}-\tau_{L2})-(\tau_{S1}-\tau_{S2})]\}. \tag{11.2.5}$$

Therefore,

$$G_{LS}^{(2)}(z_1, t_1; z_2, t_2)$$
$$= G_0^{\langle I_1\rangle\langle I_2\rangle} + G_0^{\langle\Delta I_1 \Delta I_2\rangle}\cos\{\omega_0[(\tau_{L1}-\tau_{L2})-(\tau_{S1}-\tau_{S2})]\} \tag{11.2.6}$$

Similarly, we have

$$G_{SS}^{(2)}(z_1, t_1; z_2, t_2) = G_0^{\langle I_1\rangle\langle I_2\rangle} + G_0^{\langle\Delta I_1 \Delta I_2\rangle} \tag{11.2.7}$$

and

$$G_{SL}^{(2)}(z_1, t_1; z_2, t_2)$$
$$= G_0^{\langle I_1\rangle\langle I_2\rangle} + G_0^{\langle\Delta I_1 \Delta I_2\rangle}\cos\{\omega_0[(\tau_{L1}-\tau_{L2})-(\tau_{S1}-\tau_{S2})]\}. \tag{11.2.8}$$

When scanning the optical paths of the Mach-Zehnder interferometers around $[(L_2 - S_2) - (L_1 - S_1)] \sim 0$, taking into account all possible joint photodetection events contributed from $G^{(2)}_{L_1, L_2}$, $G^{(2)}_{S_1, S_2}$, $G^{(2)}_{L_1, S_2}$, and $G^{(2)}_{S_1, L_2}$, the jointly measured photon number correlation of D_1 and D_2 is thus

$$\langle n_1 n_2 \rangle \propto 1 + \frac{1}{2} \cos\left\{ \omega_0 [(L_2 - S_2) - (L_1 - S_1)]/c \right\}. \tag{11.2.9}$$

A sinusoidal two-photon interference pattern with 50% contrast is observable from the measurement of photon number correlation when scanning the optical paths of the Mach-Zehnder interferometers around $[(L_2 - S_2) - (L_1 - S_1)] \sim 0$.

Adding only the photon number fluctuation correlations we have calculated from $G^{(2)}_{L_1, L_2}$, $G^{(2)}_{S_1, S_2}$, $G^{(2)}_{L_1, S_2}$, and $G^{(2)}_{S_1, L_2}$, the jointly measured photon number fluctuation correlation of D_1 and D_2 is therefore

$$\langle \Delta n_1 \Delta n_2 \rangle \propto 1 + \cos\left\{ \omega_0 [(L_2 - S_2) - (L_1 - S_1)]/c \right\}. \tag{11.2.10}$$

A sinusoidal two-photon interference pattern with 100% visibility is observable from the measurement of photon number fluctuation correlation when scanning the optical paths of the Mach-Zehnder interferometers around $[(L_2 - S_2) - (L_1 - S_1)] \sim 0$.

(II) Two-photon interference in single-photon state representation

In the following, we calculate the interference for the same two-photon Mach-Zehnder interferometer in single-photon state representation, under the same experimental condition of $(L - S)/c > \tau_c$. The quantum state of the fields prepared at the source can be approximated as

$$|\Psi\rangle \simeq \prod_m \left(|0\rangle + \epsilon \int d\omega\, \hat{a}_m^\dagger(\omega) |0\rangle \right)$$

$$\simeq |0\rangle + \epsilon \sum_m \int d\omega\, \hat{a}_m^\dagger(\omega) |0\rangle + \epsilon^2 \sum_{m<n} \int d\omega\, d\omega'\, \hat{a}_m^\dagger(\omega) \hat{a}_n^\dagger(\omega') |0\rangle + \dots \tag{11.2.11}$$

Under weak light condition, we may keep the necessary lowest-order terms of the state for calculating the counting rate of D_1 and D_2, respectively, and the coincidence counting rate of D_1 and D_2, jointly. The joint photodetection counting rate is proportional to the second-order coherence function $G^{(2)}(z_1, t_1; z_2, t_2)$,

$$G^{(2)}(z_1, t_1; z_2, t_2)$$

$$= \left\langle \langle \hat{E}_1^{(-)} \hat{E}_2^{(-)} \hat{E}_2^{(+)} \hat{E}_1^{(+)} \rangle_{\text{QM}} \right\rangle_{\text{En}}$$

$$\simeq \left\langle \left| \langle 0 | \sum_p [\hat{E}_p^{(+)}(\tau_{L2}) + \hat{E}_p^{(+)}(\tau_{S2})] \sum_q [\hat{E}_q^{(+)}(\tau_{L1}) + \hat{E}_q^{(+)}(\tau_{S1})] \right. \right.$$

$$\left. \left. \times \sum_{m,n} \int d\omega\, d\omega'\, \hat{a}_m^\dagger(\omega) \hat{a}_n^\dagger(\omega') |0\rangle \right|^2 \right\rangle_{\text{En}}$$

$$= \sum_{m,n} \left| \psi_m(\tau_{L1}) \psi_n(\tau_{L2}) + \psi_m(\tau_{L1}) \psi_n(\tau_{S2}) \right.$$

$$+ \psi_m(\tau_{S1}) \psi_n(\tau_{L2}) + \psi_m(\tau_{S1}) \psi_n(\tau_{S2})$$

$$+ \psi_n(\tau_{L1}) \psi_m(\tau_{L2}) + \psi_n(\tau_{L1}) \psi_m(\tau_{S2})$$

$$\left. + \psi_n(\tau_{S1}) \psi_m(\tau_{L2}) + \psi_n(\tau_{S1}) \psi_m(\tau_{S2}) \right|^2, \tag{11.2.12}$$

with

$$\psi_m(\tau_{Lj}) = \langle 0|\hat{E}_m^{(+)}(\tau_{Lj}) \int d\omega \, \hat{a}_m^\dagger(\omega)|0\rangle,$$

$$\psi_n(\tau_{Sj}) = \langle 0|\hat{E}_n^{(+)}(\tau_{Sj}) \int d\omega' \, \hat{a}_n^\dagger(\omega')|0\rangle \qquad (11.2.13)$$

the effective wavefunctions of the mth photon and the nth photon measured by D_j, where the L and S label long-path and short-path of the interferometer, respectively; and $\hat{E}_m^{(+)}(\tau_{Lj})$ and $\hat{E}_n^{(+)}(\tau_{Sj})$, $j = 1, 2$, are the field operators associated with the long-path and the short-path of the Mach-Zehnder interferometer:

$$\hat{E}_m^{(+)}(\tau_{Lj}) = \int d\omega \, \hat{a}_m(\omega)e^{i\omega(\tau_{Lj}-t_m)}$$

$$\hat{E}_n^{(+)}(\tau_{Sj}) = \int d\omega' \, \hat{a}_n(\omega')e^{i\omega'(\tau_{Sj}-t_n)}. \qquad (11.2.14)$$

Note, in Eq. (11.2.12), we have approximated $\sum_{m<n} \dots$ to $\sum_{m,n}$ by ignoring the $m = n$ terms and by renormalizing the state.

Similar to the calculation in Einstein's granularity picture, completing the sum for each term in the $|\psi_m(\tau_{L1})\psi_n(\tau_{L2}) + \dots + \psi_n(\tau_{S1})\psi_m(\tau_{S2})|^2$, we find that a number of cross terms vanish due to the experimental condition $(L - S)/c > \tau_c$. We may group the surviving terms into the following four superpositions,

$$G^{(2)}(z_1, t_1; z_2, t_2)$$
$$= \sum_{m,n} \left|\psi_m(\tau_{L1})\psi_n(\tau_{L2}) + \psi_n(\tau_{L1})\psi_m(\tau_{L2})\right|^2$$
$$+ \sum_{m,n} \left|\psi_m(\tau_{S1})\psi_n(\tau_{S2}) + \psi_n(\tau_{S1})\psi_m(\tau_{S2})\right|^2$$
$$+ \sum_{m,n} \left|\psi_m(\tau_{L1})\psi_n(\tau_{S2}) + \psi_n(\tau_{S1})\psi_m(\tau_{L2})\right|^2$$
$$+ \sum_{m,n} \left|\psi_m(\tau_{S1})\psi_n(\tau_{L2}) + \psi_n(\tau_{L1})\psi_m(\tau_{S2})\right|^2$$
$$\equiv G_{LL}^{(2)}(z_1, t_1; z_2, t_2) + G_{SS}^{(2)}(z_1, t_1; z_2, t_2)$$
$$+ G_{LS}^{(2)}(z_1, t_1; z_2, t_2) + G_{SL}^{(2)}(z_1, t_1; z_2, t_2). \qquad (11.2.15)$$

We may define and calculate $G_{L1,L2}^{(2)}$, $G_{S1,S2}^{(2)}$, $G_{L1,S2}^{(2)}$, and $G_{S1,L2}^{(2)}$, separately, as four groups of distinguishable two-photon superpositions:

$$G_{LL}^{(2)}(z_1, t_1; z_2, t_2) = \sum_{m,n} \left|\psi_m(\tau_{L1})\psi_n(\tau_{L2}) + \psi_n(\tau_{L1})\psi_m(\tau_{L2})\right|^2 \qquad (11.2.16)$$

$$G_{SS}^{(2)}(z_1, t_1; z_2, t_2) = \sum_{m,n} \left|\psi_m(\tau_{S1})\psi_n(\tau_{S2}) + \psi_n(\tau_{S1})\psi_m(\tau_{S2})\right|^2 \qquad (11.2.17)$$

$$G_{LS}^{(2)}(z_1, t_1; z_2, t_2) = \sum_{m,n} \left|\psi_m(\tau_{L1})\psi_n(\tau_{S2}) + \psi_n(\tau_{S1})\psi_m(\tau_{L2})\right|^2. \qquad (11.2.18)$$

$$G_{SL}^{(2)}(z_1, t_1; z_2, t_2) = \sum_{m,n} \left| \psi_m(\tau_{S1})\psi_n(\tau_{L2}) + \psi_n(\tau_{L1})\psi_m(\tau_{S2}) \right|^2. \qquad (11.2.19)$$

Repeating the calculations in Einstein's granularity picture, mathematically by replacing Einstein's subfields with the quantum effective wavefunctions, taking into account all possible joint photodetection events contributed from $G_{L1,L2}^{(2)}$, $G_{S1,S2}^{(2)}$, $G_{L1,S2}^{(2)}$, and $G_{S1,L2}^{(2)}$, the jointly measured photon number correlation of D_1 and D_2 is calculated to be

$$\langle n_1 n_2 \rangle \propto 1 + \frac{1}{2}\cos\frac{\omega_0}{c}[(L_2 - S_2) - (L_1 - S_1)],$$

which is the same as Eq. (11.2.9). A sinusoidal two-photon interference pattern with 50% contrast is observable from the measurement of photon number correlation when scanning the optical paths of the Mach-Zehnder interferometers around $[(L_2 - S_2) - (L_1 - S_1)] \sim 0$.

Adding only the photon number fluctuation correlations we have calculated from $G_{L1,L2}^{(2)}$, $G_{S1,S2}^{(2)}$, $G_{L1,S2}^{(2)}$, and $G_{S1,L2}^{(2)}$, the jointly measured photon number fluctuation correlation of D_1 and D_2 is therefore

$$\langle \Delta n_1 \Delta n_2 \rangle \propto 1 + \cos\frac{\omega_0}{c}[(L_2 - S_2) - (L_1 - S_1)],$$

which is the same as Eq. (11.2.10). A sinusoidal two-photon interference pattern with 100% visibility is observable from the measurement of photon number fluctuation correlation when scanning the optical paths of the Mach-Zehnder interferometers around $[(L_2 - S_2) - (L_1 - S_1)] \sim 0$.

It should be emphasized that the above two-photon interference phenomenon can only be observed when the experimental setups satisfy the conditions (A) and (B).

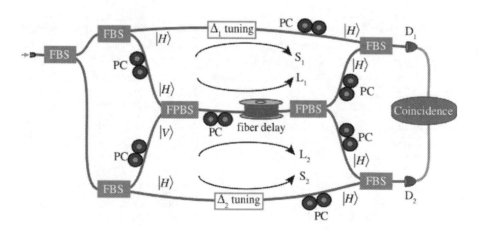

FIGURE 11.2.2 Schematic of an unbalanced two-photon Mach-Zehnder interferometer demonstrated by Ihn *et. al.* recently. A standard pseudo-thermal radiation of coherence time $\tau_c \sim 572ns$ ($l_c \sim 120m$) is generated by focusing a laser beam onto a rotating ground glass disk. The coincidence time window is set $15ns$.

(III) Experimental demonstration two-photon Mach-Zehnder interferometer

Ihn *et. al.* demonstrated such an unbalanced two-photon Mach-Zehnder interferometer recently, which is schematically shown in Fig. 11.2.2. In their experiment, a standard pseudo-thermal field is generated by focusing a laser beam onto a rotating ground glass disk. The

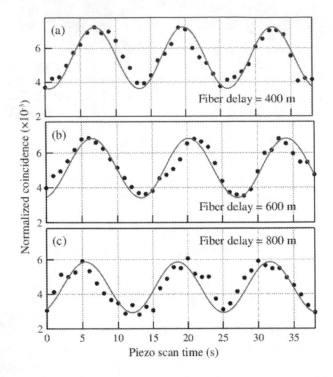

FIGURE 11.2.3 Observed two-photon interferences from the measurement of $\langle n_1 n_2 \rangle \propto \langle I_1 I_2 \rangle$ with optical fiber delays $L - S = 400m$, $600m$, and $800m$. The visibility becomes $\sim 100\%$ in the measurement of $\langle \Delta n_1 \Delta n_2 \rangle \propto \langle \Delta I_1 \Delta I_2 \rangle$. Since $l_c = 120m$, a $120m$ optical fiber delay is sufficient to completely remove the classic first-order Mach-Zehnder interferences.

laser is an external cavity diode laser operating at 780 nm, and is frequency locked to the $5S_{1/2}(F = 3) - 5P_{3/2}(F' = 4)$ transition of the 85Rb atomic energy levels. The coherence length of the pseudo-thermal field is measured approximately $120m$ ($\tau_c \sim 573ns$) in an optical fiber. To satisfy condition (b), the coincidence time window in their experiment is set $15ns$, which is much shorter than $\tau_c \sim 573ns$. In their experiment, the horizontally polarized thermal light beam is first split by a fiber beam splitter (FBS). Each beam is then sent through an unbalanced Mach-Zehnder interferometer (UMZI) with a long and a short optical fiber path. The UMZI consists of FBSs, fiber polarizing beam splitters (FPBSs), fiber polarization controllers (PCs), and optical fibers. The short paths, S1 and S2, each contain a 1-m-long optical fiber and a free-space delay line, labeled as Δ_1 tuning or Δ_2 tuning, controlled by a piezoactuator for phase modulation. The long paths, L_1 and L_2, each include a long fiber spool of length $200m$, $400m$, $600m$, or $800m$. The long paths L_1 and L_2 of the two UMZIs physically share the same fiber spool. The L_1 and L_2 paths, instead, are defined by the polarization states $|H\rangle$ and $|V\rangle$, respectively, by using PCs and FPBSs. Finally, the Δ_1 and Δ_2 delays are scanned by applying voltages to the piezoactuators while observing the single and coincidence counting rates of the two detectors D_1 and D_2.

Figure 11.2.3 reports three sets of their observed two-photon interferences from the measurement of $\langle n_1 n_2 \rangle \propto \langle I_1 I_2 \rangle$ with optical fiber delays $L - S = 400m$, $600m$, and $800m$. The visibility becomes $\sim 100\%$ in the measurement of $\langle \Delta n_1 \Delta n_2 \rangle \propto \langle \Delta I_1 \Delta I_2 \rangle$. Note that a $120m$ optical fiber delay is sufficient to completely remove the classic first-order Mach-Zehnder interferences.

11.3 TWO-PHOTON ANTI-CORRELATION OF INCOHERENT THERMAL FIELDS

In this section, we discuss a slightly different two-photon interference phenomenon: anti-correlation of incoherent thermal fields. We start from the modified double-slit interferometer of Fig. 11.3.1. In this modified double-slit interferometer, a large angular sized thermal

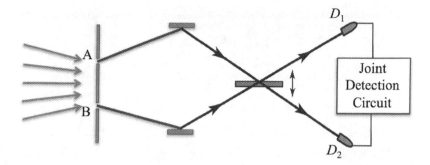

FIGURE 11.3.1 Schematic of a modified double-slit interferometer. A large angular sized thermal light source is applied. Under the experimental condition of $d \gg l_c$, E_A and E_B achieve first-order incoherent $G_{AB}^{(1)} = 0$, and second-order coherent $G_{AB}^{(2)} = $ constant, or $\langle \Delta n_A \Delta n_B \rangle \propto \langle \Delta I_A \Delta I_B \rangle = 0$. The incoherent thermal fields from slit-A and slit-B are injected from the two input ports of the beamsplitter. Two photon counting detectors D_1 and D_2 are placed at the two output ports of the beamsplitter for the measurements of $\langle n_j \rangle$ ($\langle I_j \rangle$), $j = 1, 2$, and $\langle n_1 n_2 \rangle$ ($\langle I_1 I_2 \rangle$).

light source is used to illuminate slit-A and slit-B. We keep the experimental condition of $d \gg l_c$, achieving $G_{AB}^{(1)} = 0$ and $\langle \Delta n_A \Delta n_B \rangle \propto \langle \Delta I_A \Delta I_B \rangle = 0$. The incoherent thermal fields from slit-A and slit-B are injected into the two input ports of a beamsplitter. Two photon counting detectors D_1 and D_2 are placed at the two output ports of the beamsplitter for the measurements of $\langle n_j \rangle \propto \langle I_j \rangle$, $j = 1, 2$, and $\langle n_1 n_2 \rangle \propto \langle I_1 I_2 \rangle$. If E_A and E_B are first-order coherent, i.e., $G_{AB}^{(1)} \sim 1$, the setup of Fig. 11.3.1 is equivalent a Mach-Zehnder interferometer. Classic interference patterns are expected from the single-detector counting rates of D_1 and D_2, respectively, when scanning the beamsplitter around its balanced position. However, the experimental condition of $G_{AB}^{(1)} = 0$ forces the single-detector counting rates of D_1 and D_2 constants, $\langle n_j \rangle \propto \langle I_j \rangle \sim$ constant. Under the experimental condition of $G_{AB}^{(1)} = 0$ and $\langle \Delta n_A \Delta n_B \rangle \propto \langle \Delta I_A \Delta I_B \rangle = 0$, the questions are: (1) Do we expect observing correlation or anti-correlation from the joint measurement of D_1 and D_2? (2) Do we expect observing correlation function of "peak" or anti-correlation function of "dip" by scanning the beamsplitter from its balanced position?

In the view of quantum theory of light, the probability of observing a joint photodetection event of D_1 and D_2 is proportional to the second-order coherence function

$$G^{(2)}(z_1, t_1; z, t_2)$$
$$= \left\langle \langle \hat{E}^{(-)}(z_1, t_1)\hat{E}^{(-)}(z_2, t_2)\hat{E}^{(+)}(z_2, t_2)\hat{E}^{(+)}(z_1, t_1) \rangle_{\text{QM}} \right\rangle_{\text{En}}$$
$$= \left\langle |\langle \Psi| \hat{E}^{(+)}(z_2, t_2)\hat{E}^{(+)}(z_1, t_1)|\Psi \rangle|^2 \right\rangle_{\text{En}}$$
$$= \left\langle |\langle \Psi|[\hat{E}_{A1}^{(+)} + \hat{E}_{B1}^{(+)}][\hat{E}_{A2}^{(+)} - \hat{E}_{B2}^{(+)}|\Psi \rangle|^2 \right\rangle_{\text{En}}. \tag{11.3.1}$$

The "-" sign comes from the beamsplitter. Substituting the thermal state of the randomly created and randomly paired two photons, either in the single-photon state representation or in the coherent state representation , and completing the ensemble average, we have

$$G^{(2)}(z_1, t_1; z_2, t_2)$$

$$= \sum_{m,n} \left|\psi_{mA1}\psi_{nA2} + \psi_{nA1}\psi_{mA2}\right|^2 + \sum_{m,n} \left|\psi_{mB1}\psi_{nB2} + \psi_{nB1}\psi_{mB2}\right|^2$$

$$+ \sum_{m,n} \left|\psi_{mA1}\psi_{nB2} - \psi_{nB1}\psi_{mA2}\right|^2 + \sum_{m,n} \left|\psi_{mB1}\psi_{nA2} - \psi_{nA1}\psi_{mB2}\right|^2$$

$$= G_{AA}^{(2)}(z_1, t_1; z_2, t_2) + G_{BB}^{(2)}(z_1, t_1; z_2, t_2)$$

$$+ G_{AB}^{(2)}(z_1, t_1; z_2, t_2) + G_{BA}^{(2)}(z_1, t_1; z_2, t_2). \tag{11.3.2}$$

We may calculate $G_{A_1,A_2}^{(2)}$, $G_{B_1,B_2}^{(2)}$, $G_{A_1,B_2}^{(2)}$, and $G_{B_1,A_2}^{(2)}$ separately, as four groups of distinguishable two-photon interferences. Following early sections and chapters, it is not difficult to find out that the two-photon interferences in $G_{A_1,A_2}^{(2)}$ and $G_{B_1,B_2}^{(2)}$ contribute two constants to the joint photodetection of D_1 and D_2. $G_{A_1,B_2}^{(2)}$ and $G_{B_1,A_2}^{(2)}$ may contribute a nontrivial correlation or anti-correlation. We thus exam $G_{AB}^{(2)}(z_1, t_1; z_2, t_2)$

$$G_{AB}^{(2)}(z_1, t_1; z_2, t_2)$$

$$= \sum_{m,n} \left|\psi_{mA1}\psi_{nB2} - \psi_{nB1}\psi_{mA2}\right|^2$$

$$= \sum_{m,n} \Big| \mathcal{F}_{\tau_{A1}-t_m}\{a_m(\nu)\}e^{-i\omega_0(\tau_{A1}-t_m)} \mathcal{F}_{\tau_{B2}-t_n}\{a_n(\nu)\}e^{-i\omega_0(\tau_{B2}-t_n)}$$

$$- \mathcal{F}_{\tau_{B1}-t_n}\{a_n(\nu)\}e^{-i\omega_0(\tau_{B1}-t_n)} \mathcal{F}_{\tau_{A2}-t_m}\{a_m(\nu)\}e^{-i\omega_0(\tau_{A2}-t_m)} \Big|^2 \tag{11.3.3}$$

and ask the following two questions: (1) Do we expect observing correlation or anti-correlation from the joint measurement of D_1 and D_2? (2) Do we expect observing correlation function of "peak" or anti-correlation function of "dip" by scanning the beamsplitter from its balanced position?

(1) Do we expect observing correlation or anti-correlation from the joint measurement of D_1 and D_2?

Due to the "-" sign in Eq. (11.3.3), if there is any correlation, it must be anti-correlation. Observing anti-correlation, obviously, the cross interference term of the superposition in Eq. (11.3.3) must be able to achieve a non-zero value:

$$\sum_{m,n} \mathcal{F}_{\tau_{A1}-t_m}^*\{a_m(\nu)\}e^{-i\omega_0(\tau_{A1}-t_m)} \mathcal{F}_{\tau_{B2}-t_n}^*\{a_n(\nu)\}e^{-i\omega_0(\tau_{B2}-t_n)}$$

$$\times \mathcal{F}_{\tau_{B1}-t_n}\{a_n(\nu)\}e^{-i\omega_0(\tau_{B1}-t_n)} \mathcal{F}_{\tau_{A2}-t_m}\{a_m(\nu)\}e^{-i\omega_0(\tau_{A2}-t_m)}$$

$$= \sum_{m,n} \mathcal{F}_{(t_1-t_m)-\frac{z_{A1}}{c}}^*\{a_m(\nu)\}e^{i\omega_0(\tau_{A1}-t_m)} \mathcal{F}_{(t_1-t_n)-\frac{z_{B1}}{c}}\{a_n(\nu)\}e^{-i\omega_0(\tau_{B1}-t_n)}$$

$$\times \mathcal{F}_{(t_2-t_n)-\frac{z_{B2}}{c}}^*\{a_n(\nu)\}e^{i\omega_0(\tau_{B2}-t_n)} \mathcal{F}_{(t_2-t_m)-\frac{z_{A2}}{c}}\{a_m(\nu)\}e^{-i\omega_0(\tau_{A2}-t_m)}.$$

Mathematically, this means:

(A) The conjugated wavepackets $\mathcal{F}_{\tau_{A1}-t_m}^*\{a_m(\nu)\}$ and $\mathcal{F}_{\tau_{B1}-t_n}\{a_n(\nu)\}$ $[\mathcal{F}_{\tau_{B2}-t_n}^*\{a_n(\nu)\}$ and $\mathcal{F}_{\tau_{A2}-t_m}\{a_m(\nu)\}]$ must overlap at (z_1, t_2) $[(z_2, t_2)]$.

(B) The conjugated wavepackets $\mathcal{F}_{\tau_{A1}-t_m}^*\{a_m(\nu)\}$ and $\mathcal{F}_{\tau_{A2}-t_m}\{a_m(\nu)\}$ $[\mathcal{F}_{\tau_{B2}-t_n}^*\{a_n(\nu)\}$ and $\mathcal{F}_{\tau_{B1}-t_n}\{a_n(\nu)\}]$ must overlap within the time window of the joint measurement.

If these experimental conditions are physically satisfied, the cross term turns to be

$$\int dt_m \, \mathcal{F}^*_{(t_1-t_m)-\frac{z_{A1}}{c}}\{a_m(\nu)\} \mathcal{F}_{(t_2-t_m)-\frac{z_{A2}}{c}}\{a_m(\nu)\}$$

$$\times \int dt_n \, \mathcal{F}^*_{(t_2-t_n)-\frac{z_{B2}}{c}}\{a_n(\nu)\} \mathcal{F}_{(t_1-t_n)-\frac{z_{B1}}{c}}\{a_n(\nu)\}$$

$$\simeq 1.$$

Anti-correlation is thus observable from the measurement of photon number fluctuation correlation.

(2) Do we expect observing correlation function of "peak" or anti-correlation function of "dip" by scanning the beamsplitter from its balanced position?

Due to the "-" sign, if there is any, it must be an anti-correlation "dip". For a CW thermal field, it seems impossible to observe any anti-correlation function "dip" by simply scanning the beamsplitter. If the above conditions (A) and (B) are satisfied when the beamsplitter is at its balanced position, they are still satisfied when the beamsplitter is moved away from its balanced position. The situation is, however, different for a short pulsed thermal radiation. If we can manage to creat the mth and the nth wavepackets from a short radiation pulse, $t_m \sim t_n$, we can change the degree of overlapping between the conjugats $\mathcal{F}^*_{\tau_{A1}-t_m}\{a_m(\nu)\}$ and $\mathcal{F}_{\tau_{B1}-t_n}\{a_n(\nu)\}$ $[\mathcal{F}^*_{\tau_{B2}-t_n}\{a_n(\nu)\}$ and $\mathcal{F}_{\tau_{A2}-t_m}\{a_m(\nu)\}]$ at (z_1, t_1) $[(z_2, t_2)]$ by changing the values of $z_{A1} - z_{B1}$ and $z_{A2} - z_{B2}$. When moving the beamsplitter from its balanced position, the conjugated effective wavefunctions $\mathcal{F}^*_{\tau_{A1}-t_m}\{a_m(\nu)\}$ and $\mathcal{F}_{\tau_{B1}-t_n}\{a_n(\nu)\}$ $[\mathcal{F}^*_{\tau_{B2}-t_n}\{a_n(\nu)\}$ and $\mathcal{F}_{\tau_{A2}-t_m}\{a_m(\nu)\}]$ of the pulsed thermal fields start shift away from each other. When the two conjugated wavepackets are separated completely at D_1 and D_2, the cross term of Eq. (11.3.3) vanishes, simply because $\mathcal{F}^*_{\tau_{A1}-t_m}\{a_m(\nu)\} \mathcal{F}_{\tau_{B1}-t_n}\{a_n(\nu)\}$ and $\mathcal{F}^*_{\tau_{B2}-t_n}\{a_n(\nu)\} \mathcal{F}_{\tau_{A2}-t_m}\{a_m(\nu)\}$ become zero. An anti-correlation function of "dip" is thus observable by scanning the beamsplitter from its balanced position or by other mechanisms to change $z_{A1} - z_{B1}$ $[z_{A2} - z_{B2}]$ from zero to a value that is greater than the width of the wavepackets. The width of the anti-correlation correlation is thus the same as that of the temporal coherence length of the pulsed thermal field.

(I) Experimental realization of anti-correlation "dip" of thermal fields.

In a set of recent experiments, Chen et al. observed the anti-correlation "dip" from the measurement of incoherent pulsed thermal field. The experimental setup of Chen et al. is schematically shown in Figure 11.3.2. The light source is a standard pseudo-thermal light source which consists of a CW mode-locked laser beam with ∼200 femtosecond pulses at a 78 MHz repetition rate and a fast rotating diffusing ground glass. The rotating diffusing ground glass scatters the laser beam into a large number of transversely randomly distributed point-like sub-sources. A large number of subfields scattered from these point-like sub-sources are than collected by two optical fiber tips A and B and coupled into a 50/50 optical fiber beamsplitter. This experiment measures $\langle n(z_1, t_1) n(z_2, t_2) \rangle$ or $\langle \Delta n(z_1, t_1) \Delta n(z_2, t_2) \rangle$ at the output ports of the 50/50 optical fiber beamsplitter BS2. It is BS2 in Fig. 11.3.2 introduces a "−" sign between its two input fields and produces a destructive superposition that is shown in Eqs. (11.3.2) and (11.3.3). If the two input fiber tips A and B are placed within the longitudinal temporal coherence length and the transverse coherence area of the thermal field at the output ports of the first beamsplitter BS1, i.e., $G^{(1)}(\mathbf{r}_A, t_A; \mathbf{r}_B, t_B) \neq 0$, this setup is equivalent to a Mach-Zehnder interferometer. D_1 and D_2, respectively, will each observe first-order interference as a function of the optical delay δ when scanning the fiber tip A along its longitudinal axis. Consequently, the joint photodetection of D_1 and D_2 produces an interference that is factorizable into two first-order interferences. However, in this experiment we decided to move the fiber tip A outside

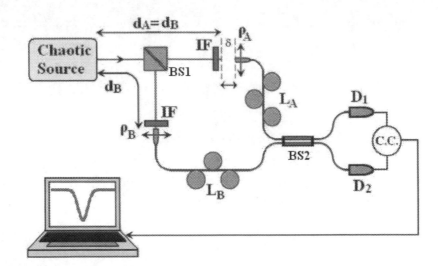

FIGURE 11.3.2 Schematic setup of an anti-correlation measurement from pulsed incoherent thermal fields by Chen *et al.* (2009).

the transverse coherence area to force $G^{(1)}(\mathbf{r}_A, t_A; \mathbf{r}_B, t_B) = 0$. Under this condition, the input radiation at the fiber tips A and B are incoherent, there would be no first-order interference observable from D_1 and D_2, respectively, even in the case of "single-exposure" or instantaneous measurement.

Although the experimental condition achieved $G^{(1)}_{AB} = 0$ and $\langle \Delta n_A \Delta n_B \rangle \propto \langle \Delta I_A \Delta I_B \rangle = 0$, an anti-correlation function was observed from pulsed incoherent thermal fields E_A and E_B in the joint photodetection counting rate of D_1 and D_2: the measurement of $\langle n_1 n_2 \rangle \propto \langle \Delta I_1 \Delta I_2 \rangle$ yielded a "dip" as function of the optical delay δ and the bandwidth of the spectral filters IF, while the counting rate of D_1 and D_2 both kept constants. Figure 11.3.3 shows two typical measured anti-correlation functions of the experiment of Chen *et al.*, corresponding to two different spectrum bandwidths of the pulsed thermal field, in photon number correlation.

The following experimental details of Chen *et al.* may be helpful for understanding the physics behind anti-correlation of thermal field.

(1) The source: the light source is a standard pseudo-thermal source that was developed since 1960's and used widely in HBT correlation measurements. The source consists of a mode-locked laser beam with ∼200 femtosecond pulses at a 78 MHz repetition rate and a fast rotating diffusing ground glass. The linearly polarized laser beam is enlarged transversely onto the ground glass with a diameter of $4.5mm$. The enlarged laser beam is scattered by the fast rotating diffusing ground into a large number of transversely randomly distributed point-like sub-sources to simulate a near-field, pseudo-thermal radiation source: a large number of randomly distributed incoherent wavepackets with random relative phases are then collected by the optical fiber tips A and B and coupled into the optical fiber beamsplitter.

(2) The interferometer: A 50/50 beam splitter (BS1) is used to split the pulsed thermal field into transmitted and reflected radiations which are then coupled into two identical polarization-controlled single-mode fibers A and B respectively. The fiber tips are located $\sim 200mm$ from the ground glass, i.e. $d_A = d_B \sim 200mm$. At this distance, the angular size of the source $\Delta\theta$ is ~ 22.5 mili-radian ($1.29°$) with respect to each input fiber tip, which satisfies the Fresnel near-field condition. Two identical narrow-band spectral filters (IF) are placed in front of the two fiber tips A and B. The transverse and longitudinal coordinates of the input fiber tips are both scannable by step-motors. The output ends of the two fibers

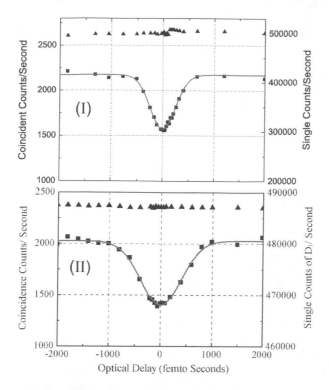

FIGURE 11.3.3 Typical observed anti-correlation "dips" in *photon number* correlation measurements with different special bandwidth: $\tau_c \sim 345fs$ for (I), $\tau_c \sim 541fs$ for (II). The temporal width of the "dip" is determined by the bandwidth of the spectral filters (IF). The counting rate of D_1 and D_2 both kept constants during the scanning of δ. The observed contrast of "dip" is $\sim 100\%$ in photon number fluctuation correlation measurement.

can be directly coupled into the photon counting detectors D_1 and D_2, respectively, for near-field HBT correlation measurements, or coupled into the two input ports of a 50/50 single-mode optical fiber beamsplitter (BS2) for the anti-correlation measurement.

(3) The measurement: two steps of measurements were made. The purpose of step-one is to confirm the pseudo-thermal light source produces thermal field by means of temporal and spatial HBT correlations. In step-one measurement, the output ends of the two fibers A and B are directly coupled into the photon counting detectors D_1 and D_2 for photon number correlation and photon number fluctuation correlation measurement, while scanning the input fiber tips longitudinally and transversely. Thermal radiation can easily be distinguished from a laser beam by examining its second-order coherence function $G^{(2)}(\mathbf{r}_A, t_A; \mathbf{r}_B, t_B)$, which is characterized experimentally by the coincidence counting rate that counts the joint photodetection events at space-time points $(\mathbf{r}_A, t_A)$ and $(\mathbf{r}_B, t_B)$. Figure 11.3.4 reports two measured second-order correlations at zero and at 1000 revolutions per minute (rpm) of the rotating ground glass. This measurement guarantees a typical HBT correlation of thermal field at rotation speeds of 1000 rpm of the ground glass, indicating the thermal nature of the light source. In this measurement, we have also located experimentally the longitudinal and transverse coordinates of the fiber tips A and B for achieving the maximum coincidence counting rate, corresponding to the maximum second-order correlation.

In step-two, we couple the 50/50 fiber beamsplitter (BS2) into the setup as shown in Fig. 11.3.2. This measurement was done in two steps. We first measured the first-order interference at $\vec{\rho}_A = \vec{\rho}_B$ by scanning the input fiber tip A longitudinally in the neighborhood

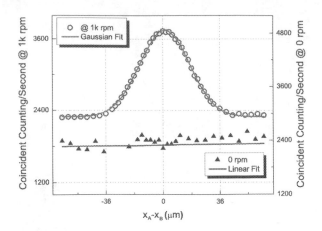

FIGURE 11.3.4 Measurement of HBT correlation of $G^{(2)}(x_A - x_B)$ on the planes of pinhole-A and pinhole-B for pseudo-thermal light (ground glass rotated at 1000 rpm) and for laser beam (ground glass stoped at 0 rpm) Here, x_A and x_B are the x-component of $\vec{\rho}_A$ and $\vec{\rho}_B$.

of $d_A \sim 200$mm. There is no surprise to have first-order interference in the counting rates of D_1 and D_2 respectively. When choosing $\vec{\rho}_A = \vec{\rho}_B$, the two input fiber tips are coupled within the spatial coherence area of the radiation field; we have effectively built a Mach-Zehnder interferometer. We then move the input fiber tip A transversely from $\vec{\rho}_A = \vec{\rho}_B$ to $|\vec{\rho}_A - \vec{\rho}_B| \gg l_c$, where l_c is the transverse coherence length of the thermal field. (In most of the measurements, $|\vec{\rho}_A - \vec{\rho}_B|$ was chosen to be $|\vec{\rho}_A - \vec{\rho}_B| \geq 40l_c$.) Then we scan the input fiber tip A again longitudinally in the neighborhood of $d_A \sim 200$mm. The optical delay between the plane $z = d_A \sim 200$mm and the scanning input fiber tip A is labeled as δ in Figure 11.3.2. Under the experimental condition of $G^{(1)}(\mathbf{r}_A, t_A; \mathbf{r}_B, t_B) = 0$ and $G^{(2)}(\mathbf{r}_A, t_A; \mathbf{r}_B, t_B) = $ constant, there is no surprise to lose first-order interference. However, it is indeed a surprise that an anti-correlation is observed in terms of the joint photodetection of D_1 and D_2 as a function of the optical delay δ that is reported in Fig. 11.3.3. In these measurements, $|\vec{\rho}_A - \vec{\rho}_B| \sim 40l_c$. The observed contrast of the "dips" are $\sim 50\%$ in photon number correlation measurements and are $\sim 100\%$ in photon number fluctuation correlation measurements. Figure 11.3.5 shows another observed photon number fluctuation anti-correlation "dip". The visibility of the "dip" is recognized as $(93.2\pm5.1)\%$.

In fact, the theoretical analysis for the experimental observation of Chen *et al.* has been given in the beginning of this section. Based on the experimental setup of Fig. 11.3.2, using the same notations $G^{(2)}_{AA}(z_1, t_1; z_2, t_2)$, $G^{(2)}_{BB}(z_1, t_1; z_2, t_2)$, $G^{(2)}_{AB}(z_1, t_1; z_2, t_2)$ and $G^{(2)}_{BA}(z_1, t_1; z_2, t_2)$, what we need is to exam the cross terms in the superposition of $G^{(2)}_{AB}(z_1, t_1; z_2, t_2)$ and $G^{(2)}_{BA}(z_1, t_1; z_2, t_2)$ by assuming $t_m \simeq t_n$. In addition, because of the use of narrow-band spectral filters (IF), we need to modify the field operators slightly by taking into account the spectral function $f(\omega)$ of the IFs. Repeating the same calculation, we found an interesting solution: when taking $t_m \simeq t_n$, the cross interference term turns into the normal square of the Fourier transform of the spectrum

$$\int d\omega \big|f(\omega)\big|^2 e^{i\omega\delta} \int d\omega' \big|f(\omega')\big|^2 e^{-i\omega'\delta} = \big|\mathcal{F}_\delta\{|f(\nu)|^2\}\big|^2. \qquad (11.3.4)$$

The photon number fluctuation correlation measurement thus takes the form of a "dip" function of the optical delay δ

$$\langle \Delta n_1 \, \Delta n_2 \rangle \propto \big[1 - e^{-\delta^2/\tau_c^2}\big] \qquad (11.3.5)$$

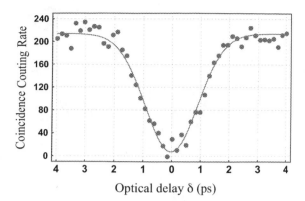

FIGURE 11.3.5 Typical observed anti-correlation function of "dip" in *photon number fluctuation* correlation measurement. The visibility of the "dip" is calculated to be $(93.2\pm5.1)\%$. The fitting curve agrees with the theoretical prediction well within the experimental error.

where we have assumed Gaussian spectrum function of $|f(\nu)|^2$. Equation (11.3.5) has been verified in the experiment of Chen *et al.*. In that experiment Chen *et al.* used a short pulsed laser beam to simulate the condition of $t_m \simeq t_n$.

(II) A few concerns about the experiment of Chen *et al.*.

After all, we may still have the following reasonable concerns:

(1) It is interesting to find that an anti-correlation function is observable under the experimental condition $d \gg l_c$, which means the radiations at pinhole-A and pinhole-B, or at the input fiber tip-A and tip-B, are "uncorrelated" with completely random photon number fluctuations: $\langle \Delta n(\mathbf{r}_A, t_A)\Delta n(\mathbf{r}_B, t_B)\rangle = 0$. A question naturally arises: How could we observe an anti-correlation "dip" in photon number fluctuation correlation measurement from two groups of completely randomly fluctuated photons? In other words, if the correlation measurement is on "uncorrelated" thermal field, instead of "correlated" thermal field, where does the anti-correlation come from?

(2) It is interesting to find that the classical statistical correlation of Gaussian field fails to explain the experimental observation of the anti-correlation "dip". Let us exam

$$\Gamma^{(2)}(\mathbf{r}_1, t_1; \mathbf{r}_2, t_2) = \langle I(\mathbf{r}_1, t_1)I(\mathbf{r}_2, t_2)\rangle \tag{11.3.6}$$
$$= \langle (E_{A1}^* + E_{B1}^*)(E_{A1} + E_{B1})(E_{A2}^* - E_{B2}^*)(E_{A2} - E_{B2})\rangle,$$

where E_{Aj} and E_{Bk} label the radiation fields from sources A and B at space-time coordinates $(\mathbf{r}_j, t_j)$ and $(\mathbf{r}_k, t_k)$, respectively. We have sixteen expectations to evaluate:

$$\langle I(\mathbf{r}_1, t_1)I(\mathbf{r}_2, t_2)\rangle$$
$$= \langle E_{A1}^* E_{A1} E_{A2}^* E_{A2}\rangle + \langle E_{A1}^* E_{A1} E_{B2}^* E_{B2}\rangle$$
$$- \langle E_{A1}^* E_{A1} E_{A2}^* E_{B2}\rangle - \langle E_{A1}^* E_{A1} E_{B2}^* E_{A2}\rangle$$
$$+ \langle E_{B1}^* E_{B1} E_{A2}^* E_{A2}\rangle + \langle E_{B1}^* E_{B1} E_{B2}^* E_{B2}\rangle$$
$$- \langle E_{B1}^* E_{B1} E_{A2}^* E_{B2}\rangle - \langle E_{B1}^* E_{B1} E_{B2}^* E_{A2}\rangle$$
$$+ \langle E_{A1}^* E_{B1} E_{A2}^* E_{A2}\rangle + \langle E_{A1}^* E_{B1} E_{B2}^* E_{B2}\rangle$$
$$- \langle E_{A1}^* E_{B1} E_{A2}^* E_{B2}\rangle - \langle E_{A1}^* E_{B1} E_{B2}^* E_{A2}\rangle$$
$$+ \langle E_{B1}^* E_{A1} E_{A2}^* E_{A2}\rangle + \langle E_{B1}^* E_{A1} E_{B2}^* E_{B2}\rangle$$
$$- \langle E_{B1}^* E_{A1} E_{A2}^* E_{B2}\rangle - \langle E_{B1}^* E_{A1} E_{B2}^* E_{A2}\rangle. \tag{11.3.7}$$

Applying the property of Gaussian field and taking the result of $\langle E_A^* E_B \rangle = 0$, it is not too difficult to find that ten terms in Equation (11.3.7) take zero value, and left six terms that may have non-zero contribution to $\langle I_1 I_2 \rangle$:

$$
\begin{aligned}
\langle I(\mathbf{r}_1, t_1) I(\mathbf{r}_2, t_2) \rangle \\
= \langle E_{A1}^* E_{A1} E_{A2}^* E_{A2} \rangle + \langle E_{B1}^* E_{B1} E_{B2}^* E_{B2} \rangle \\
+ \langle E_{A1}^* E_{A1} E_{B2}^* E_{B2} \rangle + \langle E_{B1}^* E_{B1} E_{A2}^* E_{A2} \rangle \\
- \langle E_{A1}^* E_{B1} E_{B2}^* E_{A2} \rangle - \langle E_{B1}^* E_{A1} E_{A2}^* E_{B2} \rangle.
\end{aligned}
\tag{11.3.8}
$$

The first two terms in Equation (11.3.8) correspond to the two sets of HBT correlations,

$$
\begin{aligned}
\langle E_{A1}^* E_{A1} E_{A2}^* E_{A2} \rangle \\
= \langle E_{A1}^* E_{A1} \rangle \langle E_{A2}^* E_{A2} \rangle + \langle E_{A1}^* E_{A2} \rangle \langle E_{A2}^* E_{A1} \rangle \\
= \Gamma_{A11}^{(1)} \Gamma_{A22}^{(1)} + \Gamma_{A12}^{(1)} \Gamma_{A21}^{(1)}
\end{aligned}
\tag{11.3.9}
$$

$$
\begin{aligned}
\langle E_{B1}^* E_{B1} E_{B2}^* E_{B2} \rangle \\
= \langle E_{B1}^* E_{B1} \rangle \langle E_{B2}^* E_{B2} \rangle + \langle E_{B1}^* E_{B2} \rangle \langle E_{B2}^* E_{B1} \rangle \\
= \Gamma_{B11}^{(1)} \Gamma_{B22}^{(1)} + \Gamma_{B12}^{(1)} \Gamma_{B21}^{(1)}.
\end{aligned}
\tag{11.3.10}
$$

It is clear that these two terms cannot produce the anti-correlation as a function of δ.

The next two terms in Equation (11.3.8) are the products of mean intensities, which cannot produce the anti-correlation function of δ either.

$$
\begin{aligned}
\langle E_{A1}^* E_{A1} E_{B2}^* E_{B2} \rangle \\
= \langle E_{A1}^* E_{A1} \rangle \langle E_{B2}^* E_{B2} \rangle = \Gamma_{A11}^{(1)} \Gamma_{B22}^{(1)}
\end{aligned}
\tag{11.3.11}
$$

$$
\begin{aligned}
\langle E_{B1}^* E_{B1} E_{A2}^* E_{A2} \rangle \\
= \langle E_{B1}^* E_{B1} \rangle \langle E_{A2}^* E_{A2} \rangle = \Gamma_{B11}^{(1)} \Gamma_{A22}^{(1)}
\end{aligned}
\tag{11.3.12}
$$

The last two "cross terms" may contribute non-zero values to produce an "anti-correlation", however, neither of them is a function of the optical delay δ.

$$
\begin{aligned}
\langle E_{A1}^* E_{B1} E_{B2}^* E_{A2} \rangle \\
= \langle E_{A1}^* E_{A2} \rangle \langle E_{B2}^* E_{B1} \rangle = \Gamma_{A12}^{(1)} \Gamma_{B21}^{(1)}
\end{aligned}
\tag{11.3.13}
$$

$$
\begin{aligned}
\langle E_{B1}^* E_{A1} E_{A2}^* E_{B2} \rangle \\
= \langle E_{B1}^* E_{B2} \rangle \langle E_{A2}^* E_{A1} \rangle = \Gamma_{B12}^{(1)} \Gamma_{A21}^{(1)}.
\end{aligned}
\tag{11.3.14}
$$

(3) It is interesting to find that the quantum theory of light is able to explain the observation of anti-correlation "dip" from "classically uncorrelated" thermal field. In the view of quantum theory of light, a joint photodetection between two independent point-like photodetectors D_1 and D_2 measures the probability of observing a joint photodetection event of two photons at space-time coordinates $(\mathbf{r}_1, t_1)$ and $(\mathbf{r}_2, t_2)$. If there exists more than one different yet indistinguishable alternatives for the two photons to produce that event, the probability of observing a joint photodetection event at space-time coordinates $(\mathbf{r}_1, t_1)$ and $(\mathbf{r}_2, t_2)$ is the result of the superposition among all these quantum probability-amplitudes:

$$
\begin{aligned}
G^{(2)}(\mathbf{r}_1, t_1; \mathbf{r}_2, t_2) \\
= \big\langle \langle \hat{E}^{(-)}(x_1, t_1) \hat{E}^{(-)}(x_2, t_2) \hat{E}^{(+)}(x_2, t_2) \hat{E}^{(+)}(x_1, t_1) \rangle_{\text{QM}} \big\rangle_{\text{En}} \\
\simeq \Big\langle \Big| \sum_j \Psi_j(\mathbf{r}_1, t_1; \mathbf{r}_2, t_2) \Big|^2 \Big\rangle_{\text{En}}
\end{aligned}
\tag{11.3.15}
$$

where $\Psi_j(\mathbf{r}_1, t_1; \mathbf{r}_2, t_2)$ is defined as the two-photon effective wavefunction, namely the quantum probability amplitude. It is interesting to find that the above superposition is not only applicable to entangled states, but also applicable to randomly created and randomly paired photons in thermal states, except the two-photon effective wavefunction of a randomly created and randomly paired photons in thermal state is factorizable into two single-photon effective wavefunctions. If there exists a number of distinguishable alternatives for a pair of photons, either entangled or randomly created and randomly paired, to produce a joint photodetection event, we may write $G^{(2)}(\mathbf{r}_1, t_1; \mathbf{r}_2, t_2)$ into a sum of distinguishable superpositions after completing the ensemble average,

$$G^{(2)}(\mathbf{r}_1, t_1; \mathbf{r}_2, t_2) = \sum_k G_k^{(2)}(\mathbf{r}_1, t_1; \mathbf{r}_2, t_2)$$
$$= \sum_k P_k \big| \sum_j \Psi_{kj}(\mathbf{r}_1, t_1; \mathbf{r}_2, t_2) \big|^2 \qquad (11.3.16)$$

where P_k is the probability for the two photons to produce a photodetection event in the kth distinguishable alternative way.

Either considering thermal field "correlated", "anti-correlated", or "uncorrelated", in the view of quantum theory of light, a randomly created and randomly paired photons in thermal state may interfering with the pair itself constructively or destructively

$$G_k^{(2)}(\mathbf{r}_1, t_1; \mathbf{r}_2, t_2) = \sum_{m,n} |\psi_{m1}\psi_{n2} \pm \psi_{n1}\psi_{m2}|^2, \qquad (11.3.17)$$

and is able to produce an observable "correlation peak" or an "anti-correlation dip" in a correctly designed experimental setup that satisfy a certain experimental conditions to make the two-photon constructive or destructive interference observable.

11.4 TWO-PHOTON INTERFERENCE WITH INCOHERENT ORTHOGONAL POLARIZED THERMAL FIELDS

In this section, we discuss two-photon interference with orthogonal polarized incoherent thermal fields. We start from the analysis of a simple experiment, which is schematically shown in Figure 11.4.1. The experimental setup is similar to that of the historical two-photon polarization interferometer of Alley-Shih, except the use of orthogonal polarized incoherent thermal light, instead of entangled photon pair. If the input radiation at points A and B are first-order coherent, this setup is equivalent to a classic polarization Mach-Zehnder interferometer. D_1 and D_2 will each observe first-order interference, and consequently, the joint photodetection of D_1 and D_2 outputs an interference that is the product of the two first-order interferences. To avoid this trivial effect to happen, E_A and E_B are managed mutually incoherent, $G^{(1)}(\mathbf{r}_A, t_A; \mathbf{r}_B, t_B) = 0$. The incoherent thermal radiations A and B are prepared in the same way as that in the previous experiment of Chen *et al.* by moving the fiber tip A transversely outside of the coherence area of the thermal field, which has been described in detail in previous sections. Similarly, achieving an experimental condition of $|\bar{\rho}_A - \bar{\rho}_B| \geq 40l_c$, there is no question that we will lose any first-order interferences measured by D_1 and D_2, respectively. In addition, D_1 and D_2, respectively, should observe randomly polarized light too. The incoherent $\hat{x}$ and $\hat{y}$ polarized radiations will pass A_1 and A_2 randomly according to Malus' law. The total intensity measured by D_1 and D_2, respectively, should be $\sin^2\theta_j + \cos^2\theta_j = 1$, $j = 1, 2$, in any chosen orientation of the polarizer. Again, the question is in the second-order correlation measurement. Under the experimental condition of $G^{(2)}(\mathbf{r}_A, t_A; \mathbf{r}_B, t_B) = \text{constant}$, i.e., $\langle \Delta I_A \Delta I_B \rangle = 0$, there should be no observable HBT type correlations. What can we expect from the measurement of

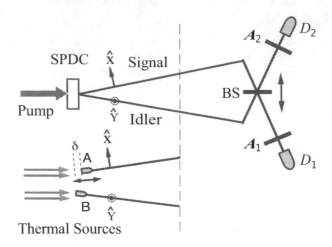

FIGURE 11.4.1 Schematic of a typical two-photon interferometer with orthogonal polarized *incoherent thermal light*. The setup is the same as the historical two-photon polarization interferometer of Alley-Shih, except the use of randomly created and randomly photon pair in thermal state, instead of entangled photon pair. The use of *incoherent radiations* E_A and E_B guarantee $G^{(1)}(\mathbf{r}_A, t_A; \mathbf{r}_B, t_B) = 0$ and $G^{(2)}(\mathbf{r}_A, t_A; \mathbf{r}_B, t_B) = $ constant. No classic Mach-Zehnder interferences are observable by D_1 and D_2, respectively. Can we expect to observe either space-time correlation or polarization correlation from the joint measurement of D_1 and D_2?

$G^{(2)}(\mathbf{r}_1, t_1; \mathbf{r}_2, t_2)$? Can we expect to observe either space-time correlation or polarization correlation from the joint measurement of D_1 and D_2?

(I) Experimental realization of Chen *et al.*

Chen *et al.* provided a positive answer to the above question in another recent experiment: two-photon interference with incoherent orthogonal polarized thermal fields produce both space-time correlation and polarization correlation from the joint measurement of D_1 and D_2. Their schematic experimental setup is illustrated in Fig. 11.4.2. Some of the experimental details are emphasized as follows:

(1) The source: the incoherent pseudo-thermal source is the same as that in the anti-correlation experiment of Chen *et al.*, which consists of a mode-locked laser beam with ~200 femtosecond pulses at a 78 MHz repetition rate and a fast rotating diffusing ground glass. The laser beam is enlarged transversely onto the ground glass with a diameter of $4.5mm$. The enlarged laser beam is scattered by the fast rotating diffusing ground into a large number of transversely randomly distributed point-like sub-sources to simulate a near-field, pseudo-thermal radiation source. A large number of randomly distributed incoherent wavepackets with random relative phases are then collected by the optical fiber tips A and B and coupled into the optical fiber beamsplitter.

(2) The polarization state: a set of polarization bases of $\hat{x}$ and $\hat{y}$ is defined by two Glen-Thompson polarizers which are coupled with fiber A and fiber B, respectively. The $\hat{x}$ polarized radiation A and the $\hat{y}$ polarized radiation B are injected onto a 50%-50% beamsplitter from its opposite input ports in a near-normal incidence configuration.[2] The beamsplitter is measured with 50% reflection and 50% transmission for both $\hat{x}$ and $\hat{y}$ polarization.

---

[2]To achieve 50%-50% reflection-transmission for both $\hat{x}$ and $\hat{y}$ polarization, near-normal incidence configuration is a better choice. This configuration is the same as that of the historical experiment of Alley and Shih which observed Bell type polarization correlation from an orthogonal polarized photon pair of SPDC. The first a few historical "dip" experiments, such as the Hong-Ou-Mandel "dip" measurment, adopted this near-normal incidence configuration from the two-photon interferometer of Alley and Shih. In fact, it is

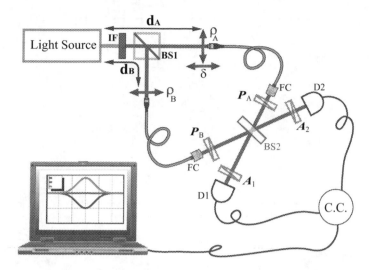

FIGURE 11.4.2 Schematic experimental setup of Chen *et al.* in which two surprises were observed from first-order and second-order incoherent orthogonal polarized thermal fields: (1) anti-correlation "dip" and correlation "peak"; (2) Bell type polarization correlation. The experimental condition of *first-order and second-oder incoherence* between radiations A and B is achieved by moving fiber tip A transversely outside of the spatial coherence area of the field. In this experiment, most of the measurements were performed at $|\bar{\rho}_A - \bar{\rho}_B| \geq 40l_c$.

(3) The polarization analyzers: two Glen-Thompson polarizers are coupled into the two output ports of the 50%-50% beamsplitter, each followed by a photon counting detector, D_1 and D_2, for the polarization correlation measurement. The polarization analyzer can be set at any angle θ relative to the $\hat{x}$. The $\hat{x}$ direction is defined as $0°$ in the above analysis.

The measurement of Chen *et al.* produces two interesting effects in one experimental setup of Fig. 11.4.2.

(1) "Unexpected" anti-correlation "dip" and correlation "peak" are observed in the joint photodetection counting rate of D_1 and D_2, $\langle n_1 n_2 \rangle$, as a function of the optical delay δ, which is defined in Fig. 11.4.1. The anti-correction "dip" is observed when P_1 and P_2 were oriented at $\theta_1 = 45°, \theta_2 = 45°$. The correlation "peak" is obtained when P_1 and P_2 are oriented at $\theta_1 = 45°, \theta_2 = 135°$.

(2) Bell type polarization correlation are observed as a function of the relative angle $(\theta_1 - \theta_2)$ of the two polarization analyzers under the condition of $\delta = 0$,

$$R_c(\theta_1, \theta_2) \propto \langle n_1(\theta_1) n_2(\theta_2) \rangle \propto \left[1 + V \sin^2(\theta_1 - \theta_2) \right], \quad (11.4.1)$$

where V is the contrast of the sinusoidal modulation, $R_c(\theta_1, \theta_2)$ is the coincidence counting rate of D_1 and D_2, which is obtained by manipulating the relative orientation of A_1 and A_2. Figure 11.4.4 is a typical observed polarization correlation reported by Chen *et al.*. Furthermore, if the measurement is on photon number fluctuation correlation $\langle \Delta n_1(\theta_1) \Delta n_2(\theta_2) \rangle \propto \Delta R_c(\theta_1, \theta_2)$, the visibility of the sinusoidal function of $\theta_1 - \theta_2$ becomes $\sim 100\%$,

$$\Delta R_c(\theta_1, \theta_2) \propto \langle \Delta n_1(\theta_1) \Delta n_2(\theta_2) \rangle \propto \sin^2(\theta_1 - \theta_2). \quad (11.4.2)$$

unnecessary to choose near-normal incidence beamsplitter for a biphoton interferometer which uses only one polarization (type-I SPDC) for "dip" measurement.

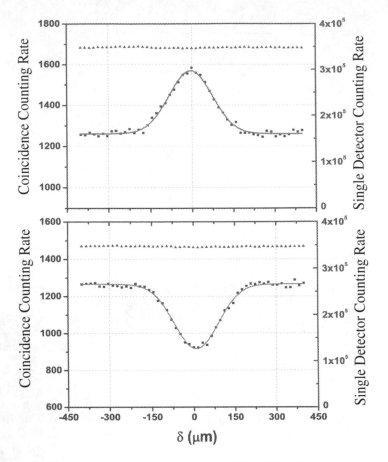

FIGURE 11.4.3 Typical observed correlation function of "peak" and anti-correlation function of "dip" as functions of the optical delay δ. The correlation "peak" is obtained when P_1 and P_2 were oriented at $\theta_1 = 45°, \theta_2 = 135°$. The anti-correction "dip" is observed when P_1 and P_2 were oriented at $\theta_1 = 45°, \theta_2 = 45°$.

(II) Theoretical analysis.

Quantum theory gives a reasonable interpretation to the observation: the observation is a two-photon interference phenomenon, the result of a superposition between indistinguishable two-photon amplitudes. Taking the result of early sections of this chapter, we start from $G^{(2)}(\delta; \theta_1, \theta_2)$ with the polarization analyzers oriented at θ_1 and θ_2, respectively,

$$
\begin{aligned}
G^{(2)}&(\theta_1, \theta_2; \delta) \\
=& \sum_{m,n} \Big| [(\hat{\theta}_1 \cdot \hat{x})\psi_{mA1} - (\hat{\theta}_1 \cdot \hat{y})\psi_{mB1}][(\hat{\theta}_2 \cdot \hat{x})\psi_{nA2} + (\hat{\theta}_1 \cdot \hat{y})\psi_{nB2}] \\
& + [(\hat{\theta}_1 \cdot \hat{x})\psi_{nA1} - (\hat{\theta}_1 \cdot \hat{y})\psi_{nB1}][(\hat{\theta}_2 \cdot \hat{x})\psi_{mA2} + (\hat{\theta}_2 \cdot \hat{y})\psi_{mB2}] \Big|^2 \\
=& \sum_{m,n} \Big| (\hat{\theta}_1 \cdot \hat{x})(\hat{\theta}_2 \cdot \hat{x})\psi_{mA1}\psi_{nA2} + (\hat{\theta}_1 \cdot \hat{x})(\hat{\theta}_2 \cdot \hat{y})\psi_{mA1}\psi_{nB2} \\
& - (\hat{\theta}_1 \cdot \hat{y})(\hat{\theta}_2 \cdot \hat{x})\psi_{mB1}\psi_{nA2} - (\hat{\theta}_1 \cdot \hat{y})(\hat{\theta}_2 \cdot \hat{y})\psi_{mB1}\psi_{nB2} \\
& + (\hat{\theta}_1 \cdot \hat{x})(\hat{\theta}_2 \cdot \hat{x})\psi_{nA1}\psi_{mA2} + (\hat{\theta}_1 \cdot \hat{x})(\hat{\theta}_2 \cdot \hat{y})\psi_{nA1}\psi_{mB2} \\
& - (\hat{\theta}_1 \cdot \hat{y})(\hat{\theta}_2 \cdot \hat{x})\psi_{nB1}\psi_{mA2} - (\hat{\theta}_1 \cdot \hat{y})(\hat{\theta}_2 \cdot \hat{y})\psi_{nB1}\psi_{mB2} \Big|^2
\end{aligned}
\tag{11.4.3}
$$

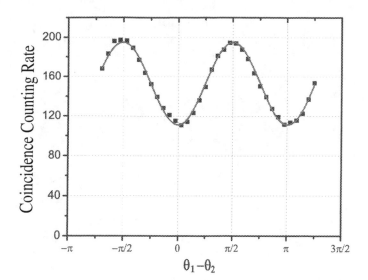

FIGURE 11.4.4 Typical observed polarization correlation $\langle n_1(\theta_1)n_2(\theta_2)\rangle$ as a function of $\theta_1 - \theta_2$. In this measurement θ_2 was fixed at $45°$ relative to $\hat{x}$ and $\hat{y}$, θ_1 was rotated at each chosen angle θ_1. If the measurement is for photon number fluctuation correlation $\langle \Delta n_1(\theta_1)\Delta n_2(\theta_2)\rangle$, the visibility of the sinusoidal function of $\theta_1 - \theta_2$ becomes $\sim 100\%$.

Completing the sum for each term in the $| \,.... \,|^2$, we find that a number of cross terms vanish due to the experimental condition $|\bar{\rho}_A - \bar{\rho}_B| \gg l_c$. We may group the surviving terms into the following four superpositions,

$$
\begin{aligned}
&G^{(2)}(\theta_1, \theta_2; \delta) \\
&= \sum_{m,n} \left| (\hat{\theta}_1 \cdot \hat{x})(\hat{\theta}_2 \cdot \hat{x})\psi_{mA1}\psi_{nA2} + (\hat{\theta}_1 \cdot \hat{x})(\hat{\theta}_2 \cdot \hat{x})\psi_{nA1}\psi_{mA2} \right|^2 \\
&+ \sum_{m,n} \left| (\hat{\theta}_1 \cdot \hat{y})(\hat{\theta}_2 \cdot \hat{y})\psi_{mB1}\psi_{nB2} + (\hat{\theta}_1 \cdot \hat{y})(\hat{\theta}_2 \cdot \hat{y})\psi_{nB1}\psi_{mB2} \right|^2 \\
&+ \sum_{m,n} \left| (\hat{\theta}_1 \cdot \hat{x})(\hat{\theta}_2 \cdot \hat{y})\psi_{mA1}\psi_{nB2} - (\hat{\theta}_1 \cdot \hat{y})(\hat{\theta}_2 \cdot \hat{x})\psi_{nB1}\psi_{mA2} \right|^2 \\
&+ \sum_{m,n} \left| (\hat{\theta}_1 \cdot \hat{y})(\hat{\theta}_2 \cdot \hat{x})\psi_{mB1}\psi_{nA2} - (\hat{\theta}_1 \cdot \hat{x})(\hat{\theta}_2 \cdot \hat{y})\psi_{nA1}\psi_{mB2} \right|^2 \\
&\equiv G_{AA}^{(2)}(\theta_1, \theta_2; \delta) + G_{BB}^{(2)}(\theta_1, \theta_2; \delta) + G_{AB}^{(2)}(\theta_1, \theta_2; \delta) + G_{BA}^{(2)}(\theta_1, \theta_2; \delta). \quad (11.4.4)
\end{aligned}
$$

The four distinguishable groups of probabilities $G_{AB}^{(2)}(\theta_1, \theta_2; \delta)$, $G_{BA}^{(2)}(\theta_1, \theta_2; \delta)$, $G_{AA}^{(2)}(\theta_1, \theta_2; \delta)$, and $G_{BB}^{(2)}(\theta_1, \theta_2; \delta)$ are schematically pictured in Fig. 11.4.5: (a) and (b) represent $G_{AB}^{(2)}(\theta_1, \theta_2; \delta)$ and $G_{BA}^{(2)}(\theta_1, \theta_2; \delta)$: the probability of triggering a joint photodetection event by two wavepackets, one coming from A and the other coming from B. (c) and (d) represent $G_{AA}^{(2)}(\theta_1, \theta_2; \delta)$ and $G_{BB}^{(2)}(\theta_1, \theta_2; \delta)$: the probability of triggering a joint photodetection event of D_1 and D_2 by two wavepackets both coming from A or both coming from B. In (a) and (b), the two wavepackets are both transmitted or both reflected at the beamsplitter. In (c) and (d), one wavepacket is transmitted, another wavepacket is reflected at the beamsplitter.

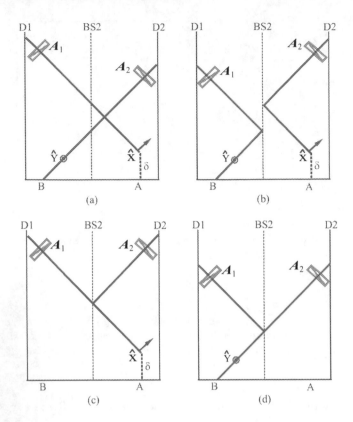

FIGURE 11.4.5 Feynman diagram of the measurement.

Examining $G_{AB}^{(2)}(\theta_1, \theta_2)$ and $G_{BA}^{(2)}(\theta_1, \theta_2)$,

$$G_{AB}^{(2)}(\theta_1, \theta_2; \delta) \tag{11.4.5}$$
$$\propto \sum_{m,n} \left| (\hat{\theta}_1 \cdot \hat{y})(\hat{\theta}_2 \cdot \hat{x}) \, \psi_{mA1}^T(\delta)\psi_{nB2}^T - (\hat{\theta}_1 \cdot \hat{x})(\hat{\theta}_2 \cdot \hat{y}) \, \psi_{nB1}^R \psi_{mA2}^R(\delta) \right|^2,$$

indicating a superposition between two-photon amplitudes (a) The mth $\hat{x}$ polarized wavepacket from A is delayed with δ and then is transmitted onto D_1; while the nth $\hat{y}$ polarized wavepacket from B is transmitted onto D_2; (b) The mth $\hat{x}$ polarized wavepacket from A is delayed by δ and then reflected onto D_2; while the nth $\hat{y}$ polarized wavepacket from B is reflected onto D_1.

$$G_{BA}^{(2)}(\theta_1, \theta_2; \delta) \tag{11.4.6}$$
$$\propto \sum_{m,n} \left| (\hat{\theta}_1 \cdot \hat{y})(\hat{\theta}_2 \cdot \hat{x})\psi_{mB1}^R \psi_{nA2}^R(\delta) - (\hat{\theta}_1 \cdot \hat{x})(\hat{\theta}_2 \cdot \hat{y})\psi_{nA1}^T(\delta)\psi_{mB2}^T \right|^2$$

indicating a superposition between two-photon amplitudes (a) The mth $\hat{y}$ polarized wavepacket from B is delayed with δ and then is reflected onto D_1; while the nth $\hat{x}$ polarized wavepacket from A is reflected onto D_2; (b) The mth $\hat{y}$ polarized wavepacket from A is delayed by δ and then transmitted onto D_2; while the nth $\hat{x}$ polarized wavepacket from A is transimitted onto D_1.

According to Eqs. (11.4.5) and (11.4.6), in the two-photon interferometer of Fig. 11.4.2 with orthogonal polarized incoherent thermal fields, we shall be able to observe two types of two-photon interference effects:

(1) Anti-correlation "dip" and correlation "peak".

In the measurement of anti-correlation "dip", we choose $\theta_1 = 45°$ and $\theta_2 = 45°$, Equation (11.4.5) becomes

$$G_{AB}^{(2)}(45°, 45°; \delta) \propto \sum_{m<n} \left| \psi_{mA1}^T(\delta)\psi_{nB2}^T - \psi_{nB1}^R \psi_{mA2}^R(\delta) \right|^2, \tag{11.4.7}$$

The relative delay between wavepackets ψ_{mA1}^* $[\psi_{nB2}]$ and ψ_{nB1} $[\psi_{mA2}]$ produces an anti-correlation "dip" as a function of the optical delay δ is thus expected in the coincidence counting rate R_{AB}.

In the measurement of correlation "peak", we choose $\theta_1 = 45°$ and $\theta_2 = 135°$, Equation (11.4.5) becomes

$$G_{AB}^{(2)}(45°, 135°; \delta) \propto \sum_{m,n} \left| \psi_{mA1}^T(\delta)\psi_{nB2}^T + \psi_{nB1}^R \psi_{mA2}^R(\delta) \right|^2, \tag{11.4.8}$$

which is the same as Equation (11.4.7), except with a "+" sign.

It is easy to see that $G_{BA}^{(2)}(45°, 45°; \delta)$ and $G_{BA}^{(2)}(45°, 135°; \delta)$ have the same contribution to the anti-correlation "dip" and the correlation "peak". The calculation of the anti-correlation "dip" and the correlation "peak" are similar to that of the anti-correlation "dip" in previous section. The experimentally observed anti-correlation "dip" and correlation "peak" of Chen *et al.* have been shown in Fig. 11.4.3.

(2) Polarization correlation.

In this measurement, we make $\delta = 0$ to achieve complete overlapping between $\psi_{mA1}^T\psi_{nB2}^T$ and $\psi_{nB1}^R\psi_{mA2}^R$. Eq. (11.4.5) becomes

$$G_{AB}^{(2)}(\theta_1, \theta_2; 0) \propto \left| (\hat{\theta}_1 \cdot \hat{y})(\hat{\theta}_2 \cdot \hat{x}) - (\hat{\theta}_1 \cdot \hat{x})(\hat{\theta}_2 \cdot \hat{y}) \right|^2$$
$$\propto \sin^2(\theta_1 - \theta_2), \tag{11.4.9}$$

corresponding to the measurement of the Bell state

$$|\Psi^{(-)}\rangle = \frac{1}{\sqrt{2}} \left[|X_1\rangle|Y_2\rangle - |Y_1\rangle|X_2\rangle \right].$$

When $\delta = 0$, we find $\psi_{mB1}^R\psi_{nA2}^R(\delta)$ and $\psi_{nA1}^T(\delta)\psi_{mB2}^T$ are also overlap completely, $G_{BA}^{(2)}(\theta_1, \theta_2; 0)$ has the same contribution to the polarization correlation measurement as that of $G_{AB}^{(2)}(\theta_1, \theta_2; 0)$.

Considering the trivial contributions from the other two terms $G_{AA}^{(2)}$ and $G_{BB}^{(2)}$, the contrast of the Bell type correlation is reduced from 100% to 50%,

$$G_{AB}^{(2)}(\theta_1, \theta_2) + G_{BA}^{(2)}(\theta_1, \theta_2) \propto \left[1 + \sin^2(\theta_1 - \theta_2) \right]. \tag{11.4.10}$$

Since the contribution of $G_{AA}^{(2)}$ and $G_{BB}^{(2)}$ can be measured experimentally by blocking the other source, in principle, the contribution of $G_{AB}^{(2)} + G_{BA}^{(2)}$ can be isolated from $G_{AA}^{(2)}$ and $G_{BB}^{(2)}$. It is interesting to find that after subtracting the contributions of R_{AA} and R_{BB} from the total joint photodetection counting rate R_c, the visibilities of the "dip" and the Bell type polarization correlation resume to ~100% in the experiment of Chen *at al.*.

(III) 100% correlation in the measurement of $\langle \Delta n_1 \Delta n_2 \rangle$.

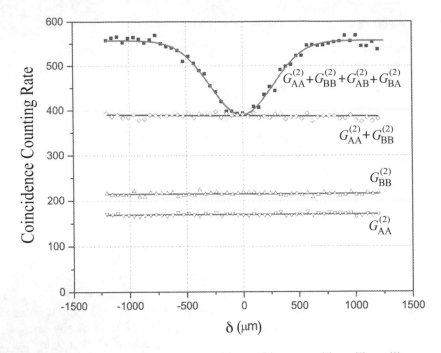

FIGURE 11.4.6 The observed $G_{AA}^{(2)} + G_{BB}^{(2)} + G_{AB}^{(2)} + G_{BA}^{(2)}$, and $G_{AA}^{(2)}$, $G_{BB}^{(2)}$. $G_{AA}^{(2)}$ and $G_{BB}^{(2)}$ are measured by blocking the other source. Subtracting the contributions of $G_{AA}^{(2)}$ and $G_{BB}^{(2)}$ the contrast of the observed anti-correlation function becomes ∼100%. The anti-correction "dip" is observed when A_1 and A_2 were oriented at $\theta_1 = \theta_2 = 45°$.

It is interesting, we also find that the visibility of the anti-correlation "dip" and the Bell type polarization correlation become 100% in the measurement of photon number fluctuation correlation $\langle \Delta n_1 \Delta n_2 \rangle$. This is easily seen by adding the cross interference terms of the superpositions in $G_{AB}^{(2)}(\theta_1, \theta_2; \delta)$, $G_{BA}^{(2)}(\theta_1, \theta_2; \delta)$, $G_{AA}^{(2)}(\theta_1, \theta_2)$ and $G_{BB}^{(2)}(\theta_1, \theta_2)$,

$$
\begin{aligned}
&\langle \Delta n_1(\theta_1, \delta) \Delta n_2(\theta_2, \delta) \rangle \\
&\propto [(\hat{\theta}_1 \cdot \hat{x})(\hat{\theta}_2 \cdot \hat{x})]^2 \psi_{mA1}^* \psi_{nA1}(\delta) \psi_{nA2}^*(\delta) \psi_{mA2}(\delta) \\
&+ [(\hat{\theta}_1 \cdot \hat{y})(\hat{\theta}_2 \cdot \hat{y})]^2 \psi_{mB1}^* \psi_{nB1} \psi_{nB2}^* \psi_{mB2} \\
&- (\hat{\theta}_1 \cdot \hat{x})(\hat{\theta}_2 \cdot \hat{y})(\hat{\theta}_1 \cdot \hat{y})(\hat{\theta}_2 \cdot \hat{x}) \psi_{mA1}^*(\delta) \psi_{nB1} \psi_{nB2}^* \psi_{mA2}(\delta) \\
&- (\hat{\theta}_1 \cdot \hat{y})(\hat{\theta}_2 \cdot \hat{x})(\hat{\theta}_1 \cdot \hat{x})(\hat{\theta}_2 \cdot \hat{y}) \psi_{mB1}^* \psi_{nA1}(\delta) \psi_{nA2}^*(\delta) \psi_{mB2}.
\end{aligned}
\tag{11.4.11}
$$

corresponding to the measurement of the Bell state

$$
|\Psi^{(-)}\rangle = \frac{1}{\sqrt{2}} \big[|X_1\rangle |Y_2\rangle - |Y_1\rangle |X_2\rangle \big].
$$

(1) 100% contrast Anti-correlation "dip" observed from the measurement of $\langle \Delta n_1 \Delta n_2 \rangle$.

Choosing $\theta_1 = \theta_2 = 45°$, Eq. (11.4.11) turns to be

$$
\begin{aligned}
&\langle \Delta n_1(45°, \delta) \Delta n_2(45°, \delta) \rangle \\
&\propto \psi_{mA1}^*(\delta) \psi_{nA1}(\delta) \psi_{nA2}^*(\delta) \psi_{mA2}(\delta) + \psi_{mB1}^* \psi_{nB1} \psi_{nB2}^* \psi_{mB2} \\
&- \psi_{mA1}^*(\delta) \psi_{nB1} \psi_{nB2}^* \psi_{mA2}(\delta) - \psi_{mB1}^* \psi_{nA1}(\delta) \psi_{nA2}^*(\delta) \psi_{mB2}.
\end{aligned}
\tag{11.4.12}
$$

The photon number fluctuation correlation $\langle \Delta n_1(45°, \delta) \Delta n_2(45°, 0) \rangle$ yields a zero value

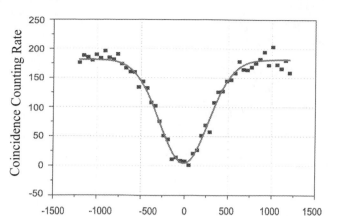

FIGURE 11.4.7 In the measurement of photo number fluctuation correlation $\langle \Delta n_1 \Delta n_2 \rangle$, the contrast of the anti-correlation is close to 100%.

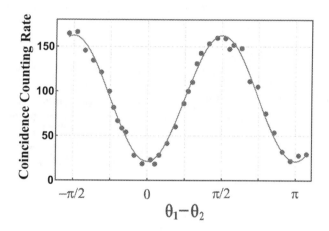

FIGURE 11.4.8 In the measurement of photo number fluctuation correlation $\langle \Delta n_1 \Delta n_2 \rangle$, the visibility of the observed polarization correlation as a function of $\theta_1 - \theta_2$ becomes $> 71\%$. Indicating the behavior of a two-photon Bell state $\Psi^{(-)}$. 71% has been considered as the border between "quantum" and "classical" in some theoretical concerns. In this measurement θ_2 was fixed at 45° relative to $\hat{x}$ and $\hat{y}$, θ_1 was rotated at each chosen angle θ_1. Either $< 71\%$ or $> 71\%$, its nonlocal two-photon interference nature never change.

when $\delta = 0$, under which all four two-photon effective wavefunctions overlap completely in space-time. However, when δ increases, the values of the first two terms of Eq. (11.4.12) have no change, while the values of the second two terms decrease because the pulsed conjugated wavepackets $\psi^*_{mA1}(\delta)$ $[\psi^*_{nB2}]$ and ψ_{nB1} $[\psi_{mA2}(\delta)]$ start shift away from each other. When the conjugated wavepackets are separated completely at D_1 and D_2, i.e., one wavepacket is delayed beyond the coherence time τ_c of the thermal field, the two cross interference terms vanish. An anti-correlation "dip" is thus observed as a function of the optical delay δ.

(2) 100% contrast Bell-type Polarization correlation observed from the measurement of $\langle \Delta n_1 \Delta n_2 \rangle$.

Choosing $\delta = 0$, under which all four two-photon effective wavefunctions overlap completely in space-time, Eq. (11.4.11) turns to be

$$
\langle \Delta n_1(\theta_1, \delta) \Delta n_2(\theta_2) \rangle
$$
$$
\propto [(\hat{\theta}_1 \cdot \hat{x})(\hat{\theta}_2 \cdot \hat{x})]^2 + [(\hat{\theta}_1 \cdot \hat{y})(\hat{\theta}_2 \cdot \hat{y})]^2
$$
$$
- (\hat{\theta}_1 \cdot \hat{x})(\hat{\theta}_2 \cdot \hat{y})(\hat{\theta}_1 \cdot \hat{y})(\hat{\theta}_2 \cdot \hat{x}) - (\hat{\theta}_1 \cdot \hat{y})(\hat{\theta}_2 \cdot \hat{x})(\hat{\theta}_1 \cdot \hat{x})(\hat{\theta}_2 \cdot \hat{y})
$$
$$
= (\cos\theta_1 \cos\theta_2)^2 + (\sin\theta_1 \sin\theta_2)^2
$$
$$
- (\sin\theta_1 \cos\theta_2 \cos\theta_1 \sin\theta_2) - (\sin\theta_1 \cos\theta_2 \cos\theta_1 \sin\theta_2)
$$
$$
= \sin^2(\theta_1 - \theta_2). \tag{11.4.13}
$$

It seems that two randomly created and randomly paired photons in thermal state can simulate the behavior of an entangled pair of photons by means of two-photon constructive and destructive interferences and by means of Bell-type polarization correlation. In fact, it is not too difficult to construct the complete set of Bell states by modifying the experimental setup of Chen *at al.* slightly. Furthermore, if the goal of the experiment is aimed for space-time correlation or Bell-type polarization correlation only, but not aimed at observing anti-correlation "dip" or correlation "peak", pulsed pseudo-thermal source may not be necessary. Anti-correlation does not have to be associated with a "dip" anyway. CW thermal light sources are able to produce anti-correlation, correlation, Bell-type polarization correlation in their photon number fluctuations under certain experimental conditions.

How could randomly created and randomly paired photons in thermal state produce the same correlation as that of an entangled state? The above analysis of the anti-correlation experiment and the Bell-type polarization correlation measurement may be helpful for us to find an answer. In fact, we have concluded the same physics in terms of two-photon interference: a pair of photons interfering with the pair itself. The superposition of two-photon amplitudes of a photon pair, either in a thermal state or in an entangled state, physically, all corresponding to different yet indistinguishable alternative ways for a pair of photons to produce a joint photodetection event.

11.5 TURBULENCE-FREE TWO-PHOTON INTERFEROMETER

It is well known that optical turbulence is harmful in optical observations. Random variations in the composition or density of the medium lead to changes in the index of refraction, known as optical turbulence, and thus vary the relative phases between different optical paths of an interferometer. These variations may "blur" the interference pattern partially or completely, thus reducing the sensitivity and effectiveness of an interferometer. Optical turbulence is particularly detrimental for extremely sensitive interferometers, interferometric spectrometers, and other interferometric sensors. For instance those used in gravitational-wave detection like the Laser Interferometer Gravitational-Wave Observatory (LIGO) which must be placed in a high-cost vacuum. In this section, we discuss the physics of turbulence-free interferometer.

We start from a recent experiment of Smith and Shih which demonstrated a turbulence-free double-slit interferometer. The experimental setup is schematically depicted in Fig. 11.5.1. This interferometer looks like a classic Young's double-slit interferometer except that it has two point-like scannable photon counting detectors, D_1 and D_2, rather than one. Together they measure the photon number fluctuation correlation $\langle \Delta n(x_1) \Delta n(x_2) \rangle$ which is proportional to the intensity fluctuation correlation $\langle \Delta I(x_1) \Delta I(x_2) \rangle$. In fact, this interferometer is able to produce three outputs corresponding to two types of measurement: (1) $\langle n(x_1) \rangle \propto \langle I(x_1) \rangle$ and $\langle n(x_2) \rangle \propto \langle I(x_2) \rangle$, corresponding to the measurement of mean intensities at D_1 and D_2, respectively; (2) $\langle \Delta n(x_1) \Delta n(x_2) \rangle \propto \langle \Delta I(x_1) \Delta I(x_2) \rangle$, corresponding

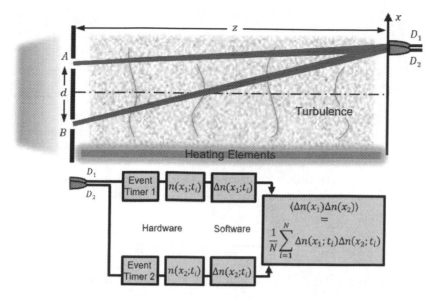

FIGURE 11.5.1   A turbulence-free double-slit interferometer. This interferometer has two types of output (1) $\langle n(x_1) \rangle \propto \langle I(x_1) \rangle$ and $\langle n(x_2) \rangle \propto \langle I(x_2) \rangle$, and (2) $\langle \Delta n(x_1) \Delta n(x_2) \rangle \propto \langle \Delta I(x_1) \Delta I(x_2) \rangle$. Due to the experimental condition of $d \gg l_c$, no interferences are observable from $\langle I(x_1) \rangle$ and $\langle I(x_2) \rangle$. However, a turbulence-free interference with 100% visibility is observed from the measurement of $\langle \Delta n(x_1) \Delta n(x_2) \rangle$. The observed interference is a two-photon phenomenon: a random pair of photons interfering with the pair itself. In the figure, the superposed two different yet indistinguishable two-photon amplitudes are indicated by red and blue colors. When the detectors are scanning in the neighborhood of $x_1 \approx x_2$, the red amplitude and the blue amplitude "overlap" which means the pair experience the same phase variations, and thus turbulence-free.

to the measurement of intensity fluctuation correlation at D_1 and D_2, jointly. In this interferometer, we managed to have the spatial coherence length, l_c, of the thermal field much smaller than the separation, d, between the upper slit (slit-A) and the lower slit (slit-B), $l_c \ll d$, where $l_c = \lambda/\Delta\theta_s$ with λ the wavelength of the monochromatic thermal radiation, and $\Delta\theta_s$ the angular diameter of the light source. Consequently, no first-order interferences are observable from $\langle I(x_1) \rangle$ and $\langle I(x_2) \rangle$. However, an interference pattern with 100% visibility is observed from the measurement of $\langle \Delta n(x_1) \Delta n(x_2) \rangle \propto \langle \Delta I(x_1) \Delta I(x_2) \rangle$ and the interference is insensitive to any index-phase variations in the optical path of the interferometer and also atmospheric vibration induced phase variations, namely, turbulence-free when scanning D_1 and D_2 in the neighborhood of $x_1 \sim x_2$.

To simplify the photon number fluctuation correlation circuit, we have employed a standard monochromatic pseudo-thermal light source consisting of a rotating ground glass and a single-frequency laser beam of wavelength $\lambda = 532$ nm. Millions of tiny diffusers within the rotating ground glass scatter the laser beam into many independent wave packets, or subfields, at the single-photon level with random relative phases, artificially simulating a natural thermal light source such as the sun. Directly following the ground glass is an adjustable pinhole used to control the transverse size of the light source, allowing us to alter the spatial coherence length of the thermal field. A double-slit with $d = 2.5$ mm and line-like slits is then placed 1.6 m after the pinhole. Using this, we simulated a thermal light source with an angular diameter of $\Delta\theta_s \approx 0.00156$ and thus obtained a spatial coherence length of $l_c = \lambda/\Delta\theta_s \approx 0.34$ mm, satisfying $d \gg l_c$. In this case, no interference is observable from $\langle n(x_1) \rangle \propto \langle I(x_1) \rangle$ and $\langle n(x_2) \rangle \propto \langle I(x_2) \rangle$. The atmospheric turbulences are introduced by heating elements of a toaster oven. These heating elements produce enough heat to rapidly vary the air density above them, thereby causing variations in the index of refraction. The resulting optical path variations "blur out" completely the classic interference pattern observed in the measurement of $\langle n(x_j)\rangle$ when $l_c > d$, where $j = 1, 2$ labels the measurements by D_1 and D_2. A photon-number fluctuation correlation (PNFC) circuit is used to measure the photon number fluctuation correlation for each chosen value of $(x_1 - x_2)$. The PNFC circuit has two synchronized event timers to record the registration times of each photodetection event of D_1 and D_2. The time axes of the event timers can be divided into a sequence of time windows Δt, each labeled by time t_i, for $i = 1, 2, ..., N$. The software first calculates the mean photon number for each detector, $\overline{n}_1$ and $\overline{n}_2$, and then calculates the photon number fluctuations for each ith time window, $\Delta n_1(t_i) = n_1(t_i) - \overline{n}_1$ and $\Delta n_2(t_i) = n_2(t_i) - \overline{n}_2$, which can either be positive or negative. The photon number fluctuation correlation is then calculated from

$$\langle \Delta n(x_1)\Delta n(x_2)\rangle = \frac{1}{N}\sum_{i=1}^{N}\Delta n_1(t_i)\Delta n_2(t_i), \tag{11.5.1}$$

where N is the total number of time windows for a data point of a chosen x_1 and x_2.

Fig. 11.5.2 reports a set of typical experimental results observed from photon number fluctuation correlation. Fig. 11.5.2a is a measurement of $\langle \Delta n(x_1)\Delta n(x_2)\rangle$ when the heating elements were powered off, i.e. no turbulence present, and Fig. 11.5.2b is the same measurement of $\langle \Delta n(x_1)\Delta n(x_2)\rangle$ when the heating elements were powered on, i.e. strong turbulence present. When D_1 and D_2 were scanned in the neighborhood of $x_1 \sim x_2$, the visibility of the interference pattern with turbulence present was 94.3±0.2%, which is consistent with the visibility without turbulence present, 94.6±0.2%.

To demonstrate that the turbulence is strong enough to blur the classic interference present in, we removed the rotating ground glass and directed the unaltered laser beam in the TEM$_{00}$ mode directly onto the double-slit. The spatial coherence length of a TEM$_{00}$ mode laser beam is as large as the transverse size of the beam itself, equivalent to having a thermal source of $\Delta\theta_s \to 0$ or $l_c \to \infty$, satisfying the condition of $l_c > d$. Fig. 11.5.3a reports a typical measured result of $\langle n(x_j)\rangle$ when the heating elements were powered off, i.e., without turbulence. Fig. 11.5.3b reports the same measurement of $\langle n(x_j)\rangle$ but now with the heating elements powered on. It is clear that the interference pattern is completely blurred out by the turbulence. This result guarantees the turbulence introduced by our heating elements is strong enough to demonstrate the turbulence-free nature of our new type of interferometer.

To see why the measurement of $\langle \Delta n(\mathbf{r}_1, t_1)\Delta n(\mathbf{r}_2, t_2)\rangle$ is turbulence-free, following previous section, we examine the second-order coherence function $G^{(2)}(x_1, t_1; x_2, t_2)$

$$G^{(2)}(x_1, t_1; x_2, t_2) = G^{(2)}_{AA}(x_1, t_1; x_2, t_2) + G^{(2)}_{BB}(x_1, t_1; x_2, t_2)$$
$$+ G^{(2)}_{AB}(x_1, t_1; x_2, t_2) + G^{(2)}_{BA}(x_1, t_1; x_2, t_2). \tag{11.5.2}$$

We first calculate $G^{(2)}_{AB}(x_1, t_1; x_2, t_2)$ in Einstein's picture,

$$G^{(2)}_{AB}(x_1, t_1; x_2, t_2)$$
$$= \sum_{m,n}|E_m g_m(x_A, t_A)g_A(x_1, t_1)E_n g_n(x_B, t_B)g_B(x_2, t_2)$$
$$+ E_m g_m(x_A, t_A)g_A(x_2, t_2)E_n g_n(x_B, t_B)g_B(x_1, t_1)|^2, \tag{11.5.3}$$

indicating a superposition between two different yet indistinguishable two-photon amplitudes: (1) the mth wavepacket passing through slit-A then propagates to D_1 while the nth wavepacket passing through slit-B then propagates to D_2; (2) the mth wavepacket passing

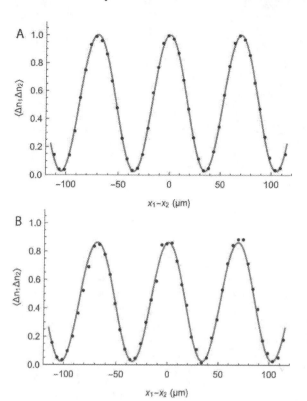

FIGURE 11.5.2 Typical measurement of turbulence-free interference. (A) Without turbulence, the measurement of photon number fluctuation correlation produces an interference pattern with 94.6±0.2% visibility, when $l_c \ll d$. (B) When turbulence was introduced, the interference pattern remained with 94.3±0.2% visibility.

through slit-A then propagates to D_2 while the nth wavepacket passing through slit-B then propagates to D_1. We may consider amplitude (1) the blue path and amplitude (2) the red path in Fig. 11.5.1, respectively. It is easy to find that the blue path and the red path overlap in the neighborhood of $x_1 \sim x_2$. In this case, the randomly created and randomly paired two wavepackets experience the same turbulences and thus the same phase variations. The phase variations associated with the two-photon amplitudes (1) and (2) cancel each other resulting in a turbulence-free two-photon interference.

The cross interference term of $G_{AB}^{(2)}(x_1, t_1; x_2, t_2)$ yields a turbulence-free sinusoidal modulation of $(x_1 - x_2)$ in the measurement of $\langle \Delta I(x_1)\Delta I(x_2)\rangle$,

$$
\begin{aligned}
\langle \Delta n_{AB}(x_1)\Delta n_{AB}(x_2)\rangle &\propto \langle \Delta I_{AB}(x_1)\Delta I_{AB}(x_2)\rangle \\
&= \sum_{m,n} E_m^* g_m^*(x_A) g_A^*(x_1) E_n^* g_n^*(x_B) g_B^*(x_2) \\
&\quad \times E_m g_m(x_A) g_A(x_2) E_n g_n(x_B) g_B(x_1) \\
&= \sum_{m\neq n} |E_m|^2 |E_n|^2 |g_m(x_A)|^2 |g_n(x_B)|^2 \\
&\quad \times g_A^*(x_1) g_B(x_1) g_A(x_2) g_B^*(x_2) \\
&= I_0^2 \cos \frac{2\pi d}{\lambda z}(x_1 - x_2)
\end{aligned}
\tag{11.5.4}
$$

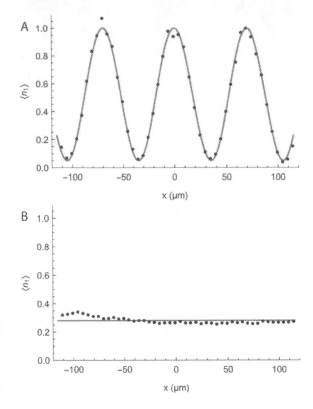

FIGURE 11.5.3 Typical measurement to confirm strong enough turbulence is present. (A) Without turbulence, the classic Young's double-slit interferometer produces an interference pattern when $l_c > d$. (B) When turbulence is present, it "blurs" the interference pattern completely.

where we have ignored the temporal variables by assuming perfect second-order temporal coherence as usual. Similar to the case of section 10.1, unlike the classic Young's double-slit interference, we find the Green's functions that propagate the fields from the source to the double-slit do not have any contributions to the two-photon interference because of $|g_m(x_A, t_A)|^2 = |g_n(x_B, t_B)|^2 = 1$.

In addition to the above alternatives for the mth and the nth subfields to produce a joint photodetection event of D_1 and D_2, the randomly created and randomly paired mth and nth subfields, or photons, may have a second superposition, $G_{BA}^{(2)}(x_1, t_1; x_2, t_2)$, to produce a joint photodetection event at (x_1, t_1) and (x_2, t_2), corresponding to the following two-photon amplitudes: (3) the nth subfield passing through slit-A then propagates to D_1 and the mth subfield passes slit-B then propagates to D_2; (4) the nth subfield passes slit-A then propagates to D_2 and the mth subfield passes slit-B then propagates to D_1. This superposition has the same mathematical expression as Eq.(11.5.3), except switching A and B. The cross terms of this superposition produces the same turbulence-free sinusoidal modulation in the measurement of $\langle \Delta I(x_1) \Delta I(x_2) \rangle$.

The randomly created and randomly paired mth and nth subfields, or photons, may have a third superposition, $G_{AA}^{(2)}(x_1, t_1; x_2, t_2)$, to produce a joint photodetection event at (x_1, t_1) and (x_2, t_2): (5) the mth subfield passes slit-A then propagates to D_1 and the nth subfield passes slit-A then propagates to D_2; (6) the mth subfield passes slit-A then propagates to

D_2 and the nth subfield passes slit-A then propagates to D_1,

$$G_{AA}^{(2)}(x_1; x_2) = \sum_{m,n} |E_m g_m(x_A) g_A(x_1) E_n g_n(x_A) g_A(x_2)$$
$$+ E_n g_n(x_A) g_A(x_1) E_m g_m(x_A) g_A(x_2)|^2 \qquad (11.5.5)$$

The cross interference term of $G_{AA}^{(2)}(x_1; x_2)$ yields a turbulence-free constant to the measurement of $\langle \Delta n(x_1) \Delta n(x_2) \rangle$,

$$\langle \Delta n_{AA}(x_1) \Delta n_{AA}(x_2) \rangle \propto \langle \Delta I_{AA}(x_1) \Delta I_{AA}(x_2) \rangle$$
$$= \sum_{m,n} |E_m^* g_m^*(x_A) g_A^*(x_1) E_n^* g_n^*(x_A) g_A^*(x_2)$$
$$\times E_n g_n(x_A) g_A(x_1) E_m g_m(x_A) g_A(x_2)|^2$$
$$= \sum_{m \neq n} |E_m|^2 |E_n|^2 |g_m(x_A)|^2 |g_n(x_A)|^2 |g_A(x_1)|^2 ||g_A(x_2)|^2$$
$$= I_0^2. \qquad (11.5.6)$$

The randomly created and randomly paired mth and nth subfields, or photons, may have a fourth superposition, $G_{BB}^{(2)}(x_1, t_1; x_2, t_2)$, to produce a joint photodetection event at (x_1, t_1) and (x_2, t_2): (7) the mth subfield passes slit-B then propagates to D_1 and the nth subfield passes slit-B then propagates to D_2; (8) the mth subfield passes slit-B then propagates to D_2 and the nth subfield passes slit-B then propagates to D_1. This superposition has the same mathematical expression as Eq.(11.5.5), except switching A and B. The cross terms of this superposition also contribute a turbulence-free constant to the measurement of $\langle \Delta I(x_1) \Delta I(x_2) \rangle$.

Adding the above four turbulence-free contributions to the intensity fluctuation correlation measurement $\langle \Delta I(x_1) \Delta I(x_2) \rangle$, we have an observable turbulence-free interference in the neighborhood of $x_1 \sim x_2$ with 100% visibility,

$$\langle \Delta n(x_1) \Delta n(x_2)$$
$$= \langle \Delta n_{AB}(x_1) \Delta n_{AB}(x_2) \rangle + \langle \Delta n_{BA}(x_1) \Delta n_{BA}(x_2) \rangle$$
$$+ \langle \Delta n_{AA}(x_1) \Delta n_{AA}(x_2) \rangle + \langle \Delta n_{BB}(x_1) \Delta n_{BB}(x_2) \rangle$$
$$= I_0^2 \left[1 + \cos \frac{2\pi d}{\lambda z} (x_1 - x_2) \right]. \qquad (11.5.7)$$

According to the quantum theory of light, there is no suppression to achieve turbulence-free in a two-photon interferometer: what we need is to make the superposed indistinguishable two-photon amplitudes experience the same turbulence so that any refraction index, length, or phase variations along the optical paths of the interferometer do not have any effect on the two-photon interference. Examining the two-photon double-slit interferometer in Fig. 11.5.1 and the superpositions of Eqs. (11.5.3) and (11.5.5), we find that all four groups of superposed two-photon amplitudes are "overlapped", respectively, when scanning D_1 and D_2 in the neighborhood of $x_1 \sim x_2$. In principle, this mechanism of achieving turbulence-free two-photon interference is able to apply to other types of interferometers and interferometric sensors, making them turbulence-free as well. To avoid atmospheric turbulence and vibrations, many interferometers and interferometric sensors, such as the gravitational wave detector, have to be maintained within complicated, high cost vacuum systems. With a turbulence-free interferometer, these complicated and expensive systems would no longer be required.

11.6 TURBULENCE INDUCED TURBULENCE-FREE TWO-PHOTON INTERFERENCE OF LASER BEAM

In section 11.5, we analyzed a turbulence-free double-slit interferometer in which incoherent thermal field was able to produce turbulence-free two-photon interference pattern from the second-order measurement of photon number fluctuation correlation $\langle \Delta n_1 \Delta n_2 \rangle$, or intensity fluctuation correlation $\langle \Delta I_1 \Delta I_2 \rangle$, while no classic interference was observable from the first-order measurement of mean intensities $\langle I_1 \rangle$ and $\langle I_2 \rangle$. In that experiment, the input light of the interferometer was in thermal state. Can we observe turbulence-free interference from an interferometer that employs coherent laser beam as the light source? The second-order coherence of laser beam has been studied since the invention of the laser. Different from thermal field, the photon number fluctuation correlation, or intensity fluctuation correlation of a laser beam is always zero, $\langle \Delta n_1 \Delta n_2 \rangle \propto \langle \Delta I_1 \Delta I_2 \rangle = 0$. Thermal field was traditionally considered as Gaussian field; and the non-trivial second-order correlation of thermal field was explained as an intrinsic property of Gaussian field. Unfortunately, the laser field is non-Gaussian and can be approximated as a coherent state.

When a coherent laser beam is incident on a double-slit, without turbulence, classic Young's double-slit interference can be easily observed from the measurement of mean photon number $\langle n \rangle$, or mean intensity $\langle I \rangle$. When optical turbulence is introduced into the interferometer, it may blur the interference pattern completely. The turbulence introduces random phase shifts following slit-A and slit-B that vary rapidly, randomly, and independently. This turns a single coherent state, representing a group of identical photons, into a mixture of two separate, distinguishable groups of identical photons in coherent states A and B with varying random relative phases from the turbulence. The incoherent superposition of coherent state A and coherent state B is unable to produce any classic interference pattern. Is it possible to observe turbulence-free second-order interference from a laser-based interferometer? Perhaps, no one would expect the same turbulence-free two-photon interference mechanism of thermal light to be applicable to a laser-based interferometer. Perhaps, no one would even expect observing nontrivial second-order correlation from a laser beam, since laser field is non-Gaussin.

Surprisingly, in a recent experiment, Smith and Shih observed turbulence-free two-photon interference from the second-order correlation measurement of photon number fluctuations, or intensity fluctuations, of an Young's double-slit interferometer which not only employed a laser beam as the light source but also was under the influence of strong turbulence. How could a measurement of photon number fluctuation correlation, or intensity fluctuation correlation, on a laser beam produce non-trivial sinusoidal function? Why is this interference pattern seemingly turbulence-free but also only present due to the turbulence itself? We address these questions in this section after describing the experimental observations of Smith and Shih.

The experimental setup of Smith and Shih is depicted in Fig. 11.6.1. A Nd:YVO$_4$ laser was used to produce a continuous wave (CW) beam in the TEM$_{00}$ spatial mode at $\lambda = 532$ nm. A beam expander with a well designed spatial filter was used to increase the diameter of the TEM$_{00}$ laser beam. The expanded beam was incident on a standard Young's double-slit interferometer with slit separation of $d = 2.5$ mm. The slits are narrow enough to be treated as lines. In this experiment, optical turbulence was introduced by a set of kilowatt heating elements beneath the optical paths of the interferometer. The heating elements heat the air and introduce random airflow and random optical index variations, i.e., optical turbulence, between the double-slit and the observation plane. The turbulence is strong enough to shorten the coherent length of the light achieving $l_c < d$ to blur the classic interference pattern. However, the turbulence is not strong enough to thermalize the coherent laser beam passing an individual line-like slit into Gaussian field. To detect the radiation at more

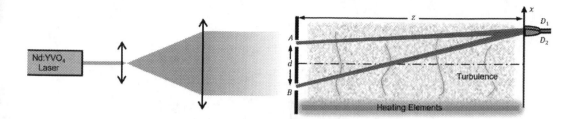

FIGURE 11.6.1 Experimental setup. Light emitted from a CW Yttrium Vanadate (Nd:YVO$_4$) laser in the TEM$_{00}$ spatial mode is incident on a double-slit with slit separation d, a beam expander is used to enlarge the laser beam to diameter D achieving $D \gg d$. Two scannable single-photon detectors, D_1 and D_2, are placed on the far-field observation plane of the double-slit interferometer. The electronics interfaced with D_1 and D_2 can simultaneously obtain mean photon number, $\langle n(x_j) \rangle$, photon number correlation, $\langle n(x_1)n(x_2) \rangle$, and photon number fluctuation correlation, $\langle \Delta n(x_1)\Delta n(x_2) \rangle$. A lab-made atmospheric turbulence is introduced between the double-slit and the photodetector. The turbulence is strong enough to shorten the coherent length of the field achieving $l_c < d$. However, the turbulence not strong enough to thermalize the coherent laser beam passing an individual line-like slit into Gaussian field.

precise spatial locations, point-like tips of single-mode optical fiber were used to interface the coherent light with or without turbulence into the single-photon counting detectors. A PNFC circuit uses a series of measurements ($\sim$ 300,000) to determine the mean photon number and photon number fluctuations for each detector, while simultaneously calculating the photon number correlation, $\langle n(x_1)n(x_2) \rangle$, and photon number fluctuation correlation, $\langle \Delta n(x_1)\Delta n(x_2) \rangle$.

No surprises were observed from the measurement of first-order classic interference. As expected, when the heating elements were powered off, the observed visibility of the classic interference pattern was $\sim$ 100% and when the heating elements were powered on, the interference pattern was blurred out by the turbulence. This confirms that the optical paths from each slit were experiencing random disturbances from the turbulence and thus introduce random phases into the two optical paths. The second-order measurements of the photon number correlation $\langle n(x_1)n(x_2) \rangle$ and the photon number fluctuation correlation $\langle \Delta n(x_1)\Delta n(x_2) \rangle$ were more interesting. When the heating elements were powered off, we observed $\sim$ 100% visibility interference in the measurement of $\langle n(x_1)n(x_2) \rangle$ (Fig. 11.6.2a) while the measurement of $\langle \Delta n(x_1)\Delta n(x_2) \rangle$ yielded a constant of $\sim$ 0, as seen in Fig. 11.6.3a. When the heating elements were powered on, interference in the measurement of $\langle n(x_1)n(x_2) \rangle$ was still present; however, the visibility was reduced (Fig. 11.6.2b), and simultaneously, an interference pattern appeared in the measurement of $\langle \Delta n(x_1)\Delta n(x_2) \rangle$, as seen in Fig. 11.6.3b. The turbulence seems to have produced interference in the measurement of photon number fluctuation correlation $\langle \Delta n(x_1)\Delta n(x_2) \rangle$. More interestingly, the interference pattern contains not only "correlation" of $\langle \Delta n_1 \Delta n_2 \rangle > 0$, but also "anti-correlation" of $\langle \Delta n_1 \Delta n_2 \rangle < 0$ within one period scanning of the interference pattern, as seen in Fig. 11.6.3b.

The observation of classic interferences from coherent laser beam without turbulence are easily understood. In the following we analyze the measurement processes of mean photon number $\langle n(x_j) \rangle \propto \langle I(x_j) \rangle$ and photon number fluctuation correlation $\langle \Delta n(x_1)\Delta n(x_2) \rangle \propto \langle \Delta I(x_1)\Delta I(x_2) \rangle$ from the turbulence disturbed double-slit interferometer of Fig. 11.6.1.

(I) Einstein's picture of light

A laser beam contains a group of large number identical photons, usually approximated

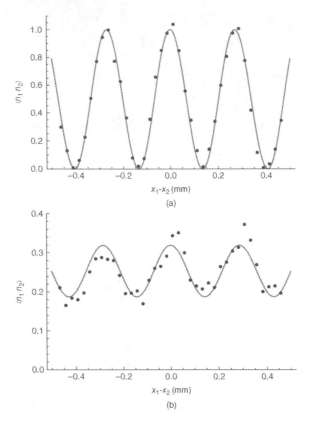

FIGURE 11.6.2 Typical measurement of photon number correlation. Each data point is estimated from $\sim 300{,}000$ measurements. (a) When the heating elements were powered off: $\sim 100\%$ visibility interference was observed. (b) When the heating elements were powered on: interference visibility was significantly reduced.

in coherent state. When turbulence is introduced into an Young's double-slit interferometer behind slit-A and slit-B, one group of identical photons is divided into two distinguishable groups of identical photons with random relative phases, if the turbulence is strong enough to shorten the coherence length of the field achieving $l_c < d$. In Einstein's picture, we may consider $E_A(x_j, t_j)$ and $E_B(x_j, t_j)$, $j = 1, 2$, distinguishable subfields,

$$E_A(x_j, t_j) = \int d\omega \, E_A(\omega) \, g_A(\omega; x_j, t_j) \, e^{-i\varphi_{Aj}(t_j)}$$

$$E_B(x_j, t_j) = \int d\omega \, E_B(\omega) \, g_B(\omega; x_j, t_j) \, e^{-i\varphi_{Bj}(t_j)} \tag{11.6.1}$$

where $g_k(\omega; x_j, t_j)$, $k = A, B$, $j = 1, 2$, is the Green's function propagating the ω mode of the kth subfield $E_k(\omega)$ from slit-k to point-like photodetector D_j, $\varphi_{kj}(t_j)$ is the turbulence induced random phase shift along path-kj

$$\varphi_{kj}(t_j) \simeq \frac{\omega}{c} \int dr_{kj} \, \delta n(r_{kj}, t_j). \tag{11.6.2}$$

The integral is along the path of k to j, the refraction index of air is approximated $n_0 \simeq 1$, $\delta n(r_{kj}, t_j)$ represents the turbulence induced variation of the refraction index along path-kj, which takes a random value from time to time and from measurement to measurement. In

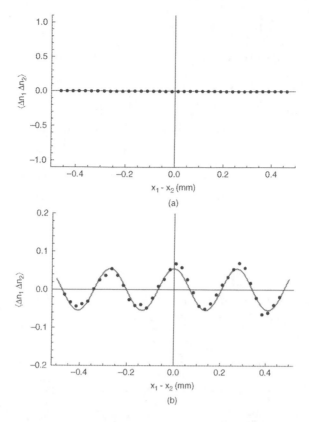

FIGURE 11.6.3 Typical measurement of photon number fluctuation correlation. The mean photon number and photon number fluctuation for each data point is estimated from ~300,000 measurements. (a) When the heating elements were powered off: No interference and correlation was observed from $\langle \Delta n(x_1) \Delta n(x_2) \rangle$. (b) When the heating elements were powered on: an interference pattern appeared in the measurement of $\langle \Delta n(x_1) \Delta n(x_2) \rangle$ with "correlation", corresponding to constructive interference, and "anti-correlation", corresponding to destructive interference, of photon number fluctuations.

the experiment of Smith and Shih, each data point is estimated from ~ 300,000 measurements, the experimentally executed time average can be treated as an ensemble average. The ensemble average of the first-order incoherent superposition of the two turbulence affected subfields gives a constant mean value for the measurement of $\langle n(x_j) \rangle$,

$$\langle n(x_j) \rangle \propto \langle I(x_j) \rangle$$
$$\propto \langle |E_A(x_j, t_j) + E_B(x_j, t_j)|^2 \rangle$$
$$= \langle |E_A(x_j, t_j)|^2 \rangle + \langle |E_B(x_j, t_j)|^2 \rangle$$
$$+ \langle E_A^*(x_j, t_j) E_B(x_j, t_j) \rangle + \langle E_A(x_1, t_j) E_B^*(x_1, t_j) \rangle$$
$$\propto n_0. \tag{11.6.3}$$

Although the fields passing through slit-A and slit-B are initially in phase, the turbulence affects each one randomly. While $[\varphi_{Aj}(t_j) - \varphi_{Bj}(t_j)]$ varies from time to time and from measurement to measurement, the time average or ensemble average of the cross terms yields $\langle E_A^*(x_j, t_j) E_B(x_j, t_j) \rangle = \langle E_A(x_1, t_j) E_B^*(x_1, t_j) \rangle = 0$ when taking into account all possible random relative phases. In reality, even for ~ 300,00 measurements, $[\varphi_{Aj}(t_j) - \varphi_{Bj}(t_j)]$ may not take all possible random values and may not vanish completely. The remaining cross terms $\langle E_A^*(x_j, t_j)E_B(x_j, t_j)\rangle + \langle E_A(x_1, t_j)E_B^*(x_1, t_j)\rangle \neq 0$ are usually treated as "noise" or "photon number fluctuations".

Now, we calculate the correlation of $\langle n(x_1)n(x_2)\rangle$ under turbulence,

$$\langle n(x_1)n(x_2)\rangle \propto \langle I(x_1)I(x_2)\rangle$$
$$= \langle E^*(x_1, t_1)E(x_1, t_1)E^*(x_2, t_2)E(x_2, t_2)\rangle$$
$$= \langle [E_A^*(x_1, t_1) + E_B^*(x_1, t_1)][E_A(x_1, t_1) + E_B(x_1, t_1)]$$
$$\times [E_A^*(x_2, t_2) + E_B^*(x_2, t_2)]^*[E_A(x_2, t_2) + E_B(x_2, t_2)]\rangle \quad (11.6.4)$$

resulting in 16 terms of expectation values to calculate. For brevity, we will drop terms that have no contribution to the measurement of $\langle n(x_1)n(x_2)\rangle$ or cannot survive the ensemble average by taking into account all possible turbulence introduced random phases along the A-path and the B-path,

$$\langle n(x_1)n(x_2)\rangle \propto \langle I(x_1)I(x_2)\rangle$$
$$\propto \langle |E_A(x_1, t_1)|^2|E_A(x_2, t_2)|^2\rangle + \langle |E_B(x_1, t_1)|^2|E_B(x_2, t_2)|^2\rangle$$
$$+ \langle |E_A(x_1, t_1)|^2|E_B(x_2, t_2)|^2\rangle + \langle |E_A(x_2, t_2)|^2|E_B(x_1, t_1)|^2\rangle$$
$$+ \langle E_A^*(x_1, t_1)E_B(x_1, t_1)E_B^*(x_2, t_2)E_A(x_2, t_2)\rangle$$
$$+ \langle E_A(x_1, t_1)E_B^*(x_1, t_1)E_B(x_2, t_2)E_A^*(x_2, t_2)\rangle$$
$$= \langle n(x_1)\rangle\langle n(x_2)\rangle + \langle \Delta n(x_1)\Delta n(x_2)\rangle \quad (11.6.5)$$

where

$$\langle \Delta n(x_1)\Delta n(x_2)\rangle \propto \langle \Delta I(x_1)\Delta I(x_2)\rangle$$
$$= \langle E_A^*(x_1, t_1)E_B(x_1, t_1)E_B^*(x_2, t_2)E_A(x_2, t_2)\rangle$$
$$+ \langle E_A(x_1, t_1)E_B^*(x_1, t_1)E_B(x_2, t_2)E_A^*(x_2, t_2)\rangle$$
$$\propto \cos\frac{2\pi d}{\lambda z}(x_1 - x_2). \quad (11.6.6)$$

when scanning D_1 in the neighborhood of D_2. It is easy to find that the above sinusoidal modulation in the measurement of $\langle \Delta n(x_1)\Delta n(x_2)\rangle$ comes from the cross interference terms of the following superposition

$$\langle \left| E_A(x_1, t_1)E_B(x_2, t_2) + E_A(x_2, t_2)E_B(x_1, t_1) \right|^2 \rangle \quad (11.6.7)$$

corresponding to two different yet indistinguishable alternatives for the two distinguishable groups of identical photons, represented by subfields $E_A(x_j, t_j)$ and $E_B(x_j, t_j)$ in Einstein's picture, to produce a join photodetection event of D_1 and D_2: (1) subfield-A is detected at D_1 and subfield-B is detected at D_2; (2) subfield-A is detected at D_2 and subfield-B is detected at D_1; namely a pair of distinguishable coherent subfields interfering with the pair itself. Note, at first glance, the cross term of this superposition would average to zero due to the random phases present. It has been discussed in previous section that by scanning D_1 in the neighborhood of D_2 such that $x_1 \sim x_2$, the optical paths of the two alternatives overlap in space-time and experience the same turbulence. The turbulence induced random phases in the intensity fluctuation correlation would have no contribution to this superposition.

It is interesting to find from Eq. (11.6.6), the interference results in a positive value of $\langle \Delta n(x_1)\Delta n(x_2)\rangle > 0$, indicating a "correlation", when the above two alternatives superpose constructively, and results in a negative value of $\langle \Delta n(x_1)\Delta n(x_2)\rangle < 0$, indicating an "anti-correlation", when the above two alternatives superpose destructively. The constructive superposition forces the measured photon number fluctuate to the same, positive-positive or negative-negative, direction, while the destructive interference forces the measured photon number fluctuate to opposite, positive-negative or negative-positive, directions: if one fluctuates positively the other one must fluctuates negatively and vice versa.

(II) Quantum coherent state representation

A laser beam of TEM$_{00}$ mode contains a group of large number of identical photons. We approximate the state of the group of identical photons a multi-mode coherent state

$$|\Psi\rangle = \prod_\omega |\alpha(\omega)\rangle. \tag{11.6.8}$$

The field operator at coordinate (x_j, t_j) of the jth photodetector

$$\begin{aligned}
\hat{E}^{(+)}(x_j, t_j) &= \int d\omega\, \hat{a}(\omega)\, g_A(\omega; x_j, t_j)\, e^{-i\varphi_{Aj}(t_j)} + \int d\omega\, \hat{a}(\omega)\, g_B(\omega; x_j, t_j)\, e^{-i\varphi_{Bj}(t_j)} \\
&= \hat{E}_A^{(+)}(x_j, t_j) + \hat{E}_B^{(+)}(x_j, t_j)
\end{aligned}$$

$$\tag{11.6.9}$$

where $g_A(\omega; x_j, t_j)$ and $g_B(\omega; x_j, t_j)$ are the Green's functions (or propagators) which propagate the ω mode of the state from slit-A and slit-B to the photodetector D_j at space-time coordinate (x_j, t_j). With the help of the quantum state and the field operators, the second-order coherence function $G^{(2)}(x_1, t_1; x_1, t_2)$ is calculated as follows

$$\begin{aligned}
&G^{(2)}(x_1, t_1; x_1, t_2) \\
&= \left\langle \langle\Psi| \hat{E}^{(-)}(x_1, t_1)\hat{E}^{(-)}(x_2, t_2)\hat{E}^{(+)}(x_2, t_2)\hat{E}^{(+)}(x_1, t_1) |\Psi\rangle \right\rangle_{\text{En}} \\
&= \left\langle \langle\Psi|[\hat{E}_A^{(-)}(x_1, t_1) + \hat{E}_B^{(-)}(x_1, t_1)][\hat{E}_A^{(-)}(x_2, t_2) + \hat{E}_B^{(-)}(x_2, t_2)] \right. \\
&\quad \left. \times [\hat{E}_A^{(+)}(x_2, t_2) + \hat{E}_B^{(+)}(x_2, t_2)][\hat{E}_A^{(+)}(x_1, t_1) + \hat{E}_B^{(+)}(x_1, t_1)]|\Psi\rangle \right\rangle_{\text{En}}. \quad (11.6.10)
\end{aligned}$$

Eq. (11.6.10) results in 16 expectations to evaluate. For brevity, we will drop terms that have no contribution to the measurement of $\langle n(x_1)n(x_2)\rangle$ or cannot survive the ensemble average by taking into account all possible turbulence introduced random phases along the A-path and the B-path,

$$\begin{aligned}
&G^{(2)}(\mathbf{r}_1, t_1; \mathbf{r}_1, t_2) \\
&= \left\langle |\Psi_A(x_1, t_1)|^2 |\Psi_A(x_2, t_2)|^2 \right\rangle_{\text{En}} + \left\langle |\Psi_B(x_2, t_2)|^2 |\Psi_B(x_1, t_1)|^2 \right\rangle_{\text{En}} \\
&\quad + \left\langle |\Psi_A(\mathbf{r}_1, t_1)|^2 |\Psi_B(\mathbf{r}_2, t_2)|^2 \right\rangle_{\text{En}} + \left\langle |\Psi_A(\mathbf{r}_2, t_2)|^2 |\Psi_B(\mathbf{r}_1, t_1)|^2 \right\rangle_{\text{En}} \\
&\quad + \left\langle \Psi_A^*(\mathbf{r}_1, t_1)\, \Psi_B^*(\mathbf{r}_2, t_2)\, \Psi_A(\mathbf{r}_2, t_2)\, \Psi_B(\mathbf{r}_1, t_1) \right\rangle_{\text{En}} \\
&\quad + \left\langle \Psi_A(\mathbf{r}_1, t_1)\, \Psi_B(\mathbf{r}_2, t_2)\, \Psi_A^*(\mathbf{r}_2, t_2)\, \Psi_B^*(\mathbf{r}_1, t_1) \right\rangle_{\text{En}} \\
&\propto \langle n(\mathbf{r}_1, t_1)\rangle\langle n(\mathbf{r}_2, t_2)\rangle + \langle \Delta n(\mathbf{r}_1, t_1)\Delta n(\mathbf{r}_2, t_2)\rangle \quad (11.6.11)
\end{aligned}$$

where $\Psi_k(x_j, t_j)$ is the effective wavefunction of the group-k identical photons, and

$$\begin{aligned}
&\langle \Delta n(x_1)\Delta n(x_2)\rangle \\
&= \left\langle \Psi_A^*(\mathbf{r}_1, t_1)\, \Psi_B^*(\mathbf{r}_2, t_2)\, \Psi_A(\mathbf{r}_2, t_2)\, \Psi_B(\mathbf{r}_1, t_1) \right\rangle_{\text{En}} \\
&\quad + \left\langle \Psi_A(\mathbf{r}_1, t_1)\, \Psi_B(\mathbf{r}_2, t_2)\, \Psi_A^*(\mathbf{r}_2, t_2)\, \Psi_B^*(\mathbf{r}_1, t_1) \right\rangle_{\text{En}} \quad (11.6.12)
\end{aligned}$$

is the cross terms of the following superposition of "two-photon" effective wavefunctions

$$\left\langle \left| \Psi_A(\mathbf{r}_1, t_1)\, \Psi_B(\mathbf{r}_2, t_2) + \Psi_A(\mathbf{r}_2, t_2)\, \Psi_B(\mathbf{r}_1, t_1) \right|^2 \right\rangle \tag{11.6.13}$$

corresponding to two different yet indistinguishable alternatives for the two distinguishable groups of identical photons to produce a join photodetection event of D_1 and D_2: (1) group A of identical photons propagate to D_1 and group B of identical photons propagate to D_2; and (2) group A of identical photons propagate to D_2 and group B of identical photons propagate to D_1; indicating the interference of two distinguishable groups of identical photons.

When scanning D_1 and D_2 in the neighborhood of $x_1 \sim x_2$, we find the measurements $\langle n(x_1)n(x_2) \rangle$ and $\langle \Delta n(x_1)\Delta n(x_2) \rangle$ yield the same results as those calculated from Einstein's picture, and agreeing well with the experimental observations of Smith and Shih. It is interesting to find that the interference results in a positive value of $\langle \Delta n(x_1)\Delta n(x_2) \rangle > 0$, indicating a "correlation", when the group-pair interferences with the pair itself constructively, and results in a negative value of $\langle \Delta n(x_1)\Delta n(x_2) \rangle < 0$, indicating an "anti-correlation", when the group-pair interferences with the pair itself destructively. The constructive superposition forces the measured photon number fluctuate to the same, positive-positive or negative-negative, direction, while the destructive interference forces the measured photon number fluctuate to opposite, positive-negative or negative-positive, directions: if one fluctuates positively the other one must fluctuates negatively and vice versa.

As discussed in early sections, photon number fluctuation correlation, or intensity fluctuation correlation, is typically observable from thermal field by means of two-photon interference: a pair of randomly created and randomly distributed photons, or a pair of distinguishable groups of identical photons, interfering with the pair itself. When the light detected is of a group of indistinguishable identical photons in coherent state, there are no "distinguishable groups" causing fluctuations and fluctuation correlation. However, under strong turbulences, variance of the turbulence introduce random phases to coherent radiations that passing through slit-A and slit-B and thus causing two distinguishable groups of identical photons in state A and state B. The observations of the experiment of Smith and Shih is the result of a pair of distinguishable groups of identical photons interfering with the group-pair itself. It should be emphasized that the turbulence-induced interference is different from the two-photon interference of thermal field. When a thermal light source is used, a randomly paired photons can pass through the same slit and contribute a constant term to the measurement of photon number fluctuation correlation of D_1 and D_2. However, when a coherent source is used and the coherent length of the turbulence is greater than the width of each single slit, all photons passing through a single slit are still in the same coherent state, preventing any photon number fluctuation correlation between D_1 and D_2 resulting from identical photons passing through a single slit. In rare cases there may exist a condition with turbulence strong enough to introduce different random phases to different sub-groups of identical photons following a single slit, but that was not the condition in Smith and Shih's experiment, and was not taken into account in the above calculation.

SUMMARY

In this chapter, we generalized the concept of two-photon interference to thermal light. The superposition of two-photon amplitudes is not restricted with entangled photon pair. Two-photon interference may occur in the joint photodetection of randomly created and randomly paired photons in a thermal state, if there exists two or more then two different yet indistinguishable alternatives for a randomly created and randomly paired photons to produce a joint photodetection event, although the jointly observed two photodetection events are just fall into the coincidence time window by chance only.

In this chapter we analyzed five typical two-photon interference experiments of incoherent thermal light, including a two-photon Young's double-slit interferometer, an unbalanced two-photon Mach-Zehnder interferometer, a two-photon anti-correlation measurement, a two-photon interference and polarization correlation experiment with first-order, second-order incoherent orthogonal polarized input pseudo-thermal light, and a turbulence-free interferometer. These experiments provided us a solid background on two-photon interferometry of thermal light. In addition, we also analyzed a turbulence induced two-photon interference experiment in which a group of identical photons in a coherent state becomes two distinguishable groups of identical photons, respectively, in coherent state A and coherent state B. This recent observation should be helpful in our understanding of the physics behind thermal field correlation and two-photon interference.

It is interesting to see the concept of two-photon interference applicable to "classical" thermal radiation. Thermal field is traditionally considered as classical. The historical HBT phenomenon has been widely accepted as the intrinsic statistical correlation of intensity fluctuations of thermal source. The two-photon anti-correlation experiments and the two-photon polarization correlation measurements as well as the two- photon interference experiments provided us undeniable evidence on the failure of the above traditional interpretation. The nontrivial second-order correlation or anti-correlation is the result of two-photon interference, i.e., a pair of photons, or a pair of distinguishable groups of identical photons, interfere with the pair itself, through the joint photodetection of two spatially separated photodetectors.

REFERENCES AND SUGGESTIONS FOR READING

[1] T. Smith and Y.H. Shih, "Turbulence-Free Double-slit Interferometer", Phys. Rev. Lett., **120**, 063606 (2018).

[2] T. Smith and Y.H. Shih, "Turbulence Induced Two-photon interference", to be published, (2020).

[3] G. Scarcelli, A. Valencia, and Y.H. Shih, "Two-photon Interference with Thermal Light", Europhysics Lett., **68**, 618 (2004).

[4] Y.S. Ihn, Y. Kim, V. Tamma, and Y.H. Kim, "Second-Order Temporal Interference with Thermal Light: Interference beyond the Coherence Time", Phys. Rev. Lett., **119**, 263603 (2017).

[5] H. Chen, T. Peng, S. Karmakar, Z.D. Xie, and Y.H. Shih, "Observation of anti-correlation in incoherent thermal light fields", Phys. Rev. A, **84**, 033835 (2011).

[6] C.O. Alley and Y.H. Shih, *Foundations of Quantum Mechanics in the Light of New Technology*, ed., M. Namiki, Physical Society of Japan, Tokyo, 47 (1986); Y.H. Shih and C.O. Alley, Phys. Rev. Lett., **61**, 2921, (1988).

[7] H. Chen, T. Peng, S. Karmaker, and Y.H. Shih, "Simulation of Bell States with Incoherent Thermal Light", *New Journal of Physics*, **13**, 083018 (2011).

CHAPTER 12

Quantum Imaging

Although questions regarding fundamental issues of quantum theory still exist, quantum entanglement has started to play important roles in practical engineering applications. Quantum imaging is one of these exciting areas. Quantum imaging has so far demonstrated three peculiar features: (1) enhancing the spatial resolution of imaging beyond the diffraction limit; (2) reproducing ghost images in a "nonlocal" manner; and (3) dispersion-cancelation or turbulence-free imaging. The enhanced spatial resolution apparently "violates" the uncertainty principle. The ghost imaging is considered as a "nonlocal" phenomenon due to a point-to-point correlation of two interaction-free photons at distance. All the above peculiar features are the results of two-photon interference, which involves the superposition of two-photon amplitudes, a nonclassical entity corresponding to different yet indistinguishable alternative ways of creating a joint-detection event. The concept of two-photon interference was introduced from the study of entangled states. It does not, however, restricted to entangled states only. Multi-photon interference is a general phenomenon occurring in multi-photon joint-detection events.

In this chapter we will focus our discussion on quantum imaging with photons in entangled state and with randomly created and randomly paired photon in thermal state. We will first discuss the physics of biphoton imaging including biphoton ghost imaging, and then discuss the ghost imaging of thermal state.

12.1 BIPHOTON IMAGING

The classic concept of optical imaging has been discussed in Chapter 5. Now, we consider a standard imaging setup as shown in Fig. 12.1.1 but replace the classical light source with entangled photon sources for joint-detection of N-fold point-like photodetectors. The entangled N-photon source will reproduce an image of the object with enhanced spatial resolution by a factor of N, despite the Rayleigh diffraction limit. Is the violation of classical diffraction limit equivalent a violation of the uncertainty principle? The answer is no! The uncertainty relation for an entangled N-photon system in N-fold joint-detection is radically different from that for N independent photons. In terms of the terminology of imaging, what we have found is that the imaging forming point-spread function $\text{somb}(x)$ has a different form in the case of entangled states. For example, an entangled biphoton system may produce an image in coincidences

$$R_c(\vec{\rho}_i) = \Big| \int_{obj} d\vec{\rho}_o \, A^2(\vec{\rho}_o) \, \text{somb}\Big[\frac{R}{s_o}\frac{2\omega}{c}\big|\vec{\rho}_o + \frac{\vec{\rho}_i}{m}\big|\Big]\Big|^2, \tag{12.1.1}$$

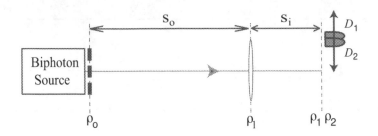

FIGURE 12.1.1 A typical setup of optical imaging. A lens of finite size is used to produce a magnified or demagnified image of an object with limited spatial resolution. Differing from classical imaging setup, the observation is in the joint-detection of two point-like photon counting detectors D_1 and D_2. Replacing classical light with an entangled biphoton system, the spatial resolution can be improved by a factor of two, despite the Rayleigh diffraction limit.

with a twice narrower point-spread function

$$\text{somb}\Big[\frac{R}{s_o}\frac{2\omega}{c}\big|\vec{\rho}_o + \frac{\vec{\rho}_i}{m}\big|\Big].$$

Comparing with the classical case, it is easy to see that the factor of 2ω yields a twice narrower image-forming function and results in a doubling spatial resolution for biphoton imaging.

Based on the setup of Fig. 12.1.1, we now calculate biphoton imaging by replacing classical light source with entangled signal-idler photon pair of SPDC and replacing the ordinary graphic film with a two-photon film which is coated with two-photon material that is sensitive to two-photon transition only. We will show that in the same experimental setup, the entangled biphoton state gives rise to, on a two-photon film or equivalently a two-fold joint-detection system as shown in Fig. 12.1.1, a point-spread function twice narrower than the one obtained in classical imaging at the same wavelength. Then, without applying shorter wavelengths, entangled biphoton states improve the spatial resolution of a *two-photon image* by a factor of 2. We will also show that the entangled biphoton system yields a peculiar Fourier transform function as if it is produced by a light source with $\lambda/2$.

To simplify the calculation, we place the biphoton source of SPDC close enough to the object plane so that the output plane of the biphoton is effectively coincide with the object plane. The above condition can be physically achieved by imaging the output plane of the SPDC onto the object plane. In this case, the signal-idler pair will be either transmitted or reflected from the same point of the object with $\vec{\rho}_s = \vec{\rho}_i$. We now propagate the field from the object plane to an arbitrary plane of $z = s_o + d$, where d may take any values for different experimental purposes:

$$
g(\vec{\kappa}_j, \omega_j; \vec{\rho}_k, z = s_o + d)
$$
$$
= \int_{obj} d\vec{\rho}_o \int_{lens} d\vec{\rho}_l \, A(\vec{\rho}_o) \, e^{i\vec{\kappa}_j \cdot \vec{\rho}_o} \Big\{ \frac{-i\omega_j}{2\pi c s_o} \, e^{i\frac{\omega_j}{c}s_o} \, G(|\vec{\rho}_o - \vec{\rho}_l|, \frac{\omega_j}{c s_o}) \Big\}
$$
$$
\times \, G(|\vec{\rho}_l|, -\frac{\omega_j}{cf}) \Big\{ \frac{-i\omega_j}{2\pi cd} \, e^{i\frac{\omega_j}{c}d} \, G(|\vec{\rho}_l - \vec{\rho}_k|, \frac{\omega_j}{cd}) \Big\}, \tag{12.1.2}
$$

where $\vec{\rho}_o$, $\vec{\rho}_l$, and $\vec{\rho}_j$ are two-dimensional vectors defined, respectively, on the (transverse) output plane of the source (which coincide with the object plane), on the transverse plane of the imaging lens and on the detection plane; and $j = s, i$, labels the signal and the idler; $k = 1, 2$, labels the photodetector D_1 and D_2. The function $A(\vec{\rho}_o)e^{i\vec{\kappa}_j \cdot \vec{\rho}_o}$ is the complex object-aperture function with a real and positive amplitude $A(\vec{\rho}_o)$ and a phase $e^{i\vec{\kappa}_j \cdot \vec{\rho}_o}$, while the terms in the first and second curly brackets of Eq. (12.1.2) describe, respectively, free propagation from the output plane of the source/object to the imaging lens, and from the imaging lens to the detection plane. Applying the properties of the Gaussian function, Eq. (12.1.2) can be simplified as

$$g(\vec{\kappa}_j, \omega_j; \vec{\rho}_k, z = s_o + d)$$

$$= \frac{-\omega_j^2}{(2\pi c)^2 s_o d} e^{i\frac{\omega_j}{c}(s_o+d)} G(|\vec{\rho}_k|, \frac{\omega_j}{cd}) \int_{obj} d\vec{\rho}_o A(\vec{\rho}_o) e^{i\vec{\kappa}_j \cdot \vec{\rho}_o} G(|\vec{\rho}_o|, \frac{\omega_j}{cs_o})$$

$$\times \int_{lens} d\vec{\rho}_l G(|\vec{\rho}_l|, \frac{\omega_j}{c}[\frac{1}{s_o} + \frac{1}{d} - \frac{1}{f}]) e^{-i\frac{\omega_j}{c}(\frac{\vec{\rho}_o}{s_o} + \frac{\vec{\rho}_k}{d}) \cdot \vec{\rho}_l}. \tag{12.1.3}$$

Substituting the Green's functions into Eq. (9.4.5), the biphoton effective wavefunction $\Psi(\vec{\rho}_1, z; \vec{\rho}_2, z)$ is thus

$$\Psi(\vec{\rho}_1, z; \vec{\rho}_2, z) = \Psi_0 \int_{obj} d\vec{\rho}_o A(\vec{\rho}_o) G(|\vec{\rho}_o|, \frac{\omega_s}{cs_o}) \int_{obj} d\vec{\rho}'_o A(\vec{\rho}'_o) G(|\vec{\rho}'_o|, \frac{\omega_i}{cs_o})$$

$$\times \int_{lens} d\vec{\rho}_l G(|\vec{\rho}_l|, \frac{\omega_s}{c}[\frac{1}{s_o} + \frac{1}{d} - \frac{1}{f}]) e^{-i\frac{\omega_s}{c}(\frac{\vec{\rho}_o}{s_o} + \frac{\vec{\rho}_1}{d}) \cdot \vec{\rho}_l}$$

$$\times \int_{lens} d\vec{\rho}'_l G(|\vec{\rho}'_l|, [\frac{\omega_i}{c}\frac{1}{s_o} + \frac{1}{d} - \frac{1}{f}]) e^{-i\frac{\omega_i}{c}(\frac{\vec{\rho}'_o}{s_o} + \frac{\vec{\rho}_2}{d}) \cdot \vec{\rho}'_l}$$

$$\times \int d\vec{\kappa}_s d\vec{\kappa}_i \, \delta(\vec{\kappa}_s + \vec{\kappa}_i) e^{i(\vec{\kappa}_s \cdot \vec{\rho}_o + \vec{\kappa}_i \cdot \vec{\rho}'_o)} \tag{12.1.4}$$

where we have absorbed all constants into Ψ_0, including the phase factor

$$e^{i\frac{\omega_s}{c}(s_o+d)} e^{i\frac{\omega_i}{c}(s_o+d)} = e^{i\frac{\omega_p}{c}(s_o+d)}$$

and

$$G(|\vec{\rho}_1|, \frac{\omega_s}{cd}) G(|\vec{\rho}_2|, \frac{\omega_i}{cd}).$$

The above phase factors have no contribution to $|\Psi(\vec{\rho}_1, z; \vec{\rho}_2, z)|^2$. Let us first complete the double integral of $d\vec{\kappa}_s$ and $d\vec{\kappa}_i$ under the EPR correlation of $\delta(\vec{\kappa}_s + \vec{\kappa}_i)$ on the transverse momentum of the photon pair. Similar to the early calculation, the double integral of $d\vec{\kappa}_s$ and $d\vec{\kappa}_i$ yields a δ-function of $\delta(\vec{\rho}_o - \vec{\rho}'_o)$. Eq. (12.1.4) is then simplified as:

$$\Psi(\vec{\rho}_1, z; \vec{\rho}_2, z)$$

$$= \Psi_0 \int_{obj} d\vec{\rho}_o A^2(\vec{\rho}_o) G(|\vec{\rho}_o|, \frac{\omega_p}{cs_o})$$

$$\times \int_{lens} d\vec{\rho}_l G(|\vec{\rho}_l|, \frac{\omega_s}{c}[\frac{1}{s_o} + \frac{1}{d} - \frac{1}{f}]) e^{-i\frac{\omega_s}{c}(\frac{\vec{\rho}_o}{s_o} + \frac{\vec{\rho}_1}{d}) \cdot \vec{\rho}_l}$$

$$\times \int_{lens} d\vec{\rho}'_l G(|\vec{\rho}'_l|, [\frac{\omega_i}{c}\frac{1}{s_o} + \frac{1}{d} - \frac{1}{f}]) e^{-i\frac{\omega_i}{c}(\frac{\vec{\rho}_o}{s_o} + \frac{\vec{\rho}_2}{d}) \cdot \vec{\rho}'_l}. \tag{12.1.5}$$

To complete Eq. (12.1.5) we divide the calculation into two parts corresponding to two different measurements:

Case (I): on the imaging plane for $\vec{\rho}_1 = \vec{\rho}_2 = \vec{\rho}$.

In this case, Eq. (12.1.5) is simplified as

$$\Psi(\vec{\rho}, z; \vec{\rho}, z) = \Psi_0 \int_{obj} d\vec{\rho}_o A^2(\vec{\rho}_o) G(|\vec{\rho}_o|, \frac{\omega_p}{cs_o}) \tag{12.1.6}$$

$$\times \int_{lens} d\vec{\rho}_l e^{-i\frac{\omega_s}{c}(\frac{\vec{\rho}_o}{s_o} + \frac{\vec{\rho}}{s_i}) \cdot \vec{\rho}_l} \int_{lens} d\vec{\rho}'_l e^{-i\frac{\omega_i}{c}(\frac{\vec{\rho}_o}{s_o} + \frac{\vec{\rho}}{s_i}) \cdot \vec{\rho}'_l}.$$

In the above double integral, we have taken advantage of the EPR correlation $\vec{\rho}_0 = \vec{\rho'}_0$ of the entangled biphoton state. We now change the double integral of $d\vec{\rho}_l$ and $d\vec{\rho'}_l$ to $d(\vec{\rho}_l - \vec{\rho'}_l)$ and $d(\vec{\rho}_l + \vec{\rho'}_l)$

$$\int_{lens} d\vec{\rho}_l \, e^{-i\frac{\omega_s}{c}(\frac{\vec{\rho}_o}{s_o} + \frac{\vec{\rho}}{s_i})\cdot\vec{\rho}_l} \int_{lens} d\vec{\rho'}_l \, e^{-i\frac{\omega_i}{c}(\frac{\vec{\rho}_o}{s_o} + \frac{\vec{\rho}}{s_i})\cdot\vec{\rho'}_l}$$

$$= \int_{lens} d(\vec{\rho}_l - \vec{\rho'}_l) \, e^{-i(\frac{\omega_s - \omega_i}{2c})(\frac{\vec{\rho}_o}{s_o} + \frac{\vec{\rho}}{s_i})\cdot(\vec{\rho}_l - \vec{\rho'}_l)}$$

$$\times \int_{lens} d(\vec{\rho}_l + \vec{\rho'}_l) \, e^{-i(\frac{\omega_p}{2c})(\frac{\vec{\rho}_o}{s_o} + \frac{\vec{\rho}}{s_i})\cdot(\vec{\rho}_l + \vec{\rho'}_l)}.$$

The integral of $d(\vec{\rho}_l - \vec{\rho'}_l)$ can be easily evaluated for degenerate monochromatic SPDC $\omega_s = \omega_i$, which contributes a trivial constant. The integral of $d(\vec{\rho}_l + \vec{\rho'}_l)$ yields a point-spread somb-function

$$\int_{lens} d(\vec{\rho}_l + \vec{\rho'}_l) \, e^{-i(\frac{\omega_p}{2c})(\frac{\vec{\rho}_o}{s_o} + \frac{\vec{\rho}}{s_i})\cdot(\vec{\rho}_l + \vec{\rho'}_l)} = \frac{2J_1\left(\frac{R}{s_o}\frac{\omega_p}{c}|\vec{\rho}_o + \frac{\vec{\rho}}{m}|\right)}{\left(\frac{R}{s_o}\frac{\omega_p}{c}|\vec{\rho}_o + \frac{\vec{\rho}}{m}|\right)}.$$

The joint-detection of a two-fold photodetector or a two-photon graphic film gives

$$G^{(2)}(\vec{\rho}, \vec{\rho}) \propto \left| \int_{obj} d\vec{\rho}_o \, A^2(\vec{\rho}_o) \, e^{i\frac{\omega_p}{2cs_o}|\vec{\rho}_o|^2} \, \text{somb}\left(\frac{R}{s_o}\frac{\omega_p}{c}|\vec{\rho}_o + \frac{\vec{\rho}}{m}|\right)\right|^2, \tag{12.1.7}$$

indicating a coherent image of the object on the image plane with magnification of $m = s_i/s_o$. In Eq. (12.1.7), the point-spread function is characterized by the pump wavelength $\lambda_p = \lambda_{s,i}/2$; hence, the point-spread function is twice narrower than that of the classical case. An entangled two-photon state thus gives an image in joint-detection with double spatial resolution when compared to what one would obtain in classical imaging.

In a more general case of a broadband SPDC, either degenerate or non-degenerate, in which $\Delta\omega_s \sim \infty$ and $\Delta\omega_i \sim \infty$, an additional integral on $d\omega_s$ and $d\omega_i$, or equivalently on $d\nu$ is necessary, where the detuning frequency ν is defined by $\omega_s = \omega_{s0} + \nu$ and $\omega_i = \omega_{i0} - \nu$ following $\omega_s + \omega_i = \omega_{s0} + \omega_{i0} = \omega_p$. The integral on $d\nu$ gives

$$\int d\nu \, e^{-i\nu(\frac{\vec{\rho}_o}{s_o c} + \frac{\vec{\rho}}{s_i c})\cdot(\vec{\rho}_l - \vec{\rho'}_l)} \simeq \delta\left[\frac{1}{s_0 c}(\vec{\rho}_o + \frac{\vec{\rho}}{m})\cdot(\vec{\rho}_l - \vec{\rho'}_l)\right],$$

where we have approximated the integral to infinity with a constant distribution function of $f(\nu)$. This result gives a further enhanced spatial resolution by means of the following $G^{(2)}(\vec{\rho}, \vec{\rho})$

$$G^{(2)}(\vec{\rho}, \vec{\rho}) \propto \left| \int_{obj} d\vec{\rho}_o \, A^2(\vec{\rho}_o) \, e^{i\frac{\omega_p}{2cs_o}|\vec{\rho}_o|^2} \frac{2J_1\left(\frac{R}{s_o}\frac{\omega_p}{c}|\vec{\rho}_o + \frac{\vec{\rho}}{m}|\right)}{\left(\frac{R}{s_o}\frac{\omega_p}{c}|\vec{\rho}_o + \frac{\vec{\rho}}{m}|\right)^2}\right|^2. \tag{12.1.8}$$

The width of the correlation now is determined by the function $2J_1(x)/x^2$, which is much narrower than that of the somb(x). In the above calculation, we have approximated the integral of $d\nu$ to infinity with a constant distribution, which is definitely unrealistic. In realty, for a certain spectrum and distribution this enhancement may not be practically helpful. The physics, however, is fundamentally interesting. In classical imaging, perhaps, we would never expect that a multi-color radiation could produce an image with enhanced spatial resolution.

Case (II): On the Fourier transform plane for $\vec{\rho}_1 = \vec{\rho}_2 = \vec{\rho}$.

The detectors are now placed in the focal plane, i.e., $d = f$. In this case, the spatial effective two-photon wavefunction $\Psi(\vec{\rho}, z; \vec{\rho}, z)$ becomes:

$$\Psi(\vec{\rho}, z; \vec{\rho}, z) \propto \int_{obj} d\vec{\rho}_o \, A^2(\vec{\rho}_o) \, G(|\vec{\rho}_o|, \frac{\omega_p}{cs_o})$$

$$\times \int_{lens} d\vec{\rho}_l \, G(|\vec{\rho}_l|, \frac{\omega_s}{cs_o}) \, e^{-i\frac{\omega_s}{c}(\frac{\vec{\rho}_o}{s_o} + \frac{\vec{\rho}}{f}) \cdot \vec{\rho}_l}$$

$$\times \int_{lens} d\vec{\rho}'_l \, G(|\vec{\rho}'_l|, \frac{\omega_i}{cs_o}) \, e^{-i\frac{\omega_i}{c}(\frac{\vec{\rho}_o}{s_o} + \frac{\vec{\rho}}{f}) \cdot \vec{\rho}'_l}. \qquad (12.1.9)$$

We will first evaluate the double integrals over the lens. To simplify the mathematics we approximate the integral to infinity. Differing from the calculation for imaging resolution, the purpose of this evaluation is to find out the Fourier transform. Thus, the approximation of an infinite sized lens is appropriate. By applying the properties of the Gaussian function, the double integrals over the lens contribute the following function of $\vec{\rho}_o$ to the integral of $d\vec{\rho}_o$ in Eq. (12.1.9):

$$C \, G(|\vec{\rho}_o|, -\frac{\omega_p}{cs_o}) \, e^{-i\frac{\omega_p}{cf} \vec{\rho}_o \cdot \vec{\rho}}$$

where C absorbs all constants including a phase factor $G(|\vec{\rho}|, -\frac{\omega_p}{cf^2/s_o})$. Replacing this result with the double integrals of $d\vec{\rho}_l$ and $d\vec{\rho}'_l$ in Eq. (12.1.9), we obtain:

$$\Psi(\vec{\rho}, z; \vec{\rho}, z) \propto \int_{obj} d\vec{\rho}_o \, A^2(\vec{\rho}_o) \, e^{-i\frac{\omega_p}{cf} \vec{\rho} \cdot \vec{\rho}_o} \propto \mathcal{F}_{[\frac{\omega_p}{cf} \vec{\rho}]} \{A^2(\vec{\rho}_o)\}, \qquad (12.1.10)$$

which is the Fourier transform of the object-aperture function, when the two photodetectors scan together (i.e., $\vec{\rho}_1 = \vec{\rho}_2 = \vec{\rho}$).

Thus, by replacing classical light with entangled biphoton source, in the double-slit setup of Fig. 12.1.1, a Young's double-slit interference-diffraction pattern with twice the interference modulation and twice narrower pattern width, compared to that of classical light at wavelength $\lambda_{s,i} = 2\lambda_p$, is observable in the joint detection. This effect has been confirmed in the "quantum lithography" experiments.

Due to the lack of two-photon sensitive material, the first experimental demonstration of quantum lithography of D'Angelo *et al.* was measured on the Fourier transform plane, instead of the image plane. Two point-like photon counting detectors were scanned jointly, similar to the setup illustrated in Fig. 12.1.1, for the observation of the biphoton interference-diffraction pattern. Fig. 12.1.2 is the published result. It is clear that the biphoton Young's double-slit interference-diffraction pattern is twice narrower with twice the interference modulation compared to that of the classical case although the wavelengths are both $916nm$.

Following linear Fourier optics, it is not difficult to see that, with the help of another lens (equivalently building a microscope), one can transform the Fourier transform function of the double-slit back onto its image plane to observe its image with twice the spatial resolution.

The key to understanding the physics of this experiment is again the entangled nature of the signal-idler biphoton system. As we have discussed earlier, the pair is always emitted from the same point on the output plane of the source, thus always passing the same slit together if the double-slit is placed close to the surface of the nonlinear crystal. There is no chance for the signal-idler pair to pass different slits in this setup. In other words, each point of the object is "illuminated" by the pair "together" and the pair "stops" on

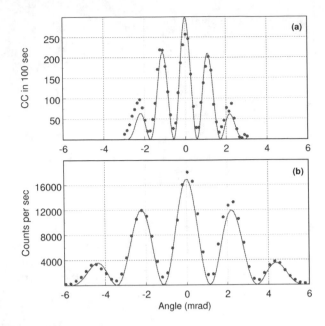

FIGURE 12.1.2 (a) Two-photon Fourier transform of a double-slit. The light source was a collinear degenerate SPDC of $\lambda_{s,i} = 916nm$. (b) Classical Fourier transform of the same double-slit. A classical light source of $\lambda = 916nm$ was used.

the image plane "together". The point-"spot" correspondence between the object plane and the image plane are based on the physics of two-photon diffraction, resulting in a twice narrower Fourier transform function in the Fourier transform plane and twice the image resolution in the image plane. The unfolded schematic setup, which is shown in Fig. 12.1.3, may be helpful for understanding the physics. It is not difficult to calculate the interference-diffraction function under the experimental condition indicated in Fig. 12.1.3. The non-classical observation is due to the superposition of the two-photon amplitudes resulting in a joint detection event, which are indicated by the straight lines connecting D_1 and D_2. The two-photon diffraction, which restrict the spatial resolution of a two-photon image, is different from that of classical light. Thus, there should be no surprise in having an improved spatial resolution even beyond the classical limit.

It is worthwhile to emphasize the following important aspects of physics in this simplified illustration:

(1) The goal of lithography is the reproduction of demagnified images of complicated patterns. The sub-wavelength interference feature does not necessarily translate into an improvement of the lithographic performances. In fact, the spatial Fourier transform argument works for *imaging setup* only; sub-wavelength interference in a temporal coherence type interferometer, like a Mach-Zehnder interferometer, for instance, does not necessarily lead to an image.

(2) In the imaging setup, it is the peculiar nature of the entangled N-photon system allows one to generate an image with N-times of spatial resolution: the entangled photons come out from one point of the object plane, undergo N-photon diffraction, and stop in the image plane within a N-times narrower spot than that of the classical imaging. The historical experiment by D'Angelo *et al*, in which the working principle of quantum lithography was first demonstrated, has taken advantage of the entangled biphoton state of SPDC: the signal-idler photon pair comes out from either the upper slit or the lower slit that is in the object plane, undergoes two-photon diffraction, and stops in the image plane within a twice

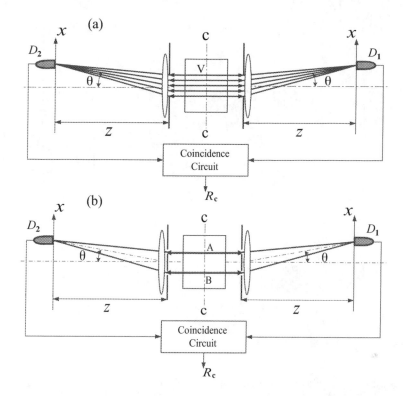

FIGURE 12.1.3 Unfolded experimental setup. The joint measurement is on the Fourier transform plane. Each point of the object is "illuminated" by the signal-idler pair "together", resulting in twice narrower interference-diffraction pattern width in the Fourier transform plane through the joint detection of the signal-idler pair, equivalent to the use of classical light of $\lambda/2$.

narrower image than that of the classical one. Although the measurement is on the Fourier transform plane, it is easy to show that a second Fourier transform, by means of a second lens to form a simple microscope, that reproduces an image on the image plane with double spatial resolution.

12.2 BIPHOTON GHOST IMAGING

The *nonlocal* position-position and momentum-momentum correlation of the entangled biphoton system of SPDC was successfully demonstrated in 1995. The experiment received an interesting name "ghost" imaging immediately in the physics community. The important physics demonstrated in that experiment, however, may not be the so called "ghost". Indeed, the original purpose of the experiment was to study the EPR correlation in position and in momentum and to test the EPR inequality of Eq. (9.1.4) for an entangled biphoton state.

The schematic setup of the ghost imaging experimental is shown in Fig. 12.2.1. An CW laser is used to pump a nonlinear crystal, which is cut for degenerate type-II phase matching to produce a pair of orthogonally polarized signal (*e*-ray of the crystal) and idler (*o*-ray of the crystal) photon. The pair emerges from the crystal as collinear, with $\omega_s \cong \omega_i \cong \omega_p/2$. The pump is then separated from the signal-idler pair by a dispersion prism, and the remaining signal and idler beams are sent in different directions by a polarization beam splitting (Thompson prism). The signal beam passes through a convex lens with a $400mm$ focal

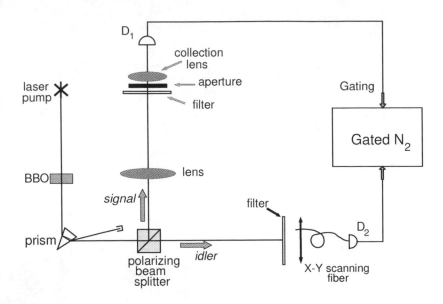

FIGURE 12.2.1 Schematic set-up of the "ghost" image experiment.

length and illuminates a chosen aperture (mask). As an example, one of the demonstrations used letters "UMBC" for the object mask. Behind the aperture is the "bucket" detector package D_1, which consists of a short focal length collection lens in whose focal spot is an avalanche photodiode. D_1 is mounted in a fixed position during the experiment. The idler beam is met by detector package D_2, which consists of an optical fiber whose output is mated with another avalanche photodiode. The input tip of the fiber is scannable in the transverse plane by two step motors. The output pulses of each detector, which are operated in photon counting mode, are sent to a coincidence circuit for counting the joint detection event of the signal-idler pair.

By recording the coincidence counts as a function of the fiber tip's transverse plane coordinates, the image of the chosen aperture (for example, "UMBC") is observed, as reported in Fig. 12.2.2, if the following experimental condition is satisfied: the focal length of the imaging lens f, the aperture's optical distance from the lens s_o, and the image's optical distance from the lens s_i (which is from the imaging lens going backward along the signal photon path to the biphoton source of SPDC crystal then going forward along the path of idler photon to the image), satisfy the Gaussian thin lens equation. When choosing $s_i/s_o = 2$, it is interesting to find that the observed image measures $7mm \times 14mm$ while the size of the "UMBC" aperture inserted in the signal beam is only about $3.5mm \times 7mm$. The observed image in the joint detection of D_1 and D_2 is magnified exactly a factor of $m = s_i/s_o = 2$. In the measurement of Fig. (12.2.2), s_o was chosen to be $s_o = 600mm$, and the twice magnified clear image was found when the fiber tip was on the plane of $s_i = 1200mm$. While D_2 was scanned on other transverse planes other than that of $s_i = 1200mm$, the images blurred out.

The first reaction of many people may be negative: "absolutely impossible!" Examine the two observers D_1 and D_2: D_1 is in the signal beam behind the object, however, it is a "bucket" detector, it does not have any ability to image the object. What D_1 can do is simply counting the signal photons; whenever a photoelectron is created by a signal photon, D_1 sends an electronic pulse to open the "gate" of the coincidence circuit for D_2. D_2, however, is placed in the idler beam, how could D_2 "see" the object that is in the signal beam?

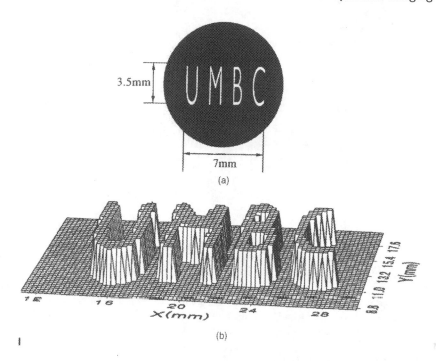

FIGURE 12.2.2 (a) A reproduction of the actual aperture "UMBC" placed in the signal beam. (b) The image of "UMBC": coincidence counts as a function of the fiber tip's transverse plane coordinates. The step size is 0.25mm. The image shown is a "slice" at the half maximum value.

The measurement of the signal and the idler subsystem themselves may further strengthen the negative reaction. In fact, the single photon counting rate of D_1 and D_2 were recorded during the scanning of the image and was found fairly constant in the entire region of the image. The counting rate of neither D_1 nor D_2 is a function of the fiber tips transverse coordinates.

The EPR-correlation δ-functions, $\delta(\vec{\rho}_s - \vec{\rho}_i)$ and $\delta(\vec{\kappa}_s + \vec{\kappa}_i)$ in transverse dimension, are the key to understand this interesting phenomenon. In degenerate SPDC, although the signal-idler photon pair has equal probability to be emitted from any points on the output surface of the nonlinear crystal, the transverse position δ-function indicates that if one of them is observed at one position, the other one must be found at the same position. In other words, the pair is always emitted from the same point on the output plane of the biphoton source. The transverse momentum δ-function, defines the angular correlation of the signal-idler pair: the transverse momenta of a signal-idler amplitude are equal but pointed in opposite directions: $\vec{\kappa}_s = -\vec{\kappa}_i$. In other words, the biphoton amplitudes are always existing at roughly equal yet opposite angles relative to the pump. This then allows for a simple explanation of the experiment in terms of "usual" geometrical optics in the following manner: we envision the nonlinear crystal as a "hinge point" and "unfold" the schematic of Fig. 12.2.1 into that shown in Fig. 12.2.3. The signal-idler biphoton amplitudes can then be represented by straight lines (but keep in mind the different propagation directions) and therefore, the image is well produced in coincidences when the aperture, lens, and fiber tip are located according to the Gaussian thin lens equation. The image is exactly the same as one would observe on a screen placed at the fiber tip if detector D_1 were replaced by a point-like light source and the nonlinear crystal by a reflecting mirror.

Following a similar analysis in geometric optics, it is not difficult to find that any geometrical "light spot" on the subject plane, which is the intersection point of all

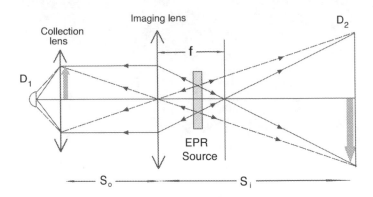

FIGURE 12.2.3 An unfolded setup of the "ghost" imaging experiment, which is helpful for understanding the physics. Since the biphoton "light" propagates along "straight-lines", it is not difficult to find that any geometrical light point on the subject plane corresponds to an unique geometrical light point on the image plane. Thus, a "ghost" image of the subject is made nonlocally in the image plane. Although the placement of the lens, the object, and detector D_2 obeys the Gaussian thin lens equation, it is important to remember that the geometric rays in the figure actually represent the biphoton amplitudes of an entangled signal-idler pair. The point to point correspondence is the result of the superposition of these biphoton amplitudes.

possible signal-idler amplitudes coming from the entangled biphoton source, corresponds to an unique geometrical "light spot" on the image plane, which is another intersection point of all the possible signal-idler amplitudes. This point to point correspondence made the "ghost" image of the subject-aperture possible. Despite the completely different physics from classical geometrical optics, the remarkable feature is that the relationship between the focal length of the lens f, the aperture's optical distance from the lens S_o, and the image's optical distance from the lens S_i, satisfy the Gaussian thin lens equation:

$$\frac{1}{s_o} + \frac{1}{s_i} = \frac{1}{f}.$$

Although the placement of the lens, the object, and the detector D_2 obeys the Gaussian thin lens equation, it is important to remember that the geometric rays in the figure actually represent the biphoton amplitudes of a signal-idler photon pair and the point to point correspondence is the result of the superposition of these biphoton amplitudes. It is the imaging lens made these biphoton paths (amplitudes) equal and thus superposed constructively at these unique pair of points on the object plane and the ghost image plane.

The signal and idler photons may propagate to any points on the object and the ghost image planes, respectively, however, if the signal is passing a point on the object plane, the idler must stopped at an unique point on the ghost image plane. In other words, the transverse coordinate uncertainty of either signal or idler is considerably greater comparing with the correlation of the transverse coordinates of the entangled signal-idler photon pair: Δx_1 (Δy_1) and Δx_2 (Δy_2) are much greater than $\Delta(x_1 - x_2)$ ($\Delta(y_1 - y_2)$). The "ghost" image is a realization of the 1935 EPR *gedankenexperiment*.

Now we calculate $G^{(2)}(\vec{\rho}_o, \vec{\rho}_i)$ for the "ghost" imaging experiment, where $\vec{\rho}_o$ and $\vec{\rho}_i$ are the transverse coordinates on the object plane and the image plane. We will show that there exists a δ-function like point-to-point relationship between the object plane and the image plane, i.e., if one observes the signal photon at a position of $\vec{\rho}_o$ on the object plane the idler photon can be observed only at a certain unique position of $\vec{\rho}_I$ on the image plane satisfying $\delta(m\vec{\rho}_o - \vec{\rho}_I)$, where $m = -(s_i/s_o)$ is the image-object magnification factor. After demonstrating the δ-function, we show how the object function of $A(\vec{\rho}_o)$ is transferred to the image plane as a magnified image $A(\vec{\rho}_i/m)$. Before showing the calculation, it is worth to emphasize again that the "straight lines" in Fig. 12.2.3 schematically represent the biphoton amplitudes all belong to a pair of signal-idler photon. A "click-click" joint measurement at $(\mathbf{r}_1, t_1)$, which is on the object plane, and $(\mathbf{r}_2, t_2)$, which is on the image plane, in the form of EPR δ-function, is the result of the coherent superposition of all these biphoton amplitudes.

We follow the unfolded experimental setup shown in Fig. 12.2.4 to establish the Green's functions $g(\vec{\kappa}_s, \omega_s, \vec{\rho}_o, z_o)$ and $g(\vec{\kappa}_i, \omega_i, \vec{\rho}_2, z_2)$. In arm-1, the signal propagates freely over a distance d_1 from the output plane of the source to the imaging lens, then passes an object aperture at distance s_o, and then is focused onto photon counting detector D_1 by a collection lens. We will evaluate $g(\vec{\kappa}_s, \omega_s, \vec{\rho}_o, z_o)$ by propagating the field from the output plane of the biphoton source to the object plane. In arm-2, the idler propagates freely over a distance d_2 from the output plane of the biphoton source to a point-like detector D_2. $g(\vec{\kappa}_i, \omega_i, \vec{\rho}_2, z_2)$ is thus a free propagator.

(I) Arm-1 (source to object):

The optical transfer function or Green's function in arm-1, which propagates the field from the source plane to the object plane, is given by:

$$g(\vec{\kappa}_s, \omega_s; \vec{\rho}_o, z_o = d_1 + s_o)$$
$$= e^{i \frac{\omega_s}{c} z_o} \int_{lens} d\vec{\rho}_l \int_{source} d\vec{\rho}_S \left\{ \frac{-i\omega_s}{2\pi c d_1} e^{i\vec{\kappa}_s \cdot \vec{\rho}_S} G(|\vec{\rho}_S - \vec{\rho}_l|, \frac{\omega_s}{c d_1}) \right\}$$
$$\times \left\{ G(|\vec{\rho}_l|, \frac{\omega_s}{cf}) \right\} \left\{ \frac{-i\omega_s}{2\pi c s_o} G(|\vec{\rho}_l - \vec{\rho}_o|, \frac{\omega_s}{c s_o}) \right\}, \tag{12.2.1}$$

where $\vec{\rho}_S$ and $\vec{\rho}_l$ are the transverse vectors defined, respectively, on the output plane of the source and on the plane of the imaging lens. The terms in the first and third curly brackets in Eq. (12.2.1) describe free space propagation from the output plane of the source to the imaging lens and from the imaging lens to the object plane, respectively. The function

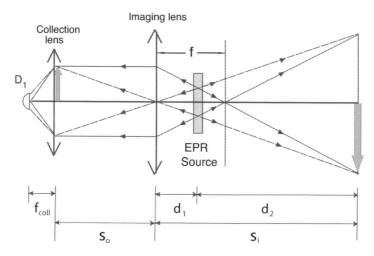

FIGURE 12.2.4 In arm-1, the signal propagates freely over a distance d_1 from the output plane of the source to the imaging lens, then passes an object aperture at distance s_o, and then is focused onto photon counting detector D_1 by a collection lens. In arm-2, the idler propagates freely over a distance d_2 from the output plane of the source to a point-like photon counting detector D_2.

$G(|\vec{\rho}_l|, \frac{\omega}{cf})$ in the second curly brackets is the transformation function of the imaging lens. Here, we treat it as a thin-lens: $G(|\vec{\rho}_l|, \frac{\omega}{cf}) \cong e^{-i\frac{\omega}{2cf}|\vec{\rho}_l|^2}$.

(II) Arm-2 (from source to image):

In arm-2, the idler propagates freely from the source to the plane of D_2, which is also the plane of the image. The Green's function is thus:

$$g(\vec{\kappa}_i, \omega_i; \vec{\rho}_2, z_2 = d_2)$$
$$= \frac{-i\omega_i}{2\pi cd_2} e^{i\frac{\omega_i}{c}d_2} \int_{source} d\vec{\rho}_S \, G(|\vec{\rho}_S - \vec{\rho}_2|, \frac{\omega_i}{cd_2}) e^{i\vec{\kappa}_i \cdot \vec{\rho}_S} \tag{12.2.2}$$

where $\vec{\rho}_S$ and $\vec{\rho}_2$ are the transverse vectors defined, respectively, on the output plane of the source, and on the plane of the photo-dector D_2.

(III) $\Psi(\vec{\rho}_o, \vec{\rho}_i)$ (object plane – image plane):

To simplify the calculation and to focus on the transverse correlation, in the following calculation we assume degenerate ($\omega_s = \omega_i = \omega$) and collinear SPDC. The transverse biphoton effective wavefunction $\Psi(\vec{\rho}_o, \vec{\rho}_2)$ is then evaluated by substituting the Green's functions $g(\vec{\kappa}_s, \omega; \vec{\rho}_o, z_o)$ and $g(\vec{\kappa}_i, \omega; \vec{\rho}_2, z_2)$ into the expression given in Eq. (9.4.5):

$$\Psi(\vec{\rho}_o, \vec{\rho}_2) \propto \int d\vec{\kappa}_s \, d\vec{\kappa}_i \, \delta(\vec{\kappa}_s + \vec{\kappa}_i) \, g(\vec{\kappa}_s, \omega; \vec{\rho}_o, z_o) \, g(\vec{\kappa}_i, \omega; \vec{\rho}_2, z_2)$$

$$\propto e^{i\frac{\omega}{c}(s_0 + s_i)} \int d\vec{\kappa}_s \, d\vec{\kappa}_i \, \delta(\vec{\kappa}_s + \vec{\kappa}_i)$$

$$\times \int_{lens} d\vec{\rho}_l \int_{source} d\vec{\rho}_S \, e^{i\vec{\kappa}_s \cdot \vec{\rho}_S} G(|\vec{\rho}_S - \vec{\rho}_l|, \frac{\omega}{cd_1})$$

$$\times \, G(|\vec{\rho}_l|, \frac{\omega}{cf}) G(|\vec{\rho}_l - \vec{\rho}_o|, \frac{\omega}{cs_o})$$

$$\times \int_{source} d\vec{\rho}_S \, e^{i\vec{\kappa}_i \cdot \vec{\rho}_S} G(|\vec{\rho}_S - \vec{\rho}_2|, \frac{\omega}{cd_2}) \tag{12.2.3}$$

where we have ignored all the proportional constants. Completing the double integral of $d\vec{\kappa}_s$ and $d\vec{\kappa}_s$

$$\int d\vec{\kappa}_s \, d\vec{\kappa}_i \, \delta(\vec{\kappa}_s + \vec{\kappa}_i) \, e^{i\vec{\kappa}_s \cdot \vec{\rho}_S} \, e^{i\vec{\kappa}_i \cdot \vec{\rho}_S'} \sim \delta(\vec{\rho}_S - \vec{\rho}_S'), \tag{12.2.4}$$

Eq. (12.2.3) becomes:

$$\Psi(\vec{\rho}_o, \vec{\rho}_2) \propto \int_{lens} d\vec{\rho}_l \int_{source} d\vec{\rho}_S \, G(|\vec{\rho}_2 - \vec{\rho}_S|, \frac{\omega}{cd_2}) G(|\vec{\rho}_S - \vec{\rho}_l|, \frac{\omega}{cd_1})$$

$$\times \, G(|\vec{\rho}_l|, \frac{\omega}{cf}) G(|\vec{\rho}_l - \vec{\rho}_o|, \frac{\omega}{cs_o}). \tag{12.2.5}$$

We then apply the properties of the Gaussian functions and complete the integral on $d\vec{\rho}_S$ by assuming a large enough transverse size of source to be treated as infinity,

$$\Psi(\vec{\rho}_o, \vec{\rho}_2) \propto \int_{lens} d\vec{\rho}_l \, G(|\vec{\rho}_2 - \vec{\rho}_l|, \frac{\omega}{cs_i}) G(|\vec{\rho}_l|, \frac{\omega}{cf}) G(|\vec{\rho}_l - \vec{\rho}_o|, \frac{\omega}{cs_o}). \tag{12.2.6}$$

Although the signal and idler propagate to different directions along two optical arms, Interestingly, the Green function in Eq. (12.2.6) is equivalent to that of a classical imaging setup, if we imagine the fields start propagating from a point $\vec{\rho}_o$ on the object plane to the lens and then stop at point $\vec{\rho}_2$ on the imaging plane. The mathematics is consistent with our previous qualitative analysis of the experiment.

The integral on $d\vec{\rho}_l$ yields a point-to-point relationship between the object plane and the image plane that is defined by the Gaussian thin-lens equation:

$$\int_{lens} d\vec{\rho}_l \, G(|\vec{\rho}_l|, \frac{\omega}{c}[\frac{1}{s_o} + \frac{1}{s_i} - \frac{1}{f}]) \, e^{-i\frac{\omega}{c}(\frac{\vec{\rho}_o}{s_o} + \frac{\vec{\rho}_i}{s_i})\cdot\vec{\rho}_l} \propto \delta(\vec{\rho}_o + \frac{\vec{\rho}_i}{m}) \tag{12.2.7}$$

where the integral is approximated to infinity and the Gaussian thin-lens equation of $1/s_o + 1/s_i = 1/f$ is applied. We have also defined $m = s_i/s_o$ as the magnification factor of the imaging system. The function $\delta(\vec{\rho}_o + \vec{\rho}_i/m)$ indicates that a point of $\vec{\rho}_o$ on the object plane corresponds to a unique point of $\vec{\rho}_i$ on the image plane. The two vectors pointed to opposite directions and the magnitudes of the two vectors hold a ratio of $m = |\vec{\rho}_i|/|\vec{\rho}_o|$.

If the finite size of the imaging lens has to be taken into account (finite diameter D), the integral yields a point-spread function of $somb(x)$:

$$\int_{lens} d\vec{\rho}_l \, e^{-i\frac{\omega}{c}(\frac{\vec{\rho}_o}{s_o} + \frac{\vec{\rho}_i}{s_i})\cdot\vec{\rho}_l} \propto somb\left(\frac{R}{s_o}\frac{\omega}{c}[\vec{\rho}_o + \frac{\vec{\rho}_i}{m}]\right) \tag{12.2.8}$$

where $somb(x) = 2J_1(x)/x$, $J_1(x)$ is the first-order Bessel function and R/s_o is named as the numerical aperture. The point-spread function turns the point-to-point correspondence between the object plane and the image plane into a point-to-"spot" relationship and thus limits the spatial resolution. This point has been discussed in detail in last section.

Therefore, by imposing the condition of the Gaussian thin-lens equation, the transverse biphoton effective wavefunction is approximated as a δ function

$$\Psi(\vec{\rho}_o, \vec{\rho}_i) \propto \delta(\vec{\rho}_o + \frac{\vec{\rho}_i}{m}) \tag{12.2.9}$$

where $\vec{\rho}_o$ and $\vec{\rho}_i$, again, are the transverse coordinates on the object plane and the image plane, respectively, defined by the Gaussian thin-lens equation. Thus, the second-order spatial correlation function $G^{(2)}(\vec{\rho}_o, \vec{\rho}_i)$ turns to be:

$$G^{(2)}(\vec{\rho}_o, \vec{\rho}_i) = |\Psi(\vec{\rho}_o, \vec{\rho}_i)|^2 \propto |\delta(\vec{\rho}_o + \frac{\vec{\rho}_i}{m})|^2. \tag{12.2.10}$$

Eq. (12.2.10) indicates a point to point EPR correlation between the object plane and the image plane, i.e., if one observes the signal photon at a position of $\vec{\rho}_o$ on the object plane, the idler photon can only be found at a certain unique position of $\vec{\rho}_i$ on the image plane satisfying $\delta(\vec{\rho}_o + \vec{\rho}_i/m)$ with $m = s_i/s_o$.

We now include an object-aperture function, a collection lens and a photon counting detector D_1 into the optical transfer function of arm-1 as shown in Fig. 12.2.1.

First, we treat the collection-lens-D_1 package as a "bucket" detector. The "bucket" detector integrates all $\Psi(\vec{\rho}_o, \vec{\rho}_2)$ that pass the object aperture $A(\vec{\rho}_o)$ as a joint photodetection event. This process is equivalent to the following convolution:

$$R_{1,2} \propto \left| \int_{obj} d\vec{\rho}_o \, A(\vec{\rho}_o) \, \Psi(\vec{\rho}_o, \vec{\rho}_i) \right|^2$$

$$\simeq \left| A(\vec{\rho}_o) \otimes \delta(\vec{\rho}_o + \frac{\vec{\rho}_2}{m}) \right|^2 = \left| A(\frac{-\vec{\rho}_2}{m}) \right|^2 \tag{12.2.11}$$

where $\otimes$ means convolution, again, D_2 is scanning in the image plane, $\vec{\rho}_2 = \vec{\rho}_i$. Eq. (12.2.11) indicates a magnified (or demagnified) image of the object-aperture function by means of the joint-detection events between distant photodetectors D_1 and D_2. The "-" sign in $A(-\vec{\rho_i}/m)$ indicates opposite orientation of the image. The model of "bucket" detector is a good and realistic approximation.

Second, we calculate the Green's function from the source to D_1 in detail by including the object-aperture function, the collection lens and the photon counting detector D_1 into arm-1. The Green's function of Eq. (12.2.1) becomes:

$$g(\vec{\kappa}_s, \omega_s; \vec{\rho}_1, z_1 = d_1 + s_o + f_{coll})$$

$$= e^{i\frac{\omega_s}{c}z_1} \int_{obj} d\vec{\rho}_o \int_{lens} d\vec{\rho}_l \int_{source} d\vec{\rho}_S \left\{ \frac{-i\omega_s}{2\pi c d_1} e^{i\vec{\kappa}_s \cdot \vec{\rho}_S} G(|\vec{\rho}_S - \vec{\rho}_l|, \frac{\omega_s}{c d_1}) \right\}$$

$$\times \ G(|\vec{\rho}_l|, \frac{\omega_s}{c f}) \left\{ \frac{-i\omega_s}{2\pi c s_o} G(|\vec{\rho}_l - \vec{\rho}_o|, \frac{\omega_s}{c s_o}) \right\} A(\vec{\rho}_o)$$

$$\times \ G(|\vec{\rho}_o|, \frac{\omega_s}{c f_{coll}}) \left\{ \frac{-i\omega_s}{2\pi c f_{coll}} G(|\vec{\rho}_o - \vec{\rho}_1|, \frac{\omega_s}{c f_{coll}}) \right\} \tag{12.2.12}$$

where f_{coll} is the focal-length of the collection lens and D_1 is placed on the focal point of the collection lens. Repeating the previous calculation, we obtain the transverse biphoton effective wavefunction:

$$\Psi(\vec{\rho}_1, \vec{\rho}_2) \propto \int_{obj} d\vec{\rho}_o \, A(\vec{\rho}_o) \, \delta(\vec{\rho}_o + \frac{\vec{\rho}_2}{m}) = A(\vec{\rho}_o) \otimes \delta(\vec{\rho}_o + \frac{\vec{\rho}_2}{m}) \tag{12.2.13}$$

which is the same as Eq. (12.2.11). Notice, in Eq. (12.2.13) we have ignored the phase factors which have no contribution to the formation of the image. The joint detection counting rate, $R_{1,2}$, between photon counting detectors D_1 and D_2 is thus:

$$R_{1,2} \propto G^{(2)}(\vec{\rho}_1, \vec{\rho}_2) \propto \left| A(\vec{\rho}_o) \otimes \delta(\vec{\rho}_o + \frac{\vec{\rho}_2}{m}) \right|^2 = \left| A(\frac{-\vec{\rho}_2}{m}) \right|^2 \tag{12.2.14}$$

where, again, $\vec{\rho}_2 = \vec{\rho}_i$.

The physical process corresponding to the above convolution can be summarized as follows. Due to the unique point-to-point correlation between the object plane and the image plane, whenever the bucket detector receives a signal photon that is either transmitted, scattered or reflected from a unique point of the object, the scanning point photodetector D_2 or a CCD element that receives the idler photon identifies the coordinate of $\vec{\rho}_o$ and the value of the aperture function $A(\vec{\rho}_o)$ for that joint photodetection event. For instance, at time t, the bucket detector receives a signal photon that is either transmitted, scattered, or reflected from a unique point $\vec{\rho}_o$ of the object plane within a coincidence time window. The joint-detection of the idler photon by the scanning point detector D_2 or a CCD element, with known coordinate $\vec{\rho}_i$, identifies $\vec{\rho}_o$ immediately for that event. At time t', the bucket detector receives another signal photon that is either transmitted, scattered, or reflected from another unique point $\vec{\rho}'_o$ of the object plane within the coincidence window, and the joint-detection of the idler photon by the scanning point detector D_2 or a CCD element, with known coordinate $\vec{\rho}'_i$, identifies $\vec{\rho}'_o$ immediately for that event. The probability of receiving a joint photodetection event at $(\vec{\rho}_o, \vec{\rho}_i)$ and at $(\vec{\rho}'_o, \vec{\rho}'_i)$ is proportional to the value of the aperture function $A(\vec{\rho}_o)$ and $A(\vec{\rho}'_o)$, respectively. Accumulating a large number of joint-detection events at each transverse coordinate on the image plane, the aperture function $A(\vec{\rho}_o)$ is thus reproduced in the joint-detection as a function of $\vec{\rho}_i$.

As we have discussed earlier, the point-to-point EPR correlation is the result of the coherent superposition of the biphoton probability amplitudes. In principle, one signal-idler pair contains all the necessary biphoton probability amplitudes that generate the ghost image. We name this kind of image as *two-photon coherent* image to distinguish the *two-photon incoherent* image of thermal light.

12.3 THERMAL LIGHT GHOST IMAGING

Ten years after the experimental demonstration of ghost imaging with entangled photon pairs, Valencia *et al.* found that the near-field, natural, non-factorizable, point-to-point image-forming correlation is not only the property of entangled photon pairs. Two-photon interference of randomly created and randomly paired photons in thermal state can produce a similar point-to-point image-forming correlation without the use of any imaging lens. This type of ghost imaging technique is commonly named lensless ghost imaging. With no lens requirement, in addition to visible light, the two-photon interference produced image-forming correlation is well suited for these radiations for which no effective imaging lenses available.

The lensless near-field ghost imaging with pseudo-thermal radiation was demonstrated by Valencia *et al.*, D'Angolo *et al.* and Scarcelli *et al.* in the years from 2005 to 2006.

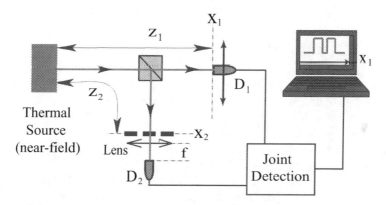

FIGURE 12.3.1 Lensless Fresnel near-field ghost imaging with pseudo-thermal light demonstrated in 2006 by Scarcelli *et al.* D_1 is a point-like photodetector that is scannable along the x_1-axis. The joint-detection between D_1 and the bucket detector D_2 is realized either by a photon-counting coincidence counter or by a standard HBT linear multiplier (RF mixer). In this measurement D_2 is fixed in the focal point of a convex lens, playing the role of a bucket detector. The counting rate or the photocurrent of D_1 and D_2, respectively, are measured to be constants. Surprisingly, an image of the 1-D object is observed in the joint-detection between D_1 and D_2 by scanning D_1 in the plane of $z_1 = z_2$ along the x_1-axis. The image is blurred out when $z_1 \neq z_2$. There is no doubt that thermal radiations propagate to any transverse plane in a random and chaotic manner. There is no lens applied to force the thermal radiation "collapsing" to a point or speckle either. What is the physical cause of the point-to-point image-forming correlation in coincidences?

Fig. 12.3.1 is the schematic experimental setup of their 2006 demonstration. A pseudo-thermal radiation with a narrow spectral bandwidth, $\Delta\omega$, and a fairly large bandwidth of spatial frequency, Δk_x, resulting from a large angular sized disk-like source, is divided into two by an optical beam-splitter[1]. The reflected light is propagated and focused onto a point-like photodetector D_2 (bucket detector) after passing through an object mask, which is a simple double-slit in Fig. 12.3.1. D_2 is fixed in the focal plane of the focusing

---

[1]A Fresnel "near-field" lensless configuration was applied in the ghost imaging experiment of Scarcelli *et al.* Note, Fresnel near-field is not "near-surface-field". In some classical treatments, "near-surface-field" is used to claim two identical copies of the "speckles" of the source plane on the object and image planes. These theories do not work for the Fresnel near-field lensless thermal light ghost imaging. Imagine we are using sun light for ghost imaging (sun has an angular size of $\sim 0.5°$ relative to earth and is in the Fresnel near-field), one cannot claim a copy of a "speckles" of the sun without the use of a lens system. Any "speckle" will become a big "spot" after its propagation, in principle: the smaller the size of the "speckle" in the source plane the greater the size of the "spot" in the Fresnel near-field planes.

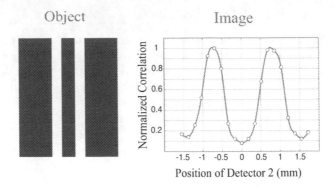

FIGURE 12.3.2 The experimentally observed ghost image of a double-slit. The lensless ghost image is observed to have equal size as that of the object. The ghost image has 50% contrast if measured by either photon-counting-coincidence circuit or by standard HBT type analog correlation circuit. If the measurement is for photon number fluctuation, or intensity fluctuation, correlation, the visibility of the ghost image is close to 100%.

convex lens. It is clear that D_2, known as a bucket detector, cannot retrieve any information about the spatial distribution, or the aperture function of the object mask. The transmitted light is freely propagated to the plane of x_1 to be detected by the scanning point-like photodetector D_1. The joint-detection between D_1 and D_2 is realized either by a photon-counting-coincidence counter or by a standard HBT type current-current linear multiplier. Although the single-detector counting rates of D_1 and D_2 are both constants during the measurement, surprisingly, an equal sized 1-D image of the object mask is observed in the joint detection when D_1 is scanned in the plane of $z_1 = z_2$ along the x_1-axis. The image contrast measures almost 50%, which is the maximum contrast we can expect from photon number correlation or intensity correlation measurements. However, if the measurement is for photon number fluctuation correlation or intensity fluctuation correlation, the image contrast can be ~100%. Figure 12.3.2 reports the measured image of the double-slit in photon number fluctuation correlation.

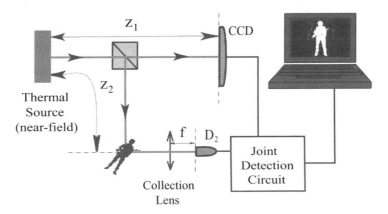

FIGURE 12.3.3 Improved near-field lensless ghost imaging of pseudo-thermal light demonstrated by Meyers *et al.*. The bucket detector collects the randomly scattered and reflected photons from an object. The CCD cannot "see" the object, but is facing the light source.

In 2008, Meyers *et al.* published a modified version of the original near-field lensless ghost imaging experiment. Figure 12.3.3 schematically illustrates their experimental setup. This

FIGURE 12.3.4 Ghost image of a toy soldier model.

experiment improves the ghost imaging experiment of Scarcelli *et al.* in two aspects: (1) the bucket detector is not triggered by the transmitted light that passing through a mask but instead trigged by the randomly scattered and reflected photons from the surface of a toy soldier; (2) no scanning point photodetector, but instead a CCD array of 2-D operated at photon counting regime is used for the joint-detection with the bucket detector D_2. A ghost image of the toy soldier is captured by the gated CCD when taking $z_1 = z_2$. The spatial resolution of the ghost image is determined by the angular size of the thermal source.

The ghost images observed from the above experiments are reproduced by a two-photon interference induced point-to-point image-forming correlation between the object plane and the ghost image plane. This two-photon interference induced diffraction limited point-to-point image-forming correlation, measurable from photon number fluctuation correlation or intensity fluctuation correlation, is part of the second-order spatial coherence function of thermal field:

$$g^{(2)}(\vec{\rho}_o, \vec{\rho}_i) \sim 1 + \delta(\vec{\rho}_o + \vec{\rho}_i)$$
$$\propto \langle I_o \rangle \langle I_i \rangle + \langle \Delta I_o \Delta I_i \rangle \propto \langle n_o \rangle \langle n_i \rangle + \langle \Delta n_o \Delta n_i \rangle \qquad (12.3.1)$$

where $\vec{\rho}_o$ and $\vec{\rho}_i$ are the transverse coordinates on the object plane and the ghost image plane, respectively. Ghost image plane is the observation plane which has equal distance to the thermal light source as that of the object plane.

The following calculation shows how the point-to-point correlation of thermal field reproduces the lensless ghost image of an aperture function $A(\vec{\rho}_{\text{obj}})$. Focusing on the second-order spatial correlation, we assume perfect second-order temporal coherence. In fact, the pseudo-thermal fields used in these experiments can be treated as single-frequency.

Examine the "unfolded" lensless ghost imaging experiment schematic setup of Fig. 12.3.5, the two-photon interference induced intensity fluctuation correlation measured by the scannable photodetector D_1 and the bucket photodetector D_2 can be easily calculated from Einstein's picture

$$\langle \Delta n(\vec{\rho}_1, z_1) \Delta n_2 \rangle$$
$$= \sum_{m,n} \left[E_m^* g_m^*(\vec{\rho}_1, z_1) \right] \left[E_n g_n(\vec{\rho}_1, z_1) \right]$$
$$\times \left[\int d\vec{\rho}_o E_n^* A^*(\vec{\rho}_o) g_n^*(\vec{\rho}_o, z_o) \right] \left[\int d\vec{\rho}'_o E_m A(\vec{\rho}'_o) g_m(\vec{\rho}'_o, z_o) \right]$$
$$= \left| \sum_m \left[E_m^* g_m^*(\vec{\rho}_1, z_1) \right] \left[\int d\vec{\rho}_o E_m A(\vec{\rho}_o) g_m(\vec{\rho}_o, z_o) \right] \right|^2$$
$$\propto \left| \int d\vec{\rho}_s \left[g_{\vec{\rho}_s}^*(\vec{\rho}_1, z_1) \right] \left[\int d\vec{\rho}_o A(\vec{\rho}_o) g_{\vec{\rho}_s}(\vec{\rho}_o, z_o) \right] \right|^2 \qquad (12.3.2)$$

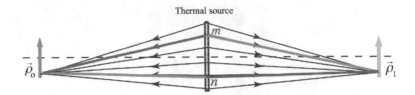

Thermal source

FIGURE 12.3.5 "Unfolded" schematic setup of lensless ghost imaging. A large number of randomly created and randomly distributed subfields are radiated from a large angular diameter disk-like thermal or pseudo-thermal light source. A "bucket" photodetector, D_2, is placed behind the object plane of $z_o = d$ to "collect" all transmitted or reflected-scattered light from the object. A scannable point-like photodetector, D_1, is placed on the ghost imaging plane of $z_1 = d$. Note: choosing $z_1 = z_o = d$, when $\vec{\rho}_1 \approx \vec{\rho}_o$, the red and blue two-photon amplitudes of the m-nth pair of subfields superpose constructively with equal path. Adding the contributions of all random pairs at $\vec{\rho}_1 \approx \vec{\rho}_o$, the intensity fluctuation correlation, or the photon number fluctuation correlation, measurement, yields a two-photon diffraction limited point-to-point correlation, which plays the role of an image-forming correlation to map the aperture function $A(\vec{\rho}_o)$ onto the ghost image plane of $z_1 = d$.

where m and n label the mth and the nth subfield emitted from the mth and the nth sub-source. In Eq. 12.3.2, we have assumed a perfect second-order temporal coherence as usual. Applying the Fresnel near field approximation to propagate the field from each sub-source to the photodetectors by means of the following Green's function

$$g_m(\omega; \vec{\rho}_j, z_j) = \frac{c_o}{z_j} e^{-i\omega\tau_j} e^{i\frac{\omega}{2cz_j}|\vec{\rho}_j - \vec{\rho}_m|^2}, \tag{12.3.3}$$

where c_0 is a normalization consistent and $\tau_j \equiv t_j - z_j/c$, $j = 1, o$. Assuming a disk-like source and randomly distributed and randomly radiated point-like sub-sources, we can approximate the sum of m into an integral of $\vec{\rho}_s$ on the source plane of $z_s = 0$. The photon number fluctuation, or intensity fluctuation, correlation measurement between D_1 and D_2 yields a diffraction limited correlation between the planes of $z_1 = d$ and $z_o = d$,

$$\langle \Delta n(\vec{\rho}_1, z_1) \Delta n_2 \rangle \big|_{z_1 = z_o}$$
$$\propto \left| \int d\vec{\rho}_s \left[g_{\vec{\rho}_s}^*(\omega; \vec{\rho}_1, z_1) \right] \left[\int d\vec{\rho}_o \, A(\vec{\rho}_o) \, g_{\vec{\rho}_s}(\omega; \vec{\rho}_o, z_o) \right] \right|^2$$
$$\simeq \left| \int d\vec{\rho}_o \, A(\vec{\rho}_o) \, \text{somb} \frac{\pi \Delta \theta_s}{\lambda} |\vec{\rho}_1 - \vec{\rho}_o| \right|^2$$
$$= \left| A(\vec{\rho}_o) \otimes \text{somb} \frac{\pi \Delta \theta_s}{\lambda} |\vec{\rho}_1 - \vec{\rho}_o| \right|^2 \tag{12.3.4}$$

where the somb-function is defined as $2J(x)/x$ and $\Delta\theta_s \approx 2R/d$ is the angular diameter of the radiation source viewed at the photodetectors. It is clear from Eq. (12.3.4) we may conclude: (1) Similar to the ghost image reproduced by entangled photon pairs, the observed lensless ghost image is the result of a convolution between the aperture function of the object, $A(\vec{\rho}_o)$, and the two-photon diffraction limited point-to-point image-forming function, $\text{somb}(\pi\Delta\theta_s/\lambda)|\vec{\rho}_1 - \vec{\rho}_o|$; (2) the spatial resolution of the lensless ghost image is determined by the wavelength of the thermal field and the angular diameter of the thermal source: the shorter the wavelength and the larger the angular size of the light source, the higher the spatial resolution of the achievable lensless ghost image.

For a large value of $\Delta\theta_s$, the point-to-"spot" sombrero function can be approximated as a delta-function, $\delta(|\vec{\rho}_1 - \vec{\rho}_o|)$,

$$\langle\Delta n(\vec{\rho}_1, z_1)\Delta n_2\rangle\big|_{z_1=z_o}$$

$$\simeq \left|\int d\vec{\rho}_o\, A(\vec{\rho}_o)\,\delta(|\vec{\rho}_1 - \vec{\rho}_o|)\right|^2$$

$$= \left|A(\vec{\rho}_o) \otimes \delta(|\vec{\rho}_1 - \vec{\rho}_o|)\right|^2 = |A(\vec{\rho}_1)|^2. \tag{12.3.5}$$

To confirm the observation is an image but not a "projection shadow", in fact, the first lensless ghost imaging demonstration of Valencia *et al.* in 2005 was observed from a secondary imaging plane of the lensless ghost image. Their unfolded experimental setup is illustrated in the upper part of Fig. 12.3.6. By using a convex imaging lens of focus

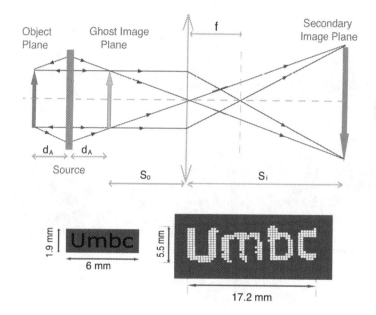

FIGURE 12.3.6 Upper: unfolded schematic experimental setup of a secondary image measurement of the primary ghost image and the measured secondary images. By using a convex lens of focal length f, the primary lensless ghost image is imaged onto a secondary image plane, which is defined by the Gaussian thin-lens equation, $1/s_o + 1/s_i = 1/f$, with magnification $m = -s_i/s_o$. This setup is useful for distant large scale ghost imaging applications. Lower: a secondary image of the primary lensless ghost image of "UMBC" with a magnification factor of $m = s_i/s_o \sim 2.9$. Notice, the secondary image is "blurred" out quickly when the scanning photodetector D_1 is moved away from the image plane.

length f, the primary lensless ghost image is mapped onto a secondary image plane, where the scanning photodetector D_1 is placed. The secondary image of the primary lensless ghost image is recorded in the joint photodetection between D_1 and D_2 by means of either photon counting coincidences or by photocurrent-photocurrent correlation. The lower part of Fig. 12.3.6 shows a secondary image of the primary lensless ghost image of "UMBC" with a magnification factor of $m = |s_i/s_o| \sim 2.9$. The secondary imaging system is useful in certain experimental conditions, especially when the size of the 1:1 lensless ghost image is either too big or too small to be captured by a CCD (scanning photodetector D_1). A magnified or demagnified secondary ghost image would be helpful for certain applications, such as satellite imaging (field is too large) and X-ray microscope (field is too small).

The following calculation confirms that the magnified or demagnified secondary ghost image is reproduced by a two-photon interference induced diffraction limited point-to-point image-forming correlation. We first calculate the point-to-point photon number fluctuation correlation between the object plane, $z_o = d_A$, and the secondary ghost image plane, $z_1 = d_A + s_o + s_i$, where s_o is the distance from the primary ghost image to the lens, s_i is the distance from the lens to the secondary ghost image, satisfying the Gaussian thin lens equation $1/s_o + 1/s_i = 1/f$ with f the focal length of the lens. Following Eq. (12.3.2), we have

$$\langle \Delta n(\vec{\rho}_o, z_o) \Delta n(\vec{\rho}_1, z_1) \rangle$$

$$\propto \left| \int d\vec{\rho}_s \left[g_{\vec{\rho}_s}(\omega; \vec{\rho}_1, z_1) \right] \left[g_{\vec{\rho}_s}^*(\omega; \vec{\rho}_o, z_o) \right] \right|^2$$

$$= \left| \int d\vec{\rho}_s \left\{ \left[g_{\vec{\rho}_s}(\omega; \vec{\rho}_{gi}, z_{gi}) \right] \left[\int d\vec{\rho}_{gi}\, g_{\vec{\rho}_{gi}}(\omega; \vec{\rho}_L, z_L) \right] \left[g_{Lens} \right] \right. \right.$$

$$\left. \left. \times \left[\int d\vec{\rho}_L\, g_L(\omega; \vec{\rho}_1, z_1) \right] \right\} \left\{ \left[g_{\vec{\rho}_s}^*(\omega; \vec{\rho}_o, z_o) \right] \right\} \right|^2$$

$$\simeq \left| \int d\vec{\rho}_{gi}\, \delta(|\vec{\rho}_o - \vec{\rho}_{GI}|) \, \mathrm{somb} \frac{\pi D}{s_o \lambda} |\vec{\rho}_{gi} - \vec{\rho}_1/\mu| \right|^2$$

$$\simeq \mathrm{somb}^2 \frac{\pi D}{s_o \lambda} |\vec{\rho}_o - \vec{\rho}_1/\mu|, \tag{12.3.6}$$

where $g_{\vec{\rho}_s}(\omega; \vec{\rho}_{gi}, z_{gi})$, $g_{\vec{\rho}_{gi}}(\omega; \vec{\rho}_L, z_L)$, g_{Lens}, and $g_{\vec{\rho}_L}(\omega; \vec{\rho}_1, z_1)$ are the Green's functions along the ghost image path that propagate the field from the source plane to the one-to-one primary ghost image plane, from the primary ghost image plane to the lens plane, from the input plane of the lens to the output plane of the lens, and from the lens plane to the secondary ghost image plane, respectively; $g_{\vec{\rho}_s}^*(\omega; \vec{\rho}_o, z_o)$ is the Green's function along the bucket detector path that propagates the field from the source plane to the object plane. It is interesting to conclude a diffraction limited *nonlocal* point-to-point correlation between the object plane of $z_o = d_A$ and the secondary ghost image plane of $z_1 = d_A + s_o + s_i$ in the presence of a lens: the sombrero-like function indicates a distant correlation between the two propagating arms of the thermal fields.

Now, we calculate the magnified or demagnified lensless ghost imaging. Examine the "unfolded" schematic setup of Fig. 12.3.6, following Eqs. (12.3.2) and (12.3.6), the two-photon interference induced intensity fluctuation correlation measured by the scannable point-like photodetector, D_1, and the "bucket" photodetector, D_2 is calculated as the follows

$$\langle \Delta n(\vec{\rho}_1, z_1) \Delta n_2 \rangle$$

$$\propto \left| \int d\vec{\rho}_s \left[g_{\vec{\rho}_s}(\omega; \vec{\rho}_1, z_1) \right] \left[\int d\vec{\rho}_o\, A(\vec{\rho}_o)\, g_{\vec{\rho}_s}^*(\omega; \vec{\rho}_o, z_o) \right] \right|^2$$

$$= \left| \int d\vec{\rho}_s \left\{ \left[g_{\vec{\rho}_s}(\omega; \vec{\rho}_{gi}, z_{GI}) \right] \left[\int d\vec{\rho}_{gi}\, g_{\vec{\rho}_{gi}}(\omega; \vec{\rho}_L, z_L) \right] \left[g_{Lens} \right] \right. \right.$$

$$\left. \left. \times \left[\int d\vec{\rho}_L\, g_L(\omega; \vec{\rho}_1, z_1) \right] \right\} \left\{ \int d\vec{\rho}_o\, A(\vec{\rho}_o) \left[g_{\vec{\rho}_s}^*(\omega; \vec{\rho}_o, z_o) \right] \right\} \right|^2$$

$$\simeq \left| \int d\vec{\rho}_o\, A(\vec{\rho}_o) \int d\vec{\rho}_{gi}\, \delta(|\vec{\rho}_o - \vec{\rho}_{GI}|) \, \mathrm{somb} \frac{\pi D}{s_o \lambda} |\vec{\rho}_{gi} - \vec{\rho}_1/\mu| \right|^2$$

$$\simeq \left| \int d\vec{\rho}_o\, A(\vec{\rho}_o) \, \mathrm{somb} \frac{\pi D}{s_o \lambda} |\vec{\rho}_o - \vec{\rho}_1/\mu| \right|^2. \tag{12.3.7}$$

In Eq. (12.3.7), D is the diameter of the lens. It is easy to find that the secondary ghost image is the result of a convolution between the aperture function $A(\vec{\rho}_o)$ and a two-photon diffraction limited sombrero-like image-forming function. Under certain experimental condition the sombrero-like function can be approximated as a delta-function of $\delta(|\vec{\rho}_o - \vec{\rho}_1/\mu|)$, we thus have

$$\langle \Delta I(\vec{\rho}_1, z_1) \Delta I_2 \rangle \simeq \left| \int d\vec{\rho}_o \, A(\vec{\rho}_o) \, \delta(|\vec{\rho}_o - \vec{\rho}_1/\mu|) \right|^2$$

$$= \left| A(\vec{\rho}_o) \otimes \delta(|\vec{\rho}_o - \vec{\rho}_1/\mu|) \right|^2 = |A(\vec{\rho}_1/\mu)|^2 \qquad (12.3.8)$$

indicating a magnified or demagnified secondary ghost image by means of a point-to-point mapping of the aperture function.

It is necessary to emphasize that the mathematics of the convolution between the aperture function and the point-to-point image-forming function has no difference in any optical imaging systems, including the traditional classic imaging, the ghost imaging of entangled states, and the ghost imaging of thermal field. The differences between different imaging systems come from different mechanisms that produce the point-to-point or point-to-"spot" image-forming function in that particular imaging system. In a classic imaging system, it is the first-order constructive-destructive interference that causes the point-to-point correspondence between the object and image planes, i.e., any radiation which is radiated (or reflected) from a point on the object plane will arrive at an unique point on the image plane. In the lensless ghost imaging system of thermal field, it is the two-photon interference that causes the nonlocal second-order correlation between the object plane and the image plane. Analogous to the ghost image of entangled photon pairs, this natural, non-factorizable, point-to-point image-forming correlation represents a nonlocal interference of a randomly created and randomly paired photons in thermal state: neither photon-one nor photon-two "knows" precisely where to go when they are created at each independent sub-source; however, if one is observed at a point on the object plane, the other one has twice the probability of arriving at a unique corresponding point on the image plane.[2]

12.4 TURBULENCE-FREE GHOST IMAGING AND CAMERA

The multi-photon interference nature of ghost imaging determines its peculiar features: (1) it is nonlocal; (2) its imaging resolution differs from that of classical. In this section, we analyze a recent ghost imaging experiment of Meyers *et al.* and a "ghost camera", which demonstrated another interesting yet peculiar feature of ghost imaging of thermal light: "turbulence-free", i.e., any index fluctuation type turbulence occur in the optical path would not affect the quality of the ghost image. As we know that atmospheric turbulence is a serious problem for classic satellite and aircraft-ground based distant imaging. This feature is thus useful for these applications.

The schematic setup of the turbulence-free ghost imaging experiment of Meyers *et al.* is shown in Fig. 12.4.1. It is a typical thermal light ghost imaging setup which captures the secondary ghost image of the primary lensless ghost image. This experiment, however, added a set of powerful heating elements underneath the optical paths to produce laboratory atmospheric turbulence. Fig. 12.4.1 illustrates the most serious situation in which turbulence occurs in all optical paths of the setup. The heating elements can be isolated to produce turbulence for any individual optical path too. Heating of the air causes temporal and spatial fluctuations on its index of refraction that makes the classic image of the object to jitter about randomly on the image plane of a classic camera.

---

[2]Similar to the far-field HBT correlation, the contrast of the Fresnel near-field point-to-point image-forming function in the measurement of photon number correlation or intensity correlation is 50%, i.e., a two to one ratio between the maximum value and the constant background. This Fresnel near-field point-to-point image-forming function turns to be 100% in the measurement of photon number fluctuation correlation or the intensity fluctuation correlation.

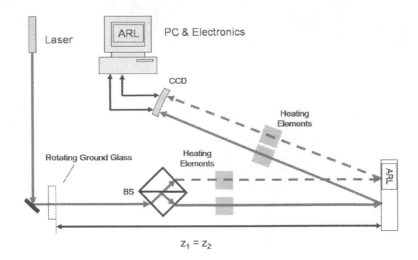

FIGURE 12.4.1 Schematic setup of a typical thermal light ghost imaging experiment which captures the secondary imaging of the primary lensless ghost image. This experiment, however, added a set of powerful heating elements underneath the optical paths to produce laboratory atmospheric turbulence. Dashed line and arrows indicate the optical path of the "bucket" detector. The solid line and arrows indicate the optical path of the ghost image arm.

Similar to their earlier demonstration of ghost imaging, the pseudo-thermal light is generated from a fairly large angular sized pseudo-thermal source and is split into two by a 50%:50% beamsplitter. One of the beams illuminates an object located at z_1, such as the letters "ARL" as shown in Fig. 12.4.1. The scattered and reflected photons from the object are collected and counted by a bucket detector which is simulated by the right-half of the photon counting CCD in Fig. 12.4.1. The other beam propagates to the ghost image plane of $z_1 = z_2$. We have learned from early analysis of thermal light ghost imaging experiments, placing a CCD array on the ghost image plane, the CCD array will capture the ghost image of the object if its exposure is gated by the bucket detector. In this experiment, the CCD array is replaced by a piece of glossy white paper. The scattered and reflect light from the glossy white paper, which contains the information of the ghost image, is then captured by the left-half of the high resolution and high sensitivity photon CCD camera, which is operated at the photon counting regime. The CCD camera is focused onto the ghost image plane and is gated by the bucket detector for the observation of the secondary ghost image. The secondary ghost image captured by the left-half CCD camera is the image of the primary lensless ghost image located at $z_1 = z_2$. In this special setup the left-half and the right-half of the CCD camera may play the poles of two independent classic cameras in their "normal" un-gated operation, and simultaneously capture the secondary ghost image in their gated joint-detection operation. The hardware circuit and the software program is designed to monitor the outputs of the left-half and the right-half of the CCDs, individually, as two independent classic cameras, and simultaneously to monitor the gated output of the left-half CCDs as a ghost camera. In the measurement, the classic image and the secondary ghost image of the object were captured and monitored simultaneously when the turbulence is introduced to each or to all optical paths.

The effect of turbulence on a classic image can be easily seen from the blurring of the images. Technically the turbulence is characterized by the refractive index structure parameter C_n^2. This experiment achieved $C_n^2 = 1.2^{-12}$ for the CCD arm and $C_n^2 = 1.5^{-12}$ for the bucket detector arm. These values correspond extremely high levels of atmospheric turbulence, causing significant temporal and spatial fluctuations of the index of refraction, as well as the blurring of the classic images. Under the same turbulence, however, the ghost images behaves differently, neither its spatial resolution nor its contrast were affected by the turbulence.

The turbulence-free ghost imaging is the result of the turbulence-free non-factorizable point-to-point image-forming correlation, which is caused by two-photon interference: superposition between two different yet indistinguishable alternative ways for a randomly created and randomly paired photons to lead a join photodetection event. We give a simple analysis in the following, staring from

$$G^{(2)}(\vec{\rho}_1, z_1; \vec{\rho}_2, z_2)$$
$$= \int d\vec{\kappa}\, d\vec{\kappa}' \left| \frac{1}{\sqrt{2}} \left[g_2(\vec{\rho}_2, z_2; \vec{\kappa}) g_1(\vec{\rho}_1, z_1; \vec{\kappa}') + g_2(\vec{\rho}_2, z_2; \vec{\kappa}') g_1(\vec{\rho}_1, z_1; \vec{\kappa}) \right] \right|^2$$

which indicates an interference between two quantum amplitudes, corresponding to two alternatives, different yet indistinguishable, which leads to a joint photodetection event. This interference involve both arms of the optical setup as well as two distant photodetection events at $(\vec{\rho}_1, z_1)$ and $(\vec{\rho}_2, z_2)$, respectively.

Optical turbulence is defined as a random change in index of refraction due to changes in composition or density of the propagation medium. These changes are related to the index of refraction along the propagation path, such that $n_j(r) = n_0 + \delta n_j(r)$, $j = 1, 2$, where r is the coordinate along the optical path. To more efficiently model this change in index of refraction, we introduce a random phase shift which is dependent on the change in index of refraction along the optical path,

$$\delta \varphi_j = \frac{2\pi}{\lambda} \int dr_j \delta n_j(r_j).$$

We thus introduce an arbitrary phase disturbance $e^{i\delta\varphi_1}$ into the ghost image arm and another phase disturbance $e^{i\delta\varphi_2}$ into the bucket-detector arm, respectively, to simulate the turbulences.

The spatial part of the second-order coherence with turbulences turns to be

$$G^{(2)}_{\text{Turb}}(\vec{\rho}_1, z_1; \vec{\rho}_2, z_2) \tag{12.4.1}$$
$$= \int d\vec{\kappa}\, d\vec{\kappa}' \left| \frac{1}{\sqrt{2}} \left[g_2(\vec{\rho}_2, z_2; \vec{\kappa}) e^{i\delta\varphi_2} g_1(\vec{\rho}_1, z_1; \vec{\kappa}') e^{i\delta\varphi_1} \right. \right.$$
$$\left. \left. + g_2(\vec{\rho}_2, z_2; \vec{\kappa}') e^{i\delta\varphi_2} g_1(\vec{\rho}_1, z_1; \vec{\kappa}) e^{i\delta\varphi_1} \right] \right|^2$$
$$= G^{(2)}(\vec{\rho}_1, z_1; \vec{\rho}_2, z_2).$$

It is easy to see that the phase turbulence has a null effect on the second-order correlation function $G^{(2)}(\vec{\rho}_1, z_1; \vec{\rho}_2, z_2)$. The normalized non-factorizable point-to-point image-forming correlation $g^{(2)}(\vec{\rho}_1; \vec{\rho}_2)$ of thermal field is thus turbulence-free. The joint photodetection counting rate between the bucket detector and the CCD array will therefore reproduce the aperture function as a turbulence-free ghost image.

$$\langle n(\vec{\rho}_1, z_1)\, n(\vec{\rho}_1, z_1) \rangle$$
$$= \langle n(\vec{\rho}_1, z_1) \rangle \langle n(\vec{\rho}_2, z_2) \rangle + \langle \Delta n(\vec{\rho}_1, z_1) \rangle \langle \Delta n(\vec{\rho}_2, z_2) \rangle$$
$$\propto G^{(2)}_{\text{Turb}}(\vec{\rho}_1, z_1; \vec{\rho}_2, z_2) \simeq n_{c0} + \left| A(\vec{\rho}_1) \right|^2$$

where n_{c0} is a constant and $A(\vec{\rho}_1)$ is the aperture function of the object.

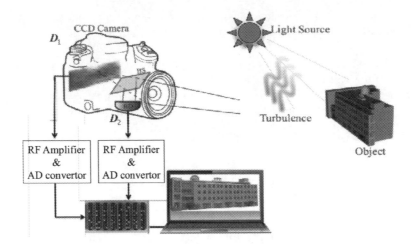

FIGURE 12.4.2 Turbulence-free camera: an image of the target object, which is under the influence of atmospheric turbulence, is produced from the photon number fluctuation correlation measurement $\langle \Delta n(\vec{\rho}_{i1}) \Delta n_2 \rangle$. The classic image observed from $\langle n(\vec{\rho}_{i1}) \rangle$ is completely "blurred" due to the influence of the atmospheric turbulence. However, a turbulence-free image is observed from the measurement of $\langle \Delta n(\vec{\rho}_{i1}) \Delta n_2 \rangle$, i.e., any atmospheric density, refraction index, or phase variations do not have any influence on this image. In this setup the turbulence may appear either in the optical paths between the camera and the object or in the optical paths between the object and the light source, or appear in both.

It should be emphasized that classical simulations of ghost imaging cannot obtain turbulence-free images. For instance, it is easy to see that the man-made factorizable speckle-to-speckle correlation

$$\gamma^{(2)}(\vec{\rho}_o; \vec{\rho}_i) \simeq \delta(\vec{\rho}_s - \vec{\rho}_o/m)\delta(\vec{\rho}_s - \vec{\rho}_i/m),$$

which is made by classically imaging the speckles of the common light source onto the object and image planes (with magnification factor m), will be "blurred" under turbulence, because each classical image of the speckle will be "blurred" independently.

In a realistic application, such as sunlight satellite imaging, it is uneasy to place a beamsplitter to produce ghost image. In fact, the real application of the turbulence-free ghost imaging of Meyers *et al.* does not have a beamsplitter (BS in Fig. 12.4.1). In that case, the lensless ghost image is right on top of the object. We may name it "ghost camera". It should be emphasized, in a real application in whcih natural light sources are used to produce ghost image, the non-perfect second-order temporal coherence has to be taken into account. Short pulse gated fast CCD is a solution.

Applying the same mechanism, a turbulence-free camera has been demonstrated recently. The turbulence-free camera is schematically illustrated in Fig. 12.4.2. It looks like a classic CCD (CMOS) camera, except (1) the image is divided into two paths, path-1 and path-2, by an optical beamsplitter in the camera. A photon counting CCD (CMOS) array, D_1, and an integrated bucket photodetector, D_2, respectively, are placed on the image plane and on the focal plane of the camera lens; (2) D_1 and D_2 measure the photon number fluctuations, $\Delta n(\vec{\rho}_{i1})$ and $\Delta n_2 = \int d\vec{\rho}_{i2} \, \Delta n(\vec{\rho}_{i2})$, where $\vec{\rho}_{i1}$ and $\vec{\rho}_{i2}$ label the transverse coordinates of the image planes of path-1 and path-2 of the camera, respectively. The photon number fluctuations $\Delta n(\vec{\rho}_{i1})$ and Δn_2 are counted, respectively, and calculated, jointly, by a novel PNFC circuit to obtain the photon number fluctuation correlation $\langle \Delta n(\vec{\rho}_{i1}) \Delta n_2 \rangle$. This circuit is especially useful for satellite imaging in which the complicated lengthy statistical

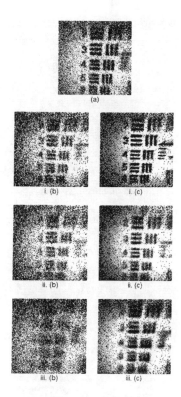

FIGURE 12.4.3 Experimental testing of a turbulence-free camera: turbulence-free image of group 0 of a 1951 USAF Resolution Testing Gauge from a demo-unit of a turbulence-free camera. (a) shows the clear classical image in the measurement of $\langle n(\vec{\rho}_{i1})\rangle$ without atmospheric turbulence; (b) shows the "blurred" first-order classic image in the presence of atmospheric turbulence. (c) shows the image in the measurement of $\langle \Delta n(\vec{\rho}_{i1})\Delta n_2 \rangle$. In (b) and (c), i labels weak turbulence, ii labels medium turbulence, and iii labels strong turbulence. In this measurement, the incoherent light source was a 6.4 mm diameter 3200 Kelvin Tungsten-Halogen white light lamp. The atmospheric turbulence between the object and the camera was simulated by a propane camp stove with variable heat settings. The level of the turbulence can be easily adjusted by varying the heat settings. Due to the use of white light thermal source, the contral of the "exposure" time of the CCD (CMOS) is critical.

calculation can be performed on the ground. The hardware located in the satellite records two sets of data only: the registration time of each photodetection events for each photo-elements (pixels) of D_1 (located at $\vec{\rho}_{i1}$) and for the bucket detector of D_2.

Fig. 12.4.3 reports a set of preliminary results of turbulence-free camera demo-unit. The camera is imaged at the group 0 of a 1951 USAF Resolution Testing Gauge and is placed 2m away from the 0.5 mm lines (0 group of the gauge). The top image (a) shows the clear classical image in the measurement of $\langle n_1(\vec{\rho}_{i1})\rangle$ without atmospheric turbulence; Column (b) shows the "blurred" first-order classic image in the presence of atmospheric turbulence. Column (c) shows the image produced by the turbulence-free camera. In columns (b) and (c), row i labels weak turbulence, row ii labels medium turbulence, row iii labels strong turbulence. It is clear that although the first-order classic images in $\langle n(\vec{\rho}_{i1})\rangle$ is "blurred" out in the presence of significant turbulence, the image in $\langle \Delta n(\vec{\rho}_{i1})\Delta n_2 \rangle$ is always an improvement.

In fact, the setup of the turbulence-free camera in Fig. 12.4.2 is the same as the ghost imaging setup of Fig. 12.3.6, except the primary lensless ghost image plane is right on top of the object plane, i.e., on top of the building of Fig. 12.4.2. The secondary ghost image of the primary lensless ghost image, which is observed from the measurement of $\langle \Delta n(\vec{\rho}_{i1}) \Delta n_2 \rangle$, is captured and resolved by the photon counting CCD (CMOS), D_1, while the bucket detector D_2 collects and integrates the scattered and reflected light from the object (the building). In addition to the secondary ghost image, the CCD (CMOS), D_1, also captures the first-order classic image of the object in its measurement of $\langle n(\vec{\rho}_{i1}) \rangle$, which may be partially or completely "blurred out" by the atmospheric turbulences.

The calculation of the image observed from the ghost camera is similar to our earlier calculation of Eq. (12.3.7), except more Green's functions are involved. Examine Figure 12.4.2, we find three different ways to reproduce a *turbulence-free* ghost image.

(1) The aperture function is in path-2 of the camera.

In this case, we assume the aperture function is in optical path-2 leading to the bucket detector and the primary lensless ghost image is in optical path-1 leading to the CCD array.

$$\langle \Delta n(\vec{\rho}_1, z_1) \Delta n_2 \rangle$$

$$\propto \left| \int d\vec{\rho}_s \left[g_{\vec{\rho}_s}(\omega; \vec{\rho}_1, z_1) \right] \left[\int d\vec{\rho}_{i2} A(\vec{\rho}_o) g_{\vec{\rho}_s}^*(\omega; \vec{\rho}_{i2}, z_{i2}) \right] \right|^2$$

$$= \left| \int d\vec{\rho}_s \left\{ \left[g_{\vec{\rho}_s}(\omega; \vec{\rho}_{gi}, z_{gi}) \right] \left[\int d\vec{\rho}_{gi} g_{\vec{\rho}_{gi}}(\omega; \vec{\rho}_L, z_L) \right] \right.$$

$$\times \left[g_{Lens} \right] \left[\int d\vec{\rho}_L g_L(\omega; \vec{\rho}_1, z_1) \right] \right\} \left\{ \int d\vec{\rho}_{i2} \left[g_{\vec{\rho}_s}^*(\omega; \vec{\rho}_o, z_o) \right] \right.$$

$$\times \left[\int d\vec{\rho}_o A^*(\vec{\rho}_o) g_{\vec{\rho}_o}^*(\omega; \vec{\rho}_L, z_L) \right] \left[g_{Lens}^* \right] \left[\int d\vec{\rho}_L g_L^*(\omega; \vec{\rho}_{i2}, z_{i2}) \right] \right\} \Bigg|^2$$

$$\simeq \left| \int d\vec{\rho}_{gi} \int d\vec{\rho}_o A^*(\vec{\rho}_o) \, \delta(|\vec{\rho}_o - \vec{\rho}_{GI}|) \, \text{somb} \frac{\pi D}{s_o \lambda} |\vec{\rho}_{gi} - \vec{\rho}_1/\mu| \right.$$

$$\times \left[\int d\vec{\rho}_{i2} \, \text{somb} \frac{\pi D}{s_o \lambda} |\vec{\rho}_o - \vec{\rho}_{i2}/\mu| \right] \Bigg|^2$$

$$\propto \left| \int d\vec{\rho}_o A^*(\vec{\rho}_o) \, \text{somb} \frac{\pi D}{s_o \lambda} |\vec{\rho}_o - \vec{\rho}_1/\mu| \right|^2 \tag{12.4.2}$$

where $g_{\vec{\rho}_s}(\omega; \vec{\rho}_{gi}, z_{GI})$, $g_{\vec{\rho}_{gi}}(\omega; \vec{\rho}_L, z_L)$, g_{Lens}, and $g_{\vec{\rho}_L}(\omega; \vec{\rho}_1, z_1)$ are the Green's functions along path-1that that propagate the field from the source plane to the one-to-one primary ghost image plane, from the primary ghost image plane to the lens plane, from the input plane of the lens to the output plane of the lens, and from the lens plane to the secondary ghost image plane, respectively; $g_{\vec{\rho}_s}^*(\omega; \vec{\rho}_o, z_o)$, $g_{\vec{\rho}_o}^*(\omega; \vec{\rho}_L, z_L)$, g_{Lens}^*, and $g_{\vec{\rho}_L}^*(\omega; \vec{\rho}_{i2}, z_{i2})$ are the Green's functions along path-2 that propagate the field from the source plane to the object plane, from the object plane to the lens plane, from the input plane of the lens to the output plane of the lens, and from the lens plane to the image plane, respectively. In Eq. (12.3.7), D is the diameter of the lens, s_o is the distance from the primary ghost image to the lens, s_i is the distance from the lens to the secondary ghost image, satisfying the Gaussian thin lens equation $1/s_o + 1/s_i = 1/f$, where f is the focal length of the lens and $\mu = s_i/s_o$ is the magnification factor of the secondary ghost image. It is easy to find that the secondary ghost image is the result of a convolution between the aperture function $A(\vec{\rho}_o)$ and a two-photon diffraction limited sombrero-like image-forming function.

(2) The aperture function is in path-1 of the camera.

In this case, we assume the aperture function is in optical path-1 leading to the CCD array and the primary lensless ghost image is in optical path-2 leading to the buck detector.

$$\langle \Delta n(\vec{\rho}_1, z_1) \Delta n_2 \rangle$$

$$\propto \left| \int d\vec{\rho}_s \left[A(\vec{\rho}_o) \, g_{\vec{\rho}_s}(\omega; \vec{\rho}_1, z_1) \right] \left[\int d\vec{\rho}_{i2} \, g_{\vec{\rho}_s}^*(\omega; \vec{\rho}_{i2}, z_{i2}) \right] \right|^2$$

$$= \left| \int d\vec{\rho}_s \left\{ \left[A(\vec{\rho}_o) \, g_{\vec{\rho}_s}(\omega; \vec{\rho}_o, z_o) \right] \left[\int d\vec{\rho}_o \, g_{\vec{\rho}_o}(\omega; \vec{\rho}_L, z_L) \right] \right. \right.$$

$$\times \left[g_{Lens} \right] \left[\int d\vec{\rho}_L \, g_L(\omega; \vec{\rho}_1, z_1) \right] \right\} \left\{ \int d\vec{\rho}_{i2} \left[g_{\vec{\rho}_s}^*(\omega; \vec{\rho}_{gi}, z_{GI}) \right] \right.$$

$$\times \left[\int d\vec{\rho}_{gi} \, g_{\vec{\rho}_{gi}}^*(\omega; \vec{\rho}_L, z_L) \right] \left[g_{Lens}^* \right] \left[\int d\vec{\rho}_L \, g_L^*(\omega; \vec{\rho}_{i2}, z_{i2}) \right] \right\} \Big|^2$$

$$\simeq \left| \int d\vec{\rho}_{gi} \int d\vec{\rho}_o \, A(\vec{\rho}_o) \, \delta(|\vec{\rho}_o - \vec{\rho}_{GI}|) \, \mathrm{somb} \frac{\pi D}{s_o \lambda} |\vec{\rho}_o - \vec{\rho}_1/\mu| \right.$$

$$\times \left[\int d\vec{\rho}_{i2} \, \mathrm{somb} \frac{\pi D}{s_o \lambda} |\vec{\rho}_{gi} - \vec{\rho}_{i2}/\mu|] \right|^2$$

$$\propto \left| \int d\vec{\rho}_o \, A(\vec{\rho}_o) \, \mathrm{somb} \frac{\pi D}{s_o \lambda} |\vec{\rho}_o - \vec{\rho}_1/\mu| \right|^2 \tag{12.4.3}$$

(3) The aperture function is in both-path.

In this case, we assume the aperture function is in both optical path-1 and path-2, so that the CCD and the bucket detector both "see" the object directly.

$$\langle \Delta n(\vec{\rho}_1, z_1) \Delta n_2 \rangle$$

$$\propto \left| \int d\vec{\rho}_s \left[A(\vec{\rho}_o) \, g_{\vec{\rho}_s}(\omega; \vec{\rho}_1, z_1) \right] \left[\int d\vec{\rho}_{i2} \, A^*(\vec{\rho}'_o) \, g_{\vec{\rho}_s}^*(\omega; \vec{\rho}_{i2}, z_{i2}) \right] \right|^2$$

$$= \left| \int d\vec{\rho}_s \left\{ \left[A(\vec{\rho}_o) \, g_{\vec{\rho}_s}(\omega; \vec{\rho}_o, z_o) \right] \left[\int d\vec{\rho}_o \, g_{\vec{\rho}_o}(\omega; \vec{\rho}_L, z_L) \right] \right. \right.$$

$$\times \left[g_{Lens} \right] \left[\int d\vec{\rho}_L \, g_L(\omega; \vec{\rho}_1, z_1) \right] \right\} \left\{ \int d\vec{\rho}_{i2} \left[A^*(\vec{\rho}'_o) \, g_{\vec{\rho}_s}^*(\omega; \vec{\rho}_{\vec{\rho}'_o}, z_{\vec{\rho}'_o}) \right] \right.$$

$$\times \left[\int d\vec{\rho}_{\vec{\rho}'_o} \, g_{\vec{\rho}_{\vec{\rho}'_o}}^*(\omega; \vec{\rho}_L, z_L) \right] \left[g_{Lens}^* \right] \left[\int d\vec{\rho}_L \, g_L^*(\omega; \vec{\rho}_{i2}, z_{i2}) \right] \right\} \Big|^2$$

$$\simeq \left| \int d\vec{\rho}_{\vec{\rho}'_o} \int d\vec{\rho}_o \, A(\vec{\rho}_o) \, A^*(\vec{\rho}'_o) \, \delta(|\vec{\rho}_o - \vec{\rho}'_o|) \, \mathrm{somb} \frac{\pi D}{s_o \lambda} |\vec{\rho}_o - \vec{\rho}_1/\mu| \right.$$

$$\times \left[\int d\vec{\rho}_{i2} \, \mathrm{somb} \frac{\pi D}{s_o \lambda} |\vec{\rho}'_o - \vec{\rho}_{i2}/\mu|] \right|^2$$

$$\propto \left| \int d\vec{\rho}_o \, |A(\vec{\rho}_o)|^2 \, \mathrm{somb} \frac{\pi D}{s_o \lambda} |\vec{\rho}_o - \vec{\rho}_1/\mu| \right|^2 \tag{12.4.4}$$

In fact, our first demonstration of *turbulence-free* camera started from a slightly different setup, in which the bucket photodetector was replaced by a carefully aligned identical CCD in path-1 of the camera. The calculation of the image is the same as above, except the measurement is for each pair of two pixels of CCD-1 and CCD-2, $\langle \Delta n(\vec{\rho}_1, z_1) \Delta n(\vec{\rho}_2, z_2) \rangle$. We may consider three similar cases in the calculation. The following is a similar calculation for case (3): assuming the aperture function is in both-path:

$$
\langle \Delta n(\vec{\rho}_1, z_1) \Delta n(\vec{\rho}_2, z_2) \rangle
$$

$$
\propto \left| \int d\vec{\rho}_s \left[A(\vec{\rho}_o)\, g_{\vec{\rho}_s}(\omega; \vec{\rho}_1, z_1) \right] \left[A^*(\vec{\rho}'_o)\, g^*_{\vec{\rho}_s}(\omega; \vec{\rho}_2, z_2) \right] \right|^2
$$

$$
= \left| \int d\vec{\rho}_s \left\{ \left[A(\vec{\rho}_o)\, g_{\vec{\rho}_s}(\omega; \vec{\rho}_o, z_o) \right] \left[\int d\vec{\rho}_o\, g_{\vec{\rho}_o}(\omega; \vec{\rho}_L, z_L) \right] \right. \right.
$$

$$
\times \left[g_{Lens} \right] \left[\int d\vec{\rho}_L\, g_L(\omega; \vec{\rho}_1, z_1) \right] \bigg\} \left\{ \left[A^*(\vec{\rho}'_o)\, g^*_{\vec{\rho}_s}(\omega; \vec{\rho}_{\vec{\rho}'_o}, z_{\vec{\rho}'_o}) \right] \right.
$$

$$
\times \left. \left[\int d\vec{\rho}_{\vec{\rho}'_o}\, g^*_{\vec{\rho}_{\vec{\rho}'_o}}(\omega; \vec{\rho}_L, z_L) \right] \left[g^*_{Lens} \right] \left[\int d\vec{\rho}_L\, g^*_L(\omega; \vec{\rho}_2, z_2) \right] \right\} \right|^2
$$

$$
\simeq \left| \int d\vec{\rho}_{\vec{\rho}'_o} \int d\vec{\rho}_o\, A(\vec{\rho}_o)\, A^*(\vec{\rho}'_o)\, \delta(|\vec{\rho}_o - \vec{\rho}'_o|)\, \mathrm{somb}\frac{\pi D}{s_o \lambda}|\vec{\rho}_o - \vec{\rho}_1/\mu| \right.
$$

$$
\times \left. \left[\mathrm{somb}\frac{\pi D}{s_o \lambda}|\vec{\rho}'_o - \vec{\rho}_2/\mu| \right] \right|^2 \tag{12.4.5}
$$

$$
\propto \left| \int d\vec{\rho}_o\, |A(\vec{\rho}_o)|^2 \left[\mathrm{somb}\frac{\pi D}{s_o \lambda}|\vec{\rho}_o - \vec{\rho}_1/\mu| \right] \left[\mathrm{somb}\frac{\pi D}{s_o \lambda}|\vec{\rho}_o - \vec{\rho}_2/\mu| \right] \right|^2.
$$

If we force $\vec{\rho}_1 = \vec{\rho}_2$ by electronic circuit, either hardware or software, Eq. (12.4.5) becomes

$$
\langle \Delta n(\vec{\rho}_1, z_1) \Delta n((\vec{\rho}_2, z_2)) \rangle
$$

$$
\propto \left| \int d\vec{\rho}_o\, |A(\vec{\rho}_o)|^2\, \mathrm{somb}^2\frac{\pi D}{s_o \lambda}|\vec{\rho}_o - \vec{\rho}_1/\mu| \right|^2
$$

$$
= \left| \int d\vec{\rho}_o\, |A(\vec{\rho}_o)|^2\, \mathrm{somb}^2\frac{\pi D}{s_o \lambda}|\vec{\rho}_o - \vec{\rho}_2/\mu| \right|^2. \tag{12.4.6}
$$

Two identical turbulence-free images are reproduced in the photon number fluctuation correlation measurement, $\langle \Delta n(\vec{\rho}_1, z_1) \Delta n(\vec{\rho}_2, z_2) \rangle$, of the two CCDs, even if the first-order classic images viewed by the two CCDs were completely blurred by the strong atmospheric turbulences.

It should be emphasized that for easy understanding of the working mechanism, the above calculation is focused on second-order spatial correlation by assuming perfect second-order temporal coherence. When dealing with a broad spectrum thermal field, such as the sunlight, (1) non-perfect second-order temporal coherence has to be taken into account: higher degree of second-order temporal coherence must be achieved experimentally; (2) the time average of slow photodetectors and correlation circuit must also be taken into account. We have given a solution (using short pulsed thermal field) in section 8.7. Although it is impossible to have a short pulsed sunlight, we can definitely "gate" our photodetectors to "expose" D_1 and D_2 within a short time window. With the help of a specially designed measurement circuit to distinguish $\langle \Delta n_1 \Delta n_2 \rangle$ from $\langle n_1 \rangle \langle n_2 \rangle$, ghost images of natural light have been observed from these turbulence-free cameras.

12.5 X-RAY GHOST MICROSCOPE

In classic imaging setups, focusing optics such as imaging lenses play a critical role in producing a diffraction-limited point-to-point relation between the object plane and the image plane, forming a magnified or demagnified image of the object. If desired, an additional lens system is then able to map the primary image onto a secondary image plane for further magnification or demagnification; notably making an optical microscope possible. Such optical microscopes are commonly used to obtain a high-resolution image a detailed surface structure of an object. Unlike visible light, X rays can pass through many materials allowing X-ray imaging devices to image the internal structure of an object. If our expectation is to obtain high-resolution images of the detailed internal structure of an object or material, an X-ray microscope is a necessary but difficult goal. The first difficulty faced was that traditional lenses are not practical to use for high-energy X rays because the refractive index is close to $n = 1$ for high energy X-rays in all known materials. To overcome this, some alternative focusing X-ray optics have been developed in recent years such as compound refractive lenses, focusing mirrors, and zone plates. While zone plates and focusing mirrors typically limited to soft X rays (< 10keV), compound refractive lenses can be designed for > 10keV. Unlike projectional radiography, which is a projection, or "shadow", of the X rays that pass through the object with different materials and thicknesses of materials causing more absorption in some areas compared to others, focusing X-ray optics take full advantage of the short wavelength of soft X rays with significantly higher resolution than current projectional radiography technology. Unfortunately, there are still many barriers in producing effective focusing devices for hard X rays. Due to this, the X-ray imaging technique most commonly used in practice is still projectional radiography. Even as focusing X-ray optics are made more readily available, it is difficult for a classic imaging device to obtain images of deeper interior structure of an object due to their limited angular resolution. To take advantage of the resolving power of short-wavelength X rays, the object plane should be near the focal point. Often the focal length of a traditional lens and focusing X-ray optics is restricted to a certain value, meaning any structure deeper than this value into the object would be unaccessible at the desired resolution.

Obviously, lensless ghost imaging is right suitable for X-ray imaging: (1) lensless imaging does not need any lens; (2) due to the "ghost" nature of the ghost image, an additional imaging device with limited angular resolution can be placed as close as possible to the selected ghost image plane, which corresponds to a "sliced" internal structure of the object, to "force" the angular separation of close neighboring points large enough to be resolvable. In addition, by scanning different lensless ghost image planes along the optical axis, which correspond to different internal cross section, or "slices", of the object, a set of slices of the internal structure of the object can be grouped together to form a magnified secondary 3-D ghost image of the object.

In this section, we apply the mechanism of two-photon ghost imaging to produce a sub-nanometer resolution, lensless, image-forming correlation between the image plane and object plane, for which X rays allow imaging of the internal structure of the object in 3-D. Through the help of a secondary imaging device, either focusing X-ray optics or a scintillator paired with a visible-light lens system, the primary lensless ghost image in sub-nanometer resolution can be mapped onto a secondary image plane with significant magnification to be resolvable by a standard CCD or CMOS; namely, an X-ray ghost microscope. In principle, once some experimental barriers are overcome, this X-ray "ghost microscope" may achieve sub-nanometer resolution and open up new capabilities that would be of interest to the fields of physics, material science, and medical imaging.

Similar to lensless ghost imaging in the visible spectrum, the X-ray lensless image-forming correlation is the result of two-photon interference: two randomly created and randomly paired X-ray photons interfering with the pair itself. It has been proved in early chapters that two-photon interference phenomena can be set up to achieve turbulence-free measurements. This is achieved when the superposed two-photon amplitudes experience the same turbulence and medium vibrations along their optical paths, meaning any composition, density, length, refractive index, or medium vibration induced random phase variations along the optical paths do not have any effect on each individual two-photon interference. The X-ray ghost microscope analyzed in this section is a typical turbulence-free imaging device. The turbulence-free nature is especially important for the extremely high resolution imaging obtainable with the X-ray ghost microscope, as we knew that vibrations would typically cause a blurred classic image.

(1) Primary lensless ghost imaging of X-ray.

To better visualize the X-ray ghost microscope it is best to start with the working mechanism of X-ray lensless ghost imaging. This simple experimental setup, schematically illustrated in Fig. 12.5.1, consists of an X-ray beamsplitter[3] that divides the X-ray beam from a disk-like source into two beams. A 2-D array of X-ray detectors, D_1, is placed in beam-one (transmitted in Fig. 12.5.1) at a selected plane of z_1 in the Fresnel near-field. Following the output of beam-two (reflected in Fig. 12.5.1), we place an object followed by a bucket X-ray photodetector, D_2, which collects all X rays transmitted from the object. This is a typical lensless ghost imaging setup. To calculate the X-ray ghost image forming

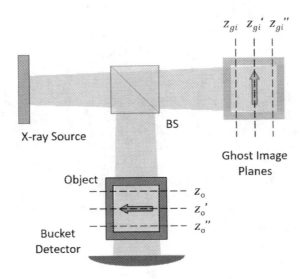

FIGURE 12.5.1 X-ray ghost imaging. A beam splitter (most likely a crystal aligned to utilize Laue diffraction which would not provide 90° separation as depicted), creates two paths for the beam, one directed at the object followed by a bucket detector and the second directed at a pixel array (CCD or CMOS). Note (1): The detectors could directly detect X rays or be paired with a scintillator to convert X-ray to visible rays. Note (2): Different ghost image plane, at distance $z = d$, $z = d'$, $z = d'' \dots$, corresponds to different "slice", or cross section, of $z = d$, $z = d'$, $z = d'' \dots$ inside the object.

correlation we apply the Fresnel near field propagator, or Green's function, to propagate the field from one space-time location, $(\vec{\rho}_m, z_m)$, to another space-time location at $(\vec{\rho}_j, z_j)$[4]

$$g_m(\vec{\rho}_j, z_j) = \frac{c_0}{|z_j - z_m|} e^{-i\omega\tau_j} e^{i\frac{\omega}{2c|z_j - z_m|}|\vec{\rho}_j - \vec{\rho}_m|^2}, \qquad (12.5.1)$$

where c_0 is a normalization constant. For the following discussion, we will assume a

---

[3]This is trivial when using visible light, but becomes more difficult with high-energy X-rays. So far, Laue diffraction of crystals have been used with success, but advancement in kinoform X-ray beam splitters may prove useful in the future

[4]Due to the use of short wavelength X rays, higher-order approximation, or numerical computation, may be necessary instead of a simple Fresnel propagator. However, it is a mathematical conclusion that higher-order approximations will produce "narrower" correlation than that of the Fresnel. To simplify the mathematics, we thus keep the Fresnel approximation in the following "prove-principle discussion".

monochromatic source to simplify the calculation as usual. Assuming a disk-like source and randomly distributed and randomly radiated point-like sub-sources, we can approximate the sum of m into an integral of $\vec{\rho}_s$ on the source plane of $z_s = z_m = 0$. Before considering the detectors, we can look at the object plane on beam-two (reflected), z_o, and the corresponding plane on beam-one (transimitted), which we will label z_{gi}. The result of X-ray intensity fluctuation correlation is

$$\langle \Delta I(\vec{\rho}_{gi}, z_{gi}) \Delta I(\vec{\rho}_o, z_o) \rangle \big|_{z_{gi}=z_o}$$

$$\propto \left\langle \left| \int d\vec{\rho}_s \left[g_s^*(\vec{\rho}_{gi}, z_{gi}) \right] \left[g_s(\vec{\rho}_o, z_o) \right] \right|^2 \right\rangle$$

$$\propto \mathrm{somb}^2 \frac{\pi \Delta \theta_s}{\lambda} |\vec{\rho}_{gi} - \vec{\rho}_o| \tag{12.5.2}$$

where the somb-function is defined as $2J(x)/x$ and $\Delta \theta_s \approx 2R/d$ is the angular diameter of the radiation source[5]. This point-to-spot correlation is similar to that of the original Hunbury Brown-Twiss experiment that started the practice of optical correlation measurements, except it is measured in the Fresnel near-field instead of the Fraunhofer far-field. This point-to-spot correlation has also been demonstrated at X-ray synchrotron sources. Due to the high energy (short wavelength) nature of X-rays, this point-to-spot correlation is much more narrower than that of the visible-light. For instance, a $\sim 0.5 \times 10^{-10}$ meter wavelength (~ 25 keV) X-ray source with angular diameter of $> 10^{-1}$ rad may achieve sub-nanometer correlation. We can approximate this point-to-spot correlation as a point-to-point correlation between $\vec{\rho}_{gi}$ and $\vec{\rho}_o$. This point-to-point correlation, or lensless ghost image forming function, identifies a unique internal plane of the object, $z_{gi} = z_o$, with the X-ray ghost image observed from the X-ray intensity fluctuation, or photon number fluctuation, correlation. Together with $z_{gi} = z_o$, we may approximate the X-ray intensity fluctuation correlation as

$$\langle \Delta I(\vec{\rho}_{gi}, z_{gi}) \Delta I(\vec{\rho}_o, z_o) \rangle \propto \delta(z_{gi} - z_o) \, \delta(|\vec{\rho}_{gi} - \vec{\rho}_o|). \tag{12.5.3}$$

We now calculate the X-ray lensless ghost image by including a 2-D photodetection array on the $z_1 = z_{gi}$ plane and a bucket detector, D_2, which integrates all possible X rays transmitted through each transverse coordinate point $\vec{\rho}_o$ of the object. The internal "aperture function" of an object cannot be simply represented as a 2-D function $A(\vec{\rho}_o, z_o)$ as typically done for visible-light imaging. In visible light imaging, $A(\vec{\rho}_o)$ usually used to represent the surface plane of an object; however, for X-ray ghost imaging is reasonable to model a 3-D aperture function representing the internal structure of an object,

$$A(\vec{\rho}_o) \simeq \int dz_o A(\vec{\rho}_o, z_o). \tag{12.5.4}$$

Assuming perfect temporal correlation is satisfied experimentally, the intensity fluctuation correlation results in

$$\langle \Delta I(\vec{\rho}_1, z_1) \Delta I_2 \rangle$$

$$= \left| \int d\vec{\rho}_o \left[\int dz_o A(\vec{\rho}_o, z_o) \right] \left[\delta(z_1 - z_o) \mathrm{somb} \frac{\pi \Delta \theta_s}{\lambda} |\vec{\rho}_1 - \vec{\rho}_o| \right] \right|^2$$

$$\simeq \left| \int d\vec{\rho}_o \left[\int dz_o A(\vec{\rho}_o, z_o) \right] \left[\delta(z_1 - z_o) \, \delta(|\vec{\rho}_1 - \vec{\rho}_o|) \right] \right|^2$$

$$\simeq \left| A(\vec{\rho}_1 = \vec{\rho}_o, z_1 = z_o) \right|^2, \tag{12.5.5}$$

---

[5]Again, due to the use of short wavelength X rays, higher-order approximation or numeral computation may be necessary. We will have a much narrower somb-function instead of $2J(x)/x$.

indicating the reproduction of an 2-D X-ray ghost image of the internal transverse cross section of $z_o = z_1$, i.e. z_{gi}. In other words, the longitudinal position of the 2-D X-ray detector array, z_1, selected an object plane, $z_o = z_1$, of the internal structure of the object to image. When the 2-D X-ray detector array is scanned from z_1 to z_1' to z_1'' along the optical axis, the selected object plane will be changed from $z_o = z_1$ to $z_o' = z_1'$ to $z_o'' = z_1''$, respectively, as illustrated in Fig. 12.3.5. By scanning the position of the 2-D X-ray detector array, a set of slices of the internal structure of the object can be grouped together to form a 3-D ghost image of the object with sub-nanometer resolution. This differs from traditional X-ray computerized tomography (CT) imaging and ghost tomography (GT) demonstrated by Kingston *et al.* which rely on rotating the object or revolving the detectors around the object 360°.

Unfortunately, the sub-nanometer resolution lensless ghost image is unresolvable by any state-of-the-art 2-D photodetector array, such as CCD or CMOS sensors unless the photodetector array technology ever advances to sub-nanometer-sized pixels. In order to resolve the two-photon X-ray ghost image with a standard 2-D photodetector array, which may have micrometer sized pixels, significant magnification is necessary.

(2) Secondary ghost image reproduced by an X-ray lens system.

One solution would be to magnify the primary X-ray ghost image to a secondary ghost image plane with X-ray lens system, such as X-ray zone plates, compound refractive lenses, or other types of X-ray lenses. Note that X-ray optics may limit the usable energy levels of the X-ray source unless the technology advances to accommodate higher energy levels. The schematic design of this type X-ray ghost microscope is shown in Fig. 12.5.2. A magnified 3-D secondary ghost image is expected from the measurement of the X-ray intensity fluctuation correlation between the bucket X-ray detector, which is placed behind the object, and the 2-D array of X-ray detecters, which is placed in the secondary ghost image plane.

To confirm the nonlocal point-to-point correlation between the object plane, z_o, and the secondary image plane, z_i (where D_1 will be placed, $z_1 = z_i$), we adjust Eq. (12.5.2) to include the lens system,

$$\langle \Delta I(\vec{\rho}_i, z_i) \Delta I(\vec{\rho}_o, z_o) \rangle$$

$$\propto \Big| \int d\vec{\rho}_s \, g_s^*(\vec{\rho}_o, z_o) g_s(\vec{\rho}_i, z_i) \Big|^2$$

$$= \Big| \int d\vec{\rho}_s \, g_{\vec{\rho}_s}^*(\vec{\rho}_o, z_o) \Big\{ g_s(\vec{\rho}_{gi}, z_{gi})$$

$$\times \Big[\int d\vec{\rho}_{gi} \, g_{gi}(\vec{\rho}_L, z_L) \Big] [g_{Lens}] \Big[\int d\vec{\rho}_L \, g_L(\vec{\rho}_i, z_i) \Big] \Big\} \Big|^2, \tag{12.5.6}$$

where $g_s(\vec{\rho}_{gi}, z_{gi})$, $g_{gi}(\vec{\rho}_L, z_L)$, g_{Lens}, and $g_L(\vec{\rho}_i, z_i)$ are the Gaussian propagator, or Green's functions propagating the field from the source plane to the one-to-one primary ghost image plane, from the primary ghost image plane to the lens plane, from the input plane of the lens to the output plane of the lens, and from the lens plane to the secondary ghost image plane, respectively.

$$\langle \Delta I(\vec{\rho}_i, z_i) \Delta I(\vec{\rho}_o, z_o) \rangle$$

$$\propto \Big\langle \Big| \int d\vec{\rho}_{gi} \, \delta(|\vec{\rho}_{gi} - \vec{\rho}_o|) \operatorname{somb} \frac{\pi D}{s_o \lambda} |\vec{\rho}_{gi} - \vec{\rho}_i/\mu| \Big|^2 \Big\rangle$$

$$\propto \operatorname{somb}^2 \frac{\pi D}{s_o \lambda} |\vec{\rho}_o - \vec{\rho}_i/\mu|, \tag{12.5.7}$$

indicating a point-to-spot correlation between the selected "slice", or internal cross section,

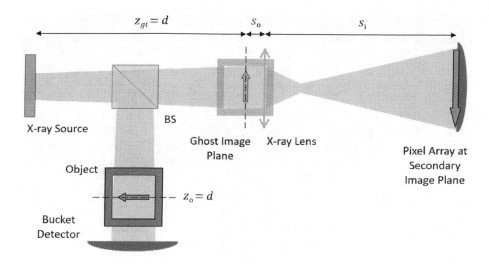

FIGURE 12.5.2 Schematic of a X-ray ghost microscope using X-ray lens system. A beam splitter (most likely a crystal aligned to utilize Laue diffraction), creates two paths for the beam, one directed at the object substance and the second directed at the primary ghost image plane. An X-ray lens system is placed behind the primary X-ray ghost image to reproduce a significantly magnified secondary ghost image that is resolvable by a standard CCD or CMOS. Note: due to the "ghost" nature of the primary ghost image, s_o of the microscope can be placed as close as possible to the primary X-ray ghost image to "force" the angular separation of nanometer scale inside the object greater than the angular resolution of the secondary imaging system; $\Delta\theta_{min} \simeq 1.22\,\lambda/D$ is the angular resolution of the secondary imaging device, where λ is the wavelength of the X-ray and D is the effective diameter of the X-ray lens.

of the object and the secondary X-ray ghost image plane, namely the secondary X-ray ghost image forming function. In Eq.(12.5.7), D is the diameter of the lens, s_o is the distance from the primary ghost image to the lens, s_i is the distance from the lens to the secondary ghost image, satisfying the Gaussian thin lens equation $1/s_o + 1/s_i = 1/f$, where f is the focal length of the lens and $\mu = s_i/s_o$ is the magnification factor of the secondary ghost image. The 2-D detector array, D_1, is now placed on the image plane of the X-ray lens (secondary ghost image plane), $z_1 = z_i$. The position of D_1 defines the position of z_{gi} and thus defines the "slice" at $z_o = z_{gi}$ of the internal cross section of the object by means of the Gaussian thin-lens equation. We then take into account the complex aperture function $\tilde{A}(\vec{\rho}_o, z_o = z_{gi})$ and the bucket detector D_2 into the calculation. A magnified secondary X-ray ghost image of the aperture function is then observed from the joint detection between the CCD (CMOS), D_1, and the bucket detector D_2,

$$
\begin{aligned}
&\langle \Delta I(\vec{\rho}_1, z_1)\Delta I_2 \rangle \\
&\propto \left\langle \left| \int d\vec{\rho}_o\, \tilde{A}(\vec{\rho}_o, z_o = z_{gi})\,\mathrm{somb}\frac{\pi D}{s_o\lambda}|\vec{\rho}_o - \vec{\rho}_1/\mu| \right|^2 \right\rangle \\
&\propto \int d\vec{\rho}_o\, |\tilde{A}(\vec{\rho}_o, z_o = z_{gi})|^2\, \mathrm{somb}^2\frac{\pi D}{s_o\lambda}|\vec{\rho}_o - \vec{\rho}_1/\mu| \\
&\simeq |\tilde{A}(\vec{\rho}_1/\mu, z_o = z_{gi})|^2.
\end{aligned}
\tag{12.5.8}
$$

Scanning D_1 from one position to another along the optical axis, or refocusing the microscope from one ghost image plane of z_{gi} to z'_{gi} etc., we obtain a magnified 3-D ghost image of the internal structure of the object.

Due to the high resolution of the primary ghost image, the result of the magnified secondary ghost image is identical to if an image from a classic X-ray microscope was obtained with the focusing X-ray optics. However, one major benefit to the ghost microscope setup is that the X-ray optics are imaging the ghost image plane, which isn't physically present, as opposed to imaging the physical object directly (hence why it is called "ghost" imaging). With an angular resolution limited by Rayleigh's criterion, the best spatial resolution one can obtain is by putting the object plane near the focal plane (with the potential for nanometer resolution). Imaging some objects with a classic X-ray microscope may not be an issue, but often it may be desirable to image the deeper internal structure (e.g. a bone) of a longer object (e.g. a body with greater longitudinal dimension) for which the focusing X-ray optics cannot physically be close enough to the internal structure to image it with proper magnification and resolution. This would not be an issue for the X-ray ghost microscope as the physical object is placed "nonlocally" on the other path following the beam splitter and is simply followed by a bucket detector.

Considering a CCD (CMOS) with pixel size of $\sim 10^{-6}$m, it would be desirable to use a microscope of $\mu > 1000$ to observe nanometer resolution primary ghost image. For X rays of $\lambda \sim 10^{-10}$m, the angular resolution of a millimeter diameter lens may achieve 10^{-7}rad, which is able to resolve a ghost image with 10^{-9}m spatial resolution, because $s_o < 10^{-2}$m to the primary ghost image plane is easily achievable to force the angular separation of nanometer scale, $\Delta\theta = 10^{-9}/s_o > 10^{-7}$rad.

(3) Secondary ghost image reproduced by scintillator-visible-light-lens assembly.

If the use of X-ray optics is not possible or not preferred, one can place a scintillator on the primary ghost image plane, z_{gi} to convert it into the visible spectrum, allowing it to then be magnified by a visible-light lens system onto a secondary image plane. This design is suitable for higher energy X-ray imaging, such as ≥ 20 keV. The schematic design of this type X-ray microscope is shown in Fig. 12.5.3. The setup is similar to that of Fig. 12.5.2, except a scintillator is placed on the primary X-ray ghost image plane to convert the X-ray ghost image into the visible spectrum. Unlike the X-ray ghost microscope shown in

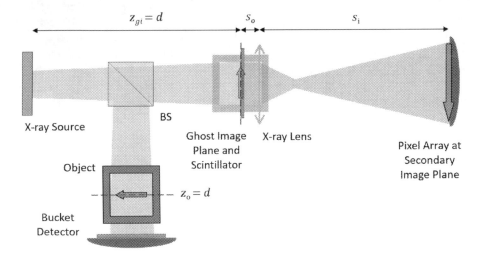

FIGURE 12.5.3 Schematic of a X-ray ghost microscope using scintillator-visible-light-lens assembly to magnify the primary X-ray ghost image. Nearly identical to the setup of Fig. 12.5.2, but now a scintillator is placed on the ghost image plane to convert the X-ray ghost image into the visible spectrum. Now an optical lens (or lens system) of visible-light produces a magnified secondary ghost image.

Fig. 12.5.2, which has a clear path for the two-photon amplitudes from the light source to the detectors, here it is not clear that the scintillator-lens system preserves the result of the two-photon interference, and thus the secondary ghost image. To understand how the secondary ghost image is preserved, we can say that the scintillator essentially "detects" the X rays, thus establishing the presence of the X-ray ghost image on the scintillator plane. Although, prior to correlation, this plane is simply a distribution of quantum speckles from interfering photon pairs. The scintillator converts these X-ray fluctuations to the visible spectrum, $\Delta I_{\mathrm{X}}(\vec{\rho}_{gi}, z_{gi}) \simeq \Delta I_{\mathrm{v}}(\vec{\rho}_{gi}, z_{gi})$. The lens system then images this distribution of fluctuations and produces a diffraction-limited magnified image of them on the secondary ghost image plane. The intensity fluctuation detected by the scintillator is the result of two-photon interference. The converted visible light from the scintillator then passes through a lens and is measured by D_1 on the secondary ghost image plane, $z_1 = z_i$

$$
\begin{aligned}
\Delta I(\vec{\rho}_i, z_i) &= \sum_{m \neq n} E_m^*(\vec{\rho}_i, z_i) E_n(\vec{\rho}_i, z_i) \\
&= \sum_{m \neq n} \int_{lens} d\vec{\rho}_l \, E_m^*(\vec{\rho}_{gi}) \left[g_{\vec{\rho}_{gi}}^*(\vec{\rho}_l, z_l) \right] \left[g_{Lens} \right] \left[g_{\vec{\rho}_l}^*(\vec{\rho}_i, z_i) \right] \\
&\quad \times \int_{lens} d\vec{\rho}'_l \, E_n(\vec{\rho}_{gi}) \left[g_{gi}(\vec{\rho}'_l, z'_l) \right] \left[g_{Lens} \right] \left[g_{l'}(\vec{\rho}_i, z_i) \right] \\
&= \sum_{m \neq n} E_m^*(\vec{\rho}_{gi}) E_n(\vec{\rho}_{gi}) \, \mathrm{somb}^2 \frac{\pi D}{s_o \lambda_{\mathrm{v}}} |\vec{\rho}_o + \vec{\rho}_i / \mu| \\
&= \Delta I(\vec{\rho}_{gi}, z_{gi}) \, \mathrm{somb}^2 \frac{\pi D}{s_o \lambda_{\mathrm{v}}} |\vec{\rho}_o + \vec{\rho}_i / \mu|.
\end{aligned}
\tag{12.5.9}
$$

where λ_{v} is the center wavelength emitted from the scintillator. It is interesting that the X-ray intensity fluctuations at $(\vec{\rho}_{gi}, z_{gi})$ of the primary ghost image plane are "propagated" to an unique point in the secondary image plane $(\vec{\rho}_i, z_i)$, where $(\vec{\rho}_i, z_i)$ is defined by the somb-function and the Gaussian thin-lens equation of the visible-light microscope. A magnified secondary ghost image is observable from the correlation measurement between the intensity fluctuations of the X-ray and the intensity fluctuations of the visible light,

$$
\begin{aligned}
&\langle \Delta I(\vec{\rho}_1, z_1) \Delta I_2 \rangle \\
&\propto \langle \left| \int d\vec{\rho}_o \, \tilde{A}(\vec{\rho}_o, z_o = z_{gi}) \, \mathrm{somb} \frac{\pi D}{s_o \lambda_{\mathrm{v}}} |\vec{\rho}_o - \vec{\rho}_1 / \mu| \right|^2 \rangle \\
&\propto \int d\vec{\rho}_o \, |\tilde{A}(\vec{\rho}_o, z_o = z_{gi})|^2 \, \mathrm{somb}^2 \frac{\pi D}{s_o \lambda_{\mathrm{v}}} |\vec{\rho}_o - \vec{\rho}'_1 / \mu|
\end{aligned}
\tag{12.5.10}
$$

Even with a new dependence on visible light, pairing the scintillator with super-resolving visible-light imaging techniques will allow for nanometer resolution. More often a standard compound microscope will be the best accessible option. Even with limited angular resolution, $\Delta \theta_{min} \simeq 1.22 \, \lambda_{\mathrm{v}}/D$, this setup would still have benefits of standard optical microscope as it allows for imaging deep internal structure of the object with limited angular resolution of the secondary imaging device.

One factor that may aid in the high resolving capabilities of the X-ray ghost microscope is the turbulence-free property of two-photon interference. It has been proved in early sections and chapters that ghost imaging and other two-photon interference phenomena can be set up to achieve turbulence-free measurements. This is achieved when the superposed two-photon amplitudes experience the same turbulence and medium vibrations along their optical paths, meaning any composition, density, length, refractive index, or medium vibration induced random phase variations along the optical paths do not have any effect on each individual two-photon interference and thus the ghost image. This also includes vibrations that would typically cause a blurred classic image.

It should be emphasized that the above calculation is focused on second-order spatial correlation by assuming perfect second-order temporal coherence. When dealing with a broad spectrum thermal field, such as X rays, (1) non-perfect second-order temporal coherence has to be taken into account: higher degree of second-order temporal coherence must be achieved experimentally; (2) the time average of slow photodetectors and correlation circuit must also be taken into account. We have given a solution in section 8.7. A short pulsed tabletop X-ray source is a good solution for X-ray microscope.

12.6 CLASSICAL SIMULATION OF GHOST IMAGING

It is always possible to replace the two-photon interference produced natural, non-factorizeable, point-to-point correlation of thermal light or entangled photon pairs by an artificial correlation made from a radiation source in which the "light knows where to go" when it is prepared at the source. There have been quite a few attempts to simulate the point-to-point ghost image-forming correlation. For instance, one may prepare two identical copies of intensity "speckles" on the object plane and on the image plane, respectively. The speckle-to-speckle correlation plays the role of spot-to-spot image-forming function. The object function $A(\vec{\rho}_{obj})$ is thus reproduced in the coincidence measurement of identical speckles. It should be emphasized: (1) the resolution and the quality of the classically simulated ghost image may not be able to achieve that of the true ghost image produced from the two-photon interference produced natural, non-factorizeable, point-to-point correlation of randomly created and randomly paired photons in thermal state or photon pairs in entangled state; (2) for a known artificial speckle distribution, it is unnecessary to use two photodetectors for joint-detection. One bucket detector is sufficient to reproduce the image of the object.

We briefly discuss three classical simulations in the following.

(I) Correlated laser beams.

In 2002, Bennink *et al.* simulated ghost imaging by two correlated laser beams. The authors intended to show that two correlated corotating laser beams can simulate similar physical effects of entangled photon pairs. Figure 12.6.1 is a schematic picture of the experi-

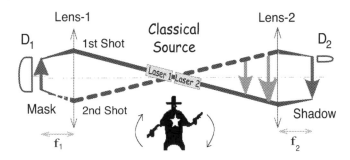

FIGURE 12.6.1 The point-to-point correlation is made shot by shot by two co-rotating laser beams. A ghost shadow can be made in coincidences by "blocking-unblocking" of the correlated laser beams, or simply by "blocking-unblocking" two correlated gun shots.

ment of Bennink *et al.* Different from ghost imaging, here the point-to-point correspondence between the object plane and the "image plane" is made artificially by two co-rotating laser beams "shot by shot". The laser beams propagated in opposite directions and focused on the object and image planes, respectively. If laser beam-1 is blocked by the object mask there would be no joint-detection between D_1 and D_2 for that "shot", while if laser beam-1 is unblocked, a coincidence count will be recorded against that angular position of the co-rotating laser beams. A shadow of the object mask is then reconstructed in coincidences by the blocking-partial blocking-unblocking of laser beam-1.

The point-to-point correlation of Bennink et al. is made shot by shot between "correlated" laser beams, which is not only different from that of ghost imaging, but also different from the standard statistical intensity fluctuation correlations. Nevertheless, the experiment of Bennink et al. obtained a ghost shadow which may be useful for certain purposes. In fact, this experiment can be considered as a good example to distinguish a man-made factorizeable classical intensity-intensity correlation from a natural, non-factorizeable second-order correlation that is caused by nonlocal two-photon interference.

(II) Correlated speckles.

Following a similar philosophy, Gatti et al. proposed a classical correlation between "speckles". The experimental setup of Gatti et al. is depicted in Fig. 12.6.2. Their experiments use either entangled photon pairs of SPDC or pseudo-thermal light for simulating ghost images in coincidences. The "ghost image" observed in coincidences comes from a

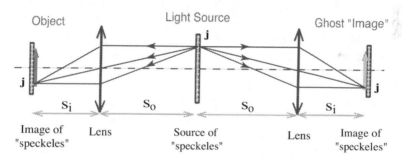

FIGURE 12.6.2 A ghost image is made by a man-made correlation of "speckles". The two identical sets of speckles are the *classical images* of the speckles of the light source. The lens, which may be part of a CCD *camera* used for the joint measurement, reconstructs classical images of the speckles of the source onto the object plane and the image plane, respectively. s_o and s_i satisfy the Gaussian thin lens equation $1/s_o + 1/s_i = 1/f$.

man-made classical speckle-to-speckle correlation. The speckles observed on the object and image planes are the classical images of the speckles of the radiation source, reconstructed by the lenses shown in the figure (the lens may be part of a CCD *camera* used for the joint measurement). Each speckle on the source, such as the jth speckle near the top of the source, has two identical images on the object plane and on the image plane. Mathematically, the speckle-to-speckle correlation is factorizeable into a product of two classical images,

$$\gamma(\vec{\rho}_1, \vec{\rho}_2) \simeq \delta(\vec{\rho}_s - \vec{\rho}_1/m)\delta(\vec{\rho}_s - \vec{\rho}_2/m), \qquad (12.6.1)$$

where $\vec{\rho}_s$ is the transverse coordinate of the light source plane, $m = s_i/s_o$ is the classical imaging magnification factor,[6] z_1 and z_2 are defined as the optical distance between the plane of the light source and the planes of the object and the ghost image. The choices of z_1 and z_2 must satisfy the Gaussian thin lens equation, respectively, see Fig. 12.6.2. It is easy to see from Fig. 12.6.2 that D_1 and D_2 will have more chance to be triggered jointly when they are in the position within the two identical speckles, such as the two jth speckles

---

[6]The original publications of Gatti et al. choose $m = 2f/2f = 1$ with $1/2f + 1/2f = 1/f$ to image the speckles of the source onto the object plane and the ghost image plane.

near the bottom of the object plane and the image plane. It is also easy to see that the size of the identical speckles determines the spatial resolution of the ghost shadow. This observation has been confirmed by quite a few experimental demonstrations. The classical simulation of Gatti *et al.* might be useful for certain applications. However, the man-made speckle-to-speckle correlation of Gatti *et al.* is fundamentally different from the natural, non-factorizeable point-to-point image-forming correlation observed in the ghost imaging experiment of Pittman *et al.* with entangled photon pairs and the lensless ghost imaging experiment of Scarcelli *et al.* with pseudo-thermal light.

(III) Computational ghost imaging.

Shapiro proposed a computational ghost imaging (CGI) experiment, which consists of a controllable (deterministic) light source, an object for imaging, and a bucket photodetector. For a nondeterministic light source, such as an entangled biphoton system or a thermal radiation source with randomly created and randomly paired photons in thermal state, a photon or subfield may be observed randomly at any arbitrary position in space, or say a photon does not "know" where to go when it is created at the light source, joint-measurements between a reference-detector (the CCD array) and a probe-detector (the bucket detector) is always necessary. The reference-detector defines the spatial coordinate of $\vec{\rho}_o$ of the observation, the probe-detector measures the value of the aperture function $|A(\vec{\rho}_o)|^2$. To reproduce a ghost image exactly and accurately, the joint-measurement must be able to distinguish the values of the aperture function, such as $|A(\vec{\rho}_o)|^2$ and $|A(\vec{\rho}_o')|^2$, in terms of each reference coordinate of the ghost image, such as $\vec{\rho}_i$ and $\vec{\rho}_i'$. However, for a deterministic light source, in which the light "knows" where to go in each shot of its operation, it is unnecessary to use two photodetectors at all. One bucket detector is good enough to reproduce the image of the object. The working principle is very simple: in each shot of the operation, the light beam propagates to a chosen "spot" of the object, or say the light knows where to go" when it is created at the source. The coordinate $\vec{\rho}_o$, which is chosen by the light source, is recorded against the counting rate of the bucket detector at that coordinate, which is proportional to $|A(\vec{\rho}_o)|^2$. The object function $|A(\vec{\rho}_o)|^2$ is thus reproduced or calculated after a large number of such records. If the purpose of the imaging is for recognizing the"shape" of the object only, the light source may not be necessary to prepare a precise spot on the target object in each of its shot-to-shot operations. Instead, the light source may prepare a known "function" of intensity speckles on the target object plane, which is randomly determined by the source from shot-to-shot. The measured counting rate of the bucket detector will be added to these coordinates in each shot of its operation. After a large number of accumulations, the shape of the object, or the shape of the mask, can be estimated statistically.

It should be emphasized that we cannot expect high quality and high resolution image from classically simulated ghost image. The main reason is the lack of a point-to-point imaging-forming function produced from two-photon interference.

SUMMARY

In this chapter, we discussed the physics of quantum imaging. Quantum imaging was stimulated in recent years by the research development of quantum entanglement. Although questions regarding fundamental issues of quantum theory still exist, quantum entanglement has started to play important roles in practical applications. Quantum imaging is one of those exciting areas. Quantum imaging, in general, reproduces the image of an object through the measurement of nontrivial second-order or higher-order ($N > 2$) coherence of entangled photon system or randomly created and randomly grouped photons in thermal state, which is the result of two-photon interference: a pair of photons, either entangled or randomly paired, interfering with the pair itself. In this chapter we started from a general model of biphoton imaging followed by a surprising biphoton ghost imaging experiment of 1995 to emphasize the physics of nonlocal two-photon interference: a pair of entangled photons interfering with the pair itself at distance. In fact, the self-interference of a pair of randomly created and randomly paired photons in thermal state can produce ghost images too. A lensless ghost imaging experiment using thermal light source was demonstrated in 2005, 10 years after the first experimental demonstration of biphoton ghost imaging. We quickly realized an unique and attractive property of thermal light ghost imaging: it can be turbulence-free. We then introduced a turbulence-free ghost imaging experiment and a turbulence-free "ghost camera" with detailed discussions on the physics behind the peculiar feature of turbulence-free. Regarding to realistic applications, perhaps, turbulence-free camera is the "easiest" candidate that can be directly adapted to satellite and long-distance imaging. In recent years, ghost imaging has also been considered for X-ray imaging, especially X-ray "ghost microscope", due to its "lensless" feature.

Quantum imaging is the result of a two-photon interference induced point-to-point imaging-forming correlation between the object plane and the imaging plane. In general, a two-photon imaging can be mathematically expressed as the convolution of the object aperture function and the second-order point-to-point image-forming correlation of a measured pair of photons

$$R_c(\vec{\rho}_i) = \int_{\mathrm{obj}} d\vec{\rho}_o \, |A(\vec{\rho}_o)|^2 \, G^{(2)}(\vec{\rho}_o, \vec{\rho}_i).$$

Ghost imaging is perhaps the most surprising and interesting observation in the family of quantum imaging. We introduced two types of ghost imaging in this chapter. Type-I ghost imaging takes advantage of a nonlocal point-to-point correlation of an entangled photon pair

$$G^{(2)}(\vec{\rho}_o, \vec{\rho}_i) \sim \delta(\vec{\rho}_o + \vec{\rho}_i/m).$$

Type-II ghost imaging utilized the two-photon inference induced photon number fluctuation correlation or intensity fluctuation correlation of thermal radiation at distance

$$G^{(2)}(\vec{\rho}_o, \vec{\rho}_i) \sim 1 + \delta(\vec{\rho}_o - \vec{\rho}_i).$$

or

$$\langle \Delta n(\vec{\rho}_o) \Delta n(\vec{\rho}_i) \rangle \sim \delta(\vec{\rho}_o - \vec{\rho}_i).$$

Quantum imaging has so far demonstrated three peculiar features: (1) enhancing the spatial resolution of imaging beyond the diffraction limit; (2) reproducing ghost images in a "nonlocal" manner; and (3) dispersion-cancelation (type-I) and turbulence-free (type-II) imaging. All these features are the results of multi-photon interference.

REFERENCES AND SUGGESTIONS FOR READING

[1] A.N. Boto et al., Phys. Rev. Lett., **85**, 27333 (2000).

[2] M.D'Angelo, M.V. Chekhova, and Y.H. Shih, Phys. Rev. Lett., **87**, 013602 (2001).

[3] T.B. Pittman, Y.H. Shih, D.V. Strekalov, and A.V. Sergienko, Phys. Rev. A **52**, R3429 (1995). This experiment was named "ghost" imaging by the physics community.

[4] D.V. Strekalov, A.V. Sergienko, D.N. Klyshko and Y.H. Shih, Phys. Rev. Lett. **74**, 3600 (1995). This experiment was named "ghost" interference by the physics community.

[5] D.N. Klyshko, Usp. Fiz. Nauk, **154**, 133 (1988); Sov. Phys. Usp, **31**, 74 (1988); Phys. Lett. A **132**, 299 (1988).

[6] M. H. Rubin, Phys. Rev. A **54**, 5349 (1996).

[7] A. Valencia, G. Scarcelli, M. D'Angelo, and Y.H. Shih, Phys. Rev. Lett. **94**, 063601 (2005).

[8] G. Scarcelli, V. Berardi, and Y. H. Shih, Phys. Rev. Lett., **96**, 063602 (2006).

[9] R. Meyers, K.S. Deacon, and Y.H. Shih, Phys. Rev. A **77**, 041801(2008).

[10] R. Meyers, K.S. Deacon, and Y.H. Shih, Appiled Phys. Lett., 98, 111115 (2011).

[11] J. Sprigg and Y.H. Shih, Sci. Rep., **6**, 38077; doi: 10.1038/srep38077 (2016).

[12] Y.H. Shih, IEEE J. of Selected Topics in Quantum Electronics, IEEE **13**, 1016 (2007).

[13] R.S. Bennink, S.J. Bentley, and R.W. Boyd, Phys. Rev. Lett. **89**, 113601 (2002); R.S. Bennink, *et al.*, Phys. Rev. Lett. **92**, 033601 (2004).

[14] A. Gatti, E. Brambilla, M. Bache and L.A. Lugiato, Phys. Rev. A **70**, 013802, (2004), and Phys. Rev. Lett. **93**, 093602 (2004).

[15] K. Wang, D. Cao, quant-ph/0404078; D. Cao, J. Xiong, and K. Wang, quant-ph/0407065 (2004).

[16] Y.J. Cai, and S.Y. Zhu, quant-ph/0407240, Phys. Rev. E, **71**, 056607 (2005).

[17] J. H. Shapiro. Phys. Rev. A **78** 061802(R) (2008).

[18] J. W. Goodman, *Introduction to Fourier Optics*, McGraw-Hill Publishing Company, New York, NY, 1968.

[19] D.N. Klyshko, *Photon and Nonlinear Optics*, Gordon and Breach Science, New York, 1988.

CHAPTER 13

Homodyne Detection and Heterodyne Detection of Light

In this chapter we introduce the concept of homodyne detection and heterodyne detection of light in the framework of electromagnetic wave theory of radiation, including Maxwell's continuous picture and Einstein's quantized granularity picture of light. Optical homodyne detection and heterodyne detection are both adapted from radio frequency modulation technology. Unlike standard photodetection, homodyne detection and heterodyne detection measure the signal radiation, which may have modulated complex amplitude, by mixing with radiation of a reference frequency, which is usually generated by a local oscillator. Roughly speaking, in homodyne detection the reference frequency equals that of the input signal radiation; in heterodyne detection, the reference light is frequency-shifted.

13.1   OPTICAL HOMODYNE AND HETERODYNE DETECTION

Figure 13.1.1 schematically shows an optical homodyne or heterodyne detection setup. The signal field and the reference field are mixed at a $50\% : 50\%$ beamsplitter and superposed at photodetectors D_1 and D_2. To simplify the notation, the following analysis will be in 1-D by focusing on one of the polarizations of the signal and reference radiation. The intensities

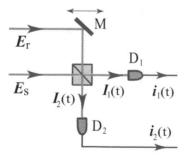

FIGURE 13.1.1   Schematic setup of a homodyne or heterodyne detection experiment.

of $I_1(t)$ and $I_2(t)$ are the results of the superposition between the input signal field E_s and the reference field E_r,

$$I_1(t) = \left|\frac{1}{\sqrt{2}}\left[E_s(t) + E_r(t)\right]\right|^2 = \frac{1}{2}\left[|E_s(t)|^2 + |E_r(t)|^2\right] + Re\left[E_s(t)E_r^*(t)\right]$$

$$I_2(t) = \left|\frac{1}{\sqrt{2}}\left[E_s(t) - E_r(t)\right]\right|^2 = \frac{1}{2}\left[|E_s(t)|^2 + |E_r(t)|^2\right] - Re\left[E_s(t)E_r^*(t)\right]. \qquad (13.1.1)$$

The common cross term $E_s(t)E_r^*(t)$ is formally written as

$$E_s(t)E_r^*(t)$$
$$= \left\{\left[\int d\nu \sum_j a_j(\nu)\, e^{i\varphi_j(\nu)} e^{-i\nu\tau_s}\right] e^{-i\omega_{s0}\tau_s}\right\}\left\{a_r\, e^{-i\varphi_r} e^{i\omega_r\tau_r}\right\} \qquad (13.1.2)$$
$$= \left\{\int d\nu\left[\sum_j a_j(\nu)\, a_r\, e^{i[\varphi_j(\nu) - \varphi_r + \omega_{s0}z_s/c - \omega_r z_r/c]}\right] e^{-i\nu\tau_s}\right\} e^{-i(\omega_{s0} - \omega_r)t},$$

where $\nu = \omega_s - \omega_{s0}$ is the detuning frequency and ω_{s0} the central frequency of the input signal radiation, $\tau_s = t - z_s/c$, $\tau_r = t - z_r/c$. In Eq. (13.1.2), we have assumed a single-mode reference radiation field

$$E_r(t) = a_r\, e^{i\varphi_r} e^{-i\omega_r(t - z_r/c)}$$

and a general multi-subfields and multi-Fourier-mode field of input signal with certain distribution of complex amplitude,

$$E_s(t) = \left[\int d\nu \sum_j a_j(\nu)\, e^{i\varphi_j(\nu)} e^{-i\nu\tau_s}\right] e^{-i\omega_{s0}\tau_s},$$

which is formally written as a wavepacket with carrier frequency ω_{s0} and modulated "envelope" corresponding to the Fourier transform of the complex amplitude.

For homodyne detection, $\omega_r = \omega_{s0} = \omega_0$, Eq. (13.1.2) can be formally written as the Fourier transform of the spectrum

$$E_s(t)E_r^*(t)$$
$$= \int d\nu\left\{\sum_j a_j(\nu)\, a_r\, e^{i[\varphi_j(\nu) - \varphi_r + \omega_0(z_s/c - z_r)/c]}\right\} e^{-i\nu\tau_s}$$
$$= \mathcal{F}_{\tau_s}\left\{\sum_j A_j(\nu)\, e^{i[\varphi_j(\nu) - \varphi_r + \omega_0(z_s - z_r)/c]}\right\}, \qquad (13.1.3)$$

where $A_j(\nu) \equiv a_j(\nu)\, a_r$, indicating an amplified amplitude of $a_j(\nu)$ in the case of a strong local oscillator. The cross term $E_s(t)E_r^*(t)$ represents the interference between the input signal radiation $E_s(t)$ and the reference field $E_r(t)$. As we have studied earlier, the relative phases $\varphi_j(\nu) - \varphi_r$ will play an important role in determining the measured values of I_1 and I_2. The interference term will contribute to the measured values of I_1 and I_2 significantly when $\varphi_j(\nu) - \varphi_r = $ constant, and will have null contribution if the relative phase $\varphi_j(\nu) - \varphi_r$ takes all possible random values from 0 to 2π. The optical path difference $z_s - z_r$ is another factor in determining the contribution of the interference term in the case of $\varphi_j(\nu) - \varphi_r = $ constant. The value of $\omega_0(z_s - z_r)/c$ determines the constructive-destructive property of the interference and consequently determines the magnitude for each and for all of the Fourier amplitudes. It is interesting to see the relative phase $\varphi_j(\nu) - \varphi_r$ and the relative phase delay $\omega_0(z_s - z_r)/c$ between the input field and the local oscillator are both included in the Fourier transform. A spectrum analyzer can retrieve this important information for certain observations. This property has been widely adapted in the studies of squeezed state and other coherent and statistical properties of light.

For heterodyne detection, taking $\omega_r \neq \omega_{s0}$, Eq. (13.1.2) can be formally written as

$$
\begin{aligned}
& E_s(t) E_r^*(t) \\
&= \left\{ \int d\nu \left[\sum_j a_j(\nu) \, a_r \, e^{i[\varphi_j(\nu) - \varphi_r + \omega_{s0} z_s/c - \omega_r z_r)/c]} \right] e^{-i\nu \tau_s} \right\} e^{-i(\omega_{s0} - \omega_r)t} \\
&= \mathcal{F}_{\tau_s} \left\{ \sum_j A_j(\nu) \, e^{i[\varphi_j(\nu) - \varphi_r + (\omega_{s0} z_s - \omega_r z_r)/c]} \right\} e^{-i\omega_d t},
\end{aligned}
\tag{13.1.4}
$$

where $\omega_d = \omega_r - \omega_{s0}$ is the frequency of beats. Eq. (13.1.4) is recognized as a modulated harmonic oscillation of frequency $\omega_d = \omega_{s0} - \omega_r$. The Fourier transform of the spectrum is the modulation function that modulates the harmonic oscillation.

13.2   BALANCED HOMODYNE AND HETERODYNE DETECTION

Figure 13.2.1 schematically shows a balanced homodyne or heterodyne detection setup. The input signal field $E_s(t)$ and the reference field $E_r(t)$ are mixed at a 50% : 50% beam-splitter. The output fields are directed and superposed at photodetectors D_1 and D_2. The photocurrent $i_1(t)$ and $i_2(t)$ are subtracted from each other in an electronic circuit. A standard spectrum analyzer follows to select, amplify and rectify a certain bandwidth of the Fourier spectral composition in the waveform of $i_1(t) - i_2(t)$, electronically. The observed output of the spectrum analyzer is a measure of the amplitude of the chosen Fourier spectral composition.

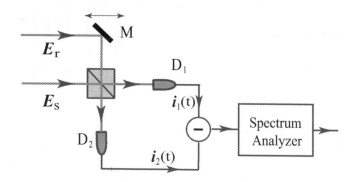

FIGURE 13.2.1   Schematic setup of a balanced homodyne and heterodyne detection experiment. In homodyne detection the reference frequency equals the central frequency of the input signal radiation, $\omega_r = \omega_{s0} = \omega_0$.

Based on the experimental setup of Fig. 13.2.1 we now calculate the expected output from the spectrum analyzer. We start from calculating $i_1(t_-) - i_2(t_-) \propto I_1(t_1) - I_2(t_2)$, where $t_\alpha = t_- - \tau_\alpha^{(e)}$, $\alpha = 1, 2$, with $\tau_\alpha^{(e)}$ the electronic delay in the cables and the electronic circuits associated with α-th photodetector, and t_- is the time for photocurrent "subtraction". To simplify the mathematics, we choose $\tau_1^{(e)} = \tau_2^{(e)}$ to achieve $t_1 = t_2 = t$. It is easy to see

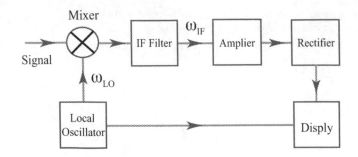

FIGURE 13.2.2 Simplified block diagram of a classic spectrum analyzer.

from Eq. (13.1.1), Eq. (13.1.3), and Eq. (13.1.4)

$$I_1(t) - I_2(t) = 2Re\left[E_s(t)E_r^*(t)\right] \tag{13.2.1}$$

$$= 2Re\,\mathcal{F}_{\tau_s}\left\{\sum_j A_j(\nu)\,e^{i[\varphi_j(\nu)-\varphi_r+\omega_0(z_s-z_r)/c]}\right\},$$

for balanced homodyne detection, and

$$I_1(t) - I_2(t) = 2Re\left[E_s(t)E_r^*(t)\right] \tag{13.2.2}$$

$$= 2Re\,\mathcal{F}_{\tau_s}\left\{\sum_j A_j(\nu)\,e^{i[\varphi_j(\nu)-\varphi_r+(\omega_{s0}z_s-\omega_r z_r)/c]}\right\}e^{-i\omega_d t},$$

for balanced heterodyne detection. We may consider homodyne detection a special case of heterodyne detection when taking $\omega_d = 0$. There is no significant difference in the modulation function, except a trivial phase factor of $\omega_0(z_s - z_r)/c$. It is the spectrum of the modulation function that will be analyzed by the spectrum analyzer. To simplify the discussion, we will focus our attention on the balanced homodyne detection in the following analysis. We will show how a spectrum analyzer works in determining the spectrum of the modulation function with the coherence and path information of the measured light. Before exploring the working mechanism of the spectrum analyzer, we estimate the expected value of its input current $i_1(t) - i_2(t) \propto I_1(t) - I_2(t)$.

The expectation value of $\langle I_1(t) - I_2(t)\rangle$ is easy to calculate by taking into account all possible values of $\varphi_j(\nu) - \varphi_r$ within the superposition. For single-mode reference field $E_r(t)$, the coherent behavior of the input signal, which is mainly determined by the phases of the sub-fields $\varphi_j(\nu)$, will determine the expectation. We discuss two extreme cases:

Case (1): Random $\varphi_j(\nu)$—pseudo-thermal light.

It is easy to see that the only surviving terms in the superposition are the terms with $\varphi_j(\nu) = \varphi_r$ when taking into account all possible values of $\varphi_j(\nu)$. Obviously, the chances of having $\varphi_j(\nu) = \varphi_r$ are quite small. The expectation value of $\langle I_1(t) - I_2(t)\rangle$ is thus effectively zero in this case. In a real measurement, however, the superposition may not take all possible values of the random phases and the interference cancelation may not be complete. These non-canceled terms of $I_1(t) - I_2(t)$ will be analyzed and displayed by the spectrum analyzer in terms of the Fourier composition of ν, which is effectively the beats frequency $\omega_s - \omega_r$ in the homodyne detection measurement,

$$Re\left\{\int d\nu\left[\sum_{\mathrm{surv}} A_j(\nu)\,e^{i[\varphi_j(\nu)-\varphi_r+\omega_0(z_s-z_r)/c]}\right]e^{-i\nu\tau_s}\right\} \tag{13.2.3}$$

where $\sum_{\text{surv}}$ represents the sum of the non-canceled surviving terms in the superposition. These surviving terms are traditionally treated as the noise or fluctuations of the radiation. The spectrum analyzer is thus considered of measuring the spectrum of the noise or the fluctuations of the radiation.

Case (2): $\varphi_j(\nu) - \varphi_r = $ constant.

In the case of $\varphi_j(\nu) - \varphi_r = \varphi_0$, where $\varphi_0 = $ constant, the intensity difference $I_1(t) - I_2(t) \propto Re\,[E_s(t)E_r^*(t)]$, i.e.,

$$Re\left\{\int d\nu \left[\sum_j A_j(\nu)\,e^{i[\varphi_0 + \omega_0(z_s - z_r)/c]}\right]e^{-i\nu\tau_s}\right\} \tag{13.2.4}$$

without interference cancelation will be received by the spectrum analyzer. The spectrum analyzer no longer measures the noise or the fluctuation of the surviving input signal, instead, it receives and analyzes the entire interference term of $E_s(t)E_r^*(t)$.

The design and working mechanism of specific spectrum analyzers can be quite different from each other. Nevertheless, the output reading of modern spectrum analyzers can be roughly divided into two categories: linear normal spectrum and nonlinear power spectrum. Linear normal spectrum simply presents the spectral amplitude of the input signal as a function of frequency. The nonlinear power spectrum provides much more detail than the normal spectrum. A power spectrum includes not only the Fourier composition of the input current $i_1(t) - i_2(t)$, but also their beats and sum-frequencies that fall within the passband of the chosen spectral filter in the heterodyne circuit of the spectrum analyzer, such as the IF filter shown in Fig. 13.2.2.

A simplified block diagram of a classic spectrum analyzer is illustrated in Fig. 13.2.2. The input signal of $i_1(t) - i_2(t)$, which is either proportional to Eq. (13.2.3) or Eq. (13.2.4), is mixed with a sinusoidal reference current of tunable RF frequency ω_{lo} in an electronic mixer. The RF current of ω_{lo} is generated from a local oscillator. The mixer has a nonlinear response to the inputs. Taking account of the first-order and the second-order response of the mixer to a good approximation, the output of the mixer contains the input signal $i_1(t) - i_2(t)$, the reference oscillation of ω_{lo}, their second-harmonics, and a cross term

$$i_{NL}$$
$$\propto \int d\nu \sum_{\text{Surv}} A_j(\nu)\cos\{\nu\tau_s - [\varphi_j(\nu) - \varphi_r] - \varphi(z_s, z_r)\}\,A_{lo}\cos\omega_{lo}t$$
$$\propto \int d\nu \sum_{\text{Surv}} A_j(\nu)A_{lo}\Big\{\cos\{(\nu + \omega_{lo})t - [\varphi_j(\nu) - \varphi_r] - \varphi(z_s, z_r)\}$$
$$+ \cos\{(\nu - \omega_{lo})t - [\varphi_j(\nu) - \varphi_r] - \varphi(z_s, z_r)\}\Big\}, \tag{13.2.5}$$

for the measurement of thermal light, and

$$i_{NL}$$
$$\propto \int d\nu \sum_j A_j(\nu)\cos\{\nu\tau_s - \varphi - \varphi(z_s, z_r)\}\,A_{lo}\cos\omega_{lo}t$$
$$\propto \int d\nu \sum_j A_j(\nu)A_{lo}\Big\{\cos\{(\nu + \omega_{lo})t - \varphi_0 - \varphi(z_s, z_r)\}$$
$$+ \cos\{(\nu - \omega_{lo})t - \varphi_0 - \varphi(z_s, z_r)\}\Big\}, \tag{13.2.6}$$

for the measurement of coherent light, where $\varphi(z_s, z_r) = \omega_{s0}(z_s - z_r)/c + \nu z_s/c$, and we have assumed a simple harmonic local oscillator of frequency ω_{lo}, $i_{lo}(t) = A_{lo}\cos\omega_{lo}t$. Eqs. (13.2.5) and (13.2.6) indicate that the nonlinear response of the mixer produces a down-converted Fourier composition $\omega_{IF} = \omega_{lo} - \nu$ and an up-converted Fourier composition $\omega_{IF} = \omega_{lo} + \nu$ in terms of each Fourier composition of the input signal. An electronic spectral filter follows after the mixer to select a narrowband RF current of frequency ω_{IF} from either the down-converted set or the up-converted set of the Fourier-modes. ω_{IF} is technically called the intermediate frequency. To simplify the mathematics, we assume the bandwidth of ω_{IF} much narrower than that of the input signal so that the selected Fourier composition of ω_{IF} can be treated as single-mode. The selected single-mode Fourier composition of ω_{IF} is then amplified by a linear amplifier and rectified by a nonlinear envelope detector, resulting in an output that is proportional to the power spectrum

$$P(\nu) \propto \left\{ \sum_{\text{Surv}} A_j(\nu)\cos\left[\varphi_j(\nu) - \varphi_r + \varphi(z_s, z_r)\right] \right\}^2, \tag{13.2.7}$$

for the measurement of incoherent thermal light, and

$$P(\nu) \propto \left\{ \sum_j A_j(\nu)\cos\left[\varphi_0 + \varphi(z_s, z_r)\right] \right\}^2, \tag{13.2.8}$$

for the measurement of coherent light.

We thus have the following results for the above two extreme cases.

(1): Measurement of incoherent thermal light.

Thermal light is statistically stationary and ergodic, by choosing an appropriate time parameter (integration time) of the spectrum analyzer, we may treat the measurement as an ensemble average

$$\begin{aligned}
\langle P(\nu) \rangle &\propto \langle \sum_{\text{Surv}} A_j(\nu)\cos\left[\varphi_j(\nu) - \varphi_r + \varphi(z_s, z_r)\right] \\
&\quad \times A_k(\nu)\cos\left[\varphi_k(\nu) - \varphi_r + \varphi(z_s, z_r)\right]\rangle \\
&\propto \langle \sum_{\text{Surv}} A_j(\nu)A_k(\nu)\left\{\cos\left[\varphi_j(\nu) - \varphi_k(\nu)\right] \right. \\
&\quad \left. + \cos\left[\varphi_j(\nu) + \varphi_k(\nu) - 2\varphi_r + 2\varphi(z_s, z_r)\right] \right\}\rangle,
\end{aligned} \tag{13.2.9}$$

where $\langle...\rangle$ denotes, again, an ensemble average by means of *taking into account all possible realizations of the field*. As we have discussed earlier, when taking into account all possible values of $\varphi_j(\nu)$, the stochastic superposition results in a non-zero value from the first cosine term, which includes all the surviving diagonal terms of $j = k$, and a zero value from the second cosine term of Eq. (13.2.9). The expected power spectrum of thermal light is thus a simple sum of the squared amplitudes

$$\langle P(\nu) \rangle \propto \sum_{\text{Surv}} A_j^2(\nu). \tag{13.2.10}$$

In reality, the radiation field may not take all possible realization within the time integral of the spectrum analyzer, the incomplete interference cancellation may still cause a random fluctuation in the neighborhood of $\langle P(\nu) \rangle$ from time to time.

Case (2): Measurement of coherent light.

Taking $\varphi_j(\nu) - \varphi_r = \varphi_0$ constant, Eq. (13.2.9) becomes

$$P(\nu) \propto \sum_{j,k} A_j(\nu) A_k(\nu) \left\{ 1 + \cos 2 \left[\varphi_0 + \varphi(z_s, z_r) \right] \right\}$$

$$\propto \left\{ \sum_{j,k} A_j(\nu) A_k(\nu) \right\} \cos^2 \left[\varphi_0 + (\omega_s z_s - \omega_r z_r)/c \right]. \tag{13.2.11}$$

It is interesting to find from Eq. (13.2.11) the power spectrum of coherent light is a sinusoidal function of $\varphi(z_s, z_r) = (\omega_s z_s - \omega_r z_r)/c \simeq \omega_{s0}(z_s - z_r)/c$. The change of the relative optical path between the signal field and the reference field, which can be realized by adjusting the position of the mirror M in Fig. 13.2.1, will produce an interference pattern as a function of $z_s - z_r$, similar to that of the interference between two individual but synchronized laser beams.

13.3 BALANCED HOMODYNE DETECTION OF INDEPENDENT AND COUPLED THERMAL FIELDS

We consider the experimental setup of Fig. 13.3.1, in which two input thermal fields E_s and E_i, either independent or coupled, are measured by two individual balanced homodyne detection setups. The balanced homodyne detection setups are similar to that of Fig. 13.2.1,

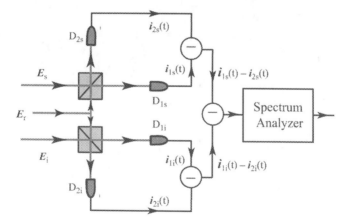

FIGURE 13.3.1 Schematic experimental setup for balanced homodyne detection of two independent or coupled thermal fields. Three subtraction circuits are used for manipulating the photocurrents $i_{1s}(t)$, $i_{2s}(t)$, $i_{1i}(t)$, and $i_{2i}(t)$. The spectrum analyzer displays the power spectrum of the final output current $[i_{1s}(t) - i_{2s}(t)] - [i_{1i}(t) - i_{2i}(t)]$.

except the two outputs of the homodyne detection are subtracted again from each other by a third subtraction circuit. The power spectrum of the final output current $[i_{1s}(t) - i_{2s}(t)] - [i_{1i}(t) - i_{2i}(t)]$ is measured by a spectrum analyzer and read out in terms of its Fourier composition.

To calculate the expected power spectrum, we start from Eq. (13.2.1) which gives the estimated values of $i_{1s}(t) - i_{2s}(t)$ and $i_{1i}(t) - i_{2i}(t)$. The output of the third subtraction circuit is therefore

$$[i_{1s}(t) - i_{2s}(t)] - [i_{1i}(t) - i_{2i}(t)]$$

$$\propto Re\,\mathcal{F}_{\tau_s}\left\{ \sum_{\text{Surv}} A_j(\nu_s)\,e^{i[\varphi_j(\nu_s) - \varphi_r + \omega_{s0}(z_s - z_r)/c]} \right\}$$

$$- Re\,\mathcal{F}_{\tau_i}\left\{ \sum_{\text{Surv}} A_k(\nu_i)\,e^{i[\varphi_k(\nu_i) - \varphi_r + \omega_{i0}(z_i - z_r)/c]} \right\}. \tag{13.3.1}$$

The output current of the third subtraction circuit is mixed with a sinusoidal reference current of tunable RF frequency ω_{lo} in the electronic mixer of the spectrum analyzer. The nonlinear response of the mixer produces the up-converted and down-converted frequencies

$$i_{NL}$$

$$\propto \int d\nu_s \sum_{\text{Surv}} A_j(\nu_s) A_{lo}\Big\{ \cos\{(\nu_s + \omega_{lo})t - [\varphi_j(\nu_s) - \varphi_r] - \varphi(z_s, z_r)\}$$

$$+ \cos\{(\nu_s - \omega_{lo})t - [\varphi_j(\nu_s) - \varphi_r] - \varphi(z_s, z_r)\}\Big\}$$

$$- \int d\nu_i \sum_{\text{Surv}} A_k(\nu_i) A_{lo}\Big\{ \cos\{(\nu_i + \omega_{lo})t - [\varphi_k(\nu_i) - \varphi_r] - \varphi(z_s, z_r)\}$$

$$+ \cos\{(\nu_i - \omega_{lo})t - [\varphi_k(\nu_i) - \varphi_r] - \varphi(z_s, z_r)\}\Big\}. \tag{13.3.2}$$

After the IF filter, the linear amplifier, and the nonlinear rectifier, the expected power spectrum is proportional to

$$\langle P(\nu) \rangle \propto \Big\langle \Big\{ \sum_{\text{Surv}} A_j(\nu_s)\cos\{\varphi_j(\nu_s) - \varphi_r + \varphi(z_s, z_r)\}$$

$$- \sum_{\text{Surv}} A_k(\nu_i)\cos\{\varphi_k(\nu_i) - \varphi_r + \varphi(z_s, z_r)\} \Big\}^2 \Big\rangle. \tag{13.3.3}$$

In the following, we discuss two measurements for different types of input radiations: independent thermal light and coupled thermal fields with $\varphi_j(\nu_s) + \varphi_k(\nu_i) = $ constant.

Case (1): Incoherent thermal fields.

Assuming the two inputs E_s and E_i are incoherent thermal fields, the phases $\varphi_j(\nu_s)$ and $\varphi_k(\nu_i)$ may take any value randomly and independently. Eq. (13.3.3) gives the following three contributions

(1) $\left\langle \Big\{ \sum_{\text{Surv}} A_j(\nu_s)\cos\big[\varphi_j(\nu_s) - \varphi_r + \varphi(z_s, z_r)\big] \Big\}^2 \right\rangle = \dfrac{1}{2} \sum_{\text{Surv}} A_j^2(\nu_s)$

(2) $\left\langle \Big\{ \sum_{\text{Surv}} A_k(\nu_i)\cos\big[\varphi_k(\nu_i) - \varphi_r + \varphi(z_i, z_r)\big] \Big\}^2 \right\rangle = \dfrac{1}{2} \sum_{\text{Surv}} A_k^2(\nu_i)$

(3) $\Big\langle \sum_{\text{Surv}} A_j(\nu_s) A_k(\nu_i)\cos\big[\varphi_j(\nu_s) - \varphi_r + \varphi(z_s, z_r)\big]$

$$\times \cos\big[\varphi_k(\nu_i) - \varphi_r + \varphi(z_i, z_r)\big] \Big\rangle$$

$$\simeq \sum_{\text{Surv}} A_j(\nu_s) A_k(\nu_i)\Big\{ \cos\big[\varphi_j(\nu_s) - \varphi_k(\nu_i)\big] \tag{13.3.4}$$

$$+ \cos\big[\varphi_j(\nu_s) + \varphi_k(\nu_i) + -2\varphi_r + \varphi(z_s, z_r) + \varphi(z_i, z_r)\big] \Big\}$$

$$= 0,$$

where we have assumed $z_s = z_i$ to simplify the notation. In Eq. (13.3.4) the expectation evaluation has taken into account all possible random values of $\varphi_j(\nu_s) - \varphi_k(\nu_i)$ and $\varphi_j(\nu_s) + \varphi_k(\nu_i)$ in the superposition. The expected power spectrum sums the above three contributions

$$\langle P(\nu) \rangle \propto \sum_{\text{Surv}} A_j^2(\nu_s) + \sum_{\text{Surv}} A_k^2(\nu_i). \tag{13.3.5}$$

This result is reasonable for the measurement of two independent thermal fields.

Case (2): Coupled thermal light with $\varphi_j(\nu_s) + \varphi_k(\nu_i) = \text{constant}$.

In case (2), we model each individual input field E_s and E_i stochastically, however, consider the system of the two coherently by means of $\varphi_j(\nu_s) + \varphi_k(\nu_i) = \varphi_p = \text{constant}$. A nonlinear optical parametric amplifier is able to generate such a state. This kind of radiation is known as "squeezed" light. To simplify the discussion, we assume the sub-fields, jth and kth, associated with a large number of sub-sources, all achieve this condition. In reality, this condition may not be achievable for all sub-fields, it is necessary to define a corresponding concept of coherence in terms of this condition. We will leave this discussion for later chapters. Due to the stochastic nature of each individual field E_s and E_i, the self-square of the two terms in Eq. (13.3.4) has the same contribution to the expected power spectrum $\langle P(\nu) \rangle$ as that of case (I). The cross term of Eq. (13.3.4), however, yields an interesting nontrivial contribution:

$$\langle \sum_{\text{Surv}} A_j(\nu_s)\cos\left[\varphi_j(\nu_s) - \varphi_r + \omega_{s0}\frac{z_s - z_r}{c} + \nu_s\frac{z_s}{c}\right]$$

$$\times \sum_{\text{Surv}} A_k(\nu_i)\cos\left[\varphi_k(\nu_i) - \varphi_r + \omega_{i0}\frac{z_i - z_r}{c} + \nu_i\frac{z_i}{c}\right] \rangle$$

$$\simeq \sum_{\text{Surv}} A_j(\nu_s)A_k(\nu_i)\left\{\cos\left[\varphi_j(\nu_s) - \varphi_k(\nu_i)\right]\right. \tag{13.3.6}$$

$$\left. + \cos\left[\varphi_p + \omega_p\frac{z_{s,i} - z_r}{c} - 2\varphi_r + (\nu_s + \nu_i)\frac{z_{s,i}}{c}\right]\right\}.$$

In Eq. (13.3.6) we have applied $\varphi_j(\nu_s) + \varphi_k(\nu_i) = \varphi_p$ and $\omega_{s0} + \omega_{i0} = \omega_p$, where φ_p and ω_p are constants corresponding to the pump phase and pump frequency of a nonlinear optical parametric amplifier. To simplify the notation, in Eq. (13.3.6) we have assumed $z_s \simeq z_i = z_{s,i}$. It is easy to find that the first cosine term yields a zero value when taking into account all possible values of $\varphi_j(\nu_s) - \varphi_k(\nu_i)$. The second cosine term has a nontrivial contribution to the power spectrum

$$\langle P(\nu) \rangle \propto \left\{\sum_j A_j^2(\nu_s) + \sum_k A_k^2(\nu_i)\right\}$$

$$+ 2\sum_{j,k} A_j(\nu_s)A_k(\nu_i)\cos\left[\omega_p\frac{z_{s,i} - z_r}{c} + \varphi_{s,i,r}\right], \tag{13.3.7}$$

where $\varphi_{s,i,r} = \varphi_p - 2\varphi_r + (\nu_s + \nu_i)z_{s,i}/c$.

Eq. (13.3.7) indicates that the expected power spectrum $\langle P(\nu) \rangle$ is a cosine function of $\omega_p(z_{s,i} - z_r)/c$. The change of the relative optical path between the input fields E_s (E_i) and the reference field E_r will produce a cosine interference modulation. This interference modulation, however, is different from a standard interference pattern, due to the relatively great contribution from the cross term. The cross term in Eq. (13.3.7) can be much greater than the sum of the two diagonal terms when taking into account a large number of sub-fields. For a large number of N sub-fields, the ratio between the number of diagonal terms and the number of cross terms is roughly $\sim 1/N$. This effect causes unavoidable difficulties for a theory in which $\langle P(\nu) \rangle$ is treated as the measure of statistical fluctuations of the radiation. Eq. (13.3.7) indicates, under certain experimental conditions, $\langle P(\nu) \rangle$ is not only able to achieve a value below "shot noise", but also able to achieve a value of negative.

SUMMARY

In this chapter we introduce the concept of homodyne detection and heterodyne detection in the framework of the classical electromagnetic wave theory of light. We analyzed the balanced homodyne and heterodyne detection in detail, including the working function of spectrum analyzer. In the last part of this chapter, we demonstrated an interesting phenomenon: for coupled thermal light with $\varphi_j(\nu_s) + \varphi_k(\nu_i) = \text{constant}$, the power spectrum turns to be

$$\langle P(\nu) \rangle \propto \left\{ \sum_j A_j^2(\nu_s) + \sum_k A_k^2(\nu_i) \right\}$$
$$+ 2 \sum_{j,k} A_j(\nu_s) A_k(\nu_i) \cos\left[\omega_p \frac{z_{s,i} - z_r}{c} + \varphi_{s,i,r} \right].$$

The sinusoidal modulation is observable when introducing optical path difference between $z_{s,i}$ and $z_{s,i}$. Furthermore, the amplitude of the sinusoidal modulation can be much greater than that of the other two constant terms, when taking into account a large number of sub-fields. This means that under certain experimental conditions, $\langle P(\nu) \rangle$ is not only able to achieve a value below "shot noise", but also able to achieve a value of negative.

REFERENCES AND SUGGESTIONS FOR READING

[1] M.I. Skolnik, *Introduction to Radar System*, McGraw-Hill, 1962.

[2] H.-A. Bachor and T.C. Ralph, *A guide to Experiments in Quantum Optics*, Wiley-VCJH, 2004.

[3] C.C. Gerry and P.L. Knight, *Introductory Quantum Optics*, Cambridge University Press, 2006.

CHAPTER 14

Optical Tests of Foundations of Quantum Theory

In the early 1950's, Bohm simplified the 1935 EPR state from continuous coordinate-momentum space to discrete spin space by introducing a singlet state of two spin 1/2 particles:

$$|\Psi\rangle = \frac{1}{\sqrt{2}} \left[|\uparrow\rangle_1 |\downarrow\rangle_2 - |\downarrow\rangle_1 |\uparrow\rangle_2 \right] \tag{14.0.1}$$

where the kets $|\uparrow\rangle$ and $|\downarrow\rangle$ represent states of spin "up" and spin "down", respectively, along an *arbitrary* direction. For the EPR-Bohm state, the spin of neither particle is determined; however, if one particle is measured to be spin up along a certain direction, the other one must be spinning down along that direction, despite the distance between the two spin 1/2 particles. We have shown in Chapter 3, Eq. (14.0.1) is independent of the choice of the spin directions. The introduction of the EPR-Bohm state simplified the physical picture and the discussion about quantum entanglement dramatically.[1]

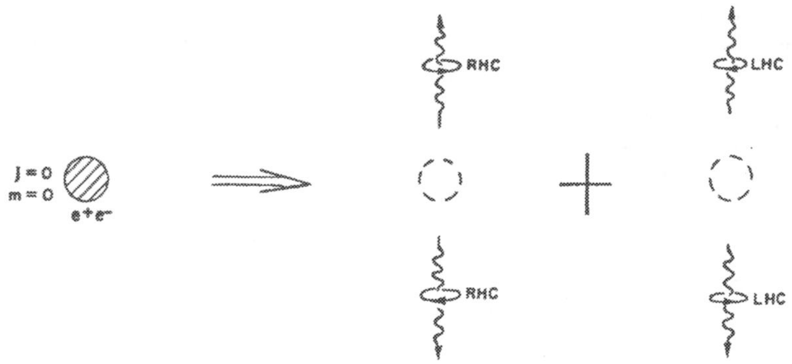

FIGURE 14.0.1 Annihilation of Positronium. Due to the conservation of angular momentum, if photon 1 is right-hand circular (RHC) polarized, photon 2 must be right-hand circular polarized. If photon 1 is left-hand circular (LHC) polarized, then photon 2 has to be left-hand circular polarized.

---

[1]But, one should keep in mind that it is not a necessary to have two "terms" in an entangled state. In fact, the first entangled state suggested by EPR in 1935 has infinite number of "terms".

A more practical example of the EPR-Bohm state concerns the polarization states of photon pairs, such as the spin-zero state of a high energy photon pair disintegrated from the annihilation of positronium. Suppose initially we have a positron and an electron in the spin-zero state with antiparallel spins. The positronium cannot exist very long: it disintegrates into two γ-ray photons within $\sim 10^{-10}$ second of its lifetime. The spin zero state is symmetric under all rotations. Therefore, the photon pair may be disintegrated into any direction in space with equal probability. The conservation of linear momentum, however, guarantees that if one of the photon is observed in a certain direction, its twin must be found in the opposite direction (with finite uncertainty $\Delta(\mathbf{p}_1 + \mathbf{p}_1) \neq 0$). The conservation of angular momentum will decide the polarization state of the photon pair. As shown in Fig. 14.0.1, in order to keep spin-zero, if photon 1 is right-hand circular polarized (RHC), photon 2 must be also right-hand circular polarized. The same argument shows that if photon 1 is left-hand circular (LHC) polarized, then photon 2 has to be left-hand circular polarized too. Therefore, the positronium may decay into two RHC photons or two LHC photons with equal probability.

What is the relationship between these two alternatives? If it is just two different ways of disintegration with equal probability, the two photon system may be described by a density matrix,

$$\hat{\rho} = \frac{1}{2} \left[\, | \, R_1 R_2 \, \rangle \langle \, R_1 R_2 \, | + | \, L_1 L_2 \, \rangle \langle \, L_1 L_2 \, | \, \right], \tag{14.0.2}$$

where $| \, R_j \, \rangle$ and $| \, L_j \, \rangle$ indicate the RHC and LHC states for photon j, respectively. The density matrix specify only the statistics: within N γ-ray photon pairs, 50% of them have RHC-RHC polarization and another 50% have LHC-LHC polarization. The physical process of positronium annihilation, however, is not that simple. The law of parity conservation must be satisfied in the disintegration: the spin-zero ground state of positronium holds an odd parity. Thus, the state of the photon pair must keep its parity odd:

$$| \Psi \rangle = \frac{1}{\sqrt{2}} \left[\, | \, R_1 \rangle | \, R_2 \rangle - | \, L_1 \rangle | \, L_2 \rangle \, \right]. \tag{14.0.3}$$

It is the conservation of parity made the two-photon state very special. The two-photon state is a non-factorizeable pure state of a special superposition between the RHC and LHC states specified with a relative phase of π. Mathematically, "non-factorizeable" means that the state cannot be written as a product state of photon 1 and photon 2. Physically, it means that photon 1 and photon 2 are not independent anymore. The two γ-ray photons are in an entangled polarization state, or spin state.

The physics behind Eq. (14.0.3) is very interesting. To describe the interesting physics of the state, however, is not an easy job. One may easily find the following popular statement about this state: the polarization state of photon 1 and photon 2 are both undefined: each has 50%:50% chance to be RHC or LHC; however, the polarization state of photon 1 can be predicted with certainty through the measurement of photon 2: whichever circular polarization state photon 2 is observed, photon 1 must be in the same state, despite the distance between them. In other words, one would always observe two RHC or two LHC polarized photons in a joint detection event:

$$| \langle R_1 R_2 | \Psi \rangle |^2 = | \langle L_1 L_2 | \Psi \rangle |^2 = 50\%$$
$$| \langle R_1 L_2 | \Psi \rangle |^2 = | \langle L_1 R_2 | \Psi \rangle |^2 = 0. \tag{14.0.4}$$

The result of Eq. (14.0.4) is interesting, but may not be surprising. In fact, a statistical ensemble of photons characterized by the density matrix of Eq. (14.0.2) would give the same

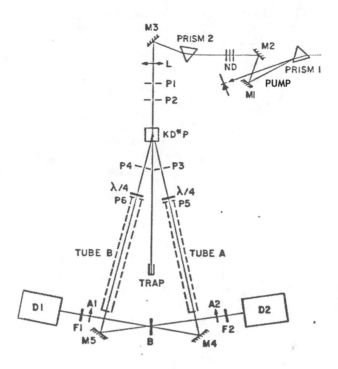

FIGURE 14.0.2   The historical two-photon polarization correlation measurement of Alley and Shih (1986). The state is prepared as that in Eq. (14.0.3) and the measurement is for linear polarization correlation.

result. The above statement, which is based on the correlation of circular polarization, has not explored the real interesting physics behind the *pure state* of Eq. (14.0.3).

Can we distinguish the pure state of Eq. (14.0.3) from the statistical mixture of Eq. (14.0.2) through a simple polarization correlation measurement? The answer is positive. Instead of measuring circular polarization, we measure linear polarization correlation. Figure 14.0.2 schematically illustrates an experimental setup of Alley and Shih which serves the purpose. The two-photon state is prepared as that in Eq. (14.0.3) and the measurement is for linear polarization correlation. Each linear polarization analyzer has two orthogonal output channels. Four photodetectors are placed at the four output ports of the two linear polarization analyzers for two-fold joint detections. There are four possible combinations: D_1 with D_2, D_1 with D'_2, D'_1 with D_2, and D'_1 with D'_2. To calculate the outcomes, it is convenient to use the conventional right-hand coordinate system as shown in Fig. 14.0.2. Suppose we choose the polarization analyzers along $|X\rangle$ and $|Y\rangle$, the measurement on the two-photon pure state of Eq. (14.0.3) yields:

$$| \langle X_1\,Y_2\,|\,\Psi\,\rangle\,|^2 = |\,\langle Y_1\,X_2\,|\,\Psi\,\rangle\,|^2 = 50\%$$
$$|\,\langle X_1\,X_2\,|\,\Psi\,\rangle\,|^2 = |\,\langle Y_1\,Y_2\,|\,\Psi\,\rangle\,|^2 = 0, \qquad (14.0.5)$$

i.e., one would always observe two orthogonal polarized photons in a joint detection event. The measurement on the statistical mixture of Eq. (14.0.2), however, gives a completely random result:

$$tr\,\hat\rho\,|\,Y_2\,X_1\rangle\langle X_1\,Y_2\,| = tr\,\hat\rho\,|\,X_2\,Y_1\rangle\langle Y_1\,X_2\,| = 25\%$$
$$tr\,\hat\rho\,|\,X_2\,X_1\rangle\langle X_1\,X_2\,| = tr\,\hat\rho\,|\,Y_2\,Y_1\rangle\langle Y_1\,Y_2\,| = 25\%. \qquad (14.0.6)$$

Furthermore, we may rotate the linear polarization analyzers to any direction in the XY plane, such as θ_1 ($\bar{\theta}_1 = \pi/2 - \theta$ in its orthogonal channel) and θ_2 ($\bar{\theta}_2 = \pi/2 - \theta$ in its orthogonal channel) for the measurement of polarization correlation, where θ_j is the angle between the polarization analyzer and the X axis. The measurement on the two-photon pure state of Eq. (14.0.3) yields

$$| \langle \theta_1 \, \theta_2 \, | \, \Psi \, \rangle |^2 = \frac{1}{2} \sin^2(\theta_1 + \theta_2) = \frac{1}{2} \sin^2 \varphi \qquad (14.0.7)$$

where $\varphi = \theta_1 + \theta_2$. Eq. (14.0.7) indicates that the pair must be always orthogonally polarized, independent of the choice of θ_j for each individual polarization analyzer. Considering the other three possible joint detections, we have the following polarization correlation functions in terms of $\varphi = \theta_1 + \theta_2$:

$$| \langle \theta_1 \, \theta_2 \, | \, \Psi \, \rangle |^2 = | \langle \bar{\theta}_1 \, \bar{\theta}_2 \, | \, \Psi \, \rangle |^2 = \frac{1}{2} \sin^2 \varphi$$

$$| \langle \theta_1 \, \bar{\theta}_2 \, | \, \Psi \, \rangle |^2 = | \langle \bar{\theta}_1 \, \theta_2 \, | \, \Psi \, \rangle |^2 = \frac{1}{2} \cos^2 \varphi \qquad (14.0.8)$$

On the other hand, the measurement on the statistical mixture of Eq. (14.0.2) gives, again, a completely random result:

$$tr \, \hat{\rho} \, | \, \theta_2 \, \theta_1 \rangle \langle \theta_1 \, \theta_2 \, | = tr \, \hat{\rho} \, | \, \bar{\theta}_2 \, \bar{\theta}_1 \rangle \langle \bar{\theta}_1 \, \bar{\theta}_2 \, | = 25\%$$

$$tr \, \hat{\rho} \, | \, \bar{\theta}_2 \, \theta_1 \rangle \langle \theta_1 \, \bar{\theta}_2 \, | = tr \, \hat{\rho} \, | \, \theta_2 \, \bar{\theta}_1 \rangle \langle \bar{\theta}_1 \, \theta_2 \, | = 25\%. \qquad (14.0.9)$$

Now we are ready to describe the interesting physics behind Eq. (14.0.3) with a better statement: the entangled two-photon polarization state of Eq. (14.0.3) has specified a peculiar two-photon system. In this system, the polarization of neither photon is defined during the course of its propagation; however, if one of the photon is measured to be in a defined polarization state, the polarization state of the other photon is determined with certainty, despite the distance between the pair and *despite the choice of the polarization vector base.*

This statement is similar to the statement we have used for the original EPR state: in an EPR two-particle system, neither *position* nor *momentum* for neither particle is defined during the course of its propagation; however, if one of the particle is measured to be in a defined *position* or *momentum*, the *position* and *momentum* of the other particle is determined with certainty, despite the distance between the two particles.

The independence of the choice for polarization vector base in Eq. (14.0.3) is equivalent to the independence of the choice of position and momentum coordinate for the original EPR state.

Regardless of whether we measure it or not, does a free propagating particle have a defined spin in the Bohm state of Eq. (14.0.1)? Does a free propagating photon have a defined polarization, either circular or linear, in the state of Eq. (14.0.3)? On one hand, the spin of neither independent particle is specified in the Bohm state of Eq. (14.0.1), and the polarization of neither independent photon is specified in the state of Eq. (14.0.3), we may have to believe that the particles do not have any defined spin, and the photons do not have any defined polarization, during the course of their propagation. On the other hand, if the spin of one particle, or the polarization of one photon, uniquely determines the spin of the other distant particle, or the polarization of the other photon, it would be hard for anyone who believes no action-at-a-distance to imagine that the spin of the two particles, or the polarization of the two photons, are not predetermined with defined values before the measurement. The Bohm state thus put us into a paradoxical situation. It seems reasonable for us to ask the same question that EPR had asked in 1935: "Can quantum-mechanical description of physical reality be considered complete?"

Is it possible to have a realistic theory which provides correct predictions of the behavior of a particle similar to quantum theory and, at the same time, respects the description of physical reality by EPR as "complete"? Bohm and his followers have attempted a "hidden variable theory" to formulate the physical reality into the wavefunction of a particle or a pair of particles. The theory seemed to be consistent with quantum mechanics and satisfied the requirements of EPR. The hidden variable theory was successfully applied to many different quantum phenomena until 1964, when Bell proved a theorem to show that an inequality, which is violated by certain quantum mechanical statistical predictions, can be used to distinguish local hidden variable statistics from quantum superpositions. Since then, the testing of Bell's inequalities became a standard instrument for the study of fundamental problems of quantum theory. The experimental testing of Bell's inequality started from the early 1970's. Most of the historical experiments concluded the violation of the Bell's inequalities and thus disproved the local hidden variable theory.[2]

14.1 HIDDEN VARIABLE THEORY AND QUANTUM CALCULATION FOR THE MEASUREMENT OF SPIN 1/2 BOHM STATE

Consider two spin 1/2 particles, in the Bohm state of Eq. (14.0.1). The particles are well separated at distance. Two groups of observers decide to measure their spins by using Stern-Gerlach analyzers. Three sets of Stern-Gerlach apparatus whose z-directions are defined by unit vectors $\hat{\mathbf{a}}$, $\hat{\mathbf{b}}$, and $\hat{\mathbf{c}}$ as shown in Fig. 14.1.1. What is the probability to find the spin

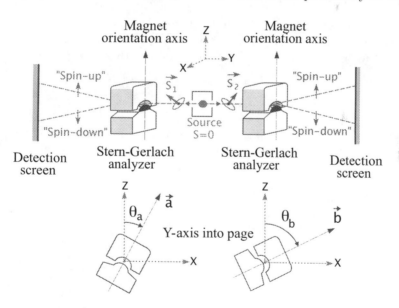

FIGURE 14.1.1 Bohm's version of the EPR *gedankenexperiment*.

of particle-1 to be $+1/2$ in the direction $\hat{\mathbf{a}}$ and the spin of particle-2 to be $+1/2$ in the direction of $\hat{\mathbf{b}}$? The following is a simple classical hidden variable model addressing this question. Before giving the detail, we first introduce the following notation P_{ab} to specify

---

[2]Although Bell's theorem and Bells' inequality involve deep philosophical concerns in general, we should keep in mind that the testing of Bell's inequality in terms of the physical measurement on a physical system is essentially a physical problem. Any proposed statistical model of classical correlation, either local or nonlocal, must obey the basic physical laws of classical theory. One cannot assume a classical correlation that is never allowed in classical theory.

the classical probability,

$$P_{ab} = P(+_a \, d_b \, d_c \,|\, d_a \, +_b \, d_c). \tag{14.1.1}$$

In this notation we chose six-elements to specify the observation. The left-side of the partition refers particle one (in channel 1) and the right-side to particle two (in channel 2). Each element represents wether a measurement of the component of the spin of the jth particle, $j = 1, 2$, gives $+$ or $-$ in the corresponding direction $\hat{\mathbf{a}}$, $\hat{\mathbf{b}}$, or $\hat{\mathbf{c}}$. d_a, d_b and d_c indicate no determination in the direction $\hat{\mathbf{a}}$, $\hat{\mathbf{b}}$ and $\hat{\mathbf{c}}$, respectively. Note, P_{ab} is defined as to have $+$ in direction $\hat{\mathbf{a}}$ which is observed in channel 1 (left-side), and to have $+$ in direction $\hat{\mathbf{b}}$ which is observed in channel 2 (right-side).

In a classical hidden variable theory, if the spin component of particle-1 is measured with $+$ in the direction $\hat{\mathbf{a}}$, then a measurement of the spin of particle-2 in the direction $\hat{\mathbf{a}}$ must be $-$ with certainty. The probability P_{ab} in which particle-1 is measured with $+$ in the direction $\hat{\mathbf{a}}$ (in channel 1), and particle-2 is measured with $-$ in the direction $\hat{\mathbf{b}}$ (in channel 2) is written as

$$P_{ab} = P(+ \ - \ d_c \,|\, - \ + \ d_c). \tag{14.1.2}$$

Similarly, we have

$$P_{bc} = P(d_a \ + \ - \,|\, d_a \ - \ +)$$
$$P_{ac} = P(+ \ d_b \ - \,|\, - \ d_b \ +). \tag{14.1.3}$$

In a classical probability theory, the undetermined spin components of particle-1 and particle-2, d_a, d_b, or d_c, must have specific values, either $+$ or $-$ in any direction. Consequently, we may write P_{ab} as a sum of probabilities for these two possibilities

$$P_{ab} = P(+ \ - \ + \,|\, - \ + \ -) + P(+ \ - \ - \,|\, - \ + \ +)$$
$$\equiv P(+ \ - \ +) + P(+ \ - \ -), \tag{14.1.4}$$

where on the second line we have dropped the redundant information involving the second particle by writing $P(+ \ - \ +)$ for $P(+ \ - \ + \,|\, - \ + \ -)$. Thus, we write Eqs. (14.1.2) to (14.1.3) as

$$P_{ab} = P(+ \ - \ +) + P(+ \ - \ -),$$
$$P_{bc} = P(+ \ + \ -) + P(- \ + \ -),$$
$$P_{ac} = P(+ \ + \ -) + P(+ \ - \ -). \tag{14.1.5}$$

It should be emphasized that the left side quantities in Eq. (14.1.5), P_{ab}, $etc.$, are measurable probabilities, but we have no procedure for measuring the separate classical probabilities $P(+ \ - \ +)$, $P(+ \ - \ -)$, $etc.$ on the right side of Eq. (14.1.5). Adding P_{ab} and P_{bc} and comparing with P_{ac}, it is easy to see that

$$P_{ab} + P_{bc} = P_{ac} + P(+ \ - \ +) + P(- \ + \ -). \tag{14.1.6}$$

Assuming the classical probabilities $P(+ \ - \ +)$ and $P(- \ + \ -)$ are positive, which is true in any classical probability theory, we then obtain the Bell-Wigner inequality

$$P_{ab} + P_{bc} \geq P_{ac}. \tag{14.1.7}$$

The seemingly obvious inequality of Eq. (14.1.7) has been experimentally violated.

We next present a simple quantum mechanical calculation for the same measurement. Let us rewrite the state vector of Eq. (14.0.1) into the following form

$$|\Psi\rangle = \frac{1}{\sqrt{2}}\left[\begin{pmatrix} 1 \\ 0 \end{pmatrix}_1 \begin{pmatrix} 0 \\ 1 \end{pmatrix}_2 - \begin{pmatrix} 0 \\ 1 \end{pmatrix}_1 \begin{pmatrix} 1 \\ 0 \end{pmatrix}_2\right]. \tag{14.1.8}$$

Recalling that for a spin 1/2 particle in a state $|\psi\rangle$ the probability of observing spin up in a direction at angle θ to the z-axis is

$$|\langle\theta|\Psi\rangle\}^2 = \langle\Psi|\theta\rangle\langle\theta|\Psi\rangle = \langle\Psi|\hat{\pi}_\theta|\Psi\rangle, \tag{14.1.9}$$

where $\hat{\pi}_\theta = |\theta\rangle\langle\theta|$ is the projection operator, and $|\theta\rangle = e^{-(i/2)\,\sigma_y\theta\,|\uparrow\rangle}$. The projection operator for spin up in the direction of a unit vector $\hat{\mathbf{r}}$ at angle θ to the z-axis is

$$\hat{\pi}_\theta = |\theta\rangle\langle\theta| = \frac{1}{2}(1 + \sigma_z\cos\theta + \sigma_x\sin\theta) \equiv \frac{1}{2}(\mathbf{I} + \vec{\sigma}\cdot\hat{\mathbf{r}}), \tag{14.1.10}$$

where $\vec{\sigma} = (\sigma_x, \sigma_y, \sigma_z)$ are the Pauli matrices and $\mathbf{I}$ is the identity matrix. We can also explicitly rewrite $\hat{\pi}_\theta$ as a matrix,

$$\hat{\pi}_\theta = \frac{1}{2}\begin{pmatrix} 1 + \cos\theta & \sin\theta \\ \sin\theta & 1 - \cos\theta \end{pmatrix} \tag{14.1.11}$$

The physical observable P_{ab} is defined as the joint probability for particle-1 to be spin up along an SGA oriented in direction $\hat{\mathbf{a}}$ and particle-2 to be spin up along another SGA oriented in direction $\hat{\mathbf{b}}$

$$P_{ab} = \langle\Psi|\hat{\pi}_{1\hat{\mathbf{a}}}\,\hat{\pi}_{2\hat{\mathbf{b}}}|\Psi\rangle, \tag{14.1.12}$$

where each projector has two subscripts defining the particles (and channel) and the measurement direction in the $x - z$ plane. Similarly,

$$P_{bc} = \langle\Psi|\hat{\pi}_{1\hat{\mathbf{b}}}\,\hat{\pi}_{2\hat{\mathbf{c}}}|\Psi\rangle,$$
$$P_{ac} = \langle\Psi|\hat{\pi}_{1\hat{\mathbf{a}}}\,\hat{\pi}_{2\hat{\mathbf{c}}}|\Psi\rangle. \tag{14.1.13}$$

Evaluation of P_{ab}, P_{bc}, and P_{ac} using the operator in Eq. (14.1.11) and the state in Eq. (14.1.8) gives

$$P_{ab} = \frac{1}{4}\left[1 - \cos\left(\theta_a - \theta_b\right)\right] = \frac{1}{4}\left[1 - \hat{\mathbf{a}}\cdot\hat{\mathbf{b}}\right],$$
$$P_{bc} = \frac{1}{4}\left[1 - \cos\left(\theta_b - \theta_c\right)\right] = \frac{1}{4}\left[1 - \hat{\mathbf{b}}\cdot\hat{\mathbf{c}}\right],$$
$$P_{ac} = \frac{1}{4}\left[1 - \cos\left(\theta_a - \theta_c\right)\right] = \frac{1}{4}\left[1 - \hat{\mathbf{a}}\cdot\hat{\mathbf{c}}\right]. \tag{14.1.14}$$

It is easy to find that the Bell-Wigner inequality in Eq. (14.1.7) achieves its maximum violation when $\theta_a = 0$, $\theta_b = \pi/3$, and $\theta_c = 2\pi/3$. The experimental observations have confirmed the violation.

14.2 BELL'S THEOREM AND BELL'S INEQUALITY

Bell first proved his theorem in 1964 under idealized experimental condition of 100% detection efficiency. It was soon realized that the Bell's inequality under idealized condition cannot be tested experimentally. Any realistic particle detector cannot have quantum efficiency of 100%. Bell improved his theorem and inequality of 1964 in 1972 by including realistic non-perfect particle detection.

(I) Bell's theorem of 1964.

In his 1964 paper, based on the EPR-Bohm state of Eq. (14.0.1) and a hidden parameter λ, Bell derived an inequality to distinguish quantum mechanics from local realistic probability theory of hidden variable. In his pioneer work Bell introduced a "more complete specification effected by means of parameter λ" with probability distribution $\rho(\lambda)$ for the classical statistical estimation of the expectation value of the joint measurement $\langle \Psi | (\vec{\sigma}_1 \cdot \hat{\mathbf{a}}) (\vec{\sigma}_2 \cdot \hat{\mathbf{b}}) | \Psi \rangle$ of particle-1 and particle-2 in the directions $\hat{\mathbf{a}}$ and $\hat{\mathbf{b}}$, simultaneously and respectively. The quantum mechanical result of this measurement gives

$$E_{ab} = \langle \Psi | (\vec{\sigma}_1 \cdot \hat{\mathbf{a}}) (\vec{\sigma}_2 \cdot \hat{\mathbf{b}}) | \Psi \rangle = -\hat{\mathbf{a}} \cdot \hat{\mathbf{b}}. \tag{14.2.1}$$

A special case of this result contains the determinism implicit in this idealized system, When the SGA analyzers are parallel, we have

$$E_{ab} = \langle \Psi | (\vec{\sigma}_1 \cdot \hat{\mathbf{a}}) (\vec{\sigma}_2 \cdot \hat{\mathbf{a}}) | \Psi \rangle = -1 \tag{14.2.2}$$

for all λ and all $\hat{\mathbf{a}}$. Thus, we can predict with certainty the result B by obtaining the result of A. Since the quantum mechanical state $|\Psi\rangle$ does not determine the result of an individual measurement, this fact (via EPR's argument) suggests that there exits a more complete specification of the state by a single symbol λ it may have many dimensions, discrete and/or continuous parts, and different parts of it interacting with either apparatus, *etc.* Let Λ be the space of λ for an ensemble composed of a very large number of the particle systems. Bell represented the distribution function for the state λ on the space Λ by the symbol $\rho(\lambda)$ and take $\rho(\lambda)$ to be normalized

$$\int_\Lambda \rho(\lambda)\, d\lambda = 1. \tag{14.2.3}$$

In a deterministic hidden variable theory the observable $[A(\hat{\mathbf{a}})B(\hat{\mathbf{b}})]$ has a defined value $[A(\hat{\mathbf{a}})B(\hat{\mathbf{b}})](\lambda)$ for the state λ.

The locality is defined as follows: a deterministic hidden variable theory is local if for all $\hat{\mathbf{a}}$ and $\hat{\mathbf{b}}$ and and all $\lambda \in \Lambda$

$$[A(\hat{\mathbf{a}})B(\hat{\mathbf{b}})](\lambda) = A(\hat{\mathbf{a}}, \lambda)\, B(\hat{\mathbf{b}}, \lambda). \tag{14.2.4}$$

This is, once λ is specified and the particle have separated, measurements of A can depend only upon λ and $\hat{\mathbf{a}}$ but not $\hat{\mathbf{b}}$. Likewise measurements of B depend only upon λ and $\hat{\mathbf{b}}$. Any reasonable physical theory that is realistic and deterministic and that denies action-at-a-distance is local in this sense. For such theories the expectation value of $[A(\hat{\mathbf{a}})B(\hat{\mathbf{b}})]$ is given by:

$$\begin{aligned} E(\hat{\mathbf{a}}, \hat{\mathbf{b}}) &= \int_\Lambda d\lambda\, \rho(\lambda)\, [A(\hat{\mathbf{a}})B(\hat{\mathbf{b}})](\lambda) \\ &= \int_\Lambda d\lambda\, \rho(\lambda)\, A(\hat{\mathbf{a}}, \lambda)\, B(\hat{\mathbf{b}}, \lambda), \end{aligned} \tag{14.2.5}$$

where $E(\hat{\mathbf{a}}, \hat{\mathbf{b}}) \equiv E_{ab}$, corresponding to our previous notation. It is clear that Eq. (14.2.2) can hold if only if

$$A(\hat{\mathbf{a}}, \lambda) = -B(\hat{\mathbf{b}}, \lambda) \tag{14.2.6}$$

hold for all $\lambda \in \Lambda$.

Using Eq. (14.2.6) we calculate the following expectation values, which involves three different orientations of the SGA analyzers:

$$
\begin{aligned}
& E(\hat{\mathbf{a}}, \hat{\mathbf{b}}) - E(\hat{\mathbf{a}}, \hat{\mathbf{c}}) \\
& = \int_\Lambda d\lambda \, \rho(\lambda) \left[A(\hat{\mathbf{a}}, \lambda) \, B(\hat{\mathbf{b}}, \lambda) - A(\hat{\mathbf{a}}, \lambda) \, B(\hat{\mathbf{c}}, \lambda) \right] \\
& = - \int_\Lambda d\lambda \, \rho(\lambda) \left[A(\hat{\mathbf{a}}, \lambda) \, A(\hat{\mathbf{b}}, \lambda) - A(\hat{\mathbf{a}}, \lambda) \, A(\hat{\mathbf{c}}, \lambda) \right] \\
& = - \int_\Lambda d\lambda \, \rho(\lambda) \, A(\hat{\mathbf{a}}, \lambda) \, A(\hat{\mathbf{b}}, \lambda) \left[1 - A(\hat{\mathbf{b}}, \lambda) \, A(\hat{\mathbf{c}}, \lambda) \right]. \quad (14.2.7)
\end{aligned}
$$

Since $A(\hat{\mathbf{a}}, \lambda) = \pm 1$, $A(\hat{\mathbf{b}}, \lambda) = \pm 1$, this expression can be written as

$$
\left| E(\hat{\mathbf{a}}, \hat{\mathbf{b}}) - E(\hat{\mathbf{a}}, \hat{\mathbf{c}}) \right| \leq \int_\Lambda d\lambda \, \rho(\lambda) \left[1 - A(\hat{\mathbf{b}}, \lambda) \, A(\hat{\mathbf{c}}, \lambda) \right], \quad (14.2.8)
$$

and consequently,

$$
\left| E(\hat{\mathbf{a}}, \hat{\mathbf{b}}) - E(\hat{\mathbf{a}}, \hat{\mathbf{c}}) \right| \leq 1 + E(\hat{\mathbf{b}}, \hat{\mathbf{c}}). \quad (14.2.9)
$$

This inequality is the first of a family of inequalities which are collectively called "Bell's inequalities".

It is easy to find a disagreement between the quantum mechanics prediction of Eq. (14.2.1) and the inequality of Eq. (14.2.9). When we choose $\hat{\mathbf{a}}$, $\hat{\mathbf{b}}$ and $\hat{\mathbf{c}}$ to be coplanar with $\hat{\mathbf{c}}$ making an angle of $2\pi/3$ with $\hat{\mathbf{a}}$, and $\hat{\mathbf{b}}$ making an angle of $\pi/3$ with both $\hat{\mathbf{a}}$ and $\hat{\mathbf{c}}$, the quantum prediction gives

$$
\left| \left[E(\hat{\mathbf{a}}, \hat{\mathbf{b}}) - E(\hat{\mathbf{a}}, \hat{\mathbf{c}}) \right]_{QM} \right| = 1, \quad (14.2.10)
$$

while

$$
1 + \left[E(\hat{\mathbf{b}}, \hat{\mathbf{c}}) \right]_{QM} = \frac{1}{2}. \quad (14.2.11)
$$

It does not satisfy inequality of Eq. (14.2.9).

In summary, Bell's theorem has proved that no deterministic hidden variable theory satisfying Eq. (14.2.2) and Eq. (14.2.4) can agree with all of the predications of quantum mechanics concerning the spin of a pair of spin-1/2 particles in the singlet state of Eq. (14.0.1).

Since the space of Λ in Eq. (14.2.5) is spanned into four regions with classical probabilities P_{ab}, P_{-ab}, P_{a-b}, P_{-a-b} in which A and B have values ± 1, the expectation value evaluation of Eq. (14.2.5) can be explicitly calculated as

$$
\begin{aligned}
E_{ab} &= (+1)(+1)P_{ab} + (-1)(+1)P_{-ab} + (+1)(-1)P_{a-b} + (-1)(-1)P_{-a-b} \\
&= P_{ab} - P_{-ab} - P_{a-b} + P_{-a-b}. \quad (14.2.12)
\end{aligned}
$$

Consider three directions, $\hat{\mathbf{a}}$, $\hat{\mathbf{b}}$, $\hat{\mathbf{c}}$, using the results of Eq. (14.1.5), Eq. (14.2.12) gives

$$
\begin{aligned}
E_{ab} = \; & P(+ \; - \; +) + P(+ \; - \; -) - P(- \; - \; +) - P(- \; - \; -) \\
& - P(+ \; + \; +) - P(+ \; + \; -) + P(- \; + \; +) + P(- \; + \; -).
\end{aligned}
$$

Together with the normalization condition

$$\sum P(\pm\ \pm\ \pm) = 1,$$

E_{ab} becomes

$$E_{ab} = 2\big[P(+\ -\ +) + P(+\ -\ -) + P(-\ +\ +) + P(-\ +\ -)\big] - 1.$$

Similarly, we have

$$E_{bc} = 2\big[P(+\ +\ -) + P(-\ +\ -) + P(+\ -\ +) + P(-\ -\ +)\big] - 1.$$

$$E_{ac} = 2\big[P(+\ +\ -) + P(+\ -\ -) + P(-\ +\ +) + P(-\ -\ +)\big] - 1.$$

From the above three equations, it is not difficult to find that

$$E_{ab} - E_{ac} = 1 + E_{bc} - 4P(+\ +\ -) - 4P(-\ -\ +)$$
$$E_{ac} - E_{ab} = 1 + E_{bc} - 4P(-\ +\ -) - 4P(+\ -\ +). \tag{14.2.13}$$

Considering the positive values of the probabilities, Eq. (14.2.13) implies

$$E_{ab} - E_{ac} \leq 1 + E_{bc}$$
$$E_{ac} - E_{ab} \leq 1 + E_{bc}, \tag{14.2.14}$$

and consequently we obtain the original Bell's inequality,

$$\big|E_{ab} - E_{ac}\big| \leq 1 + E_{bc}. \tag{14.2.15}$$

(II) Bell's theorem of 1971.

It was soon realized that the Bell's inequality of Eq. (14.2.9) cannot be tested in a real experiment. Because Eq. (14.2.2) cannot be hold exactly in an realistic measurement. Any real detector cannot have a perfect quantum efficiency of 100%, and any real analyzer cannot have a perfect distinguish ration between orthogonal channels. In 1971, Bell proved a new inequality which includes these concerns by assuming the outcomes of measurement A or B may take one of the following possible results

$$A(\hat{\mathbf{a}}, \lambda)\,\text{or}\,B(\hat{\mathbf{b}}, \lambda) = \begin{cases} +1 & \text{``spin-up''} \\ -1 & \text{``spin-down''} \\ 0 & \text{particle\ \ not\ \ detected} \end{cases} \tag{14.2.16}$$

For a given state λ, we define the measured values for these quantities by the symbols $\bar{A}(\hat{\mathbf{a}}, \lambda)$ and $\bar{B}(\hat{\mathbf{b}}, \lambda)$, which satisfy

$$\big|\bar{A}(\hat{\mathbf{a}}, \lambda)\big| \leq 1 \ \ \text{and} \ \ \big|\bar{B}(\hat{\mathbf{b}}, \lambda)\big| \leq 1. \tag{14.2.17}$$

Following the same definition of locality, the expectation value of $A(\hat{\mathbf{a}})B(\hat{\mathbf{b}})$ is calculated as

$$E(\hat{\mathbf{a}}, \hat{\mathbf{b}}) = \int_\Lambda d\lambda\,\rho(\lambda)\,\bar{A}(\hat{\mathbf{a}}, \lambda)\,\bar{B}(\hat{\mathbf{b}}, \lambda). \tag{14.2.18}$$

Consider a measurement which involves $E(\hat{\mathbf{a}}, \hat{\mathbf{b}})$ and $E(\hat{\mathbf{a}}, \hat{\mathbf{b}}')$

$$E(\hat{\mathbf{a}}, \hat{\mathbf{b}}) - E(\hat{\mathbf{a}}, \hat{\mathbf{b}}')$$
$$= \int_\Lambda d\lambda\,\rho(\lambda)\big[\bar{A}(\hat{\mathbf{a}}, \lambda)\,\bar{B}(\hat{\mathbf{b}}, \lambda) - \bar{A}(\hat{\mathbf{a}}, \lambda)\,\bar{B}(\hat{\mathbf{b}}', \lambda)\big], \tag{14.2.19}$$

which can be written in the following form:

$$E(\hat{\mathbf{a}}, \hat{\mathbf{b}}) - E(\hat{\mathbf{a}}, \hat{\mathbf{b}}')$$

$$= \int_\Lambda d\lambda \, \rho(\lambda) \, \bar{A}(\hat{\mathbf{a}}, \lambda) \, \bar{B}(\hat{\mathbf{b}}, \lambda) \big[1 \pm \bar{A}(\hat{\mathbf{a}}', \lambda) \, \bar{B}(\hat{\mathbf{b}}', \lambda) \big]$$

$$- \int_\Lambda d\lambda \, \rho(\lambda) \, \bar{A}(\hat{\mathbf{a}}, \lambda) \, \bar{B}(\hat{\mathbf{b}}', \lambda) \big[1 \pm \bar{A}(\hat{\mathbf{a}}', \lambda) \, \bar{B}(\hat{\mathbf{b}}, \lambda) \big]. \qquad (14.2.20)$$

Using inequality in Eq. (14.2.17), we then have

$$\big| E(\hat{\mathbf{a}}, \hat{\mathbf{b}}) - E(\hat{\mathbf{a}}, \hat{\mathbf{b}}') \big| \leq \int_\Lambda d\lambda \, \rho(\lambda) \, \big[1 \pm \bar{A}(\hat{\mathbf{a}}', \lambda) \, \bar{B}(\hat{\mathbf{b}}', \lambda) \big]$$

$$+ \int_\Lambda d\lambda \, \rho(\lambda) \, \big[1 \pm \bar{A}(\hat{\mathbf{a}}', \lambda) \, \bar{B}(\hat{\mathbf{b}}, \lambda) \big], \qquad (14.2.21)$$

or

$$\big| E(\hat{\mathbf{a}}, \hat{\mathbf{b}}) - E(\hat{\mathbf{a}}, \hat{\mathbf{b}}') \big| \leq \pm \big[E(\hat{\mathbf{a}}', \hat{\mathbf{b}}') + E(\hat{\mathbf{a}}', \hat{\mathbf{b}}) \big] + 2 \int_\Lambda d\lambda \, \rho(\lambda). \qquad (14.2.22)$$

We thus derive a measurable inequality

$$-2 \leq E(\hat{\mathbf{a}}, \hat{\mathbf{b}}) - E(\hat{\mathbf{a}}, \hat{\mathbf{b}}') + E(\hat{\mathbf{a}}', \hat{\mathbf{b}}) + E(\hat{\mathbf{a}}', \hat{\mathbf{b}}') \leq 2. \qquad (14.2.23)$$

The quantum mechanical prediction of the Bhom state in a realistic measurement with in-perfect detectors, analyzers *etc.*, can be written as

$$\big[E(\hat{\mathbf{a}}, \hat{\mathbf{b}}) \big]_{QM} = \mathrm{C} \, \hat{\mathbf{a}} \cdot \hat{\mathbf{b}} \qquad (14.2.24)$$

where $|\mathrm{C}| \leq 1$. Suppose we take $\hat{\mathbf{a}}$, $\hat{\mathbf{a}}'$, $\hat{\mathbf{b}}$, $\hat{\mathbf{b}}'$ to be coplanar with $\phi = \pi/4$, we can easily find a disagreement between the quantum mechanics prediction and the inequality of Eq. (14.2.23):

$$\big[E(\hat{\mathbf{a}}, \hat{\mathbf{b}}) - E(\hat{\mathbf{a}}, \hat{\mathbf{b}}') + E(\hat{\mathbf{a}}', \hat{\mathbf{b}}) + E(\hat{\mathbf{a}}', \hat{\mathbf{b}}') \big]_{QM} = 2\sqrt{2} \, \mathrm{C}. \qquad (14.2.25)$$

Although Bell derived his inequalities based on the measurement of EPR-Bohm state of Eq. (14.0.1), Eq. (14.2.9) and Eq. (14.2.23) are not restricted to the measurement of spin 1/2 particle pairs. In fact, most of the historical experimental testing have been the polarization measurements of photon pairs. The photon pairs are prepared in similar states which have been called EPR-Bohm-Bell states, or Bell states in short. Most of the experimental observations violated Bell's inequalities which may have different forms and have their violation occur at different orientations of the polarization analyzers. However, the physics behind the violations are all similar to that of Bell's theorem.

14.3 BELL STATES

We have given an example of EPR-Bohm state in earlier discussions. The high energy γ-ray photon pair disintegrated from the annihilation of positronium is a good example to explore the physics of the EPR-Bohm state, however, the γ-ray photon pairs are difficult to handle experimentally: (1) There is no effective polarization analyzers available for the high energy γ-rays; (2) The uncertainty in momentum correlation, $\Delta(\mathbf{p}_1 + \mathbf{p}_2)$, has considerable large value, resulting in a "pair collection efficiency loophole" in Bell inequality measurements, i.e., one may never have $\sim 100\%$ chance to "collect" a pair for joint photodetection measurement. Fortunately, the two-photon state of Eq. (14.0.3) is also observed in atomic cascade decay with visible-ultraviolet wavelengths and we have plenty of high efficiency polarization analyzers available in that wavelengths. Thus, most of the early EPR-Bohm-Bell experiments demonstrated in 70's and early 80's used two-photon source of atomic cascade decay. These experiments, unfortunately, still experienced the difficulties in the momentum uncertainty. The "pair collection" efficiency is as low as that of the annihilation of positronium. It was in the middle of 1980's, Alley and Shih introduced the nonlinear spontaneous parametric down-conversion to the preparation of entangled states. The entangled signal-idler photon pair can be easily prepared in visible-infrared wavelengths, and very importantly, the uncertainty in momentum correlation was improved significantly. The "pair collection efficiency loophole" was finally removed.

The following set of four two-photon spin states can be generated by manipulating the signal-idler pair of SPDC with beamsplitters, half and/or quarter wave-plates:

$$| \Psi_{RL}^{\pm} \rangle = \frac{1}{\sqrt{2}} \big[| R_1 \rangle | R_2 \rangle \pm | L_1 \rangle | L_2 \rangle \big]$$

$$| \Phi_{RL}^{\pm} \rangle = \frac{1}{\sqrt{2}} \big[| R_1 \rangle | L_2 \rangle \pm | L_1 \rangle | R_2 \rangle \big]. \tag{14.3.1}$$

The four states in Eq. (14.3.1), named as EPR-Bohm-Bell states, or simply "Bell states", form a complete orthonormal basis in two-particle vector space. In principle, any arbitrary two-photon polarization state can be expressed as an appropriate superposition of Bell states. Bell states have been extensively involved in studying fundamental issues of quantum theory as well as in practical applications of quantum entanglement. In the field of quantum information, Bell states are usually represented in the following form of "qubit":

$$| \Psi_{01}^{(\pm)} \rangle = \frac{1}{\sqrt{2}} \big[| 0_1 1_2 \rangle \pm | 1_1 0_2 \rangle \big]$$

$$| \Phi_{01}^{(\pm)} \rangle = \frac{1}{\sqrt{2}} \big[| 0_1 0_2 \rangle \pm | 1_1 1_2 \rangle \big] \tag{14.3.2}$$

where $|0\rangle$ and $|1\rangle$ represent two arbitrary orthogonal polarization bases. The most popular Bell states are prepared from SPDC in linear polarization representation:

$$| \Psi_{XY}^{\pm} \rangle = \frac{1}{\sqrt{2}} \big[| X_1 \rangle | Y_2 \rangle \pm | Y_1 \rangle | X_2 \rangle \big]$$

$$| \Phi_{XY}^{\pm} \rangle = \frac{1}{\sqrt{2}} \big[| X_1 \rangle | X_2 \rangle \pm | Y_1 \rangle | Y_2 \rangle \big] \tag{14.3.3}$$

where $|X\rangle$ and $|Y\rangle$, respectively, are defined by the polarization of the o-ray and the e-ray of the nonlinear crystal of SPDC. Treating the circular polarization state as the superposition of the linear polarization states,

$$| R \rangle = | X \rangle + i | Y \rangle$$
$$| L \rangle = - | X \rangle + i | Y \rangle, \tag{14.3.4}$$

we may find the following states exchangeable between the linear polarization representation and the circular polarization representation:

$$\begin{aligned} | \Psi_{XY}^{+} \rangle = | \Psi_{RL}^{-} \rangle, \qquad & | \Psi_{XY}^{-} \rangle = | \Phi_{RL}^{-} \rangle; \\ | \Phi_{XY}^{+} \rangle = | \Phi_{RL}^{+} \rangle, \qquad & | \Phi_{XY}^{-} \rangle = | \Psi_{RL}^{+} \rangle. \end{aligned} \tag{14.3.5}$$

To simplify the notation, we will ignore the subscription for the linear polarization representation to replace $| \Psi_{XY}^{\pm} \rangle$ and $| \Phi_{XY}^{\pm} \rangle$ with $| \Psi^{\pm} \rangle$ and $| \Phi^{\pm} \rangle$, respectively. The polarization correlation functions, in terms of chosen orientations of the polarization analyzers, θ_1 and θ_2, for the Bell states of Eq. (14.3.3) can be easily calculated:

$$| \langle \theta_1 \, \theta_2 \, | \, \Psi^\pm \, \rangle |^2 = \frac{1}{2} \sin^2(\theta_1 \pm \theta_2)$$

$$| \langle \theta_1 \, \theta_2 \, | \, \Phi^\pm \, \rangle |^2 = \frac{1}{2} \cos^2(\theta_1 \mp \theta_2). \tag{14.3.6}$$

The other correlation functions in terms of $\bar{\theta}$s or in terms of θ and $\bar{\theta}$ can be derived accordingly based on Eq. (14.3.6).

It is noticed that the condition for achieving maximum correlation, in terms of θ_1 and θ_2, in $| \, \Psi^+ \, \rangle$ $(| \, \Psi^- \, \rangle)$ and in $| \, \Phi^+ \, \rangle$ $(| \, \Phi^- \, \rangle)$ are different: $\theta_1 + \theta_2 = \pi/2$ against $\theta_1 - \theta_2 = \pi/2$ $(\theta_1 - \theta_2 = 0$ against $\theta_1 + \theta_2 = 0)$. Thus, if one of them implies orthogonal (parallel) polarization of the photon pair, the other one must imply a different geometrical relationship between the two polarization.

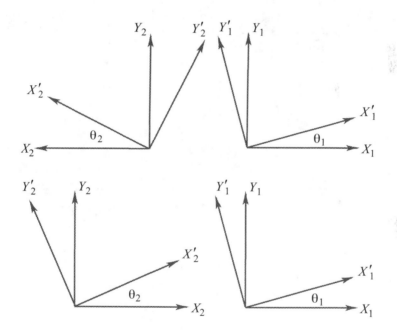

FIGURE 14.3.1 Upper: right hand coordinate system with anti-parralle propagation. The Z_1 (Z_2) axis is point out from (into) the paper. The relative angle between X_1-Y_1 and X_2-Y_2 are $\theta_2 + \theta_1$ after the rotation. Lower: right hand coordinate system with collinear or near collinear propagation. Both Z_1 and Z_2 axes are point out the paper. The relative angle between X_1-Y_1 and X_2-Y_2 are $\theta_2 - \theta_1$ after the rotation.

Now, we are ready to discuss the rotation symmetry of Bell states. We have mentioned earlier that while preparing Bell states in SPDC, the polarization vector base of $|X\rangle$ and $|Y\rangle$ are usually defined by the chosen orientation of the nonlinear crystal and the associated wave-plates. What happen if we rotate $|X\rangle$-$|Y\rangle$ vector base to $|X'\rangle$-$|Y'\rangle$ with angle θ? It is easy to find that in each set of Eq. (14.3.1) and Eq. (14.3.3), only two of the states reserve the symmetry of rotation. In a right-hand coordinate system with anti-parallel propagation, i.e., the pair propagate to opposite directions, which is illustrated in the upper part of Fig. 14.3.1, the two Bell states which reserve rotation symmetry are:

$$|\Psi^+\rangle = \frac{1}{\sqrt{2}}\left[|X_1'\rangle|Y_2'\rangle + |Y_1'\rangle|X_2'\rangle\right]$$

$$|\Phi^-\rangle = \frac{1}{\sqrt{2}}\left[|X_1'\rangle|X_2'\rangle - |Y_1'\rangle|Y_2'\rangle\right]. \tag{14.3.7}$$

In a right-hand coordinate system with collinear or near collinear propagation, i.e., the pair propagate to the same direction, which is illustrated in the lower part of Fig. 14.3.1, the two Bell states which reserve rotation symmetry are:

$$|\Psi^-\rangle = \frac{1}{\sqrt{2}}\left[|X_1'\rangle|Y_2'\rangle - |Y_1'\rangle|X_2'\rangle\right]$$

$$|\Phi^+\rangle = \frac{1}{\sqrt{2}}\left[|X_1'\rangle|X_2'\rangle + |Y_1'\rangle|Y_2'\rangle\right]. \tag{14.3.8}$$

It might be easier to figure out the results in Eqs. (14.3.7) and (14.3.8) from the circular polarization based states of Eq. (14.3.1). If the rotation from X-Y to X'-Y' means a rotation of $\theta_1 = \theta$ in arm-1 and $\theta_2 = -\theta$ in arm-2, which is illustrated in the upper part of Fig. 14.3.1, $|\Psi^\pm_{RL}\rangle$ will reserve rotation symmetry:

$$
\begin{aligned}
|\Psi^\pm_{RL}\rangle &= \frac{1}{\sqrt{2}}\left[|R_1\rangle|R_2\rangle \pm |L_1\rangle|L_2\rangle\right] \\
&= \frac{1}{\sqrt{2}}\left[e^{i(\theta_1+\theta_2)}|R_1'\rangle|R_2'\rangle \pm e^{-i(\theta_1+\theta_2)}|L_1'\rangle|L_2'\rangle\right] \\
&= \frac{1}{\sqrt{2}}\left[|R_1'\rangle|R_2'\rangle \pm |L_1'\rangle|L_2'\rangle\right],
\end{aligned}
\tag{14.3.9}
$$

corresponding to Eq. (14.3.7). However, if the rotation from X-Y to X'-Y' means a rotation of $\theta_1 = \theta_2 = \theta$ in both arm-1 and arm-2, which is illustrated in the lower part of Fig. 14.3.1, $|\Phi^\pm_{RL}\rangle$ will reserve rotation symmetry:

$$
\begin{aligned}
|\Phi^\pm_{RL}\rangle &= \frac{1}{\sqrt{2}}\left[|R_1\rangle|L_2\rangle \pm |L_1\rangle|R_2\rangle\right] \\
&= \frac{1}{\sqrt{2}}\left[e^{i(\theta_1-\theta_2)}|R_1'\rangle|L_2'\rangle \pm e^{-i(\theta_1-\theta_2)}|L_1'\rangle|R_2'\rangle\right] \\
&= \frac{1}{\sqrt{2}}\left[|R_1'\rangle|L_2'\rangle \pm |L_1'\rangle|R_2'\rangle\right],
\end{aligned}
\tag{14.3.10}
$$

corresponding to Eq. (14.3.8).

14.4   BELL STATE PREPARATION

SPDC has been one of the most convenient two-photon sources for the preparation of Bell state since the early 1980's. Although Bell state is for polarization (or spin), the space-time part of the state cannot be ignored. One important "preparation" is to make the two biphoton wavepackes, corresponding to the first and the second terms in the Bell state, completely "overlap" in space-time, or indistinguishable for the joint detection event. This is especially important for type-II SPDC. Type-II SPDC seems easier to use for the preparation of Bell state due to the orthogonal polarization of the signal and idler.

A very interesting situation for type-II SPDC is that of "noncollinear phase matching". The signal-idler pair are emitted from a SPDC crystal, such as BBO, cut in type-II phase

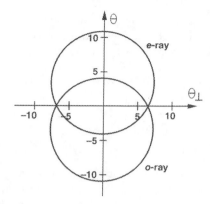

FIGURE 14.4.1 Type-II noncolinear phase matching: a cross section view of the degenerate 702.2nm cones. The 351.1nm pump beam is in the center. The numbers along the axes are in degrees.

matching, into two cones, one ordinarily polarized, the other extraordinarily polarized, see Fig. 14.4.1. Along the intersection, where the cones overlap, two pinholes numbered 1 and 2 are used for defining the direction of the **k** vectors of the signal-idler pair. It is very reasonable to consider the polarization state of the signal-idler pair as

$$|\Psi\rangle = \frac{1}{\sqrt{2}}(|o_1 e_2\rangle + |e_1 o_2\rangle) = |\Psi_{XY}^+\rangle \qquad (14.4.1)$$

where o_j and e_j, $j = 1, 2$, are ordinarily and extraordinarily polarization, respectively. It seems straightforward to realize an EPR-Bohm-Bell measurement by simply setting up a polarization analyzer in series with a photon counting detector behind pinholes 1 and 2, respectively, and to expect observe the polarization correlation of Eq. (14.3.6). This is, however, *incorrect*! One can never observe the EPR–Bohm-Bell polarization correlation unless a "compensator" is applied. The "compensator" is a piece of birefringent material. For example, one may place another piece of nonlinear crystal behind the SPDC. It could be the same type of crystal as that of the SPDC, with the same cutting angle, except having half the length and a 90° rotation in respect to that of the SPDC crystal.

What is the role of the "compensator"? There have been naive explanations about the compensator. One suggestion was that the problem comes from the longitudinal "walk-off" of the type-II SPDC. For example, if one uses a type II BBO, which is a negative uni-axis crystal, the extraordinary-ray propagates faster than the ordinary-ray inside the BBO. *Suppose* the $o - e \leftrightarrow e - o$ pair is generated in the *middle* of the crystal, the e-polarization will trigger the detector earlier than the o-polarization by a time $\Delta t = (n_o - n_e)L/2c$. This implies that D_2 would be fired first in $|o_1 e_2\rangle$ term; but D_1 would be fired first in $|e_1 o_2\rangle$ term. If Δt is greater than the coherence length of the signal-idler field, one would be able to distinguish which amplitude gave rise to the "click-click" coincidence event. One may compensate the "walk-off" by introducing an additional piece of birefringent material, like the compensator we have suggested above, to delay the e-ray relative to the o-ray by the same amount of time, Δt. If, however, the signal-idler pair is generated in the *front face* or the *back face* of the SPDC, the delay time would be very different: $\Delta t = (n_o - n_e)L/c$ for the *front face* and $\Delta t = 0$ for the *back face*. One can never satisfy all the pairs which are generated at different places along the SPDC crystal. Nevertheless, since SPDC is a *coherent* process, the signal-idler pair is generated in such a way that it is impossible to know the birth place of the pair. So, how is the delay time Δt determined?

Quantum mechanics has provided the correct answer. It is the overlap or indistinguishability of the biphoton wavepackets in space-time.

The Bell state of Eq. (14.4.1) as well as Eqs. (14.3.1) and (14.3.3) are written in the simplified Schrödinger representation. It is sufficient for certain types of discussions of physics. The simplified form, however, has ignored the space-time aspect of the state, which is important here.

Adopting our earlier result, we may rewrite the state of the signal-idler pair of Fig. 14.4.1 in the following form:

$$|\Psi\rangle = \sum_{\mathbf{k}_o, \mathbf{k}_e} \delta(\omega_o + \omega_e - \omega_p) \Phi(\Delta_k L) \; \hat{\mathbf{o}} \; \hat{a}_o^\dagger(\omega(\mathbf{k}_o)) \; \hat{\mathbf{e}} \; \hat{a}_e^\dagger(\omega(\mathbf{k}_e)) \; |0\rangle \qquad (14.4.2)$$

where $\hat{\mathbf{o}}$ and $\hat{\mathbf{e}}$ are unit vectors along the o-ray and the e-ray polarization direction of the SPDC crystal, and $\Delta_k = k_o + k_e - k_p$. In $\Phi(\Delta_k L)$, the finite length of the nonlinear crystal has been taken into account. Suppose the polarizers of the detectors D_1 and D_2 are set at angles θ_1 and θ_2, relative to the polarization direction of the o-ray of the SPDC crystal, respectively, the field operators can be written as

$$E_j^{(+)}(t_j, r_j) = \int d\omega \; \hat{\theta}_j \; \hat{a}(\omega) e^{-i[\omega t_j - k(\omega) r_j]}$$

where $j = 1, 2$, $\hat{\theta}_j$ is the unit vector along the ith analyzer direction. Following the standard Glauber formula, the joint detection counting rate of D_1 and D_2 is thus:

$$R_c \propto \int_T dt_1 dt_2 \, |(\hat{\theta}_1 \cdot \hat{\mathbf{o}})(\hat{\theta}_2 \cdot \hat{\mathbf{e}}) \; \Psi(\tau_1^o, \tau_2^e) + (\hat{\theta}_1 \cdot \hat{\mathbf{e}})(\hat{\theta}_2 \cdot \hat{\mathbf{o}}) \; \Psi(\tau_1^e, \tau_2^o)|^2$$
$$= cos^2\theta_1 \sin^2\theta_2 \; + \; sin^2\theta_1 \cos^2\theta_2$$
$$+ \; cos\theta_1 \, sin\theta_2 \, sin\theta_1 \, cos\theta_2 \int_T dt_1 dt_2 \, \Psi^*(\tau_1^o, \tau_2^e) \, \Psi(\tau_1^e, \tau_2^o) \qquad (14.4.3)$$

where $\Psi(\tau_1^o, \tau_2^e)$ and $\Psi(\tau_1^e, \tau_2^o)$ are the effective two-photon wavefunctions with the following normalization:

$$\int_T dt_1 dt_2 \, | \, \Psi(\tau_1, \tau_2) \, |^2 = 1.$$

The third term of Eq. (14.4.3) determines the degree of two-photon coherence. Considering degenerate CW laser pumped SPDC, the biphoton wavepacket of Eq. (10.1.9) can be simplified as:

$$\Psi(\tau_1, \tau_2) = \Psi_0 \, e^{-i\omega_p(\tau_1 + \tau_2)/2} \mathcal{F}_{\tau_-} \{ f(\Omega) \}.$$

The coefficient of $cos\theta_1 \, sin\theta_2 \, sin\theta_1 \, cos\theta_2$ in the third term of Eq. (14.4.3) is thus

$$e^{-i\omega_p(\Delta\tau_1 - \Delta\tau_2)/2} \mathcal{F}_{\tau_1^o - \tau_2^e} \{ f(\Omega) \} \otimes \mathcal{F}_{\tau_1^e - \tau_2^o} \{ f(\Omega) \}.$$

Therefore, two important factors will determine the result of the polarization correlation measurement: (1) the phase of $e^{-i\omega_p(\Delta\tau_1 - \Delta\tau_2)/2}$; and (2) the value of the convolution between biphoton wavepackets $\Psi^*(\tau_1^o, \tau_2^e)$ and $\Psi(\tau_1^e, \tau_2^o)$. Examining the two wavepackets associated with the $o_1 - e_2$ and $e_1 - o_2$ terms, we found that the convolution of the two dimensional biphoton wavepackets of type II SPDC gives a null result, due to the *asymmetrical* rectangular function of $\Pi(\tau_1 - \tau_2)$ as indicated in Fig. 14.4.2. In order to make the two wavepackets overlap, we may either (1) move both wavepackets a distance of $DL/2$ (case I), or (2) move one of the wavepackets a distance of DL (case II). In both case I and case II, the convolution obtains its maximum value of one. The use of "compensator" is for

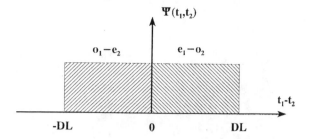

FIGURE 14.4.2 Without "compensator", the two dimensional wavepackets of $\Psi(\tau_1^o, \tau_2^e)$ and $\Psi(\tau_1^e, \tau_2^o)$ do not overlap along $\tau_1 - \tau_2$ axis.

this purpose. After compensating the two asymmetrical function of $\Pi(\tau_1 - \tau_2)$, we need to further manipulate the phase of $e^{-i\omega_p(\Delta\tau_1 - \Delta\tau_2)/2}$ to finalize the desired Bell states. This can be done by means of a retardation plate to introduce phase delay of 2π $(+1)$ or π (-1) between the o-ray and the e-ray in either arm 1 or arm 2. The EPR-Bohm-Bell polarization correlation

$$R_c \propto \sin^2(\theta_1 \pm \theta_2)$$

is expected only when the above two conditions are satisfied. We can simplify the polarization state of the signal-idler photon pair in the form of Bell states $|\Psi_{XY}^{(\pm)}\rangle$ in this situation only.

The biphoton wavepacket looks very different in the case of femtosecond laser pumped SPDC. As an example, a calculation of type-II biphoton wavepacket is given in the following.

We start with the quantum state of type-II SPDC in the following format without losing generality:

$$|\Psi\rangle = \int dk_o \int dk_e \int d\omega_p \int_0^L dz\, e^{-4\ln 2\left[\frac{\omega_p - \Omega_p}{\sigma_p}\right]^2} e^{i\Delta z} \delta(\omega_o + \omega_e - \omega_p) a_o^\dagger a_e^\dagger |0\rangle \quad (14.4.4)$$

where $\Delta = k_p - k_o - k_e$, L is the length of the nonlinear crystal. We have considered a Gaussian like pump pulse with bandwidth σ_p.

To generalize the result to a realistic experimental setup, we consider placing a Gaussian like spectral filter in front of the detector $D_j, j = 1, 2$. The field operator is written as

$$E_j^{(+)}(t_j, r_j) = \int d\omega\, e^{-4\ln 2\left[\frac{\omega - \Omega_j}{\sigma_j}\right]^2} a_o(\omega) e^{-i[\omega t_j - k(\omega)r_j]} \quad (14.4.5)$$

Define

$$\omega_o = \Omega_o + \nu_o,$$
$$\omega_e = \Omega_e + \nu_e,$$
$$\omega_p = \Omega_p + \nu_p,$$

where $\Omega_o \equiv \Omega_1$ and $\Omega_e \equiv \Omega_2$. ν_j is the detuning from the central frequency Ω_j. Note that $\Omega_o + \Omega_e = \Omega_p$ and $\nu_o + \nu_e = \nu_p$. The type-II phase mismatch Δ is now expanded to the first leading order. Assume $k(\nu) = [\Omega + \nu]n(\Omega + \nu)/c$, where $n(\Omega + \nu)$ is the index of refraction of the crystal at frequency $\Omega + \nu$. $k(\nu)$ can then be expanded to

$$k_j = K_j + \frac{\nu_j}{u_j(\Omega_j)} \quad (14.4.6)$$

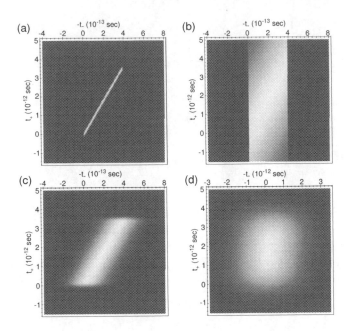

FIGURE 14.4.3 Type-II biphoton as a function of the pump bandwidth and the filter bandwidth. $\Pi(\tau_+, \tau_-)$ is shown in a density plot. (a) With no spectral filters, pump pulse duration $\delta t_p = 80fs$. (b) With no spectral filters, $\delta t_p = 3ps$. (c) With $10nm$ FWHM spectral filters, $\delta t_p = 80fs$. (d) With $1nm$ FWHM spectral filters, $\delta t_p = 80fs$.

where the subscripts $j = o, e, p$ and $K_j = \frac{n_j \Omega_j}{c}$, and where n_j is the index of refraction inside the nonlinear crystal. Therefore, Δ can now be written as

$$\Delta = k_p - k_o - k_e$$
$$= \frac{\nu_p}{u_p(\Omega_p)} - \frac{\nu_o}{u_o(\Omega_o)} - \frac{\nu_e}{u_e(\Omega_e)}$$
$$= -\nu_p D_+ - \frac{1}{2}\nu_- D, \tag{14.4.7}$$

where $\nu_- = \nu_o - \nu_e$, $D_+ \equiv \frac{1}{2}\left[\frac{1}{u_o(\Omega_o)} + \frac{1}{u_e(\Omega_e)}\right] - \frac{1}{u_p(\Omega_p)}$, and $D \equiv \frac{1}{u_o(\Omega_o)} - \frac{1}{u_e(\Omega_e)}$. To make the calculation easier, we assume $\sigma_1 = \sigma_2 = \sigma$; i.e., the same bandwidths for both filters. This assumption makes the ν_- and ν_+ integrals uncoupled. We also define $\Omega_o = \Omega_p/2 + \Omega_d$ and $\Omega_e = \Omega_p/2 - \Omega_d$. Note that in the case of degenerate type-II, $\Omega_d = 0$.

By combining Eq. (14.4.5), Eq. (14.4.4), and Eq. (14.4.7), the biphoton wavefunction $\Psi(\tau_+, \tau_-)$ is given by

$$\Psi(\tau_+, \tau_-) = e^{-i\Omega_p \tau_+} e^{-i\Omega_d \tau_-} \Pi(\tau_+, \tau_-) \tag{14.4.8}$$

where

$$\Pi(\tau_+, \tau_-) = \int d\nu_p \int d\nu_- \int_0^L dz \, e^{-[\nu_p/\delta]^2} e^{-2\ln 2[\nu_-/\sigma]^2}$$
$$\times e^{-i\nu_p \tau_+} e^{-\frac{i\nu_- \tau_-}{2}} e^{-i[\nu_p D_+ + \frac{\nu_-}{2}D]z} \tag{14.4.9}$$

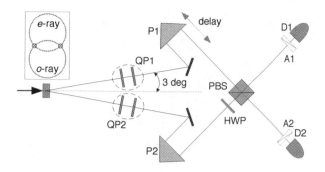

FIGURE 14.4.4 Bell states generation by ultrashort laser pulse. The non-overlapping $o_1 - e_2$ and $e_1 - o_2$ amplitudes are generated via non-collinear type-II SPDC. The use of half-waveplate (HWP) and polarization beamsplitter (PBS) make the reflection-reflection and transmission-transmission two-photon wavepackets "tilted" in the same orientation. QP1 and QP2 are quartz plates for fine tuning of the relative phase between the biphoton amplitudes.

with $\frac{1}{\delta^2} = \frac{2\ln 2}{\sigma^2} + \frac{4\ln 2}{\sigma_p^2}$. If the bandwidths of the filters are taken to be infinite, Eq. (14.4.9) can be simplified as

$$\Pi(\tau_+, \tau_-) = \begin{cases} e^{-\sigma_p^2[\tau_+ - [D_+/D]\tau_-]^2/16\ln 2} & 0 < \tau_- < DL \\ 0 & \text{otherwise.} \end{cases}$$

The Π function for pulsed type-II SPDC looks quite different from a CW pumped type-II SPDC. In a CW pumped type-II SPDC, the Π function is approximately independent of τ_+. However, in short pulse pumped type-II SPDC, it must be treated as a function of both τ_- and τ_+.

As an example, we consider a type-II BBO crystal with $2mm$ thickness pumped by a laser pulse which has a $400nm$ central wavelength, and for the degenerate case, $\lambda_o = \lambda_e = 800nm$. Fig. 14.4.3(a) shows the biphoton wavepacket generated by a $80fs$ pump pulse with no spectral filtering. Strong dependence on τ_+ is observed. The Π function is "tilted" on the τ_- - τ_+ plane. In Fig. 14.4.3(b), the biphoton wavepacket is generated by $3ps$ pump pulse with no spectral filtering. Note that the biphoton is now stretched in τ_+ direction and starts to resemble the CW pumped type-II biphoton. In Fig. 14.4.3(c) and Fig. 14.4.3(d), we show the effects of spectral filtering. The pump pulse duration is fixed at $80fs$, but the bandwidth of the spectral filters are varied. In Fig. 14.4.3(c), we use $10nm$ FWHM spectral filters and in Fig. 14.4.3(d), a $1nm$ FWHM spectral filter is applied. As expected, spectral filtering broadens the biphoton in both τ_+ and τ_- directions. The effects in τ_- is greater than that in τ_+.

Femtosecond pulse pumped type-II SPDC has a very peculiar space-time structure. The "tilted" biphoton wavepacket requires special efforts for preparing Bell states. Differing from CW pumped SPDC, applying a "compensator" in this situation is not enough. Although we could shift the $e - o$ and $o - e$ terms to make them approach each other, the "tilted" wavepackets cannot be overlapped due to the different orientation. To overcome this problem, basically, we need to either (1) use a narrow band spectral filter to broaden the wavepacket as we have learned in Fig. 14.4.3(d); or (2) prepare the two wavepackets in the same "tilted" orientation, i.e., making e-ray reach detector D_1 and o-ray reach detector D_2, or vise versa, for both amplitudes.

A clever experimental scheme developed by Kim *et al.* is shown in Fig. 14.4.4. This design uses a non-collinear type-II SPDC. The use of half-waveplate (HWP) and polarization beamsplitter (PBS) makes both reflection-reflection and transmission-transmission two-photon wavepackets "tilted" in the same orientation (see Fig. 14.4.5). The e-rays and

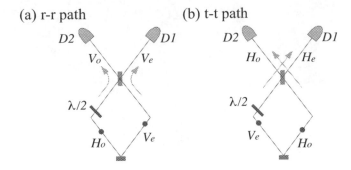

FIGURE 14.4.5 Two alternative ways for a photon pair trigging a joint detection event. This figure further clarifies the role of the half-waveplate (HWP) and the polarization beamsplitter (PBS): how to make the reflection-reflection and transmission-transmission two-photon wavepackets "tilted" in the same orientation. The e-rays and o-rays are always go to detectors D_1 and D_2, respectively. Bell state measurements were tested for CW and femtosecond pump. In both cases, high interference visibilities were observed.

o-rays, respectively, are always go to detectors D_1 and D_2. The Bell state measurements have been tested for both CW and Femtosecond pumps. In both cases, high visibilities of interferences were achieved, indicating high degree overlapping of the reflection-reflection and transmission-transmission two-photon amplitudes.

14.5 SCULLY'S QUANTUM ERASER

Quantum eraser, proposed by Scully and Drühl in 1982, is another thought experiment challenging the "basic mystery" of quantum mechanics: wave-particle duality. So far, several quantum eraser experiments have been demonstrated with interesting results supporting the ideas of Scully and Drühl.

A double-slit type quantum eraser experiment, closing to the original Scully-Drühl thought experiment of 1982, is illustrated in Fig. 14.5.1. A pair of entangled photons, photon 1 and photon 2, is excited by a weak laser pulse either from atom A, which is located in slit A, or from atom B, which is located in slit B. Photon 1, propagates to the right, is registered by detector D_0, which can be scanned by a step motor along its x_0-axis for the examination of interference fringes. Photon 2, propagating to the left, is injected into a beamsplitter. If the pair is generated in atom A, photon 2 will follow the A path meeting BSA with 50% chance of being reflected or transmitted. If the pair is generated in atom B, photon 2 will follow the B path meeting BSB with 50% chance of being reflected or transmitted. In view of the 50% chance of being transmitted by either BSA or BSB, photon 2 is detected by either detector D_3 or D_4. The registration of D_3 or D_4 provides which-path information (path A or path B) on photon 2 and in turn provides which-path information for photon 1 because of the entanglement nature of the two-photon state generated by atomic cascade decay. Given a reflection at either BSA or BSB photon 2 will continue to follow its A or B path to meet another 50-50 beamsplitter BS and then be detected by either detectors D_1 or D_2.

The experimental condition was arranged in such a way that no interference is observable in the single counting rate of D_0., i.e., the distance between A and B is large enough to be "distinguishable" for D_0 to learn which-path information of photon 1. When the states of a photon associated with slit-A and slit-B are distinguishable, the photon carries which-path

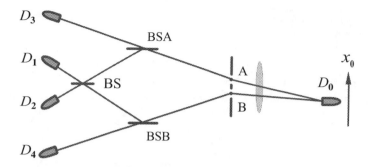

FIGURE 14.5.1 Quantum erasure: a thought experiment of Scully-Drühl. A pair of entangled photons is emitted from either atom A or atom B by atomic cascade decay. The experimental condition guarantees no interference fringes is observable in the single detector counting rate of D_0. The "clicks" at D_1 or D_2 erase the which-path information, thus helping to restore the interference even after the "click" of D_0. On the other hand, the "clicks" at D_3 or D_4 record which-slit information. Thus, no observable interference is expected with the help of these "clicks".

information. If the states of a photon associated with slit-A and slit-B are indistinguishable, the photon carries both-paths information[3].

However, the "clicks" at D_1 or D_2 will erase the which-path information of photon 1 and help to restore the interference. On the other hand, the "clicks" at D_3 or D_4 record which-path information. Thus, no observable interference is expected with the help of these "clicks". It is interesting to note that both the "erasure" and "recording" of the which-path information can be made as a "delayed choice": the experiment is designed in such a way that L_0, the optical distance between atoms A and B and detector D_0, is much shorter than L_A (L_B), which is the optical distance between atoms A and B and the beamsplitter BSA (BSB) where the "which-path" or "both-paths" "choice" is made randomly by photon 2. Thus, after the annihilation of photon 1 at D_0, photon 2 is still on its way to BSA (BSB), i.e., "which-path" or "both-path" choice is "delayed" compared to the detection of photon 1. After the annihilation of photon 1, we look at these "delayed" detection events of D_1, D_2, D_3, and D_4 which have constant time delays, $\tau_i \simeq (L_i - L_0)/c$, relative to the triggering time of D_0. L_i is the optical distance between atoms A, B and detectors D_1, D_2, D_3, and D_4, respectively. It was predicted that the "joint-detection" counting rate R_{01} (joint-detection rate between D_0 and D_1) and R_{02} will show an interference pattern as a function of the position of D_0 on its x-axis. This reflects the wave nature (both-path) of photon 1. However, no interference fringes will be observable in the joint detection counting events R_{03} and R_{04} when scanning detector D_0 along its x-axis. This is as would be expected because we have now inferred the particle (which-path) property of photon 1. It is important to emphasize that all four joint detection rates R_{01}, R_{02}, R_{03}, and R_{04} are recorded at the same time during one scanning of D_0. That is, in the present experiment, we "see" both wave (interference) and which-path (particle-like) with the same measurement apparatus.

It should be mentioned that (1) the "choice" in this experiment is not actively switched

---

[3]In the early times of quantum theory, Einstein and Bohr had serious debates based on the double-slit interference phenomenon. Now a days, "which-path" and/or "both-paths" information is still a topic of debate. What do we mean which-path and both-paths information? For physicists, at least for Einstein and Bohr, when a photon passes through slit-A and slit-B with distinguishable states, the photon carries which-path information. If its states associated with slit-A and slit-B are indistinguishable, the photon carries both-paths information. In quantum optics, when the separation between slit-A and slit-B is greater than the transverse coherence length of the field, the states of a photon associated with slit-A and slit-B are distinguishable; therefore, the photon carries which-path information. The idea of delayed choice quantum eraser is to erase this information after the annihilation of the photon itself by restoring the interference.

by the experimentalist during the measurement. The "delayed choice" associated with either the wave or particle behavior of photon 1 is "randomly" made by photon 2. The experimentalist simply looks at which detector D_1, D_2, D_3, or D_4 is triggered by photon 2 to determine either wave or particle properties of photon 1 after the annihilation of photon 1; (2) the photo-detction event of photon 1 at D_0 and the delayed choice event of photon 2 at BSA (BSB) are space-like separated events. The "coincidence" time window is chosen to be much shorter than the distance between D_0 and BSA (BSB). Within the joint-detection time window, it is impossible to have the two events "communicating".

14.5.1 Random Delayed Choice Quantum Eraser One

Kim *et. al.* realized the above random delayed choice quantum eraser in 2000. The schematic diagram of the experimental setup of Kim *et al.* is shown in Fig. 14.5.2. Instead of atomic cascade decay, SPDC is used to prepare the entangled two-photon state. In the experiment,

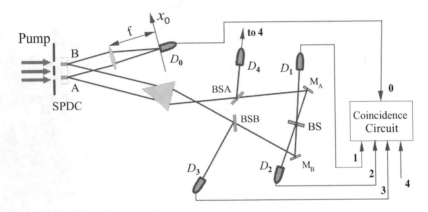

FIGURE 14.5.2 Delayed choice quantum eraser: Schematic of an actual experimental setup of Kim *et al..* Pump laser beam is divided by a double-slit and makes two regions A and B inside the SPDC crystal. A pair of signal-idler photons is generated either from the A or B region. The "delayed choice" to observe either wave or particle behavior of the signal photon is made randomly by the idler photon about $7.7ns$ after the detection of the signal photon.

a $351.1nm$ Argon ion pump laser beam is divided by a double-slit and directed onto a type-II phase matching nonlinear crystal BBO at regions A and B. A pair of $702.2nm$ orthogonally polarized signal-idler photon is generated either from region A or region B. The width of the region is about $a = 0.3mm$ and the distance between the center of A and B is about $d = 0.7mm$. A Glen-Thompson prism is used to split the orthogonally polarized signal and idler. The signal photon (photon 1, coming either from A or B) propagates through lens LS to detector D_0, which is placed on the Fourier transform plane of the lens. The use of lens LS is to achieve the "far field" condition, but still keep a short distance between the slit and the detector D_0. Detector D_0 can be scanned along its x-axis by a step motor for the observation of interference fringes. The idler photon (photon 2) is sent to an interferometer with equal-path optical arms. The interferometer includes a prism PS, two 50-50 beamsplitters BSA, BSB, two reflecting mirrors M_A, M_B, and a 50-50 beamsplitter BS. Detectors D_1 and D_2 are placed at the two output ports of the BS, respectively, for erasing the which-path information. The triggering of detectors D_3 and D_4 provides which-path information for the idler (photon 2) and, in turn, which-path information for the signal (photon 1). The detectors are fast avalanche photodiodes with less than $1ns$ rise time and about $100ps$ jitter. A constant fractional discriminator is used with each of the detectors to

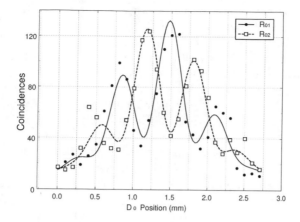

FIGURE 14.5.3 Joint detection rates R_{01} and R_{02} against the x coordinates of detector D_0. Standard Young's double-slit interference patterns are observed. Note the π phase shift between R_{01} and R_{02}. The solid line and the dashed line are theoretical fits to the data.

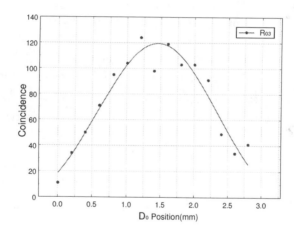

FIGURE 14.5.4 Joint detection counting rate of R_{03}. Absence of interference is clearly demonstrated. The solid line is a sinc-function fit.

register a single photon whenever the leading edge of the detector output pulse is above the threshold. Coincidences between D_0 and D_j ($j = 1, 2, 3, 4$) are recorded, yielding the joint detection counting rates R_{01}, R_{02}, R_{03}, and R_{04}.

In the experiment, the optical delay $(L_{A,B} - L_0)$ is chosen to be $\simeq 2.3m$, where L_0 is the optical distance between the output surface of BBO and detector D_0, and L_A (L_B) is the optical distance between the output surface of the BBO and the beamsplitter BSA (BSB). This means that any information (which-path or both-path) one can infer from photon 2 must be at least $7.7ns$ later than the registration of photon 1. Compared to the $1ns$ response time of the detectors, $2.3m$ delay is thus enough for "delayed erasure". Although there is an arbitrariness about when a photon is detected, it is safe to say that the "choice" of photon 2 is delayed with respect to the detection of photon 1 at D_0 since the entangled photon pair is created simultaneously.

Figs.14.5.3, reports the joint detection rates R_{01} and R_{02}, indicating the regaining of standard Young's double-slit interference pattern. An expected π phase shift between the two interference patterns is clearly shown in the measurement. The single detector counting rates of D_0 and D_1 are recorded simultaneously. Although interference is observed in the joint detection counting rate, there is no significant modulation in any of the single detector counting rate during the scanning of D_0. R_0 is a constant during the scanning of D_0. The absence of interference in the single detector counting rate of D_0 is simply because the separation between slits-A and slit-B is much greater than the coherence length of the single field.

Fig. 14.5.4 reports a typical R_{03} (R_{04}), joint detection counting rate between D_0 and "which-path detector" D_3 (D_4). An absence of interference is clearly demonstrated. The fitting curve of the experimental data indicates a sinc-function like envelope of the standard Young's double slit interference-diffraction pattern. Two features should bring to our attention that (1) there is no observable interference modulation as expected, and (2) the curve is different from the constant single detector counting rate of D_0.

The experimental result is surprising from a classical point of view. The result, however, is easily explained in the contents of quantum theory. In this experiment, there are two kinds of very different interference phenomena: single-photon interference and two-photon interference. As we have discussed earlier, single-photon interference is the result of the superposition between single-photon amplitudes, and two-photon interference is the results of the superposition between two-photon amplitudes. Quantum mechanically, single-photon amplitude and two-photon amplitude represent very different measurements and, thus, very different physics.

In this regard, we analyze the experiment by answering the following questions:

(1)Why is there no observable interference in the single-detector counting rate of D_0?

This question belongs to single-photon interferometry. The absence of interference in single-detector counting rate of D_0 is very simple: the separation between slit-A and slit-B is much greater than the spatial coherence length of the the signal field. The single-photon states of the signal photon associated with slit-A and slit-B are distinguishable. Therefore, the signal photon carries "which-slit information. The idea of delayed choice quantum eraser is to erase this information after the annihilation of the signal photon it self.

(2)Why is there observable interference in the joint detection counting rate of R_{01} and R_{02}?

This question belongs to two-photon interferometry. Two-photon interference is very different from single-photon interference. Two-photon interference involves the addition of different yet indistinguishable two-photon amplitudes. The coincidence counting rate R_{01}, again, is proportional to the probability P_{01} of joint detecting the signal-idler pair by detectors D_0 and D_1,

$$R_{01} \propto P_{01} = \langle \Psi | E_0^{(-)} E_1^{(-)} E_1^{(+)} E_0^{(+)} | \Psi \rangle = \left| \langle 0 | \ E_1^{(+)} E_0^{(+)} | \Psi \rangle \right|^2. \qquad (14.5.1)$$

To simplify the mathematics, we use the following "two-mode" expression for the state, bearing in mind that the transverse momentum δ-function will be taken into account.

$$| \Psi \rangle = \epsilon \, [a_s^\dagger a_i^\dagger e^{i\varphi_A} + b_s^\dagger b_i^\dagger \ e^{i\varphi_B}] \, | 0 \rangle$$

where ϵ is a normalization constant that is proportional to the pump field and the nonlinearity of the SPDC crystal, φ_A and φ_B are the phases of the pump field at A and B, and $a_j^\dagger$ ($b_j^\dagger$), $j = s, i$, are the photon creation operators for the lower (upper) mode in Fig. 14.5.2.

In Eq. (14.5.1), the fields at the detectors D_0 and D_1 are given by:

$$E_0^{(+)} = a_s\, e^{ikr_{A0}} + b_s\, e^{ikr_{B0}}$$
$$E_1^{(+)} = a_i\, e^{ikr_{A1}} + b_i\, e^{ikr_{B1}}$$

(14.5.2)

where r_{Aj} (r_{Bj}), $j = 0, 1$, are the optical path lengths from region A (B) to the jth detector. Substituting the biphoton state and the field operators into Eq. (14.5.1),

$$R_{01} \propto \left| e^{i(kr_A + \varphi_A)} + e^{i(kr_B + \varphi_B)} \right|^2 = |\Psi_A + \Psi_B|^2$$
$$= 1 + \cos\left[k(r_A - r_B)\right] \simeq \cos^2\left(x_0 \pi d/\lambda z_0\right)$$

(14.5.3)

where, $r_A = r_{A0} + r_{A1}$, $r_B = r_{B0} + r_{B1}$; Ψ_A and Ψ_B are the two-photon effective wave functions of path A and path B, representing the two different yet indistinguishable probability amplitudes to produce a joint photodetection event of D_0 and D_1, indicating a two-photon interference. In Dirac's language: a signal-idler photon pair interferes with the pair itself.

To calculate the diffraction effect of a single-slit, again, we need an integral of the effective two-photon wavefunction over the slit width (the superposition of infinite number of probability amplitudes results in a click-click joint detection event):

$$R_{01} \propto \left| \int_{-a/2}^{a/2} dx_{AB}\, e^{-ik\, r(x_0,\, x_{AB})} \right|^2 \cong \text{sinc}^2(x_0 \pi a/\lambda z_0)$$

(14.5.4)

where $r(x_0, x_{AB})$ is the distance between points x_0 and x_{AB}, x_{AB} belongs to the slit's plane, and the far-field condition is applied.

Repeating the above calculations, the combined interference-diffraction joint detection counting rate for the double-slit case is given by:

$$R_{01} \propto \text{sinc}^2(x_0 \pi a/\lambda z_0) \cos^2(x_0 \pi d/\lambda z_0).$$

(14.5.5)

If the finite size of the detectors are taken into account, the interference visibility will be reduced.

(3) Why is there no observable interference in the joint detection counting rate of R_{03} and R_{04}?

This question belongs to two-photon interferometry. From the view of two-photon physics, the absence of interference in the joint detection counting rate of R_{03} and R_{04} is obvious: only one two-photon amplitude contributes to the joint detection events.

In fact, the two-photon states of the signal-idler pair associated with slit-A and slit-B are "indistinguishable and thus carries "both-paths information. The "both-paths information can be read by means of the joint photodetection events between D_0 and D_1 as well as between D_0 and D_2. These joint photodetection events are the results of the superposition between the indistinguishable two-photon amplitudes. The "both-paths information can be erased by means of the joint photodetection events between D_0 and D_3 as well as between D_0 and D_4 which contains only one two-photon amplitude.

14.5.2 Random Delayed Choice Quantum Eraser Two

Now we ask, what would happen if we replace the entangled photons with a randomly created and randomly paired photons, or wavepackets, in thermal state? Can a randomly paired photons in thermal state erase the which path information? The answer is positive. A random delayed choice quantum eraser with thermal light has been demonstrated by Peng et. al. recently.

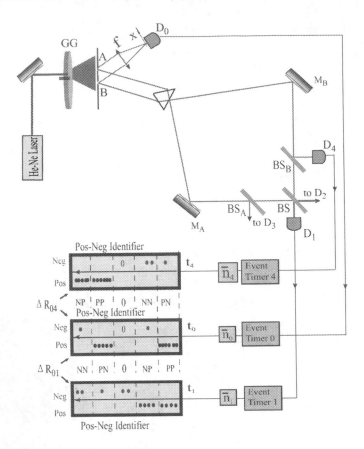

FIGURE 14.5.5 Schematic of a random delayed choice quantum eraser. The He-Ne laser beam spot on the rotating ground glass (GG) has a diameter of $\sim 2mm$. A double-slit, with slit-width $150\mu m$ and slit-separation $d = 0.7mm$, is placed $\sim 25cm$ away from the GG. The spatial coherence length of the pseudo-thermal field on the double-slit plane is $l_c = \lambda/\Delta\theta \sim 160\mu \ll d$, which guarantees the two fields E_A and E_B are spatially incoherent. The states of the photons associated with slit-A and slit-B are distinguishable and thus carries "which-slit" information. All beamsplitters are 50/50 non-polarizing beamsplitters. The two fields from the two slits may propagate to detector D_0 which is transversely scanned on the focal plan of lens f for observing the interference patten of the double-slit interferometer; and may also pass along a Mach-Zehnder-like interferometer and finally reach at D_1 or D_4 (D_2 or D_3). A photon number fluctuation correlation (PNFC) measurement circuit is followed to evaluate the photon number fluctuation correlations measured by D_0-D_1 and D_0-D_4 (or D_0-D_2 and D_0-D_3).

The schematic of the experiment of Peng et. al. is illustrated in Fig.14.5.5. The experimental setup is almost the same as that of the experiment of Kim et. al. of 2000, except (1) the entangled signal-idler photon pair is replaced by two randomly created and randomly paired photons in a thermal state; (2) the joint photodetection measures the photon number fluctuation correlation, or the intensity fluctuation correlation, resulting from two-photon interferences.

The experimental setup in Fig.14.5.5 can be divided into four parts: a thermal light source, a Young's double-slit interferometer, a Mach-Zehnder-like interferometer, and a photon number fluctuation correlation (PNFC) measurement circuit. (1) The light source is a standard pseudo-thermal source which consists of a He-Ne laser beam ($\sim$2mm diameter) and a rotating ground glass (GG). Within the $\sim$2mm diameter spot, the ground glass contains millions of tiny diffusers. A large number of randomly distributed sub-fields, or wavepacket, are scattered from millions of randomly distributed tiny diffusers with random phases. The pseudo-thermal field then passes a double-slit which is about 25cm away from the GG. (2) The double-slit has a slit-width $150\mu m$, and a slit-separation $d = 0.7mm$ (distance between the center of two slits). The spatial coherence length of the pseudo-thermal field on the double slit plane is $l_c = \lambda/\Delta\theta \sim 160\mu \ll d$, which guarantees the two fields E_A and E_B are spatially incoherent. Under this experimental condition, (a) no first-order classic Young's interference is observable; (b) the photon number fluctuates correlatively only within slit-A or slit-B. Therefore, the states of the photons associated with slit-A and slit-B are distinguishable and carries "which-slit" information. We may learn the which-slit information from either a photon number measurement or from a photon number-fluctuation correlation measurement. A lens, f, is placed following the double-slit. On the focal-plane of the lens a scannable point-like photodetector D_0 is used to learn the which-slit information or to observe the Young's double-slit interference pattern. (c) The Mach-Zehnder-like interferometer and the photodetectors D_1, D_2, are used to "erase" the which-slit information. Simultaneously, the joint-detection between D_0 and D_3 or D_4 are used to "read" the which-slit information. All five photodetectors are photon-counting detectors working at single-photon level. The Mach-Zehnder-like interferometer has three beamsplitters, BS, BS_A, and BS_B, all of them are 50/50 non-polarizing beamsplitters. Moreover, the detectors are fast avalanche photodiodes with rise time less than 1 ns, and the path delay between BS_A or BS_B, and D_0 is $\approx 1.5m$ which ensure that, at each joint-detection measurement, when a photon chooses to be reflected (read which-way) or transmitted (erase which-way) at BS_A or BS_B, it is already 5 ns later than the annihilation of its partner at D_0. Comparing the 1 ns rise time, we are sure this is a "delayed choice" made by that photon. (4) The PNFC circuit consists of five synchronized "event-timers" which record the registration times of D_0, D_1, D_2, D_3, and D_4. A positive-negative fluctuation identifier follows each event-timer to distinguish "positive-fluctuation" Δn^+, from "negative-fluctuation" Δn^-, for each photodetector within each coincidence time window. The photon number fluctuation correlations of D_0-D_1: $\Delta R_{01} = \langle \Delta n_0 \Delta n_1 \rangle$ and D_0-D_4: $\Delta R_{04} = \langle \Delta n_0 \Delta n_4 \rangle$ are calculated, accordingly and respectively, based on their measured positive-negative fluctuations. The detailed description of the PNFC circuit can be found in the references.

The experimental observation of ΔR_{04} is reported in Fig. 14.5.6. The data excludes any possible existing interferences. This measurement means the coincidences that contributed

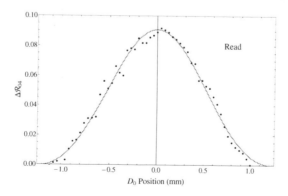

FIGURE 14.5.6 The measured ΔR_{04} by scanning D_0 on the observation plane of the Young's double-slit interferometer. The black dots are experimental data, the red line is the theoretical fitting with Eq.(14.5.13).

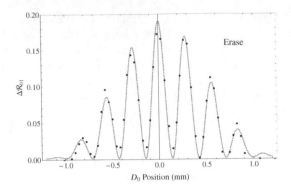

FIGURE 14.5.7 The measured ΔR_{01} as a function of the transverse coordinate of D_0. The black dots are experimental data, the red line is the theoretical fitting with Eq.(14.5.14).

to ΔR_{04} must have passed through slit-B. Fig. 14.5.7 reports a typical experimental result of ΔR_{01}: a typical double-slit interference-diffraction pattern. The 100% visibility of the sinusoidal modulation indicates complete erasure of the which-slit information.

Assuming a random pair of sub-fields at single-photon level, such as the mth and nth wavepackets, is scattered from the mth and the nth sub-sources located at transverse coordinates $\vec{\rho}_{0m}$ and $\vec{\rho}_{0n}$ of the ground glass and fall into the coincidence time windows of D_0-D_1 and D_0-D_4, the mth wavepacket may propagate to the double-slit interferometer and the nth wavepacket may pass through the Mach-Zehnder, or vice versa. Under the experimental condition of spatial incoherence between E_A and E_B, the which-slit information is learned from the photon number fluctuation correlation measurements $\Delta R_{04} = \langle \Delta n_0 \Delta n_4 \rangle = \langle \Delta n_{B0} \Delta n_{B4} \rangle$ of D_0-D_4, and no interference is observable by scanning D_0. It is interesting that the which-slit information are erasable in the photon number fluctuation correlation measurements of $\Delta R_{01} = \langle \Delta n_0 \Delta n_1 \rangle$ of D_0-D_1, resulting in a reappeared interference pattern as a function of the scanning coordinate of D_0.

The field operator at detector D_0 can be written in the following form in terms of the subfields:

$$\hat{E}^{(+)}(\mathbf{r}_0, t_0) = \hat{E}_A^{(+)}(\mathbf{r}_0, t_0) + \hat{E}_B^{(+)}(\mathbf{r}_0, t_0)$$
$$= \sum_m \left[\hat{E}_{mA}^{(+)}(\mathbf{r}_0, t_0) + \hat{E}_{mB}^{(+)}(\mathbf{r}_0, t_0) \right] \quad (14.5.6)$$
$$= \sum_m \int d\mathbf{k} \, \hat{a}_m(\mathbf{k}) \left[g_m(\mathbf{k}; \mathbf{r}_A, t_A) g_A(\mathbf{k}; \mathbf{r}_0, t_0) + g_m(\mathbf{k}; \mathbf{r}_B, t_B) g_B(\mathbf{k}; \mathbf{r}_0, t_0) \right].$$

where $g_m(\mathbf{k}; \mathbf{r}_s, t_s)$ is a Green's function which propagates the mth subfield from the mth sub-source to the sth slit ($s = A, B$). $g_s(\mathbf{k}; \mathbf{r}_0, t_0)$ is another Green's function that propagates the field from the sth slit to detector D_0. It is easy to notice that, although there are two ways a photon can be detected at D_0, due to the first order incoherence of E_A and E_B, there should be no interference at the detection plane.

D_4 (D_3) in the experiment can only receive photons from slit-B (slit-A), so the field operator is then:

$$\hat{E}^{(+)}(\mathbf{r}_4, t_4) = \sum_m \hat{E}_{mB}^{(+)}(\mathbf{r}_4, t_4)$$
$$= \sum_m \int d\mathbf{k} \, \hat{a}_m(\mathbf{k}) g_m(\mathbf{k}; \mathbf{r}_B, t_B) g_B(\mathbf{k}; \mathbf{r}_4, t_4). \quad (14.5.7)$$

The detector D_1 (D_3), however, can receive photons from both slit-A and slit-B through the Mach-Zehnder-like interferometer, so the field operator has two terms:

$$\hat{E}^{(+)}(\mathbf{r}_1, t_1) = \sum_m \left[\hat{E}_{mA}^{(+)}(\mathbf{r}_1, t_1) + \hat{E}_{mB}^{(+)}(\mathbf{r}_1, t_1) \right] \tag{14.5.8}$$

$$= \sum_m \int d\mathbf{k}\, \hat{a}_m(\mathbf{k}) \left[g_m(\mathbf{k}; \mathbf{r}_A, t_A) g_A(\mathbf{k}; \mathbf{r}_1, t_1) + g_m(\mathbf{k}; \mathbf{r}_B, t_B) g_B(\mathbf{k}; \mathbf{r}_1, t_1) \right].$$

Based on the state, either in the single-photon state representation or in the coherent state representation, and the field operators of Eq.(14.5.6–14.5.8), we apply the Glauber-Scully theory to calculate the photon number fluctuation correlation or the second-order coherence function $G^{(2)}(\mathbf{r}_0, t_0; \mathbf{r}_\alpha, t_\alpha)$ from the coincidence measurement of D_0 and D_α,($\alpha = 1, 2, 3, 4$):

$$G^{(2)}(\mathbf{r}_0, t_0; \mathbf{r}_\alpha, t_\alpha)$$

$$= \left\langle \left\langle \Psi \middle| E^{(-)}(\mathbf{r}_0, t_0) E^{(-)}(\mathbf{r}_\alpha, t_\alpha) E^{(+)}(\mathbf{r}_\alpha, t_\alpha) E^{(+)}(\mathbf{r}_0, t_0) \middle| \Psi \right\rangle \right\rangle_{Es}$$

$$= \left\langle \left\langle \Psi \middle| \sum_m E_m^{(-)}(\mathbf{r}_0, t_0) \sum_n E_n^{(-)}(\mathbf{r}_\alpha, t_\alpha) \sum_q E_q^{(+)}(\mathbf{r}_\alpha, t_\alpha) \sum_p E_p^{(+)}(\mathbf{r}_0, t_0) \middle| \Psi \right\rangle \right\rangle_{Es}$$

$$= \sum_m \psi_m^*(\mathbf{r}_0, t_0) \psi_m(\mathbf{r}_0, t_0) \sum_n \psi_n^*(\mathbf{r}_\alpha, t_\alpha) \psi_n(\mathbf{r}_\alpha, t_\alpha)$$

$$+ \sum_{m,n} \psi_m^*(\mathbf{r}_0, t_0) \psi_n(\mathbf{r}_0, t_0) \psi_n^*(\mathbf{r}_\alpha, t_\alpha) \psi_m(\mathbf{r}_\alpha, t_\alpha)$$

$$= \langle n_0 \rangle \langle n_\alpha \rangle + \langle \Delta n_0 \Delta n_\alpha \rangle. \tag{14.5.9}$$

Here $\psi_m(\mathbf{r}_\alpha, t_\alpha)$ is the effective wavefunction of the mth subfield at $(\mathbf{r}_\alpha, t_\alpha)$. In the case of $\alpha = 1, 2$

$$\psi_m(\mathbf{r}_\alpha, t_\alpha) = \psi_{mA\alpha} + \psi_{mB\alpha} \tag{14.5.10}$$

$$= \int d\mathbf{k}\, \alpha_m(\mathbf{k}) \left[g_m(\mathbf{k}; \mathbf{r}_A, t_A) g_A(\mathbf{k}; \mathbf{r}_\alpha, t_\alpha) + g_m(\mathbf{k}; \mathbf{r}_B, t_B) g_B(\mathbf{k}; \mathbf{r}_\alpha, t_\alpha) \right].$$

This shows that the measured effective wavefunction $\psi_m(\mathbf{r}_\alpha, t_\alpha)$ is the result of a superposition between two alternative amplitudes in terms of path-A and path-B, $\psi_{m\alpha} = \psi_{mA\alpha} + \psi_{mB\alpha}$. When $\alpha = 4$ (or $\alpha = 3$), the effective wavefunction has only one amplitude

$$\psi_m(\mathbf{r}_4, t_4) = \psi_{mB4} = \int d\mathbf{k}\, \alpha_m(\mathbf{k}) g_m(\mathbf{k}; \mathbf{r}_B, t_B) g_B(\mathbf{k}; \mathbf{r}_4, t_4). \tag{14.5.11}$$

From Eq.(14.5.9) and the measurement circuit in Fig.14.5.5, it is easy to find that what we measure in this experiment is the photon number fluctuation correlation:

$$\langle \Delta n_0 \Delta n_\alpha \rangle = \sum_{m,n} \psi_m^*(\mathbf{r}_0, t_0) \psi_n(\mathbf{r}_0, t_0) \psi_n^*(\mathbf{r}_\alpha, t_\alpha) \psi_m(\mathbf{r}_\alpha, t_\alpha). \tag{14.5.12}$$

We thus obtain:

$$\Delta R_{04} \propto \langle \Delta n_0 \Delta n_4 \rangle = \sum_{n \neq m} \psi_{mB0}^* \psi_{nB0} \psi_{nB4}^* \psi_{mB4} \propto \mathrm{sinc}^2(x\pi a / \lambda f), \tag{14.5.13}$$

indicating a diffraction pattern which agrees with the experimental observation of Fig.14.5.6.

In the case of $\alpha = 1, 2$, we obtain:

$$\Delta R_{01} \propto \langle \Delta n_0 \Delta n_1 \rangle$$

$$\propto \sum_{n \neq m} \left[\psi_{mA0}^* \psi_{nA0} \psi_{nA1}^* \psi_{mA1} + \psi_{mB0}^* \psi_{nB0} \psi_{nB1}^* \psi_{mB1} \right.$$

$$\left. + \psi_{mA0}^* \psi_{nB0} \psi_{nB1}^* \psi_{mA1} + \psi_{mB0}^* \psi_{nA0} \psi_{nA1}^* \psi_{mB1} \right]$$

$$\propto \mathrm{sinc}^2(x\pi a/\lambda f) \cos^2(x\pi d/\lambda f), \tag{14.5.14}$$

which agrees with the experimental observation in Fig.14.5.7.

14.6 POPPER'S EXPERIMENT

In 1934, one year before the 1935 paper of Einstein-Podolsky-Rosen, Popper published a thought experiment to probe the foundation of quantum theory according to his philosophy of realism. Popper's original thought experiment is schematically shown in Fig. 14.6.1. A point source S, positronium as Popper suggested, is placed at the center of the experimental arrangement from which entangled pair of particle 1 and particle 2 are emitted in opposite directions along the respective positive and negative x-axes toward two screens A and B. There are slits on both screens parallel to the y-axis and the slits may be adjusted by varying their widths Δy. Beyond the slits on each side stand an array of Geiger counters for the joint measurement of the particle pairs as shown in the figure. The entangled pair could be emitted to any direction in 4π solid angles from the point source. However, if particle 1 is detected in a certain direction, particle 2 is then known to be in the opposite direction due to the momentum conservation of the quanta pair.

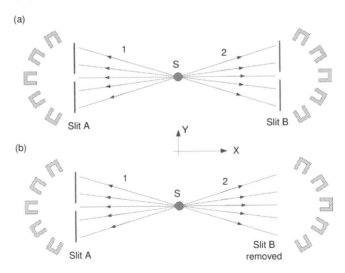

FIGURE 14.6.1 Popper's thought experiment. An entangled pair of particles are emitted from a point source with momentum conservation. A narrow slit on screen A is placed in the path of particle 1 to provide the precise knowledge of its position on the y-axis and this also determines the precise y-position of its twin, particle 2, on screen B. (a) Slits A and B are both adjusted very narrowly. (b) Slit-A is kept very narrow and slit-B is left wide open.

First, let us imagine the case in which slits A and B are both adjusted very narrowly. In this circumstance, particle 1 and particle 2 experience diffraction at slit-A and slit-B, respectively, and exhibit greater Δp_y for smaller Δy of the slits. There seems to be no disagreement in this situation between Copenhagen and Popper.

Next, suppose we keep slit-A very narrow and leave slit-B wide open. The main purpose of the narrow slit-A is to provide the precise knowledge of the position y of particle 1 and this subsequently determines the precise position of its twin (particle 2) on side B through quantum entanglement. Now, Popper asks, in the absence of the physical interaction with an actual slit, does particle 2 experience a greater uncertainty in Δp_y due to the precise knowledge of its position? Based on his beliefs, Popper provides a straightforward prediction: *particle 2 must not experience a greater Δp_y unless a real physical narrow slit-B is applied.* However, if Popper's conjecture is correct, this would imply the product of Δy and Δp_y of particle 2 could be smaller than h ($\Delta y \, \Delta p_y < h$). This may pose a serious difficulty for Copenhagen and perhaps for many of us. On the other hand, if particle 2 going to the right does scatter like its twin, which has passed though slit-A, while slit-B is wide open, we are then confronted with an apparent *action-at-a-distance!*

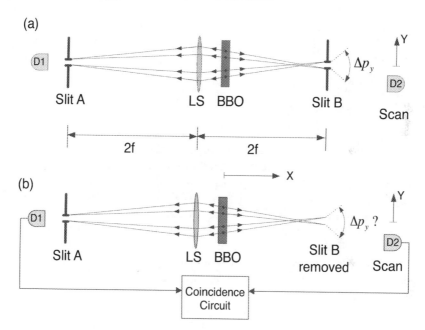

FIGURE 14.6.2 Modified version of Popper's experiment. An entangled signal-idler photon pair is generated from the nonlinear optical process of SPDC of a BBO crystal. A lens and a narrow slit-A are placed in the path of photon 1 to provide the precise knowledge of its position on the y-axis and also to determine the precise y-position of its twin, photon 2, on screen B by means of biphoton ghost imaging. Photon counting detectors D_1 and D_2 are used to scan in y-directions for joint detections. (a) Slits A and B are both adjusted very narrowly. (b) Slit-A is kept very narrow and slit-B is left wide open.

The use of a *point source* in Popper's proposal has been criticized historically as the fundamental error Popper made. It is true that a point source can never produce a pair of entangled particles which preserves EPR correlation in momentum as Popper expected. However, notice that a *point source* is *not* a necessary requirement for Popper's experiment. What is required is a precise position-position EPR correlation: if the position of particle 1 is precisely known, the position of particle 2 is 100% determined. Ghost imaging is a perfect tool to achieve this.

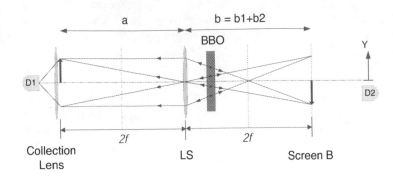

FIGURE 14.6.3 An unfolded schematic of ghost imaging, which is helpful for understanding Kim and Shih's realization of Popper's experiment. We assume the entangled signal-idler photon pair holds a perfect momentum correlation $\delta(\mathbf{k}_s + \mathbf{k}_i) \sim 0$. The locations of the slit-A, the imaging lens LS, and the ghost image must be governed by the Gaussian thin lens equation. In this experiment, we have chosen $s_o = s_i = 2f$. Thus, the ghost image of slit-A is expected to be the same size as that of slit-A.

14.6.1 Popper's Experiment One

In 1999, Popper's experiment was realized by Kim and Shih with the help of biphoton ghost imaging. Fig. 14.6.2 is a unfolded schematic diagram of their experiment, which is helpful for comparison with the original Popper's thought experiment. It is easy to see that this is a typical ghost imaging experimental setup. An entangled signal-idler photon pair, generated from the nonlinear optical process of SPDC of a BBO crystal, is used to image slit-A onto its distant ghost image plane of "screen" B. In the setup, s_o is chosen to be twice the focal length of the imaging lens LS, $s_o = 2f$. According to the Gaussian thin lens equation, an equal size "ghost" image of slit-A appears on the two-photon image plane at $s_i = 2f$. The use of slit-A provides a precise knowledge of the position of photon 1 (signal) on the y-axis and also determines the precise y-position of its twin, photon 2 (idler), on screen B by means of the biphoton ghost imaging. The experimental condition specified in Popper's experiment is then achieved: when slit-A is adjusted to a certain narrow width and slit-B is wide open, slit-A provides precise knowledge about the position of photon 1 on the y-axis up to an accuracy Δy which equals the width of slit-A, and the corresponding ghost image of pinhole A at screen B determines the precise position y of photon 2 to within the same accuracy Δy. Δp_y of photon 2 can be independently studied by measuring the width of its "diffraction pattern" at a certain distance from "screen" B. This is obtained by recording coincidences between detectors D_1 and D_2 while scanning detector D_2 along its y-axis, which is behind screen B at a certain distance.

Figure 14.6.3 is a conceptual diagram to connect the modified Popper's experiment with biphoton ghost imaging. In this unfolded ghost imaging setup, we assume the entangled signal-idler photon pair holds a perfect EPR correlation in momentum with $\delta(\mathbf{k}_s + \mathbf{k}_i) \sim 0$, which can be easily realized in a large transverse sized SPDC. In this experiment, we have chosen $s_o = s_i = 2f$. Thus, an equal size ghost image of slit-A is expected to appear on the image plane of screen B.

The detailed experimental setup is shown in Fig.14.6.4 with indications of the various distances. A CW Argon ion laser line of $\lambda_p = 351.1nm$ is used to pump a $3mm$ long beta barium borate (BBO) crystal for type-II SPDC to generate an orthogonally polarized signal-idler photon pair. The laser beam is about $3mm$ in diameter with a diffraction limited divergence. It is important not to focus the pump beam so that the phase-matching condition, $\mathbf{k}_s + \mathbf{k}_i = \mathbf{k}_p$, is well reinforced in the SPDC process, where $\mathbf{k}_j$ ($j = s, i, p$) is the

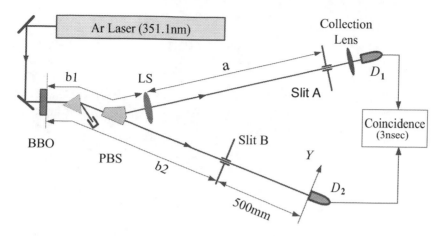

FIGURE 14.6.4 Schematic of the experimental setup of Kim and Shih. The laser beam is about $3mm$ in diameter. The "phase-matching condition" is well reinforced. Slit-A ($0.16mm$) is placed $1000mm = 2f$ behind the converging lens, LS ($f = 500mm$). The one-to-one ghost image ($0.16mm$) of slit-A is located at B. The optical distance from LS in the signal beam taken as back through PBS to the SPDC crystal ($b_1 = 255mm$) and then along the idler beam to "screen B" ($b_2 = 745mm$) is $1000mm = 2f$ ($b = b_1 + b_2$).

wavevectors of the signal (s), idler (i), and pump (p) respectively. The collinear signal-idler beams, with $\lambda_s = \lambda_i = 702.2nm = 2\lambda_p$ are separated from the pump beam by a fused quartz dispersion prism, and then split by a polarization beam splitter PBS. The signal beam (photon 1) passes through the converging lens LS with a $500mm$ focal length and a $25mm$ diameter. A $0.16mm$ slit is placed at location A which is $1000mm$ ($s_o = 2f$) behind the lens LS. A short focal length lens is used with D_1 for collecting all the signal beam that passes through slit-A. The point-like photon counting detector D_2 is located $500mm$ behind "screen B". "Screen B" is the image plane defined by the Gaussian thin equation. Slit-B, either adjusted as the same size as that of slit-A or opened completely, is placed to coincide with the ghost image. The output pulses from the detectors are sent to a coincidence circuit. During the measurements, the bucket detector D_1 is fixed behind slit-A while the point-like detector D_2 is scanned on the y-axis by a step motor.

Measurement 1: this measurement studied the case in which both slits A and B were adjusted to be $0.16mm$. The y-coordinate of D_1 was chosen to be 0 (center) while D_2 was allowed to scan along its y-axis. The circled dot data points in Fig. 14.6.5 show the *coincidence* counting rates against the y-coordinates of D_2. It is a typical single-slit diffraction pattern with $\Delta y \Delta p_y = h$. Nothing is special in this measurement except that we have learned the width of the diffraction pattern for the $0.16mm$ slit and this represents the minimum uncertainty of Δp_y. We should emphasize at this point that the *single* detector counting rate of D_2 as a function of its position y is basically the same as that of the coincidence counts except for a higher counting rate.

Measurement 2: the same experimental conditions were maintained except that slit-B was left wide open. This measurement is a test of Popper's prediction. The y-coordinate of the bucket detector D_1 was chosen to be 0 (center) while D_2 was allowed to scan along its y-axis. Due to the entangled nature of the signal-idler photon pair and the use of coincidence measurement circuit, only those twins which have passed through slit-A and the "ghost image" of slit-A at screen B with an uncertainty of $\Delta y = 0.16mm$ (which is the same width as the real slit-B we have used in measurement 1) would contribute to the coincidence

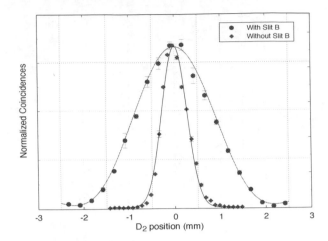

FIGURE 14.6.5 The observed coincidence patterns. The y-coordinate of D_1 was chosen to be 0 (center) while D_2 was allowed to scan along its y-axis. Circled dot points: slit-A $=0.16mm$, slit-B $= 0.16mm$. Diamond dot points: slit-A $= 0.16mm$, slit-B wide open. The width of the sinc-function curve fitted by the circled dot points is a measure of the minimum Δp_y determined by a $0.16mm$ slit. The fitting curve for the diamond dots is numerical result of Eq. (14.6.1), indicating a *blurred* ghost image of silt-A.

counts through the joint detection of the bucket detector D_1 and the point-like detector D_2. The diamond dot data points in Fig. 14.6.5 report the measured coincidence counting rates against the y coordinates of D_2. The measured width of the pattern is narrower than that of the diffraction pattern shown in measurement 1. It is also interesting to notice that the single detector counting rate of D_2 keeps constant in the entire scanning range, which is very different from that in measurement 1. The experimental data has provided a clear indication of $\Delta y \, \Delta p_y < h$ in the joint measurements of the entangled photon pairs.

Given that $\Delta y \, \Delta p_y < h$, is this a violation of the uncertainty principle? Does quantum mechanics agree with this peculiar experimental result? If quantum mechanics does provide a solution with $\Delta y \, \Delta p_y < h$ for photon 2. We would indeed be forced to face a paradox as EPR had pointed out in 1935.

Quantum mechanics does provide a solution that agrees with the experimental results. We now examine the experimental results from the view of quantum mechanics:

(1) When slit-A $= 0.16mm$, slit-B $= 0.16mm$.

This is the experimental condition for measurement one: slit-B is adjusted to be the same as slit-A. There is nothing surprise for this measurement. The measurement simply provide us the knowledge of Δp of photon 2 after the diffraction coursed by slit-B of $\Delta y = 0.16mm$. The experimental data shown in Fig. 14.6.5 agrees with the calculation. Notice that slit-B is about $745mm$ far away from the $3mm$ two-photon source, the angular size of the light source is roughly the same as $\lambda/\Delta y$, $\Delta\theta \sim \lambda/\Delta y$, where $\lambda = 702nm$ is the wavelength and $\Delta y = 0.16mm$ is the width of the slit. The calculated diffraction pattern is very close to that of the "far-field" Fraunhofer diffraction of a $0.16mm$ single-slit.

(2) When slit-A $= 0.16mm$, slits-B $\sim \infty$ (wide open).

Now we remove slit-B (wide open) from the ghost image plane. The calculation of the transverse effective two-photon wavefunction and the second-order correlation is the same as that of the ghost image except the observation plane of D_2 is moved from the image plane a distance of $500mm$ behind. The two-photon image of slit-A is located at a distance $s_i = 2f = 1000mm$ $(b_1 + b_2)$ from the imaging lens, in this measurement D_2 is placed at $d = 1500mm$ from the imaging lens. The measured pattern is simply a "blurred" two-photon image of slit-A. The "blurred" two-photon image can be calculated from Eq. (14.6.1)

$$
\begin{aligned}
\Psi(\vec{\rho}_o, \vec{\rho}_2) &\propto \int_{lens} d\vec{\rho}_l \, G(|\vec{\rho}_2 - \vec{\rho}_l|, \frac{\omega}{cd}) \, G(|\vec{\rho}_l|, \frac{\omega}{cf}) \, G(|\vec{\rho}_l - \vec{\rho}_o|, \frac{\omega}{cs_o}) \\
&\propto \int_{lens} d\vec{\rho}_l \, G(|\vec{\rho}_l|, \frac{\omega}{c}[\frac{1}{s_o} + \frac{1}{d} - \frac{1}{f}]) \, e^{-i\frac{\omega}{c}(\frac{\vec{\rho}_o}{s_o} + \frac{\vec{\rho}_i}{d}) \cdot \vec{\rho}_l}
\end{aligned}
\qquad (14.6.1)
$$

where d is the distance between the imaging lens and D_2. In this measurement, D_2 was placed $500mm$ behind the image plane, i.e., $d = s_i + 500mm$. The numerically calculated "blurred" image, which is narrower then that of the diffraction pattern of the $0.16mm$ slit-B, agrees with the measured result of Fig 14.6.5 within experimental error.

The measurement does show a result of $\Delta y \, \Delta p_y < h$. The measurement, however, has nothing to do with the uncertainty relation that governs the behavior of photon 2 (the idler). Popper and EPR were correct in the prediction of the outcomes of their experiments. Popper and EPR, on the other hand, made the same error by applying the physics of two-particle system to the explanation of the behavior of an individual particle.

In both the Popper and EPR experiments, the measurements are *joint detection* between two detectors applied to entangled states. Quantum mechanically, an entangled two-particle state only provides *the precise knowledge of the correlations of the pair*. The behavior of *photon 2* observed in the joint measurement is conditioned upon the measurement of its twin. A quantum must obey the uncertainty principle but the *conditional behavior* of a quantum in an entangled biparticle system is different in principle. We believe paradoxes are unavoidable if one insists that the *conditional behavior* of a particle is the *behavior* of the particle. This is the central problem in the rationale behind both Popper and EPR. $\Delta y \, \Delta p_y \geq h$ is not applicable to the conditional behavior of either *photon 1* or *photon 2* in the experiments of Popper and EPR.

The behavior of photon 2 being conditioned upon the measurement of photon 1 is well represented by the two-photon amplitudes. Each of the *straight lines* in Fig 14.6.3 corresponds to a two-photon amplitude. Quantum mechanically, the superposition of these two-photon amplitudes are responsible for a "click-click" measurement of the entangled pair. A "click-click" joint measurement of the two-particle entangled state projects out certain two-particle amplitudes, and only these two-particle amplitudes are featured in the quantum formalism. In the above analysis we never consider photon 1 or photon 2 *individually*. Popper's question about the momentum uncertainty of photon 2 is then inappropriate. The correct question to ask in these measurements should be: what is the uncertainty of Δp_y for the signal-idler *pair* which are localized within $\Delta y = 0.16mm$ at "screen" A with and without slit-B? This is indeed the central point for Popper's experiment.

Once again, the demonstration of Popper's experiment calls our attention to the important message: the physics of the entangled two-particle system must inherently be very different from that of individual particles.

14.6.2 Popper's Experiment Two

In fact, the nonfactorizable, point-to-point image-forming correlation is not only the property of entangled photon pairs; it can also be realized in the joint-detection of two randomly created and randomly paired photons in thermal state. In 2005, ten years after the first ghost imaging experiment, a near-field lensless ghost imaging experiment that used pseudo-thermal radiation source, was demonstrated by Valencia et. al.. This experiment opened a door for the realization of Popper's thought experiment through the joint measurement of randomly created and randomly paired photons in thermal state.

With the help of a novel joint detection scheme, namely the photon number fluctuation correlation (PNFC) circuit, which distinguishes the positive and negative photon number fluctuations measured by two single-photon counting detectors, and calculates the correlation between them, we were able to produce the ghost image of an object at a distance with 100% visibility. By modifying the Kim-Shih experiment of 1999 with a different light source and a lensless configuration, Peng *et al.* realized Popper's thought experiment again in 2015. Fig. 14.6.6 is an unfolded schematic, in which a large enough angular sized thermal source produces an equal-sized ghost image of slit-A at the plane $d_B = d_A$. The ghost image of slit-A can be verified by scanning the point-like photodetector D_B in the plane of slit-B. This ghost image provides the value of Δy through the correlation measurement. Again, the question of Popper is: Do we expect to observe a diffraction pattern that satisfies $\Delta p_y \Delta y > h$? To answer this question, we again make two measurements following Popper's suggestion. Measurement (1) is illustrated in the upper part of Fig. 14.6.6. In this measurement, we place slit-B, which has the same width as that of slit-A, coincident with the 1:1 ghost image of slit-A and measure the diffraction pattern by scanning D_B along the y-axis in the far-field of the ghost image. In this measurement, we learn the value of Δp_y due to the diffraction of a real slit of Δy. Measurement (2) is illustrated in the lower part of Fig. 14.6.6. Here, we open slit-B completely, scanning D_B again along the same y-axis to measure the "diffraction" pattern of the 1:1 ghost image with the same width as slit-A. By comparing the observed pattern width in measurement (2) with that of measurement (1), we can examine Popper's prediction.

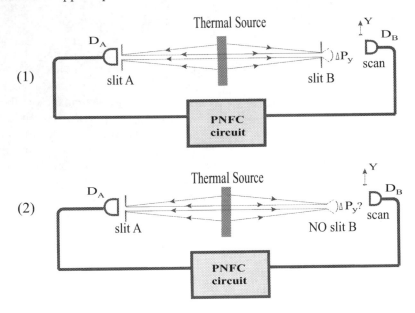

FIGURE 14.6.6 Unfolded schematic of Popper's experiment with thermal light. The lensless ghost imaging setup with PNFC protocol produces an equal sized 100% visibility ghost image of slit-A at the position of slit-B. Detector D_B is scanning transversely in the y direction to measure the photon number fluctuation correlation with D_A when (1) Slit-A and slit-B are adjusted both very narrowly, and (2) Slit-A is kept very narrow and slit-B is left wide open.

The experimental details are shown in Fig. 14.6.7. The light source is a standard pseudo-thermal source, consisting of a He-Ne laser beam and a rotating ground glass(GG). A 50/50 beamsplitter (BS) is used to split the pseudo-thermal light into two beams. One of the beams illuminates a single slit, slit-A, of width $D = 0.15mm$ located $d_A \sim 400mm$ from the source. A "bucket" photodetector D_A is placed right behind slit-A. An equal-sized ghost image of slit-A is then observable from the positive-negative photon number fluctuation correlation measurement between the "bucket" detector D_A and the transversely scanning point-like photodetector D_B, if D_B is scanned on the ghost image plane located at $d_B = d_A = 400mm$. In this experiment, however, D_B is scanned on a plane that is located $d'_B \sim 900mm$ behind the ghost image plane, to measure the "diffraction" pattern of the ghost image. The output pulses from the two single-photon counting detectors are then sent to a PNFC circuit, which starts from two Pos-Neg identifiers follow two event-timers distinguish the "positive-fluctuation" Δn^+, from the "negative-fluctuation" Δn^-, measured by D_A and D_B, respectively, within each coincidence time window. The photon number fluctuation-correlations of D_A-D_B: $\Delta R_{AB} = \langle \Delta n_A \Delta n_B \rangle$ is calculated, accordingly and respectively, based on their measured positive-negative fluctuations.

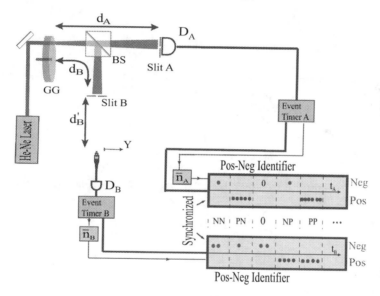

FIGURE 14.6.7 Schematic of the experimental setup of Peng *et al.* A rotating ground glass (GG) is employed to produce pseudo-thermal light. BS is a 50/50 non-polarizing beam splitter. After BS, the transmitted beam passes through slit-A($0.15mm$) and collected by a "bucket" detector D_A which is put right after the slit. The reflected beam passes slit-B, which can be adjusted to be the same width as that of slit-A or wide open, and then reaches the scanning detector D_B. The distances from slit-A and slit-B to the source are the same($d_A = d_B = 400mm$). The distance from the scanning fiber tip of D_B to the plane of slit-B is $d'_B = 900mm$. A PNFC protocol is followed to evaluate the photon number fluctuation correlations from the coincidences between D_A and D_B.

The experiment was performed in two steps after confirming the 1:1 ghost image of slit-A. In measurement (1), we place slit-B ($D = 0.15mm$) coincident with the ghost image and move D_B to a plane at $d'_B \sim 900mm$ to measure the diffraction pattern of slit-B. In measurement (2), we keep the same experimental condition as that of measurement (1), except slit-B is set wide open.

Fig. 14.6.8 reports the experimental results. The circles show the normalized photon number fluctuation correlation from the PNFC protocol against the position of D_B along the y-axis for Popper's measurement (1). As expected, we observed a typical single-slit diffraction pattern giving us the uncertainty in momentum, Δp_y^{real}. The squares show the experimental observation from the PNFC for Popper's measurement (2), when slit-B is wide open. The measured curves agree well with our theoretical fittings. We found the width of the curve representing no physical slit is much narrower than that of the real diffraction pattern, which agrees with Popper's prediction.

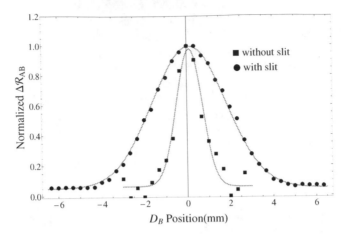

FIGURE 14.6.8 The observed diffraction patterns. Circles: slit-A and slit-B are both adjusted for $0.15mm$. Squares: slit-A is $0.15mm$, slit-B is wide open. The width of the curve without the slit is almost three times narrower than that of the curve with slit, agreeing well with the theoretical predictions from Eq.(14.6.7) and Eq.(14.6.9).

We give a simple analysis in the following. Similar to our early analysis in quantum eraser, we chose coherent state representation for the calculation of the joint photodetection counting rate of D_A and D_B which is proportional to the second-order coherence function $G^{(2)}_{AB}$:

$$
\begin{aligned}
G^{(2)}_{AB} = \big\langle \langle \hat{E}^{(-)}(\vec{\rho}_A, z_A, t_A) \hat{E}^{(-)}(\vec{\rho}_B, z_B, t_B) \\
\times \hat{E}^{(+)}(\vec{\rho}_A, z_B, t_B) \hat{E}^{(+)}(\vec{\rho}_A, z_A, t_A) \rangle_{\mathrm{QM}} \big\rangle_{\mathrm{Es}},
\end{aligned}
\tag{14.6.2}
$$

where $E^{(+)}(\vec{\rho}_j, z_j, t_j)$ $(E^{(-)}(\vec{\rho}_j, z_j, t_j))$ is the positive (negative) field operator at space-time coordinate $(\vec{\rho}_j, z_j, t_j)$, $j = A, B$, with $(\vec{\rho}_j, z_j, t_j)$ the transverse, longitudinal, and time coordinates of the photodetection event of D_A or D_B. Note, in the Glauber-Scully theory, the quantum expectation and classical ensemble average are evaluated separately, which allows us to examining the two-photon interference picture before ensemble averaging.

The field at each space-time point is the result of a superposition among a large number of subfields propagated from a large number of independent, randomly distributed and randomly radiating sub-sources of the entire thermal source,

$$
\begin{aligned}
\hat{E}^{(\pm)}(\vec{\rho}_j, z_j, t_j) &= \sum_m \hat{E}^{(\pm)}(\vec{\rho}_{0m}, z_{0m}, t_{0m}) g_m(\vec{\rho}_j, z_j, t_j) \\
&\equiv \sum_m \hat{E}^{(\pm)}_m(\vec{\rho}_j, z_j, t_j),
\end{aligned}
\tag{14.6.3}
$$

where $\hat{E}^{(\pm)}(\vec{\rho}_{0m}, z_{0m}, t_{0m})$ is the mth subfield at the source coordinate $(\vec{\rho}_{0m}, z_{0m}, t_{0m})$, and $g_m(\vec{\rho}_j, z_j, t_j)$ is the optical transfer function that propagates the mth subfield from coordinate $(\vec{\rho}_{0m}, z_{0m}, t_{0m})$ to $(\vec{\rho}_j, z_j, t_j)$. We can write the field operators in terms of the annihilation and creation operators:

$$
\hat{E}^{(+)}_m(\vec{\rho}_j, z_j, t_j) = C \int d\mathbf{k} \, \hat{a}_m(\mathbf{k}) \, g_m(\mathbf{k}; \vec{\rho}_j, z_j, t_j),
\tag{14.6.4}
$$

where C is a normalization constant, $g_m(\mathbf{k}; \vec{\rho}_j, z_j, t_j)$, $j = A, B$, is the optical transfer function for mode $\mathbf{k}$ of the mth subfield propagated from the mth sub-source to the jth detector, and $\hat{a}_m(\mathbf{k})$ is the annihilation operator for the mode $\mathbf{k}$ of the mth subfield.

Substituting the field operators and the state, in the multi-mode coherent representation, into Eq.(14.6.2), we then write $G_{AB}^{(2)}$ in terms of the superposition of a large number of effective wavefunctions, or wavepackets:

$$
\begin{aligned}
&G^{(2)}(\vec{\rho}_A, z_A, t_A; \vec{\rho}_B, z_B, t_B) \\
&= \Big\langle \sum_{m,n,p,q} \psi_m^*(\vec{\rho}_A, z_A, t_A)\psi_n^*(\vec{\rho}_B, z_B, t_B)\psi_p(\vec{\rho}_B, z_B, t_B)\psi_q(\vec{\rho}_A, z_A, t_A) \Big\rangle_{\text{Es}} \\
&= \Big\langle \sum_{m,n} \big|\psi_m(\vec{\rho}_A, z_A, t_A)\psi_n(\vec{\rho}_B, z_B, t_B) + \psi_n(\vec{\rho}_A, z_A, t_A)\psi_m(\vec{\rho}_B, z_B, t_B)\big|^2 \Big\rangle_{\text{Es}} \\
&= \Big\langle \sum_m \big|\psi_m(\vec{\rho}_A, z_A, t_A)\big|^2 \sum_n \big|\psi_n(\vec{\rho}_B, z_B, t_B)\big|^2 \\
&\quad + \sum_{m \neq n} \big[\psi_m^*(\vec{\rho}_A, z_A, t_A)\psi_m(\vec{\rho}_B, z_B, t_B)\psi_n(\vec{\rho}_A, z_A, t_A)\psi_n^*(\vec{\rho}_B, z_B, t_B)\big] \Big\rangle_{\text{Es}} \\
&\equiv \langle n_A\rangle\langle n_B\rangle + \langle \Delta n_A \Delta n_B\rangle.
\end{aligned}
\tag{14.6.5}
$$

with

$$
\psi_s(\vec{\rho}_j, z_j, t_j) = \int d\mathbf{k}\, \alpha_s(\mathbf{k}) e^{i\varphi_{0s}} g_s(\mathbf{k}; \vec{\rho}_j, z_j, t_j),
$$

where $s = m, n, p, q$, $j = A, B$, and the phase factor $e^{i\varphi_{0s}}$ represents the random initial phase of the mth subfield. In Eq.(14.6.5), we have completed the ensemble average in terms of the random phases of the subfields, i.e. φ_{0s}, and kept the nonzero terms only. Eq.(14.6.5) indicates the second-order coherence function, is the result of a sum of a large number of subinterference patterns, each subpattern indicates an interference in which a random pair of wave packets interfering with the pair itself. For example, the mth and the nth wave packets have two different yet indistinguishable alternative ways to produce a joint photodetection event, or a coincidence count, at different space-time coordinates: (1) the mth wavepacket is annihilated at D_A and the nth wavepacket is annihilated at D_B; (2) the mth wavepacket is annihilated at D_B and the nth wavepacket is annihilated at D_A. In quantum mechanics, the joint detection probability of D_A and D_B is proportional to the normal square of the superposition of the above two probability amplitudes. We name this kind of superposition "nonlocal interference". The superposition of the two amplitudes for each random pair results in an interference pattern, and the addition of these large number of interference patterns yields the nontrivial correlation of the thermal light.

The cross interference term in Eq.(14.6.5) indicates the photon number fluctuation correlation $\langle \Delta n_A \Delta n_B\rangle$:

$$
\begin{aligned}
&\langle \Delta n_A(\vec{\rho}_A, z_A, t_A)\Delta n_B(\vec{\rho}_B, z_B, t_B)\rangle_{\text{Es}} \\
&= \Big\langle \sum_{m \neq n} \big[\psi_m^*(\vec{\rho}_A, z_A, t_A)\psi_n(\vec{\rho}_A, z_A, t_A)\big]\big[\psi_m(\vec{\rho}_B, z_B, t_B)\psi_n^*(\vec{\rho}_B, z_B, t_B)\big] \Big\rangle_{\text{Es}} \\
&\simeq \Big\langle \sum_m \psi_m^*(\vec{\rho}_A, z_A, t_A)\psi_m(\vec{\rho}_B, z_B, t_B) \sum_n \psi_n(\vec{\rho}_A, z_A, t_A)\psi_n^*(\vec{\rho}_B, z_B, t_B) \Big\rangle_{\text{Es}}. \tag{14.6.6}
\end{aligned}
$$

Measurement (1): slit-A $= 0.15mm$, slits-B $= 0.15mm$.

In measurement (1), the optical transfer functions that propagate the fields from the source to D_A and D_B are

$$g_m(\vec{\kappa}, \omega; \vec{\rho}_A, z_A = d_A) = \frac{-i\omega e^{i(\omega/c)z_A}}{2\pi c d_A} \int d\vec{\rho}_s f(\vec{\rho}_s) e^{i\vec{\kappa}\cdot\vec{\rho}_s} G(|\vec{\rho}_s - \vec{\rho}_o|)_{[\omega/(cd_A)]},$$

and

$$g_n(\vec{\kappa}, \omega; \vec{\rho}_B, z_B = d_B + d'_B)$$
$$= \frac{-\omega^2 e^{i(\omega/c)z_B}}{(2\pi c)^2 d_B d'_B} \int d\vec{\rho}_s \int d\vec{\rho}_i \, f(\vec{\rho}_s) e^{i\vec{\kappa}\cdot\vec{\rho}_s} G(|\vec{\rho}_s - \vec{\rho}_i|)_{[\omega/(cd_B)]} t(\vec{\rho}_i) G(|\vec{\rho}_i - \vec{\rho}_B|)_{[\omega/(cd'_B)]},$$

where $\vec{\rho}_s$ is defined on the output plane of the source and $f(\vec{\rho}_s)$ denotes the aperture function of the source. We also assumed a perfect "bucket" detector D_A, which is placed at the object plane of slit-A ($\vec{\rho}_A = \vec{\rho}_o$), in the following calculation. $\vec{\rho}_i$ is defined on the ghost image plane, which is coincided with the plane of slit-B, and $\vec{\rho}_B$ is defined on the detection plane of D_B, $t(\vec{\rho}_i)$ is the aperture function of slit-B. The function $G(|\alpha|)_{[\beta]}$ is the Gaussian function $G(|\alpha|)_{[\beta]} = e^{i\frac{\beta}{2}|\alpha|^2}$. The measured fluctuation correlation can be calculated from Eq. (14.6.6)

$$\Delta R_{AB} = \int d\vec{\rho}_o \, |t(\vec{\rho}_o)|^2 \text{sinc}^2\left[\frac{\omega_0 D\vec{\rho}_B}{2cd'_B}\right] \equiv C' \times \text{sinc}^2\left[\frac{\omega_0 D\vec{\rho}_B}{2cd'_B}\right], \tag{14.6.7}$$

where $t(\vec{\rho}_o)$ is the aperture function of slit-A. The above calculation indicates a product between a constant C', which is from the integral on the "bucket" detector D_A, and a first order diffraction pattern of slit-B. With our experimental setup, the width of the diffraction pattern is estimated to be $\sim 4mm$, which agrees well with the experimental observation, as shown in Fig. 14.6.8.

Measurement (2): slit-A $= 0.15mm$, slits-B $\sim \infty$ (wide open).

In measurement (2), with slit-B wide open, the field at D_B becomes

$$g_n(\vec{\kappa}, \omega; \vec{\rho}_B, z_B) = \frac{-i\omega e^{i(\omega/c)z_B}}{2\pi c z_B} \int d\vec{\rho}_s f(\vec{\rho}_s) e^{i\vec{\kappa}\cdot\vec{\rho}_s} G(|\vec{\rho}_s - \vec{\rho}_B|)_{[\omega/(cz_B)]}.$$

We first check if a ghost image of slit-A is present when scanning D_B in the ghost image plane of $d_B = d_A$. The photon number fluctuation correlation is calculated to be:

$$\Delta R_{AB} = \int d\vec{\rho}_o |t(\vec{\rho}_o)|^2 \text{sinc}^2\left[\frac{\omega_0 a}{cd_A}|\vec{\rho}_o - \vec{\rho}_B|\right]$$
$$= |t(\vec{\rho}_o)|^2 \otimes \text{sinc}^2\left[\frac{\omega_0 a}{cd_A}|\vec{\rho}_o - \vec{\rho}_B|\right] \approx |t(\vec{\rho}_B)|^2. \tag{14.6.8}$$

Note, we have placed D_A right behind slit-A and thus $\vec{\rho}_A = \vec{\rho}_o$. This suggests an equal-sized 100% visibility ghost image on the plane of $d_B = d_A$.

When we move D_B away from the ghost image plane to the far-field plane of $d_B + d_{B'}$, the photon number fluctuation correlation becomes:

$$\Delta R_{AB} = \int d\vec{\rho}_o |t(\vec{\rho}_o)|^2 \tilde{\mathcal{F}}_s^2 (m\vec{\rho}_o - \vec{\rho}_B) = |t(\vec{\rho}_o)|^2 \otimes \tilde{\mathcal{F}}_s^2 (m\vec{\rho}_o - \vec{\rho}_B), \tag{14.6.9}$$

where $\tilde{\mathcal{F}}_s$ is the Fourier transform of the defocused pupil function $\mathcal{F}_s = f(\vec{\rho}_s)e^{-i(\omega_0/2c\mu)\vec{\rho}_s^2}$ and μ, m are defined as $1/\mu = 1/d_A - 1/(d_B + d_{B'})$, $m = (d_B + d_{B'})/d_A$, respectively. The measured result of measurement (2) is thus a convolution between the aperture function of slit-A, $t(\vec{\rho}_o)$, and the correlation function $\tilde{\mathcal{F}}_s(m\vec{\rho}_o - \vec{\rho}_B)$, resulting in a "blurred" image of slit-A. With our experimental setup, the width of the "diffraction" pattern is estimated to be $\sim 1.4mm$, which is almost three times narrower than the diffraction pattern of measurement (1) and agrees well with the experimental observation, as shown in Fig. 14.6.8. Compared with the Kim-Shih experimental result, we can see that although the number varies due to different experimental parameters, we have obtained a very similar result: the measured width of the "diffraction pattern" in measurement (2) is much narrower than that of the diffraction pattern in measurement (1).

The above analysis indicates that the experimental observations are reasonable from the viewpoint of the quantum coherence theory of light. The important physics we need to understand is to distinguish the first-order coherent effect and the second-order coherent effect, even if the measurement is for thermal light. In measurement (1), the fluctuation correlation is the result of first-order coherence. The joint measurement can be "factorized" into a product of two first-order diffraction patterns. After the integral of the "bucket" detector, which turns the diffraction pattern of slit-A into a constant, the joint measurement between D_A and D_B is a product between a constant and the standard first-order diffraction pattern of slit-B. There is no question the measured width of the diffraction pattern satisfies $\Delta p_y \Delta y \geq h$. In measurement (2) when slit-B is wide open or removed, the measurement can no longer be written as a product of single-photon detections but as a non-separable function, i.e., a convolution between the object aperture function and the photon number fluctuation correlation function of randomly paired photons, or the second-order coherence function of the thermal field. We thus consider the observation of $\Delta p_y \Delta y < h$ the result of the second-order coherence of thermal field which is caused from nonlocal two-photon interferences: a randomly paired photon interferes with the pair itself at a distance by means of a joint photodetection event between D_A and D_B. The result of nonlocal two-photon interference does not contradict the uncertainty principle that governs the behavior of single photons. Again, the observation of this experiment is not a violation of the uncertainty principle. The observation of $\Delta p_y \Delta y < h$ from thermal light, however, may reveals a concern about nonlocal interference as we have mentioned earlier.

SUMMARY

In this chapter, we studied three types of optical experiments to probe the foundations of quantum theory: (1) EPR-Bohm-Bell correlation and Bells inequality; (2) Scully's quantum eraser; (3) Poppers experiment. The results of these experiments are very interesting. On one hand, the experimental observations confirm the predictions of EPR-Bell, Scully, and Popper which lead to their questions to the quantum theory. On the other hand, the calculations from quantum theory perfectly agree with the experimental data. Moreover, apparently, the experimental observations do not lead to any "violations" of the principles of quantum mechanics. One important conclusion we may draw from these optical tests is that all the observations are the results of multi-photon interference: a group of photons interferes with the group itself at distance. The nonlocal multi-photon interference phenomena may never be understood in classical theory, however, it is legitimate in quantum mechanics. The superposition principle of quantum theory supports the superposition of multi-photon amplitudes, whether the photons are entangled or randomly created and randomly grouped, and despite the distances between these individual photodetection events. Perhaps we must accept the probabilistic nature of the "wavefunction" associated with a quantum or a group of quanta. Although a photon does not have a "wavefunction", we have developed the concept of an effective wavefunction for a photon and for a group of photons which have similar physical meanings as that of the wavefunction of a particle or the wavefunction of a group of particles. In terms of the superposition, although the effective wavefunction plays the same role as that of the electromagnetic wave, apparently, the effective wavefunction is different from the electromagnetic field in nature. Any efforts attempting to physically equal the two concepts would trap us in the question posed by Einstein: how long does it take for the energy on the other side of the two-lightyear diameter sphere to arrive at the detector?

In this chapter, we especially reviewed the Bell's theorem and Bell's inequality. Bell's theorem addresses two fundamental problems on quantum mechanics: *Reality* and *Locality*, both originated from the 1935 article of Einstein, Podolsky, and Rosen. In that paper, EPR formally raised a question: "Can quantum-mechanical description of physical reality be considered complete?" After giving a simple yet carefully stated criterion that defines their locality and reality: "If, without in any way disturbing a system, we can predict with certainty the value of a physical quantity, then there exist an element of physical reality corresponding to this quantity", EPR proposed an entangled two-particle state in a *gedankenexperiment* to show that it is possible to determine the position and momentum of a particle with certainty through the distant measurement of the position and momentum of its twin. They are confident on their beliefs of locality that the measurement of one particle does not disturbing its twin at distance.

Does a particle in the EPR state, or in the EPR-Bohm state, have a defined position and momentum, or polarization, during the course of its propagation despite if we measure it or not? Quantum theory answers NO. In the view of quantum mechanics, a particle or a pair of particles may take any or all possible values of position and momentum or polarization if the particle or the pair of particles is in the state of a superposition. It is also an experimental evidence that even if each photon in a pair is prepared with a defined polarization, an appropriate two-photon superposition may result in a state in which no polarization is specified anymore for either photon as a sub-system of the two-photon state. As a realistic physicist, Einstein felt very uncomfortable about the quantum superposition. Einstein insisted that the quantum mechanical wavefunction should be and must be able to specify physical reality that was defined in their 1935 paper. It was Bohm and his followers attempted a hidden variable theory and formulated the physical reality into the wavefunction of a particle or a pair of particles. Bohm's theory demonstrated its success and consistency with quantum mechanics in almost every aspect of physics until Bell proved a theorem and derived an inequality to show a quantitative difference between the local hidden variable theory and quantum mechanics. In 1964, Bell constructed a general statistical model in a hidden variable parameter space with his mathematically formulated "locality" to show that an inequality can be used to distinguish quantum superposition from classical statistics. Although concerns and criticisms about Bell's theorem and Bell's inequality still exist, a large number of experimental violations of a variety of Bell's inequalities have been reported in favor of quantum superposition.

REFERENCES AND SUGGESTIONS FOR READING

[1] A. Einstein, B. Podolsky, and N. Rosen, Phys. Rev., **47**, 777 (1935).

[2] D. Bohm, *Quantum theory*, Prentice-Hall, Yew York, 1951.

[3] R.F. Feynman, R.B. Leighton, and M.L. Sands, *Lectures on Physics*, Addison-Wesley, Reading, MA, 1965.

[4] J.A. Wheeler, and W.H. Zurek, *Quantum Theory and Measurement*, Princeton University Press, Princeton, New Jersey, 1983.

[5] J.S. Bell, Physics, **1**, 195 (1964).

[6] J.S. Bell. *Speakable and unspeakable in quantum mechanics*, Cambridge University Press, 1993.

[7] D.M. Greenberger, M. Horne, and A. Zeilinger, *Bell's theorem, quantum theory, and conceptions of the universe*, ed. M. Kafatos, Kluwer Academic, Dordrecht, The Netherlands, 1989.

[8] D.M. Greenberger, M. Horne, A. Shimony, and A. Zeilinger, Am. J. Phys., **58**, 1131 (1990).

[9] M.O. Scully, N. Erez, and E.S. Fry, Phys Lett. A, **347** 56 (2005).

[10] An excellent review on Bell measurement before the introduction of SPDC: J.F. Clauser and A. Shimony, *Reports on Progress in Physics* **41**, 1881 (1978).

[11] S.J. Freedman and J.F. Clauser, Phys. Rev. Lett., **28**, 938 (1972).

[12] J.F. Clauser, Phys. Rev. Lett., **36**, 1223 (1976).

[13] E.S. Fry and P.C. Thompson PC, Phys. Rev. Lett., **37**, 465 (1976).

[14] A. Aspect, P. Grangier, and G. Roger, Phys. Rev. Lett., **47**, 460 (1981).

[15] A. Aspect, P. Grangier, and G. Roger, Phys. Rev. Lett., **49**, 91 (1982).

[16] C.O. Alley and Y.H. Shih, *Foundations of Quantum Mechanics in the Light of New Technology*, edited by Namiki M *et al.* Physical Society of Japan, Tokyo, Japan, 1986; Y.H. Shih and C.O. Alley, Phys. Rev. Lett., **61**, 2921 (1988).

[17] P.G. Kwiat, K. Mattle, H. Weinfurter, A. Zeilinger, A.V. Sergienko, and Y.H. Shih, Phys. Rev. Lett., **75**, 4337 (1995).

[18] M.O. Scully and H. Druhl, Phys. Rev. A, **25**, 2208 (1982).

[19] Y.H. Kim, S.P. Yu, S.P. Kulik, Y.H. Shih, and M.O. Scully, Phys. Rev. Lett., **84**, 1 (2000).

[20] T. Peng, H. Chen, Y.H. Shih YH, and M.O. Scully, Phys. Rev. Lett., **112**, 180401 (2014).

[21] K. Popper, *Naturwissenschaften*, **22**, 807 (1934).

[22] Y.H. Kim and Y.H. Shih, *Foundations of Physics*, **29**, 1849 (1999).

[23] T. Peng, J. Simon, H. Chen, R. French, and Y.H. Shih, Euro Phys. Lett., **109**, 14003 (2015).

Index